Introduction to Automata Theory, Formal Languages and Computation

Introduction to Automata Theory, Formal Languages and Computation

Shyamalendu Kandar
Assistant Professor, Computer Science and Engineering
Haldia Institute of Technology
Haldia, West Bengal

Pearson

Assistant Editor–Acquisitions: Neha Goomer
Assistant Editor–Production: Sonam Arora

ISBN 978-81-317-9351-0

First Impression, 2013

Published by Pearson India Education Services Pvt. Ltd, CIN: U72200TN2005PTC057128.

Head Office: 15th Floor, Tower-B, World Trade Tower, Plot No. 1, Block-C, Sector 16,
Noida 201 301, Uttar Pradesh, India.

Registered Office: 4th floor, Software Block, Elnet Software City, TS 140 Block 2 & 9,
Rajiv Gandhi Salai, Taramani, Chennai - 600 113, Tamil Nadu, Fax: 080-30461003,
Phone: 080-30461060, Website: in.pearson.com Email: companysecretary.india@pearson.com

Compositor: White Lotus Infotech Pvt. Ltd, Puducherry

Digitally Printed in India by Repro Books Limited, Thane in the year of 2019.

Dedication

The students

The future architect of the nation

and

to my parents…

Arun Kumar Kandar and Hirabati Kandar

Contents

Foreword

The subject of automata theory is an important component of computer science and engineering. Automata is known as the backbone of computer science and it is the link between digital logic and computer algorithm, it is the base of compiler design. Thus, without any hesitation I decisively believe that it is a core part of the subject of computer science and engineering.

Many books on automata theory are available and all these books have obvious attractions and drawbacks. The current book is authored by one of my beloved student, Shyamalendu. He is students' friendly and informative. As a teacher, I believe the book by Shyamalendu will add new dimensions and flavours to the readership.

I wish to congratulate Shyamalendu for his credible academic achievement and hope to receive many more such academic outputs from him in the days to come.

Professor Chandan Tilak Bhunia
Director, National Institute of Technology
Arunachal Pradesh

Preface

What, How and Why—these three words are related to education. 'What' only fulfills the basic knowledge, 'How' is related to engineering and 'Why' makes the knowledge complete.

Bachelors of Computer Science and Engineering has to not only learn some application software and some programming languages but also they must learn how a programming language works, how a program is compiled, how input is converted to output from the machine hardware level. Theoretical part of computer science includes complexity analysis, compiler construction, verifying correctness of circuits and protocols, etc. Automata theory is one of the core courses in the curriculum of Bachelors of Technology of Computer Science or Information Technology under any university.

Automata theory lies on digital electronics and extends to algorithm and computational complexity. It is the base of compiler design. So, the knowledge of automata theory is very important for graduate and post-graduate students of computer science or information technology.

During my teaching career I have seen that students are frightened of this subject and have realized that they need a book with detailed discussion on few topics presented lucidly. They want more numbers of solved problems where solving is done step by step. Now-a-days multiple choice questions are essential as in different university question papers and competitive exams, these types of questions come. Keeping all this in mind, this book is framed. It contains more than 450 solved problems which include question papers from UPTU, WBUT, JNTU, Pune University, Gujarat Technical University, etc. With each chapter a number of MCQ and fill in the blanks are added. One of the important features of the book is that it contains 20 years' (1992–2012) GATE solved questions on automata theory. This will help the students a lot in preparing for GATE, NET, etc. In addition, each chapter contains exercises for practice. The difficult questions have clues to solve.

The book contains 14 chapters which covers preliminaries, grammar and languages, finite automata, finite state machine, regular expression, context-free grammar, push down automata, Turing machine, variations of Turing machine, undecidability, recursive function, computational complexity, basics of compiler design and some advance topics related to automata. The Index will help to easily search any topic included in the book.

The book is written mainly according to the syllabus of major technical universities like Jadavpur University, ISM, BESU, WBUT, UPTU, Andhra University, JNTU, Pune University, Gujarat Technical University and different NITs like NIT Durgapur, NIT Rourchella, NIT, Trichi, NIT, Shilchar, etc.

The readers are welcomed to send suggestions, advice for further improvement of the book. The mail address is shyamalenduk@yahoo.com.

My dream of writing this book will be successful if the students are benefited from this book.

Shyamalendu Kandar

Acknowledgements

At starting of my career as a lecturer I was offered two subjects. One of them is automata theory. At the very beginning I used to prepare notes, solve examples, etc., on plain paper. Those notes helped me a lot in teaching this subject.

November 2008 steered my career to a new direction. On 3 November 2008, Professor C. T. Bhunia left our institute. He used to inspire me in good teaching, research and in different creative activities. His departure from our institute made me roofless. I got a huge time beyond my academic activities. I started making hand written notes in soft copies which started the journey of writing the book. In January 2012 the Express Learning book on Automata was published by Pearson Education. Since then there was a desire of writing a text book on this subject which can cover the syllabuses of most of the universities of India.

Through this work I want to pay respect to Professor C. T. Bhunia, Director NIT Arunachal, whose blessings are always with me and I consider him as my Guru in the field of technical education.

I want to pay respect to Bibhas Ch. Dhara, Head of the Department, IT, Jadavpur University and my Ph.D. guide. He wanted all things to be perfect. He pointed out some mistakes in the Express Learning book as well as some useful suggestions on Automata theory to make it a quality book.

My friend Bidesh Chakrabory and Arindam Giri, Assistant Professor, Computer Science helped me a lot by providing needful information and consistent inspiration and support. A number of my students also helped me a lot in selecting problems, solving problems, proof checking, etc. Among them Satyakam Shailesh, Akash Ranjan, Nishant Kumar, Nijhar Bera and Somnath Kayal deserve special mention.

Finally I want to express gratitude to my colleagues especially Mrinmoy Sen, Sabyasachi Samanta, Ashish Bera, Anupam Pattanayak, Milan Bera, Mrityunjay Maity and Tanuka Sinha, Bapida, who are an important part of our department.

I express my thanks to parents Arun Kumar Kandar and Hirabati Kandar for their constant support and encouragement during the preparation of this book.

Neha Goomer of Pearson Education deserves a special mention. My sincere thanks to Neha for making my dream of publishing the book come true.

Shyamalendu Kandar

About the Author

Shyamalendu Kandar is working as an Assistant Professor of Computer Science and Engineering department of Haldia Institute of Technology. He served as a co-ordinator of M.Tech (IT) of HIT (centre distance mode) with Javadpur University. Professor Kandar has done his M.Tech in IT from Jadavpur University in 2006. He has a number of research papers in reputed international conferences and journals. His subject of interests are Automata Theory, Compiler Designing, Algorithm, Object oriented programming, Web Technology, etc. His areas of research interest are cryptography, secret sharing, Image processing, etc.

Basic Terminology 1

Introduction

In the automata theory, we have to deal with some mathematical preliminaries. As examples in finite automata and finite state machine the knowledge of set theory is necessary, in grammar and language section we need the basic knowledge of alphabet, string, and substring, and in the regular expression chapter we need the concept of prefix, suffix, etc. The knowledge of the basic operations on a set such as union, intersection, difference, Cartesian product, power set, and concatenation product are required throughout the syllabus of formal language and automata theory.

For this reason, in this chapter, we shall discuss some basic terminologies related to mathematics which are required for automata theory.

1.1 Basics of String

A string has some features as follows:

- **Symbol:** A symbol is a user-defined entity.
- **Alphabet:** An alphabet is a finite set of symbols denoted by Σ in automata. An alphabet is a set of symbols used to construct a language. As an example, {0, 1} is a binary alphabet, and {A……, Z, a…….z} is an alphabet set for the English language.
- **String:** A string is defined as a sequence of symbols of finite length. A string is denoted by w in automata. As an example, 000111 is a binary string.
  (The length of a string w is denoted by |w|. For the previous case, |w| = |000111| = 6.)
- **Prefix:** A prefix of a string is the string formed by taking any number of symbols of the string.
  Example: Let us take a string w = 0111. For the particular string, λ, 0, 01, 011, and 0111 are prefixes of the string 0111. For a string of length n, there are n + 1 number of prefixes.
- **Proper prefix:** For a string, any prefix of the string other than the string itself is called as the proper prefix of the string.
  Example: For the string w = 0111, the proper prefixes are λ, 0, 01, and 011.
- **Suffix:** A suffix of a string is formed by taking any number of symbols from the end of the string.
  Example: Let us take a string w = 0110. For the particular string, λ, 0, 10, 110, and 0110 are suffixes of the string 0110. For a string of length n, there are n + 1 number of suffixes.
- **Proper suffix:** For a string, any suffix of the string other than the string itself is called as the proper suffix of the string.
  Example: For the string w = 0110, the proper suffixes are λ, 0, 10, and 110.

- **Substring:** A substring of a string is defined as a string formed by taking any number of symbols of the string.
  Example: For the string w = 012, the substrings are λ, 0, 1, 2, 01, 12, and 012.

1.2 Basics of Set Theory

A *set* is a well-defined collection of objects. The objects used for constructing a set are called elements or the members of the set.

A set has some features as follows:

- A set is a collection of objects. This collection is regarded as a single entity.
- A set is comprised of distinct elements. If an element, say 'a', is in set S, then it is denoted as $a \in S$.
- A set has a well-defined boundary. If S is a set and 'a' is any element, then depending on the properties of 'a', it can be said whether $a \in S$ or $a \notin S$.
- A set is characterized by its property. In general, if p is the defined property for the elements of S, then S is denoted as S = {a : a has the property p}.

Example:

- The set of all integers is denoted as S = {a: a is an integer}.
  Here, $7 \in S$ but $1/7 \notin S$.
- The set of all odd numbers denoted as S = {a: a is not divisible by 2}.
  Here, $7 \in S$ but $8 \notin S$.
- The set of prime numbers less than 100 is denoted as S = {a: a is prime and less than 100}.
  Here, $23 \in S$ but 98 or $101 \notin S$.

1.2.1 Subset

Let there be two sets S and S_1. S_1 is said to be a subset of S if every element S_1 is an element of S. Symbolically, it is denoted as $S_1 \subset S$.

The reverse of a subset is the *superset*. In the previous example, S is the superset of S_1.

Example:

- Let Z be the set of all integers. E is the set of all even numbers. All even numbers are natural numbers. So, it can be denoted as $E \subset Z$.
- Let S be the set of the numbers divisible by 6, where T is the set of numbers divisible by 2. Property says that if a number is divisible by 6, it must be divisible by 2 and 3. So, it can be denoted as $S \subset T$.

1.2.2 Finite and Infinite Set

A set is said to be finite if it contains no element or a finite number of elements. Otherwise, it is an infinite set.

Example:

- Let S be the set of one digit integers greater than 1. S is finite as its number of elements is 8.
- Let P be the set of all prime numbers. T is infinite, as the number of prime numbers is infinite.

1.2.3 Equal

Two sets S and S_1 are said to be equal if S is a subset of S_1 and S_1 is a subset of S.

1.2.4 Algebraic Operations on Sets

- **Union:** If there are two sets A and B, then their union is denoted by $A \cup B$. Let A = {2, 3, 4} and B = {3, 5, 6}. Then, $A \cup B$ = {2, 3, 4, 5, 6}. In general, $A \cup B = \{x \mid x \in A \text{ or } x \in B\}$.

  Diagrammatically, the union operation on two sets can be represented as shown in Fig. 1.1. This diagrammatic representation of sets is called the Venn diagram.

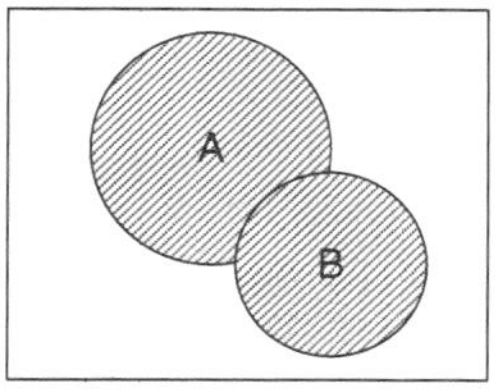

Fig. 1.1 $A \cup B$

- **Intersection:** If there are two sets A and B, then their intersection is denoted by $A \cap B$. Let A = {2, 3, 4} and B = {3, 4, 5, 6}. Then, $A \cap B$ = {3, 4}. In general, $A \cap B = \{x \mid x \in A \text{ and } x \in B\}$.

  The Venn representation of the intersection operation on two sets can be represented as shown in Fig. 1.2.

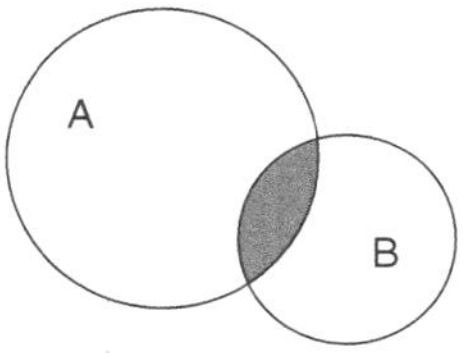

Fig. 1.2 $A \cap B$

- **Difference:** If there are two sets A and B, then their difference is denoted by A − B. Let A = {2, 3, 4, 5} and B = {3, 4}. Then, A − B = {2, 5}. In general, $A - B = \{x \mid x \in A \text{ and } x \notin B\}$.

  The Venn representation of the difference operation on two sets can be represented as shown in Fig. 1.3.

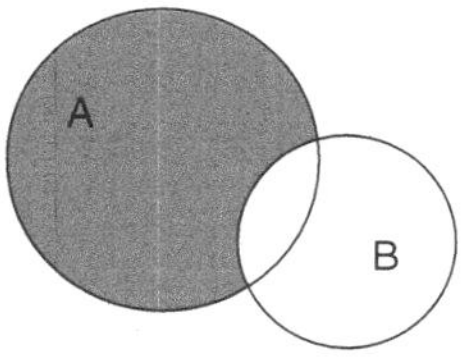

Fig. 1.3 $A - B$

- **Complementation:** The complement of a set A, which is a subset of a large set U is denoted by A^C or A', defined by $A' = \{x \in U: x \notin A\}$.

  The Venn representation of the complement operation is given in Fig. 1.4.

- **Cartesian product:** If there are two sets A and B, then their Cartesian product is denoted by A × B. Let A = {2, 3, 4, 5} and B = {3, 4}. Then, A × B = {(2, 3), (2, 4), (3, 3), (3, 4),(4, 3), (4, 4), (5, 3), (5, 4)}. In general, $A \times B = \{(a, b) \mid a \in A \text{ and } b \in B\}$.

- **Power set:** The power set of a set A is the set of all possible subsets of A. Let A = {a, b}. Then, the power set of A is {(∅), (a), (b), (a, b)}. For a set of elements n, the number of elements of the power set of A is 2^n.

- **Concatenation:** If there are two sets A and B, then their concatenation is denoted by A.B. Let A = {a, b} and B = {c, d}. Then, A.B = {ac, ad, bc, bd}. In general, A.B = {ab|a is in A and b is in B}.

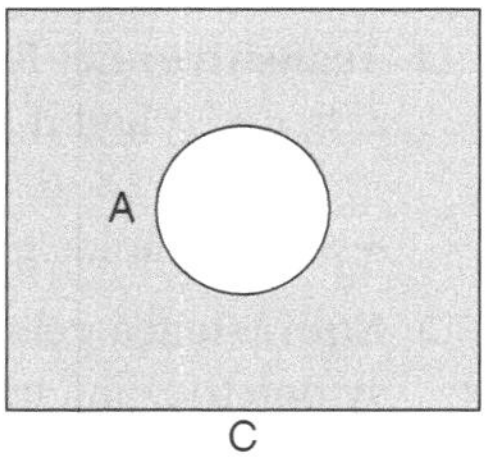

Fig. 1.4 A^C

1.2.5 Properties Related to Basic Operation

Some properties related to basic operations on set are as follows:

- $A \cup \varnothing = A$, $A \cap \varnothing = A$ (∅ is called null set)
- $A \cup U = U$, $A \cap U = A$ (where $A \subset U$)
- $A \cup A = A$, $A \cap A = A$ (idempotent law)

- $A \cup B = B \cup A$, $A \cap B = B \cap A$ (commutative property)
- $A \cup (B \cup C) = (A \cup B) \cup C$, $A \cap (B \cap C) = (A \cap B) \cap C$ (associative property)
- $A \cup (B \cap C) = (A \cup B) \cap (B \cup C)$, $A \cap (B \cup C) = (A \cap B) \cup (B \cap C)$ (distributive property)
- $A \cup A^C = U$ (U is the universal set)
- $A \cap A^C = \varnothing$
- $(A^C)^C = A$
- $(A \cup B)^C = A^C \cap B^C$, $(A \cap B)^C = A^C \cup B^C$ (D'Morgan's Law)
- $A - (B \cup C) = (A - B) \cap (A - C)$
- $A - (B \cap C) = (A - B) \cup (A - C)$

1.3 Relation on Set

- **Relation:** Let there be two non-empty sets S and T. A relation R, defined between S and T, is a subset of $S \times T$. If the ordered pair $(s, t) \in R$, then the element $s \in S$ is said to be related to $t \in T$ by the relation R.

  Let S = {2, 4, 6, 8} and T = {12, 16, 18, 19}. A relation R between S and T is defined as an element s in S is related to an element t of T if t is divided by s. Here, R = {(2, 12), (2, 16), (2, 18), (4, 12), (6, 12), (6, 18)}. But $(4, 18) \notin R$ as 18 is not divisible by 4.

- **Reflexive:** A relation R is said to be reflexive on a non-empty set R if every element of A is related to itself by that relation R.

  Let A = {B, C, D}, where B, C, and D are brothers. Then, the relation brotherhood is a reflexive relation on the set A as {B, C}, {B, D}, and {C, D} are all sets of brother.

- **Symmetric:** A relation R is said to be symmetric, if for two elements 'a' and 'b' in X, if 'a' is related to 'b' then b is related to a.

  Let A be a set of all students in a class. Let R be a relation called classmate. Then we can call R as a symmetric relation. If a and b are two students belonging to the set, then a is a classmate of b and b is a classmate of a.

- **Transitive:** Let R be a relation defined on a set A; then the relation R is said to be transitive if for $a, b, c \in A$ and if aRb, bRc holds good then aRc also holds good.

  Let A = {8, 6, 4}, where R is a relation called greater than. If 8 > 6 and 6 > 4 hold good, then 8 > 4 also holds good. So, 'greater than' is a transitive relation.

- **Equivalence relation:** A relation R is called as an equivalence relation on 'A' if R is reflexive, symmetric, and transitive.

- **Right invariant**: An equivalence relation R on a strings of symbols from some alphabet Σ is said to be a right invariant if for all $x, y \in \Sigma^*$ with x R y and all $w \in \Sigma^*$ we have that xw R yw. This definition states that an equivalence relation has the right invariant property if two equivalent strings (x and y) that are in the language still are equivalent if a third string (w) is appended to the right of both of them.

- **Closure:** A set is closed (under an operation) if and only if the operation on two elements of the set produces another element of the set. If an element outside the set is produced, then the operation is not closed.

  Closure is a property which describes when we combine any two elements of the set; the result is also included in the set.

If we multiply two *integer numbers*, we will get another integer number. Since this process is always true, it is said that the integer numbers are 'closed under the operation of multiplication'. There is simply no way to escape the set of integer numbers when multiplying.

Let S = {1, 2, 3, 4, 5, 6, 7, 8, 9, 10....} be a set of integer numbers.

$$1 \times 2 = 2$$
$$2 \times 3 = 6$$
$$5 \times 2 = 10$$

All are included in the set of integer numbers.

So, we can say that integer numbers are closed under the operation of multiplication.

Example 1.1 A relation R is defined on set (I) of all integers, by aRb, if and only if ab > 0, for all a, b ∈ I. Examine if R is (a) reflexive, (b) symmetric, and (c) transitive.

Solution:

a) Let a ∈ I. Then a.a > 0 holds good if a ≠ 0. If a = 0, then a.a = 0. Therefore, aRa does not hold for all a in I. So, R is not reflexive.

b) Let a, b ∈ I. If ab > 0, then ba >0 also. That is, aRb ⇒ bRa. So, R is symmetric.

c) Let a, b, c ∈ I. Let a Rb and bRc both hold good. It can be said that ab > 0 and bc > 0. If we multiply these two, then (ab)(bc) > 0. $ab^2c > 0$. We know b^2 is always > 0. It can be said clearly that ac > 0.

Thus, aRb and bRc ⇒ aRc. So, R is transitive.

Example 1.2 A relation R is defined on a set of integers (I) by aRb if a – b is divisible by 3, for a, b ∈ I. Examine if R is (a) reflexive (b) symmetric, and (c) transitive.

Solution:

a) Let a ∈ I. Then, a – a is divisible 3. Therefore, aRa holds good for all a ∈ I. So the relation R is reflexive.

b) Let a, b ∈ I and aRb hold good. It means a– b is divisible by 3. If it is true, then b – a is also divisible by 3.

(Let a = 6, b = 3, a– b = 3, divisible by 3, b – a = – 3, which is also divisible by 3). So, bRa holds good. Therefore, R is symmetric.

c) Let a, b, c ∈ I and aRb, bRc hold good. It means a – b and b – c both are divisible by 3. Therefore, (a – c) = (a – b) + (b – c) is divisible by 3.

(Let a = 18, b = 12, c = 6, a – b = 6, b – c = 6. Both are divisible by 3. a – c = 12 is also divisible by 3.)

So, aRc holds good. Therefore, R is transitive.

Example 1.3 Consider the relation R = {(a,b), where len(a) = len(b)}. Show that it is an equivalence relation.

Solution: Here, len(a) means the length of string a

a) It is reflexive: as len(a) = len(a).

b) It is symmetric: if len(a) = len(b), then len(b) = len(a) holds good.

c) It is transitive: if len(a) = len(b) and len(b) = len(c), then len(a) = len(c).
Thus, R is an equivalence relation.

Example 1.4 Check whether the set of numbers divisible by 5 is closed under the addition operation.

Solution: Let the set S = {0, 5, 10, 15…}.

$$0 + 5 = 5, \text{ divisible by } 5.$$
$$5 + 10 = 15, \text{ divisible by } 5.$$
$$10 + 15 = 25, \text{ divisible by } 5.$$

All the results are included in the same set S.
So, the set of numbers divisible by 5 is closed under the addition operation.

Example 1.5 Prove that a set of numbers divisible by n is closed under the addition and multiplication operations.

Solution: The set of numbers divisible by n can be represented as S = {0, n, 2n, 3n, 4n,……}.

Addition:

$$0 + n = n$$
$$n + 2n = 3n$$
$$0 + 2n = 2n$$
$$\ldots\ldots\ldots\ldots$$

In general, $(m - p)n + (m - q).n = n(2m - p - q)$. This is divisible by n.
So, a set of numbers divisible by n is closed under addition.

Multiplication:

$$0 \times n = 0$$
$$n \times 2n = 2n^2$$
$$2n \times 3n = 6n^2$$
$$\ldots\ldots\ldots\ldots$$

In general, $(m - p)n \times (m - q).n = (m - p)(m - q)n^2$. This is divisible by n.
So, a set of numbers divisible by n is closed under multiplication.

1.4 Graph and Tree

1.4.1 Graph

A graph is an abstract representation of a set of objects where some pairs of the objects are connected by directed or undirected links. The interconnected objects are called *vertices*, and the links that connect some pairs of vertices are called *edges*. A graph G = (V, E) consists of a set of vertices $V = \{v_1, v_2, v_3, \ldots..\}$ and a set of edges $E = \{e_1, e_2, e_3, \ldots\ldots\}$. The vertices associated with an edge e are called the end vertices of that edge.

As an example, the graph in Fig. 1.5 has 5 vertices (A, B, C, D, and E) and 6 edges (AB, BC, CD, DA, DE, and EA).

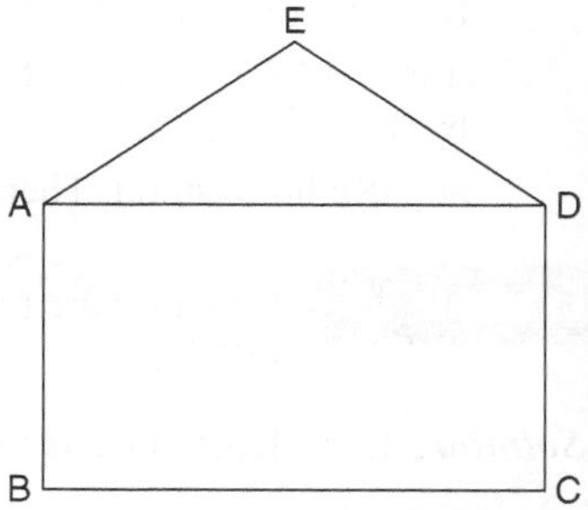

Fig. 1.5 *Example of Graph*

1.4.2 Incident and Degree of a Vertex

The edges meet at a common vertex are called incident on that common vertex

The number of edges incident on a vertex is called the degree of the vertex. If the edge is a self-loop, it will be counted twice.

As an example, for the graph in Fig. 1.5, the edges AE, AB, and AD are incident on the vertex A. The degree of the vertex is d(A) = 3.

It can be proved that the number of odd degree vertices in a graph is always even.

1.4.3 Isolated Vertex, Pendant Vertex

A vertex of a graph having no incident edge is called an isolated vertex.

A vertex of degree one is called a pendant vertex. Sometimes, the pendant vertex is called an end vertex.

In the graph shown in Fig. 1.6, G is called the isolated vertex and F is called the pendant vertex, as it has degree 1.

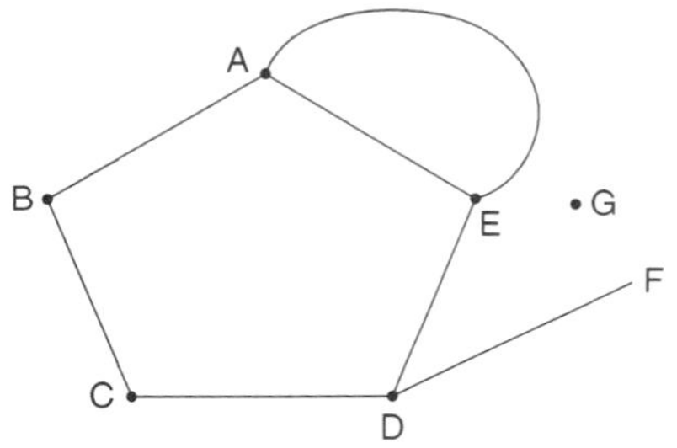

Fig. 1.6 *Example of Isolated and Pendant Vertex*

1.4.4 Walk

Starting from a vertex, traversing some edges and at last ends on some vertex forms a walk. It may be defined as a finite alternating sequence of vertices and edges, beginning and ending with vertices. One important condition of a walk is that no edges are traversed more than once in a single walk, whereas a vertex may be traversed more than once.

For the graph in Fig. 1.6, B(BC)C(CD)D(DF)F is a walk. The label in () represents edge.

If the beginning and end vertices of a walk are the same, then it is called closed walk or else open walk.

1.4.5 Path and Circuit

An open walk in which no vertex appears more than once is called a path.

A closed walk where no vertex appears more than once is called a circuit.

1.4.6 Connected and Disconnected Graph

A graph is called a connected graph if between every pair of vertices there exists at least one path. Else, the graph is called a disconnected graph.

1.4.7 Tree

A tree is defined as an undirected connected graph with no circuit.

A tree has a specially designed vertex called a root. The rest of the nodes could be partitioned into t disjoint sets ($t \geq 0$), where each set represents a tree T_i. i = 0 to t. These are called *sub-trees*.

The number of sub-trees of a node is called the *degree* of the node.

The vertex of a tree having degree 0 is called the *leaf node* or the terminal node. The nodes other than the leaf nodes are called the non-terminal nodes.

In a rooted tree, the *parent* of a vertex is the vertex connected to it on the path to the root; every vertex except the root has a unique parent. A *child* of a vertex v is a vertex of which v is the parent.

The children of a same parent node are called *siblings*.

The *depth* of a node is the length of the path from the root of the tree to the node. The set of all nodes at a given depth is called the *level* of the tree. The root node has depth 0.

The *height* of a tree is the length of the path from the root to the deepest node in the tree.

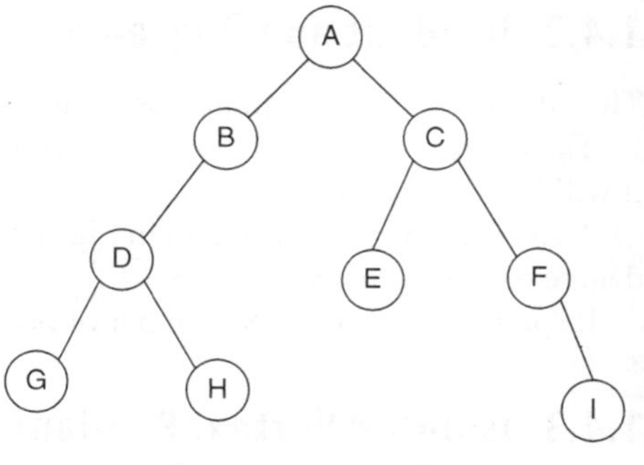

Fig. 1.7 *Example of Tree*

A tree is given in Fig. 1.7. The root node is 'A'. The node C has two sub-trees. So, the node C has a degree.

In the tree, the leaf nodes are G, H, E, and I. For the node E, the parent node is C. E and F are siblings as they are from the same parent C. The height of the three is three. The highest level of the tree is 3.

A *binary tree* is a tree in which each node has at most two children.

A *full binary tree* is a tree in which every node other than the leaves has two children.

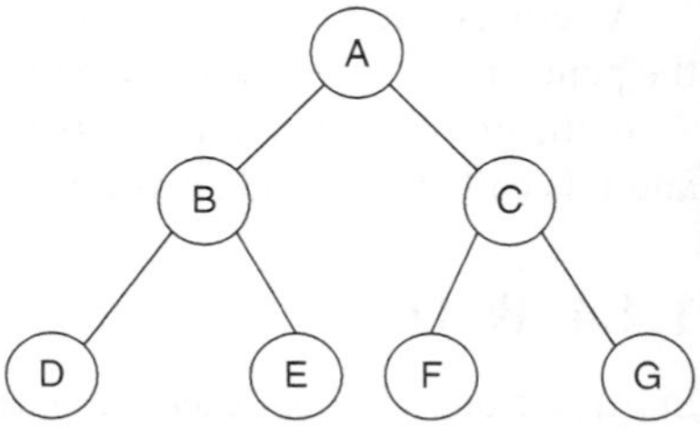

Fig. 1.8 *Full Binary Tree*

The number of nodes in a full binary tree is $n = 2^{h+1} - 1$, where h is the height of the tree.

A full binary tree is given in Fig. 1.8. The height of the tree is 2. The number of nodes is $2^3 - 1 = 7$. If you count the number of nodes, it is 7.

A *complete binary tree* is a binary tree in which every level, except leaves, is completely filled.

The number of nodes of a complete binary tree is minimum: $n = 2^h$ and maximum: $n = 2^{h+1} - 1$, where h is the height of the tree.

1.5 Basics of Digital Electronics

This is the age of electronics. From toy to radio, television, and music player, all are electronics components. The computer hardware is also a complex electronics circuit. Electronics is a branch of technology concerned with the design of circuits using different gates, transistors, microprocessor chips, etc., and with the behaviour and movement of electrons in a semiconductor, conductor, vacuum, or gas. By definition, circuits are closed paths used to describe many situations. An electronics circuit is a closed path consisting of a series of electronic components through which electricity flows to perform some operations.

Signals are of two types, analog and digital. An analog signal is a fluctuating voltage which can have any numerical value, from tiny fractions of a volt to a hundreds of volts. It may be a constant voltage or rapidly changing. It is a continuous signal and provides a true reflection of the real world signals.

In contrast to the analog signal, a digital signal can only have one of the two possible values. The exact value of these voltages varies from circuit to circuit. In common, + 5 V is referred to as the digital high and 0 V as the digital low. Though this voltage level is not fixed, it has a certain tolerance.

The electronic gate receives input in the form of voltage and produces output also in the form of voltage. As discussed earlier, the value of the voltage varies from circuit to circuit and from device to device. As gates are digital components, all these voltage ranges are restricted to two values 'high' and 'low'. In digital electronics, 'high' is represented as '1' and low is represented as '0'. The electronic gate acts as a switching device which either permits flow of current or blocks it. The basic gates are AND, OR, NOT, and XOR.

The **AND gate** has two or more inputs and a single output (Fig. 1.9). If all of the inputs are '1' the output is '1'; otherwise, the output is '0'. For a two-input AND gate, with input labels 'A' and 'B', the output function is written as T = AB. A two-input AND gate is represented by the following symbol and the truth table is as follows.

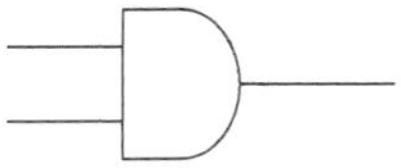

A	B	O/P
0	0	0
0	1	0
1	1	1
1	0	0

Fig. 1.9 *The AND Gate with Truth Table*

The **OR gate** also has two or more inputs and a single output (Fig. 1.10). If any of the inputs is '1', the output is '1'; otherwise, it is '0'. For a two-input OR gate, with input labels 'A' and 'B', the output function is written as T = A + B. A two-input OR gate is represented by the following symbol and the truth table is as follows.

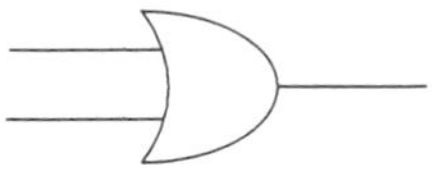

A	B	O/P
0	0	0
0	1	1
1	1	1
1	0	1

Fig. 1.10 *The OR Gate with Truth Table*

The **NOT gate** has a single input and a single output (Fig. 1.11). The function of the NOT gate is to reverse the input. The output function of the NOT gate is represented as T = A, where 'A' is the input. The symbol and truth table of the NOT gate is given as follows.

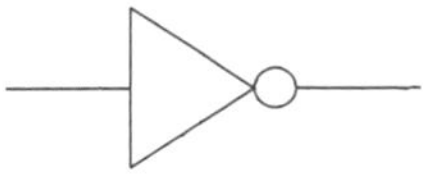

A	O/P
0	1
1	0

Fig. 1.11 *The NOT Gate with Truth Table*

The exclusive-OR, in short **XOR,** gate is a complex gate (Fig. 1.12). A two-input XOR gate gives the output '1' when its two inputs are different and the output '0' when the inputs are the same. For two inputs 'A' and 'B', the output functions are represented by T = A ⊕ B = AB + AB.

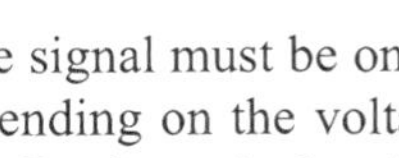

A	B	O/P
0	0	0
0	1	1
1	1	0
1	0	1

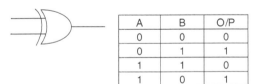

Fig. 1.12 *The XOR Gate with Truth Table*

1.6 Digital Circuit

A digital circuit is a circuit using logic gates where the signal must be one of two discrete levels. Each level is interpreted as one of two different states depending on the voltage level (on/off, 0/1 or true/false). The digital circuit is operated by the logic of the Boolean algebra. This logic is the foundation of digital electronics and computer processing.

Depending on the output function, digital circuits are divided into two groups:

1. **Combinational circuits:** The circuits where the output depends only on the present input, i.e., output is the function of only the present input, are called combinational circuits.

   O/P = Func.(Present I/P)

2. **Sequential circuit:** The circuits where the output depends on the external input and the stored information at that time, i.e., output is the function of external input and the present stored information, are called sequential circuits.

   O/P = Func.(External I/P and Present stored information)

The difference between the sequential and combinational circuit is that sequential circuit has memory in the form of flip flop, whereas combinational circuit does not have the memory element. A general block diagram of sequential circuit is shown in Fig. 1.13.

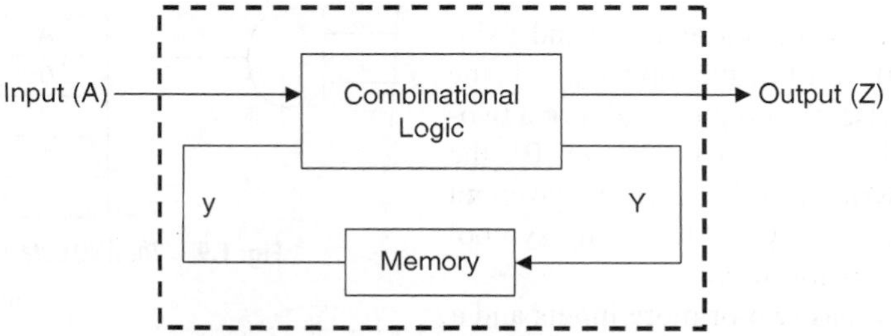

Fig. 1.13 *Block Diagram of Sequential Circuit*

Sequential circuits fall into two classes: synchronous and asynchronous. Synchronization is usually achieved by some timing device, such as clock. A clock produces equally spaced pulses. These pulses are fed into a circuit in such a way that various operations of the circuit take place with the arrival of appropriate clock pulses. Generally, the circuits, whose operations are controlled by clock pulses, are called synchronous circuit.

The operation of an asynchronous circuit does not depend on clock pulses. The operations in an asynchronous circuit are controlled by a number of completion and initialization signals. Here, the completion of one operation is the initialization of the execution of the next consecutive operation. The following (Fig. 1.14) block diagram is that of a synchronous sequential circuit.

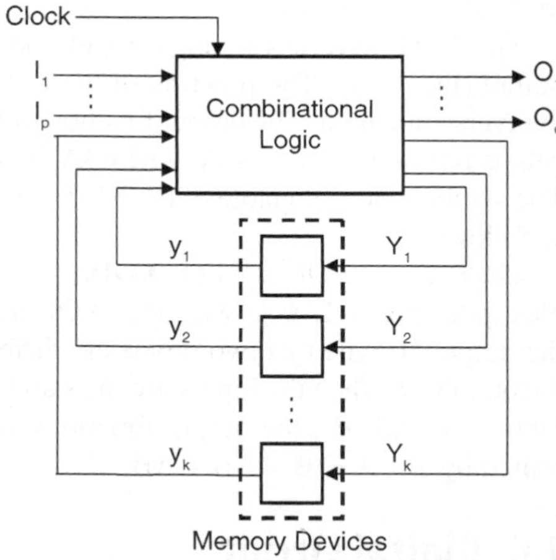

Fig. 1.14 *Synchronous Sequential Circuit*

A synchronous sequential machine has finite number of inputs. If a machine has n number of input variables, the input set consists of 2^n distinct inputs called input alphabet I.

In the figure, the input alphabet is $I = \{I_1, I_2, \ldots\ldots, I_p\}$.

The number of outputs of a synchronous sequential machine is also finite.

A synchronous sequential circuit can be designed by the following process:

- From the problem description, design a state table and a state diagram (whichever is first applicable).
- Make the state table redundant by machine minimization. This removes some states and makes the table redundant.
- Perform state assignment by assigning the states to binary numbers. n binary numbers can assign 2^n states.
- After performing the state assignment, derive a transition table and an output table.
- Derive the transitional function and the output function from the transitional table and the output table.
- Draw the circuit diagram.

The following examples design some synchronous sequential circuit.

Example 1.6 Design a sequential circuit which performs the following:

A	B	O/P	Carry
0	0	0	0
0	1	1	0
1	0	1	0
1	1	0	1

The carry is added with the I/P's in the next clock pulse.

Solution: Let us take two input strings $X_1 = 0111$ and $X_2 = 0101$.

Here, the output at time t_i is a function of the inputs X_1 and X_2 at the time t_i and of the carry generated for the input at t_{i-1}.

O/P = func.(I/P at t_i and the carry generated for the input at t_{i-1})

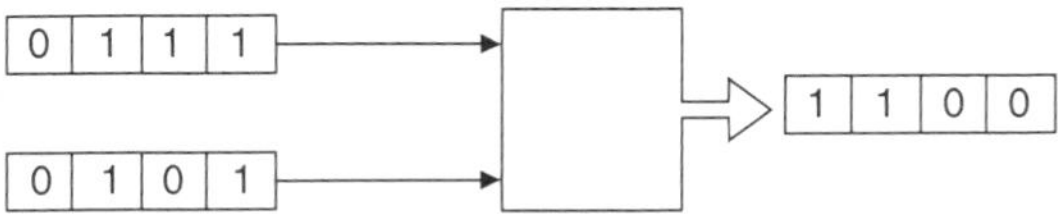

Fig. 1.15

Therefore, this is a sequential circuit (Fig. 1.15).

If we look into the previous table, we will see two types of cases arisen there. These are

1. The producing carry '0'
2. The producing carry '1'.

We have to consider this as the O/P depends of the carry also.

Let us take the cases as states. So, we can consider two states, A for (1) and B for (2). If we construct a table for the inputs X_1 and X_2 by considering the states, it will become

	Next State, O/P(Z)			
Present State	$X_1X_2 = 00$	$= 01$	$= 11$	$= 10$
A	A, 0	A, 1	B, 0	A, 1
B	A, 1	B, 0	B, 1	B, 0

This type of table is called the state table.

We can make this tabular form more clear by a graphical notation (Fig. 1.16).

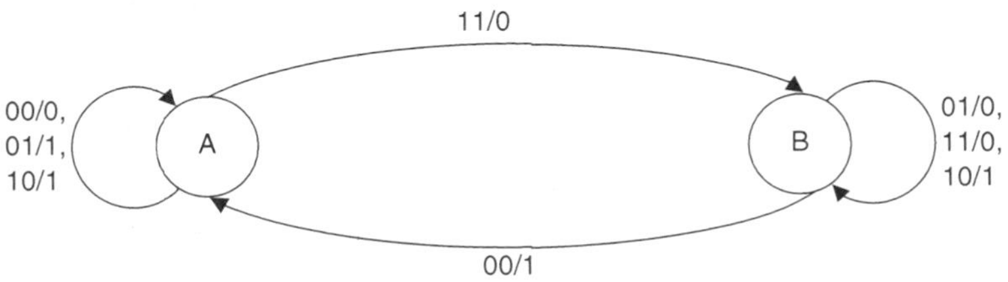

Fig. 1.16 *State Diagram for the Sequential Circuit*

This type of graph is called state graph or state diagram.

For designing a circuit, we need only '0' and '1', i.e., Boolean values. So, the states A and B must be assigned to some Boolean number. As there are only two states A and B, only one-digit Boolean value is sufficient. Let us represent A as '0' and B as '1'.

By assigning these Boolean values to A and B, the modified table becomes

	Next State, (Y)				O/P(Z)			
Present State (y)	$X_1X_2 = 00$	$= 01$	$= 11$	$= 10$	$= 00$	$= 01$	$= 11$	$= 10$
0	0	0	1	0	0	1	0	1
1	1	1	1	1	1	0	1	0

The function for next state $\quad Y = X_1X_2 + X_{1y} + X_{2y}$

The function for output $\quad Z = X_1'X_{2y}' + X_1'X_{2y'} + X_1X_{2y'}' + X_1X_{2y}$

$$= X_1 \oplus X_2 \oplus y.$$

From this function, the digital circuit is designed as denoted in Fig.1.17.

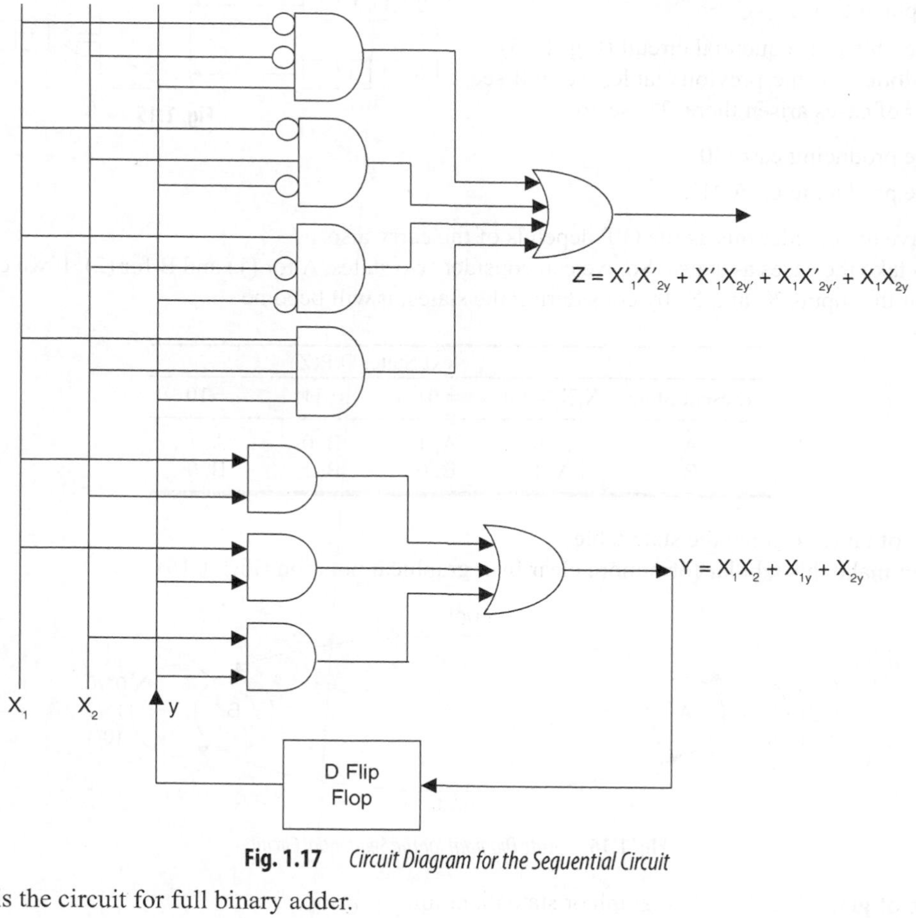

Fig. 1.17 *Circuit Diagram for the Sequential Circuit*

This is the circuit for full binary adder.

Example 1.7 Design a full binary substractor.

Solution: The truth table for a substractor is given in the following table:

A	B	O/P	Borrow
0	0	0	0
0	1	1	1
1	0	1	0
1	1	0	0

Let us take two input strings $X_1 = 1010$ and $X_2 = 0111$.

Here, the output at time t_i is a function of the inputs X_1 and X_2 at the time t_i and of the borrow generated for the input at t_{i-1}.

O/P = func.(I/P at t_i and the carry generated for the input at t_{i-1})

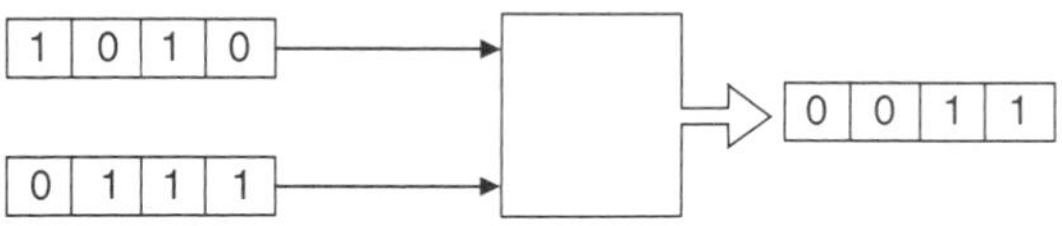

Fig. 1.18

Therefore, this is a sequential circuit (Fig. 1.18).

If we look into the previous table, we will see two types of cases arisen there. These are

1. The producing borrow '0'
2. The producing borrow '1'.

We have to consider this, as the O/P depends on the borrow also.

Let us take the cases as states. So, we can consider two states, S_1 for (1) and S_2 for (2). If we construct a table for the inputs X_1 and X_2 by considering the states, it will become.

	Next State (Y), O/P(Z)			
Present State	$X_1X_2 = 00$	$= 01$	$= 11$	$= 10$
S_1	S_1, 0	S_2, 1	S_1, 0	S_1, 1
S_2	S_2, 1	S_2, 0	S_2, 1	S_1, 0

This is the state table for binary substractor.

The state graph for the binary substractor is shown in Fig. 1.19.

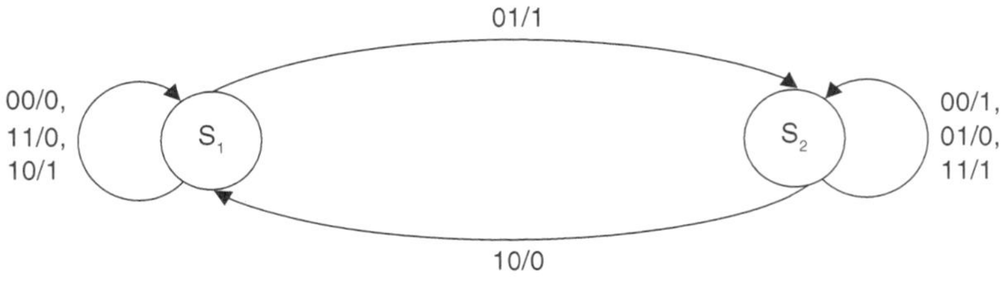

Fig. 1.19 *State Diagram for the Binary Substractor*

State Assignment: For designing a digital circuit, we need to convert S_1 and S_2 into some digital number. There are only two states, and so a 1-bit digital number which can produce two types of digital values 0 or 1 is sufficient to represent S_1 and S_2.

Let us represent S_1 as 0 and S_2 as 1.

The modified state table is

	Next State, (Y)				O/P(Z)			
Present State (y)	$X_1X_2 = 00$	= 01	= 11	= 10	= 00	= 01	= 11	= 10
0	0	1	0	0	0	1	0	1
1	1	1	1	0	1	0	1	0

The function for next state $\quad Y = X_1'X_2 + X_1'y + X_2y$

The function for output $\quad Z = X_1'X_2'y + X_1'X_2y' + X_1X_2'y' + X_1X_2y$

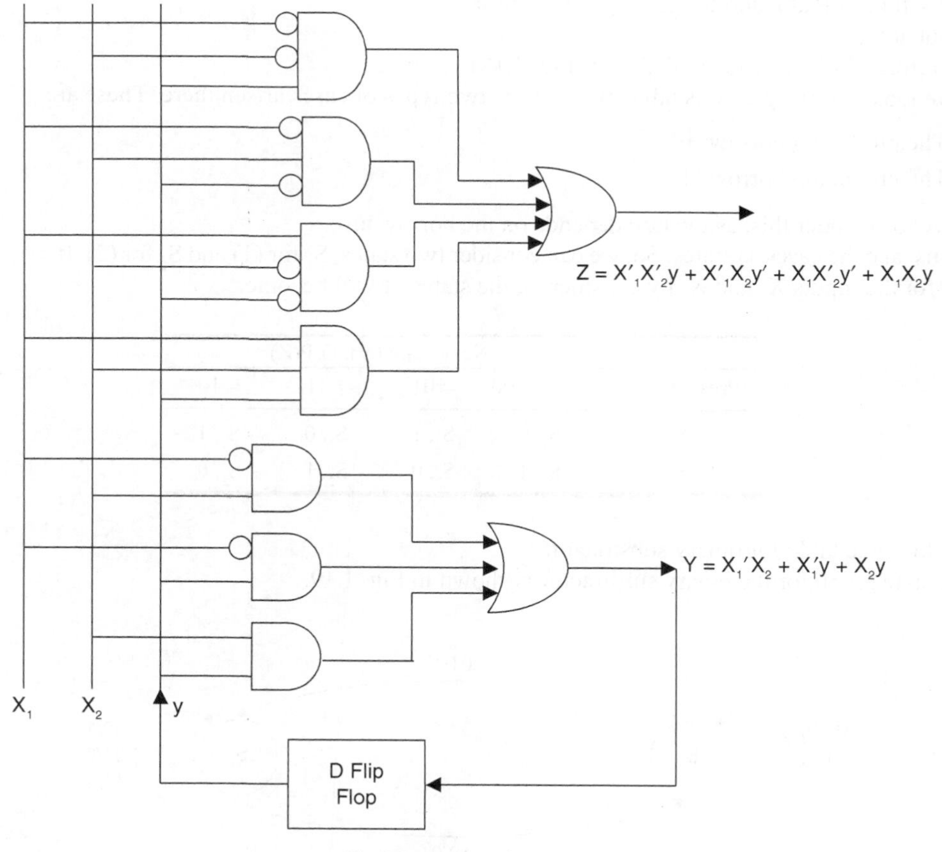

Fig. 1.20 *Circuit Diagram*

1.7 Basics of Automata Theory and Theory of Computation

The term 'automata' has some typical pronounceable symmetry with the term 'automatic'. In a computer, all the processes appear to be done automatically. A given input is being processed in the CPU, and an output is generated. We are not concerned about the internal operation in the CPU. We are only concerned about the given input and the received output. However, in reality, the input is converted to '0' and '1' and assigned to the process for which the input was given. It then performs some internal operation in its electronic circuit and generates the output in '0' and '1' format. The output generated in the '0' and '1' format is converted to a user-understandable format and we get the output. From the discussion, it is clear that the CPU performs machine operations internally. In automata, we shall learn about how to design such machines. To get a clear idea about automata theory, we need to travel back to the history of computer science. The engineering branch 'electrical engineering' stands on the base of physics. Electronics engineering came from electrical engineering. The hardware of a computer is a complex electronics circuit. The circuit understands only binary command, which is impossible for a human user to memorize. Then, mathematics extends its hand for developing the logic. Mathematics and electronics have jointly given birth to a new branch of engineering called 'computer science and engineering'.

The name of the subject is formal language and automata theory. We have a basic idea about the automata theory. Now what is formal language? Let us discuss what language is. Language is a communication medium through which two persons communicate. For every nation, there is some language to communicate such as Hindi, English, Bengali, etc. It is the same for communicating with a computer, wherein the user needs some language called the programming language. C, C++, and Java are some examples of programming languages. The characteristics of these types of languages are that they are similar to the English language and are easily understandable by the user. However, the computer does not understand English-like statements. It understands only binary numbers. Therefore, the computers have a compiler that checks the syntax and acts as a converter between English-like statements to binary numbers and vice versa. But, to design the compiler, some logic is needed. That logic can be designed by using mathematics. For each language, there is a grammar. Grammar is the constructor for any language. Similarly, the languages that are used for computer programming purpose have some grammar to construct them. These rules and grammar and the process to convert these grammar and languages to machine format are the main objectives of automata theory.

Any problem which can be modeled to be solved by a computer is called a computational problem. On the contrary, non-computational problems cannot be solved by any computer. As an example, 'what will be the result of the forthcoming cricket match?' In computer science, we consider only computational problems.

1.8 History of Automata Theory

Theoretical computer science or automata theory is the study of abstract machines and the computational problems related to these abstract machines. These abstract machines are called automata. The origin of the word 'automata' is from a Greek word (Αυτόματα) which means that something is doing something by itself.

Around 1936, when there were no computers, *Alen Turing* proposed a model of abstract machine called the Turing machine which could perform any computational process carried out by the present day's computer. A formal version of the finite automata model was proposed in 1943 in the *McCulloch-Pitts* neural network models. Finite automata and finite state machines got a complete form by the efforts of *George H. Mealy* at the Bell Labs and *Edward F. Moore* in IBM around 1960. The machines are named as the Mealy and Moore machines, respectively. In mid-1950, *Noam Chomsky* at the Harvard University started working on formal languages and grammars. Grammar is a set of rules that are performed to produce sentences in the language. Chomsky structured a hierarchy of formal languages based on the properties of the grammars required to generate the languages. The Chomsky hierarchy drives the automata theory one step forward. In 1971, *Stephen Arthur Cook* in his research paper 'The Complexity of Theorem Proving Procedures' formalized the notions of polynomial-time reduction and NP-completeness. He separated those problems that can be solved efficiently by the computer from the set of problems that can be solved but taking much time. These second classes of problems are called NP-hard.

This subject is sometimes called the 'Theory of Computation' because it includes all of these rules for constructing a computer language and converts them into machine format. That is the theory of computer science. Basically, formal language and automata theory and theory of computation are different names for a single subject that covers all the aspects of the theoretical part of computer science.

1.9 Use of Automata

It is unfruitful to study a subject without knowing its usefulness. The subject 'automata theory' is called one of the core subjects of computer science. Some of the uses of this subject in different fields of computer science and engineering them are.

- The main usefulness of the automata theory is in the compiler design. The first phase of compiler called lexical analyser is designed by the finite automata. In syntax analysis part, a common error called syntax error is checked by the rules of context-free grammar.
- The working principle of the software used for checking the behaviour of digital circuits is based on the automata theory.
- The software for natural language processing (NLP) takes the help of the automata theory.
- Automata is useful for designing the software for counting the occurrence of a particular word, pattern, etc., in a large text.

What We Have Learned So Far

1. A string is defined as a sequence of symbols of finite length.
2. A set is a well-defined collection of objects.
3. The operations union, intersection, difference, complement, Cartesian product etc. can be performed on set.
4. A relation R is called as an equivalence relation on 'A' if R is reflexive, symmetric, and transitive.
5. A set is closed (under an operation) if and only if the operation on two elements of the set produces another element of the set.
6. A graph is an abstract representation of a set of objects where some pairs of the objects are connected by directed or undirected links.
7. A tree is an undirected connected graph with no circuit.

8. A binary tree is a tree in which each node has at most two children.
9. A full binary tree is a tree in which every node other than the leaves has two children.
10. The operations of synchronous circuit are controlled by clock pulses. The operations in an asynchronous circuit are controlled by a number of completion and initialization signals. Here, the completion of one operation is the initialization of the execution of the next consecutive operation.
11. The circuit whose output depends on the present state only is called a combinational circuit., i.e., O/P = Func.(Present I/P).
12. If the output is the function of the external input and the present stored information, then the circuit is called a sequential circuit., i.e., O/P = Func.(External I/P and Present stored information).

Solved Problems

1. Prove that $A \cup (B \cap C) = (A \cup B) \cap (A \cup C)$.

Solution: Let $P = A \cup (B \cap C)$ and $Q = (A \cup B) \cap (A \cup C)$.
Consider an element $a \in P$ and $b \in Q$.
As $P = A \cup (B \cap C)$ and $a \in P$, it implies that

$$\begin{aligned} & a \in A \text{ or } a \in (B \cap C) \\ \Rightarrow\ & a \in A \text{ or } \{a \in B \text{ and } a \in C\} \\ \Rightarrow\ & \{a \in A \text{ or } a \in B\} \text{ and } \{a \in A \text{ or } a \in C\} \\ \Rightarrow\ & a \in (A \cup B) \text{ and } a \in (A \cup C) \\ \Rightarrow\ & a \in (A \cup B) \cap (A \cup C) \end{aligned}$$

It is proved that P is a subset of Q.
As $Q = (A \cup B) \cap (A \cup C)$ and $b \in Q$, it implies that

$$\begin{aligned} & b \in (A \cup B) \text{ and } b \in (A \cup C) \\ \Rightarrow\ & \{b \in A \text{ or } b \in B\} \text{ and } \{b \in A \text{ or } b \in C\} \\ \Rightarrow\ & b \in A \text{ or } \{b \in B \text{ and } b \in C\} \\ \Rightarrow\ & b \in A \text{ or } b \in (B \cap C) \end{aligned}$$

It is proved that Q is a subset of P.
As P is a subset of Q and Q is a subset of P, it implies that $P \equiv Q$.
Thus $A \cup (B \cap C) = (A \cup B) \cap (A \cup C)$ is proved.

2. Prove the following identity:
$r(s + t) = rs + rt$

Solution: Let $x \in r(s + t)$,

$$\begin{aligned} \Rightarrow\ & x \in r \text{ and } x \in (s + t) \\ \Rightarrow\ & x \in r \text{ and } (x \in s \text{ or } x \in t) \\ \Rightarrow\ & (x \in r \text{ and } x \in s) \text{ or } (x \in r \text{ and } x \in t) \\ \Rightarrow\ & x \in (rs + rt) \end{aligned}$$

That is $r(s + t) \subseteq$ (is a subset of) $(rs + rt)$.

Let $x \in (rs + rt)$

$$\Rightarrow (x \in rs \text{ or } x \in rt)$$
$$\Rightarrow (x \in r \text{ and } x \in s) \text{ or } (x \in r \text{ and } x \in t)$$
$$\Rightarrow x \in r \text{ and } (x \in s \text{ or } x \in t)$$
$$\Rightarrow x \in r(s + t)$$

That is $(rs + rt) \subseteq r(s + t)$.

So, $r(s + t) = rs + rt$ (proved).

3. Prove that
$(A \cup B) \cap (B \cup C) \cap (C \cup A) = (A \cap B) \cup (B \cap C) \cup (C \cap A)$.

Solution: From the distributive property, we know $A \cup (B \cap C) = (A \cup B) \cap (B \cup C)$.

$$\rightarrow (B \cup A) \cap (B \cup C) \rightarrow B \cup (A \cap C) \rightarrow B \cup (C \cap A)$$

LHS $(A \cup B) \cap (B \cup C) \cap (C \cup A)$

$$\Rightarrow [B \cup (C \cap A)] \cap (C \cup A)$$
$$\Rightarrow [B \cap (C \cup A)] \cup [(C \cap A) \cup (C \cup A)] \quad [\text{As } A \cap A = A]$$
$$\Rightarrow (B \cap C) \cup (B \cap A) \cup (C \cap A)$$
$$\Rightarrow (A \cap B) \cup (B \cap C) \cup (C \cap A) = \text{RHS (proved)}.$$

4. Let R be an equivalence relation in $\{0\}^*$ with the following equivalence classes:
$[\ ]R = \{0\}^0$
$[0]R = \{0\}^1$
$[00]R = \{0\}^2 \cup \{0\}^3 \cup \{0\}^4 \ldots\ldots\ldots\ldots$
Show that R is right invariant.

Solution: From the definition of right invariant, we know that if R is right invariant then for all x, y, and z,

$$xRy \Rightarrow xzRyz.$$

From the given equivalence classes,

$[\ ]_R$ means a null string.

$[0]_R$ has only string 0.

$[00]_R$ has strings 00, 000, 0000, ……, etc.

Three cases may occur:

i) If x and y are null string, i.e., if $x, y \in [\ ]_R$. $xz = z$ and $yz = z$ and zRz holds good.

ii) If $x, y \in [0]_R$, and $z \in [\ ]_R$, then $xz = x$ and $yz = y$. So, $xRy = xzRyz$.
If $z \in [0]_R$, then $xz = 00$ and $yz = 00$, Hence, $xz = yz$, which implies that xzRyz holds good.
If $z \in [00]_R$, then z has at least 2 zeros. This means that xz and yz have at least 3 zeros which belongs to $[00]_R$. This means that xzRyz holds good.

iii) If $x, y \in [00]_R$, then x and y each have at least 2 zeros. z may belong to any one of the three, but xy and yz produce strings of 2 or more zeros. That is, both of them belong to $[00]_R$. This means that xzRyz holds good.

It is clear from the previous discussion that for all the cases xRy $\Rightarrow$ xzRyz. Thus, R is a right invariant.

5. Check whether a set of numbers divisible by n is closed under subtraction and division (except division by 0) operation.

   Solution: A set of numbers divisible by n can be represented as S = {0, n, 2n, 3n, 4n,......}.

   Subtraction:

$$n - 0 = n$$
$$n - 2n = -n$$
$$3n - n = 2n$$
$$\ldots\ldots\ldots\ldots$$

   In general, (m – p)n – (m – q).n = n (q – p). This is divisible by n. So, a set of numbers divisible by n is closed under subtraction.

   Division:

$$2n/n = 2$$
$$3n/2n = 3/2$$
$$\ldots\ldots\ldots\ldots$$

   In general, (m – p)n/(m–q).n = (m–p)/(m–q). This may or may not be divisible by n. So, a set of numbers divisible by n is not closed under division.

Fill in the Blanks

1. For a string, any prefix of the string other than the string itself is called as the ____________ of the string.
2. For a string, any suffix of the string other than the string itself is called as the ____________ of the string.
3. If there are two sets A and B, then their intersection is denoted by ____________.
4. For a set of elements n, the number of elements of the power set of A is ____________.
5. A relation R is said to be ____________, if for two elements 'a' and 'b' in X, if a is related to b then b is related to a.
6. A relation R is said to be ____________, if for a, b, c $\in$ A and if aRb, bRc hold good then aRc also holds good.
7. A relation R is called as an ____________ on 'A', if R is reflexive, symmetric, and transitive.

Answers:

1. proper prefix
2. proper suffix
3. $A \cap B$
4. 2^n
5. symmetric
6. transitive
7. equivalence relation

Exercise

1. Using Venn diagram prove that $A \cup (B \cap C) = (A \cup B) \cap (A \cup C)$.
2. Prove that (a) $A \cup (A^C \cap B) = A \cup B$ and (b) $A \cap (A^C \cup B) = A \cap B$.
3. A relation R is defined as a R b if $a^2 - b^2$ is divisible by 5, where a and b are integers. Prove that R is an equivalence relation.
4. A relation R is defined on set N (natural numbers) such that $R = \{(a, b) \in N \times N$ if 'a' is a divisor of 'b'$\}$. Examine if R is reflexive, symmetric, and transitive.
5. Check whether the relation power defined as a^b, where a and b are natural numbers, is closed or not.

Language and Grammar

2

Introduction

To express ourselves to someone, i.e., to communicate with someone, we need some medium. That medium is language. Hindi, English, Bengali, etc., are all used by people to communicate among themselves. So, these are all languages.

For constructing a language, there are some rules. Without the rules, a language cannot exist. These rules are called the grammar for that language. Without grammar, a language cannot exist.

Formal language is an abstraction of the general characteristics of a programming language.

In computer science, to communicate with the computer hardware, the user needs some languages for programming purposes, such as C, C++, and Java. These are called programming languages. These languages are used for communicate between the computer and the user. So, they can be called as languages. For constructing the languages, there are some rules to be followed. The rules are called the grammar for that programming language.

2.1 Grammar

Grammar is a set of rules to define a valid sentence in any language. A language is not complete without grammar.

Grammar consists of four touples:

$$G = \{V_N, \Sigma, P, S\},$$

where

V_N : Set of non-terminals
Σ : Set of terminals
P : Set of production rule
S : Start symbol

(*Non-terminal* symbols are those symbols which can be replaced multiple times. Terminal symbols are those symbols which cannot be replaced further.)

Let us take a sentence in English:

Bikash goes to college.

This sentence can be represented in grammatical form like this:

<Sentence> → <Subject><Predicate>
<Subject> → <Noun>/<Pronoun>

<Predicate> → <Verb Phase>
<Verb phase> → <Verb>.<Preposition>.<Noun>
<Verb> → goes/eats/runs
<Preposition> → to/at
<Noun> → Ram/Hari/Bikash/College/School
<Pronoun> → I/We/He/She/They

Here,

V_N : {<Sentence>, <Subject>, <Predicate> … .}
Σ : {Ram, Hari, Bikash, He, She … etc.}
P : The production rules given previously
S : <Sentence>

A language is generated from the rules of a grammar. A language contains only terminal symbols ε ∑. Let a language L be generated from a grammar G. The language is written as L(G) and read as the language generated by the grammar G. The grammar is called G(L) and read as the grammar for the language L.

The production rules of a grammar consist of two parts: left hand side (LHS) and right hand side (RHS). The LHS of a production rule mainly contains the non-terminal symbols to be replaced. It may contain terminal and non-terminal both for some cases but at least one non-terminal. The RHS of a production rule may contain terminal, non-terminal, or any combination of terminal and non-terminal even null. To be a valid grammar, at least one production must contain the start symbol S at its LHS.

In most of the cases, a grammar is represented by only the production rules because the production rules contain a set of non-terminals, a set of terminals, and obviously the production rules and the start symbol. Sometimes in production rules, two or more rules are grouped into one. For those cases, the LHS of the production rules must be same. The RHS rules are grouped separated by '/' under the same LHS. As an example,

$$A \rightarrow aA$$
$$A \rightarrow bCa$$
$$A \rightarrow ab$$

are grouped as A → aA/bCa/ab.

Example 2.1 Construct the language generated from the given grammar:

$$S \rightarrow aS/\varepsilon$$

Solution: The grammar consists of one non-terminal symbol S and one terminal symbol a. (ε is a null alphabet and not treated as terminal.) The start symbol S generates two strings ε and aS. Between these, ε belongs to the language set generated by the grammar as it does not contain any non-terminal symbol.

S in aS is a non-terminal, and so it can be replaced by any of ε or aS as there is a production rule **S → aS/ε**. This replacing generates two strings, namely aaS and a (a ε means a). 'a' is only a terminal symbol, and so it also belongs to the language set generated by the grammar. By this process, aa, aaa, aaaa.... also belong to the language set. Let this iteration continue for n steps. Up to (n – 1) step, the string was $a^{(n-1)}S$. In the nth step, S is replaced by ε producing a^n.

In the general form, the language generated by the grammar is a^n, where $n \geq 0$. (If n = 0, the null string is produced.)

$$L(G) = a^n, n \geq 0$$

The process is described in Fig. 2.1.

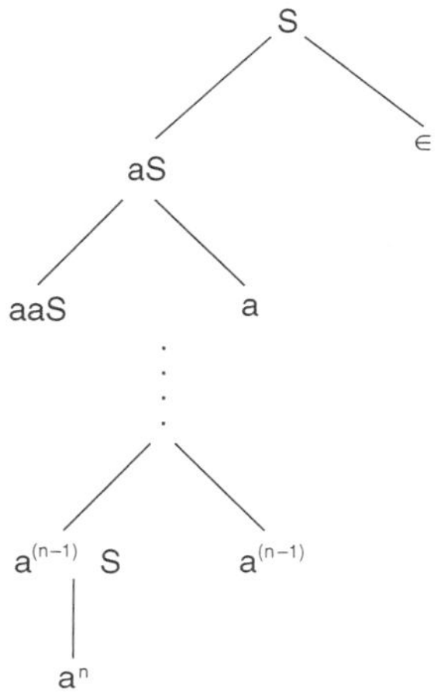

Fig. 2.1

Note: *a^n means n number of 'a'. $(ab)^n$ means n number of 'ab'. a^nb^n means n number of 'a' followed by n number of 'b'. $(ab)^n$ and a^nb^n are not the same because we are dealing here with strings.*

Example 2.2 Construct the language generated from the given grammar:

$$S \rightarrow aSb/\varepsilon$$

Solution: The grammar consists of two terminal symbols a and b. Here $S \rightarrow aSb$ and $S \rightarrow \varepsilon$ are combined into a single production. S is the start symbol. S is also a non-terminal symbol. So, S can be replaced. S can be replaced by two types of strings, aSb and ε. In aSb, there is a non-terminal symbol S. So, aSb is not in the language set produced from the grammar. ε will be in the language set.

In aSb, there is a non-terminal S. So, S can be replaced by either ε or aSb. If S is replaced by ε, it will be ab, which consists of only terminals. So, ab belongs to the language set.

By this process, the strings aabb, aaabbb will be in the language set. After the nth iteration, the language will be a^nb^n, where the value of n will be 0,1,2,….n.

[n = 0 means a^0b^0 means there is no a and no b in the string, which means a null string (Λ).]

So, the language generated from the grammar is

$$L(G) = a^nb^n \ n \geq 0.$$

This is described in Fig. 2.2.

(If the grammar was $S \rightarrow aSb/ab$, the language will be $L(G) = a^nb^n$, $n > 0$.)

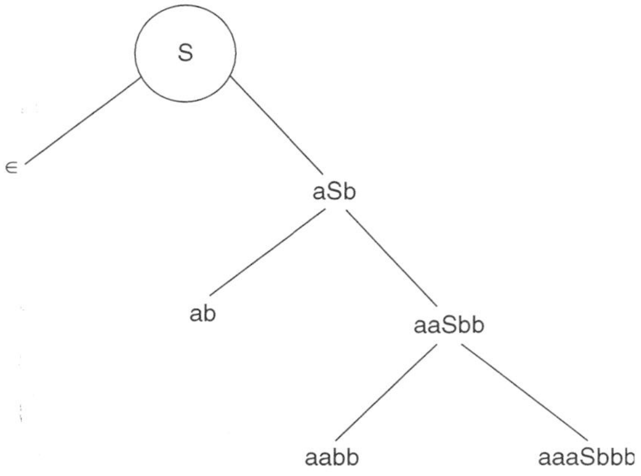

Fig. 2.2

Example 2.3 Construct the language generated by the grammar:

$$S \rightarrow aCa$$
$$C \rightarrow aCa/b$$

Solution: In the grammar, there are two terminal symbols 'a' and 'b'. From the start symbol S, the control reaches to C and, then at the last step, C is replaced by b to form the language. S is replaced only one time and C can be replaced (n – 1) times producing $a^nC\ a^n$. At the last step, C is replaced by 'b'. The language generation process is shown in Fig. 2.3.

$$S \Rightarrow aCa \Rightarrow aaCaa$$
$$\vdots$$
$$\Rightarrow a^nCa^n$$
$$\Rightarrow a^nba^n$$

Fig. 2.3

In the language generated by the grammar, there will be at least two 'a' (the two a's are coming from the first replacement of S). Null does not belong to the language set.

The language generated by the grammar $L(G) = a^nba^n$, $n > 0$.

Example 2.4 Construct the language generated by the grammar:

$$S \rightarrow AbB$$
$$A \rightarrow aA/a$$
$$B \rightarrow aB/a$$

Solution: The start symbol S generates another two non-terminal symbols, A and B, separated by a terminal symbol b. A can be replaced n times whereas B can be replaced m times as they are independent on each other. The language generation process is described in Fig. 2.4.

$$S \Rightarrow AbB \Rightarrow aAbB \Rightarrow aaAbB$$
$$\vdots$$
$$\Rightarrow a^nbB \Rightarrow a^nbaB \Rightarrow a^nbaaB$$
$$\vdots$$
$$\Rightarrow a^nba^m$$

Fig. 2.4

In the language set generated by the grammar, there will be at least one 'a' in both the sides of 'b'. But it may or may not be possible that in both the sides of 'b' A and B are replaced in same number because they are different non-terminals.

So, the language generated by the grammar becomes

$$L(G) = a^nba^m,\ m, n > 0$$

Example 2.5 Construct the language generated from the given grammar:

$$S \rightarrow aS/bS/\varepsilon$$

Solution: In this production rule, there are three productions, $S \rightarrow aS$, $S \rightarrow bS$, and $S \rightarrow \varepsilon$. In ε, there is no non-terminal, and so in the language set there is a null string (Λ). S can be replaced by aS, bS, or ε. If S is replaced by aS in aS, it will be aaS. If S is replaced by bS, it will be abS. If S is replaced by aS in bS, it will be baS. If S is replaced by bS in bS, it will be bbS. If S is replaced by ε in aS and bS it will be a and b, respectively, which will belong to the language set. By this process, we will get aa, bb, ab, ba, …, aba, bab....abaaba, …. etc.

In a single statement, we can represent the language set as 'any combination of a, b including null'.
In mathematics, it is represented as $L(G) = \{a, b\}^*$
(Any combination of a, b excluding null is represented by $\{a, b\}^+$)

Example 2.6 Construct the language generated from the given grammar:

$$S \rightarrow aSa/bSb/C \text{ (C is also a terminal symbol)}$$

Solution: From the start symbol, C is generated which is a terminal symbol. So, C belongs to the language set. From S, two another production rules are generated, namely aSa and bSb. In aSa, there is a non-terminal S. So, S can be replaced by aSa or bSb or C. If S is replaced by these production rules, the strings will be aaSaa, abSba, and aCa, respectively. Similarly, for bSb, the strings will be baSab, bbSbb, and bCb. Among these, aCa and bCb will be in the language set generated from the grammar.

By this process, we will get aaCaa, abCba, baCab, bbCbb...abaCaba,...., ababbCbbaba and so on.

If we look into the string, we will see that before C there is any combination of 'a' and 'b' including ε (For the string C, there is no a and no b), and after C the string is the reverse of the previous string. If we take the string before C as W, the string after C will be W^R.

The language generated from the grammar will be

$$L(G) = WCW^R, \text{ where } W \varepsilon (a, b)^*$$

The process is described in Fig. 2.5.

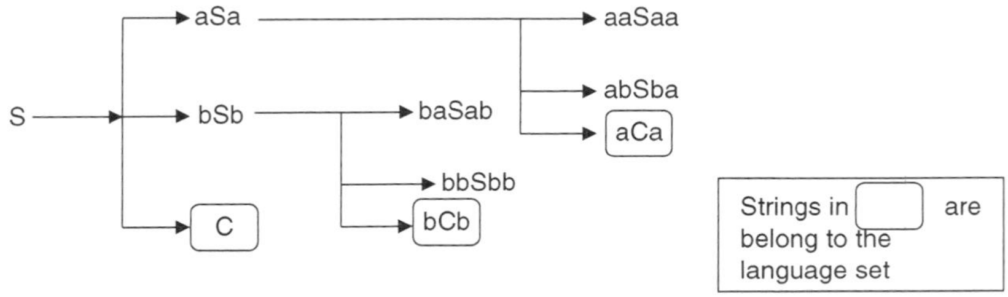

Fig. 2.5

Example 2.7 Construct the language generated by the given grammar:

$$S \rightarrow aSAc/abc$$
$$bA \rightarrow bb$$
$$cA \rightarrow Ac$$

Solution: The start symbol S generates two strings; among them, abc is the string of terminals. So, abc is the smallest string that belongs to the language set generated by the grammar. Another production is **S → aSAc**. The production's RHS contains two non-terminal symbols S and A. S can be replaced (n – 1) times producing $a^{(n-1)}S\ (Ac)^{(n-1)}$. At the last step, if S is replaced by abc, it will produce $a^n bc(Ac)^{(n-1)}$.

Here, we can easily apply cA → Ac (n–1) times. This generates $a^n bA^{(n+1)}c^n$.

In this phase, the production **bA → bb** can easily be applied (n – 1) times producing $a^n b^n c^n$. The minimum length string is abc. So, the language generated by the grammar is

$$L(G) = a^n b^n c^n, n \geq 1$$

This is described in Fig. 2.6.

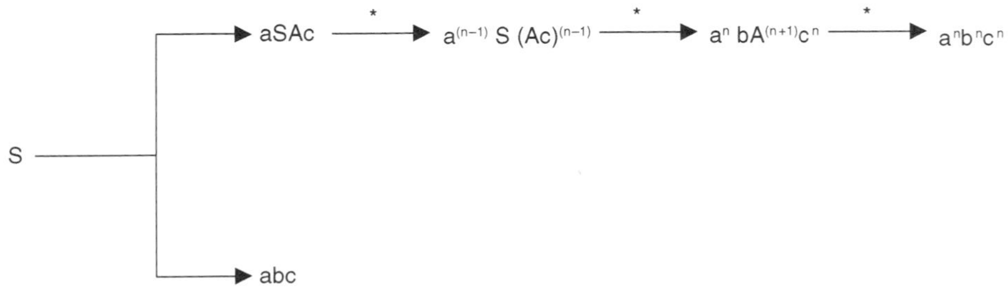

Fig. 2.6

Example 2.8 Find the language generated by the given grammar:

$$S \rightarrow aS/A$$
$$A \rightarrow bB/b$$
$$B \rightarrow cC/\varepsilon$$
$$C \rightarrow \varepsilon$$

Solution: The grammar consists of three terminal symbols, namely 'a', 'b', and 'c'. The start symbol S is replaced by aS or A. If it is replaced by aS, then S can again be replaced by aS or A. If S is once replaced by A, then the control goes to the production of A. So, $S \rightarrow A$ is the window to switch control from S to A. After the nth iteration of $S \rightarrow aS$, the string becomes a^nS.

In this stage, S is replaced by A with the two productions $A \rightarrow bB/b$ generating a^nbB or a^nb. The last one consists of only the terminal symbol, and so it belongs to the language set generated by the grammar. Here, it can be written as $L(G) = a^nb, n \geq 0$. (If S is replaced by A in the first step and A is replaced by 'b', the least length string becomes b.)

There is one branch in the left which consists of a non-terminal B. B can be replaced by either cC or ε, which generates a^nbcC or a^nb. The last one is already taken. 'C' in the first one that can be replaced by ε producing a^nbc. It is a string which consists of only terminal symbols. So, it belongs to the language set generated by the grammar. It can be written as $L(G) = a^nb, n \geq 0$. As G generates two strings, the language is written as

$$L(G) = \{a^nb, n \geq 0\} \cup \{a^nbc, n \geq 0\}$$

Example 2.9 Construct the grammar for the language $a^n, n > 0$.

Solution: The language consists of any number of 'a'. If n is replaced by 1, 2, 3, and so on, the strings are a, aa, aaa.........$a^{(n-1)}$, a^n.

From the previous list, it is clear that in the language set there is at least one 'a'.

At the time of constructing the rules, it must be kept in mind that the rules must generate all the strings generated by the language set. In this case, it is [a, aa, aaa.........$a^{(n-1)}$, a^n]. The least length string is a. So, there must be a production $S \rightarrow a$. Now, there is a chance to generate other strings such as aa, aaa, etc. If it is thought that a non-terminal is attached with S, which is replaced again and again and at last replaced by 'a' to generate the string, then the construction process will be easier. Take the non-terminal as S, and so S can be replaced multiple times. Hence, the grammar for the language is

$$S \rightarrow aS/a$$

In the form of $\{V_N, \Sigma, P, S\}$, the grammar is

$$V_N : \{S\}, \Sigma : \{a\}, P : \{S \rightarrow aS/a\}, S : \{S\}$$

But, writing the production rules are sufficient in describing a grammar.

(Note: If the language is $a^n, n \geq 0$, then the production rules are $S \rightarrow aS/\varepsilon$.)

Example 2.10 Construct the grammar for the language $(ab)^n, n > 0$.

Solution: The language consists of any number of ab (here we can think of ab as a single character).

The grammar for the language is

$$S \rightarrow abS/ab$$

Example 2.11 Construct the grammar for the language a^nb^n, $n > 0$.

Solution: From the string, we can say that

1. The language consists of a and b.
2. In the language, a will appear before b.
3. The number of 'a' and of 'b' are the same in any string that belongs to the language set.
4. In the language, there is no null string.

The value of n is greater than 0. The lowest value of n is 1 here. If n = 1, the lowest length string that is produced from the language is ab, which is produced from the start symbol S. So, in the grammar, there will be a production $S \rightarrow ab$.

In the language set, there are strings such as aabb, aaabbb….., a^nb^n.

Every time, one 'a' and one 'b' is added in the string for n = 1, 2, 3….. It can be thought like this—in the middle there is a non-terminal, replacing which it is adding one a at the left and one b at the right. And, at last replacing that non-terminal by ab, it is producing aabb, aaabbb …….. .

So, the another production is

$$S \rightarrow aSb$$

So, the grammar becomes

$$S \rightarrow aSb/ab,$$

where $V_N : \{S\}, \Sigma : \{a, b\}, P : S \rightarrow aSb/ab, S : \{S\}$

Example 2.12 Construct a grammar for the language a^nb^{n+1}, $n > 0$.

Solution: The number of 'b' is one more than the number of 'a'. As n > 0, in the language set, there is at least one 'a' and at least two 'b'(as the number of 'b' is one more than the number of 'a'). The least string is abb. For $\mathbf{a^nb^n}$, the production rule was $S \rightarrow aSb/ab$. The modified production $S \rightarrow aSb/abb$ generates the language.

The grammar is $\mathbf{S \rightarrow aSb/abb}$.

This can be constructed by importing one extra non-terminal as described in the following.

For $\mathbf{a^nb^n}$, the production rule was $S \rightarrow aSb/ab$. If one extra 'b' is added in the first step, then it will be prepared for generating $\mathbf{a^nb^{n+1}}$. But, here, the problem is in 'S' at the RHS. Each replacement of S at the RHS adds one extra 'b' to the language set; but the extra 'b' must be limited to the first replacement only. If, in the first production's RHS, one extra non-terminal, say 'A', is fetched and the production for A is made $A \rightarrow aAb/\varepsilon$, then there is no problem.

The grammar for the language is

$$S \rightarrow aAbb$$
$$A \rightarrow aAb/\varepsilon$$

The grammar can be written in another way

$$S \rightarrow aSb/aAb$$
$$A \rightarrow aAb/b$$

Example 2.13 Construct the grammar for the language $a^n b^{n+1}$, $n \geq 0$.

Solution: The number of 'b' is one more than the number of 'a'. As $n \geq 0$, the number of 'a' may be zero. If the number of 'a' is zero, then the number of 'b' is one. That means, in the language set, there exists at least one 'b'.

The grammar for the language is

$$S \rightarrow aAbb/b$$
$$A \rightarrow aAb/\varepsilon$$

The grammar can be written in another way as

$$S \rightarrow aSb/b$$

Example 2.14 Construct the grammar for the language $a^n\ c^i\ b^n$ $n, i \geq 0$.

Solution: From the string, we can say that

1. The language consists of a, b, and c.
2. In the language set, a comes before c, and c comes before b.
3. The number of a and the number of b are the same; the number of c may be different.
4. A null string may exist in the language set.

In the language set, the number of a and the number of b are same. That means, each time a and b come simultaneously. For $a^n b^n$, we have seen that there is a production rule $S \rightarrow aSb$. In the string, c appears in between a and b by using the production rule $S \rightarrow Sc/\varepsilon$. If c directly comes from S, then c may come after b, if someone does like this

$$S \rightarrow Sc \rightarrow aSbc \rightarrow abc$$

But abc does not belong to the language set.

From S, c will not come directly. We have to introduce another non-terminal A which generates c^i between a and b. If $n = 0$, the language set consists of only c. The production rule $S \rightarrow A$ opens the provision of generating c only.

The production rules are

$$S \rightarrow aSb/A$$
$$A \rightarrow Ac/\varepsilon$$

Example 2.15 Construct the grammar for the language $a^n\ c^i\ b^n$ $n > 0, i \geq 0$.

Solution: In the language set, c may be null. But there will be at least one a and one b. The production $S \rightarrow aAb$ opens the possibility of replacing A by a string of c and fulfills the condition of a and b becoming not null.

The grammar is

$$S \rightarrow aSb/aAb$$
$$A \rightarrow Ac/\varepsilon$$

Example 2.16 Construct the grammar for the language $a^n\ c^i\ b^n$ $n \geq 0, i > 0$.

Solution: In the language set, a and b may be null but there will be at least one c. The production $A \rightarrow Ac/c$ generates strings of c fulfilling the condition that the number of c is not null.

The grammar is

$$S \rightarrow aSb/A$$
$$A \rightarrow Ac/c$$

Example 2.17 Construct the grammar for the language $a^n c^i b^n$ $n > 0, i > 0$

Solution: A string that belongs to the language generated by the grammar contains at least one a, b, and c.

The grammar for the language is

$$S \rightarrow aSb/aAb$$
$$A \rightarrow Ac/c$$

Example 2.18 Construct a grammar generating the language $a^n b^n c^i$, where $n \geq 1, i \geq 0$.

Solution: The language consists of two parts: a^nb^n, where $n \geq 1$, and c^i, where $i \geq 0$. From the start symbol S, we need to take two non-terminals A and B, where A generates a^nb^n with $n \geq 1$ and B generates c^i with $i \geq 0$.

The grammar for the language is

$$S \rightarrow AB$$
$$A \rightarrow aAb/ab$$
$$B \rightarrow Bc/\varepsilon$$

Example 2.19 Construct a grammar generating the language $a^i b^n c^n$, where $n \geq 1, i \geq 0$.

Solution: The language consists of two parts: b^nc^n, where $n \geq 1$, and a^i, where $i \geq 0$. So, from the start symbol S, we need to take two non-terminals A and B, where A generates a^i with $i \geq 0$ and B generates b^nc^n with $n \geq 1$.

The grammar for the language is

$$S \rightarrow AB$$
$$A \rightarrow Aa/\varepsilon$$
$$B \rightarrow bBc/bc$$

Example 2.20 Construct the grammar for the language $a^l b^m c^n$, where one of l, m, or n = 1 and the remaining two are equal.

Solution: In the language set, at least one of 'a', 'b', or 'c' will exist in single. 'a' and 'c' are terminal elements. If in the language set there is single 'a', the production rule is $S \rightarrow aS_1$ and $S_1 \rightarrow bS_1c/\varepsilon$. If in the language set there is single 'c', the production rule is $S \rightarrow S_2c$ and $S_2 \rightarrow aS_2b/\varepsilon$. 'b' is the middle element; for a single 'b' in the middle, the production rule is $S \rightarrow aS_3c$ and $S_3 \rightarrow aS_3c/b$.

Combining all these production rules, the grammar for the language is

$$S \rightarrow aS_1/S_2c/aS_3c$$
$$S_1 \rightarrow bS_1c/\varepsilon$$
$$S_2 \rightarrow aS_2b/\varepsilon$$
$$S_3 \rightarrow aS_3c/b$$

Example 2.21 Construct the grammar for the language $a^l b^m c^n$, where $l + m = n$.

Solution: The total number of 'a' and 'b' are equal to the total number of 'c'. If the number of 'a' or the number of 'b' is zero and the non-zero element is 1, then the number of 'c' is 1. If the number of 'a' and the number of 'b' are zero, then the number of 'c' is also zero.

So, there are productions $S \rightarrow ac/bc/\varepsilon$.

If, in the language, there is no 'a', then the number of 'b' is equal to the number of 'c'. The same thing is applicable for a language without 'b'. The production rules are $S \rightarrow aSc/bSc$. But in this case, there is a chance for 'a' to occur after 'b'. The modified production is $S \rightarrow aSc/bS_1c$.

The total number of 'a' and 'b' are equal to the total number of 'c'. This can be solved, if in each occurrence of 'a' or 'b', one 'c' is added to the language. The production rules for this case are $S_1 \rightarrow bS_1c/bc/\varepsilon$.

Combining all the production rules, the grammar for the language is

$$S \rightarrow aSc/bS_1c/ac/bc/\varepsilon$$
$$S_1 \rightarrow bS_1c/bc/\varepsilon$$

Example 2.22 Construct the grammar for palindrome of binary numbers.

Solution: A string which when read from left to right and right to left gives the same result is called palindrome.

A palindrome can be of two types: odd palindrome and even palindrome.

Odd palindromes are those types of palindromes where the number of characters is odd.

Even palindromes are those types of palindrome where the number of characters is even.

A null string is also a palindrome. The palindrome will consist of '0' and '1'.

A string will be a palindrome if its first and last characters are same.

The character in the 'n'th position from the beginning will be the same as the character from the 'n'th position from the last.

The palindrome can start with 0 or can start with 1. 0 can come after 0, or 0 can come after 1. 1 can come after 1, or 1 can come after 0.

From the previous discussion, the grammar for an even palindrome is

$$S \rightarrow 0S0/1S1/\varepsilon \text{ (null string is also a palindrome).}$$

The grammar for an odd palindrome is

$$S \rightarrow 0S0/1S1/0/1$$

By combining the two grammars for even and odd palindromes, the final grammar will be

$$S \rightarrow 0S0/1S1/0/1/\varepsilon$$

Example 2.23 Construct the grammar for the language L = {Set of all string over 0,1 containing twice as many '0' than '1'}.

Solution: The language consists of '0' and '1'. In the language set, the number of '0' is twice that of the number of '1'. '0' and '1' can occur in any pattern it must be confirmed that number of '0' = 2 × number of 1.

The least strings generated by the grammar are 001, 100, and 010. (considering null string is not in the language set).

The grammar for the language is

$$S \rightarrow A1A0A0A/A0A0A1A/A0A1A0A$$
$$A \rightarrow A1A0A0A/A0A0A1A/A0A1A0A/\varepsilon$$

Example 2.24 Construct a grammar consisting of equal number of '0' and '1'.

Solution: $|0| = |1|$ in any string generated by the grammar. If the string is null, $|0| = |1| = 0$, which also fulfills the condition.

The grammar for the language is

$$S \rightarrow S0S1S/S1S0S/\varepsilon$$

Example 2.25 Find the grammar for the language a^nb^m where $n \geq 2$, $m \geq 3$.

Solution: Each string in the language set generated by the grammar has at least two a's and three b's. It is not always true that the number of 'b' is one more than the number of 'a', as m and n are different numbers. In the language set, all the 'a' will come before 'b', and so we need to introduce two different non-terminals A and B to generate a string of 'a' and a string of 'b', respectively, with the conditions fulfilled.

As $n \geq 2$ and $m \geq 3$, two 'a' and three 'b' are added in the production rule with the start symbol.

The grammar for the language is

$$S \rightarrow aaABbbb$$
$$A \rightarrow aA/\varepsilon$$
$$B \rightarrow bB/\varepsilon$$

Example 2.26 Find the grammar for the language 0^m1^n where $m \neq n$, $m, n \geq 1$.

Solution: In the language set, there is at least one '0' and one '1'. The number of '0' is not equal to the number of '1', i.e., the number of '0' may be more than the number of '1' or vice versa.

By introducing a production $S \rightarrow 0S1$, the condition $m, n \geq 1$ is satisfied. For introducing more '0' than '1', a new non-terminal A is introduced. B is introduced for the reverse case.

The grammar for the language is

$$S \rightarrow 0S1/0A/B1$$
$$A \rightarrow 0A/\varepsilon$$
$$B \rightarrow B1/\varepsilon$$

Alternatively, the grammar can be in the following form:

$$S \rightarrow S_1/S_2$$
$$S_1 \rightarrow 0S_11/0A$$
$$A \rightarrow 0A/\varepsilon$$
$$S_2 \rightarrow 0S_21/B1$$
$$B \rightarrow B1/\varepsilon$$

Example 2.27 Construct the grammar for the language 0^m1^n, where $m \geq 1$ and $n \geq 1$ and $m < n$.

Solution: In any string that belongs to the language set, the number of '0' and the number of '1' are greater than or equal to '1', and the number of '0' is always less than the number of '1'. The least value of m is 1 here. So, the number of 1 must be at least 2.

By introducing a production S → 0S1, the condition m, n ≥ 1 is satisfied. Introducing more '1', S is replaced by another non-terminal A.

The grammar for the language is

$$S \rightarrow 0S1/A1/1$$
$$A \rightarrow A1/\varepsilon$$

Example 2.28 Construct the grammar for generating all positive integer numbers.

Solution: Positive integer numbers are from 0 to infinity and of any length. 0 is an integer but the starting character '0' of an integer has no value. Single-digit integers are generated from the production S→ 0/1/2/3/4/5/6/7/8/9.

If the production is made as S → 1S/2S/3S/4S/5S/6S/7S/8S/9S, then it can generate any integer number.

Combining these, the production becomes

$$S \rightarrow 1S/2S/3S/4S/5S/6S/7S/8S/9S/0/1/2/3/4/5/6/7/8/9$$

Example 2.29 Construct a grammar which generates all even integers up to 998.

Solution: In even numbers, the right most number is one of '0', '2', '4', '6', '8'. The middle and left-most number is one of '0', '1', '2', '3', '4', '5', '6', '7', '8', '9'. We have to generate even numbers up to 998, i.e., 3 digits. So, we have to consider all even numbers of one and two digits also.

The production S → 0/2/4/6/8 generates one-digit even numbers.

For generating two-digit numbers, another non-terminal A is introduced making the productions S → AS and A → 0/1/2/3/4/5/6/7/8/9. But here the problem is 'S' in RHS, which can be replaced by AS again making the string of an infinite length. To check this, the production is modified to $S \rightarrow AS_1$, and $S_1 \rightarrow 0/2/4/6/8$.

The production rules are

$$S \rightarrow AS1/AAS1/0/2/4/6/8$$
$$A \rightarrow 0/1/2/3/4/5/6/7/8/9$$
$$S_1 \rightarrow 0/2/4/6/8$$

Example 2.30 Construct the grammar for generating real numbers greater than 0.

Solution: Real numbers greater than 0 consist of two parts: (i) integer part (ii) real part. Let us take two non-terminals A and B to generate a real number where A generates the integer part and B generates the real part. The construction of the integer number is already described.

After the decimal point, 0 has value if there is number after 0.

The production B → 0B/1B/2B/3B/4B/5B/6B/7B/8B/9B generates a real part of any length. At last, step B is replaced by any of 1, 2, 3, 4, 5, 6, 7, 8, or 9.

From the previous discussion, the grammar is

$$S \rightarrow A.B$$
$$A \rightarrow 1A/2A/3A/4A/5A/6A/7A/8A/9A/0/1/2/3/4/5/6/7/8/9$$
$$B \rightarrow 0B/1B/2B/3B/4B/5B/6B/7B/8B/9B/1/2/3/4/5/6/7/8/9$$

Example 2.31 Construct the grammar for WW^R, where $W \in (a, b)^*$.

Solution: The string W consists of a and b. W can be a null string also. W^R is the reverse string of W. The number of characters in the string WW^R will always be even, except the null string.

So, the string is nothing but an even palindrome of a and b including the null string.

The grammar is

$$S \rightarrow aSa/bSb/\varepsilon$$

Example 2.32 Construct the grammar for the language $L = a^n b^n c^m d^m \quad m, n > 0$.

Solution: The total string can be divided into two parts: (i) string of a and b, where the number of a and the number of b are the same and (ii) string of c and d, where the number of c and d is the same.

From the start symbol, there will be two non-terminals A and B, where A will generate $\mathbf{a^n b^n}$ and B will generate $\mathbf{c^m d^m}$. In the language set, there will be no null string.

So the grammar is

$$S \rightarrow AB$$
$$A \rightarrow aAb/ab$$
$$B \rightarrow cBd/cd$$

Example 2.33 Construct the grammar for the language $L = a^n c^m d^m b^n$ m, n > 0.

Solution: In between a and b, there is an equal number of c and d. The number of a and the number of b are the same.

So, we have to construct $a^n b^n$ with a non-terminal in between a^n and b^n. From that non-terminal, $c^m d^m$ will be produced.

The grammar constructed according to the previous discussion is

$$S \rightarrow aSb/aAb$$
$$A \rightarrow cAd/cd$$

2.2 The Chomsky Hierarchy

The Chomsky hierarchy is an important contribution in the field of formal language and automata theory. In the mid-1950, *Noam Chomsky*, at the Harvard University, started working on formal languages and grammars. He structured a hierarchy of formal languages based on the properties of the grammars required to generate the languages. The Chomsky hierarchy drives automata theory one step forward.

Chomsky classified the grammar into four types depending on the production rules.

These are as follows:

Type 0: Type 0 grammar is a phase structure grammar without any restriction. All grammars are type 0 grammar.

For type 0 grammar, the production rules are in the format of

$$\{(L_c)(NT)(R_c)\} \rightarrow \{\text{String of terminals or non-terminals or both}\}$$
$$L_c : \text{Left context } R_c : \text{Right context NT : Non-terminal}$$

Type 1: Type 1 grammar is called context-sensitive grammar.

For type 1 grammar, all production rules are context-sensitive if all rules in *P* are of the form

$$\alpha A\beta \rightarrow \alpha\gamma\beta,$$

where A ε NT (i.e., *A* is a single non-terminal), α, β ε (NT U Σ)* (i.e., α and β are strings of non-terminals and terminals), and γ ε $(NT \cup \Sigma)^+$ (i.e., γ is a non-empty string of non-terminals and terminals).

Type 2: Type 2 grammar is called context-free grammar. In the LHS of the production, there will no left or right context.

For type 2 grammar, all the production rules are in the format of

$$(NT) \rightarrow \alpha, \text{ where } |NT| = 1 \text{ and } \alpha \; \varepsilon \; (NT \cup T)^*$$
$$\text{NT : Non-terminal T : Terminal}$$

Type 3: Type 3 grammar is called regular grammar. Here, all the productions will be in the following forms:

$$A \rightarrow \alpha \text{ or } A \rightarrow \alpha B, \text{ where A, B } \varepsilon \text{ NT and } \alpha \; \varepsilon \text{ T}$$

The Chomsky classification is called the Chomsky hierarchy. This is represented diagrammatically in Fig. 2.7

From this diagrammatical representation, we can say that all regular grammar is context-free grammar, all context-free grammar is context-sensitive grammar, and all context-sensitive grammar is unrestricted grammar.

The following table shows the different machine formats for different languages.

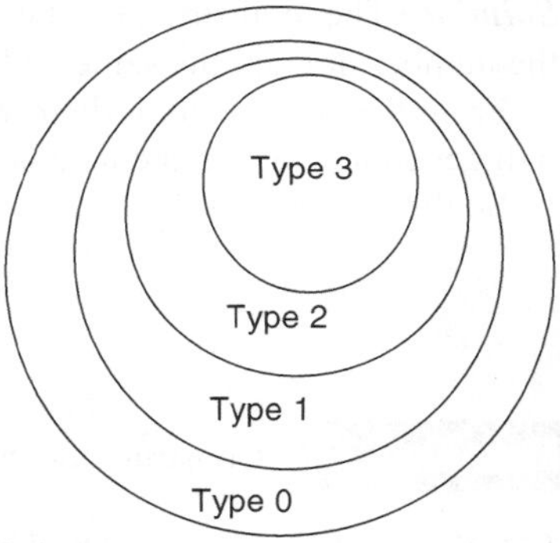

Fig. 2.7 *The Chomsky Hierarchy*

Grammar	Language	Machine Format
Type 0	Unrestricted language	Turing machine
Type 1	Context-sensitive language	Linear bounded automata
Type 2	Context-free language	Push down automata
Type 3	Regular expression	Finite automata

(Null alphabet is represented by ε. Null string is represented by Λ. Null set is represented by Φ.)

The following are some examples from context-sensitive grammar.

Example 2.34 Construct a grammar for the language $a^n b^n c^n$, where n ≥ 1.

Solution: We construct it by the following process:

1. First, we construct $a^n \alpha^n$.
2. Then, from α^n we construct $b^n c^n$.

From α, generating b^nc^n is difficult, From α^n, $(bc)^n$ can be constructed. But $(bc)^n$ is not equal to b^nc^n. Introduce a new non-terminal B.

$$S \to Abc/ABSc$$
$$BA \to AB$$
$$Bb \to bb$$
$$A \to a$$

Example 2.35 Construct a grammar for the language xx, where x ∈ (a, b)*

Solution:

$$S \to aAS/bBS$$
$$S \to aAZ/bBZ$$
$$Aa \to aA$$
$$Bb \to bB$$
$$AZ \to Za$$
$$BZ \to Zb$$
$$Z \to \lambda$$

Example 2.36 Construct a grammar for the language $\{a^nb^nc^nd^n, n \geq 1\}$.

Solution:

$$S \to abcd$$
$$S \to aXbcd$$
$$Xb \to bX$$
$$Xc \to bYc$$
$$Yc \to cY$$
$$Yd \to Rcdd$$
$$cR \to Rc$$
$$bR \to Rb$$
$$aR \to aaX \mid aa$$

In the Chomsky classification, the different types of grammar and their production rules are described. From the production rules, it is possible to find the highest type of the grammar. The following are some examples of finding the highest type of grammar from the production rules.

Example 2.37 Find the highest type of the following grammar.

$$A \to aA/bB$$
$$B \to aB/bB/a/b$$

Solution: The grammar is context-free as the LHS of each production contains only one non-terminal.

The RHS of each production contains a single terminal or a single terminal followed by non-terminal. This matches with regular grammar. So, the grammar is type 3 grammar.

Example 2.38 Find the highest type of the following grammar.

$$S \rightarrow a/aAS$$
$$A \rightarrow SS/SbA/ba$$

Solution: The grammar is context-free as the LHS of each production contains only one non-terminal.

Now, it is needed to check whether it is regular or not. The RHS of all productions does not contain a single terminal or a single terminal followed by non-terminal. So, the grammar is not regular. Hence, it is type 2 grammar.

Example 2.39 Find the highest type of the following grammar.

$$S \rightarrow aS/A$$
$$aS \rightarrow aa$$
$$A \rightarrow a$$

Solution: The LHS of the second production rule contains a terminal and a non-terminal (left context). So, the grammar is context-sensitive, and hence it is type 1.

What We Have Learned So Far

1. The set of rules for constructing a language is called the grammar for that language.
2. For every programming language such as C, C++, or Java, there is a grammar.
3. Grammar consists of four touples. $G = \{V_N, \Sigma, P, S\}$, where V_N : set of non-terminals, Σ : set of terminals, P : set of production rules, S : start symbol.
4. Chomsky, a linguist, classified the grammar into four types depending on the productions. The types are type 0, i.e., unrestricted grammar, type 1, i.e., context-sensitive grammar, type 2, i.e., context-free grammar, type 3, i.e., regular grammar.
5. For each of the grammar, there is a specific language namely unrestricted language, context-sensitive language, context-free language, and regular expression.
6. Unrestricted language is accepted by the Turing machine, context-sensitive language is accepted by linear bounded automata, context-free language is accepted by push down automata, and regular language is accepted by finite automata.
7. For type 0 grammar, the production rules are in the format of $\{(L_c)(NT)(R_c)\} \rightarrow$ {String of terminals or non-terminals or both}, where L_c : Left context, R_c : Right context, and NT: Non-terminal
8. For type 1 grammar, all production rules are context-sensitive if all rules in *P* are of the form $\alpha A\beta \rightarrow \alpha\gamma\beta$, where $A \in NT$ (i.e., A is a single non-terminal), $\alpha, \beta \in (NT \cup \Sigma)^+$ (i.e., α and β are strings of non-terminals and terminals), and $\gamma \in (NT \cup \Sigma)^+$ (i.e., γ is a non-empty string of non-terminals and terminals).
9. Type 2 grammar is called context-free grammar. In the LHS of the production, there will be no left or right context. For type 2 grammar, all the production rules are in the format of $(NT) \rightarrow \alpha$, where $|NT| = 1$ and $\alpha \in (NT \cup T)^*$, NT : Non terminal, and T: Terminal.
10. Type 3 grammar is called regular grammar. Here, all the productions will be in the following forms: $A \rightarrow \alpha$ or $A \rightarrow \alpha B$, where $A, B \in NT$ and $\alpha \in T$.

Solved Problems

1. Find the languages generated by the following grammars.

   a) $S \rightarrow aSa/aba$
   b) $S \rightarrow aSb/aAb, A \rightarrow bAa/ba$
   c) $S \rightarrow 0S1/0A1, A \rightarrow 1A/1$
   d) $S \rightarrow 0A/0/1B, A \rightarrow 1A/1, B \rightarrow 0B/1S$

   Solution:

   a) S is replaced by aSa or aba. aba is a string of terminals. S in aSa can again be replaced by aSa or aba. By this process, the language is

   $$S \rightarrow aSa \rightarrow aaSaa \;\ldots\ldots\; \rightarrow a^{(n-1)}\, Sa^{(n-1)} \rightarrow a^n ba^n$$
   $$L(G) = a^n ba^n,\ n > 0$$

   b) S can be replaced by aSa or aAb. $S \rightarrow aAb$ shifts the control from S to A. Replacing S (n–1) times produces $a^{n-1}Sb^{b-1}$. Replacing S by aAb produces a^nAb^n.

   A is replaced by bAa or ba. As S and A are separate non-terminals, there is no guarantee of replacing S and A at the same time. Let A be replaced (m–1) times, and then finally replaced by ba. This generates the string $a^nb^ma^mb^n$.

   $$S \rightarrow aSb \rightarrow aaSbb \;\ldots\; \rightarrow a^{n-1}Sb^{b-1} \rightarrow a^nAb^n \rightarrow a^nbAab^n \;\ldots\; \rightarrow a^nb^{m-1}\, A\, a^{m-1}b^n \rightarrow a^nb^ma^mb^n$$

   The language generated by the grammar is $L(G) = a^nb^ma^mb^n$, where m, n > 0.

   c) $S \rightarrow 0S1 \rightarrow 00S11 \ldots \rightarrow 0^{m-1}A1^{m-1} \rightarrow 0^mA1^m \rightarrow 0^m1A1^m \ldots \rightarrow 0^m1^{n-1}A1^m \rightarrow 0^m1^n1^m \rightarrow 0^m1^{m+n}$
   The language generated by the grammar is $L(G) = 0^m1^{m+n}$ where m, n > 0.

   d) First, let us try to find the expressions generated by $B \rightarrow 0B/1S$ and $A \rightarrow 1A/1$. Then, replace the expressions generated by A and B into $S \rightarrow 0A/1B$.

   $$B \rightarrow 0B \rightarrow 00B \;\ldots\ldots\; \rightarrow 0^*B \rightarrow 0^*1S$$
   $$A \rightarrow 1A \rightarrow 11A \;\ldots\ldots \rightarrow 1^*A \rightarrow 1^*1$$
   $$S \rightarrow 0A/0/1B$$
   $$\rightarrow 01^*1/0/10^*1S \rightarrow (01^*1/0)(10^*1)^*$$

2. Generate a CFG for the language L = Alternating sequence of ‘a’ and ‘b’.

   Solution: The language may start with ‘a’ or start with ‘b’. If the string starts with ‘a’, then the language is $(ab)^n$. If the string starts with ‘b’, then the language is $(ba)^n$. If the string starts with ‘a’, then the production rule is $S \rightarrow aA$. If the string starts with ‘b’, then the production rule is $S \rightarrow bB$. For alternating sequences of ‘a’ and ‘b’, the production rules are $A \rightarrow bB$, $B \rightarrow aA$, $A \rightarrow b$, $B \rightarrow a$.

   The final grammar is

   $$S \rightarrow aA/bB$$
   $$A \rightarrow bB/b$$
   $$B \rightarrow aA/a$$

3. Find the grammar for the language $L = \{a^nb^m$, where n + m is even $\}$ [UPTU 2003]

   Solution: This may happen in three cases if

   i) both ‘a’ and ‘b’ are odd
   ii) both ‘a’ and ‘b’ are even
   iii) any one of ‘a’ and ‘b’ are even and the other is zero.

If the first two are constructed, then the third will be fulfilled. (0 can be considered as even.)

The grammar is

$$S \rightarrow \{\text{odd number of 'a'}\}\ S\ \{\text{odd number of 'b'}\}/$$
$$\{\text{even number of 'a'}\}\ S\ \{\text{even number of 'b'}\}/\varepsilon$$

It can be constructed another way by taking two productions from S; among them, one generates both 'a' and 'b' odd and another generates both 'a' and 'b' even.

$$S \rightarrow AB/CD$$
$$A \rightarrow aaA/a$$
$$B \rightarrow bbB/b$$
$$C \rightarrow aaA/\varepsilon$$
$$D \rightarrow bbB/\varepsilon$$

4. Find the grammar for the following language [UPTU 2004]

$$L = a^n b^n c^k, \text{ where } k \geq 3$$

Solution: In the language, there is at least three 'c'. The number of 'a' and number of 'b' are the same and may be null.

The grammar is

$$S \rightarrow AcccB$$
$$A \rightarrow aAb/\varepsilon$$
$$B \rightarrow cB/\varepsilon$$

5. Construct a grammar which generates all odd integers up to 999.

Solution: The odd numbers from 0 to 9 are 1, 3, 5, 7, 9. The grammar for generating 1, 3, 5, 7, 9 are $S \rightarrow 1/3/5/7/9$.

For an odd number of length 2, the grammar is $S \rightarrow AS_1$, where $A \rightarrow 0/1........7/8/9$, $S_1 \rightarrow 1/3/5/7/9$.

For an odd number of length 3, the grammar is $S \rightarrow BAS_1$, where $A \rightarrow 0/1........7/8/9$, $B \rightarrow 0/1........7/8/9$ $S_1 \rightarrow 1/3/5/7/9$.

Combining all these rules, the grammar becomes

$$S \rightarrow AS_1/BAS_1/1/3/5/7/9$$
$$A \rightarrow 0/1/2/3/4/5/6/7/8/9$$
$$S_1 \rightarrow 1/3/5/7/9$$

6. Construct a grammar which generates $\{(01)^n \cup (10)^n\}$, where $n > 0$

Solution: It consists of two languages $(01)^n$ and $(10)^n$. Both are connected by union. Take A and B as two non-terminals, which generate $(01)^n$ and $(10)^n$, respectively. From the start symbol S it goes to A and B. The grammar is

$$S \rightarrow A/B$$
$$A \rightarrow abA/ab$$
$$B \rightarrow baB/ba$$

7. Find the grammar for the language $L = \{a^n b^m, \text{ where } n \neq m\}$.

Solution: Here, two cases are possible (i) The number of 'a' is more than the number of 'b' (ii) The number of 'b' is more than the number of 'a'.

For case (i) the grammar is

$$A \rightarrow X_1Y_1$$
$$X_1 \rightarrow aX_1/a$$
$$Y_1 \rightarrow aY_1b/\varepsilon$$

For case (ii) the grammar is

$$B \rightarrow X_2Y_2$$
$$Y_2 \rightarrow bY_2/b$$
$$X_2 \rightarrow aX_2b/\varepsilon$$

The complete grammar is

$$S \rightarrow A/B$$
$$A \rightarrow X_1Y_1$$
$$X_1 \rightarrow aX_1/a$$
$$Y_1 \rightarrow aY_1b/\varepsilon$$
$$B \rightarrow X_2Y_2$$
$$Y_2 \rightarrow bY_2/b$$
$$X_2 \rightarrow aX_2b/\varepsilon$$

8. Find a grammar generating $L = \{a^nb^nc^f \mid n \geq 1, f \geq 0\}$. [WBUT 2010(IT)]

Solution: The language consists of two parts: a^nb^n, where $n \geq 1$, and c^f, where $f \geq 0$. So, from the start symbol S, we need to take two non-terminals A and B, where A generates a^nb^n with $n \geq 1$ and B generates c^f with $f \geq 0$.

The grammar for the language is

$$S \rightarrow AB$$
$$A \rightarrow aAb/ab$$
$$B \rightarrow Bc/\varepsilon$$

9. Construct a grammar for the language

$$L = (0 + 1)^*111(0 + 1)^*$$

Solution: It can be thought of as a language set having any combination of 0 or 1 in both the sides of 111. We know that the grammar for **(0 + 1)*** is $A \rightarrow 0A/1A/0/1$.

From the previous discussion, the grammar is

$$S \rightarrow A111A$$
$$A \rightarrow 0A/1A/0/1$$

10. Find the grammar for $L = \{a^nb^mc^m \mid n \geq 0, m \geq 1\}$. [WBUT 2003]

Solution: There are two parts of the language, a^n and $\mathbf{b^mc^m}$. Take two non-terminals A and B to generate these two parts.

$$S \rightarrow AB$$
$$A \rightarrow aA/\varepsilon$$
$$B \rightarrow bBc/bc$$

11. Find the grammar for the language $L = \{a^nb^m \mid n + m \text{ is even}\}$. [UPTU 2004]

Solution: There are two conditions for n + m to become even.

a) both n and m are even
b) both n and n are odd

For case (a), the language is (aa)*(bb)*. For case (b), the language is a(aa)*(bb)*b.
For case (a), the grammar is

$$S_1 \rightarrow aaS_1bb/\varepsilon$$

For case (b), the grammar is

$$S_2 \rightarrow aS_1b$$
$$S_1 \rightarrow aaS_1bb/\varepsilon$$

Thus, the final grammar is

$$S \rightarrow S_1/S_2$$
$$S_1 \rightarrow aaS_1bb/\varepsilon$$
$$S_2 \rightarrow aS_1b$$

12. Construct a grammar for the language (i) $a^i\ b^2\ i$ for $i > 0$ and (ii) a^nba^n for $n > 0$. [Anna University 06]

Solution:

i) For a single 'a', two 'b's are added. The grammar is

$$S \rightarrow aSbb/abb$$
or
$$S \rightarrow aSbb/aAbb$$
$$A \rightarrow \varepsilon$$

ii) In the language, there is only one 'b'. The grammar is

$$S \rightarrow aSa/aAa$$
$$A \rightarrow b$$

Multiple Choice Questions

1. Which is correct
   a) $a^+ = a^*. a^*$ b) $a^* = a^+. a^+$
   c) $a^+ = a^*. a$ d) $a^* = a^+. a^*$

2. (a, b)* means
   a) Any combination of a, b including null
   b) Any combination of a, b excluding null
   c) Any combination of a, b, but 'a' will come first
   d) None of these

3. (a, b)⁺ means
   a) Any combination of a, b including null
   b) Any combination of a, b excluding null
   c) Any combination of a, b, but 'a' will come first
   d) None of these

4. What is the language generated by the grammar $S \rightarrow aSb$, $S \rightarrow A$, $A \rightarrow aA$
   a) a^mb^m b) $\varnothing$ c) a^nb^m d) a^mb^m

5. Which type of grammar is the following in particular $S \rightarrow aSb$ $S \rightarrow ab$
   a) Unrestricted
   b) Context-sensitive grammar
   c) Context-free grammar
   d) Regular grammar

6. Which type of grammar is the following in particular $S \rightarrow aS/bA$ $A \rightarrow aA/a$
   a) Unrestricted
   b) Context-sensitive grammar
   c) Context-free grammar
   d) Regular grammar

7. The language genarated by the grammar $S \rightarrow 0S0$, $S \rightarrow 1S1$, $S \rightarrow 0$, $S \rightarrow 1$ is
   a) Even palindrome of 0 and 1
   b) Odd palindrome of 0 and 1
   c) Any combination of 0 and 1
   d) None of these
8. What is the highest type number to the grammar given by the following production rules $S \rightarrow Aa$, $A \rightarrow c|Ba$, $B \rightarrow abc$. [WBUT 2008]
   a) Zero b) One c) Two d) Three
9. The language $\{a^m b^n c^m + n \mid m, n \geq 1\}$ is
   a) Regular
   b) Context-free but not regular
   c) Context-sensitive but not context-free
   d) Type 0 but not context-sensitive
10. Which type of grammar is the following in particular $A \rightarrow aB$, $B \rightarrow ac$, $C \rightarrow bC/aD$, $D \rightarrow bA/b$
    a) Context-sensitive grammar
    b) Context-free grammar
    c) Context-free but regular in particular
    d) Not type0 but context-free
11. The machine format of context-sensitive grammar is [WBUT 2009]
    a) Finite automata
    b) Push down automata
    c) Linear bounded automata
    d) All of the above
12. Which of the following grammar generates strings with any number of 1's? [WBUT 2010]
    a) $S \rightarrow 1A, A \rightarrow \varepsilon$ b) $S \rightarrow 1S, S \rightarrow \varepsilon$
    c) $S \rightarrow S1, S \rightarrow \varepsilon$ d) (b) & (c)
13. Which of the following is true?
    a) $(ab)^2 = a^2b^2$ b) $(ab)^2 = abab$
    c) $(ab)^2 = aabb$ d) None of these
14. If $\Sigma = \{1\}$, then $\Sigma^* - \Sigma^+$ is
    a) 1^+ b) $\{1\}$ c) $\{\wedge\}$ d) $\{\wedge, 1, 11.....\}$
15. A language set contains
    a) Alphabet b) String
    c) Set d) None of these
16. In automata, why it is called formal language?
    a) Some alphabet forms a string
    b) Some strings form a language
    c) Well-defined use of symbols is significant
    d) Only the form of the string generated by alphabets is significant.
17. The language constructed from the grammar $S \rightarrow aSbb/aAbb$, $A \rightarrow a$ is
    a) $a^{n+1} b^{2n}$ b) $a^n b^{2n}$
    c) $a^n b^{2n-1}$ d) $a^{n+1} b^{2n+1}$
18. The language generated by the grammar $S \rightarrow aSa/aBa$, $A \rightarrow Ba/b$ is
    a) $a^n b a^n$ b) $a^n b a^{n+1}$ c) a^{2n} d) Φ

Answers:

1. c	2. a	3. b	4. b	5. c	6. d	7. b	8. c	9. b
10. c	11. c	12. d	13. b	14. c	15. b	16. d	17. a	18. d

GATE Questions

1. In the given context-free grammar, S is the start symbol, a and b are terminals, and ε denotes the empty string.

$$S \rightarrow aSa/bSb/a/b/\varepsilon$$

Which of the following strings is NOT generated by the grammar?
a) aaaa b) baba c) abba d) babaaabab

2. In the given context-free grammar, S is the start symbol, a and b are terminals, and ε denotes the empty string.

$$S \rightarrow aSAb/\varepsilon$$
$$A \rightarrow bA/\varepsilon$$

The grammar generates the language
a) $((a + b)^*b)^*$ b) $\{a^m b^n \mid m \leq n\}$ c) $\{a^m b^n \mid m = n\}$ d) a^*b^*

3. The two grammars given generate a language over the alphabet (x, y, z)

$$G1: S \rightarrow x|z|xS|zS|yB$$
$$B \rightarrow y|\ z|\ yB|\ zB$$
$$G2: S \rightarrow y|z|yS|zS|xB$$
$$B \rightarrow y|\ yS$$

Which of the following choices describe the properties satisfied by the strings in these languages?
a) G1: No y appears before any x
   G2: Every x is followed by at least one y
b) G1: No y appears before any x
   G2: No x appears before any y
c) G1: No y appears after any x
   G2: Every x is followed by at least one y.
d) G1: No y appears after any x
   G2: Every y is followed by at least one x.

4. Consider the given grammar

$$S \rightarrow xB/yA$$
$$A \rightarrow x/xS/yAA$$
$$B \rightarrow y/yS/xBB$$

Consider the following strings. Which are accepted by the grammar?
i) xxyyx   ii) xxyyxy   iii) xyxy   iv) yxxy   v) yxx   vi) xyx

a) (i), (ii) and (iii)   b) (ii), (v) and (vi)
c) (ii), (iii) and (iv)   d) (i), (iii) and (iv)

5. Consider the following grammars. Names representing terminals have been specified in capital letters.

G1: stmnt → WHILE (expr) stmnt
stmnt → OTHER
expr → ID

G2: stmnt → WHILE (expr) stmnt
stmnt → OTHER
expr → expr + expr
expr → expr * expr
expr → ID

Which of the following statements is true?
a) G1 is context-free but not regular and G2 is regular
b) G2 is context-free but not regular and G1 is regular

c) Both G1 and G2 are regular
d) Both G1 and G2 are context-free but neither of them is regular.

6. Consider a grammar with the following productions

$$S \rightarrow a\alpha b/b\alpha c/aB$$
$$S \rightarrow S/b$$
$$S \rightarrow \alpha bb/ab$$
$$S\alpha \rightarrow bdb/b$$

a) Context-free b) Regular c) Context-sensitive d) LR(k)

7. Which of the following definitions generate the same language as L, where L = $\{x^n y^n$ such that $n \geq 1\}$?
   i) $E \rightarrow xEy/xy$
   ii) $xy/(x^+ xyy^+)$
   iii) $x^+ y^+$

   a) I only b) I and II c) II and III d) II only

8. Consider the grammar

$$S \rightarrow bSe/PQR$$
$$P \rightarrow bPc/\varepsilon$$
$$Q \rightarrow cQd/\varepsilon$$
$$R \rightarrow dRe/\varepsilon$$

   The string generated by the grammar is $b^i c^j d^k e^m$ for some i, j, k, m ≥ 0. What is the relation among i, j, k, and m?

   a) $i = j = k = m$ b) $i + k = j + m$ c) $i = j$ but $k \neq m$ d) $i + j + k = m$

9. Consider the grammar $S \rightarrow aSAb/\varepsilon$
$A \rightarrow bA/\varepsilon$

   The grammar generates the strings of the form $a^i b^j$ for some i, j ≥ 0. What are the conditions of the values of i and j?

   a) $i = j$ b) $j \leq 2i$ c) $j \geq 2i$ d) $i \leq j$

10. The language $\{a^m b^n c^{m+n} \mid m, n \geq 1\}$ is
   a) regular
   (b) context-free but not regular
   c) context-sensitive but not context-free
   (d) type-0 but not context-sensitive

11. Consider the following grammar G:

$$S \rightarrow bS/aA/b$$
$$A \rightarrow bA/aB$$
$$B \rightarrow bB/aS/a$$

   Let $N_a(w)$ and $N_b(w)$ denote the number of a's and b's in a string w, respectively. The language $L(G) \subseteq \{a, b\}^+$ generated by G is

   a) $\{w \mid N_a(w) > 3N_b(w)\}$
   b) $\{w \mid N_b(w) > 3N_a(w)\}$
   c) $\{w \mid N_a(w) = 3k, k \in \{0, 1, 2, \ldots\}\}$
   d) $\{w \mid N_b(w) = 3k, k \in \{0, 1, 2, \ldots\}\}$

12. Let $L_1 = \{0^{n+m}1^n0^m \mid n, m \geq 0\}$, $L_2 = \{0^{n+m}1^{n+m}0^m \mid n, m \geq 0\}$, and $L_3 = \{0^{n+m}1^{n+m}0^{n+m} \mid n, m \geq 0\}$. Which of these languages are NOT context-free?

    a) L_1 only b) L_3 only c) L_1 and L_2 d) L_2 and L_3

13. Which one of the following grammars generate the language $L = \{a^ib^j \mid i \neq j\}$?

    a) $S \to AC \mid CB$
    $C \to aCb \mid a \mid b$
    $A \to aA \mid \varepsilon$
    $B \to Bb \mid \varepsilon$

    b) $S \to aS \mid Sb \mid a \mid b$

    c) $S \to AC \mid CB$
    $C \to aCb \mid \varepsilon$
    $A \to aA \mid \varepsilon$
    $B \to Bb \mid \varepsilon$

    d) $S \to AC \mid CB$
    $C \to aCb \mid \varepsilon$
    $A \to aA \mid a$
    $B \to Bb \mid b$

14. In the correct grammar above, what is the length of the derivation (number of steps starting from S) to generate the string a^lb^m with $l \neq m$?

    a) $\max(l, m) + 2$ b) $l + m + 2$ c) $l + m + 3$ d) $\max(l, m) + 3$

15. $S \to aSa \mid bSb \mid a \mid b$

    The language generated by the above grammar over the alphabet {a, b) is the set of

    a) all palindromes.
    b) all odd length palindromes.
    c) strings that begin and end with the same symbol.
    d) all even length palindromes.

16. Consider the languages $L_1 = \{0^i1^j \mid i \neq j\}$, $L_2 = \{0^i1^j \mid i = j\}$, $L_3 = \{0^i1^j \mid i = 2j + 1\}$, and $L4 = \{0^i1^j \mid i \neq 2j\}$. Which one of the following statements is true?

    a) Only L_2 is context-free
    b) Only L_2 and L_3 are context-free
    c) Only L_1 and L_2 are context-free
    d) All are context-free

Answers:

1. b	2. b	3. a	4. c	5. b	6. c	7. a	8. b	9. d
10. b	11. c	12. d	13. d	14. a	15. b	16. d		

Hints:

2. The grammar is derived as $S \to aSAb \to aaSAbAb$ - - - $\to a^mS(Ab)^m$. If S and A are replaced by ε, them m = n. If A is replaced by bA, then $m \leq n$.

3. $S \to xB \to xy/xyS$

4. $S \to xB \to xxBB \to xxySB \to xxyyAB \to xxyyxy$
   $S \to xB \to xyS \to xyxB \to xyxy$
   $S \to yA \to yxS \to yxxB \to yxxy$

5. G_2 is not regular for the productions
   expr → expr + expr
   expr → expr * expr

8. $S \to bSe$ - - $\to b^aSe^a \to b^aPQRe^a$ - - $\to b^a \underbrace{b^xc^x}_{P} \underbrace{c^yd^y}_{Q} \underbrace{d^ze^z}_{R} e^a$

9. Same as 2
11. It accepts b (k = 0) aaa, $b^+ ab^+ ab^+ a$ (k = 1)
14. $S \to AC \to Aa\underline{C}b \to Aaa\underline{C}bb \to \underline{A}aabb \to aaabb$ a = 3 b = 2
    max(3, 2) + 2 = 5 = Number of derivations
16. L_1 is CFG according to 14.
    The grammar of L_3 is
    $S \to 00S1/0$

Fill in the Blanks

1. Grammar consists of four touples—Set of non-terminals, __________, set of production rule, and __________.
2. According to the Chomsky hierarchy, there are __________ types of grammars.
3. Type 1 grammar is called __________.
4. Type 2 grammar is called __________.
5. According to the Chomsky hierarchy, regular grammar is type __________ grammar.
6. All languages are accepted by __________.
7. The machine format of context-free language is __________.
8. Linear bounded automata is the machine format of __________.
9. The machine format of type 3 language is __________.
10. Grammar where production rules are in the format $\alpha A\beta \to \alpha\gamma\beta$ is __________ grammar
11. In a context-free grammar at the left hand side, there is __________ non-terminal.
12. Type 3 language is called __________.
13. $a^nb^nc^n$ is an example of __________ language in particular.
14. The grammar $S \to aSb/A$, $A \to Ac/c$ is an example of __________ grammar in particular.
15. The grammar $S \to Abc/ABSc$, $BA \to AB$, $Bb \to bb$, $A \to a$ is an example of __________ grammar in particular.
16. The grammar $A \to aA/bB/a/b$, $B \to bB/b$ is an example of __________ grammar in particular.
17. The language $a^*(a + b)b^*$ is an example of __________ language in particular.
18. L = Alternating sequence of 'a' and 'b' is an example of __________ language in particular.
19. Regular expression is accepted by type __________ grammar.

Answers:

1. set of terminals, start symbol
2. Four
3. Context-sensitive grammar
4. Context-free grammar
5. Three
6. Turing machine
7. Push down automata
8. Context-sensitive grammar
9. Finite automata
10. Context-sensitive
11. Single
12. Regular expression
13. Context-sensitive
14. Context-free
15. Context-sensitive
16. Regular grammar
17. Type 4
18. Context-free
19. Three

Exercise

1. Define grammar. Is any grammar possible without a start symbol?
2. Is it possible to guess the type of language from a given grammar? Is the reverse true? Give reasons in support of your answer.
3. What is the Chomsky hierarchy? What is the usefulness of it?
4. Find the languages generated by the following grammars.
   a) $S \rightarrow aSb/A, A \rightarrow Ac/c$
   b) $S \rightarrow aSb/aAb, A \rightarrow Ac/\varepsilon$
   c) $S \rightarrow aSb/aAb, A \rightarrow bA/b$
   d) $S \rightarrow S_1/S_2, S_1 \rightarrow 0S_11/0A, A \rightarrow 0A/, S_2 \rightarrow 0S_21/B1, B \rightarrow B1/\varepsilon$
   e) $S \rightarrow AB/CD, A \rightarrow aA/a, B \rightarrow bB/bC, C \rightarrow cD/d, D \rightarrow aD/AD$

      Justify your answer for this.
   f) $S \rightarrow AA, A \rightarrow BS, A \rightarrow b, B \rightarrow SA, B \rightarrow a$
   g) $E \rightarrow E + E| E–E| E * E| E/E|id$
5. Construct a grammar for the following languages.
   a) $L = \{\Phi\}$
   b) $L = (a, b)^*$, where all 'a' appears before 'b'
   c) $L = (a, b)^*$, where all 'b' appears before 'a'
   d) $L = (a, b)^*$, where there are equal number of 'a' and 'b'
   e) $L = (a, b)^*$, where ab and ba appear in an alternating sequence.
   f) $L = (a, b)^*$, where the number of 'b' is one more than the number of 'a'
   g) $L = (a, b)^*aa(a, b)^*$
6. Construct a grammar for the following languages.
   a) $L = a^mb^n$, where $m \neq n$.
   b) $L = a^xb^yc^z$, where $y = x + z$
   c) $L = a^xb^yc^z$, where $z = x + y$
   d) $L = a^xb^yc^z$, where $x = y + z$
   e) L = Set of all string over a, b containing aa or bb as substring
   f) L = Set of all string over a, b containing at least two 'a'
   g) L = Set of all string over 0, 1 containing 011 as substring
7. Construct a grammar for the following languages.
   a) $\{ a^nb^n| n > 0\} \cup \{c^md^m \mid m \geq 0 \}$
   b) $\{ a^nb^n| n > 0\} \cup \{a^mb^m \mid m \geq 0 \}$
   c) $\{ a^xb^yc^z$, where $x = y + z \} \cup \{ L = a^xb^yc^z$, where $z = x + y\}$
8. Construct a grammar for the following languages and find the type of the grammar in particular.
   a) $L = (0 + 1)^*\ 11\ (11)^*$
   b) L = (Set of all string of 'a', 'b' beginning and ending with 'a')
   c) $L = a^{2n+1}$, where $n > 0$

Finite Automata

3

Introduction

The term 'automata' has some typical pronounceable symmetry with the term 'automatic'. In a computer, all processes appear to be done automatically. A given input is processed in the CPU, and an output is generated. We are not concerned about the internal operation in the CPU, but only about the given input and the received output. However, in reality, the input is converted to '0' and '1' and assigned to the process for which the input was given. It then performs some internal operation in its electronic circuit and generates the output in '0' and '1' format. The output generated in the '0' and '1' format is converted to a user-understandable format and we get the output. From the discussion, it is clear that the CPU performs machine operations internally. In automata, we shall learn about how to design such machines. To get a clear idea about the automata theory, we need to travel back to the history of computer science. The engineering branch 'electrical engineering' stands on the base of physics. Electronics engineering came from electrical engineering. The hardware of a computer is a complex electronics circuit. The circuit understands only binary command, which is impossible for a human user to memorize. Then, mathematics extends its hand for developing the logic. Mathematics and electronics have jointly given birth to a new branch of engineering called 'computer science and engineering'.

The name of the subject is formal language and automata theory. We have a basic idea about the automata theory. Now what is formal language? Let us discuss what language is. Language is a communication medium through which two persons communicate. For every nation, there is some language to communicate such as Hindi, English, Bengali, etc. It is the same for communicating with a computer, wherein the user needs some language called the programming language. C, C ++, and Java are some examples of programming languages. The characteristics of these types of languages are that they are similar to the English language and are easily understandable by the user. However, the computer does not understand English-like statements. It understands only binary numbers. Therefore, the computers have a compiler that checks the syntax and acts as a converter between English-like statements to binary numbers and vice versa. But, to design the compiler, some logic is needed. That logic can be designed using mathematics. For each language, there is a grammar. Grammar is the constructor for any language. Similarly, the languages that are used for computer programming purpose have some grammar to construct them. These rules and grammar and the process of converting these grammar and languages to machine format are the main objectives of this subject.

3.1 History of the Automata Theory

Theoretical computer science or automata theory is the study of abstract machines and the computational problems related to these abstract machines. These abstract machines are called automata. The origin of the word 'automata' is from a Greek word (Αυτόματα) which means that something is doing something by itself.

Around 1936, when there were no computers, *Alen Turing* proposed a model of abstract machine called the Turing machine which could perform any computational process carried out by the present day's computer. A formal version of the finite automata model was proposed in 1943 in the *McCulloch–Pitts* neural network models. Finite automata and finite state machines got a complete form by the efforts of *George H. Mealy* at the Bell Labs and *Edward F. Moore* in IBM around 1960. The machines are named as Mealy and Moore machines, respectively. In mid-1950s, *Noam Chomsky* at the Harvard University started working on formal languages and grammars. Grammar is a set of rules that are performed to produce sentences in the language. Chomsky structured a hierarchy of formal languages based on the properties of the grammar required to generate the languages. The Chomsky hierarchy drives the automata theory one step forward. In 1971, *Stephen Arthur Cook* in his research paper 'The Complexity of Theorem Proving Procedures' formalized the notions of polynomial-time reduction and NP-completeness. He separated those problems that can be solved efficiently by the computer from the set of problems that can be solved but with so much time-taking. These second classes of problems are called NP-hard.

This subject is sometimes called the 'Theory of Computation' because it includes all of these rules for constructing a computer language and converts them into machine format. That is the theory of computer science. Basically formal language and automata theory and the theory of computation are different names for a single subject that covers all the aspects of the theoretical part of computer science.

3.2 Use of Automata

It is unfruitful to study a subject without knowing its usefulness. The subject 'automata theory' is called one of the core subjects of computer science. Some of the uses of this subject in different fields of computer science and engineering are as follows:

- The main usefulness of the automata theory is in the compiler design. The first phase of compiler called lexical analyser is designed by the finite automata. In syntax analysis part, a common error called syntax error is checked by the rules of context-free grammar.
- The working principle of the software used for checking the behaviour of digital circuits is based on the automata theory.
- The software for natural language processing (NLP) takes the help of the automata theory.
- Automata is useful for designing the software for counting the occurrence of a particular word, pattern, etc., in a large text.

3.3 Characteristics of Automaton

The word 'automata' is plural and the singular is 'automaton'. A Finite Automaton, in short FA, provides the simplest model of a computing device. It is called finite because it has a finite number of states. The state of the system memorizes the information concerning the past input. It is necessary to determine the future behaviour of the system.

Definition: An automaton is a system where materials, energy, or information are transformed and transmitted for performing some operation without the direct participation of a human.

Any automated machine can be given as an example of automaton. A model of finite automata is given in Fig. 3.1.

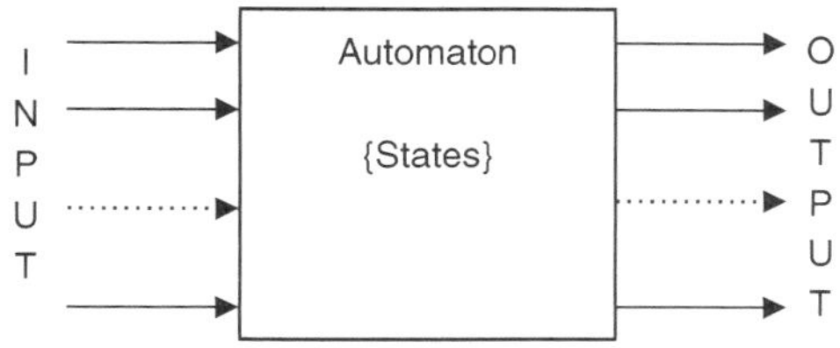

Fig. 3.1 *Block Diagram of Finite Automata*

3.3.1 Characteristics

- **Input (I/P):** The input is taken in each clock pulse. For every single instance of time t_1, t_2, t_3 .... t_n, the inputs are taken as I_1, I_2, I_3 ...... I_n. As there are a number of input lines, n number of inputs will be taken in each single time instance. The input for each input line is finite and taken from a set called the set of input alphabets Σ.
- **Output (O/P):** The output is generated in each clock pulse. For every single instance of time t_1, t_2, t_3 .... t_m, the outputs are generated as O_1, O_2, O_3 ...... O_m. The output generated from each output line is finite and belongs to a set called the output alphabet set.
- **State:** At any discrete instance of time, the automaton can be in one of the states q_1, q_2, q_3 .... q_n. The state belongs to a set called 'State' Q.
- **State transition:** At any instance of time, the automaton must be in one of the states that belong to the set Q. By getting an input in a clock pulse, the automaton must reside in a state. The state in which the automaton resides by getting that particular input is determined by state transition. The state transition is a function of the present state and the present input, which produces the next state. The function is represented as δ.
- **Output relation:** Similar to the state transition for state, there is a relation for output. The output depends either on the present state and present input or on the present state only depending on the type of machine.

3.4 Finite Automata

Definition: Finite automata (singular: automaton) are the machine formats of regular expression, which is the language format of type 3 grammar. An FA is defined as

$$M = \{Q, \Sigma, \delta, q_0, F\}$$

where Q : finite non-empty set of states

Σ : finite non-empty set of input symbols

δ : transitional function

q_0 : beginning state

F : finite non-empty set of final states

Finite automata are one type of the finite state machine. It has a finite number of states. Finite automata can be thought of as a finite state machine without output.

Mechanically, finite automata can be described as an input tape containing the input symbols (Σ) with a reading head scanning the inputs from left to right. The inputs are fed to finite control which contains the transitional functions (δ). According to the transitional functions, the state (Q) change occurs.

The mechanical diagram of finite automata is given in Fig. 3.2.

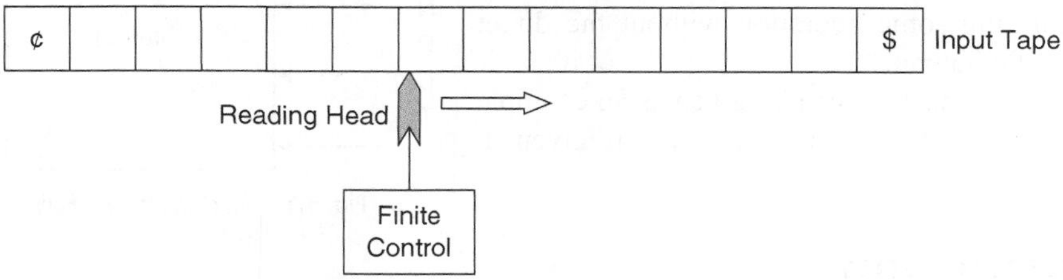

Fig. 3.2 *Mechanical Diagram of Finite Automata*

- **Input tape:** The input tape contains the input symbol. It is divided into several squares, which contain single characters of the input alphabet. Both the left and right ends of the input tape contain end markers. Between two end markers, the input string is placed. This string is needed to be processed from left to right.
- **Reading head:** The head scans each square in the input tape and reads the input from the tape. The head can move from left to right or right to left. But, in most of the cases, the head moves from left to right. In two-way finite automata and the Turing machine, the head can move in both directions.
- **Finite control:** Finite control can be considered as the control unit of an FA. An automaton always resides in a state. The reading head scans the input from the input tape and sends it to finite control. In this finite control, it is decided that 'the machine is in this state and it is getting this input, so it will go to this state'. The state transition relations are written in this finite control.

3.5 Graphical and Tabular Representation of FA

Finite automata can be represented in two ways: (i) graphical and (ii) tabular.

1. In the graphical representation, a state is represented as 

   A beginning state is represented as 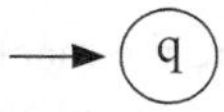

   A final state is represented as 

   An FA is represented in graphical format in Fig. 3.3.

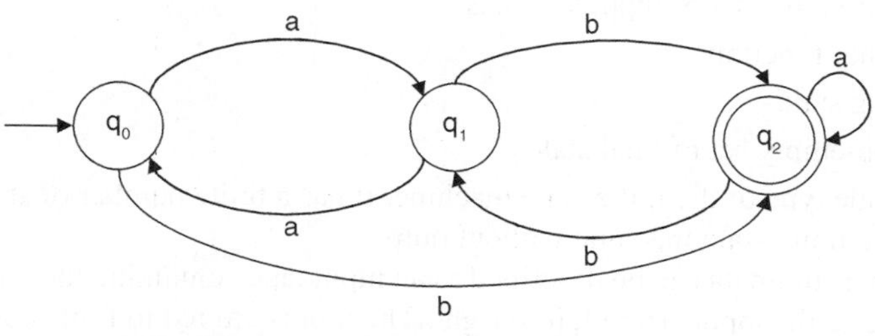

Fig. 3.3 *Graphical Representation of Finite Automata*

Here

$Q : \{q_0, q_1, q_2\}$
$\Sigma : \{a, b\}$
$\delta : \delta(q_0, a) \to q_1$
$\delta(q_0, b) \to q_2$
$\delta(q_1, a) \to q_0$
$\delta(q_1, b) \to q_2$
$\delta(q_2, a) \to q_2$
$\delta(q_2, b) \to q_1$
$q_0 : \{q_0\}$
$F : \{q_2\}$

2. In the tabular format, a state is represented by the name of the state.
   The beginning state is represented as $\longrightarrow q_n$
   The final state is represented as $\textcircled{q_n}$
   For the previous finite automata, the tabular format is

Present State	Next State a	Next State B
$\rightarrow q_0$	q_1	q_2
q_1	q_0	q_2
$\textcircled{q_2}$	q_2	q_1

Here

$Q : \{q_0, q_1, q_2\}$
$\Sigma : \{a, b\}$
$\delta : \delta(q_0, a) \to q_1$
$\delta(q_0, b) \to q_2$
$\delta(q_1, a) \to q_0$
$\delta(q_1, b) \to q_2$
$\delta(q_2, a) \to q_2$
$\delta(q_2, b) \to q_1$
$q_0 : \{q_0\}$
$F : \{q_2\}$

3.6 Transitional System

A transitional system (sometimes called the transitional graph) is a finite-directed graph in which each node (or vertex) represents a state, and the directed arc indicates the transition of the state. The label of the arc indicates the input or output or both.

A transitional function has two properties.

1. **Property I:** $\delta(q, \Lambda) \to q$. It means if the input is given null for a state, the machine remains in the same state.

2. **Property II:** For all string X and input symbol a ε Σ,

$$\delta(q, Xa) \rightarrow \delta(\delta(q, X), a)$$
$$\delta(q, aX) \rightarrow \delta(\delta(q, a), X)$$

3.6.1 Acceptance of a String by Finite Automata

There are two conditions for declaring a string to be accepted by a finite automaton. The conditions are

Condition I: The string must be totally traversed.

Condition II: The machine must come to a final state.

In short, it can be said that if $\delta(q_0, W) = q_n$, where W is the string given as input to the FA, q_0 is the beginning state, and q_n belongs to the set of final states, then the string W can be said to be accepted by the FA.

If these two conditions are fulfilled, then we can declare a string to be accepted by an FA.

If any of the conditions are not fulfilled, then we can declare a string to be not accepted by an FA.

The following examples (Examples 3.1 and 3.2) describe this.

Example 3.1 Check whether the string 011001 is accepted or not by the following FA.

State	Input 0	Input 1
→ q_0	q_0	q_1
q_1	q_2	q_3
q_2	q_0	q_3
(q_3)	q_1	q_3

Solution: For the given FA, $Q = \{q_0, q_1, q_2, q_3\}$ $\Sigma = \{0, 1\}$. The beginning state is q_0 and the final state is q_3.

The transitional functions are given in the table.

For checking whether a string is accepted by an FA or not, we will assume that the input string is given input in the beginning state q_0. But only a single input is given in each clock pulse. So, at first, the left most character is given the input to the beginning state. From the transitional function given in the table, the next state is determined. The next character of the input string is treated as the input to the state just achieved. And it will process like this till the string is finished or such a condition has arrived so that there is no transitional function mentioned in the table.

If the string is finished and the state achieved is the final state, then the string will be declared accepted by the machine.

If it does not happen, then the string will be declared not accepted.

$$\begin{aligned}\delta(q_0, 011001) &\rightarrow \delta(q_0, 11001)\\ &\rightarrow \delta(q_1, 1001)\\ &\rightarrow \delta(q_3, 001)\\ &\rightarrow \delta(q_1, 01)\\ &\rightarrow \delta(q_2, 1)\\ &\rightarrow q_3\end{aligned}$$

q_3 is the final state of the FA. So, the two conditions for acceptability of a string by an FA is fulfilled. So, the string 011001 is accepted by the given finite automata.

Example 3.2 Test whether the strings 010010 and 01010 are accepted by the finite automata given in Fig. 3.4 or not.

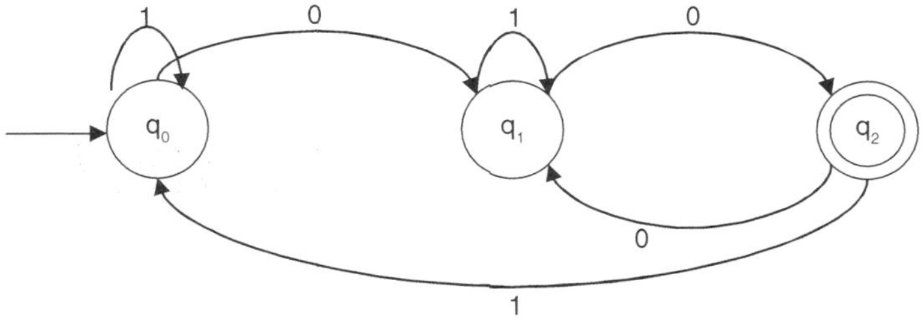

Fig. 3.4

Solution:

$$\begin{aligned}\delta(q_0, 010010) &\rightarrow \delta(q_1, 10010)\\ &\rightarrow \delta(q_1, 0010)\\ &\rightarrow \delta(q_2, 010)\\ &\rightarrow \delta(q_1,10)\\ &\rightarrow \delta(q_1,0)\\ &\rightarrow q_2\end{aligned}$$

q_2 is the final state of the FA. So, the string 010010 is accepted by the FA.

$$\begin{aligned}\delta(q_0, 01010) &\rightarrow (q_1, 1010)\\ &\rightarrow (q_1, 010)\\ &\rightarrow (q_2, 10)\\ &\rightarrow (q_0, 0)\\ &\rightarrow q_1\end{aligned}$$

q_1 is not the final state of the FA. The string is finished, but the machine has not come to the final state. So, the string is not accepted by the FA.

3.7 DFA and NFA

Finite automata are of two types: deterministic finite automata or DFA and non-deterministic finite automata or NFA.

DFA is a finite automata where, for all cases, when a single input is given to a single state, the machine goes to a single state, i.e., all the moves of the machine can be uniquely determined by the present state and the present input symbol.

A DFA can be represented as

$$M_{DFA} = \{Q, \Sigma, \delta, q_0, F\},$$

where δ is a transitional function mapping $Q \times \Sigma \rightarrow Q$, where $| Q | = 1$. The meanings of all other notations are the same as finite automata.

Non-deterministic finite automata or NDFA or NFA is a finite automaton where, for some cases, when a single input is given to a single state, the machine goes to more than one states, i.e., some of the moves of the machine cannot be uniquely determined by the present state and the present input symbol.

An NFA can be represented as

$$M_{NFA} = \{Q, \Sigma, \delta, q_0, F\},$$

where δ is a transitional function mapping $Q \times \Sigma \rightarrow 2^Q$, where 2^Q is the power set of Q.

(A power set means the set of all subsets of the set. For a set with n number of elements, the number of subsets or power set is 2^n. For example, for the set {1, 2, 3}, the subsets are { }, {1}, {2}, {3}, {1, 2}, {2, 3}, {1, 3}, and {1, 2, 3}. A total of 8, i.e., 2^3.)

As an example, consider the following machine.

Present State	Next State 0	Next State 1
→ q_0	q_0, q_1	q_2
q_1	q_2	q_1
(q_2)	q_1	q_2

Here, for the state q_0 for input 0, it can go to q_0 or q_1. So, it is an example of NDFA or NFA.

For a DFA, it can be easily determined whether a string is accepted by it or not. However, for an NFA, it is difficult.

(In real life, an NFA can be thought of driving through a path with multiple wing paths. An inexperienced driver can make mistake in choosing the wings to reach to the destination. If he/she chooses the right wing, he/she will be able to reach, but choosing the wrong route will cost failure.)

Consider Example 3.3 to clarify this.

Example 3.3 Test whether the string '00101' is accepted by the following finite automata given in Fig. 3.5 or not.

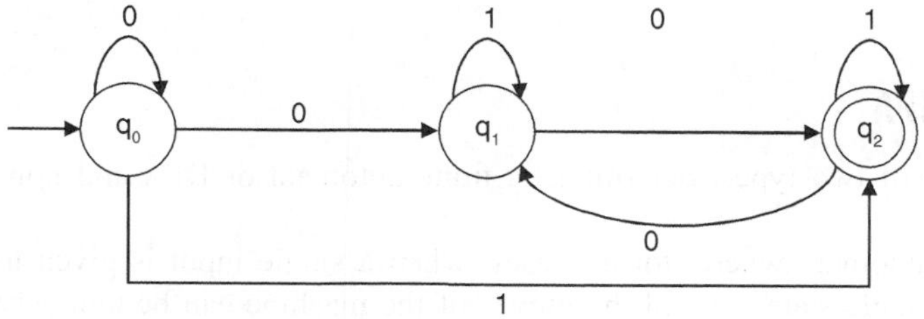

Fig. 3.5

Solution: In the previous given FA, we see that from q_0 for input 0 it can go to q_0 and q_1. That is, for q_0 for input 0 there are two paths.

Let us draw a transaction diagram like Fig. 3.6 to make the understanding clear.

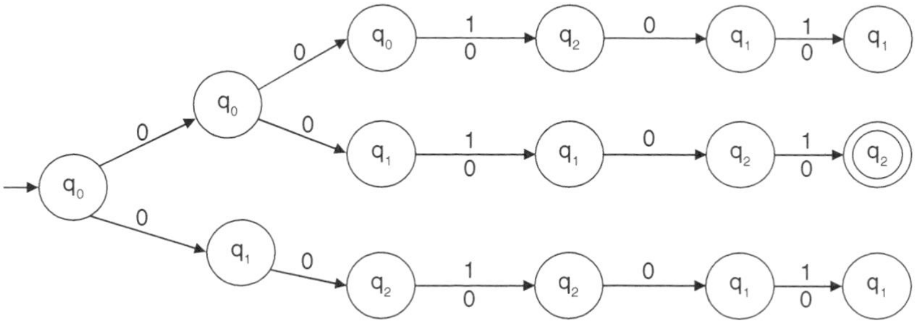

Fig. 3.6 *Transitional Path for Testing the Acceptability of 00101*

From the transaction diagram, from q_0 for input string 00101, we get three paths. For traversing the string 00101, we get the last states as q_1, q_2, and q_1, respectively. Among them, q_2 is the final state. That is, for the string 00101 from q_0 we can reach the final state. So, the string 00101 is accepted by the FA.

But, if someone follows path 1 and path 3, it will not be possible to reach the final state. Then, it can be declared that the string is not accepted by the FA. That means it depends on the path which is being followed by the FA.

A string $w \in \Sigma^*$ is accepted by an NFA if $\delta(q_0, w)$ contains some final state.

So, the string 00101 is accepted by the FA.

3.8 Conversion of an NFA to a DFA

For a DFA, δ is a transitional function mapping $Q \times \Sigma \rightarrow Q$, where $|Q| = 1$

For an NFA, δ is a transitional function mapping $Q \times \Sigma \rightarrow 2^Q$, where 2^Q is the power set of Q.

For an NFA, for a single input given to a single state, the machine can go to more than one state. So, for an NFA, it is hard to decide for a string to be accepted by the NFA or not. If an NFA can be converted to a DFA, then this type of problem will not arise.

That means, we have to convert $Q \times \Sigma \rightarrow 2^Q$ to $Q' \times \Sigma \rightarrow Q'$, where Q is the finite set of states for an NFA, and Q′ is the set of states for the converted DFA.

3.8.1 Process of Converting an NFA to a DFA

- Start from the beginning state of the NFA. Take the state within [ ].
- Place the next states for the beginning state for the given inputs in the next state columns. Put them also in [ ].
- If any new combination of state appears in next state column, which is not yet taken in the present state column, then take that combination of state in the present state column.
- If in the present state column more than one state appears, then the next state for that combination will be the combination of the next states for each of the states.
- If no new combination of state appears, which is not yet taken in the present state column, stop the process.
- The beginning state for the constructed DFA will be the beginning state of the NFA.
- The final state or final states for the constructing DFA will be the combination of states containing at least one final state.

Consider the following examples (Examples 3.4 to 3.6).

Example 3.4 Construct a DFA from the given NFA.

Present State	Next State 0	Next State 1
→ q_0	q_0, q_1	q_2
q_1	q_2	q_1
(q_2)	q_1	q_2

Solution:

Present State	Next State 0	Next State 1
$[q_0]$	$[q_0, q_1]$	$[q_2]$
$[q_0, q_1]$	$[q_0, q_1, q_2]$	$[q_1, q_2]$
$[q_1, q_2]$	$[q_1, q_2]$	$[q_1, q_2]$
$[q_0, q_1, q_2]$	$[q_0, q_1, q_2]$	$[q_1, q_2]$
$[q_2]$	$[q_1]$	$[q_2]$
$[q_1]$	$[q_2]$	$[q_1]$

Let us replace $[q_0]$ as A, $[q_0, q_1]$ as B, $[q_1, q_2]$ as C, $[q_0, q_1, q_2]$ as D, $[q_2]$ as E, and $[q_1]$ as F.

Then, the constructed finite automata is

Present State	IP	Next State 0	Next State 1
A		B	E
B		D	C
C		C	C
D		D	C
E		F	E
F		E	F

In the finite automata, we are seeing that, for a single state, for a single input the machine can go to only one state. So it is a DFA.

The beginning state is $[q_0]$, i.e., A. The final states are $[q_1, q_2]$, i.e., C, $[q_0, q_1, q_2]$, i.e., D, and $[q_2]$, i.e., E.

The transaction diagram for the DFA is given in Fig. 3.7.

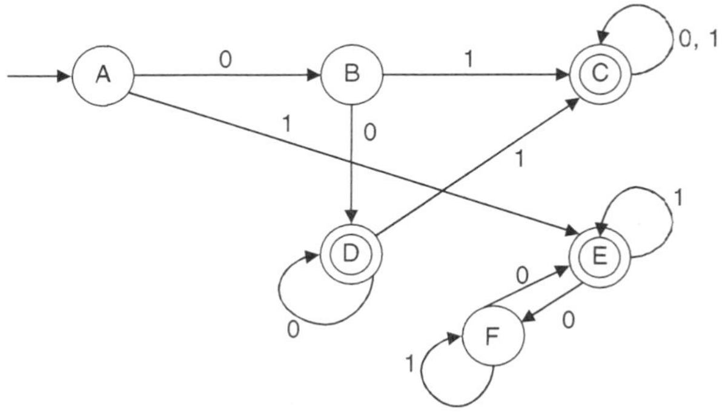

Fig. 3.7

Example 3.5 Construct a DFA from the given NFA.

Present State	Next State 0	Next State 1
→ q_0	q_0, q_1	q_0
q_1	q_2	q_1
q_2	q_3	q_3
(q_3)	–	q_2

Solution:

Present State	Next State 0	Next State 1
$[q_0]$	$[q_0, q_1]$	$[q_0]$
$[q_0, q_1]$	$[q_0, q_1, q_2]$	$[q_0, q_1]$
$[q_0, q_1, q_2]$	$[q_0, q_1, q_2, q_3]$	$[q_0, q_1, q_3]$
$[q_0, q_1, q_3]$	$[q_0, q_1, q_2]$	$[q_0, q_1, q_2]$
$[q_0, q_1, q_2, q_3]$	$[q_0, q_1, q_2, q_3]$	$[q_0, q_1, q_2, q_3]$

Let us replace $[q_0]$ as A, $[q_0, q_1]$ as B, $[q_0, q_1, q_2]$ as C, $[q_0, q_1, q_3]$ as D, and $[q_0, q_1, q_2, q_3]$ as E.

Present State IP	Next State 0	Next State 1
→ A	B	A
B	C	B
C	E	D
(D)	C	C
(E)	E	E

In the finite automata, we are seeing that, for a single state, for a single input the machine can go to only one state. So it is a DFA.

The beginning state is A. The final states are D and E.

The transaction diagram for the DFA is given in Fig. 3.8.

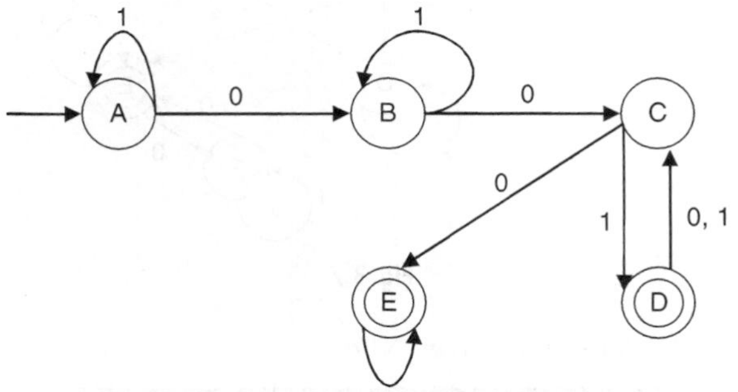

Fig. 3.8

Example 3.6 Construct an equivalent DFA for the given NFA.

Present State	IP	Next State a	Next State b
q_0		q_0, q_1	q_2
q_1		q_0	q_1
(q_2)		–	q_0, q_1

Solution:

Present State	IP	Next State a	Next State b
[q_0]		[q_0, q_1]	[q_2]
[q_0, q_1]		[q_0, q_1]	[q_1, q_2]
[q_1, q_2]		[q_0]	[q_0, q_1]
[q_2]		[–]	[q_0, q_1]
[–]		[–]	[–]

Here, for the state q_2 for input 'a' in the given NFA, no state was mentioned. So, we have taken it as a new state [–], when converting from NFA to DFA. For the new state [–], as no next state is mentioned in the NFA, we have taken the same state [–] for both the inputs a and b.

(Soon we shall learn that these types of states are called dead state.)

Let us replace [q_0] by A, [q_0, q_1] by B, [q_1, q_2] by C, [q_2] by D, and [–] by E. The tabular representation of the DFA is as follows.

	Next State	
Present State IP	a	b
A	B	D
B	B	C
C	A	B
D	E	B
E	E	E

In the finite automata, we are seeing that, for a single state, for a single input the machine can go to only one state. So, it is a DFA.

The beginning state is A. The final states are C and D as both of them contain q_2.

The transaction diagram for the DFA is given in Fig. 3.9.

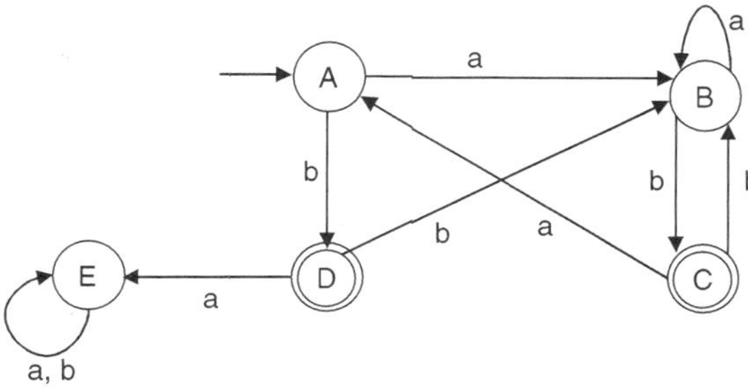

Fig. 3.9

3.9 NFA with ε (null) Move

If any finite automata contains any ε (null) move or transaction, then that finite automata is called NFA with ε moves. As an example,

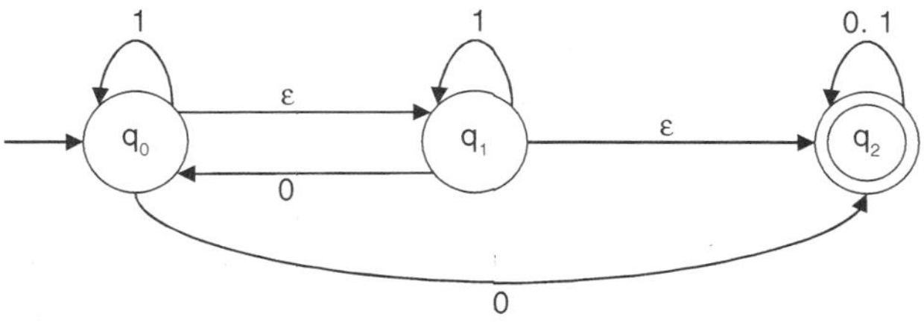

Fig. 3.10

The finite automata in Fig. 3.10 contain two null moves. So, the previous FA is NFA with null move according to the definition.

Now, in this stage, the question comes—why is this called an NFA? From the definition of an NFA, we know that for an NFA from a single state for a single input the machine can go to more than one state, i.e., $Q \times \Sigma \rightarrow 2^Q$, where 2^Q is the power set of Q.

From a state by getting ε input, the machine is confined into that state (Ref. *Property of transitional system*). An FA with null move must contain at least one ε move. For this type of finite automata, for input ε, the machine can go to more than one state. (One is that same state and another is the ε-transaction next state). So, a finite automata with ε move can be called as an NFA.

For the previous finite automata in Fig. 3.10.

$$\delta(q_0, \varepsilon) \rightarrow q_0 \text{ and } \delta(q_0, \varepsilon) \rightarrow q_1$$
$$\delta(q_1, \varepsilon) \rightarrow q_1 \text{ and } \delta(q_1, \varepsilon) \rightarrow q_2$$

An NFA with ε move can be defined as

$$M_{\text{NFA, null}} = \{Q, \Sigma, \delta, q_0, F\}$$

where Σ: set of input alphabets including ε, and δ is a transitional function mapping. $Q \times \Sigma \rightarrow 2^Q$, where 2^Q is the power set of Q.

3.9.1 Usefulness of NFA with ∈ Move

If we want to construct an FA which accepts a language, sometimes it becomes very difficult or seems to be impossible to construct a direct NFA or DFA. But if NFA with ε moves is used, then the transitional diagram can be constructed and described easily.

Consider the following example.

Example 3.7 Let us assume that we are given to construct a DFA that accepts any combinations of string of a, b, c. That means the string can start with a or b or c. After any symbol, any of the symbols can come in any combination and all strings are accepted. The DFA for this is given in Fig. 3.11.

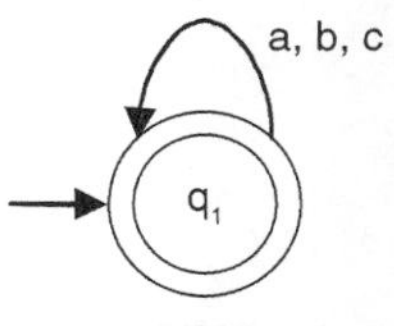

Fig. 3.11

Solution: But it is not clearly understandable as to how any combinations of a, b, c can come. In this case, an NFA with ε move plays its role as represented by the transitional diagram in Fig. 3.12.

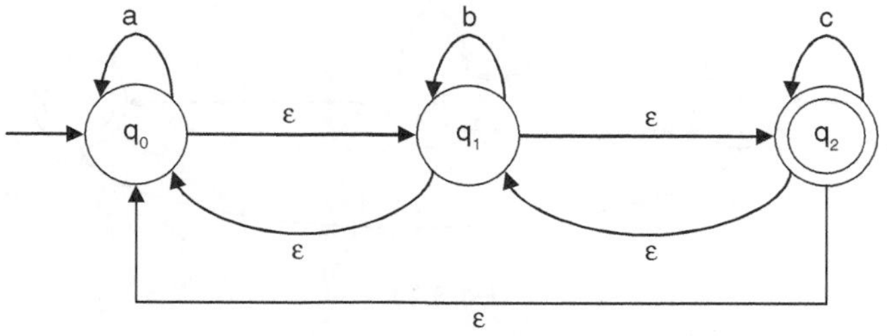

Fig. 3.12

Now it is clearly understandable that the string can start with a or b or c. And for the null transitions from any state, we can move to any other state with no input. So, if c comes after 'a', we can go from q_0 to q_2 easily with no input. The same thing can occur for all the other states.

3.9.2 Conversion of an NFA with ε Moves to DFA without ε Move

Let us assume that in the NFA we want to remove the ε move which exists between the states S_1 and S_2. This can be removed in the following way

- Find all the edges (transitions) those start from the state S_2.
- Duplicate all these transitions starting from the state S_1, keeping the edge label the same.
- If S_1 is the initial state, also make S_2 as the initial state.
- If S_2 is the final state, also make S_1 as the final state.

Consider the following examples.

Example 3.8 Convert the following NFA with null move in Fig. 3.13 to an equivalent DFA without ε move.

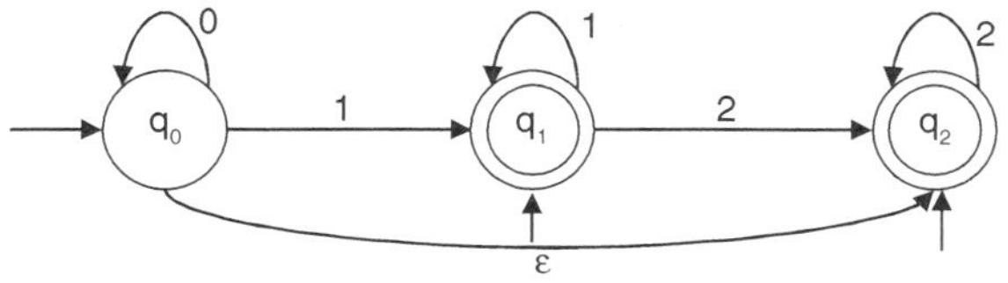

Fig. 3.13

Solution: Three null moves exist in the previous transitional diagram. The transitions are

1. from q_0 to q_1
2. from q_1 to q_2
3. from q_0 to q_2

These null transitions are removed step by step.

Step I:

1. Between the states q_0 and q_1, there is a null move. If we want to remove that null transition, we have to find all the edges starting from q_1.The edges are q_1 to q_1 for input 1 and q_1 to q_2 for input ε.
2. All these transitions starting from the state q_0 are duplicated, keeping the edge label the same. q_0 is the initial state, so make q_1 also an initial state. The modified transitional diagram is given in Fig. 3.14.

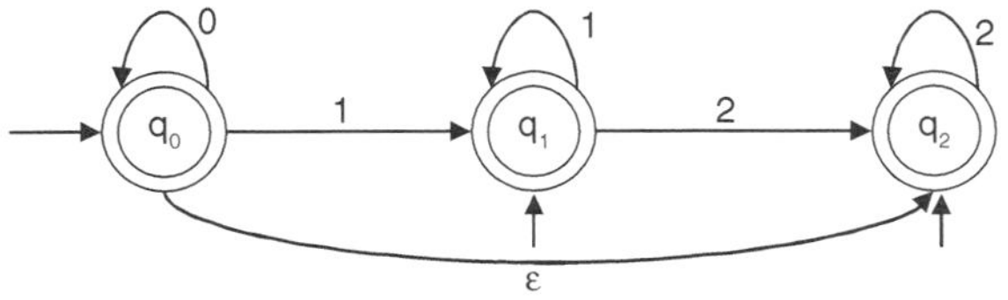

Fig. 3.14

Step II:

1. Between q_1 and q_2, there is a null transaction. To remove that null transition, find all the edges starting from q_2. The edge is q_2 to q_2 for input 2.

2. Duplicate the transition starting from the state q_1, keeping the edge label the same. q_1 is the initial state, so make q_2 also an initial state. q_2 is the final state, so make q_1 also a final state. The modified transitional diagram is given in Fig. 3.15.

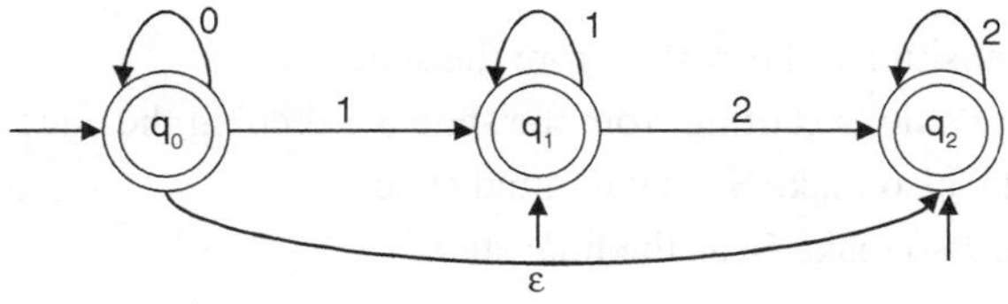

Fig. 3.15

Step III:

1. Between q_0 and q_2, there is a null transaction. To remove that null transition, find all the edges starting from q_2. The edge is q_2 to q_2 for input 2.
2. Duplicate the transition starting from the state q_0, keeping the edge label the same. q_0 is the initial state, so make q_2 also an initial state. q_2 is the final state, so make q_0 also a final state. The modified transitional diagram is given in Fig. 3.16.

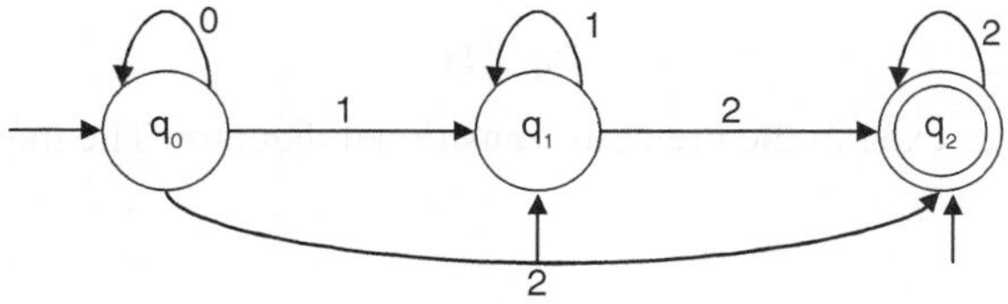

Fig. 3.16

This is the equivalent NFA without ε move obtained from the NFA with null transaction.

Example 3.9 Convert the following NFA with null move to an equivalent NFA without ε move as given in Fig. 3.17.

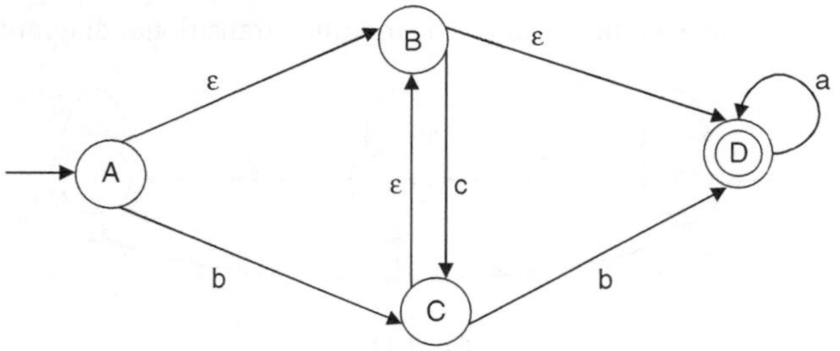

Fig. 3.17

Solution: In the given NFA with ε move, there are three ε transitions: from A to B, from B to D, and from B to C. The three ε transitions are removed step by step.

Step I: There is a ε transition from A to B. To remove that ε transition, we have to find all the edges starting from B. The edges are B to C for input c and B to D for input ε.

Duplicate the transition starting from the state A, keeping the edge label the same. A is the initial state, so make B also an initial state. The modified transaction diagram is shown in Fig. 3.18.

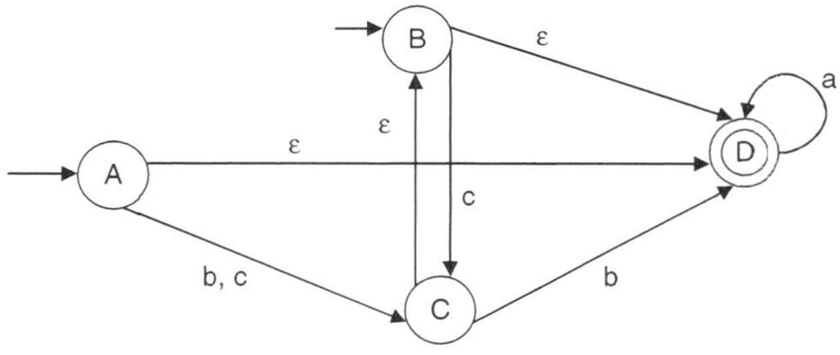

Fig. 3.18

Step II: Again there are three ε transactions in the NFA. These are from A to D, from B to D, and from C to B.

Let us remove the ε transaction from B to D. The edge starting from D is D to D for input a. Duplicate the transition starting from the state B, keeping the edge label the same. As B is the initial state, D will also be the initial state. As D is the final state, B will also be the final state. The modified transaction diagram will be as shown in Fig. 3.19.

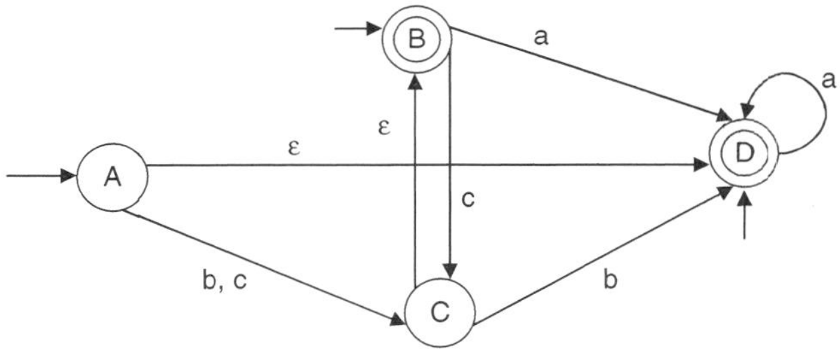

Fig. 3.19

Step III: Now we are going to remove the ε transaction from A to D. The edge starting from D is D to D for input a. Duplicate the transition starting from the state A, keeping the edge label the same. As A is the initial state, D will also be the initial state. As D is the final state, A will also be the final state. The modified transaction diagram will be as shown in Fig. 3.20.

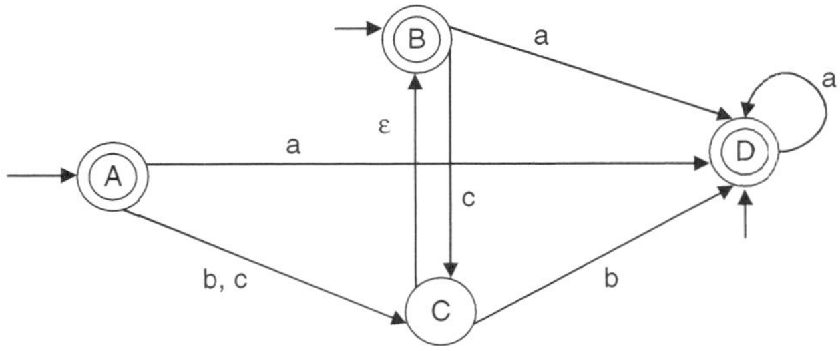

Fig. 3.20

Step IV: There is only one ε transaction from C to B. To remove this, find the edges starting from B. There are two: from B to D for input a and from B to C for input c. Start these edges from C. B is the final state, so make C as the final state. The modified transitional diagram will be as shown in Fig. 3.21.

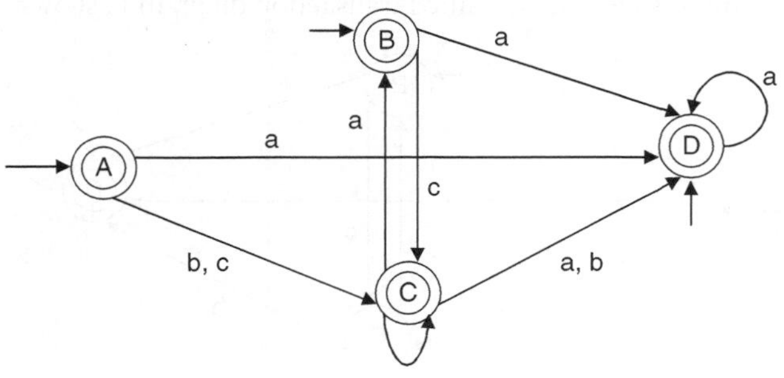

Fig. 3.21

This is the equivalent NFA without ε move obtained from the NFA with null transaction.

3.10 Equivalence of DFA and NFA

A DFA can simulate the behaviour of an NFA by increasing its number of states. The DFA does not contain any null move and any transitional functions of a state and input reaching to more than one states. In this section, we shall prove the following:

'If a language L is accepted by an NDFA, then there exists a DFA that accepts L'.

Proof: Let M be an NDFA denoted by $\{Q, \Sigma, \delta, q_0, F\}$ which accepts L.

We construct a DFA $M' = \{Q', \Sigma, \delta', q'_0, F'\}$, where Q′ contains the subsets of Q, i.e., $Q' = 2^Q$.

Q′ may be denoted by $[q_1, q_2, \ldots.. q_n]$ as a single state where $q_1, q_2, \ldots.. q_n \in Q$

Initial state $q'_0 = [q_0]$

Final state F′ = set of all states in Q′ containing at least one final state of M.

Transitional function δ′ is defined as

$$\delta'([q_1, q_2, \ldots.. q_n], a) = [P_1, P_2, \ldots.. P_k] \text{ where } q_1, q_2, \ldots.. q_n \in Q, a \in \Sigma \text{ and}$$
$$\delta(\{q_1, q_2, \ldots.. q_n\}, a) = \delta(q_1, a) \cup \delta(q_2, a) \cup \ldots.. \cup \delta(q_n, a)$$
$$= P_1 \cup P_2 \cup \ldots.. \cup P_k$$

This is the case for a single input 'a'. Now, we shall prove it for some input string x. x may be of length 0 to length n. We prove this by induction.

Let $| x | = 0$, i.e., x is a null string.

$$\delta'(q'_0, x) = [q_0] \text{ as } q'_0 = [q_0] \text{ and } \delta(q_0, x) = \delta(q_0, \varepsilon) = q_0.$$

Thus, the induction has a base condition.

Let us assume that it is true for each string of length n. Now we need to prove that the result is true for any string of length (n + 1).

Let $\quad S = xa.$

So, $\quad | S | = | xa | = | x | + | a | = n + 1.$

$$\delta'(q'_0, S) = \delta'(q'_0, xa) = \delta'(\delta'(q'_0, x)a).$$

By induction, it is proved that $\delta'(q'_0, x) = [P_1, P_2, \ldots. P_k]$ if and only if

$$\delta(q_0, x) = \{P_1, P_2, \ldots. P_k\}$$

So, $\delta'(\delta'(q'_0, x)a) = \delta'([P_1, P_2, \ldots. P_k], a) = [r_1, r_2, \ldots. r_k]$ if and only if

$$\delta(\{P_1, P_2, \ldots. P_n\}, a) = \delta(P_1, a) \cup \delta(P_2, a) \cup \ldots. \cup \delta(P_n, a)$$
$$= r_1 \cup r_2 \cup \ldots. \cup r_k$$

Thus $\delta'(q'_0, xa) = \delta'(\delta'(q'_0, x)a) = [r_1, r_2, \ldots. r_k]$

It is proved that the result is true for a string of length n + 1, if it is true for a string of length n. Now $\delta'(q'_0, x) \in F'$ when $\delta(q_0, x)$ to a state of Q in F.

Therefore, it is proved that L(M) = L(M′).

3.11 Dead State

The name dead state was mentioned in Example 3.6 of converting an NFA to an equivalent DFA. Dead state is a state where the control can enter and be confined till the input ended, but there is no way to come out from that state. (The string is dead or finished on entering the state.)

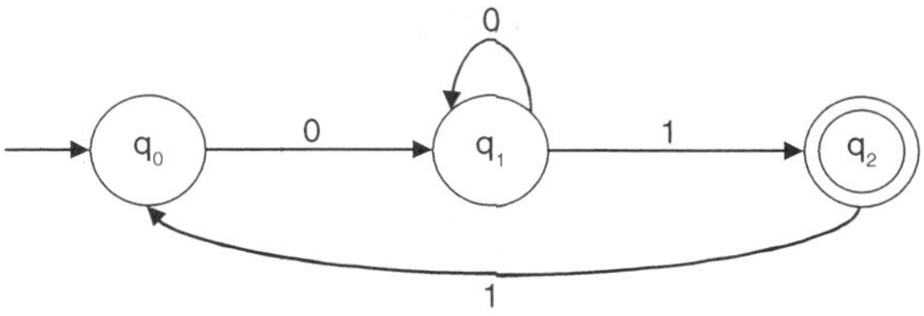

Fig. 3.22

An example for the previous finite automata is shown in Fig. 3.22; from q_0 for input 0, there is a path to go to q_1. But there is no path mentioned from q_0 for input 1. The same thing happens also for q_2 for input 0. Now, let a string 101 be given as input to the FA to test the acceptability of the string by the finite automata.

According to the condition of the acceptability, the first condition, i.e., the string will be totally traversed, is not fulfilled. Obviously, the second condition is also not fulfilled.

We have to decide that the string is not accepted by the finite automata without traversing the string totally!

For making the string totally traversed, we have to include an extra state, say q_f, where the control will go from the states for which there is only one path for one input. In the state q_f, for the inputs, there will be a self-loop. Here, q_f is the dead state.

By including the dead state, the finite automata becomes as shown in Fig. 3.23.

Dead state makes the string totally traversed.

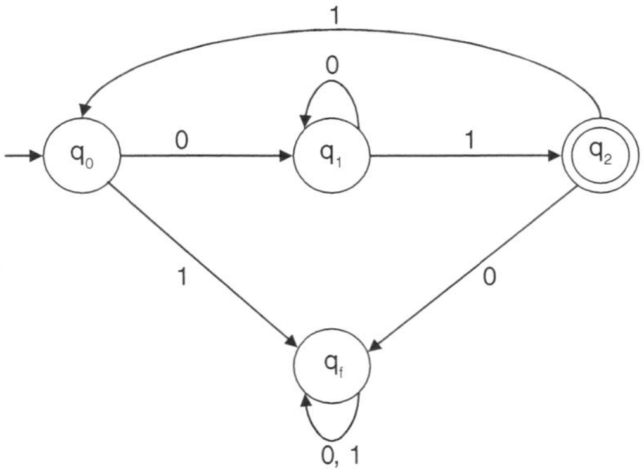

Fig. 3.23

3.12 Finite Automata with Output

Till now, we were discussing machines with no output. But two things, output and output relations, are omitted. This was mentioned in the characteristics of the automaton. As FA is a machine, it must produce output. In this section, we shall discuss finite automata with output. Finite automata with output can be divided into two types:

1. The Mealy machine
2. The Moore machine

3.12.1 The Mealy Machine

The Mealy machine was proposed by *George H. Mealy* at the Bell Labs in 1960.

The Mealy machine is one type of finite automata with output, where the output depends on the present state and the present input.

The Mealy machine consists of six touples

$$M = (Q, \Sigma, \Delta, \delta, \lambda, q_0),$$

where

Q : finite non-empty set of states
Σ : set of input alphabets
Δ : set of output alphabets
δ : transitional function mapping $Q \times \Sigma \rightarrow Q$
λ : output function mapping $Q \times \Sigma \rightarrow \Delta$
q_0 : beginning state

3.12.2 The Moore Machine

The Moore machine was proposed by *Edward F. Moore* in IBM around 1960.

The Moore machine is one type of finite automata where output depends on the present state only, but the output is independent of the present input.

The Moore machine consists of six touples

$$M = (Q, \Sigma, \Delta, \delta, \lambda, q_0),$$

where

Q : finite non-empty set of states
Σ : set of input alphabets
Δ : set of output alphabets
δ : transitional function mapping $Q \times \Sigma \rightarrow Q$
λ : output function mapping $Q \rightarrow \Delta$
q_0 : beginning state

3.12.3 Tabular and Transitional Representation of the Mealy and Moore Machines

3.12.3.1 Tabular

The output of the Mealy machine depends on both the present state and the present input. So, for the mealy machine, there will be n number of output columns if there are n number of input.

The tabular format of the Mealy machine is as follows.

	I/P = 0		I/P = 1	
Present State	Next State	O/P	Next State	O/P
A	D	0	B	0
B	A	1	D	0
C	B	1	A	1
D	D	1	C	0

The output of the Moore machine depends on the present state only. As in a machine, there is only one present state column, so there is only one output column.

The tabular format of the Moore machine is given in the following table.

	Next State		
Present State	I/P = 0	I/P = 1	O/P
A	A	B	0
B	A	C	0
C	A	C	1

3.12.3.2 *Transitional*

The Mealy machine is one type of finite automata with output. The transitional diagram is like finite automata, but the output is mentioned. As we know, the output of a Mealy machine depends on the present state and the present input. So, for a Mealy machine, the transitional arc will be labelled with both input and output like the example in Fig. 3.24.

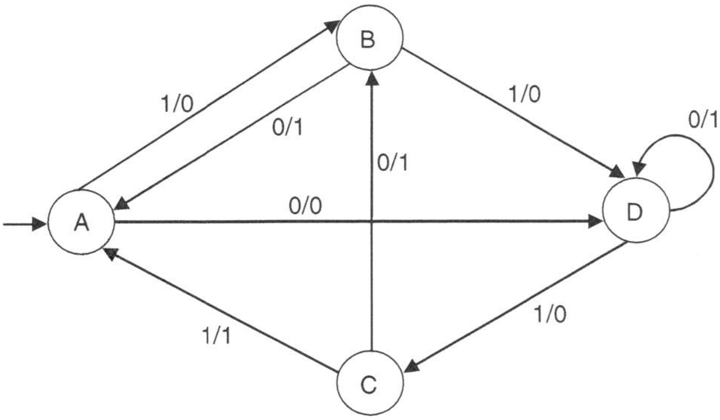

Fig. 3.24

The output of the Moore machine depends only on the present state. For a Moore machine, the states are labelled with the state name and the output like the example shown in Fig. 3.25.

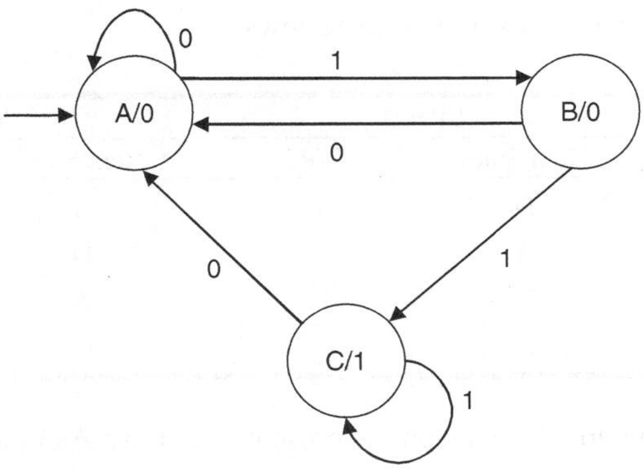

Fig. 3.25

The following are some examples of finite automata with output.

Example 3.10 Design a Mealy machine to get complement of binary number.

Solution: For input 0, the machine gives the output 1, and for input 1 the machine will give the output 0. The Mealy machine is constructed as shown in Fig. 3.26.

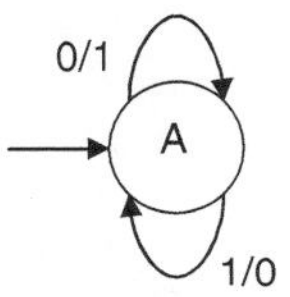

Fig. 3.26

Example 3.11 Design a Moore machine which determines the residue mod-3 for each binary string treated as binary integer.

Solution: Binary string consists of two types of characters 0 and 1. Binary integer means the decimal equivalent of a binary string as integer number. Residue mod-3 means the reminder received when the decimal equivalent of the binary number is divided by 3.

If a decimal number is divided by 3, only three types of reminders 0 or 1 or 2 are generated. The output of the Moore machine depends only on the present state. So, for three types of reminders (output), three states are needed.

Let us assume that state q_0 is producing output 0, state q_1 is producing output 1, and state q_2 is producing output 2. Now, it is needed to construct the transitional arcs. Before construction, take a table containing binary equivalent of decimal numbers from 0 to 9 and the reminders for each of them, when divided by 3.

Decimal	Binary	Reminders
0	0	0
1	1	1
2	10	2
3	11	0
4	100	1
5	101	2
6	110	0
7	111	1
8	1000	2
9	1001	0

Input 0, output is 0. The transitional function is $\delta(q_0, 0) \rightarrow q_0$, because q_0 state produces output 0. Input 1, output 1, the transitional function is $\delta(q_0, 1) \rightarrow q_1$, because q_1 state produces output 1. Input 10, output is 2. So, the transitional function is $\delta(q_1, 0) \rightarrow q_2$, because q_2 state produce output 2. Input 11, output is 0. So, the transitional function is $\delta(q_1, 1) \rightarrow q_0$. By this process, all the transitional functions can be produced.

The Moore machine is given in Fig. 3.27 for the given problem.

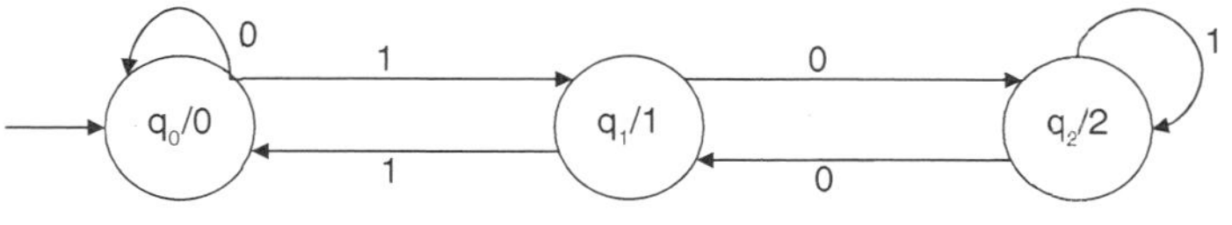

Fig. 3.27

Example 3.12 Design a Mealy machine which determines the residue mod-3 for each binary string treated as a binary integer.

Solution: The machine is a Mealy machine, and so the outputs depend on the present state and the present input.

For input 0, output is 0. The transitional function is $\delta(q_0, 0) \rightarrow q_0$ and output function λ is $q_0 \times 0 \rightarrow$ 0. For input 1, output is 1. The transitional function becomes $\delta(q_0, 1) \rightarrow q_1$ and the output function λ is $q_0 \times 1 \rightarrow 1$. For input 10, output is 2. So, the transitional function is $\delta(q_1, 0) \rightarrow q_2$, and the output function λ is $q_1 \times 0 \rightarrow 2$. For input 11, output is 0, which makes the transitional function as $\delta(q_1, 1) \rightarrow q_0$, and the output function λ is $q_1 \times 0 \rightarrow 0$. By this process, all the transitional functions are constructed. The final Mealy machine for the problem is given in Fig. 3.28.

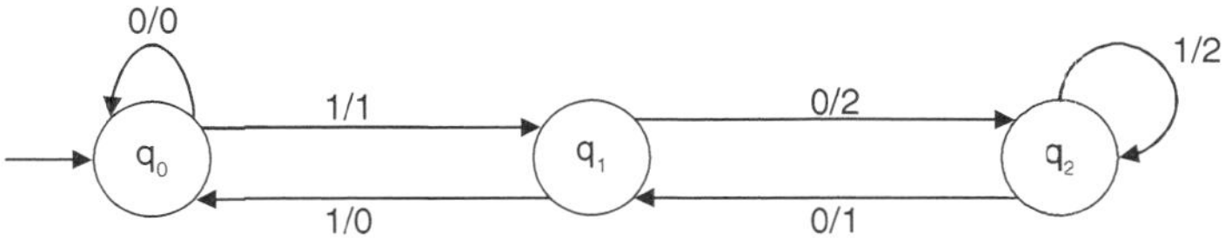

Fig. 3.28

Example 3.13 Design a Moore machine which counts the occurrence of substring aba in a given input string.

Solution: It is one type of sequence detector. When the string aba occurs as a substring, then output 1 is produced. For all the other cases, the output 0 is produced (overlapping sequences are also accepted).

We have to design a Moore machine, i.e., the output depends only on the present state. For a string length of 3, we need four states if from a state with single input the machine goes to another new state. Let us name the states as q_0, q_1, q_2, and q_3. Except q_3, all the states produce output 0. In state q_0, we can get two types of input, a or b. If we get 'a', then there is a chance to get 'aba'. So, it moves forward to q_1. If we get 'b', then there is no chance to get 'aba', so it confines to the same state. In q_1, again two types of input can occur, 'a' and 'b'. If we get 'a', then by considering this 'a' there is a chance to get 'aba'. So, it confines to the same state. If we get b, then we are two steps forward to get 'aba'. So, the machine moves forward to state q_2. In q_2, if we get 'b' then there is no chance to get 'aba' except starting from the beginning. So, it traverses to q_0. If we get 'a', we have got the string aba. The machine will go to state q_3. In q_3, if we get 'a', then again there is a chance to get 'aba'. From q_0, by getting 'a', the machine has moved forward to q_1. So, from q_2 by getting 'a' the machine will go to q_1. If we get 'b', then by considering the previous 'a' again there is a chance to get 'aba', so the machine moves to state 'q_2'.

From the earlier discussions, the transitional functions can easily be derived. As it is a Moore machine, the outputs are included with the state (output depends only on the present state and not on the present input), and so there is no need to derive the output functions. The constructed Moore machine for the earlier problem is given in Fig. 3.29.

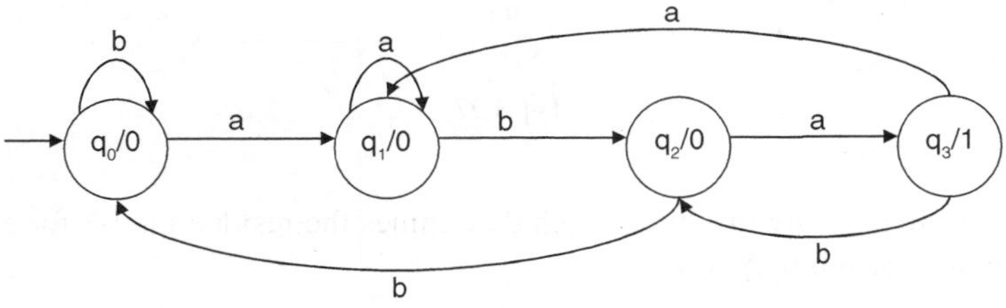

Fig. 3.29

Example 3.14 Design a Mealy machine which counts the occurrence of substring aab in a given input string.

Solution: The machine will be a Mealy machine, that is, the output depends on the present state and the present input. (Discussions will be like previous.)

The machine is given in Fig. 3.30.

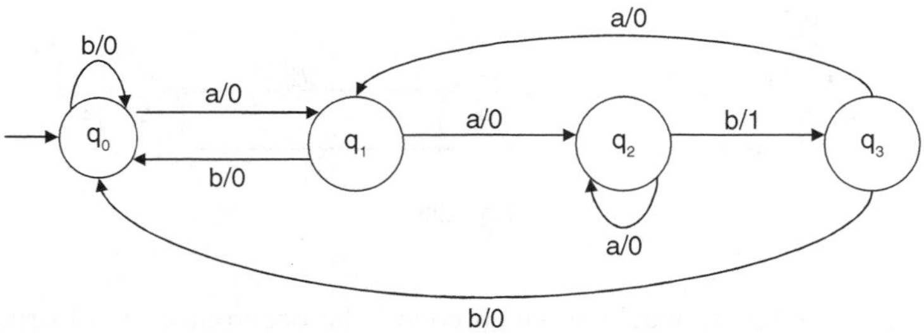

Fig. 3.30

3.13 Conversion of One Machine to Another

Mealy and Moore are two different machines. In the Mealy machine, the output depends on the present state and the present input. For this reason, the length of the output is the same as the given input string. But in the Moore machine, the output depends only on the present state. One output is counted as the traversal starts from the initial state without taking any input character. For this reason, the length of the output is one more than the length of the input string.

It can be proved that if $M = (Q, \Sigma, \Delta, \delta, \lambda, q_0)$ is a Moore machine, then there exists an equivalent Mealy machine $M' = (Q, \Sigma, \Delta, \delta, \lambda', q_0)$, where $\lambda'(Q, a) = \lambda(\delta(Q, a))$.

The following section covers how to convert a Moore machine to a Mealy machine and vice versa.

3.13.1 Tabular Format

3.13.1.1 Conversion of a Moore Machine to a Mealy Machine

- Draw the tabular format of a Mealy machine.
- Put the present states of the Moore machine in the present state column of the constructing Mealy machine.
- Put the next states for the corresponding present states and input of the Moore machine in the next state column for those present states and input, respectively, of the corresponding Mealy machine.
- For the output, look into the present state column and the output column of the Moore machine. The output for Q_{Next} (next state for the present state $Q_{Present}$ and input I_P of the constructing Mealy machine) will be the output for that state (Q_{Next}) as a present state in the given Moore machine.

The following examples (Examples 3.15 to 3.17) describe the previous method.

Example 3.15 Construct a Mealy machine equivalent to the given Moore machine.

Present State	Next State I/P = 0	Next State I/P = 1	O/P
→ q_0	q_3	q_1	0
q_1	q_1	q_2	1
q_2	q_2	q_3	0
q_3	q_3	q_3	0

Solution:

	I/P = 0		I/P = 1	
Present State	Next State	O/P	Next State	O/P
→ q_0	q_3	0	q_1	1
q_1	q_1	1	q_2	0
q_2	q_2	0	q_3	0
q_3	q_3	0	q_3	0

In the Moore machine, for q_1 as a present state, the output is 1. So, in the constructing Mealy machine, for q_1 as a next state the output will be 1.

In the Moore machine, for q_2 as a present state, the output is 0. So, in the constructing Mealy machine, for q_2 as a next state the output will be 0.

Example 3.16 Construct a Mealy machine equivalent to the given Moore machine.

Present State	Next State I/P = 0	Next State I/P = 1	O/P
→ q_0	q_1	q_2	1
q_1	q_3	q_2	0
q_2	q_2	q_1	1
q_3	q_0	q_3	1

Solution:

Present State	I/P = 0 Next State	I/P = 0 O/P	I/P = 1 Next State	I/P = 1 O/P
→ q_0	q_1	0	q_2	1
q_1	q_3	1	q_2	1
q_2	q_2	1	q_1	0
q_3	q_0	1	q_3	1

Example 3.17 Construct a Mealy machine equivalent to the given Moore machine.

Present State	Next State I/P = 0	Next State I/P = 1	O/P
→ q_0	q_1	q_0	0
q_1	q_1	q_2	0
q_2	q_3	q_0	0
q_3	q_1	q_2	1

Solution:

Present State	I/P = 0 Next State	I/P = 0 O/P	I/P = 1 Next State	I/P = 1 O/P
→ q_0	q_1	0	q_0	0
q_1	q_1	0	q_2	0
q_2	q_3	1	q_0	0
q_3	q_1	0	q_2	0

3.13.1.2 Conversion of a Mealy Machine to a Moore Machine

- Draw the tabular format of a Mealy Machine first.
- Look into the next state and output columns of the given Mealy machine.
- If for the same next state in the next state column, the output differs in the output column, break the state q_i into different number of states. The number is equal to the number of different outputs associated with q_i.
- Put the states of the present state column in the new table. Put the broken states in the place of those states that are broken into a number of different states.
- Change the next states in the next state columns according to the new set of states.
- Put the output from the output column of the original Mealy machine in the new machine also.
- Draw the tabular format of a Moore machine.
- Put the present states and next states from the newly constructed Mealy machine into the constructed Moore machine.

- For output, look into the next state and output column of the newly constructed Mealy machine. For a state as a next state (let q_i) in the new constructed Mealy Machine if output is 0 then for the state (q_i) as a present state in the constructing Moore Machine; the output will be 0.
- For the Moore machine, the output depends only on the present state. This means that from the beginning state for Λ input, we can get an output. If the output is 1, the newly constructed Moore machine can accept zero length string, which was not accepted by the given Mealy machine. To make the Moore machine not to accept Λ string, we have to add an extra state q_b (new beginning state), whose state transactions will be identical with those of the existing beginning state, but with output as 0.

(For a Mealy machine with m number of states and n number of outputs, the Moore machine will be of n number of outputs but not more than mn + 1 number of states.)

The following examples (Example 3.18 to 3.20) describe the previous method.

Example 3.18 Convert the given Mealy machine to an equivalent Moore machine.

	I/P = 0		I/P = 1	
Present State	Next State	O/P	Next State	O/P
→ q_0	q_0	1	q_1	0
q_1	q_3	1	q_3	1
q_2	q_1	1	q_2	1
q_3	q_2	0	q_0	1

Solution: Look into the next state and output columns of the given Mealy machine. For I/P 0 for q_1 as a next state, the output is 1. For I/P 1 for q_1 as a next state, the output is 0. The same thing happens for q_2 as a next state for input 0 and input 1. So the state q_1 is broken as q_{10} and q_{11}, and the state q_2 is broken as q_{20} and q_{21}. After breaking, the modified Mealy machine becomes

	I/P = 0		I/P = 1	
Present State	Next State	O/P	Next State	O/P
→ q_0	q_0	1	q_{10}	0
q_{10}	q_3	1	q_3	1
q_{11}	q_3	1	q_3	1
q_{20}	q_{11}	1	q_{21}	1
q_{21}	q_{11}	1	q_{21}	1
q_3	q_{20}	0	q_0	1

For the present state q_0 for input 1, the next state is q_{10}, because there is no q_1 in the modified Mealy machine. It has been broken into q_{10} and q_{11} depending on the output 0 and 1, respectively. For the present state q_0 for input 1, the output is 0. So the next state is q_{10}. The same cases occur for the others also. For the broken states, the next states and outputs are the same as the original, from where the broken states have come.

From this, the Moore machine becomes

Present State	Next State I/P = 0	Next State I/P = 1	O/P
→ q_0	q_0	q_{10}	1
q_{10}	q_3	q_3	0
q_{11}	q_3	q_3	1
q_{20}	q_{11}	q_{21}	0
q_{21}	q_{11}	q_{21}	1
q_3	q_{20}	q_0	1

For Moore machine the beginning state is q_0 and the corresponding output is 1. That means, with null length input (no input), we are getting an output of 1. That is, the Moore machine accepts 0 length sequence [because here output depends only on the present state] which is not acceptable for the Mealy machine. To overcome this situation, we must add a new beginning state q_b with the same transactions as q_0, but with output as 0. By including the new state, the Moore machine is

Present State	Next State I/P = 0	Next State I/P = 1	O/P
→ q_b	q_0	q_{10}	0
q_0	q_0	q_{10}	1
q_{10}	q_3	q_3	0
q_{11}	q_3	q_3	1
q_{20}	q_{11}	q_{21}	0
q_{21}	q_{11}	q_{21}	1
q_3	q_{20}	q_0	1

Example 3.19 Convert the given Mealy machine to an equivalent Moore machine.

	I/P = 0		I/P = 1	
Present State	Next State	O/P	Next State	O/P
→ q_0	q_2	z_0	q_1	z_1
q_1	q_0	z_0	q_2	z_0
q_2	q_0	z_1	q_2	z_1

Solution: For q_0 for input 0, the output differs. For q_2 for inputs 0 and 1, the output differs. So the states are broken as q_{00}, q_{01} and q_{20}, q_{21}. According to the new states, the modified Mealy machine becomes

	I/P = 0		I/P = 1	
Present State	Next State	O/P	Next State	O/P
→ q_{00}	q_{20}	z_0	q_1	z_1
q_{01}	q_{20}	z_0	q_1	z_1

	I/P = 0		I/P = 1	
Present State	Next State	O/P	Next State	O/P
q_1	q_{00}	z_0	q_{20}	z_0
q_{20}	q_{01}	z_1	q_{21}	z_1
q_{21}	q_{01}	z_1	q_{21}	z_1

From this modified Mealy machine, the Moore machine is

	Next State		
Present State	I/P = 0	I/P = 1	O/P
→ q_{00}	q_{20}	q_1	z_0
q_{01}	q_{20}	q_1	z_1
q_1	q_{00}	q_{20}	z_1
q_{20}	q_{01}	q_{21}	z_0
q_{21}	q_{01}	q_{21}	z_1

Example 3.20 Convert the given Mealy machine to an equivalent Moore machine.

	I/P = 0		I/P = 1	
Present State	Next State	O/P	Next State	O/P
→ q_0	q_2	0	q_1	0
q_1	q_0	1	q_3	0
q_2	q_1	1	q_0	1
q_3	q_3	1	q_2	0

Solution: In the next state column of the given Mealy machine, the output differs for q_1 and q_3 as the next state. So, the states are divided as q_{10}, q_{11} and q_{30}, q_{31}, respectively. After dividing the states, the modified Mealy machine becomes

	I/P = 0		I/P = 1	
Present State	Next State	O/P	Next State	O/P
→ q_0	q_2	0	q_{10}	0
q_{10}	q_0	1	q_{30}	0
q_{11}	q_0	1	q_{30}	0
q_2	q_{11}	1	q_0	1
q_{30}	q_{31}	1	q_2	0
q_{31}	q_{31}	1	q_2	0

In the next state column of the modified Mealy machine, when q_0 is a next state, the output is 0. So, in the constructing Moore machine, for the present state q_0, the output is also 0. Similarly, for the present state q_2, the output is 0. For the divided states like q_{10}, q_{11}, there is no need to mention the output as they were divided according to the distinguished output. So, the constructing Moore machine is

Present State	Next State		O/P
	I/P = 0	I/P = 1	
→ q_0	q_2	q_{10}	1
q_{10}	q_0	q_{30}	0
q_{11}	q_0	q_{30}	1
q_2	q_{11}	q_0	0
q_{30}	q_{31}	q_2	0
q_{31}	q_{31}	q_2	1

To get rid of the problem of the occurrence of null string, we need to include another state, q_a, with same transactions as of q_0 but with output 0.

The modified final Moore machine equivalent to the given Mealy machine becomes

Present State	Next State		O/P
	I/P = 0	I/P = 1	
→ q_a	q_2	q_{10}	0
q_0	q_2	q_{10}	1
q_{10}	q_0	q_{30}	0
q_{11}	q_0	q_{30}	1
q_2	q_{11}	q_0	0
q_{30}	q_{31}	q_2	0
q_{31}	q_{31}	q_2	1

3.13.2 Transitional Format

From the transitional diagram of a Moore or Mealy machine, conversion can be done easily. That process is described here.

3.13.2.1 Moore Machine to Mealy Machine

Let us assume that a Moore machine M_o is to be converted to an equivalent Mealy machine M_C. There are certain steps for this conversion.

Step I: For a Moore machine, each state is labelled with the output, because for a Moore machine, output depends only on the present state. For the conversion, let us take a state S, which is labelled with output O. Look into the incoming edges to the state S. These incoming edges are labelled by the input alphabets of the Moore machine. These incoming edges are relabelled by the input alphabet as well as the output of the state S. The output for the state must be removed.

Step II: Keep the outgoing edges from the state S as it was. (The outgoing edge of a state must be the incoming edge of some other state.)

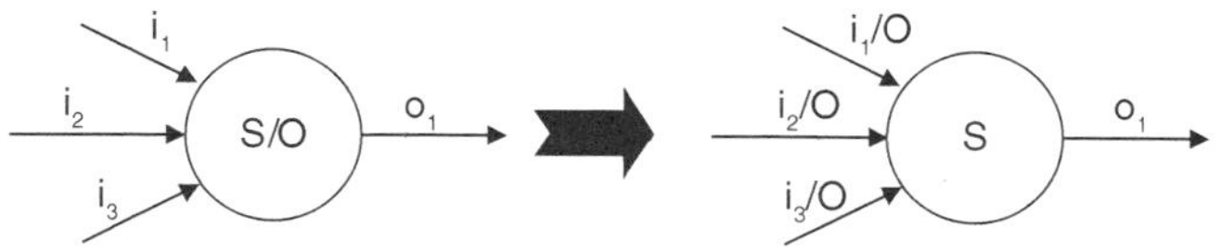

Fig. 3.31 *Process of Conversion of a Moore Machine to a Mealy*

Step III: Repeat these steps for each of the states. By this process, the Mealy machine-equivalent Moore machine is constructed. The process is described in Fig.3.31.

Consider the following examples.

Example 3.21 Convert the following Moore machine into an equivalent Mealy machine as given in Fig. 3.32.

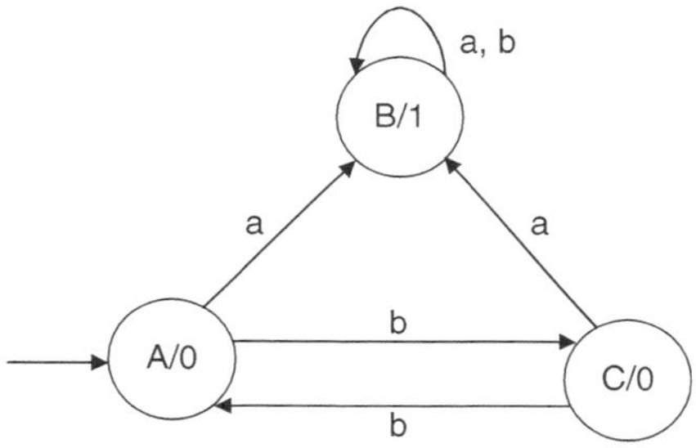

Fig. 3.32

Solution:

1. In this machine, A is the beginning state. So, start from A. For A, there are two incoming arcs, from C to A with input b and another in the form of the start-state indication with no input. State A is labelled with output 0. As the start-state indication contains no input, it is useless and, therefore, keep it as it is.

   Modify the label of the incoming edge from C to B including the output of state A. So, the label of the incoming state is C to A with label b/0.
2. State B is labelled with output 1. The incoming edges to the state B are from A to B with input a, B to B with inputs a and b, and C to B with input a.

   Modify the labels of the incoming edges including the output of state B. So, the labels of the incoming states become A to B with label a/1, B to B with labels a/1 and b/1, and C to B with label a/1.
3. State C is labelled with output 0. There is only one incoming edge to this state, from A to C with input b. The modified label is b/0.

   The converted Mealy machine is given in Fig. 3.33.

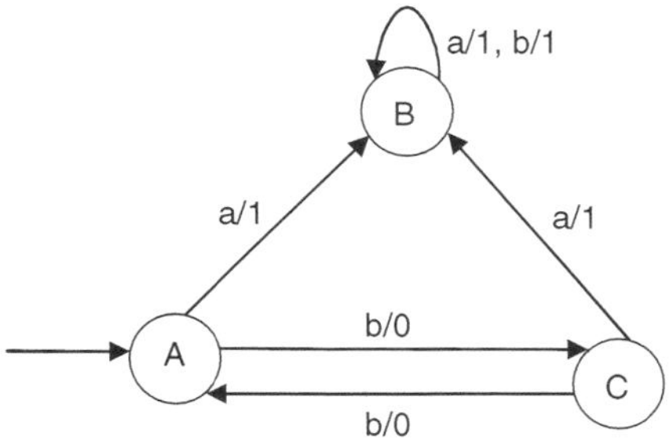

Fig. 3.33

Example 3.22 Convert the following Moore machine into an equivalent Mealy machine as given in Fig. 3.34.

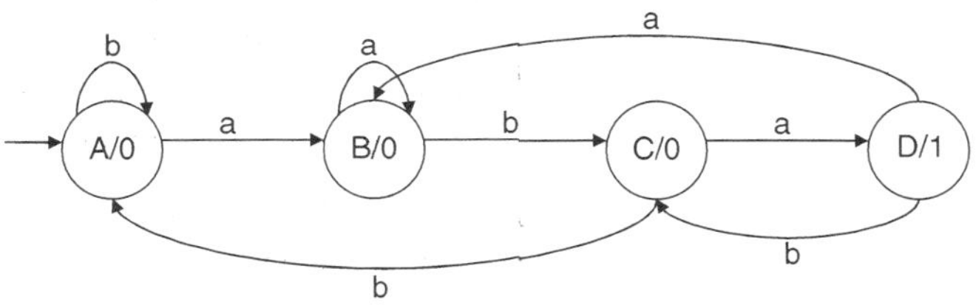

Fig. 3.34

Solution: State A is labelled with output 0. The incoming edges to this state are C to A for input b, A to A for input b, and one as start-state indication with no input. Ignore the edge with start-state indication.

Modify the labels of the incoming edges by including the output of state A. So, the modified labels become A to A with label b/0 and C to A with label b/0.

State B is labelled with output 0. The incoming edges to this state are A to B for input a, B to B for input a, and D to B for input a. The modified labels are A to B with label a/0, B to B with label a/0, and D to B with label a/0.

State C is labelled with output 0. The incoming edges to this state are B to C for input b, and D to C for input b. The modified labels are B to C with label b/0 and D to C with label b/0.

State D is labelled with output 1. The incoming edge to this state is C to D for input a. The modified label is from C to D with label a/1.

The Mealy machine equivalent to the given Moore machine is given in Fig. 3.35.

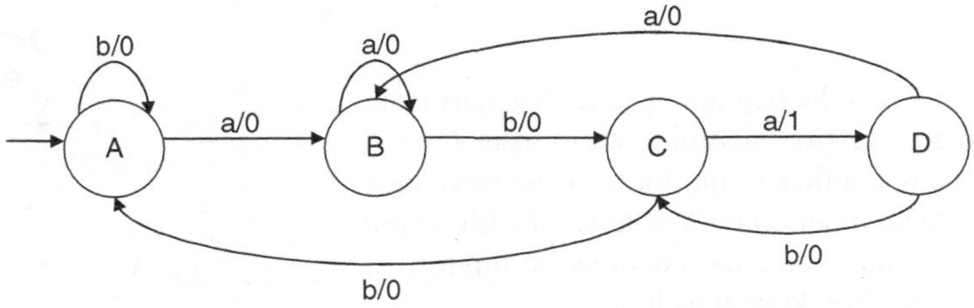

Fig. 3.35

3.13.2.2 Mealy Machine to Moore Machine

For a given Mealy machine M_C, there is an equivalent Moore machine M_o. These can be constructed in several steps.

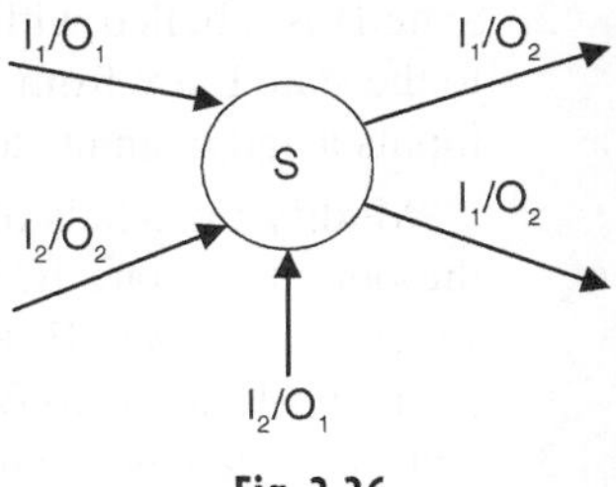

Fig. 3.36

Step I: In a Mealy machine, the output depends on the present state and the present input. So, in the transactional diagram of a Mealy machine, the transactional edges are labelled with input as well as output. For a state, there two types of edges, namely, incoming edges and outgoing edges. For incoming edges, it may happen that the output differs for two incoming edges like the example shown in Fig. 3.36.

In the transitional diagram for state S, the incoming edges are labelled as I_1/O_1, I_2/O_2, and I_2/O_1, and the outgoing edges are labelled as I_1/O_2 and I_2/O_2. For state S for incoming edges, we get two types of output O_1 and O_2. The state S is divided into n number of parts, where n = number of different outputs for the incoming edges to the state. The output edges are repeated for all the divided states. The transitional diagram for this case is shown in Fig. 3.37.

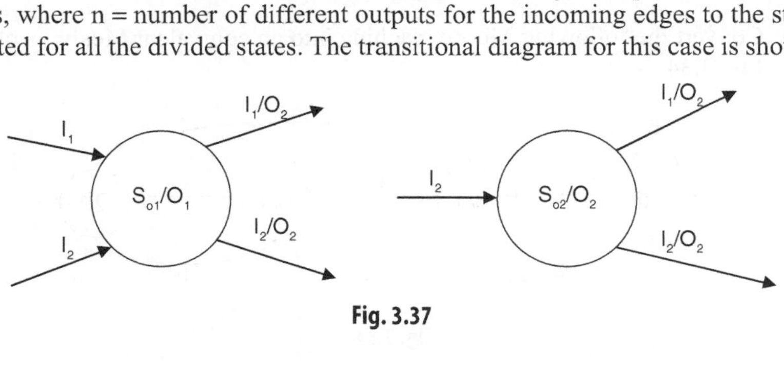

Fig. 3.37

Step II: If a state has a loop like Fig. 3.38, that state also needs to be divided as shown in Fig. 3.39.

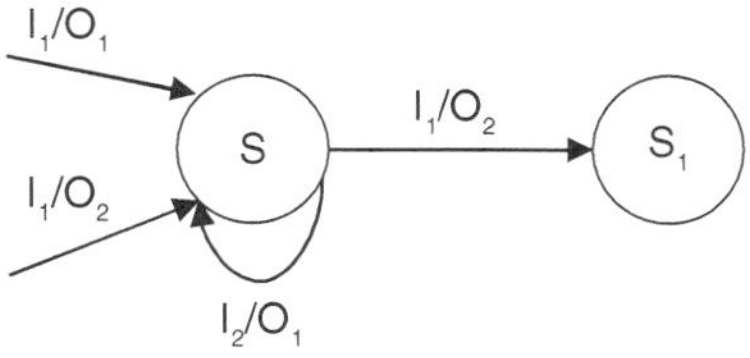

Fig. 3.38

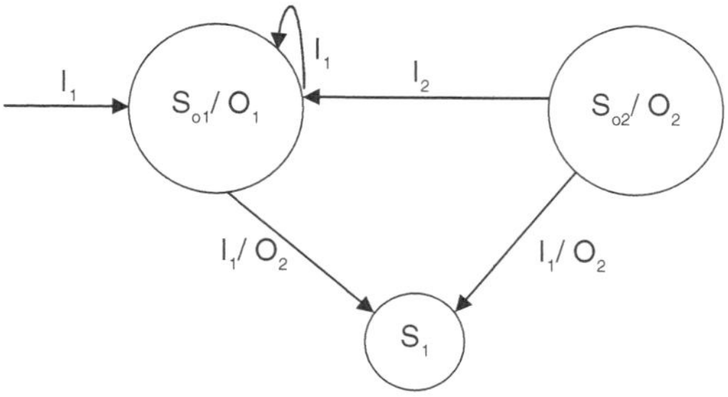

Fig. 3.39

[Here, for input I_1, the output is O_1 as well as O_2. So, the state S needs to be divided. S has a loop with input I_2 and output O_1.]. The state S will be divided into two states S_{O1} and S_{O2}. As the loop for input I_2 is labelled with output O_1, there will be a loop on state S_{O1}. But this loop is not possible on S_{O2}, because it produces the output O_2. So, there will be a transition from S_{O2} to S_{O1} with input label I_2.

Step III: Repeat the steps I and II. By this, the Moore machine equivalent to the Mealy machine is constructed.

The following examples (Examples 3.23 and 3.24) describe the process.

Example 3.23 Convert the following Mealy machine to an equivalent Moore machine as given in Fig. 3.40.

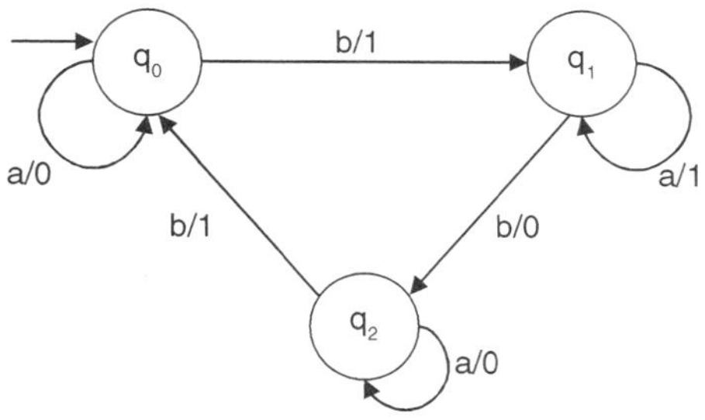

Fig. 3.40

Solution: For state q_0, there are two incoming states, q_0 to q_0 with label a/0 and q_2 to q_0 with label b/1. Two incoming edges are labelled with two different outputs 0 and 1. So, the state q_0 needs to be divided into two states as q_{00} and q_{01}. A loop for input 'a' on q_{00} is constructed. A transition is made from q_{01} to q_{00} with input label 'a'. From both the states, transitions are made to the state q_1 with label b/1. The modified machine is given in Fig. 3.41.

For all the other states q_1 and q_2 the outputs for all the incoming edges are 1 and 0 respectively. So there is no need to divide the states. The final Moore machine is as given in Fig. 3.42.

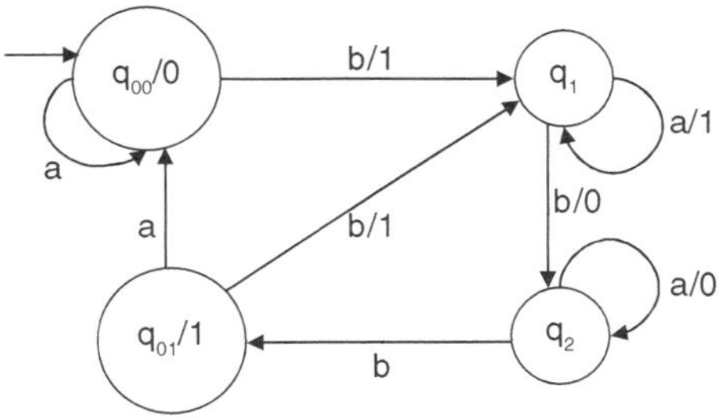

Fig. 3.41

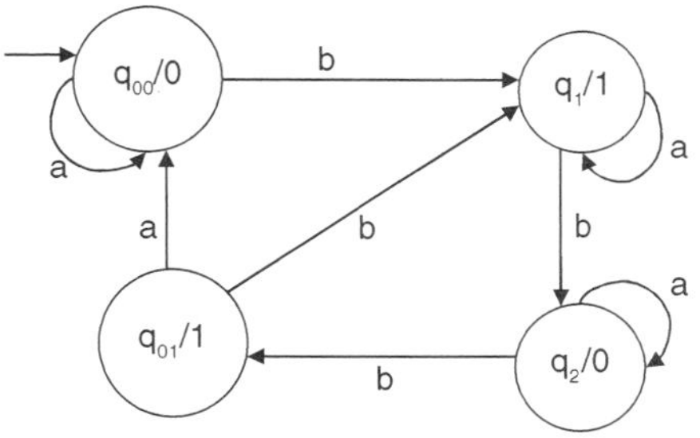

Fig. 3.42

Example 3.24 Convert the following Mealy machine as given in Fig. 3.43 to an equivalent Moore machine.

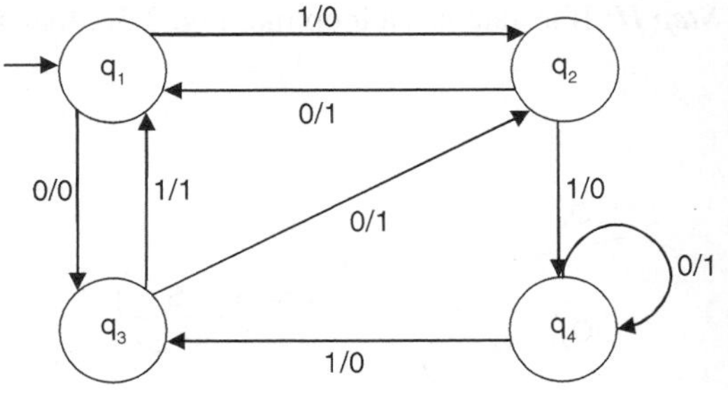

Fig. 3.43

Solution: The machine contains four states. Let us start from the state q_1. The incoming edges to this state are from q_2 to q_1 with label 0/1 and from q_3 to q_1 with label 1/1. There is no difference in the outputs of the incoming edges to this state and, therefore, in the constructing Moore machine, the output for this state is 1. The modified machine is as shown in Fig. 3.44.

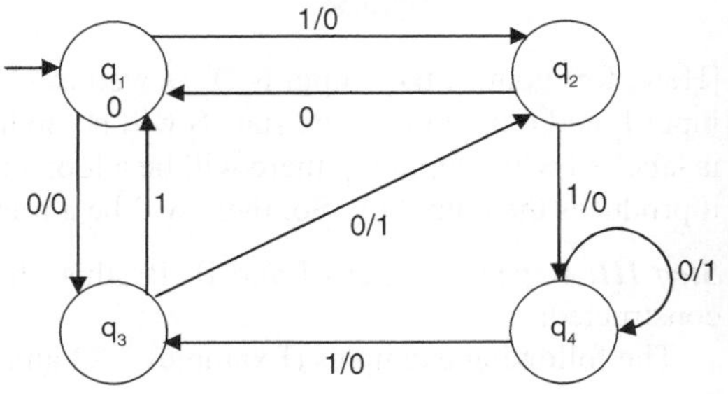

Fig. 3.44

Consider the state q_2. This state contains two incoming edges: from q_1 to q_2 with label 1/0 and q_3 to q_2 with label 0/1. We get two different outputs for the two incoming edges (q_1 to q_2 output 0, q_3 to q_2 output 1). So, the state q_2 is divided into two, namely, q_{20} and q_{21}. The outgoing edges are duplicated for both the states generated from q_2 as shown in Fig. 3.45.

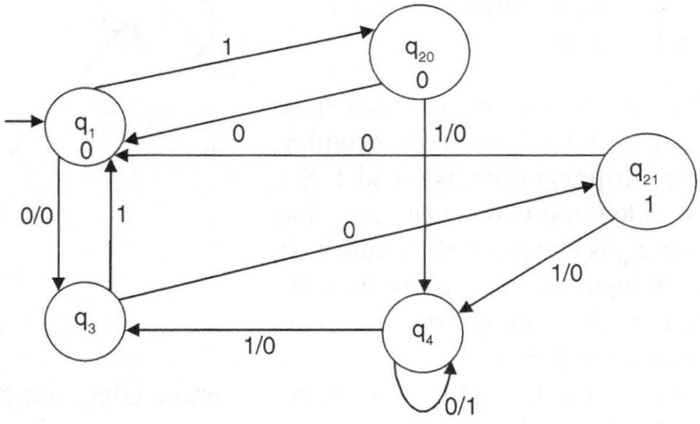

Fig. 3.45

Consider the state q_3. The incoming edges to this state are from q_1 to q_3 with label 0/0 and from q_4 to q_3 with label 1/0. There is no difference in the outputs of the incoming edges to this state and, therefore, in the constructing Moore machine, the output for this state is 0. The modified machine is given in Fig. 3.46.

Consider the state q_4. This state contains three incoming edges, from q_{20} to q_4 with label 1/0, from q_{21} to q_4 with label 1/0, and from and q_4 to q_4 with label 0/1. We get two different outputs for the three incoming edges (q_{20} to q_4 output 0, q_{21} to q_4 output 0, and q_4 to q_4 output 1). So, the state q_4 will be divided into two, namely, q_{40} and q_{41}.

The modified Moore machine is given in Fig. 3.47.

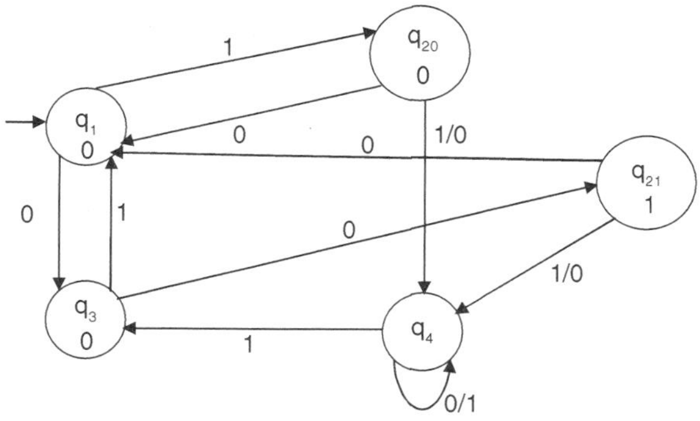

Fig. 3.46

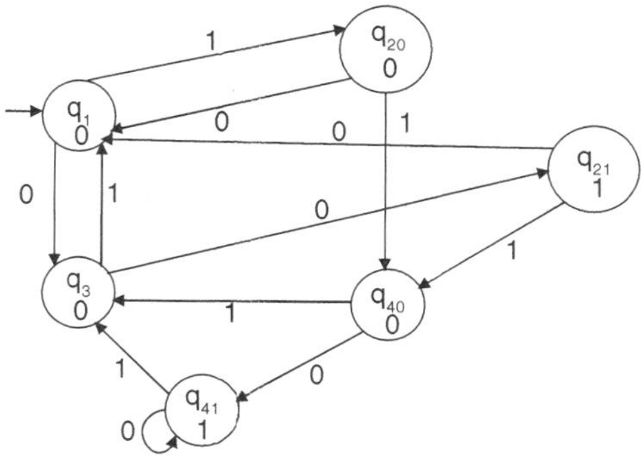

Fig. 3.47

3.14 Minimization of Finite Automata

The language (regular expression) produced by a DFA is always unique. But the reverse, i.e., a language produces a unique DFA, is not true. For this reason, there may be different DFAs in a given language. By minimizing, we can get a minimized DFA with minimum number of states and transitions which produces that particular language. The DFA determines how computers manipulate regular languages (expressions). The DFA size determines the space/time efficiency. So, a DFA with minimized states needs less time to manipulate a regular expression.

Before describing the process, we need to know the definitions of dead state, inaccessible state, equivalent state, distinguishable state, and k-equivalence in relation with finite automata.

- **Dead State:** A state q_i is called a dead state if q_i is not a final state and for all the inputs to this state, the transitions are confined to that state. In mathematical notation, we can denote $q_i \notin F$ and $\delta(q_i, \Sigma) \rightarrow qi$.
- **Inaccessible State:** The states which can never be reached from the initial state are called inaccessible states.

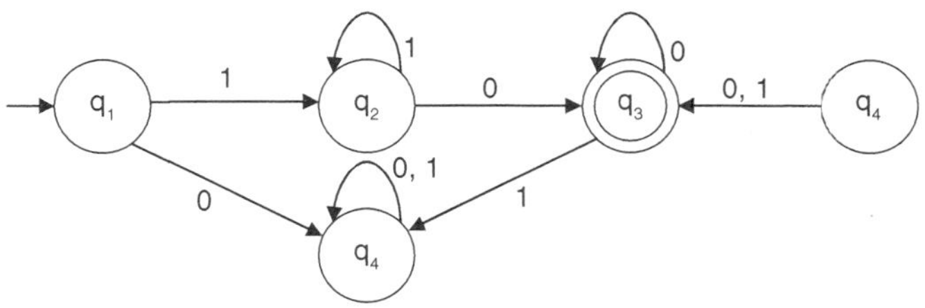

Fig. 3.48

Here, q_d is a dead state and q_a is an inaccessible state as shown in Fig. 3.48.

- **Equivalent state:** Two states q_i and q_j of a finite automata M are called equivalent if both $\delta(q_i, x)$ and $\delta(q_j, x)$ produce final states or both of them produce non-final states for all $x \in \Sigma^*$. It is denoted by $q_i \equiv q_j$.
- **Distinguishable state:** Two states q_i and q_j of a finite automata M are called distinguishable if, for a minimum length string x, for $\delta(q_i, x)$ and $\delta(q_j, x)$, one produces final state and another produces non-final state or vice versa for all $x \in \Sigma^*$.

- **K-equivalent:** Two states q_i and q_j of a finite automata M are called k-equivalent (k > = 0) if both $\delta(q_i, x)$ and $\delta(q_j, x)$ produce final states or both of them produce non-final states for all $x \in \Sigma^*$ of length k or less.

3.14.1 Process of Minimizing

1. All the states are '0' equivalent. Mark this as S_0.
2. Divide the set of states into two subsets: set of final states and set of non-final states. Mark this as S_1.
3. Apply all the inputs separately on the two subsets and find the next state combinations. If it happens that, for applying input on one set of states, the next states belong to different subsets, then separate the states which produce next states belonging to different subsets.
4. Continue step (3) for n + 1 times.
5. If S_n and S_{n+1} are the same, then stop and declare S_n as an equivalent partition.
6 Mark each subset of S_n as a different state and construct the transitional table accordingly. This is the minimum automata.

Consider the following examples to make the minimization process clear.

Example 3.25 Construct a minimum state automaton from the transitional table given below.

Present State	Next State I/P = 0	Next State I/P = 1
→ q_0	q_1	q_2
q_1	q_2	q_3
q_2	q_2	q_4
(q_3)	q_3	q_3
(q_4)	q_4	q_4
(q_5)	q_5	q_4

Solution: In the finite automata, the states are $\{q_0, q_1, q_2, q_3, q_4, q_5\}$. Name this set as S_0.

$$S_0: \{q_0, q_1, q_2, q_3, q_4, q_5\}$$

All of the states are 0 equivalents.

In the finite automata, there are two types of states: final state and non-final states. So, divide the set of states into two parts, Q_1 and Q_2.

$$Q_1 = \{q_0, q_1, q_2\} \quad Q_2 = \{q_3, q_4, q_5\}$$

$$S1: \{\{q_0, q_1, q_2\}, \{q_3, q_4, q_5\}\}$$

The states belonging to the same subset are 1-equivalent because they are in the same set for string length 1. The states belonging to different subsets are 1-distinguishable.

For input 0 and 1, q_0 goes to q_1 and q_2, respectively. Both of the states belong to the same subset. For q_1 and q_2 with input 0 and 1, the next states are q_2, q_3 and q_2, q_4, respectively. For both of the states, for input 0, the next state belongs to one subset, and for input 1, the next state belongs to another subset. So, q_0 can be distinguished from q_1 and q_2.

The next states with input 0 and 1 for states q_3, q_4, and q_5 belong to the same subset. So, they cannot be divided.

$$S_2: \{\{q_0\}, \{q_1, q_2\}, \{q_3, q_4, q_5\}\}$$

q_0 is the single state in the subset. So, it cannot be divided.

For states q_1 and q_2 with input 0 and 1, for both of the cases, one state belongs to one subset and another state belongs to another subset. So, they cannot be divided.

The next states with input 0 and 1 for states q_3, q_4, and q_5 belong to the same subset. So, they cannot be divided.

So, in the next step,

$$S_3: \{\{q_0\}, \{q_1, q_2\}, \{q_3, q_4, q_5\}\}$$

S_2 and S_3 are equivalent.

As step (n–1) and step n are the same, there is no need of further advancement.

In the minimized automata, the number of states is 3.

The minimized finite automata is presented in tabular format as follows:

State	Next State	
	I/P = 0	I/P = 1
$\{q_0\}$	$\{q_1\}$	$\{q_2\}$
$\{q_1, q_2\}$	$\{q_2\}$	$\{q_3, q_4\}$
$\{q_3, q_4, q_5\}$	$\{q_3, q_4, q_5\}$	$\{q_3, q_4, q_5\}$

But $\{q_1\}$, $\{q_2\}$, and $\{q_3, q_4\}$ do not exist under the column of present state. They are not states of the minimized finite automata, but they are subset of the states. In the next state columns, by replacing the subsets by proper state, the modified table becomes

State	Next State	
	I/P = 0	I/P = 1
$\{q_0\}$	$\{q_1, q_2\}$	$\{q_1, q_2\}$
$\{q_1, q_2\}$	$\{q_1, q_2\}$	$\{q_3, q_4, q_5\}$
$\{q_3, q_4, q_5\}$	$\{q_3, q_4, q_5\}$	$\{q_3, q_4, q_5\}$

As q_0 is the beginning state of the original finite automata, $\{q_0\}$ will be the beginning state of minimized finite automata. As q_3, q_4, and q_5 are the final states of the original finite automata, the set of the states containing any of the states as element is final state. Here, all the states are contained in a single set $\{q_3, q_4, q_5\}$ and, therefore, it is the final state. By replacing $\{q_0\}$ as A, $\{q_1, q_2\}$ as B, and $\{q_3, q_4, q_5\}$ as C, the modified minimized finite automata becomes

State	Next State I/P = 0	I/P = 1
→A	B	B
B	B	C
(C)	C	C

The transitional diagram of the minimized finite automata is given in Fig. 3.49.

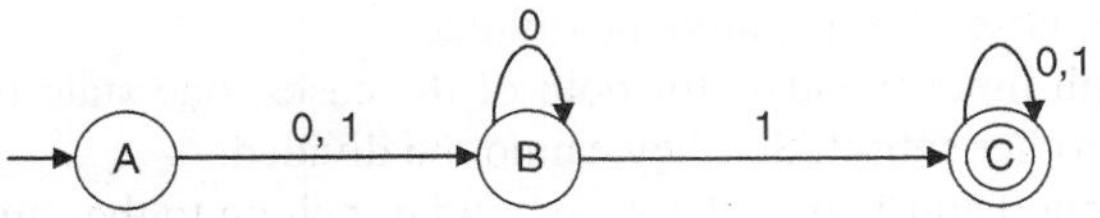

Fig. 3.49

Example 3.26 Construct a minimum state automaton from the following transitional table.

Present State	Next State I/P = 0	I/P = 1
A	F	B
B	C	G
C	C	A
D	G	C
E	F	H
F	G	C
G	E	G
H	C	G

A is initial state and C is final state

Solution: In the Finite automata, the states are {A, B, C, D, E, F, G, H}. Name this set as S_0.

S_0: {A, B, C, D, E, F, G, H}

All of the states are 0 equivalents.

In the finite automata, there are two types of states: final state and non-final state. The set of states is divided into two parts, namely, Q_1 and Q_2.

Q_1 = {C} Q_2 = {A, B, D, E, F, G, H}

S_1: {{C} {A, B, D, E, F, G, H}}

The states belonging to the same subset are 1-equivalent because they are in the same set for string length 1. The states belonging to different subsets are 1-distinguishable.

C is a single state, and so it cannot be divided. Among the states {A, B, D, E, F, G, H} for {B, D, F, H}, for an input of either 0 or 1, the next state belongs to {C} which is a different subset (from B with input 0 goes to C, from D with input 1 goes to C, from F with input 1 goes to C, and from H with input 0 goes to C).

For the other states {A, E, G}, for an input of either 0 or 1, the next state belongs to Q_2.

This divides Q_2 into two parts: {A, E, G} and {B, D, F, H}. Let us name them as Q_3 and Q_4.

The divided sets are

$$S_2: \{\{C\}, \{A, E, G\}, \{B, D, F, H\}\}.$$

Consider the subset of states {A, E, G}. A and E with input 0 and 1 both go to F, B and F, H, respectively, i.e., to the subset {B, D, F, H}. G with input 0 and 1 goes to E, G, i.e., the same subset. Here, {A, E, G} is divided into two subsets: {A, E} and {G}.

The subset {B, D, F, H} can be divided depending on the input and the next state combination. B and H produce the next states C and G for input 0 and 1, respectively.

D and F produce the next states G and C for input 0 and 1, respectively. So, the set {B, D, F, G} is divided into two subsets: {B, H} and {D, F}.

The divided sets are

$$S_3: \{\{C\}, \{A, E\}, \{G\}, \{B, H\}, \{D, F\}\}.$$

The subsets cannot be divided further making these as the states of minimized DFA. Let us rename the subsets as q_0, q_1, q_2, q_3, and q_4. The initial state was A, and so here the initial state is {A, E}, i.e., q_1. The final state was C, and so here the final state is {C}, i.e., q_0. The tabular representation of minimized DFA is

	Next State	
Present State	I/P = 0	I/P = 1
(q_0)	q_0	q_1
→ q_1	q_4	q_3
q_2	q_1	q_2
q_3	q_0	q_2
q_4	q_2	q_0

3.15 Myhill–Nerode Theorem

John Myhill and *Anil Nerode* of the University of Chicago proposed a theorem in 1958 which provides a necessary and sufficient condition for a language to be regular. This theorem can also be used to minimize a DFA. But before going into the details of the theorem statement, we need to know some definitions related to the theorem.

3.15.1 Equivalence Relation

- A relation R in set S is reflexive if xRx for every x in S.
- A relation R in set S is symmetric if for x, y in S, yRx whenever xRy.
- A relation R in set S is transitive if for x, y, and z in S, xRz whenever xRy and yRz.

A relation R in set S is called an equivalence relation if it is reflexive, symmetric, and transitive.

3.15.1.1 Right Invariant

An equivalence relation R on strings of symbols from some alphabet Σ is said to be right invariant if for all x, y ∈ Σ* with xRy and all w ∈ Σ* we have that xw R yw. This definition states that an equivalence relation has the right invariant property if two equivalent strings (x and y) that are in the language still are equivalent if a third string (w) is appended to the right of both of them.

3.15.2 Statement of the Myhill–Nerode Theorem

The Myhill–Nerode theorem states that the following three statements are equivalent.

1. The set L, a subset of Σ*, is accepted by a DFA, i.e., L is a regular language.
2. There is a right-invariant equivalence relation R of finite index such that L is the union of some of the equivalence classes of R.
3. Let equivalence relation R_L be defined as xR_Ly, if and only if for all z in Σ*, xz is in L exactly when yz is in L then R_L is of finite index.

3.15.3 Myhill–Nerode Theorem in Minimizing a DFA

Step I: Build a two-dimensional matrix labelled by the states of the given DFA at the left and bottom side. The major diagonal and the upper triangular parts are shown as dashes.

Step II: One of the three symbols, X, x, or 0 are put in the locations where there is no dash.

1. Mark X at p, q in the lower triangular part such that p is the final state and Q is the non-final state.
2. Make distinguished pair combination of the non-final states. If there are n number of non-final states, there are nC_2 number of distinguished pairs.
   Take a pair (p, q) and find (r, s), such that r = δ(p, a) and s = δ(q, a). If in the place of (r, s) there is X or x, in the place of (p, q), there will be x.
3. If (r, s) is neither X nor x, then (p, q) is 0.
4. Repeat (2) and (3) for final states also.

Step III: The combination of states where there is 0, they are the states of the minimized machine.

Consider the following examples to get the earlier discussed method.

Example 3.27 Minimize the following DFA using the Myhill–Nerode theorem.

	Next State	
Present State	I/P = a	I/P = b
→A	B	E
B	C	D
C	H	I
D	I	H
E	F	G
F	H	I
G	H	I
H	H	H
I	I	I

Here C, D, F, G are final states.

Solution:

Step I: Divide the states of the DFA into two subsets: final (F) and non-final (Q–F).

$$F = \{C, D, F, G\},\ Q\text{–}F = \{A, B, E, H, I\}$$

Make a two-dimensional matrix as shown in Fig. 3.50 labelled at the left and bottom by the states of the DFA.

A	—	—	—	—	—	—	—	—	—
B		—	—	—	—	—	—	—	—
C			—	—	—	—	—	—	—
D				—	—	—	—	—	—
E					—	—	—	—	—
F						—	—	—	—
G							—	—	—
H								—	—
I									—
	A	B	C	D	E	F	G	H	I

Fig. 3.50

Step II:

1. The following combinations are the combination of the beginning and final state.
   (A, C), (A, D), (A, F), (A, G), (B, C), (B, D), (B, F), (B, G), (E, C), (E, D), (E, F), (E, G), (H, C), (H, D), (H, F), (H, G), (I, C), (I, D), (I, F), (I, G).

   Put X in these combinations of states. The modified matrix is given in Fig. 3.51.

A	—	—	—	—	—	—	—	—	—
B		—	—	—	—	—	—	—	—
C	X	X	—	—	—	—	—	—	—
D	X	X		—	—	—	—	—	—
E			X	X	—	—	—	—	—
F	X	X			X	—	—	—	—
G	X	X			X		—	—	—
H			X	X		X	X	—	—
I			X	X		X	X		—
	A	B	C	D	E	F	G	H	I

Fig. 3.51

2. The pair combination of non-final states are (A, B), (A, E), (A, H), (A, I), (B, E), (B, H), (B, I), (E, H), (E, I), and (H, I).

r = δ(A, a) → B s = δ(B, a) → C

in the place of (B, C) there is X. So, in the place of (A, B), there will be x.

Similarly,

(r, s) = δ((A, E), a) → (B, F) (there is X). In the place of (A, E), there will be x.
(r, s) = δ((A, H), a) → (B, H) (neither X nor x). In the place of (A, H), there will be 0.
(r, s) = δ((A, I), a) → (B, I) (neither X nor x]. In the place of (A, I), there will be 0.
(r, s) = δ((B, E), a) → (C, F) (neither X nor x). In the place of (B, E), there will be 0.
(r, s) = δ((B, H), a) → (C, H) (there is X). In the place of (B, H), there will be x.
(r, s) = δ((B, I), a) → (C, I) (there is X). In the place of (B, I), there will be x.
(r, s) = δ((E, H), a) → (F, H) (there is X). In the place of (E, H), there will be x.
(r, s) = δ((E, I), a) → (F, I) (there is X). In the place of (E, I), there will be x.
(r, s) = δ((H, I), a) → (H, I) (neither X nor x). In the place of (H, I), there will be 0.

3. The pair combination of final states are (C, D), (C, F), (C, G), (D, F), (D, G), and (F, G).

(r, s) = δ((C, D), a) → (H, I) (neither X nor x). In the place of (C, D), there will be 0.
(r, s) = δ((C, F), a) → (H, H) (there is dash, neither X nor x). In the place of (C, F), there will be 0.
(r, s) = δ((C, G), a) → (H, H) (neither X nor x). In the place of (C, G), there will be 0.
(r, s) = δ((D, F), a) → (I, H) (neither X nor x). In the place of (D, F), there will be 0.
(r, s) = δ((D, G), a) → (I, H) (neither X nor x). In the place of (D, G), there will be 0.
(r, s) = δ((F, G), a) → (H, H) (neither X nor x). In the place of (F, G), there will be 0.

The modified matrix becomes as shown in Fig. 3.52.

B	x							
C	X	X						
D	X	X	0					
E	x	0	X	X				
F	X	X	0	0	X			
G	X	X	0	0	X	0		
H	x	x	X	X	x	X	X	
I	x	x	X	X	x	X	X	0
	A	B	C	D	E	F	G	H

Fig. 3.52

The combination of entries 0 are the states of the modified machine. The states of the minimized machine are [A], [B, E], [C, D], [C, F], [C, G], [D, F], [D, G], [F, G], and [H, I].

For the minimized machine M′

Q′ = ({A}, {B, E}, {C, D, F, G}, {H, I}). [C, D], [C, F], [C, G], [D, F], [D, G], and [F, G] are combined to a new state [CDFG].

Σ = {a, b} δ′: (given in the following table)

Present State	Next State	
	a	b
{A}	{B, E}	{B, E}
{B, E}	{C, D, F, G}	{C, D, F, G}
{C, D, F, G}	{H, I}	{H, I}
{H, I}	{H, I}	{H, I}

q'_0: {A}

F′ : {C, D, F, G}

Example 3.28 Minimize the following finite automata by the Myhill–Nerode theorem.

Present State	Next State	
	I/P = a	I/P = b
→A	B	F
B	A	F
C	G	A
D	H	B
E	A	G
F	H	C
G	A	D
H	A	C

Here F, G, H are final states.

Solution:

Step I: Divide the states of the DFA into two subsets: final (F) and non-final (Q–F).

F = {E, F, G}, Q–F = {A, B, C, D}

Make a two-dimensional matrix (Fig. 3.53) labelled at the left and bottom by the states of the DFA.

A	—	—	—	—	—	—	—	—
B		—	—	—	—	—	—	—
C			—	—	—	—	—	—
D				—	—	—	—	—
E					—	—	—	—
F						—	—	—
G							—	—
H								—
	A	B	C	D	E	F	G	H

Fig. 3.53

Step II:

1. The following combinations are the combination of the beginning and final states.
   (A, E), (A, F), (A, G), (B, E), (B, F), (B, G), (C, E), (C, F), (C, G), (D, E), (D, F), (D, G)
   Put X in these combinations of states. The modified matrix is given in Fig. 3.54.

B							
C							
D							
E							
F	X	X	X	X	X		
G	X	X	X	X	X		
H	X	X	X	X	X		
	A	B	C	D	E	F	G

Fig. 3.54

2. The pair combination of non-final states are (A, B), (A, C), (A, D), (A, E), (B, C), (B, D), (B, E), (C, D), (C, E), and (D, E).
   r = δ(A, a) → B s = δ(B, a) → A, in the place of (A, B), there is neither X nor x. So, in the place of (A, B), there will be 0.
   Similarly,

   (r, s) = δ((A, C), a) → (B, G) (there is X). In the place of (A, C), there will be x.
   (r, s) = δ((A, D), a) → (B, H) (there is X). In the place of (A, D), there will be x.
   (r, s) = δ((A, E), a) → (B, A) (there is neither X nor x). In the place of (A, E), there will be 0.
   (r, s) = δ((B, C), a) → (A, G) (there is X). In the place of (B, C), there will be x.
   (r, s) = δ((B, D), a) → (A, H) (there is X). In the place of (B, D), there will be x.
   (r, s) = δ((B, E), a) → (A, A) (there is neither X nor x, only dash). In the place of (B, E), there will be 0.
   (r, s) = δ((C, D), a) → (G, H) (there is neither X nor x). In the place of (C, D), there will be 0.
   (r, s) = δ((C, E), a) → (G, A) (there is X). In the place of (C, E), there will be x.
   (r, s) = δ((D, E), a) → (H, A) (there is X). In the place of (D, E), there will be x.

3. The pair of combinations of final states are (F, G), (F, H), and (G, H).

   (r, s) = δ((F, G), a) → (A, H) (there is X). In the place of (F, G), there will be x.
   (r, s) = δ((F, H), a) → (H, A) (there is X). In the place of (F, H), there will be x.
   (r, s) = δ((G, H), a) → (A, A) (there is neither X nor x, there is only dash). In the place of (G, H), there will be 0.

   The modified table is given in Fig. 3.55.

B	0						
C	x	x					
D	x	x	0				
E	0	0	x	x			
F	X	X	X	X	X		
G	X	X	X	X	X	x	
H	X	X	X	X	X	x	0
	A	B	C	D	E	F	G

Fig. 3.55

The combination of entries 0 are the states of the modified machine. The states of the minimized machine are (A, B), (A, E), (B, E), (C, D), (G, H), i.e., (A, B, E), (C, D), (G, H), and (F) (As F is a final state of the machine, it is left in the state combinations).

(A, B, E) for input 'a' gives the output (A, B, A) and for input 'b' gives the output (F, F, G), where (F, F) belongs to one set and (G) belongs to another set. So, it will be divided into (A, B), (E).

The states of the minimized machines are (A, B),(E), (C, D), (G, H), and (F). Let us name them as S_1, S_2, S_3, S_4, and S_5.

For the minimized machine M′,

$$Q = \{S_1, S_2, S_3, S_4, S_5\}$$
$$\Sigma = \{a, b\}$$

State Table (transitional function δ)

Present State	Next State	
	a	b
S_1	S_1	S_5
S_2	S_1	S_4
S_3	S_4	S_1
S_4	S_1	S_3
S_5	S_4	S_3

3.16 Two-way Finite Automata

Till now, we have studied automata where the reading head moves only in one direction, from left to right.

Two-way finite automata are machines which can traverse (read) an input string in both directions (left and right).

A two-way DFA (2DFA) consists of five touples M = {Q, Σ, δ, q_0, F} where Q, Σ, q_0, and F are defined like one-way DFA, but here the transitional function δ is a map from Q × Σ to Q × (L, R). L means left and R means right. Block diagram of a two-way finite automata is shown in Fig. 3.56.

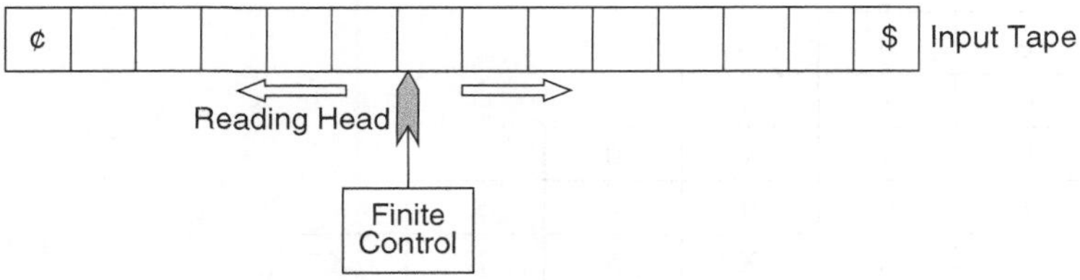

Fig. 3.56 *Two-way Finite Automata*

Consider the following example of a two-way DFA; M is given in the following table.

Present State	Next State 0	Next State 1
→ A	A, R	B, R
(B)	B, R	C, L
C	A, R	C, L

Let us give a string 101001 to check whether it is accepted by the 2DFA or not.

$$(A, 101001) \rightarrow (B, 01001R) \rightarrow (B, 1001R) \rightarrow (C, 01001L) \rightarrow (A, 1001R) \rightarrow (B, 001R) \rightarrow (B, 01R) \rightarrow (B, 1R) \rightarrow (C, 01L) \rightarrow (A, 1R) \rightarrow B$$

We have reached the final state, and the string is finished. So, the string 101001 is accepted by the 2DFA.

3.17 Applications of Finite Automata

Finite automata can be applied in different fields of computer science and in different engineering fields. Some of them are spelling checkers and advisers, multilanguage dictionaries, minimal perfect hashing, and text compression. Perhaps, the most traditional application is found in compiler construction where such automata can be used to model and implement efficient lexical analysers.

A typical example is given in Fig. 3.57 (transitional diagram for relational operators in C).

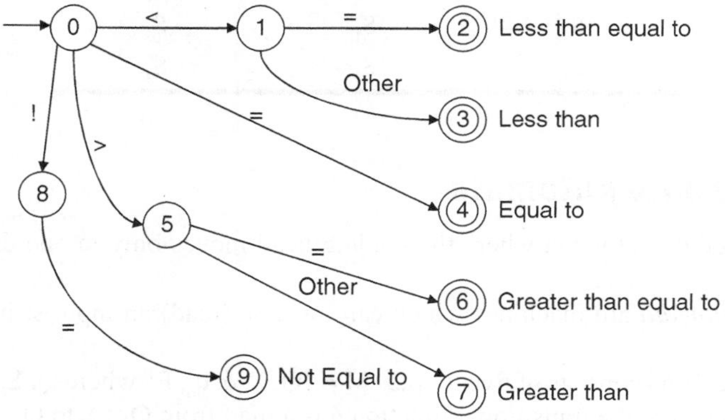

Fig. 3.57 *Transitional Diagram for Relational Operators in C.*

(*Source:* "Compilers: Principles, Techniques and tools" Aho, Sethi, Ullman, Pearson Education.)

3.18 Limitations of Finite Automata

An FA is a finite state machine without considering the output. Thus, it has finite number of states. So, it cannot perform operations where infinite amount of memory is needed. As an example, let us check a string in the form a^nb^n, where $n > 0$. As n is a positive integer from 0 to any number, it is not possible to design a general finite automata for checking a string in the form a^nb^n. It is possible to design separate finite automata for ab, a^2b^2, a^3b^3 ..., but general finite automata for a^nb^n is not possible.

What We Have Learned So Far

1. An FA has a finite number of states.
2. An FA has five characteristics: (i) input, (ii), state, (iii) state transition, (iv) output, and (v) output relation.
3. Finite automata M is represented as $M = \{Q, \Sigma, \delta, q_0, F\}$.
4. A string is declared accepted by an FA if the string is finished and the machine has reached a final state.
5. Finite automata can be of two types (i) deterministic finite automata (DFA) and (ii) non-deterministic finite automata (NDFA).
6. For a DFA, for all cases for a single input given to a single state, the machine goes to a single state, i.e., $Q \times \Sigma \rightarrow Q$.
7. For an NFA, for some cases for a single input given to a single state, the machine goes to more than one states, i.e., $Q \times \Sigma \rightarrow 2^Q$, where 2^Q is the power set of Q.
8. NFA can be converted to an equivalent DFA.
9. An FA is called an NFA with null move if there exists a null transaction, i.e., $Q \times \varepsilon \rightarrow Q$.
10. A dead state is a state where the control can enter and be confined till the input ended, but there is no way to come out from that state.
11. In a Mealy machine, the output depends on the present state and the present input.
12. In a Moore machine, the output depends only on the present state.
13. The Myhill–Nerode theorem is used to minimize finite automata.
14. Two-way finite automata are machines which can traverse (read) an input string in both directions (left and right).
15. Finite automata are used in designing lexical analysers.

Solved Problems

1. Test whether the following strings are accepted by the following finite automata or not: (i) 0001101, (ii) 00000, (iii) 01010.

Present State	Next State 0	Next State 1
→ q_0	q_2	q_3
q_1	q_0	q_2
q_2	q_1	q_3
(q_0)	q_3	q_1

Solution:

i)

$$\begin{aligned}\delta(q_0, 0001101) &\rightarrow \delta(q_2, 001101)\\ &\rightarrow \delta(q_1, 01101)\\ &\rightarrow \delta(q_0, 1101)\\ &\rightarrow \delta(q_3, 101)\\ &\rightarrow \delta(q_1, 01)\\ &\rightarrow \delta(q_0, 1)\\ &\rightarrow q_3\end{aligned}$$

As q_3 is the final state, the string is accepted by the given finite automata.

ii)

$$\begin{aligned}\delta(q_0, 00000) &\rightarrow \delta(q_2, 0000)\\ &\rightarrow \delta(q_1, 000)\\ &\rightarrow \delta(q_0, 00)\\ &\rightarrow \delta(q_2, 0)\\ &\rightarrow q_1\end{aligned}$$

As q_1 is the non-final state, the string is not accepted by the given finite automata.

iii)

$$\begin{aligned}\delta(q_0, 01010) &\rightarrow \delta(q_2, 1010)\\ &\rightarrow \delta(q_3, 010)\\ &\rightarrow \delta(q_3, 10)\\ &\rightarrow \delta(q_1, 0)\\ &\rightarrow q_0\end{aligned}$$

As q_0 is a non-final state, the string is not accepted by the given finite automata.

2. Test whether the following strings are accepted by the following finite automata or not: (i) 0111100, (ii) 11111, (iii) 11010.

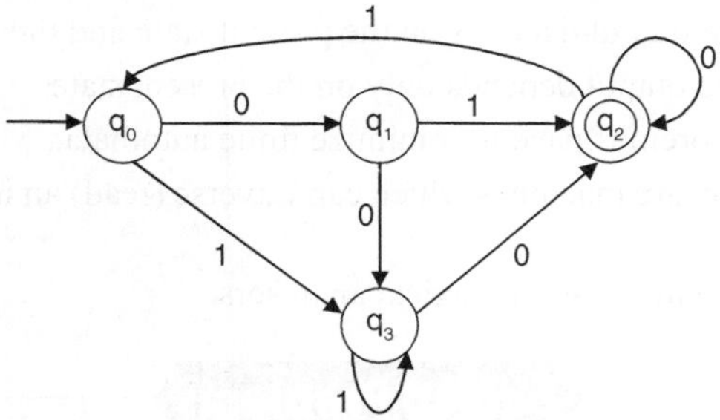

Solution:

i)

$$\begin{aligned}\delta(q_0, 0111100) &\rightarrow \delta(q_1, 111100)\\ &\rightarrow \delta(q_2, 11100)\\ &\rightarrow \delta(q_0, 1100)\\ &\rightarrow \delta(q_3, 100)\\ &\rightarrow \delta(q_3, 00)\\ &\rightarrow \delta(q_2, 0)\\ &\rightarrow q_2\end{aligned}$$

As q_2 is the final state, the string is accepted by the given finite automata.

ii)

$$\delta(q_0, 11111) \rightarrow \delta(q_3, 1111)$$
$$\rightarrow \delta(q_3, 111)$$
$$\rightarrow \delta(q_3, 11)$$
$$\rightarrow \delta(q_3, 1)$$
$$\rightarrow q_3$$

As q_3 is a non-final state, the string is not accepted by the FA.

iii)

$$\delta(q_0, 11010) \rightarrow \delta(q_3, 1010)$$
$$\rightarrow \delta(q_3, 010)$$
$$\rightarrow \delta(q_2, 10)$$
$$\rightarrow \delta(q_0, 0)$$
$$\rightarrow q_1$$

As q_1 is a non-final state, the string is not accepted by the given finite automata.

3. Test whether the string 10010 is accepted by the following NFA or not.

	Next State	
Present State	a = 0	a = 1
→ A	B	A, C
B	B	C
C	A, D	B
(D)	D	B

Solution:

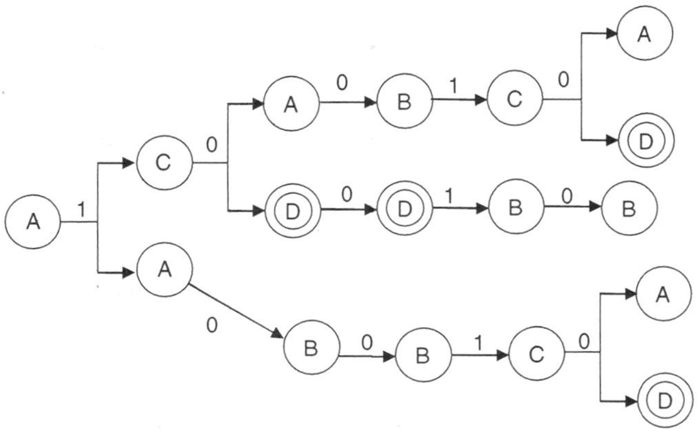

For the string 10010, we get the terminal state as D which is a final state for two cases. Therefore, the string is accepted by the NFA.

4. Convert the following NFA into an equivalent DFA.

Present State	Next State a = 0	Next State a = 1
→ q_0	q_0, q_1	q_0, q_2
q_1	q_3	–
q_2	–	q_3
(q_3)	q_3	q_3

Solution:

Present State	Next State a = 0	Next State a = 1
$[q_0]$	$[q_0, q_1]$	$[q_0, q_2]$
$[q_0, q_1]$	$[q_0, q_1, q_3]$	$[q_0, q_2]$
$[q_0, q_2]$	$[q_0, q_1]$	$[q_0, q_2, q_3]$
$[q_0, q_1, q_3]$	$[q_0, q_1, q_3]$	$[q_0, q_2, q_3]$
$[q_0, q_2, q_3]$	$[q_0, q_1, q_3]$	$[q_0, q_2, q_3]$

For simplification, let us replace $[q_0]$ by A, $[q_0, q_1]$ by B, $[q_0, q_2]$ by C, $[q_0, q_1, q_3]$ by D, and $[q_0, q_2, q_3]$ by E. Here, A is the initial state, and D and E are the final states as they contain the state q_3.

The simplified DFA is

Present State	Next State a = 0	Next State a = 1
→ A	B	C
B	D	C
C	B	E
(D)	D	E
(E)	D	E

5. Design an equivalent DFA corresponding to the following NFA. [WBUT 2005]

NFA M = < $\{q_0, q_1, q_2, q_3, q_4\}$, $\{0, 1\}$, δ, q_0, $\{q_2, q_4\}$ > where δ is as follows.

Present State	Next State 0	Next State 1
q_0	$\{q_0, q_3\}$	$\{q_0, q_1\}$
q_1	Φ	q_2
q_2	q_2	q_2
q_3	q_4	Φ
q_4	q_4	q_4

Solution:

Present State	Next State 0	Next State 1
$[q_0]$	$[q_0, q_3]$	$[q_0, q_1]$
$[q_0, q_3]$	$[q_0, q_3, q_4]$	$[q_0, q_1]$
$[q_0, q_3, q_4]$	$[q_0, q_3, q_4]$	$[q_0, q_1, q_4]$
$[q_0, q_1, q_4]$	$[q_0, q_3, q_4]$	$[q_0, q_1, q_4]$
$[q_0, q_1]$	$[q_0, q_3]$	$[q_0, q_1]$

For simplification, let us replace $[q_0]$ by A, $[q_0, q_3]$ by B, $[q_0, q_3, q_4]$ by C, $[q_0, q_1, q_4]$ by D, and $[q_0, q_1]$ by E. Here, A is the initial state, and C and D are the final states as they contain the state q_4. The simplified DFA is

Present State	Next State 0	Next State 1
A	B	E
B	C	E
C	C	D
D	C	D
E	B	E

6. Convert the following NFA to an equivalent DFA.

Σ States	0	1
q_0	q_0	q_0, q_1
q_1	q_2	q_2
q_2	–	q_2

(q_0 is the initial state and q_1 is the final state)

Solution: Conversion is done in the following ways:

Σ States	0	1
$[q_0]$	$[q_0]$	$[q_0, q_1]$
$[q_0, q_1]$	$[q_0, q_2]$	$[q_0, q_1, q_2]$
$[q_0, q_1, q_2]$	$[q_0, q_2]$	$[q_0, q_1, q_2]$
$[q_0, q_2]$	$[q_0]$	$[q_0, q_1, q_2]$

Rename $[q_0]$ as A, $[q_0, q_1]$ as B, $[q_0, q_1, q_2]$ as C, and $[q_0, q_2]$ as D. The beginning state is A, and final states are B and C.

Σ States	0	1
A	A	B
B	D	C
C	D	C
D	A	C

7. Convert the following NFA to an equivalent DFA. [UPTU 2005]

Σ States	0	1
p	{q, s}	{q}
q	{r}	{q, r}
r	{s}	{p}
s	∅	{p}

where p is the initial state and q and s are the final states.

Solution:

Σ States	0	1
{p}	{q, s}	{q}
{q}	{r}	{q, r}
{r}	{s}	{p}
{s}	∅	{p}
{q, r}	{r, s}	{p, q, r}
{r, s}	{s}	{p}
{p, q, r}	{q, r, s}	{p, q, r}
{q, r, s}	{r, s}	{p, q, r}
{q, s}	{r}	{p, q, r}
{∅}	{∅}	{∅}

Here {p} is the beginning state and {q}, {s}, {q, r}, {r, s}, {p, q, r}, {q, r, s}, and {q, s} are the final states. {∅} is the dead state.

8. Construct a DFA equivalent to the following NDFA given in the following figure. [UPTU 2004]

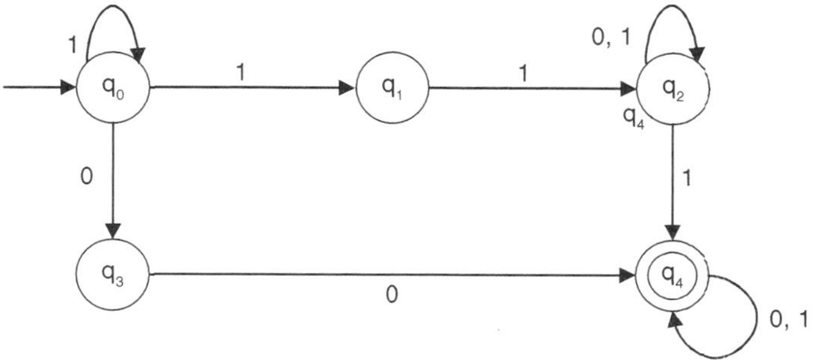

Solution: The tabular representation of the NDFA is

Present State	Next State 0	1
q_0	q_3	$\{q_0, q_1\}$
q_1	Φ	q_2
q_2	q_2	$\{q_2, q_4\}$
q_3	q_4	Φ
q_4	q_4	q_4

(q_0 is the initial state and q_4 is the final state)
The corresponding DFA is

Present State	Next State 0	1
$\{q_0\}$	$\{q_3\}$	$\{q_0, q_1\}$
$\{q_3\}$	$\{q_4\}$	$\{\Phi\}$
$\{q_4\}$	$\{q_4\}$	$\{q_4\}$
$\{q_0, q_1\}$	$\{q_2\}$	$\{q_2, q_4\}$
$\{q_2\}$	$\{q_2\}$	$\{q_2, q_4\}$
$\{q_2, q_4\}$	$\{q_2, q_4\}$	$\{q_2, q_4\}$
$\{\Phi\}$	$\{\Phi\}$	$\{\Phi\}$

Here $\{q_0\}$ is the beginning state, and $\{q_4\}$, and $\{q_2, q_4\}$ are the final states.
(Draw a transitional diagram to complete the answer.)

9. Find the minimal DFAs for the language $L = \{a^n b^m, n \geq 2, m \geq 1\}$ [Andhra University 2008]

Solution: All 'a' will appear before 'b'. There is atleast 2 'a' and 1 'b'. The DFA is the following.

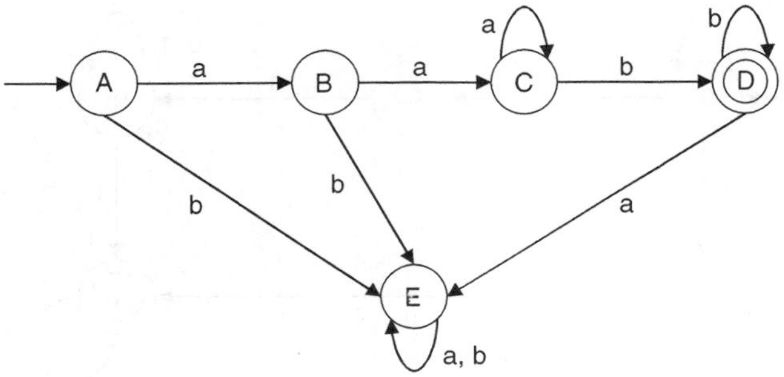

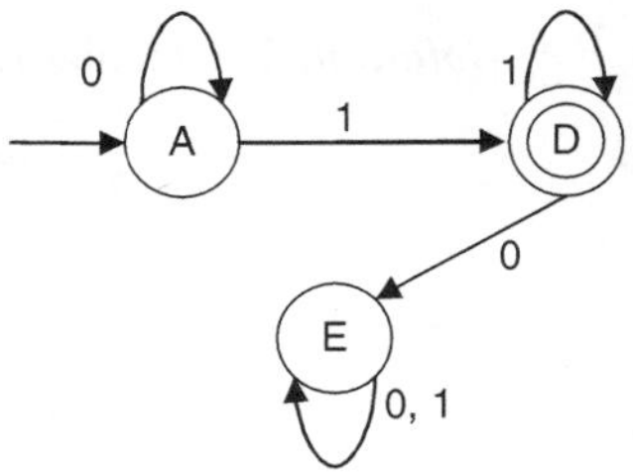

Fig. 3.58

10. Design a DFA for the language L = $\{0^m1^n, m \geq 0, n \geq 1\}$ [JNTU 2007]

Solution: All '0's will appear before '1'. There is at least one '1', but the number of '0's may be zero. The DFA is shown in Fig. 3.58.

11. Construct a DFA which accepts the set of all binary strings that, interpreted as the binary representation of an unsigned decimal integer, is divisible by 5. [WBUT 2008]

Solution: For Mod 5, the remainders are 0, 1, 2, 3, and 4. We can assign the states as q_0, q_1, q_2, q_3, and q_4.

For any binary string, if we add a bit at LSB, then the previous value becomes doubled. (Let the string be 101. The decimal value is 5. If we add another 1 at LSB, the string becomes 1011. The decimal value of the previous 101 becomes 10.)

In general, we can write that 'n' becomes 2n + b, where n is the previous number and b is the added bit.

$$(2n + b) \bmod 5 = 2n \bmod 5 + b \bmod 5$$

As b is either 0 or 1, b mod 5 = b.

2n mod 5 is any one of 0, 1, 2, 3, or 4, i.e., 2 X (state number) + a.

For this machine, the input alphabets are 0 and 1.

$$\delta(q_0, 0) \rightarrow 2 \times 0 + 0 = 0 \text{ means } q_0$$
$$\delta(q_0, 1) \rightarrow 2 \times 0 + 1 = 1 \text{ means } q_1$$
$$\delta(q_1, 0) \rightarrow 2 \times 1 + 0 = 2 \text{ means } q_2$$
$$\delta(q_1, 1) \rightarrow 2 \times 1 + 1 = 3 \text{ means } q_3$$
$$\delta(q_2, 0) \rightarrow 2 \times 2 + 0 = 4 \text{ means } q_4$$
$$\delta(q_2, 1) \rightarrow 2 \times 2 + 1 = 5\%5 = 0 \text{ means } q_0$$
$$\delta(q_3, 0) \rightarrow 2 \times 3 + 0 = 6\%5 = 1 \text{ means } q_1$$
$$\delta(q_3, 1) \rightarrow 2 \times 3 + 1 = 7\%5 = 2 \text{ means } q_2$$
$$\delta(q_4, 0) \rightarrow 2 \times 4 + 0 = 8\%5 = 3 \text{ means } q_3$$
$$\delta(q_4, 1) \rightarrow 2 \times 4 + 1 = 9\%5 = 4 \text{ means } q_4$$

The transitional diagram is

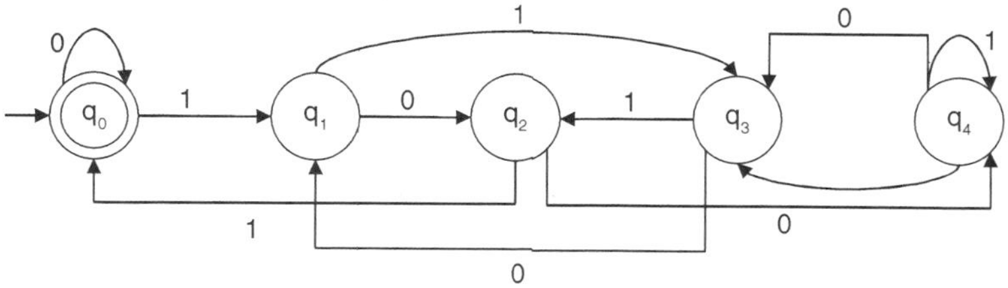

12. Convert the following NFA with ε move to an equivalent DFA.

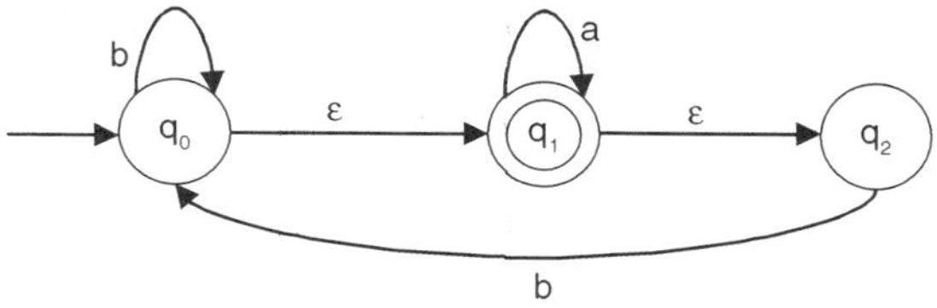

Solution: In the given automata, there are two ε moves: from q_0 to q_1 and from q_1 to q_2. If we want to remove the first ε move, from q_0 to q_1, then we have to find all the edges starting from q_1.The edges are q_1 to q_1 for input a and q_1 to q_2 for input ε.

Duplicate all these transitions starting from the state q_0, keeping the edge label the same. q_0 is the initial state, and so make q_1 also an initial state. q_1 is the final state, and so make q_0 as the final state The modified transitional diagram will be

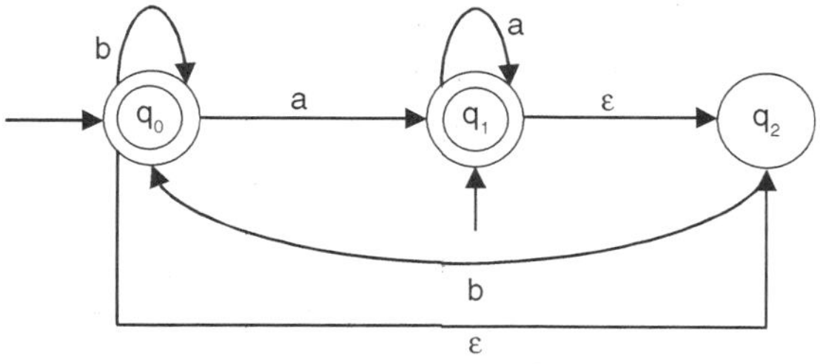

Again, there are two ε moves, from q_0 to q_2 and from q_1 to q_2. If we want to remove the null transition from q_0 to q_2, we have to find all the edges starting from q_2.The edges are q_2 to q_0 for input b. Duplicate this transition starting from the state q_0, keeping the edge label the same. The modified transitional diagram will be as follows. (As in q_0 there is a loop with label b, we need not make another loop with the same label).

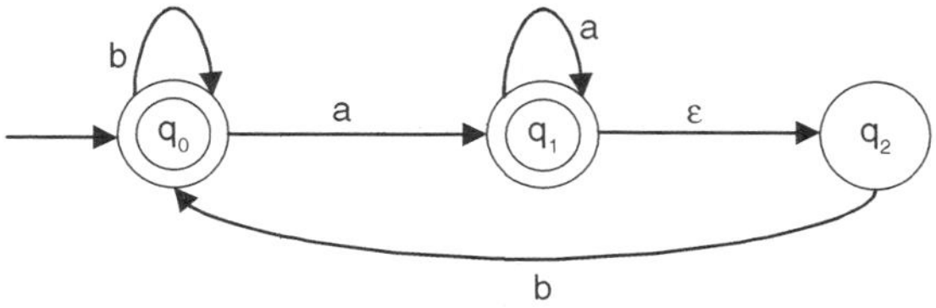

If we want to remove the null transition from q_1 to q_2, we have to find all the edges starting from q_2.The edges are q_2 to q_0 for input b. Duplicate this transition starting from the state q_1, keeping the edge label the same. q_1 is the initial state, and so make q_2 also an initial state. The modified transitional diagram will be

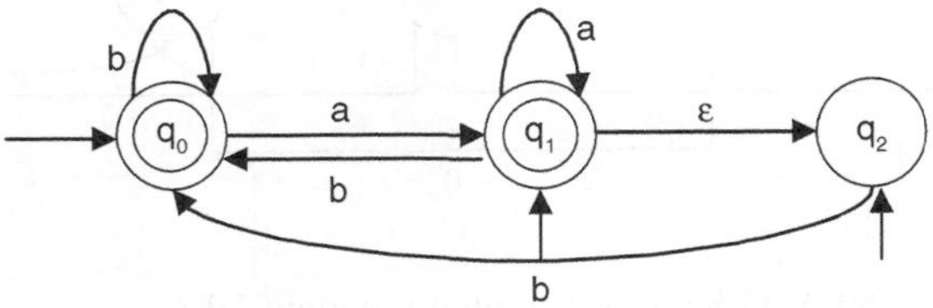

13. Convert the following Moore machine to an equivalent Mealy machine by the tabular format.

Present State	Next State I/P = 0	Next State I/P = 1	O/P
→ q_0	q_2	q_1	0
q_1	q_0	q_3	1
q_2	q_3	q_4	1
q_3	q_4	q_1	0
q_4	q_4	q_2	1

Solution: The equivalent Mealy machine is

Present State	I/P = 0 Next State	I/P = 0 O/P	I/P = 1 Next State	I/P = 1 O/P
→ q_0	q_2	1	q_1	1
q_1	q_0	0	q_3	0
q_2	q_3	0	q_4	1
q_3	q_4	1	q_1	1
q_4	q_4	1	q_2	1

14. Convert the following Moore machine to an equivalent Mealy machine by the tabular format.

Present State	Next State I/P = 0	Next State I/P = 1	O/P
→ q_0	q_1	q_3	1
q_1	q_0	q_2	0
q_2	q_4	q_1	0
q_3	q_4	q_3	1
q_4	q_3	q_1	0

Solution: The equivalent Mealy machine is

	I/P = 0		I/P = 1	
Present State	Next State	O/P	Next State	O/P
→ q_0	q_1	0	q_3	1
q_1	q_0	1	q_2	0
q_2	q_4	0	q_1	0
q_3	q_4	1	q_1	1
q_4	q_3	1	q_1	0

15. Convert the following Mealy machine to an equivalent Moore machine by the tabular format.

	I/P = 0		I/P = 1	
Present State	Next State	O/P	Next State	O/P
→ q_0	q_1	1	q_2	1
q_1	q_3	0	q_0	1
q_2	q_4	0	q_3	1
q_3	q_1	0	q_4	0
q_4	q_2	1	q_4	0

Solution: In the next state column of the given Mealy machine, the output differs for q_1 and q_3 as the next states. So, the states will be divided as q_{10}, q_{11} and q_{30}, q_{31}, respectively. After dividing the states, the modified Mealy machine becomes

	I/P = 0		I/P = 1	
Present State	Next State	O/P	Next State	O/P
→ q_0	q_{11}	1	q_2	1
q_{10}	q_{30}	0	q_0	1
q_{11}	q_{30}	0	q_0	1
q_2	q_4	0	q_{31}	1
q_{30}	q_{10}	0	q_4	0
q_{31}	q_{10}	0	q_4	0
q_4	q_2	1	q_4	0

The converted Moore machine is

	Next State		
Present State	I/P = 0	I/P = 1	O/P
→ q_0	q_{11}	q_2	1
q_{10}	q_{30}	q_0	0

Present State	Next State I/P = 0	Next State I/P = 1	O/P
q_{11}	q_{30}	q_0	1
q_2	q_4	q_{31}	1
q_{30}	q_{10}	q_4	0
q_{31}	q_{10}	q_4	1
q_4	q_2	q_4	0

To get rid of the problem of occurrence of a null string, we need to include another state, q_a, with the same transactions as that of q_0 but with output 0.

The modified final Moore machine equivalent to the given Mealy machine is

Present State	Next State I/P = 0	Next State I/P = 1	O/P
q_a	q_{11}	q_2	0
q_0	q_{11}	q_2	1
q_{10}	q_{30}	q_0	0
q_{11}	q_{30}	q_0	1
q_2	q_4	q_{31}	1
q_{30}	q_{10}	q_4	0
q_{31}	q_{10}	q_4	1
q_4	q_2	q_4	0

16. Convert the following Mealy machine to an equivalent Moore machine by the tabular format.

Present State	I/P = 0 Next State	I/P = 0 O/P	I/P = 1 Next State	I/P = 1 O/P
→ q_0	q_0	1	q_1	0
q_1	q_3	1	q_3	1
q_2	q_1	1	q_2	1
q_3	q_2	0	q_0	1

Solutuion: In the next state column of the given Mealy machine, the output differs for q_1 and q_2 as the next states. So, the states will be divided as q_{10}, q_{11} and q_{20}, q_{21}, respectively. After dividing the states, the modified Mealy machine will be

Present State	I/P = 0 Next State	I/P = 0 O/P	I/P = 1 Next State	I/P = 1 O/P
→ q_0	q_0	1	q_{10}	0
q_{10}	q_3	1	q_3	1
q_{11}	q_3	1	q_3	1

Present State	I/P = 0 Next State	O/P	I/P = 1 Next State	O/P
q_{20}	q_{11}	1	q_{21}	1
q_{21}	q_{11}	1	q_{21}	1
q_3	q_{20}	1	q_0	1

The converted Moore machine is

Present State	Next State I/P = 0	Next State I/P = 1	O/P
→ q_0	q_0	q_{10}	1
q_{10}	q_3	q_3	0
q_{11}	q_3	q_3	1
q_{20}	q_{11}	q_{21}	0
q_{21}	q_{11}	q_{21}	1
q_3	q_{20}	q_0	1

To get rid of the problem of occurrence of a null string, we need to include another state, q_a, with the same transactions as that of q_0 but with output 0.

The modified final Moore machine equivalent to the given Mealy machine is

Present State	Next State I/P = 0	Next State I/P = 1	O/P
→ q_a	q_0	q_{10}	0
q_0	q_0	q_{10}	1
q_{10}	q_3	q_3	0
q_{11}	q_3	q_3	1
q_{20}	q_{11}	q_{21}	0
q_{21}	q_{11}	q_{21}	1
q_3	q_{20}	q_0	1

17. Convert the following Mealy Machine to a Moore Machine. [WBUT 2008]

Present State	Next State I/P = 0 State	I/P = 0 Output	Next State I/P = 1 State	I/P = 1 Output
Q_1	Q_2	1	Q_1	0
Q_2	Q_3	0	Q_4	1
Q_3	Q_1	0	Q_4	0
Q_4	Q_3	1	Q_2	1

Solution: Q_3 and Q_4 as next states produce outputs 0 and 1, and so the states are divided into Q_{30}, Q_{31} and Q_{40}, Q_{41}. Thus, the constructing Moore machine contains six states. The Moore machine becomes

Present State	Next State I/P = 0	Next State I/P = 1	Output
Q_1	Q_2	Q_1	0
Q_2	Q_{30}	Q_{41}	1
Q_{30}	Q_1	Q_{40}	0
Q_{31}	Q_1	Q_{40}	1
Q_{40}	Q_{31}	Q_2	0
Q_{41}	Q_{31}	Q_2	1

18. From the following Mealy machine, find the equivalent Moore machine. Check whether the Mealy machine is a minimal one or not. Give proper justification to your answer. [WBUT 2007]

Present State	I/P = 0 Next State	I/P = 0 O/P	I/P = 1 Next State	I/P = 1 O/P
S_1	S_2	0	S_1	0
S_2	S_2	0	S_3	0
S_3	S_4	0	S_1	0
S_4	S_2	0	S_5	0
S_5	S_2	0	S_1	1

Solution:

i) In the Mealy machine, S_1 as the next state produces output 0 for some cases and produces output 1 for one case. For this reason, the state S_1 is divided into two parts: S_{10} and S_{11}. All the other states produce output 0.
To get rid of the problem of occurrence of a null string, we need to include another state, S_a, with the same transactions as that of S_{10} but with output 0.
The modified final Moore machine equivalent to the given Mealy machine will be as follows.
The converted Moore machine is

Present State	Next State I/P = 0	Next State I/P = 1	O/P
S_a	S_2	S_{10}	0
S_{10}	S_2	S_{10}	0
S_{11}	S_2	S_{10}	1
S_2	S_2	S_3	0
S_3	S_4	S_{10}	0
S_4	S_2	S_5	0
S_5	S_2	S_{11}	0

ii) All the states are 0 equivalents.

$$P_0 = \{S_1S_2S_3S_4S_5\}$$

For string length 1, all the states produce output 0 except S_5.

$$P_1 = \{S_1S_2S_3S_4\}\{S_5\}$$

The next states of all the states (belong to the first subset) for all inputs belong to one set except S_4. The modified partition is

$$P_2 = \{S_1S_2S_3\}\{S_4\}\{S_5\}$$

By this process, $P_3 = \{S_1S_2\}\{S_3\}\{S_4\}\{S_5\}$

$$P_4 = \{S_1\}\{S_2\}\{S_3\}\{S_4\}\{S_5\}$$

The machine is a reduced machine as the number of subsets of the machine is the same as the number of states of the original Mealy machine. Hence, the machine is a minimal machine.

19. Convert the following Moore machine into an equivalent Mealy machine by the transitional format.

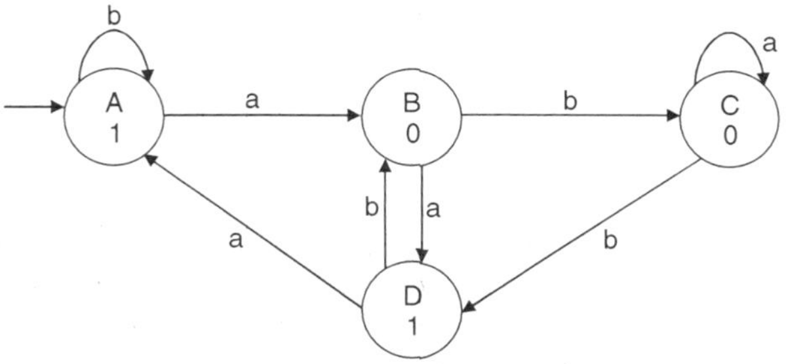

Solution:

i) In this machine, A is the beginning state. So start from A. For A, there are three incoming arcs, from A to A with input b, one in the form of start-state indication with no input, and the last is from D to A with input a. State A is labelled with output 1. As the start-state indication contains no input, it is useless and, therefore, keep it as it is.

Modify the label of the incoming edge from D to A and from A to A including the output of state A. So, the label of the incoming state will be D to A with label a/1 and A to A with label b/1.

ii) State B is labelled with output 0. The incoming edges to the state B are from A to B with input a and from D to B with input b.

Modify the labels of the incoming edges including the output of state B. So, the labels of the incoming states will be A to B with label a/0 and from D to B with label b/0.

iii) State C is labelled with output 0. There are two incoming edges to this state, from B to C with input b and from C to C with input a.

The modified label will be B to C with label b/0 and C to C with label a/0.

iv) State D is labelled with output 1. There are two incoming edges to this state, from B to D with input a and from C to D with input b.

The modified label will be B to D with label a/1, and C to D with label b/1.

The converted Mealy machine will be

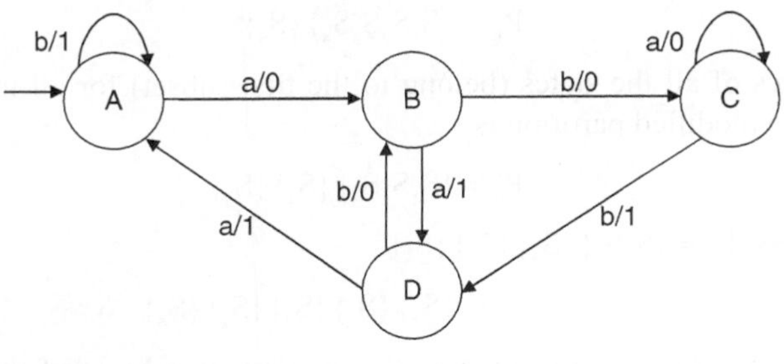

20. Convert the following Mealy machine into an equivalent Moore machine by the transitional format.

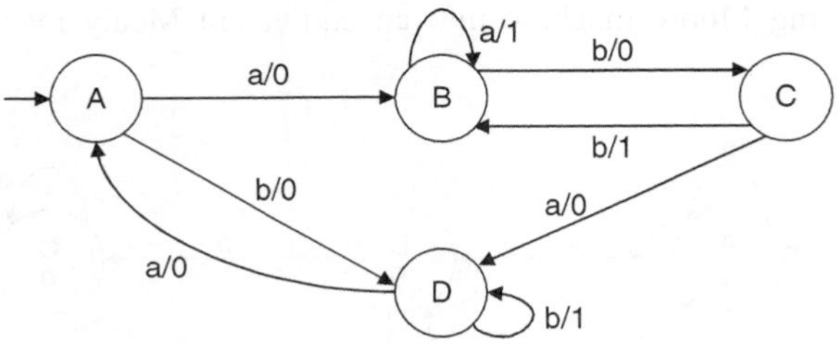

Solution: The machine contains four states. Let us start from the state A. The incoming edges to this state are from D to A with label a/0. There is no difference in the outputs of the incoming edges to this state, and so in the constructing Moore machine the output for this state will be 0.

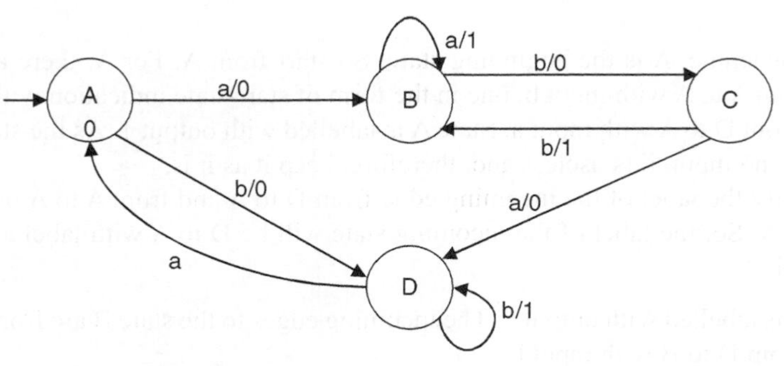

For the state B, the incoming edges are B to B with label a/1, from A to B with label a/0, and from C to B with label b/1.

We get two different outputs for two incoming edges (B to B output 1, A to B output 0). So, the state B will be divided into two, namely, B_0 and B_1. The outgoing edges are duplicated for both the states generated from B. The modified machine is

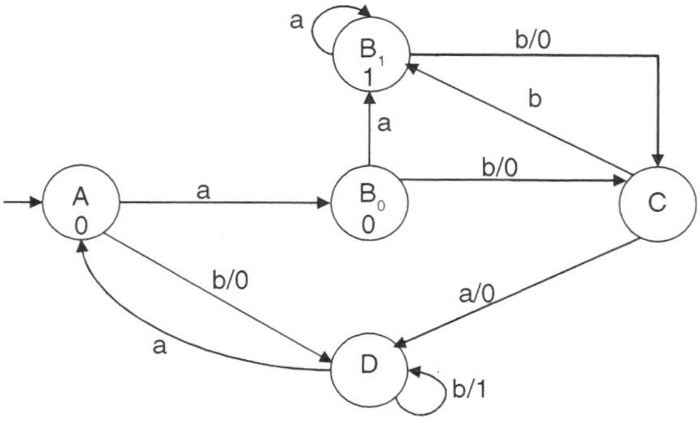

For the state A, the incoming edges to this state are from B_0 to C with label b/0 and B_1 to C with label b/0. There is no difference in the outputs of the incoming edges to this state, and so in the constructing Moore machine the output for this state will be 0.

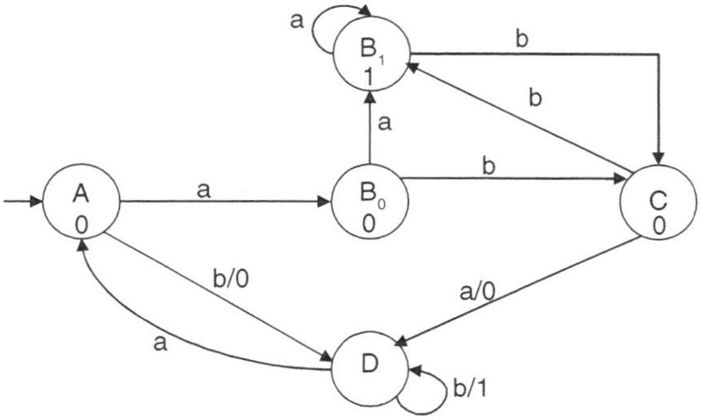

For the state D, the incoming edges are A to D with label b/0, from C to D with label a/0, and from D to D with label b/1.

We get two different outputs for two incoming edges (D to D output 1, C to D output 0). So, the state D will be divided into two, namely, D_0 and D_1. The outgoing edges are duplicated for both the states generated from D. The modified machine is

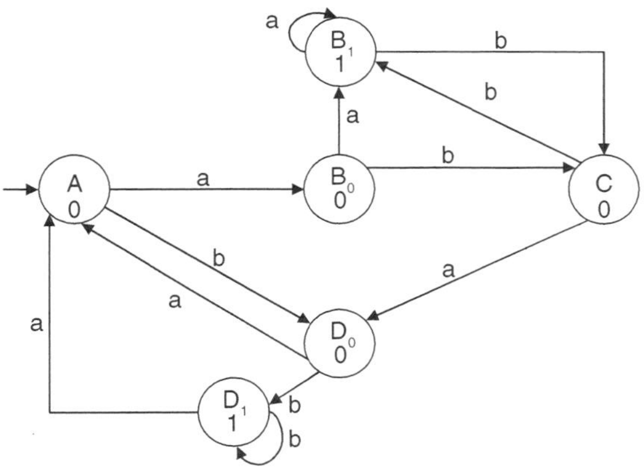

21. Convert the following Mealy machine into an equivalent Moore machine. [UPTU 2004]

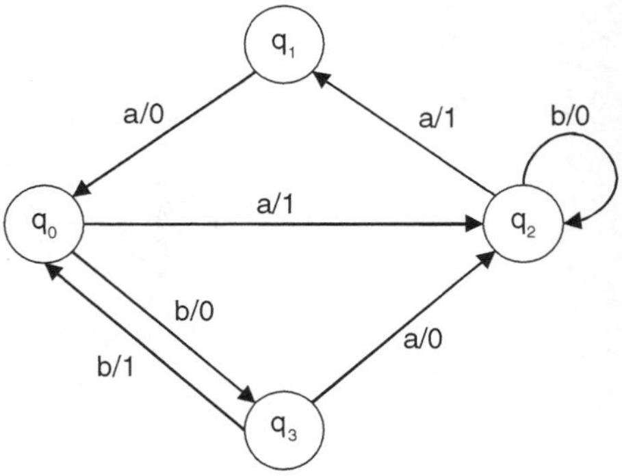

Solution: The state q_0 has two incoming edges: from q_1 with label a/0 and from q_3 with label b/1. As there is a difference in output, the state q_0 is divided into q_{00} and q_{01} with outputs 0 and 1, respectively. The states q_1 and q_3 have only one incoming edge each, and so there is no need of division. The state q_2 has three incoming edges; among those, two are of output '0' and another is of output '1'. Thus, it is divided into q_{20} and q_{21} with outputs 0 and 1, respectively.

From q_1 input with label 'a' ends on q_{00}, and from q_3 input with label 'b' ends on q_{01}. The outputs from old q_1 state are duplicated from q_{00} and q_{01}.

The state q_1 and q_3 are not divided. q_1 gets output '1' and q_3 gets output '0'.

Dividing the state q_0 and placing q_1 and q_3, the intermediate machine becomes as follows

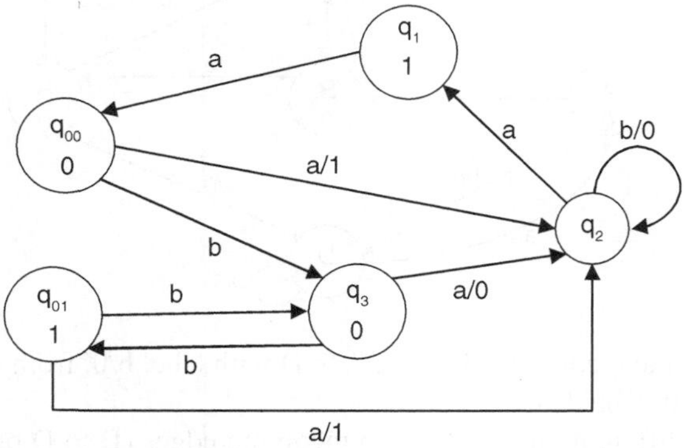

The state q_2 is divided into q_{20} and q_{21}. From q_{00} and q_{01} input with label 'a' ends on q_{21}. From q_3 input with label 'a' ends on q_{20}. There is a loop on q_2. That loop will be on q_{20} with label 'b'. Another transition with label 'b' is drawn from q_{21} to q_{20}. The final Moore machine is as follows

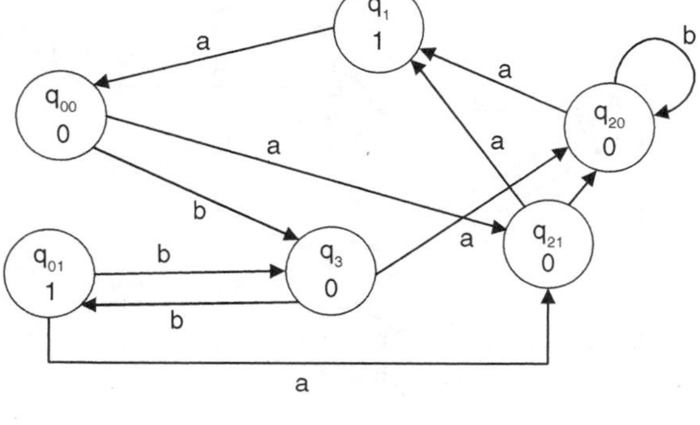

22. Minimize the following finite automata.

Present State	Next State I/P = a	I/P = b
→A	B	F
B	A	F
C	G	A
D	H	B
E	A	G
F	H	C
G	A	D
H	A	C

Here F, G, and H are the final states.

Solution: In the finite automata, the states are {A, B, C, D, E, F, G, H}. Name this set as S_0.

$$S_0\text{: \{A, B, C, D, E, F, G, H\}}$$

All of the states are 0 equivalents.

In the finite automata, there are two types of states: final state and non-final states. So, divide the set of states into two parts, Q_1 and Q_2.

$$Q_1 = \{F, G, H\} \quad Q_2 = \{A, B, C, D, E\}$$

$$S_1\text{: \{\{F, G, H\} \{A, B, C, D, E\}\}}$$

The states belonging to same subset are 1-equivalent because they are in the same set for string length 1. The states belonging to different subsets are 1-distinguishable.

The next states of F are H and C. The next states of G and H are A, D and A, C, respectively.

A, D and A, C belong to the same subset but H and C belong to a different subset. So, F, G, and H are divided into {F}, {G, H}.

For input 0, the next states of A, B, C, D, and E are B, A, G, H, and A, respectively. For input 1, the next states of A, B, C, D, and E are F, F, A, B, and G, respectively. So, the set {A, B, C, D, E} is divided into {A, B, E} and {C, D}.

$$S_2\text{: \{\{F\} \{G, H\} \{A, B, E\} \{C, D\}\}}$$

By the same process, {A, B, E} is divided into {A, B}, {E}.

$$S_3\text{: \{\{F\} \{G, H\} \{A, B\} \{E\} \{C, D\}\} = \{\{A, B\}, \{C, D\}, \{E\}, \{F\}, \{G, H\}\}}$$

The set is not dividable further. So, these are the states of minimized DFA. Let us rename the subsets as q_0, q_1, q_2, q_3, and q_4. The initial state was A, and so here the initial state is {A, B}, i.e., q_0. The final state was F, G, and H, and so here the final states are {F}, i.e., q_3 and {G, H}, i.e., q_4. The tabular representation of minimized DFA is

Present State	Next State	
	I/P = 0	I/P = 1
→ q_0	q_0	q_3
q_1	q_4	q_0
q_2	q_0	q_4
(q_3)	q_4	q_1
(q_4)	q_0	q_1

23. Design a Mealy and Moore machine for detecting a sequence 1010 where overlapping sequences are also accepted. Convert the Moore machine that you have got into a Mealy machine. Are there any differences? How will you prove that the two Mealy machines are equivalent?

Solution: The Mealy machine is

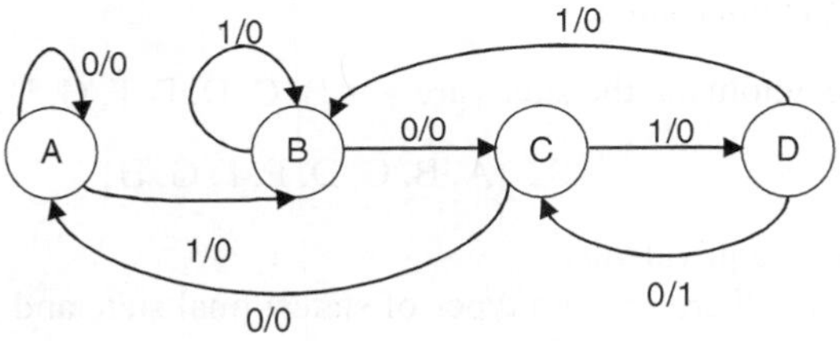

The Moore machine is

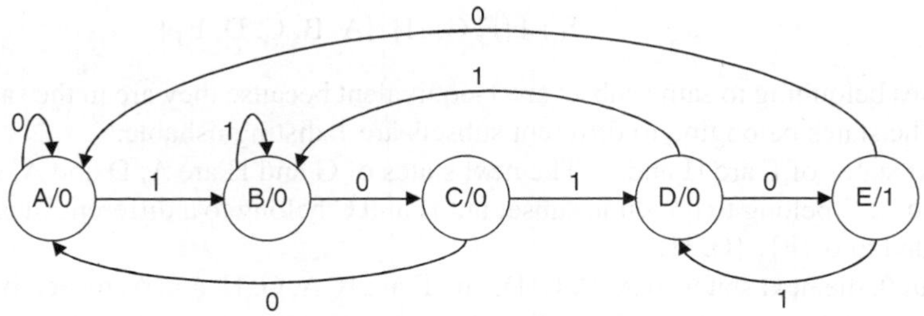

The converted Mealy machine from the given Moore machine is (by using the transactional format)

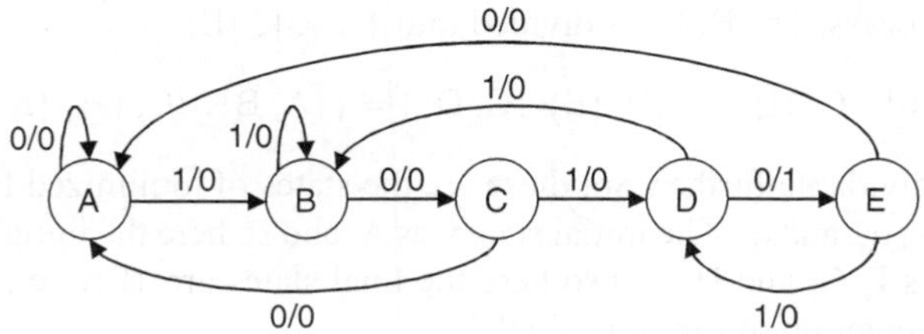

This is surely different from the previously constructed Mealy machine in the number of states and in transactions.

But these two Mealy machines are equivalent. This can be proved by finding the equivalent partitions of the second machine.

The tabular format of the previous machine is

Present State	Next State, z	
	x = 0	x = 1
A	A, 0	B, 0
B	C, 0	B, 0
C	A, 0	D, 0
D	E, 1	B, 0
E	A, 0	D, 0

P_0 = {A, B, C, D, E}

P_1 = {A, B, C, E} {D}

P_2 = {A, B} {C, E} {D}

P_3 = {A} {B} {C, E} {D}

P_4 = {A} {B} {C, E} {D}

Rename the states as S_1, S_2, S_3, and S_4.

Present State	Next State, z	
	x = 0	x = 1
S_1(A)	S_1, 0	S_2, 0
S_2(B)	S_3, 0	S_2, 0
S_3(C, E)	S_1, 0	S_4, 0
S_4 (D)	S_3, 1	S_2, 0

The machine is same as the first Mealy machine.

24. Minimize the following DFA M using the Myhill–Nerode theorem.

Present State	Next State	
	I/P = a	I/P = b
A	A	D
B	C	F
C	D	E
D	A	F
E	A	G
F	B	E
G	B	D

A is the initial state, and G is the final state. [WBUT 2003]

Solution:

Step I: Divide the states of the DFA into two subsets: final (F) and non-final (Q–F).

F = {G}, Q–F = {A, B, C, D, E, F}

Make a two-dimensional matrix labelled at the left and bottom by the states of the DFA.

A	—	—	—	—	—	—	—
B		—	—	—	—	—	—
C			—	—	—	—	—
D				—	—	—	—
E					—	—	—
F						—	—
G							—
	A	B	C	D	E	F	G

Step II:

i) The following combinations are the combination of the beginning and final states. (A, G), (B, G), (C, G), (D, G), (E, G), (F, G). Put X in these combinations of states.
The modified matrix is

A	—	—	—	—	—	—	—
B		—	—	—	—	—	—
C			—	—	—	—	—
D				—	—	—	—
E					—	—	—
F						—	—
G	X	X	X	X	X	X	—
	A	B	C	D	E	F	G

ii) The pair combination of non-final states are (A, B), (A, C), (A, D), (A, E), (A, F), (B, C), (B, D), (B, E), (B, F), (C, D), (C, E), (C, F), (D, E), (D, F), and (E, F).
r = δ(A, b) → D s = δ(B, b) → F, in the place of (DF), there is neither X nor x. So, in the place of (A, B), there will be 0.
Similarly,

(r, s) = δ((A, C), b) → (D, E) (neither X nor x). In the place of (A, C), there will be 0.

(r, s) = δ((A, D), b) → (DF) (neither X nor x). In the place of (A, D), there will be 0.

(r, s) = δ((A, E), b) → (DG) (there is X). In the place of (AE), there will be x.

All these processes will result in 0, except (AE), (BE), (CE), (DE) and (EF) (these are x).

The modified matrix is

A	—	—	—	—	—	—	—
B	0	—	—	—	—	—	—
C	0	0	—	—	—	—	—
D	0	0	0	—	—	—	—
E	x	x	x	x	—	—	—
F	0	0	0	0	x	—	—
G	X	X	X	X	X	X	—
	A	B	C	D	E	F	G

The combination of entries 0 are the states of the modified machine. The states of the minimized machine are [AB], [AC], [AD], [AF], [BC], [BD], [BF], [CD], [CF], and [DF], which are simplified to [ABCDF]. It denotes that the machine is already minimized.

25. Use the Myhill–Nerode theorem to minimize the following FA:

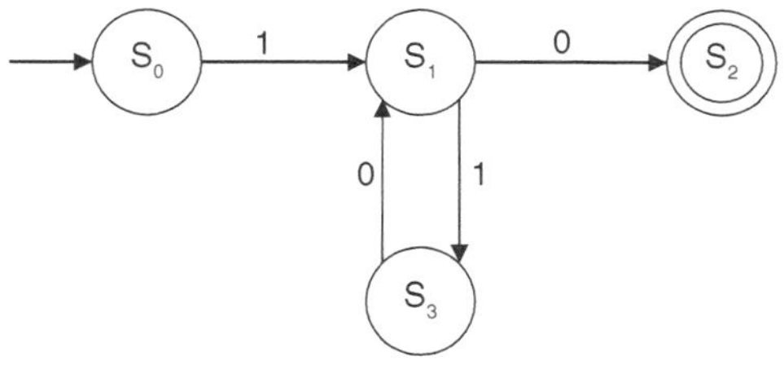

[WBUT 2006]

Solution: The tabular format of the finite automata is

Present State	Next State	
	i = 0	i = 1
S_0	–	S_1
S_1	S_3	S_2
S_2	S_1	–
S_3	–	–

Step I: Divide the states of the DFA into two subsets, final (F) and non-final (Q-F).

$$F = \{S_3\}, Q–F = \{S_0, S_1, S_2\}$$

Make a two-dimensional matrix labelled at the left and bottom by the states of the DFA.

S_0	–	–	–	–
S_1		–	–	–
S_2			–	–
S_3				–
	S_0	S_1	S_2	S_3

Step II:

i) The following combinations are the combination of the beginning and final states (S_0, S_3), (S_1, S_3), (S_2, S_3).
Put X in these combinations of states. The modified matrix becomes

S_0	–	–	–	–
S_1		–	–	–
S_2			–	–
S_3	X	X	X	–
	S_0	S_1	S_2	S_3

The pair combination of non-final states are (S_0, S_1), (S_0, S_2), (S_1, S_2).

$r = \delta(S_0, 0) \rightarrow$ not mentioned $s = \delta(S_1, 0) \rightarrow S_3$, in the place of all combination of S_3, there is X. So, in the place of (S_0, S_1), there will be x.

$r = \delta(S_0, 0) \rightarrow$ not mentioned $s = \delta(S_2, 0) \rightarrow S_1$, as S_1 is a non-final state, we shall consider that $\delta(S_0, 0)$ will produce a non-final state. We are considering that (S_0, S_1) is filled by 0.

$r = \delta(S_1, 0) \rightarrow S_3$ $s = \delta(S_2, 1) \rightarrow S_1$, in the place of (S_1, S_3), there is X. So, in the place of (S_1, S_2), there will be x.

The modified matrix becomes

S_0	–	–	–	–
S_1	x	–	–	–
S_2	0	x	–	–
S_3	X	X	X	–
	S_0	S_1	S_2	S_3

The combination (S_0, S_2) is filled by 0. But another two states S_1 and S_3 are left. The states of the minimized machine are (S_0, S_2), S_1, S_3, where the final state is S_3. The minimized DFA becomes

Present State	Next State	
	i = 0	i = 1
(S_0, S_2)	S_1	S_1
S_1	S_3	(S_0, S_2)
S_3	–	–

26. Prove that the following two FA are equivalent.

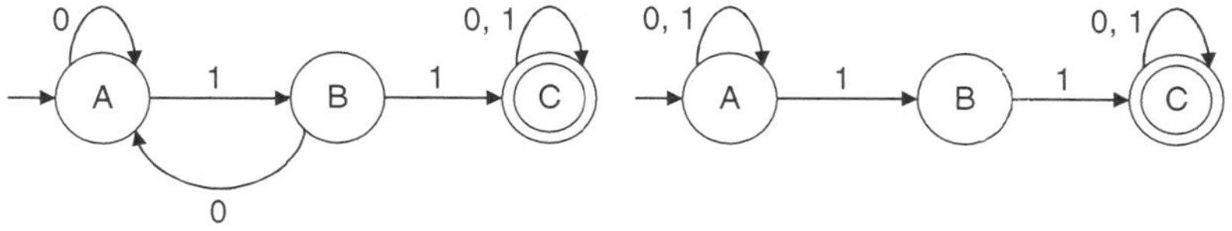

Solution: The second one is an NFA. The tabular representation of the FA is

Present State	Next State i = 0	i = 1
A	A	A, B
B	–	C
C	C	C

The DFA from the given NFA is

Present State	Next State i = 0	i = 1
[A]	[A]	[A, B]
[A, B]	[A]	[A, B, C]
[A, B, C]	[A, C]	[B, C]
[B, C]	[B, C]	[B, C]
[B, C]	[A, C]	[C]
[C]	[C]	[C]

Simplifying this, the DFA becomes

Present State	Next State i = 0	i = 1
S_1	S_1	S_2
S_2	S_1	S_3
S_3	S_4	S_5
S_4	S_5	S_5
S_5	S_4	S_6
S_6	S_6	S_6

Here, S_1 is the initial and S_3, S_4, S_5, and S_6 are the final states.

Now try to minimize the DFA.

$$P_0 = (S_1\ S_2\ S_3\ S_4\ S_5\ S_6)$$

$$P_1 = (S_1\ S_2)(S_3\ S_4\ S_5\ S_6)$$

$$P_2 = (S_1)\ (S_2)\ (S_3\ S_4\ S_5\ S_6)$$

Rename (S_1) as A, (S_2) as B, and (S_3 S_4 S_5 S_6) as C. The minimized FA is

Present State	Next State	
	i = 0	i = 1
A	A	B
B	A	C
C	C	C

where A is the initial state and C is the final state.

It is proved that the two DFA are equivalent.

27. Draw the state transition of a deterministic finite state automaton which accepts all strings from the alphabet (a, b), such that no string has three consecutive occurrences of the letter b.

[GATE 1993]

Solution:

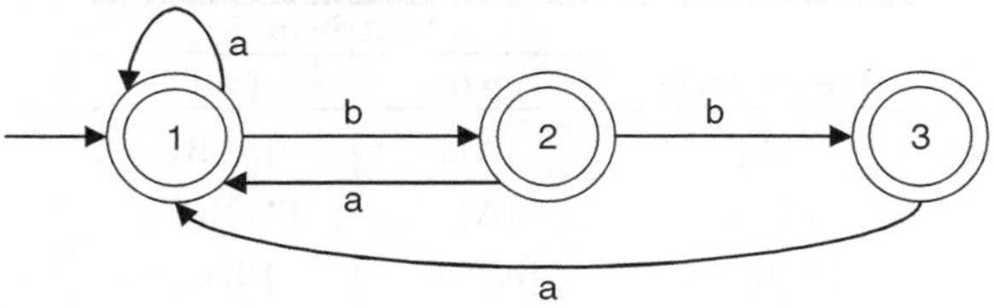

28. Construct a finite state machine with minimum number of states, accepting all strings over (a, b) such that the number of a' s is divisible by two and the number of b's is divisible by three.

[GATE 1997]

Solution:

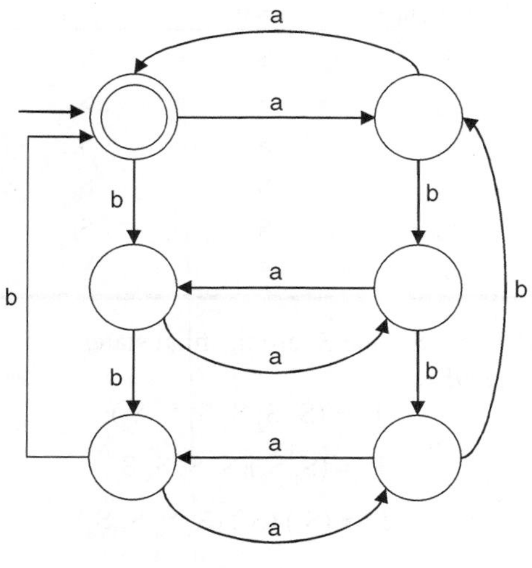

Multiple Choice Questions

1. A language L from a grammar G = {V_N, Σ, P, S} is
   a) Set of symbols over V_N
   b) Set of symbols over Σ
   c) Set of symbols over P
   d) Set of symbols over S
2. Which is true for δ(q, ab)
   a) δ(q, a) ∪ δ(q, b)
   b) δ(δ(q, a), b)
   c) δ(q, a), b
   d) δ(q, a) ∩ δ(q, b)
3. The transitional function of a DFA is
   a) $Q \times \Sigma \to Q$ b) $Q \times \Sigma \to 2^Q$
   c) $Q \times \Sigma \to 2^n$ d) $Q \times \Sigma \to Q^n$
4. The transitional function of an NFA is
   a) $Q \times \Sigma \to Q$ b) $Q \times \Sigma \to 2^Q$
   c) $Q \times \Sigma \to 2^n$ d) $Q \times \Sigma \to Q^n$
5. The maximum number of states of a DFA converted from an NFA with n states is
   a) n b) n^2
   c) 2^n d) None of these
6. A string after full traversal is called not accepted by an NFA if it results in
   a) Some non-final states
   b) All non-final states
   c) A single non-final state
   d) Some final states
7. An NFA with a set of states Q is converted to an equivalent DFA with a set of states Q′. Find which is true.
   a) $Q' = Q$ b) $Q' \subseteq Q$
   c) $Q \subseteq Q'$ d) None of these
8. The basic limitations of a finite state machine is
   a) It cannot remember arbitrarily large amount of information
   b) It cannot remember state transitions
   c) It cannot remember grammar for a language
   d) It cannot remember language generated from a grammar
9. The string WW^R is not recognized by any FSM because
   a) An FSM cannot remember arbitrarily large amount of information
   b) An FSM cannot fix the mid-point
   c) An FSM cannot match W with W^R
   d) An FSM cannot remember the first and last inputs.
10. A finite automata recognizes
    a) Any language
    b) Context sensitive language
    c) Context-free language
    d) Regular language
11. Which is true for a dead state?
    a) It cannot be reached anytime
    b) There is no necessity of the state
    c) If control enters, there is no way to come out from the state
    d) If control enters, FA is dead
12. The language accepted by the given FA is

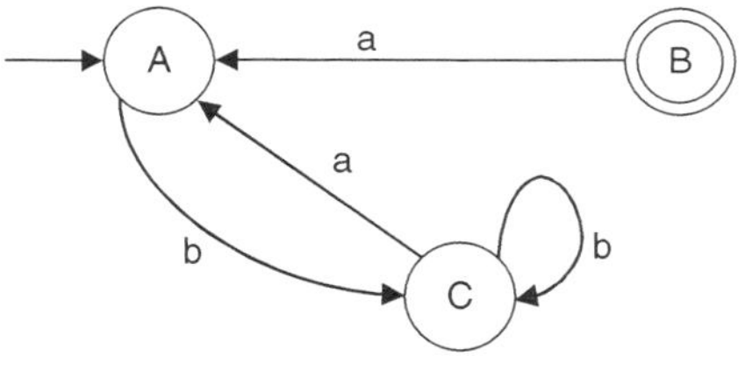

    a) (ab)* b) bb*a
    c) b(ba)*a d) Null
13. In the previous FA, B is called
    a) Dead state
    b) Inaccessible state
    c) Both a and b
    d) None of these
14. Which is true for a Moore machine?
    a) Output depends on the present state
    b) Output depends on the present input
    c) Output depends on the present state and the present input
    d) Output depends on the present state and the past input

15. Which is true for the Mealy machine?
    a) Output depends on the present state
    b) Output depends on the present input
    c) Output depends on the present state and the present input
    d) Output depends on the present state and the past input

16. Which is true for the inaccessible state?
    a) It cannot be reached anytime
    b) There is no necessity of the state
    c) If control enters, there is no way to come out from the state
    d) If control enters, FA is dead

17. In Mealy Machine, O/P is a function of
    a) Present state only
    b) Next state only
    c) Present state and Input
    d) Input only

18. In Moore Machine, O/P is associated with
    a) Present state only
    b) Next state only
    c) Present state and Input
    d) Input only

19. Which type of string is accepted by the following finite automata?

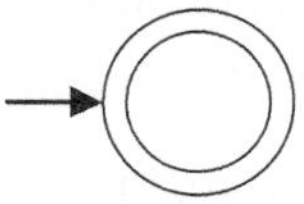

    a) All string
    b) Null string
    c) No string
    d) All of the above

Answers:

1. b	2. b	3. a	4. b	5. c	6. b	7. d	8. a	9. b	10. d
11. c	12. d	13. b	14. a	15. c	16. a	17. b	18. a	19. b	

GATE Questions

1. Consider the strings u = abbaba, v = bab, and w = aabb. Which of the following statement is true for the given transitional system?

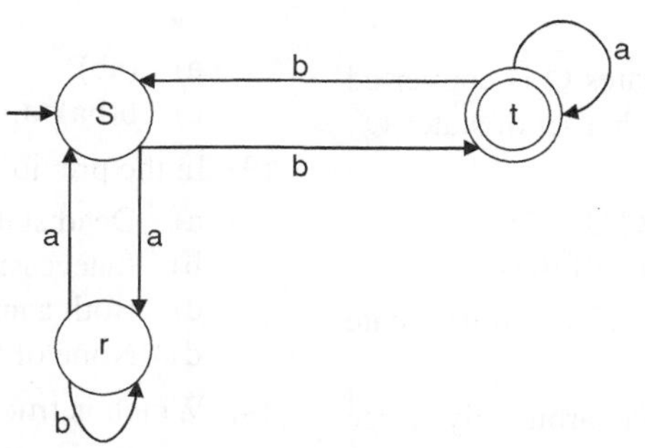

    a) The automaton accepts u and v but not w.
    b) The automaton accepts each of u, v, and w.
    c) The automaton rejects each of u, v, and w.
    d) The automaton accepts u but rejects v and w.

2. Consider the transitional system.

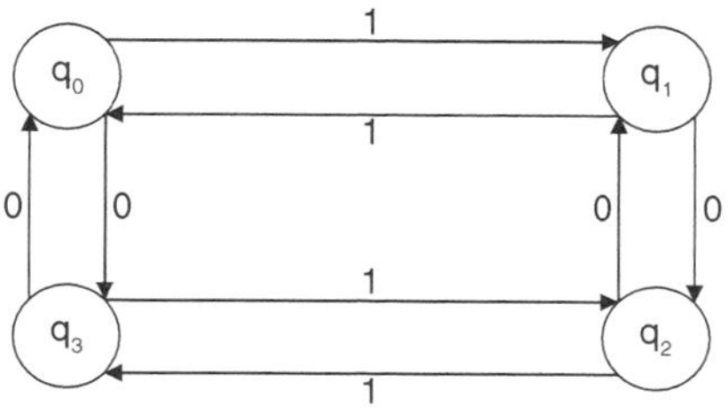

Which of the states are to be marked as starting state and final state, respectively, so as to turn the above system into a DFA that accepts all strings having odd number of zeros and even number of 1's?

a) q_0, q_2 b) q_0, q_1 c) q_1, q_2 d) None of these

3. Consider the following DFA in which S_0 is the start state and S_1 and S_3 are the final states. Which one is true?

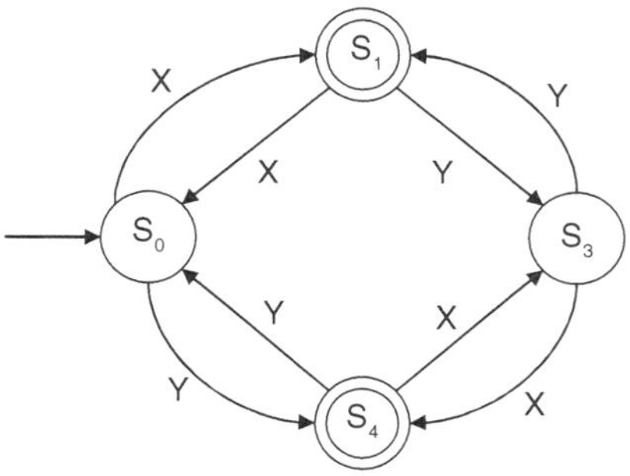

a) All strings of x and y.
b) All strings of x and y which have either an even number of x and even number of y or an odd number of x and odd number of y.
c) All strings of x and y which have an equal number of x and y.
d) All strings of x and y which have either an even number of x and odd number of y or an odd number of x and even number of y.

4. Let N be an NFA with n states and let M be the minimized DFA with m states recognizing the same language. Which of the following is NECESSARILY true?

a) $m \leq 2^n$ b) $n \leq m$ c) M has one accept state d) $m = 2^n$

5. If the final state and non-final states in the following DFA are interchanged, then which of the following languages over the alphabet (a, b) will be accepted by the new DFA?

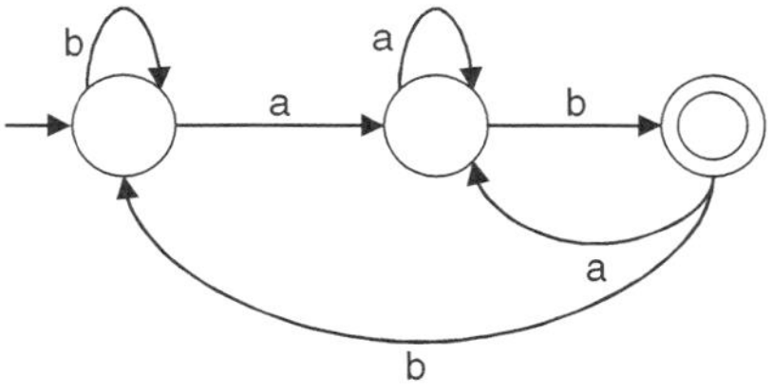

a) Set of all strings that do not end with ab
b) Set of all strings that begin with either an a or a b.
c) Set of all strings that do not contain the substring ab.
d) The set described by the regular expression b*aa*(ba)*b*

6. A finite state machine with the following state table has a single input x and a single output z.

Present State	Next State, z X = 1	Next State, z X = 0
A	D, 0	B, 0
B	B, 1	C, 1
C	B, 0	D, 1
D	B, 1	C, 0

If the initial state is unknown, then the shortest input sequence to reach the final state C is:

a) 01 b) 10 c) 101 d) 110

7. Which of the following sets can be recognized by a deterministic finite state automaton?
   a) The numbers 1, 2, 4, 8, ... 2^n written in binary.
   b) The numbers 1, 2, 4,... 2^n written in unary.
   c) The set of binary string in which the number of zeros is the same as the number of ones.
   d) The set {1, 101, 11011, 1110111, ...}

8. Consider the regular expression (0 + 1)(0 + 1) ..... N times. The minimum state FA that recognizes the language represented by this regular expression contains
   a) n states b) (n + 1) states c) (n + 2) states d) None of the above

9. What can be said about a regular language L over {a} whose minimal finite state automation has two states?
   a) L must be $\{a^n \mid n \text{ is odd}\}$ b) L must be $\{a^n \mid n \text{ is even}\}$
   c) L must be $\{a^n \mid n \geq 0\}$ d) Either L must be $\{a^n \mid n \text{ is odd}\}$ or L must be $\{a^n \mid n \text{ is even}\}$

10. The smallest FA which accepts the language {x | length of x is divisible by 3} has
    a) 2 states b) 3 states c) 4 states d) 5 states

11. Consider the following deterministic finite state automaton M.

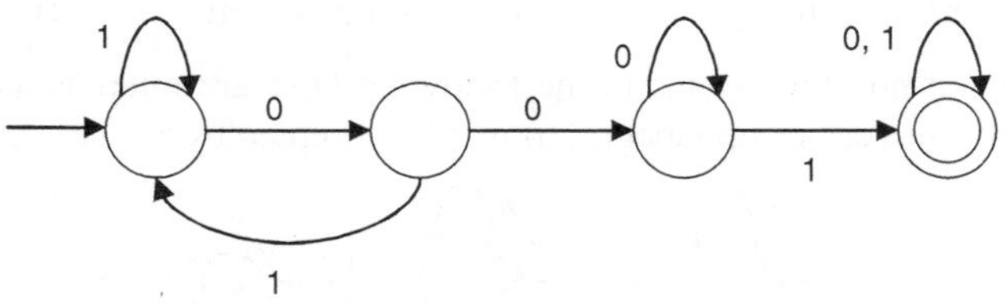

Let S denote the set of seven-bit binary strings in which the first, fourth, and last bits are 1. The number of strings in S that are accepted by M is

(a) 1 b) 5 c) 7 d) 8

12. The following finite state machine accepts all those binary strings in which the number of 1's and 0's are, respectively,

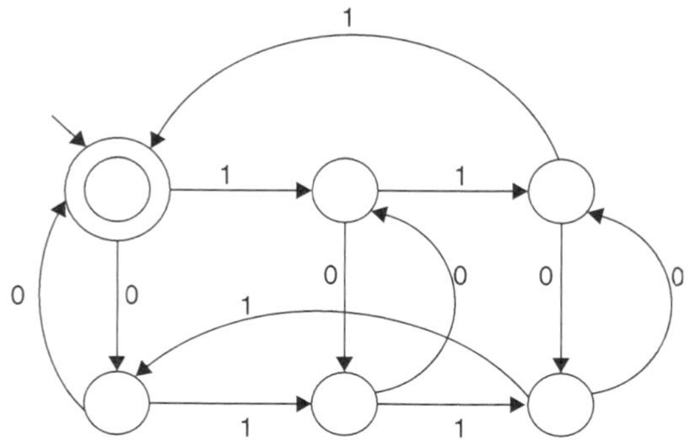

a) divisible by 3 and 2
b) odd and even
c) even and odd
d) divisible by 2 and 3

13. Consider the machine M.

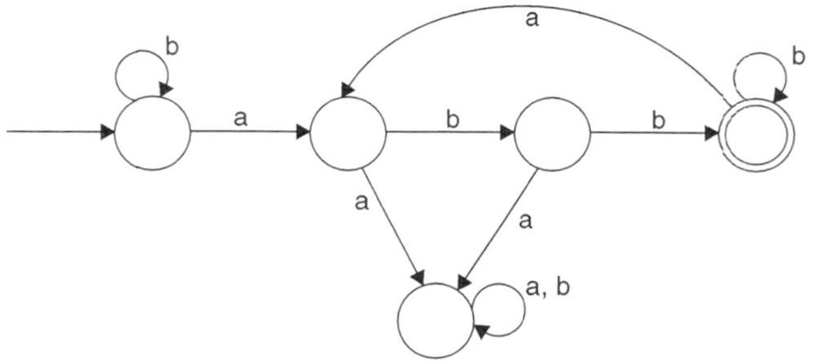

The language recognized by M is:

a) {w ε{a, b}* | every a in w is followed by exactly two b's}
b) {w ε{a, b}* | every a in w is followed by at least two b's}
c) {w ε{a, b}* | w contains the substring 'abb'}
d) {w ε{a, b}* | w does not contain 'aa' as a substring}

14. A minimum state deterministic FA accepting the language L = {w | w ε {0, 1}*} where number of 0's and 1's in w are divisible by 3 and 5, respectively, has

a) 15 states b) 11 states c) 10 states d) 9 states

15. Following are two finite automata (→ indicates the start state and F indicates the final state)

Y:

	a	b
→ 1	1	2
2(F)	2	1

Z:

	a	b
→ 1	2	2
2(F)	1	1

Which of the following represents the product automaton Z × Y?

a)

	a	b
→P	S	R
Q	R	S
R(F)	Q	P
S	Q	P

b)

	a	b
→P	S	Q
Q	R	S
R(F)	Q	P
S	P	Q

c)

	a	b
→P	Q	S
Q	R	S
R(F)	Q	P
S	Q	P

d)

	a	b
→P	S	Q
Q	S	R
R(F)	Q	P
S	Q	P

16.

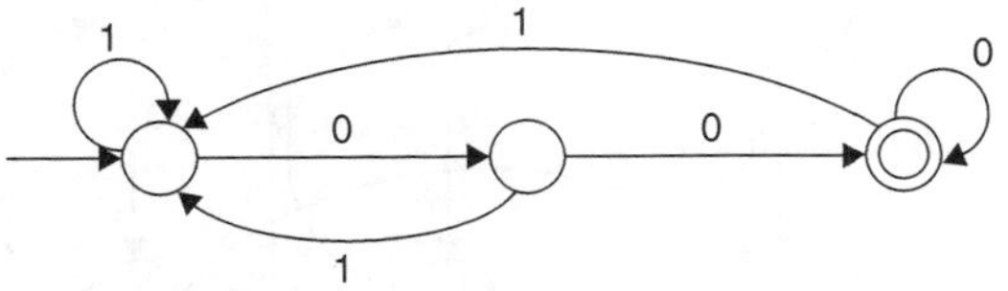

The given DFA accepts the set of all strings over {0, 1} that

a) begin either with 0 or 1.
b) end with 0
c) end with 00.
d) contain the substring 00.

17. Let w be any string of length n in $\{0, 1\}^*$. Let L be the set of all substrings of w. What is the minimum number of states in a non-deterministic FA that accepts L?

a) $n - 1$ b) n c) $n + 1$ d) 2^{n-1}

18. Definition of a language L with alphabet {a} is given as following $\{a^{nk} \mid k > 0$ and n is a positive integer constant$\}$
What is the minimum number of states needed in a DFA to recognize L?

a) $k + 1$ b) $n + 1$ c) 2^{n+1} d) 2^{k+1}

19. A deterministic finite automation (DFA) D with alphabet $\Sigma = \{a, b\}$ is given as follows:

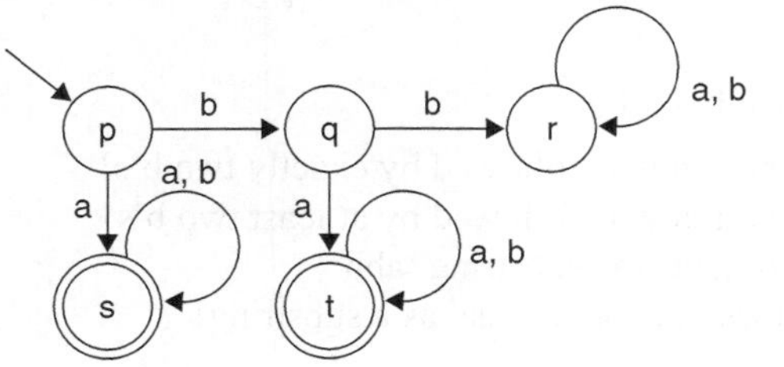

Which of the following finite state machines is a valid minimal DFA which accepts the same language as D?

a)

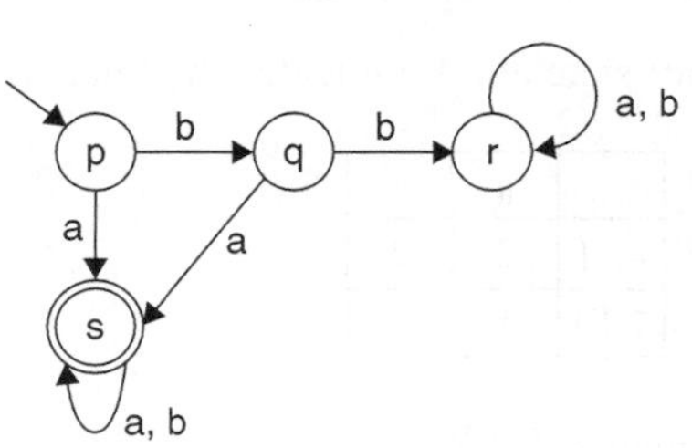

b)

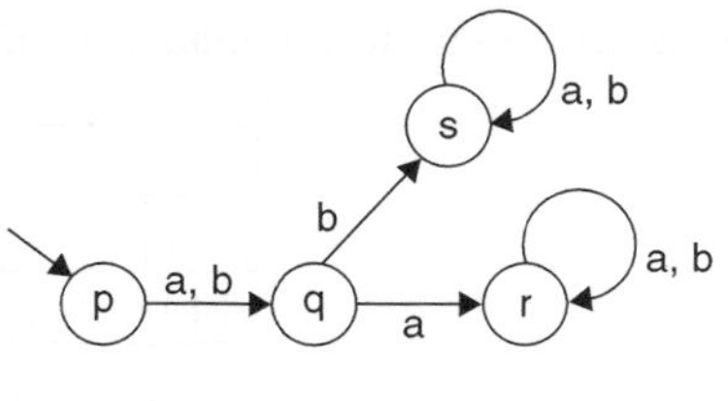

c)

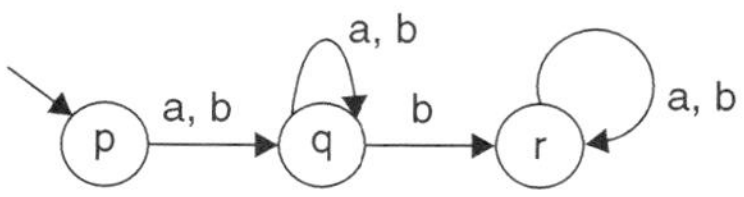

d)

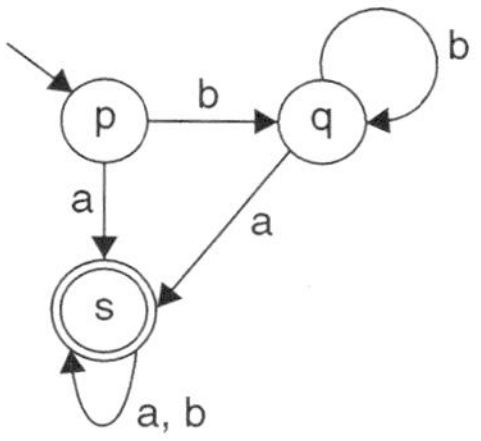

Answers:

1. d	2. c	3. d	4. a	5. a	6. b	7. a	8. b	9. a	10. b
11. a	12. a	13. a	14. a	15. a, b	16. c	17. c	18 b	19. a	

Hints:

3. If S_1 is the final state, then it accepts an odd number of x and even number of y. If S_3 is the final state, then an even number of x and odd number of y.
4. The highest possibility is to be 2^n. May be less than that.
5. The diagram is

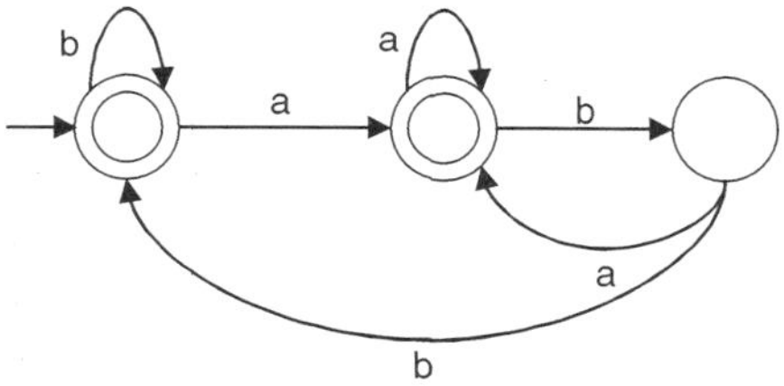

6. Consider the initial state as A or B.
7.

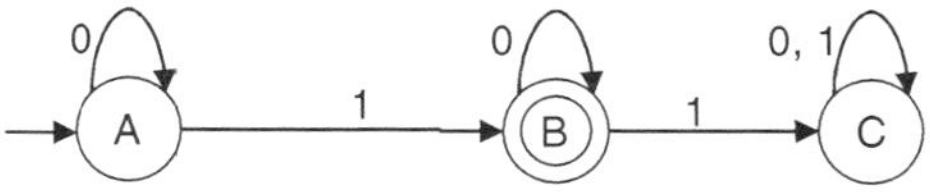

9.

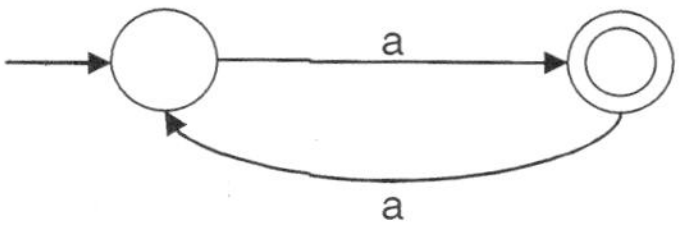

10.

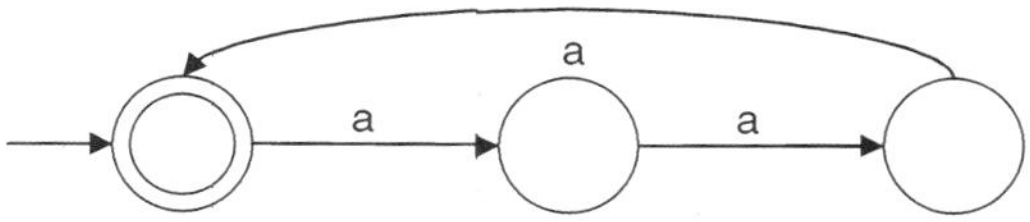

11. The string is 1001001.

15. It came in the 2008 GATE CSE. All answers are wrong!
 a is true if δ(S, a) → P and δ(S, b) → Q. b is true if δ(P, b) → R.
 The cross-product of two DFA $\{Q_1, \Sigma, \delta_1, q_{01}, F_1\}$, $\{Q_2, \Sigma, \delta_2, q_{02}, F_2\}$ is $\{Q, \Sigma, \delta, q_0, F\}$ where $q_0 = (q_{01}, q_{02})$, $F = \{F_1, F_2\}$ and
 $\delta(Q_1, Q_2, \text{i/p } \varepsilon\ \Sigma) = \delta(Q_1, \text{i/p } \varepsilon\ \Sigma) \cup \delta(Q_2, \text{i/p } \varepsilon\ \Sigma)$.
 For the example, q_0 = (1, 1)(P) and F = (2, 2)(R)

	A	b
(1, 1)(P)	(1, 2)	(2, 2)
(2, 1)(Q)	(2, 2)	(1, 2)
(2, 2)(R)	(2, 1)	(1, 1)
(1, 2)(S)	(1, 1)	(2, 1)

16. a is wrong as 10 is not accepted. b is wrong as 10 is not accepted. d is wrong as 1001 is not accepted.
17. Let w = 01. The FA is

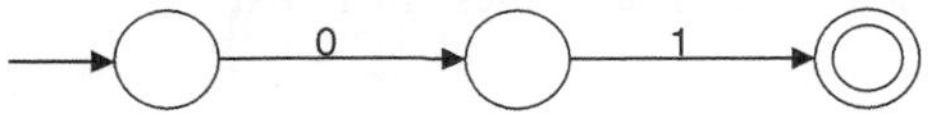

18.

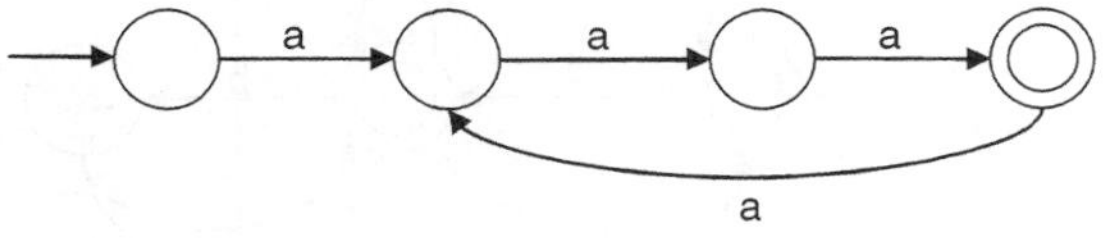

19. Minimize it.

Fill in the Blanks

1. Finite automata is defined as M = {___, ___, ___, ___, ___}.
2. Finite automata is one type of __________ state machine.
3. In the graphical format representation of finite automata, by double circle __________ state is represented.
4. In the tabular format representation of finite automata, by single circle __________ state is represented.
5. In the block diagram of finite automata, the input from input tape is read by __________.
6. Finite control can be considered as a __________ unit of a finite automaton.
7. If the input is given null for a state of finite automata, the machine goes to the __________.
8. There are two conditions for declaring a string accepted by a finite automaton. The conditions are __________ and __________.
9. For DFA, the transitional function δ is represented by $Q \times \Sigma \to$ __________.
10. DFA is a finite automata where for all cases for a single input given to a single state, the machine goes to a __________ state.

11. In NFA, δ is a transitional function mapping $Q \times \Sigma \rightarrow$ __________.
12. If any finite automata contains any ε (null) move or transaction, then that finite automata is called __________ with ε moves.
13. A state where the control can enter and be confined till the input ended, but there is no way to come out from that state is called __________.
14. The Mealy machine and Moore machine are example of finite automata with __________.
15. For the Mealy machine, the output depends on __________ and __________.
16. For the Moore machine, the output depends on __________.
17. The states which can never be reached from the initial state are called __________ states.
18. A relation R in set S is called equivalence relation if it is __________, __________, and __________.
19. A relation R in set S is __________ if for x, y in S, yRx whenever xRy.
20. A relation R in set S is __________ if for x, y, and z in S, xRz whenever xRy and yRz.
21. R is __________ then for all x, y, and z, xRy $\Rightarrow$ xzRyz.
22. The Myhill–Nerode theorem is used for __________.
23. A two-way finite automata is like finite automata but can traverse in __________.
24. The lexical analyser is designed by __________.
25. The maximum number of states of a DFA converted from an NFA with n states is __________.
26. A finite automata recognizes __________ language.

Answers:

1. $Q, \Sigma, \delta, q_0, F$	2. finite	3. final
4. final	5. reading head	6. control
7. same state	8. string must be totally traversed, the machine must come to a final state	9. Q
10. single	11. 2^Q	12. NFA
13. dead state	14. outputs	15. present state, present input
16. present state	17. inaccessible	18. reflexive, symmetric, transitive
19. symmetric	20. transitive	21. right invariant
22. minimization of FA	23. both directions	24. finite automata
25. 2^n	26. regular	

Exercise

1. Define finite automata. Can an FA have multiple initial and final states? Can an FA have no final state? Give reason in support of your answer.
2. Is it possible that a single string is accepted by two different finite automata? Is it true for a single regular expression to be accepted by two non-equivalent finite automata? Give reason.
3. Test whether the following strings are accepted by the given finite automata or not.

   a) 01010 b) 1010100 c) 1110011

Present State	Next State a = 0	a = 1
→ q_0	q_1	q_2
q_1	q_3	q_2
q_2	q_2	q_3
(q_3)	q_3	q_0

4. Test whether the following strings are accepted by the given finite automata or not.
   a) 100010 b) 0001101 c) 1000

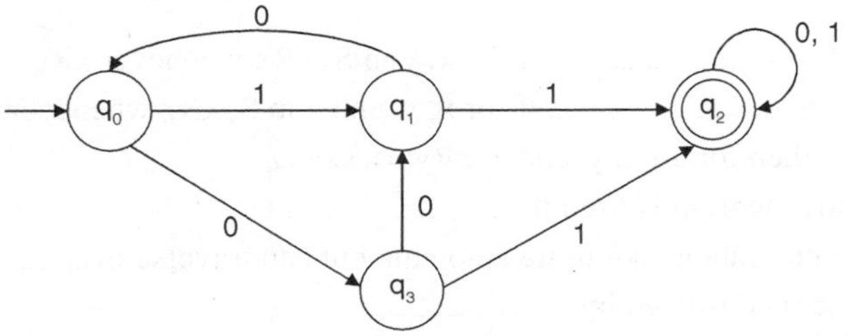

5. Test whether the following string 101001 is accepted by the given finite automata or not taking all possible combinations of the transitional path.

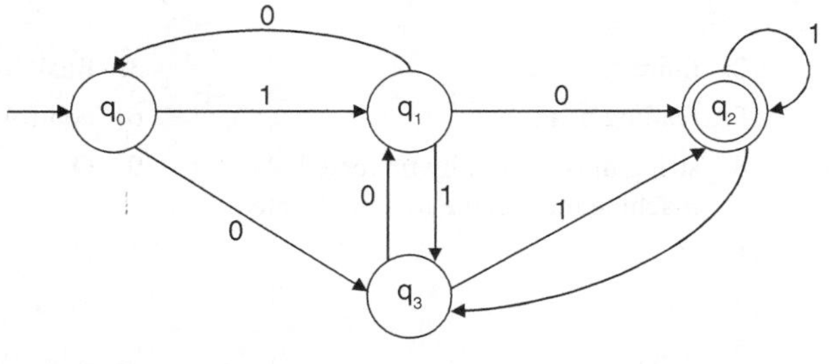

   Why are you getting different results? Explain.
6. Define NFA. Mention the differences between NFA and DFA. What is the necessity of converting an NFA to a DFA?
7. Convert the following NFAs into equivalent DFAs.

a)

Present State	Next State a = 0	a = 1
→ A	B	A, C
B	B	C
C	A, D	B
D	D	B

D is final.

b)

Present State	Next State a = 0	a = 1
→ q_0	q_0	q_0, q_1
(q_1)	–	q_2
q_2	q_2	q_2

c)

Present State	Next State a = 0	Next State a = 1
→ A	A	A, B
B	C	–
C	–	D
D	D	D

A, B, C, and D are final states.

8. Why is an FA with ε transition called NFA? What is the necessity of NFA with ε transition? If in an FA a self-loop is made on a state with label ε, will it be called an NFA with ε move? Give reason.

9. Convert the following NFAs with null move to equivalent DFAs.

a)

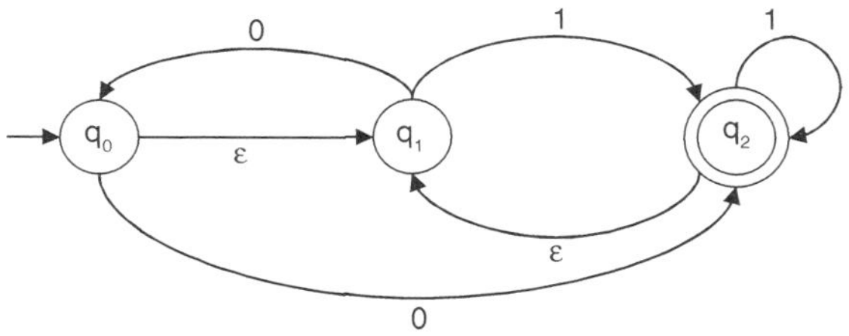

b)

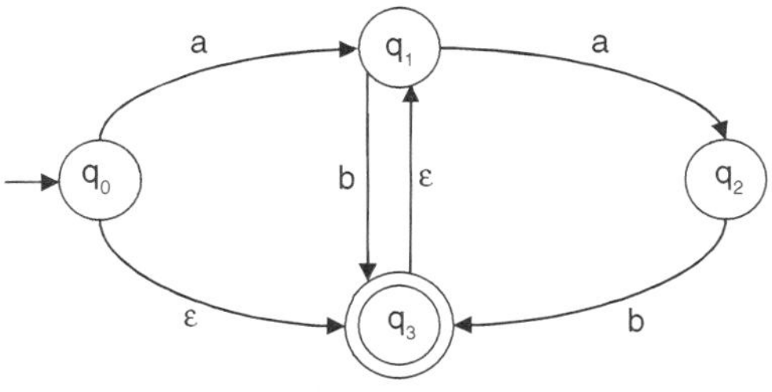

10. Define Mealy machine and Moore machine. If a sequence detector is designed for a string using Mealy and Moore machine, then the Moore is converted to a Mealy machine. Will the two Mealy machines be the same always?
    a) A Mealy machine is converted to an equivalent Moore Machine. A string is given input to both of the machines. Will the output be same always?
    b) Convert the following Moore machines into equivalent Mealy machines.

i)

Present State	Next State I/P = 0	Next State I/P = 1	O/P
→ q_0	q_3	q_1	1
q_1	q_1	q_2	0
q_2	q_2	q_3	1
q_3	q_3	q_3	1

ii)

Present State	Next State I/P = 0	Next State I/P = 1	O/P
→ A	D	B	1
B	A	E	0
C	A	E	1
D	C	A	0
E	F	D	0
F	F	D	1

11. Convert the following Mealy machines into equivalent Moore machines.

a)

	I/P = 0		I/P = 1	
Present State	Next State	O/P	Next State	O/P
A	E	0	C	0
B	F	0	C	1
C	E	0	A	0
D	F	0	A	1
E	A	0	D	0
F	D	0	E	1

b)

	I/P = 0		I/P = 1	
Present State	Next State	O/P	Next State	O/P
A	B	0	C	0
B	D	0	E	1
C	A	0	A	1
D	E	1	E	1
E	E	1	C	0

12. Convert the following Moore machine into an equivalent Mealy machine by the transitional format.

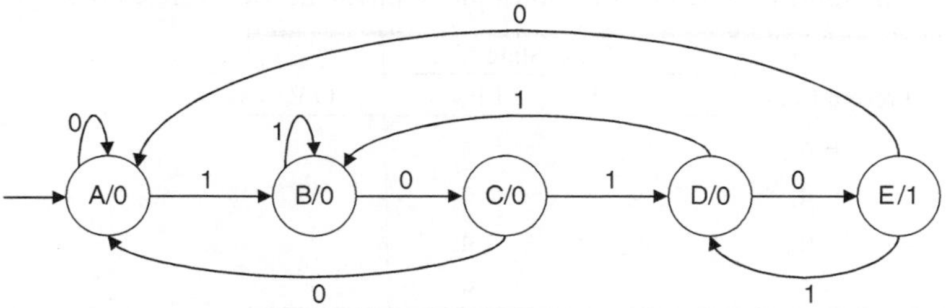

13. Convert the following Mealy machine into an equivalent Moore machine by the transitional format.

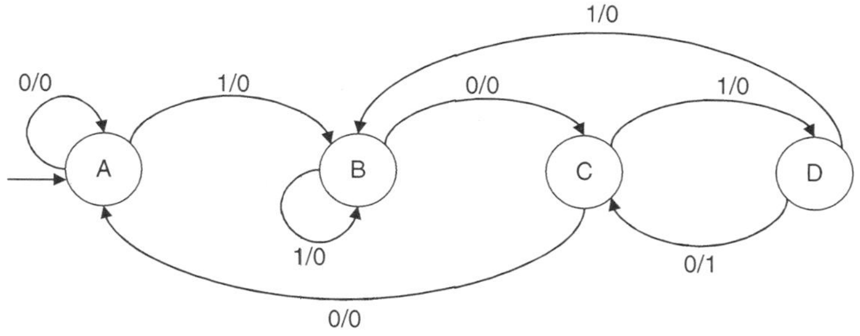

14. Why is minimization of finite automata necessary? Define equivalent state and distinguishable state.

15. Minimize the following finite automata by using the equivalent state method.

a)

Present State	Next State I/P = 0	Next State I/P = 1
→ A	E	C
B	C	A
C	B	G
D	G	A
E	F	B
F	E	D
(G)	D	G

b)

Present State	Next State I/P = 0	Next State I/P = 1
→ A	B	H
B	F	D
C	D	E
D	C	F
E	D	C
(F)	C	C
G	C	D
H	C	A

16. Minimize the following finite automata by the Myhill–Nerode theorem.

Present State	Next State	
	I/P = 0	I/P = 1
→A	B	F
B	G	C
(C)	A	C
D	C	G
E	H	F
F	C	G
G	G	E
H	G	C

Finite State Machine

4

Introduction

Chapter 3 already discusses the Mealy and Moore machines. These machines fall in the category of automata with output. Finite state machine, in short FSM, is a machine with a finite number of states. As all finite automata contain finite number of states, they are FSMs also. In this chapter, we shall discuss elaborately the different features of an FSM, some application such as sequence detector, incompletely specified machine, where some states and/or some outputs are not mentioned, definite machine, finite machine, information lossless machine, minimization of machines, etc.

4.1 Sequence Detector

The sequence detector detects and counts a particular sequence from a long input. Let us assume that, in a circuit, a long input of '0' and '1' is fed. Let us also assume that, from there, it needs to count the occurring of a particular sequence where overlapping sequences are also counted. The circuit generates '1' if the particular output pattern is generated; otherwise, '0' is generated.

The sequence detector can be designed either in a Mealy or Moore machine format.

The following examples describe the design of a sequence detector elaborately.

Example 4.1 Design a two-input two-output sequence detector which generates an output '1' every time the sequence 1001 is detected. And for all other cases, output '0' is generated. Overlapping sequences are also counted.

Solution: Before designing this circuit, some clarification regarding overlapping sequence is needed.

Let the input string be 1001001.

We have to design the circuit in such a way that it will take one input at a time. The input can be either '0' or '1' (two types of input). The output will also be of two types, either '0' or '1'. The circuit can store (memorize) the input string up to four clock pulses from t_{i-3} to t_i.

If the input string is placed according to clock pulses, the output will become

	t_1	t_2	t_3	t_4	t_5	t_6	t_7
I/P	1	0	0	1	0	0	1
O/P	0	0	0	1	0	0	1

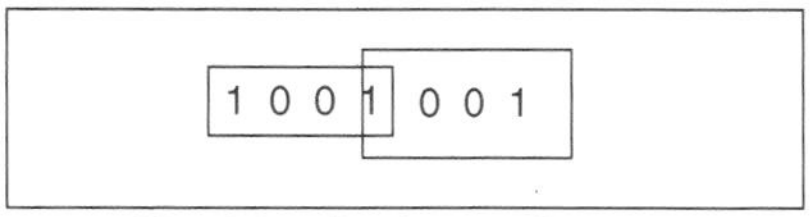

Fig. 4.1 *Overlapping Sequence*

The first input at t_1 is 1, and as there is no input from t_{i-3} to t_i, the input sequence does not equal 1001. So, the output will be 0. The same cases occur for the inputs up to t_3.

But at time t_4, the input from t_{i-3} to t_i becomes 1001, and so the output '1' is produced at time t_4. At time t_5 and t_6, the input string from t_{i-3} to t_i are 0010 and 0100, respectively. So, the output '0' is produced. But at t_7 clock pulse, the input string is 1001, and so the output '1' is produced. As the '1' at t_4 is overlapped from t_1 to t_4 and from t_4 to t_7, this is called overlapping condition as shown in Fig. 4.1.

For this case, we have to first draw the state diagram given in Fig. 4.2.

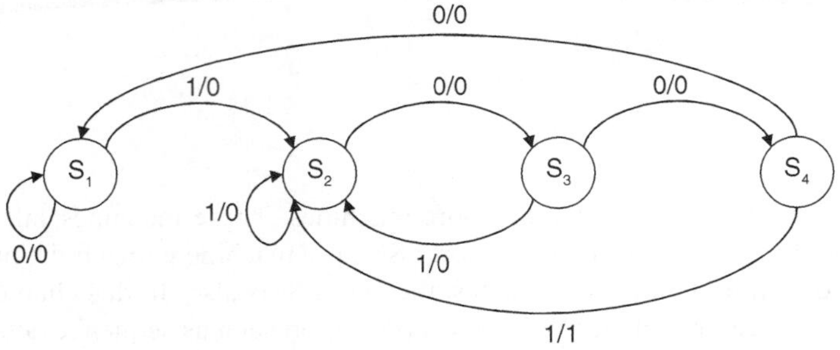

Fig. 4.2 *State Diagram of the Sequence Detector*

In state S_1, the input may be either '0' or '1'. If the input is given '0', there is no chance to get 1001. The machine has to wait for the next input. So, '0' becomes a loop on S_1 with output '0'.

If the input is '1', there is a chance to get 1001 by considering that input, and so by getting input '1' the machine moves to S_2 producing output '0'.

In S_2, again the input may be either '0' or '1'. If it is '1', the input becomes 11. There is no chance to get 1001 by taking the previous inputs from t_{i-1} to t_i. But there is a chance to get 1001 by considering the given input '1' at t_i. So, it will be in state S_2. (If it goes to S_1, then there will be a loss of clock pulse, which means again from S_1 by taking input '1', it has to come to S_2, i.e., one extra input, which means one clock pulse is needed, and for this the output will not be in right pattern.) If the input is '0', the input becomes 10, by considering the previous input. As there is a chance to get 1001, it will move to S_3.

In S_3, if it gets input '0', the input becomes 100 by considering the previous input. As it has a chance to get 1001, it shifts to S_4. But if it gets '0', it has no chance to get 1001 considering the previous input, but there is a chance to get 1001 by considering the given input '1'. So, it will shift to S_2 as we know by getting '1' in S_1 the machine comes to S_2.

In S_4, if it gets '0', the input will become 1000, but it does not match with 1001. So it has to start from the beginning, i.e., S_1. Getting '1', the string becomes the desired 1001. The overlapping part is the last '1' as given in Fig. 4.1. From the last '1' of 1001, if it gets 001 only, it will give an output '1'. So, it will go to S_2.

State Table: From the previous state diagram, a state graph can be easily constructed.

	Next State, O/P	
Present State	X = 0	= 1
S_1	S_1, 0	S_2, 0
S_2	S_3, 0	S_2, 0
S_3	S_4, 0	S_2, 0
S_4	S_1, 0	S_2, 1

State Assignment: For making a digital circuit, the states must be assigned to some binary numbers. This is called state assignment. As the number of states is 4, only two-digit is sufficient to represent the four states ($2^2 = 4$).

Let S_1 be assigned to 00, S_2 be assigned to 01, S_3 be assigned to 11, S_4 be assigned to 10.

After doing this state assignment, the state table becomes

Present State (y_1y_2)	Next State, (Y_1Y_2) X = 0	= 1	O/P (z) = 0	= 1
00	00	01	0	0
01	11	01	0	0
11	10	01	0	0
10	00	01	0	1

From this state assignment table, the digital function can easily be derived.

Y_1

X / y_1y_2	0	1
00	0	0
01	1	0
11	1	0
10	0	0

Y_2

X / y_1y_2	0	1
00	0	1
01	1	1
11	0	1
10	0	1

z

X / y_1y_2	0	1
00	0	0
01	0	0
11	0	0
10	0	1

$$Y_1 = X'y_2$$
$$Y_2 = X + y_1'y_2$$
$$z = Xy_1y_2'$$

Y_1 and Y_2 are the next states, which are the memory elements. These will be feedbacked to the input as states y_1 and y_2 with some delay by D flip flop.

The circuit diagram for this sequence detector is given in Fig. 4.3.

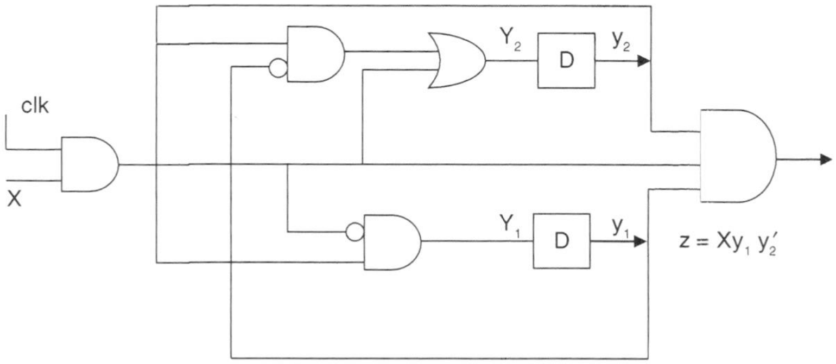

Fig. 4.3 *Digital Circuit for the Sequence Detector*

Example 4.2 Design a two-input two-output sequence detector which generates an output '1' every time the sequence 1010 is detected. And for all other cases, output '0' is generated. Overlapping sequences are also counted.

Solution: The input string 1010100 is placed according to the clock pulses, and it looks like the following.

	t_1	t_2	t_3	t_4	t_5	t_6	t_7
I/P	1	0	1	0	1	0	0
O/P	0	0	0	1	0	1	0

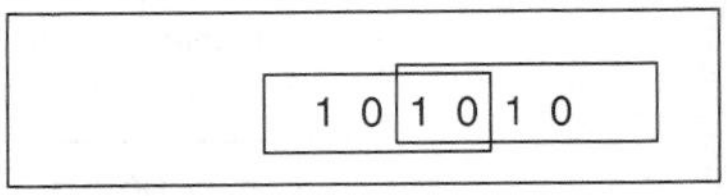

Fig. 4.4 *Overlapping Sequence*

And the output becomes as given earlier. Overlapping portion is 10, as shown in Fig. 4.4. The state diagram is given in Fig. 4.5.

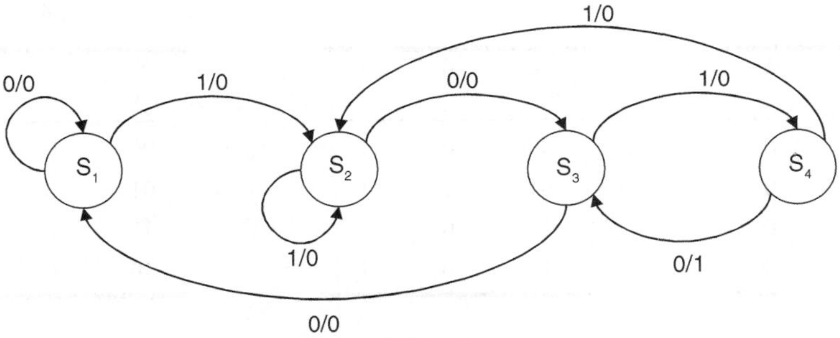

Fig. 4.5 *State Diagram*

State Table: From the previous state diagram, a state table as follows can easily be constructed.

Present State	Next State, O/P	
	X = 0	= 1
S_1	S_1, 0	S_2, 0
S_2	S_3, 0	S_2, 0
S_3	S_1, 0	S_4, 0
S_4	S_3, 1	S_2, 0

State Assignments: For making a digital circuit, the states must be assigned to some binary numbers to make a digital circuit. This is called state assignment. As the number of states is 4, two-digit is sufficient to represent the four states ($2^2 = 4$).

Let S_1 be assigned to 00, S_2 be assigned to 01, S_3 be assigned to 11, S_4 be assigned to 10.

After doing this state assignment, the state table becomes

Present State (y_1y_2)	Next State, (Y_1Y_2) X = 0	= 1	O/P (z) = 0	= 1
00	00	01	0	0
01	11	01	0	0
11	00	10	0	0
10	11	01	1	0

From this state assignment table, the digital function can easily be derived as follows.

Y_1

X / y_1y_2	0	1
00	0	0
01	1	0
11	0	1
10	1	0

Y_2

X / y_1y_2	0	1
00	0	1
01	1	1
11	0	0
10	1	1

z

X / y_1y_2	0	1
00	0	0
01	0	0
11	0	0
10	1	0

$$Y_1 = X'y_1'y_2 + Xy_1y_2 + X'y_1y_2'$$
$$Y_2 = y_1'y_2 + y_1'X + y_1y_2'$$
$$z = X'y_1y_2'$$

Y_1 and Y_2 are the next states, which are the memory elements. These will be feedbacked to the input as states y_1 and y_2 with some delay by D flip flop. The circuit diagram is shown in Fig. 4.6.

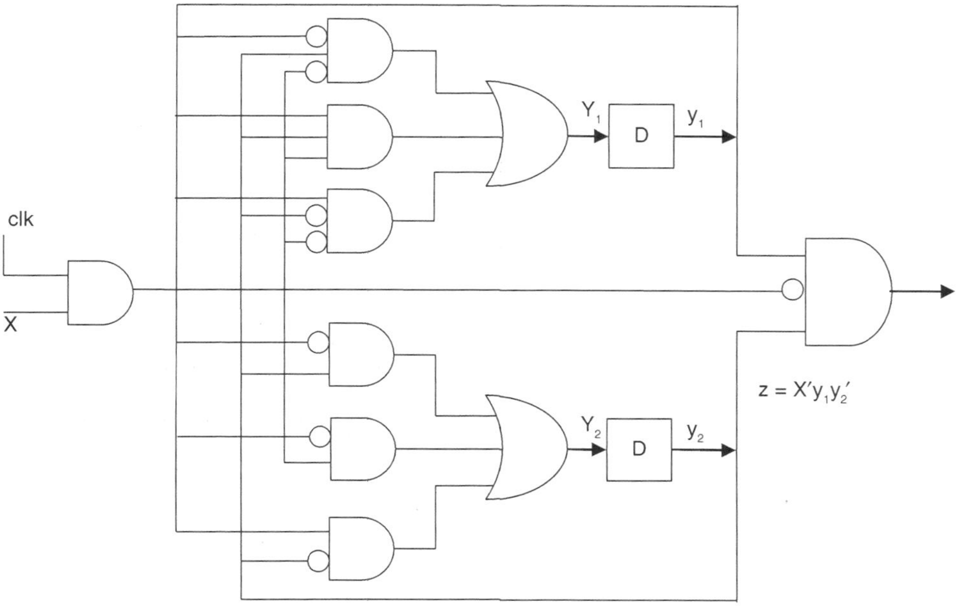

Fig. 4.6 *Digital Circuit Diagram*

4.2 Binary Counter

The binary counter counts in binary.

Example 4.3 Design a Modulo 3 binary counter.

Solution: A Modulo 3 binary counter can count up to 3. The binary representation of 3 is 11. It can count 00, 01, 10, and 11. There will be an external input x, which will act as a control variable and determine when the count should proceed. After counting 3, if it has to proceed, then it will come back to 00 again.

The state diagram for a Mod 3 binary counter is given in Fig. 4.7.

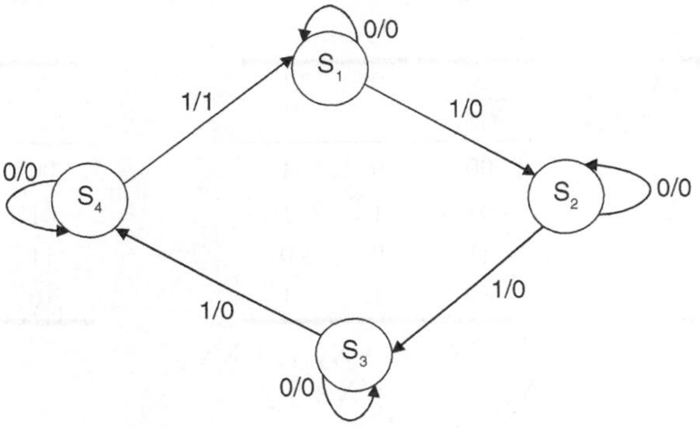

Fig. 4.7 *State Diagram of a Mod 3 Binary Counter*

The state table for Mod 3 binary counter is

Present State	Next State, O/P	
	X = 0	X = 1
S_1	S_1, 0	S_2, 0
S_2	S_2, 0	S_3, 0
S_3	S_3, 0	S_4, 0
S_4	S_4, 0	S_1, 1

There are four states in the machine. Two bits are sufficient to assign four states into the binary number.

Let us assign S_1 to 00, S_2 to 01, S_3 to 10, and S_4 to 11.

After doing this state assignment, the state table becomes

Present State	Next State, (Y_1Y_2)		O/P (z)	
(y_2y_1)	X = 0	= 1	= 0	= 1
00	00	01	0	0
01	01	10	0	0
10	01	11	0	0
11	11	00	0	1

4.2.1 Designing Using Flip Flop (T Flip Flop and SR Flip Flop)

The excitation table for T flip flop is given in the following:

Circuit From	Changed To	T
0	0	0
0	1	1
1	0	1
1	1	0

In state assignment, 00 is changed to 00 for input 0. Here, y_1 is changed from 0 to 0, and so T_1 will be 0. y_2 is changed from 0 to 1, and so T_1 will be 0. 00 is changed to 01 for input 1. Here, y_1 is changed from 0 to 1, and so T_1 will be 1. y_2 is changed from 0 to 0, and so T_1 will be 0. By this process, the excitation table of the counter using T flip flop is given in the following table.

Present State	T_2T_1	
(y_2y_1)	X = 0	X = 1
00	00	01
01	00	11
10	00	01
11	00	11

$$T_1 = X$$
$$T_2 = Xy_1$$
$$z = Xy_1y_2$$

The circuit diagram for this is presented in Fig. 4.8.

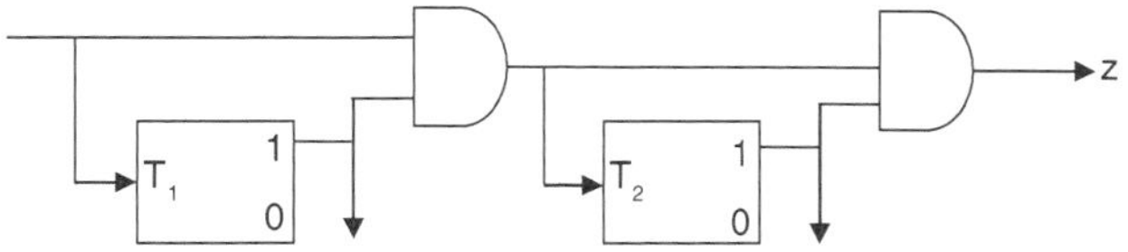

Fig. 4.8 *Circuit Diagram Using T Flip Flop*

The excitation table for SR flip flop is denoted in the following table.

Circuit From	Changed To	S	R
0	0	0	–
0	1	1	0
1	0	0	1
1	1	–	0

In state assignment, 00 is changed to 00 for input 0. Here, y_1 is changed from 0 to 0, and so R_1 will be don't care and S_1 will be 0. y_2 is changed from 0 to 0, and so R_2 will be don't care and S_2 will be 0. In the state assignment table, 00 is changed to 01 for input 1. Here, y_1 is changed from 0 to 1, and so R_1 will be 0 and S_1 will be 1. y_2 is changed from 0 to 0, and so R_2 will be don't care and S_2 will be 0. By this process, the excitation table of the counter using SR flip flop is given as follows.

Present State	X = 0		X = 1	
(y_2y_1)	S_1R_1	S_2R_2	S_1R_1	S_2R_2
00	0 –	0 –	1 0	0 –
01	– 0	0 –	0 1	1 0
10	0 –	– 0	1 0	– 0
11	– 0	– 0	0 1	0 1

$$S_1 = Xy_1' \quad R_1 = Xy_1$$
$$S_2 = Xy_1y_2' \quad R_2 = Xy_1y_2$$

The circuit diagram for this is presented in Fig. 4.9.

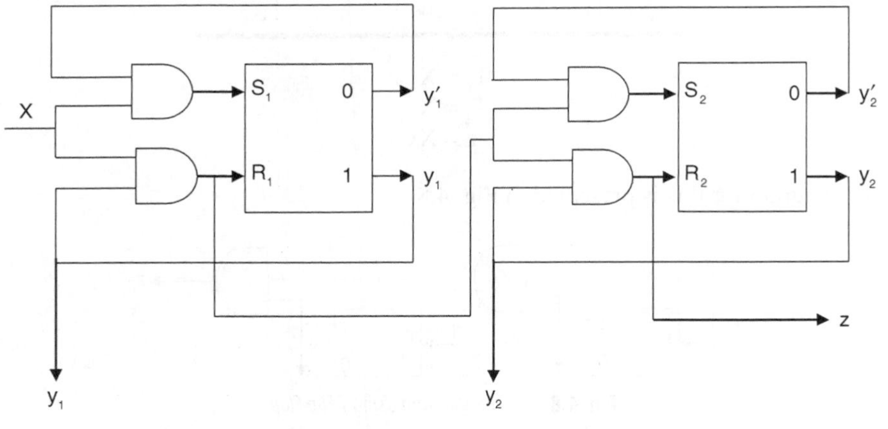

Fig. 4.9 *Circuit Diagram Using SR Flip Flop*

Example 4.4 Design a Modulo 8 binary counter

Solution: A Modulo 8 binary counter can count up to 8 from 000 to 111. There will be an external input x, which will act as a control variable and determine when the count should proceed. After counting 8, if it has to proceed, then it will come back to 000 again.

The state diagram for a Mod 8 binary counter is given in Fig. 4.10.

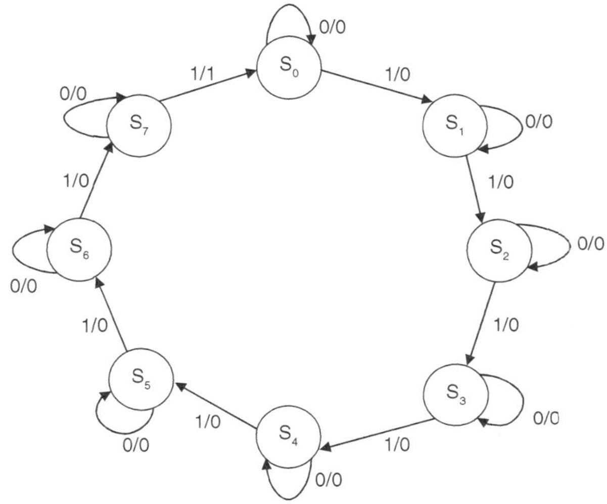

Fig. 4.10 *A Modulo 8 Binary Counter*

The state table for a Mod 8 binary counter is

Present State	Next State, O/P	
	X = 0	X = 1
S_0	S_0, 0	S_1, 0
S_1	S_1, 0	S_2, 0
S_2	S_2, 0	S_3, 0
S_3	S_3, 0	S_4, 0
S_4	S_4, 0	S_5, 0
S_5	S_5, 0	S_6, 0
S_6	S_6, 0	S_7, 0
S_7	S_7, 0	S_0, 1

State Assignment: There are eight states in the machine. Three bits are sufficient to assign eight states into binary number ($2^3 = 8$).

Let us assign S_1 to 000, S_2 to 001, S_3 to 010, S_4 to 011, S_5 to 100, S_6 to 101, S_7 to 101, and S_8 to 111.

Present State ($y_3y_2y_1$)	Next State, (Y_1Y_2)		O/P (z)	
	X = 0	= 1	= 0	= 1
000	000	001	0	0
001	001	010	0	0
010	010	011	0	0
011	011	100	0	0

Present State $(y_3y_2y_1)$	Next State, (Y_1Y_2) X = 0	= 1	O/P (z) = 0	= 1
100	100	101	0	0
101	101	101	0	0
101	101	111	0	0
111	111	000	0	1

The excitation table of the counter using T flip flop is as follows.

Present State $(y_3y_2y_1)$	$T_3T_2T_1$ X = 0	X = 1
000	000	001
001	000	011
010	000	001
011	000	111
100	000	001
101	000	011
101	000	001
111	000	111

$$T_1 = X \quad T_2 = Xy_1 \quad T_3 = Xy_1y_2 \quad z = Xy_1y_2y_3$$

The circuit diagram for this is shown in Fig. 4.11.

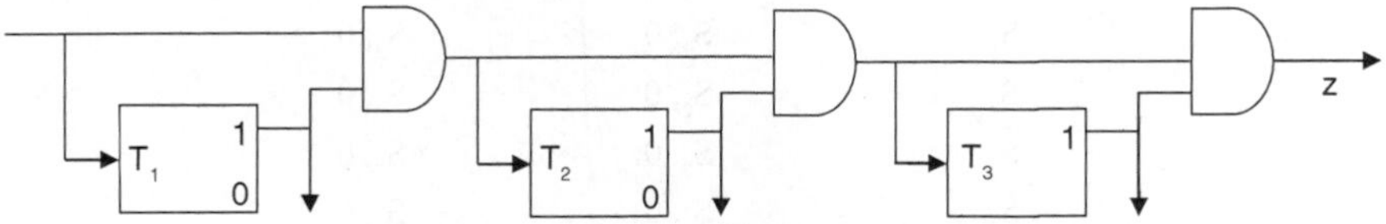

Fig. 4.11 *Circuit Diagram Using T Flip Flop*

The excitation table of the counter using SR flip flop is given in the following table.

Present State $(y_3y_2y_1)$	X = 0 S_1R_1	S_2R_2	S_3R_3	X = 1 S_1R_1	S_2R_2	S_3R_3
000	0 –	0 –	0 –	1 0	0 –	0 –
001	– 0	0 –	0 –	0 1	1 0	0 –
010	0 –	– 0	0 –	1 0	– 0	0 –
011	– 0	– 0	0 –	0 1	0 1	1 0
100	0 –	0 –	– 0	1 0	0 –	– 0
101	– 0	0 –	– 0	0 1	1 0	– 0
110	0 –	– 0	– 0	1 0	– 0	– 0
111	– 0	– 0	– 0	0 1	0 1	0 1

$$S_1 = Xy_1' \qquad R_1 = Xy_1$$
$$S_2 = Xy_1y_2' \qquad R_2 = Xy_1y_2$$
$$S_3 = Xy_1y_2y_3' \qquad R_3 = Xy_1y_2y_3$$

The circuit diagram for this is presented in Fig. 4.12.

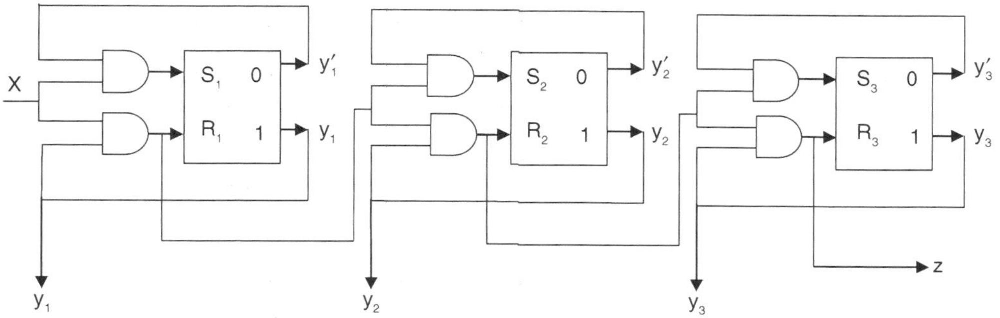

Fig. 4.12 *Circuit Diagram Using SR Flip Flop*

4.3 Finite State Machine

An FSM can be defined as a type of machine whose past histories can affect its future behaviour in a finite number of ways.

Need some clarification.... Take the example of a binary full adder. Its output depends on the present input and the carry generated from the previous input.

It may have a large number of previous input histories, but they can be divided into two types: (i) input combination that produces carry and (ii) input combination that produces no carry.

This means that the past histories can affect the future behaviour in a finite number of ways (here 2).

4.3.1 Capabilities and Limitations of Finite-State Machine

Let us assume that an FSM has n states. Let a long sequence of input be given to the machine. The machine will progress starting from its beginning state to the next states according to the state transitions as described in Fig. 4.13. But after some time, the input string may be longer than 'n', the number of states. But as there are only 'n' states in the machine, it must come to a state it was previously been in, and from this phase if the input remains the same the machine will function in a periodically repeating fashion. From here, a conclusion that 'for an 'n' state machine, the output will become periodic after a number of clock pulses less than equal to 'n' can be drawn.

The states are memory elements. As for an FSM, the number of states is finite, and so a finite number of memory elements are required to design a finite state machine.

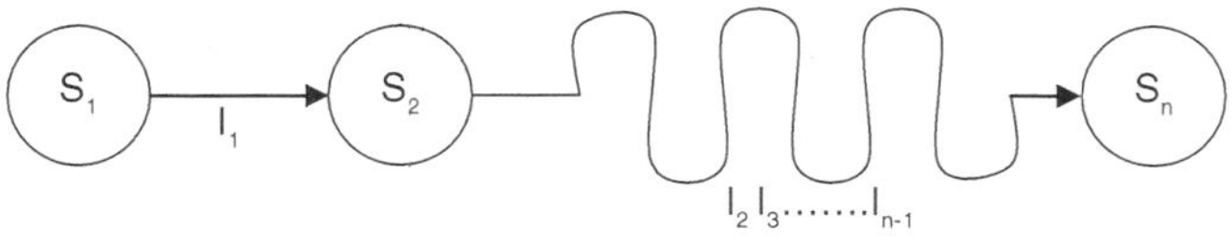

Fig. 4.13

4.3.1.1 Limitations

No FSM Can be Produced for an Infinite Sequence: Let us design an FSM which receives a long sequence of 1. The machine will produce an output 1, when the input string length will be equal to [p(p + 1)]/2, where p = 1, 2, 3, ..., and 0 for all other cases.

That means for p = 1, [p(p + 1)]/2 = 1. In first place, there will be o/p 1.
For p = 2, [p(p + 1)]/2 = 3. In third place, there will be o/p 1.
For p = 3, [p(p + 1)]/2 = 6. In sixth place, there will be o/p 1.

For this type of machine, the input–output form will be

Time	t_1	t_2	t_3	t_4	t_5	t_6	t_7	t_8	t_9	t_{10}	t_{11}	t_{12}	t_{13}	t_{14}	t_{15}..................
I/P =	1	1	1	1	1	1	1	1	1	1	1	1	1	1	1...........
O/P =	1	0	1	0	0	1	0	0	0	1	0	0	0	0	1..........

Here, the output does not become eventually periodic after a certain number of clock pulses. So, from this type of sequence, no FSM can be produced.

No FSM Can Multiply Two Arbitrary Large Binary Numbers: Let us multiply two binary numbers that are given input serially to an FSM for multiplication. The inputs are given to the machine with the least significant bit (LSB) first and then the other bits. Suppose we want to multiply $2^m \times 2^m$, where m > n ('n' is the number of states of the machine). The result will be 2^{2m}.

2^m is represented by one 1 followed by m number of '0's (like $2^3 = 1000$). So, the inputs are given to the machine from t_1 to t_{m+1} time. Throughout the time, the machine produces '0'. At t_{m+2} times, the input stops and the machine produces the output 0 followed by 1 from t_{m+2} to t_{2m} time.

In the time period t_{m+1} to t_{2m}, no input is given to the machine, but the machine produces outputs. As m > n, according to the definition of FSM, the output must be periodic and the period must be < m.

As we are not getting any repeating output sequence, the binary multiplication of two arbitrary, large binary numbers is not possible to be designed by using FSM.

t_{2m+1}	t_{2m}	.....t_{m+1}	t_m..................	t_2	t_1 = time	
		1	0	0	0	1st Number
		1	0	0	0	2nd Number
1	0......	0	0.....................	0	0	Result

FSM Cannot Check Whether a String is Palindrome or Not: It is not able to find where a string ends and its reverse string starts or the middle of the string.

4.4 State Equivalence and Minimization of Machine

In chapter 3, we have already learnt how to minimize finite automata. In this section, we shall learn how to minimize an FSM.

The state of a machine means memory elements. A machine with more states means more memory elements are required. For an N state machine, $\log_2 N$ state variables are needed for state assignment. More state variables means designing the machine becomes more complex. Machine minimization reduces the number of states, which consequently reduces the complexity of designing the machine.

Two states S_i and S_j of machine M are said to be *equivalent* if they produce the same output sequences for an input string applied to the machine M, considering S_i and S_j as initial states.

Two states S_i and S_j of machine M are said to be *distinguishable* if there exists a minimum length input sequence which when applied to the machine M, considering S_i and S_j as the initial states, produce different output sequence. (The input string is always applied on initial state.)

The sequence that distinguishes those two states is called the *distinguishing sequence* for the pair S_i and S_j.

If two states S_i and S_j are distinguished for the input string of length 'k', then S_i and S_j are called k-distinguishable. 'k' is the minimum number of the length of the string for which the states produce different output. If two states are k-distinguishable, then they are (k − 1) equivalent for k = k to 1.

From the following state table, the concept will be clear.

Present State	Next State, z	
	X = 0	X = 1
A	E, 0	C, 0
B	F, 0	C, 1
C	E, 0	A, 0
D	F, 0	A, 1
E	A, 0	D, 0
F	D, 0	E, 1

Take the previous example.

A and C give the same output (here 0) for the input string of length 1 (either 0 or 1). So, A and C are 1-equivalent.

A and B give different outputs for input string of length 1. (For input string length 0, i.e., for no input—the outputs are same; but for 1, they produce different outputs.) So, A and B are 1-distinguishable.

Let us check for string length 2. String length 2 means that it gives four types of combinations 00, 01, 11, and 10.

Present State	00	01	11	10
A	00 (A)	00 (A)	00 (A)	00 (E)
B	00 (D)	01 (E)	10 (A)	10 (A)
C	00 (E)	00 (D)	00 (C)	00 (E)
D	00 (D)	01 (E)	10 (C)	10 (C)
E	00 (E)	00 (C)	01 (A)	00 (F)
F	00 (D)	01 (A)	10 (D)	10 (A)

A and E are 2-distinguishable, as they produce different outputs for 11. The distinguishing sequence for A and E is 11.

The *equivalent partition* of a machine M can be defined as a set of maximum number subsets of states where states which reside in same subset are equivalent for any input given to the states. The states which reside in different subsets are distinguishable for some input.

Definition: *The equivalent partition of an FSM is unique.*

Proof: Suppose for a machine M there exist two equivalent partitions P_1 and P_2, where $P_1 \neq P_2$. As $P_1 \neq P_2$, there must exist at least two states S_i and S_j, which are in the same block of one partition (say, P_1)

and in different blocks in other partition (say, P_2). As they are in different blocks in P_2, there must exist an input sequence which distinguishes S_i and S_j. So, they cannot be in the same block of P_1. Therefore, our assumption is wrong. For a single machine, there cannot exist two equivalent partitions.

So, from here we can conclude that equivalent partition is unique, i.e., $P_1 \equiv P_2$.

The following are some examples of finding an equivalent partition of some machines and minimizing those machines.

Example 4.5 Find the equivalent partitions and minimize the following finite state machine.

Solution:

Present State	Next State, z	
	X = 0	X = 1
A	E, 0	C, 0
B	F, 0	C, 1
C	E, 0	A, 0
D	F, 0	A, 1
E	A, 0	D, 0
F	D, 0	E, 1

For string length 0 (i.e., for no input), there is no output. So, all the states are equivalent.

$$P_0 = (ABCDEF)$$

Consider for string length 1 (i.e., two inputs 0 or 1). For 0, for all states, the output is 0. For 1, we get different outputs for (ACE) and (BDF). A and B are 1-distinguishable. A and C are 1-equivalent.

So,

$$P_1 = ((ACE)\ (BDF))$$

Consider for string length 2 (i.e., four types of input 00, 01, 11, and 10).

Present State	00	01	11	10
A	00 (A)	00 (A)	00 (A)	00 (E)
B	00 (E)	00 (D)	00 (C)	00 (E)
C	00 (E)	00 (C)	01 (A)	00 (F)
D	00 (D)	01 (E)	10 (A)	10 (A)
E	00 (D)	01 (E)	10 (C)	10 (C)
F	00 (D)	01 (A)	10 (D)	10 (A)

For A and C, the outputs are the same for 00, 01, 11, and 10. But E has a different output for 11. For B, D, F, the outputs are the same for all inputs.

$$P_2 = ((AC)(E)(BDF))$$

Consider for string length 3 (i.e., eight types of input combinations). So, it will be difficult to find the output sequences and the equivalent partitions, as the length of input string increases.

It will be easier to check for the next states.

We know

$$P_1 = ((ACE)\ (BDF))$$

Let us rename (ACE) as set S_1 and (BDF) as set S_2.

For ACE with input 0, the next states are (EEA). Both A and E belong to set S_1. For input 1, the next states are (CAD). Here, A and C belong to set S_1, but D belongs to set S_2. So, the set (ACE) will be divided as (AC) and (E) for input string length 2.

For (BDF) with input 0, the next states are (FFD). Both F and D belong to the same set S_2. For input 1, the next states are (CAE). All of these belong to same set S_1. So, (BDF) cannot be divided for input string length 2.

The partition of states become

$$P_2 = (AC)(E)(BDF) \text{ (the same result is obtained with considering output).}$$

Let us check for input string length 3.

For (AC) with input 0, the next states are (EE). Both the next states belong to a single set. For input 1, the next states are (AC). Both the next states belong to a single set. So, (AC) cannot be divided for input string length 3.

(E) is a set of single state. So, (E) cannot be divided.

For (BDF) with input 0, the next states are (FFD). All of them belong to a single set. With input 1, the next states are (CAE). C and A belong to one set in P_2 and E belongs to another set in P_2. So, (BDF) will be divided as (BD) and (F) for input string length 3.

The partition of states become

$$P_3 = (AC)(E)(BD)(F)$$

Let us check for input string length 4.

For (AC) with input 0, the next states are (EE), belonging to a single set. For input 1, the next states are (AC), belonging to a single set. So, (AC) cannot be divided for input string length 4.

(E) is a set of single state. So, (E) cannot be divided.

For (BD) with input 0 and 1, the next states are (FF) and (AC), respectively. (FF) belong to a single set and (AC) also belong to a single set. So, (BD) cannot be divided. As (F) is a single state, it cannot be divided.

The partition of states become

$$P_4 = (AC)(E)(BD)(F)$$

As P_3 and P_4 consist of the same partitions, $P_3 = (AC)(E)(BD)(F)$ is the equivalent partition for the machine M.

Minimization: We know that equivalent partition is unique. So, $P_3 = (AC)(E)(BD)(F)$ is the unique combination. Here, every single set represents one state of the minimized machine.

Let us rename these partitions for simplification.

Rename (AC) as S_1, (E) as S_2, (BD) as S_3, and (F) as S_4.

AC with input 0 goes to (EE) with output 0, and so there will be a transaction from S_1 to S_2 with output 0. E with input 0 goes to A producing output 0. A belongs to set S_1 in the minimized machine, and so there will be a transaction from S_2 to S_1 with output 0. By this process, the whole table of the minimized machine is constructed.

The minimized machine becomes

Present State	Next State, z	
	X = 0	X = 1
S_1(AC)	S_2, 0	S_1, 0
S_2(E)	S_1, 0	S_3, 0
S_3(BD)	S_4, 0	S_1, 1
S_4(BD)	S_3, 0	S_2, 1

Example 4.6 Find equivalent partitions and minimize the following finite state machine.

Solution:

Present State	Next State, z	
	X = 0	X = 1
A	B, 0	H, 1
B	C, 0	G, 1
C	B, 0	F, 1
D	F, 1	C, 1
E	B, 1	C, 1
F	B, 1	B, 1
G	C, 1	B, 1
H	D, 1	A, 1

For string length 0 (i.e., for no input), there is no output. So, all the states are equivalent. It is called 0-equivalent. We can write this as

$$P_0 = (ABCDEFGH)$$

For string length 1, there are two types of inputs—0 and 1. The states A, B, and C give output 0 for input 0, and the states D, E, F, G, and H give output 1 for input 0. All of the states give output 1 for input 1.

So, the states in the set P_0 are divided into (ABC) and (DEFGH). We can write this as

$$P_1 = ((ABC)\ (DEFGH))$$

Here, A and F are 1-distinguishable because they produce different outputs for input string length 1.

For input string length 2, check the distinguishability by the next state combination.

The states A, B, and C for input 0 produce next states B, C, and B, respectively, and produce next states H, G, F for input 1. The states BCB belong to the same set, and HGF also belong to the same set. So, ABC cannot be partitioned for input string length 2.

The states DEFGH for input 0 produce next states F, B, B, C, D, respectively, and produce next states C, C, B, B, A for input 1. The states B, B, C belong to the same set, but F, D belong to different sets. So, the set (DEFGH) is partitioned into (DH) and (EFG). The new partition becomes

$$P_2 = ((ABC)\ ((DH)\ (EFG)))$$

The states D and G are 2-distinguishable, because they produce different outputs for input string length 2.

The states A, B, C for input 0 produce next states B, C, B, respectively, and produce next states H, G, F for input 1. The states BCB belong to the same set, but H and G, F belong to different sets. So, the set (ABC) is partitioned into (A) and (BC).

The states B and H produce next states C and D, respectively, for input 0 and produce next states G and A for input 1. C and D belong to different sets, and so the set (BH) is divided into (B) and (H).

The states E, F, G produce next states B, B, and C for input 0 and next states C, B, B for input 1. Both B and C belong to the same set, and so the set (EFG) cannot be partitioned.

The new partition becomes

$$P_3 = (((A)\ (BC))\ (((D)\ (H))\ (EFG)))$$

Here, A and B are 3-distinguishable, because they produce different outputs for input string length 3.

By this process, we will get P_4 also as

$$P_4 = (((A)\ (BC))\ (((D)\ (H))\ (EFG)))$$

As P_3 and P_4 consist of the same partitions,

$$P_3 = (((A)(BC))\ (((D)(H))\ (EFG)))$$

is the equivalent partition for the machine M.

Minimization: We know that equivalent partition is unique. So, $P_3 = (((A)(BC))\ (((D)(H))\ (EFG)))$ is the unique combination. Here, every single set represents one state of the minimized machine.

Let us rename these partitions for simplification.

Rename (A) as S_1, (BC) as S_2, (D) as S_3, (H) as S_4, and (EFG) as S_5(A) with input 0 goes to (B), and so there will be a transaction from S_1 to S_2 with input 0. (A) with input 1 goes to (H), and so there will be a transaction from S_1 to S_4 with input 1. (BC) with input 0 goes to (BC) for input 0. There will be a transaction from S_2 to S_2 for input 0. (BC) with input 1 goes to (FG). There will be a transaction from S_2 to S_5 for input 1.

By this process, the whole table of the minimized machine is constructed.

The minimized machine becomes

Present State	Next State, z	
	X = 0	X = 1
S_1(A)	S_2, 0	S_4, 1
S_2(BC)	S_2, 0	S_5, 1
S_3(D)	S_5, 1	S_2, 1
S_4(H)	S_3, 1	S_1, 1
S_5(EFG)	S_2, 1	S_2, 1

4.5 Incompletely Specified Machine, Minimal Machine

In real life, for all states and for all inputs, the next state or outputs or both are not mentioned. Those types of machines, where for all states and for all inputs, the next state, or output, or both are not mentioned, are called incompletely specified machine.

In the following machine, for state A and for 00 input, no next state and outputs are specified. So, the previous machine is an example of an incompletely specified machine.

Present State	00	01	11	10
A	_, _	B, 0	C, 1	_, _
B	A, 0	B, 1	_, 0	C, _
C	D, 1	A, _	B, 0	D, 1
D	_, 0	_, _	C, 0	B, 1

4.5.1 Simplification

An incompletely specified machine can be simplified by the following steps:

- If the next state is not mentioned for a state, for a given input, put a temporary state T in that place.
- If the output is not mentioned, make it blank.
- If the next state and output are not mentioned, put a temporary state T in the place of the next state and nothing in the place of output.
- Add the temporary state T in the present state column, putting T as the next state and no output for all inputs.

By following the previous steps, the simplification of the previous incompletely specified machine will be

	Next State, z			
Present State	00	01	11	10
A	T, _	B, 0	C, 1	T, _
B	A, 0	B, 1	T, 0	C, _
C	D, 1	A, _	B, 0	D, 1
D	T, 0	T, _	C, 0	B, 1
T	T, _	T, _	T, _	T, _

Here 'T' is the same as the dead state in finite automata.

Example 4.7 Simplify the following incompletely specified machine.

Solution:

	Next State, z		
Present State	I_1	I_2	I_3
A	C, 0	E, 1	_ _
B	C, 0	E, _	_ _
C	B, _	C, 0	A, _
D	B, 0	C, _	E, _
E	_ _	E, 0	A, _

Put a temporary state T in the place of the next state, where the next states are not specified. If the output is not mentioned, there is no need to put any output.

As the temporary state T is considered, put T in the present state column with the next state T for all inputs with no output.

The simplified machine becomes

Present State	Next State, z		
	I_1	I_2	I_3
A	C, 0	E, 1	T, _
B	C, 0	E, _	T, _
C	B, _	C, 0	A, _
D	B, 0	C, _	E, _
E	T, _	E, 0	A, _
T	T, _	T, _	T, _

Minimal Machine: Is the minimum of the machines obtained by minimizing an incompletely specified machine.

In an incompletely specified machine, for all states and for all inputs, the next state, or output, or both are not mentioned. At the time of minimizing the incompletely specified machine, different persons can take the unmentioned next states or outputs according to their choice. Therefore, there is a great possibility to get different equivalent partitions for a single machine. But, we know that equivalent partition is unique for a given machine. It is not possible to find a u nique minimized machine for a given incompletely specified machine most of the times. Therefore, our aim must be to find a reduced machine which not only covers the original machine but also has a minimal (least of the minimum) number of states. This type of machine is called *minimal machine*, i.e., it is the minimum of the machines obtained by minimizing an incompletely specified machine.

Let us consider the following incompletely specified machine

Present State	Next State, z	
	X = 0	X = 1
A	E, 1	D, 0
B	E, 0	C, 1
C	A, 0	B, _
D	A, 0	D, 1
E	A, _	B, 0

In this machine, for C with input 1, the output is not specified and the same for E with input 0. There are two types of outputs that can occur in the machine. So, the unspecified outputs can be any of the following:

(a) (B, 0 A, 0)
(b) (B, 0 A, 1)
(c) (B, 1 A, 1)
(c) (B, 1 A, 0).

If it is (a), then the machine and its equivalent partition is

Present State	Next State, z X = 0	X = 1
A	E, 1	D, 0
B	E, 0	C, 1
C	A, 0	B, 0
D	A, 0	D, 1
E	A, 0	B, 0

$P_0 = (ABCDE)$
$P_1 = (A)(BD)(CE)$
$P_2 = (A)(B)(D)(CE)$

If it is (b), then the machine and its equivalent partition is

Present State	Next State, z X = 0	X = 1
A	E, 1	D, 0
B	E, 0	C, 1
C	A, 0	B, 0
D	A, 0	D, 1
E	A, 1	B, 0

$P_0 = (ABCDE)$
$P_1 = (AE)(BD)(C)$
$P_2 = (AE)(B)(D)(C)$
$P_3 = (A)(E)(B)(D)(C)$

If it is (c), then the machine and its equivalent partition is

Present State	Next State, z x = 0	x = 1
A	E, 1	D, 0
B	E, 0	C, 1
C	A, 0	B, 1
D	A, 0	D, 1
E	A, 1	B, 0

$P_0 = (ABCDE)$
$P_1 = (AE)(BCD)$

If it is (d), then the machine and its equivalent partition is

Present State	Next State, z x = 0	x = 1
A	E, 1	D, 0
B	E, 0	C, 1
C	A, 0	B, 1
D	A, 0	D, 1
E	A, 0	B, 0

$P_0 = (ABCDE)$
$P_1 = (A)(BCD)(E)$
$P_2 = (A)(B)(CD)(E)$
$P_3 = (A)(B)(C)(D)(E)$

In cases (b) and (d), the number of equivalent partitions are 5, which is equal to the number of states of the machine, i.e., the machine constructed from the equivalent partitions obtained in cases (b) and (d) and the original machines for cases (b) and (d) are the same.

4.6 Merger Graph

In the previous section, it was discussed that minimizing an incompletely specified machine and finding a minimal machine from there is difficult. Sometimes, it is a time-taking assignment if the number of unmentioned next states and outputs are many. For minimizing an incompletely specified machine there is a technique which consist of constructing **merger graph** and **compatible graph** and from there finding **minimal closed covering**.

A merger graph of a machine M of 'n' states is an undirected graph defined as follows:

1. A merger graph consists of n number of vertices, where 'n' is the number of states of the machine. That is, in other words, each states of the machine represent one vertex.
2. There is an undirected arc between a pair of vertices (states) if the outputs do not conflict for those pair of states.
   (a) The arc will be an uninterrupted arc if the next states of the two states (vertices) do not conflict.
   (b) The arc will be an interrupted arc if the next states of the states (vertices) conflict. The conflicting next states will be placed in the interrupted portions.
3. There will be no arc between the two vertices if the outputs of the pair of states conflict.

Consider the following examples to construct a merger graph.

Example 4.8 Construct the merger graph for the following machine.

Solution:

	Next State, z			
Present State	I_1	I_2	I_3	I_4
A	E, 1	B, 0	_, _	E, 0
B	_, _	_, _	_, _	C, 0
C	D, 1	F, 0	A, 0	_, _
D	_, _	B, 1	_, _	_, _
E	_, _	_, _	A, 1	A, 0
F	B, 1	C, 1	E, 1	_, _

In this machine, there are 6 states. So, the number of vertices of the merger graph is 6, namely A, B, C, D, E, and F.

Consider two vertices A and B. In the state table for A and B for input I_1, the outputs are 1 and don't care. It is treated as the same output (don't care can be either 0 or 1). Similarly, for I_2, I_3, and I_4, the outputs do not conflict. (If the outputs are don't care, consider it as the same with the other.) So, an undirected arc is drawn between A and B.

For input I_4, A produces the next state E, and B produces the next state C. So, the undirected arc between A and B is an interrupted arc and EC is placed in the interrupted portion.

Consider A and C. The outputs do not conflict for inputs I_1, I_2, I_3, and I_4. An undirected arc is drawn between A and C. The next states produced by A and C for input I_1 are D and E, i.e., conflicting. For input I_2, the next states are B and F—also conflicting. The arc is an interrupted arc, and (BF) and (DE) are placed in the interrupted portion.

Consider A and D. For input I_2, A and D produce conflicting outputs—0 for A and 1 for B. So, no arc can be drawn between A and D.

Consider A and E. For all the inputs, they produce the same outputs. An undirected arc is drawn between A and E. A and E produce the next states E and A, respectively, for input I_4. So, the arc is an interrupted arc and (EA) is placed in the interrupted portion.

Consider A and F. A and F produce conflicting outputs (0 for A, 1 for F) for input I_2. So, no arc can be drawn between A and F.

By this process, an uninterrupted arc is drawn between B and C.

An uninterrupted arc is drawn between B and D.

An interrupted arc is drawn between B and E, and AC is placed in the interrupted portion.

An uninterrupted arc is drawn between B and F.

For input I_2, C and D produce conflicting outputs. No arc can be drawn between C and D.

For input I_3, C and E produce conflicting outputs. No arc can be drawn between C and E.

For input I_3, C and F produce conflicting outputs. No arc can be drawn between C and F.

D and E produce the same outputs and the same next states for all the inputs. An uninterrupted arc is drawn between D and E.

D and F produce the same outputs for all the inputs but conflicting next states (B and C) for input I_2. An interrupted arc is drawn between D and F, and BC is placed in the interrupted portion.

E and F produce the same output for all the inputs but conflicting next states (A and E) for input I_3. An interrupted arc is drawn between E and F, and AE is placed in the interrupted portion.

The final merger graph is shown in Fig. 4.14.

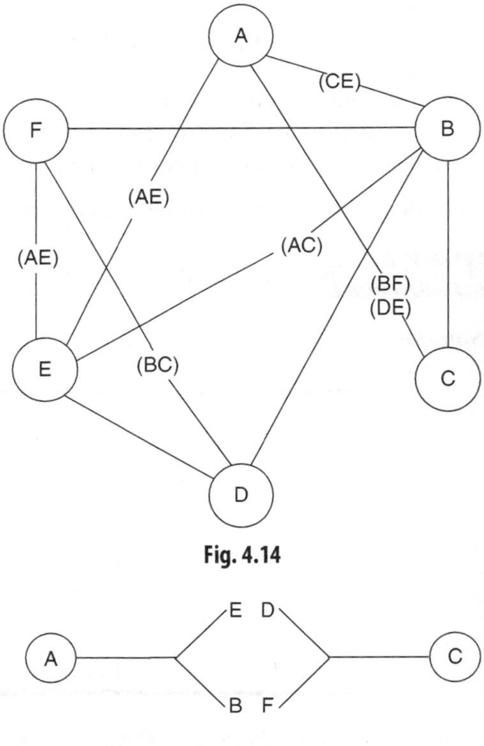

Fig. 4.14

A and B are connected by an uninterrupted, undirected arc.

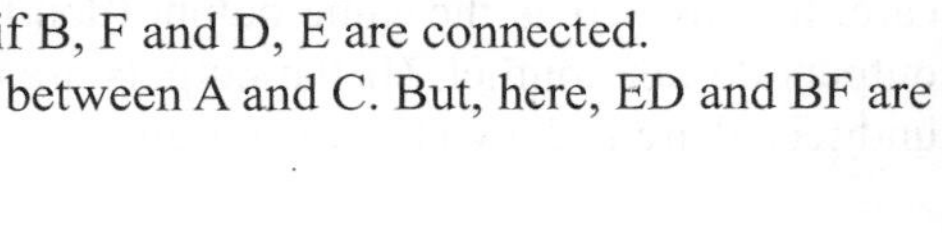

In the interrupted portion, CE is placed. The connection between A and B will be uninterrupted if C and E are connected by an uninterrupted arc. But there is no arc between C and E. So, there will be no arc between A and B. CE is crossed, therefore.

A and C are connected by an uninterrupted undirected arc. In the interrupted portion, DE and BF are placed. A and C will be connected by an uninterrupted arc if B, F and D, E are connected.

If either ED or BF is not connected, there will be no arc between A and C. But, here, ED and BF are connected.

Example 4.9 Develop the merger graph for the following machine.

Solution:

Present State	Next State, z		
	I_1	I_2	I_3
A	E, 0	B, _	C, _
B	_, _	D, _	B, 0
C	E, _	D, _	C, 0
D	C, 0	_, _	B, 1
E	C, 0	_, _	B, 1

The number of states of the machine is 5. So, the number of vertices in the merger graph is 5. Let us name them by the name of the states.

Consider two vertices A and B. The states A and B produce the same outputs for all the inputs but different next states for input I_2 (conflicting states BD) and for input I_3 (Conflicting states BC). So, an interrupted undirected arc is drawn between A and B, and (BC)(BD) is placed in the interrupted portion.

Consider A and C. The two states produce the same outputs for all the inputs. But A and C produce conflicting next states (BD) for input I_2. Therefore, between A and C an undirected interrupted arc is drawn, and (BD) is placed in the interrupted portion.

Consider A and D. The states produce the same outputs for all the inputs. But they produce conflicting next states (BC) and (EC) for inputs I_1 and I_3, respectively. Therefore, an undirected interrupted arc is drawn between A and D and (BC)(EC) is placed in the interrupted portion.

Consider A and E. The states produce the same outputs for all the inputs. But they produce conflicting next states for input I_1 and I_3. The conflicting next states are (CE) and (BC). An interrupter arc is drawn between A and E, and (CE)(BC) is placed in the interrupted portion.

Consider B and C. They produce the same outputs for all the inputs but conflicting next state pair (BC) for input I_3. So, an interrupted arc is drawn between B and C placing (BC) in the interrupted portion.

Consider B and D. No arc is drawn between B and D as they produce different outputs for input I_3.

Consider B and E. No arc is drawn between B and E as they produce different outputs for input I_3.

Consider C and D. These states produce different outputs for input I_3. Therefore, no arc is drawn between C and D.

Consider C and E. These states produce different outputs for input I_3. Therefore, no arc is drawn between C and E.

Consider D and E. They produce the same outputs and the same next state combination for all the inputs. Therefore, an uninterrupted arc is drawn between D and E.

The merger graph for the machine is given in Fig. 4.15.

Between C and E, there is no arc. So, the combinations (EC)(BC) placed between A, D and A, E are crossed off. Now, there is no connection between A and E and A and D. There is no arc between B and D. So, the next state combinations (BC)(BD) and (BD) placed between A, B and A, C, respectively, are crossed off.

Fig. 4.15

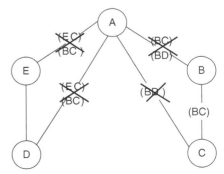

4.7 Compatibility Graph and Minimal Machine Construction

A compatibility graph is constructed from the merger graph. Before constructing a compatibility graph, we have to know two definitions, namely, compatible pair and implied pair. These are obtained from the merger graph. Without finding the compatible pair and implied pair, a compatible graph cannot be constructed.

4.7.1 Compatible Pair

Two states, say S_i and S_j, are said to be compatible, if both of them give the same output strings for all input strings applied to the machine separately, considering S_i and S_j as the initial states.

(In the sense of a merger graph, two states are said to be compatible if there exists an uninterrupted arc between the two states.)

4.7.2 Implied Pair

Two states S_i and S_j are said to be implied on S_p and S_q if and only if (S_p, S_q) is compatible and then only (S_i, S_j) is compatible. The compatibility of (S_i, S_j) is dependent on the compatibility of (S_p, S_q). (S_p, S_q) is said to be the implied pair of (S_i, S_j).

4.7.3 Compatibility Graph

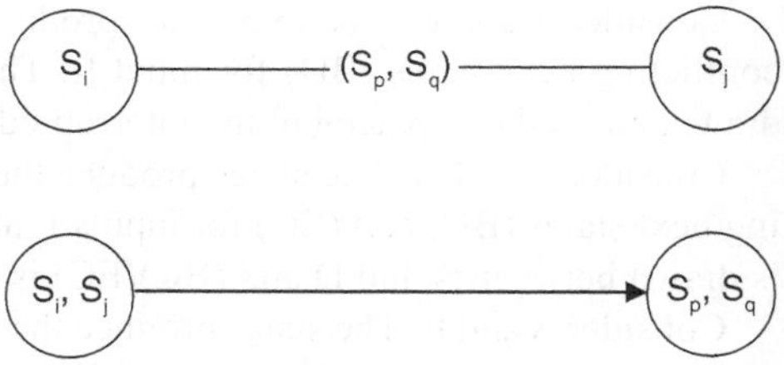

A compatibility graph is a directed graph constructed as follows:

- The number of vertices of the compatibility graph corresponds to the number of compatible pairs obtained from the machine.
- A directed arc is drawn between two compatible pairs (vertices), say from (S_i, S_j) to (S_p, S_q) [where $(S_i, S_j) \neq (S_p, S_q)$] if (S_p, S_q) is the implied pair for (S_i, S_j).

The following example describes how to construct a compatible graph.

Example 4.10 Consider the following finite state machine whose compatible graph is to be constructed.

Solution:

Present State	Next State, z			
	I_1	I_2	I_3	I_4
A	E, 1	C, 1	_, _	B, 1
B	_, _	_, _	D, 0	D, 1
C	_, _	_, _	E, 0	_, _
D	C, 0	B, 1	B, 0	_, _
E	A, 0	F, 1	_, _	D, 0
F	C, 1	_, _	_, _	_, _

To find the compatible pairs, we need to construct the merger graph of the machine first. The merger graph of the machine is given in Fig. 4.16.

The pair of states which are connected by undirected arcs (not crossed in the interrupted portion) are compatible pairs.

The compatible pairs of the given machine are (AB), (AC), (AF), (BC), (BD), (BF), (CE), (CF), and (DE). Therefore, in the compatibility graph there are nine vertices.

In the given machine,

(BD) is the implied pair for (AB). So, a directed arc is drawn from (AB) to (BD).

(CE) is the implied pair for (AF). So, a directed arc is drawn from (AF) to (CE).

(DE) is the implied pair for (BC). So, a directed arc is drawn from (BC) to (DE).

(BD) is the implied pair for (BD). As $(S_i, S_j) \neq (S_p, S_q)$, no arc is drawn from (BD) to (BD).

(AC) and (BF) are implied pairs for (DE). So, two directed arcs are drawn from (DE) to (AC) and (BF).

The compatibility graph for the given machine is given in Fig. 4.17.

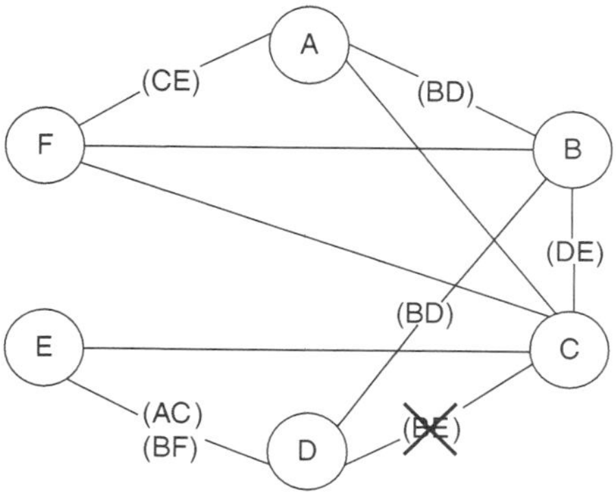

Fig. 4.16 *Merger Graph*

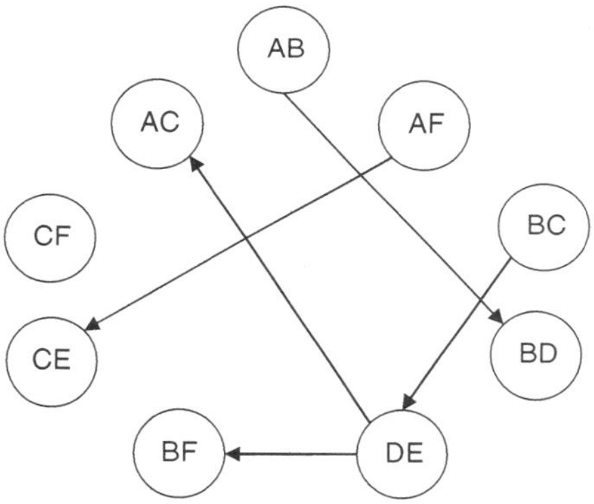

Fig. 4.17 *Compatible Graph*

4.7.4 Minimal Machine Construction

To construct the minimal machine, a compatible graph construction is an essential part. After this, we have to find closed compatible, closed covering, and from there minimal closed covering.

A subgraph of a compatibility graph is called *closed* if, for every vertex in the subgraph, all outgoing arcs and the terminal vertices of the arcs also belong to the subgraph. The pair of states that belong to the subgraph as terminal vertices are called *closed compatible*.

If a subgraph is closed, and if every state of the machine is covered by at least one vertex of the subgraph of a compatibility graph, then the subgraph forms a *closed covering* of the machine.

To find a *minimal machine*, we need to find the subgraphs which closed cover the machine. Then, we need to find the subgraph which contains less number of vertices and which can generate a simpler machine. The states of the minimal machine are the vertices of the subgraph. From these states, the transition functions are constructed. By this process, a minimal machine of the given machine is constructed.

The minimal machine is constructed by the following process for the machine given as an example in the compatibility graph section.

4.7.5 Minimal Machine

The subgraphs (AC), (DE), (BF) or (AB), (BD), (AF), (CE) or (AF), (CE), (BC), (DE) are closed subgraphs of the compatibility graph and forms a closed cover of the machine (here every state of the machine is covered by at least one vertex of the subgraph). Among them, the subgraph (AC)(DE)(BF) contains less number of vertices.

In the minimal machine, the states are (AC), (BF), and (DE). Let us rename them as S_1, S_2, S_3, respectively. The minimal machine becomes

Present State	Next State, z			
	I_1	I_2	I_3	I_4
S_1(AC)	S_3, 1	S_1, 1	S_3, 0	S_2, 1
S_2(BF)	S_1, 1	_, _	S_3, 0	S_3, 1
S_3(DE)	S_1, 0	S_2, 1	S_2, 0	S_3, 0

Example 4.11 Draw a compatible graph for the following machine. Hence, find the minimal machine.

Solution:

Present State	Next State, z			
	I_1	I_2	I_3	I_4
A	_, _	D, 0	D, 0	C, _
B	A, 1	_, _	_, _	D, _
C	B, 1	C, 0	E, 0	_, _
D	E, 1	C, 0	_, _	D, 0
E	_, _	_, _	_, _	A, 1

To find the compatible pairs of a machine, we need to construct the merger graph first. According to the rules for the construction of the merger graph, the following graph as shown in Fig. 4.18 is constructed.

The pair of states which are connected by undirected arcs (not crossed in the interrupted portion) are compatible pairs.

The compatible pairs of the given machine are (AB), (AD), (BC), (BE), (CD), and (CE). Therefore, in the compatibility graph of the previous machine, there are nine vertices.

(CD) is the implied pair for (AB). A directed arc is drawn from (AB) to (CD).

(CD) is the implied pair for (AD). A directed arc is drawn from (AD) to (CD).

(AB) is the implied pair for (BC). A directed arc is drawn from (BC) to (AB).

(AD) is the implied pair for (BE). A directed arc is drawn from (BE) to (AD).

(BE) is the implied pair for (CD). A directed arc is drawn from (CD) to (BE).

The compatibility graph is given in Fig. 4.19.

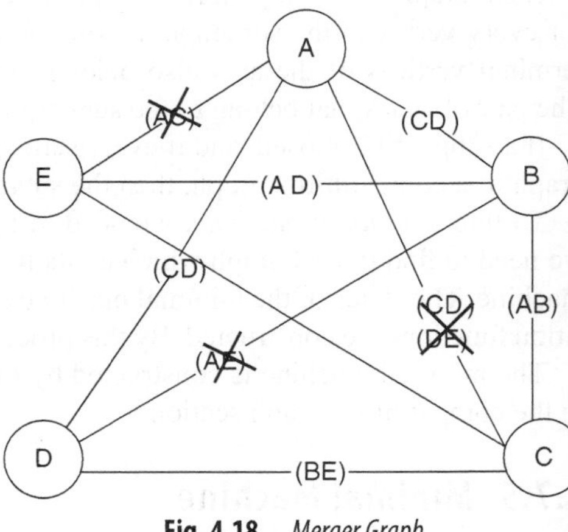

Fig. 4.18 *Merger Graph*

Minimal Machine: In the previous compatibility graph, the subgraph (AD), (CD), (BE) forms a closed covering of the given machine.

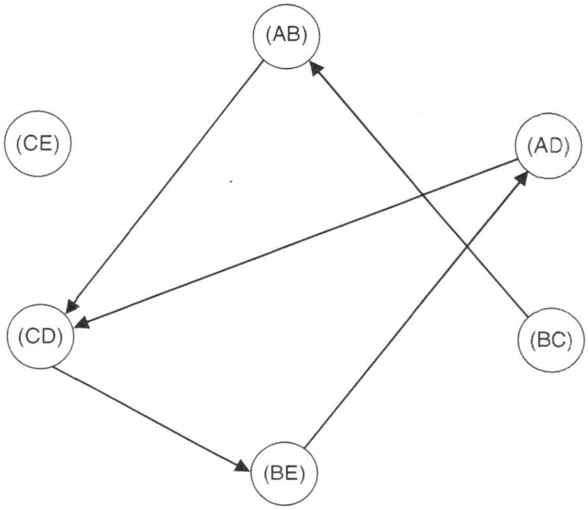

Fig. 4.19 *Compatible Graph*

The states of the minimal machine are (AD), (CD), and (BE). Let us rename them as S_1, S_2, and S_3, respectively. The minimal machine of this machine is

Present State	Next State, z			
	I_1	I_2	I_3	I_4
S_1(AD)	S_3, 1	S_2, 0	S_1/S_2, 0	S_2, 0
S_2(CD)	S_3, 1	S_2, 0	S_3, 0	S_1/S_2, 0
S_3(BE)	S_1, 1	_, _	_, _	S_1, 1

4.8 Merger Table

The merger table is a substitute application of the merger graph. Similar to the merger graph from the merger table, we can also get the compatible pairs and implied pairs.

If the number of states of a machine increases, the number of combination pair increases. For a machine of n states, the number of two state combinations are $^nC_2$, i.e., n(n−1)/2. If n = (n−1), the number of combinations are (n−1) (n−2)/2. The number of combinations increases to (n−1) if the number of states increases from (n−1) to n. It is difficult to connect two states by arcs if the number of states increases. In substitute of that, the merger table is an easier process to find compatible pairs and implied pairs.

4.8.1 Construction of a Merger Table

A merger table can be constructed by the following way.

- Make a table of (n−1) × (n−1), where the left hand side is labelled by the 2nd state to the nth state and the right hand side is labelled by the 1st state to the (n−1)th state.
- Each box represents a pair of state combination.

- Put a cross in the box if the outputs conflict for the pair of states.
- Put a right sign in the box if the outputs as well as next states do not conflict for the pair of states.
- If the outputs do not conflict but the next states conflict for the pair of states, put the conflicting next states in the box.

The following examples describe the process in detail.

Example 4.12 Construct a merger table of the following machine and find the compatible pairs.

Solution:

Present State	Next State, z		
	I_1	I_2	I_3
A	E, 0	B, _	C, _
B	_, _	D, _	B, 0
C	E, _	D, _	C, 0
D	C, 0	_, _	B, 1
E	C, 0	_, _	B, _

Make a table like the following:

B				
C				
D				
E				
	A	B	C	D

Consider the state combination (AB). Here, the outputs do not conflict but the next states for I_2 and I_3 conflict. So, the conflicting next state combinations, (BD) and (BC), are placed in the box labelled (AB).

Consider the state combination (AC). Here, the outputs do not conflict but the next states for I_2 conflict. So, the conflicting next state combination, (BD), is placed in the box labelled (AC).

Consider the state combination (AD). The outputs do not conflict, but the next states for I_1 and I_3 conflict. The conflicting next state combinations (CE) and (BC) are placed in the box labelled (AD).

Consider the state combination (AE). The outputs for the states do not conflict, but the next states for input I_1 and I_3 conflict. (CE) and (BC) are placed in the box labelled (AE).

Consider the state combination (BD). Here, the outputs for input I_3 conflict. There will be an × (cross) in the box labelled (BD).

Consider state combination (DE). Here, both the outputs and the next state combination are the same for all the inputs. So a √ (tick) is placed in the box labelled (DE).

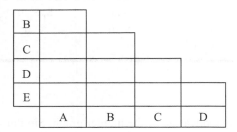

By this process, the merger table is

B	(BD) (BC)			
C	(BD)	(BC)		
D	(CE) (BC)	√	×	
E	(CE) (BC)	√	(CE) (BC)	√
	A	B	C	D

The boxes which are not crossed are compatible pairs. So, the compatible pairs are (AB), (AC), (AD), (AE), (BC), (BD), (BE), (CD), and (DE).

Example 4.13 Construct a merger table of the following machine and find the compatible pairs.

Solution:

Present State	Next State, z			
	I_1	I_2	I_3	I_4
A	C, _	_, _	_, _	_, _
B	_, _	C, _	D, _	E, _
C	_, _	F, 0	B, _	_, _
D	E, _	_, 1	_, _	A, _
E	_, _	B, _	_, _	C, _
F	B, _	_, _	E, _	_, _

For the states AB, for all the inputs, next states and output do not conflict. So a √ (tick) is placed in the box labelled AB.

For states AC, next states and outputs do not conflict for all the inputs. So a √ (tick) is placed in the box labelled AC.

For states AD, the outputs do not conflict, but the next states for input I_1 conflict. So, the conflicting next state pair (CE) is placed in the box labelled AD.

In the box labelled (AE), a √ (tick) is placed.

In the box (AF), the conflicting next state pair (BC) is placed.

In the box (BC), the conflicting next state pairs (CF) and (BD) are placed.

In the box (BD), the conflicting next state pair (AE) is placed.

In the box (BE), the conflicting next state pairs (BC) and (CE) are placed.

In the box (BF), the conflicting next state pair (DE) is placed.

The outputs for I_2 for the state (CD) conflict. So a × is placed in the box (CD).

In the box (CE), the conflicting next state pair (BF) is placed.

In the box (CF), the conflicting next state pair (BE) is placed.

By this process, the constructed merger table is

B	√				
C	√	(CF) (BD)			
D	(CE)	(AE)	×		
E	√	(BC) (CE)	(BF)	(AC)	
F	(BC)	(DE)	(BE)	(BE)	√
	A	B	C	D	E

The compatible pairs are (AB), (AC), (AD), (AE), (AF), (BC), (BD), (DE), (DF), (CE), (CF), (DE), (DF), and (EF).

4.9 Finite Memory and Definite Memory Machine

If we recall the definition of an FSM, it is told that an FSM is a machine whose past histories can affect its future behaviour in a finite number of ways. It means that the present behaviour of the machine is dependent on its past histories. To memorize the past histories, an FSM needs memory elements. The amount of past input and corresponding output information is necessary to determine the machine's future behaviour. This is called the memory span of the machine.

Let us assume that a machine is deterministic (for a single state with single input, only one next state is produced) and completely specified. For this type of a machine, if the initial state and the input sequence are known, one can easily find the output sequence and the corresponding final state. One interesting thing is that, this output sequence and the final state are unique. But the reverse is not always true. If the final state and the output sequence are known, it is not always possible to determine uniquely the input sequence. This section describes the minimum amount of past input–output information required to find the future behaviour of the machine and the condition under which the input to the machine can be constructed from the output produced.

4.9.1 Finite Memory Machine

An FSM M is called a finite memory machine of order μ if μ is the least integer so that the present state of the machine M can be obtained uniquely from the knowledge of last μ number of inputs and the corresponding μ number of outputs.

There are two methods to find whether a machine is finite or not

1. Testing table and testing graph for finite memory
2. Vanishing connection matrix.

4.9.1.1 Testing Table and Testing Graph for Finite Memory Method

The testing table for finite memory is divided into two halves. The upper half contains a single state input–output combination. If, in a machine, there are two types of inputs and two types of outputs, say

0 and 1, the input–output combinations are 0/0, 0/1, 1/0, and 1/1. Here, 0/0 means 0 input and 0 outputs, that is, for the cases we are getting output 0 for input 0, and 0/1 means 0 input and 1 output, that is, for the cases we are getting output 1 for input 0.

The lower half of the table contains all the combinations of the present states taking two into combination. For four present states, (say, A, B, C, and D) there are ${}^4C_2$, which is six, combinations: AB, AC, AD, BC, BD, and CD.

The table is constructed according to the machine given.

The pair of the present state combination is called the uncertainty pair. And its successor is called the implied pair.

In the testing graph for finite memory,

1. The number of nodes will be the number of present state combination taking two into account.
2. There will be a directed arc with a label of input–output combination, from S_iS_j [$i \neq j$] to S_pS_q [$p \neq q$], if S_pS_q is the implied pair of S_iS_j.

If the testing graph is loop-free, the machine is of finite memory. The order of finiteness is the length of the longest path in the testing Graph (l) + 1, i.e., $\mu = l + 1$.

4.9.1.2 Vanishing Connection Matrix Method

If the number of states increases, then it becomes difficult to find the longest path in the Testing graph for finite memory. There is an easy method to determine whether a machine is finite or not, and if finite, to find its order of finiteness. The process is called vanishing connection matrix method.

4.9.2 Constructing the Method of Connection Matrix

1. The number of rows will be equal to the number of columns ($p \times p$ matrix).
2. The rows and columns will be labelled with the pair of the present state combinations. The labels associated with the corresponding rows and columns will be identical.
3. In the matrix, the (i, j)th entry will be 1 if there is an entry in the (S_aS_b) and (S_pS_q) combination in the corresponding testing table. Otherwise, the entry will be 0.

4.9.3 Vanishing of Connection Matrix

1. Delete all the rows having 0's in all positions and delete the corresponding columns also.
2. Repeat this step until one of the following steps is achieved
   (a) No row having 0's in all positions left
   (b) The matrix vanishes, which means there are no rows and columns left.

If the condition 2(a) arrives, the machine is not of finite memory.

If the condition 2(b) arrives, the machine is of finite memory and the number of steps required to vanish the matrix is the order of finiteness of the machine.

The following examples describe the processes in detail.

Example 4.14 Test whether the following machine is of finite memory or not by using testing table–testing graph and vanishing matrix method.

Solution:

Present State	Next State, z X = 0	X = 1
A	D, 1	A, 1
B	D, 0	A, 1
C	B, 1	B, 1
D	A, 1	C, 1

- **Testing Table and Testing Graph for Finite Memory Method:** A table which is divided into two halves is constructed. The machine has two inputs and two outputs. There are four input–output combinations namely 0/0, 0/1, 1/0, and 1/1. The upper half of the machine contains single state input–output combination and the lower half contains two state input–output combinations. There are four states, and so six combination pairs are made. The testing table becomes

Present State	0/0	0/1	1/0	1/1
A	D	—	—	A
B	—	D	—	A
C	—	B	—	B
D	—	A	—	C
AB	—	—	—	AA
AC	—	—	—	AB
AD	—	—	—	AC
BC	—	BD	—	AB
BD	—	AD	—	AC
CD	—	AB	—	BC

In the testing table, there are six present states combinations. So, in the testing table there are six nodes. There is a directed arc with a label of input–output combination, from S_iS_j $[i \neq j]$ to S_pS_q $[p \neq q]$, if S_pS_q is the implied pair of S_iS_j. The testing graph for finite memory is given in Fig. 4.20.

(There will be no arc from AB to AA as AA, is the repetition of same state 'A'.)

The testing graph is loop-free. The longest path in the testing graph is 5 (CD → BC → BD → AD → AC → AB), and so the order of definiteness $\mu = 5 + 1 = 6$.

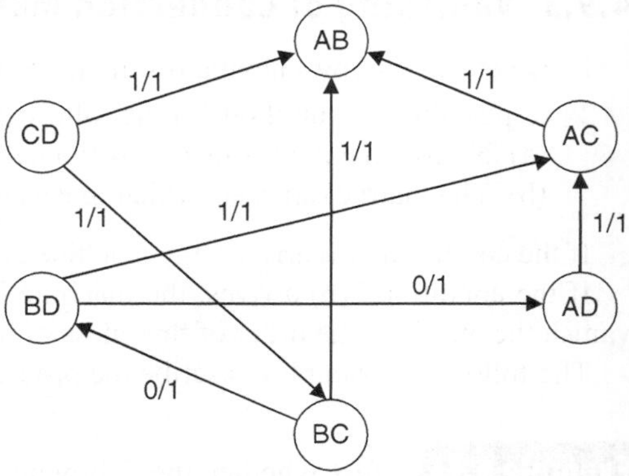

Fig. 4.20 *Testing Graph for Finite Memory*

- **Vanishing Connection Matrix Method:** According to the rule of the construction of the connection matrix, a table is constructed with six rows and six columns labelled with the present state combinations. In the testing table, (AB) is the next state combination for (AC) for input 1/1. So, an entry 1 is placed in the box labelled (AB)(AC). All the other entries are 0.

By this process, the following vanishing connection matrix is generated.

	AB	AC	AD	BC	BD	CD
AB	0	0	0	0	0	0
AC	1	0	0	0	0	0
AD	0	1	0	0	0	0
BC	1	0	0	0	1	0
BD	0	1	1	0	0	0
CD	1	0	0	1	0	0

The row labelled AB contains all the entries as '0'. So, the row labelled AB and the corresponding column are crossed.

	AB	AC	AD	BC	BD	CD
~~AB~~	~~0~~	~~0~~	~~0~~	~~0~~	~~0~~	~~0~~
AC	1	0	0	0	0	0
AD	0	1	0	0	0	0
BC	1	0	0	0	1	0
BD	0	1	1	0	0	0
CD	1	0	0	1	0	0

Now, the table does not contain the row and column labelled AB. The row labeled AC contains all '0'. So, the row labelled AC and the corresponding column are crossed.

	AC	AD	BC	BD	CD
~~AC~~	~~0~~	~~0~~	~~0~~	~~0~~	~~0~~
AD	1	0	0	0	0
BC	0	0	0	1	0
BD	1	1	0	0	0
CD	0	0	1	0	0

Now, the table does not contain the row and column labelled AC. The row labelled AD contains all '0'. So, the row labeled AD and the corresponding column are crossed.

	AD	BC	BD	CD
~~AD~~	~~0~~	~~0~~	~~0~~	~~0~~
BC	0	0	1	0
BD	1	0	0	0
CD	0	1	0	0

The table does not contain the row and column labelled AD. The row labelled BD contains all '0'. So, the row labelled BD and the corresponding column are crossed.

	BC	BD	CD
BC	0	1	0
~~BD~~	~~0~~	~~0~~	~~0~~
CD	1	0	0

The table does not contain the row and column labelled BD. The row labelled BC contains all '0'. So, the row labeled BC and the corresponding column are crossed.

	BC	CD
~~BC~~	~~0~~	~~0~~
CD	1	0

The table does not contain the row and column labelled BC. The row labelled CD contains all '0'. So, the row labelled CD and the corresponding column are crossed, which results in the vanishing of the matrix.

	CD
~~CD~~	~~0~~

The number of steps required to vanish the matrix is 6. So, the order of finiteness $\mu = 6$.

Example 4.15 Check whether the following machine is of finite memory or not by the testing table–testing graph and vanishing connection matrix method.

Solution:

	Next State, z	
Present State	X = 0	X = 1
A	B, 1	D, 0
B	A, 1	C, 1
C	D, 1	A, 0
D	B, 1	A, 1

- **Testing Table and Testing Graph for Finite Memory Method:** A table which is divided into two halves is made. The machine has two inputs and two outputs. There are four input–output combinations namely 0/0, 0/1, 1/0, and 1/1. The upper half of the machine contains single state input–output combination and the lower half contains two state input–output combinations. There are four states, and so six combination pairs are made. The testing table becomes

Present State	0/0	0/1	1/0	1/1
A	—	B	D	—
B	—	A	—	C
C	—	D	A	—
D	—	B	—	A
AB	—	AB	—	—
AC	—	BD	AD	—
AD	—	BB	—	—
BC	—	AD	—	—
BD	—	AB	—	AC
CD	—	BD	—	—

In the testing table, there are six present states combinations. So, in the testing table, there are six nodes. There is a directed arc with a label of input–output combination, from S_iS_j [i ≠ j] to S_pS_q [p ≠ q], if S_pS_q is the implied pair of S_iS_j. The testing graph for finite memory is given in Fig. 4.21.

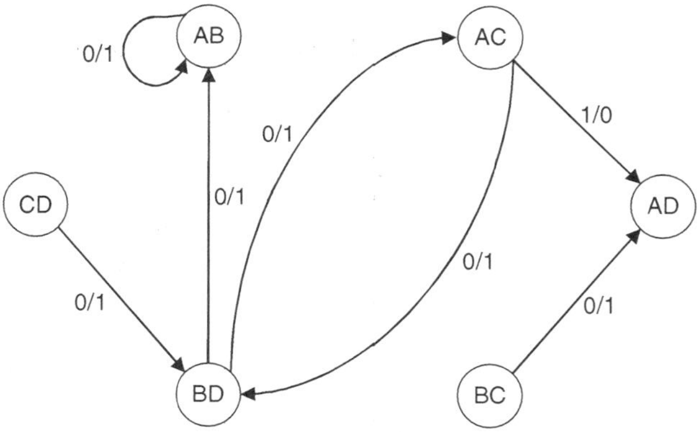

Fig. 4.21 *Testing Graph for Finite Memory*

The testing graph for finite memory contain two loops AB to AB with label 0/1 and AC to BD with label 0/1, BD to AC with label 0/1. As the testing graph is not loop-free, the machine is not of finite memory.

- **Test by Vanishing Connection Matrix Method:** In the testing table, six present state pairs are formed. Therefore, the connection matrix consists of six rows and six columns. The rows and columns are labelled with a pair of present state combinations.

In the matrix, the (i, j)th entry will be 1 if there is an entry in (S_aS_b) (a ≠ b) and (S_pS_q) (p ≠ q) combination in the corresponding testing table. Otherwise, the entry will be 0.

The connection matrix of the previous machine is as follows.

	AB	AC	AD	BC	BD	CD
AB	1	0	0	0	0	0
AC	0	0	1	0	1	0
AD	0	0	0	0	0	0
BC	0	0	1	0	0	0
BD	1	1	0	0	0	0
CD	0	0	0	0	1	0

In the previous matrix, the row AD contains all '0'. So, the row labelled AD and the corresponding column vanish.

	AB	AC	~~AD~~	BC	BD	CD
AB	1	0	~~0~~	0	0	0
AC	0	0	~~1~~	0	1	0
~~AD~~	~~0~~	~~0~~	~~0~~	~~0~~	~~0~~	~~0~~
BC	0	0	~~1~~	0	0	0
BD	1	1	~~0~~	0	0	0
CD	0	0	~~0~~	0	1	0

In the modified matrix, there is no row and column labelled AD. The row labelled BC contains all '0'. So, the row labelled BC and the corresponding column vanish.

	AB	AC	~~BC~~	BD	CD
AB	1	0	~~0~~	0	0
AC	0	0	~~0~~	1	0
~~BC~~	~~0~~	~~0~~	~~0~~	~~0~~	~~0~~
BD	1	1	~~0~~	0	0
CD	0	0	~~0~~	1	0

In the modified matrix, there is no row and column labelled BC. The modified matrix becomes

	AB	AC	BD	CD
AB	1	0	0	0
AC	0	0	1	0
BD	1	1	0	0
CD	0	0	1	0

The matrix does not contain any row containing all '0'. So, the matrix does not vanish. Therefore, the machine is not of finite memory.

4.9.4 Definite Memory Machine

A sequential machine M is called definite if there exists a least integer μ, so that the present state of the machine M can be uniquely obtained from the past μ number of inputs to the machine M. μ is called the order of definiteness of the machine.

There are three methods to find whether a machine is definite or not.

1. Synchronizing tree method
2. Contracted table method
3. Testing table, testing graph for definiteness method.

4.9.4.1 Synchronizing Tree Method

For a machine with definite memory, only inputs play a role. There is no role of outputs. So, the present state combinations can be divided into two parts for two types of inputs, 0 and 1. Those divided combinations of states can again be divided for input 0 and 1.

If the leaf nodes are of single state, stop constructing the tree. The order of definiteness is the maximum label of the synchronizing tree.

Else, the machine is not of definite memory.

4.9.4.2 Contracted Table Method

i) Find those present states whose next states are identical.
ii) Represent those present states as single states. (Those present states are equivalent.)
iii) Obtain the contracted table by replacing only one of the present states from the equivalent state set and modify the table according to it.
iv) Repeat steps from (i) to (iii) until new contractions are possible.

By this process, if at last a machine with single state is received, the machine is definite. Its order is the number of steps required to obtain the single state machine.

Else, the machine is not definite.

4.9.4.3 Testing Table, Testing Graph For Definiteness Method

The testing table for definiteness is divided into two halves. The upper half contains the input and the next state combination. The lower half of the table contains all combination of present states taking two into combination. For four present states (say, A, B, C, and D), there are $^4C_2$, which is six, combinations: AB, AC, AD, BC, BD, and CD.

The table is constructed according to the machine given.

For these present state combinations, find the next state combinations for input 0 and input 1.

The pair of present state combination is called the uncertainty pair. And its successor is called the implied pair.

In the testing graph for definite memory,

- The number of nodes will be the number of present state combination taking two into account.
- There will be a directed arc with a label of input–output combination, from S_iS_j $[i \neq j]$ to S_pS_q $[p \neq q]$, if S_pS_q is the implied pair of S_iS_j.

If the testing graph is loop-free, the machine is of definite memory. The order of definiteness is the length of the longest path in the testing graph (l) + 1, i.e., $\mu = l + 1$.

The following examples describe the methods to check definiteness.

Example 4.16 Test whether the following machine is definite or not by any of the methods. If definite, find the order of definiteness.

Solution:

Present State	Next State X = 0	Next State X = 1
A	C	B
B	A	D
C	C	B
D	C	D

- **Synchronizing Tree Method:** (ABCD) is the present state combination of the machine. It has two types of inputs, 0 and 1. (ABCD) with input 0 produces the next state combination (CACC). It can be grouped into two distinct states (AC). (ABCD) with input 1 produces the next state combination (BDBD) which can be grouped into (BD).

In the next level, (AC) with input 0 produces a single next state C. With input 1, (AC) produces the single next state B. As C and B are single states, there is no need to approach further to the next level. The state combination (BD) with input 0 produces the next state combination (AC). With input 1, the state combination (BD) produces a single state D. In the next level, the state combination (AC) produces C and B, respectively, for input 0 and 1.

As the leaf nodes are of single state, stop constructing the tree. The order of definiteness is the maximum label of the synchronizing tree in Fig. 4.22. In this case, the order is 3.

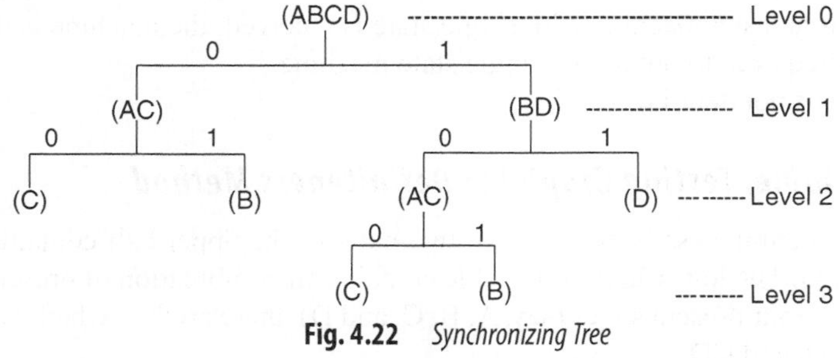

Fig. 4.22 *Synchronizing Tree*

- **Contracted Table Method**

Present State	Next State X = 0	Next State X = 1
A	C	B
B	A	D
C	C	B
D	C	D

In the previous machine, A and C have the same next state combination (both cases C and B). Therefore, A and C are equivalent states. Among A and C, let us take only A. This means that all the C in the state table are replaced by A. The contracted table is

	Next State	
Present State	X = 0	X = 1
A	A	B
B	A	D
D	A	D

In the table, B and D produce the same next state combination for input 0 and 1. So, B and D are equivalent states. D is replaced by B. The contracted table becomes

	Next State	
Present State	X = 0	X = 1
A	A	B
B	A	B

The table contains two states A and B with the same next state combination for input 0 and 1. So they are equivalent states. B is replaced by A. The contracted table becomes

	Next State	
Present State	X = 0	X = 1
A	A	A

A machine with a single state is received, and the machine is definite. Its order is the number of steps required to obtain the single state machine, here 3.

- **Testing Table Testing Graph Method:** A table with two halves is made. The upper half contains the input and next state combinations. The lower half of the table contains all the combination of present states taking two into combination. Here, for four present states (say, A, B, C, and D), there are ${}^4C_2$, which is six, combinations: AB, AC, AD, BC, BD, and CD. The table is

Present State	0	1
A	C	B
B	A	D
C	C	B
D	C	D
AB	AC	BD
AC	CC	BB
AD	CC	BD
BC	AC	BD
BD	AC	DD
CD	CC	BD

The lower half of the testing table contains six present state combination pairs. The testing graph is given in Fig. 4.23.

The testing graph is loop-free. So, the machine is of definite memory. The longest path of the testing graph is 2 (AB → BD → AC or AD → BD → AC). So, the order of definiteness of the machine is 2 + 1 = 3.

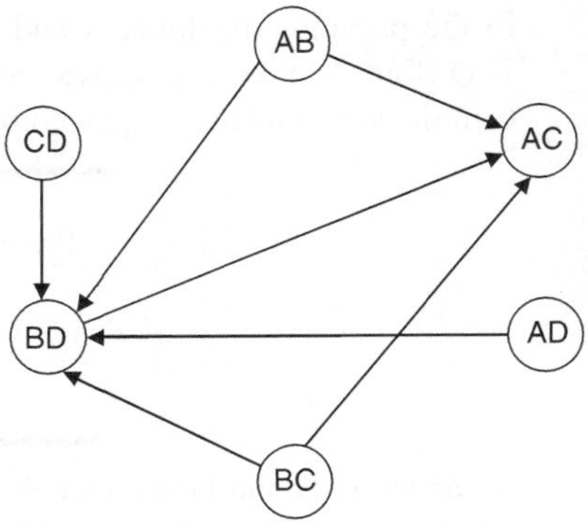

Fig. 4.23 *Testing Graph for Definiteness*

Example 4.17 Test whether the following machine is definite or not by any of the methods (synchronizing tree or contracted table). If definite, find the order of definiteness.

Solution:

Present State	Next State X = 0	Next State X = 1
A	D	E
B	A	B
C	C	B
D	C	B
E	A	B

- **Synchronizing Tree Method:** The synchronizing tree thus obtained contains the leaf nodes of a single state. So, the machine is of definite memory. The number of levels of the tree, as given in Fig. 4.24, is 3. Therefore, the order of definiteness of the machine is 3.

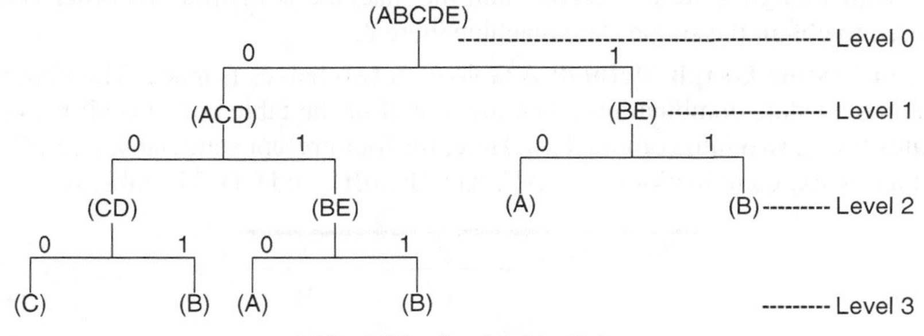

Fig. 4.24 *Synchronizing Tree*

- **Contracted Table Method**

Present State	Next State X = 0	Next State X = 1
A	D	E
B	A	B
C	C	B
D	C	B
E	A	B

B and E are producing the same next state combination for input 0 and 1. So, they are equivalent. E is replaced by B. C and D are producing the same next state combination, and D is replaced by C. The contracted table becomes

	Next State	
Present State	X = 0	X = 1
A	C	B
B	A	B
C	C	B

States A and C are producing the same next state combination for input 0 and 1. So, A and C are equivalent. C is replaced by A. The contracted table becomes

	Next State	
Present State	X = 0	X = 1
A	A	B
B	A	B

Both of the states are producing the same next state combination, and so they are equivalent. B is replaced by A.

	Next State	
Present State	X = 0	X = 1
A	A	A

The machine becomes a single state machine. It is of definite memory. The number of steps required to make it a single state machine is 3. Therefore, the order of definiteness of the machine is 3.

4.10 Information Lossless Machine

The main problem of coding and information transmission theory is to determine the conditions for which it is possible to regenerate the input sequence given to a machine from the achieved output sequence. Let us assume that the information used for a coding device in a machine be the input and the coded message be the output achieved, and let the initial and final states be known. It this case, *information losslessness* of the machine guarantees that the coded message (output) can always be deciphered.

A machine is called *information lossless* if its initial state, final state, and output string are sufficient to determine uniquely the input string.

A machine is said to be (information) lossless of order μ (ILF-μ) if the knowledge of the initial state and the first μ output symbols is sufficient to determine uniquely the first input symbol.

Let us take a machine M as follows

	Next State, z	
Present State	X = 0	X = 1
A	A, 1	B, 0
B	B, 0	A, 1

Let the initial state of the machine be A, the final state achieved be A and the coded message (output) be 01. We have to find the information given as input.

Now we are constructing the machine according to the output achieved.

	Previous State, I/P	
Next State	z = 0	z = 1
A	–	A, 0 B, 1
B	A, 1 B, 0	–

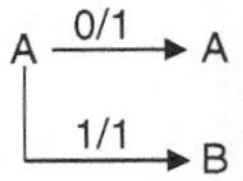

An output of 01 means that from the reverse side it is 10. The initial state is A and final state is A. So, the diagram will become

The output 1 with the next state A is produced for two cases—previous state A with input 0 or previous state B with input 1.

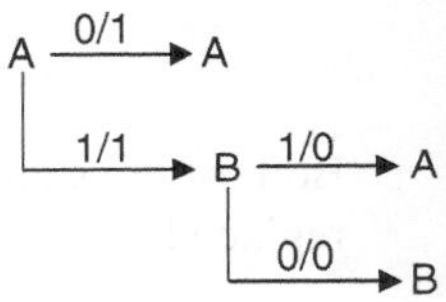

The output 0 with next state A is produced for no cases. The output 0 with next state B is produced for two cases—previous state A with input 1 and previous state B with input 0.

The initial state is A. In the previous case, A is produced for input 11. Therefore, the input is 11.

Lossy Machine: A machine which is not lossless is called lossy.

Consider the following state table

	Next State, z	
Present State	X = 0	X = 1
A	A, 1	B, 0
B	B, 1	A, 1

If the testing table obtained from the given machine contains repeated entry of states, the machine is lossy. As an example, for the previous machine, the testing table is as follows

	Next State	
Present State	z = 0	z = 1
A	B	A
B	–	AB
AB	–	AA, AB

The testing table contains repeated entry of states in the form of AA. So, the machine is lossy.

Two states $S_{i,}$ S_j of a machine M are said to be *output compatible* if there exists some state S_{p}, such that both S_i and S_j are its O_k successor or there exists a compatible pair of states $S_{a,}$ S_b such that S_i S_j are its O_k successor.

Present State	Next State, z	
	X = 0	X = 1
A	A, 1	B, 0
B	B, 1	A, 1

If we construct an output successor table for it, the table will be

Present State	Previous State, I/P	
	z = 0	z = 1
A	B, 1	A, 0
B	–	B, 0 A, 1

Let us apply input 1 and 0, respectively, on states B and A. The next state will be A for both the cases, and the output will be 1 for both the cases. Therefore, we can say A and B are output compatible as there exists a state A such that both A and B are its 1 successor.

4.10.1 Test for Information Losslessness

The process of testing whether a machine is information lossless or not is done by constructing a testing table and a testing graph.

The testing table is constructed in the following way.

- The testing table for checking information losslessness is divided into two halves. The upper half contains the present states and its output successors.
- The lower half of the table is constructed in the following way
  - Every compatible pair appearing in the output successor table is taken into the present state column. The successors of these pairs are constructed from the original table. Here, if any compatible pair appears, then that pair is called the implied pair for the compatible pair in the present state column.
  - That new pair is again taken in the present state column if it is not taken.
  - The process terminates when all compatible pairs have been taken in the present state column.

The machine is information lossless if and only if its testing table does not contain any compatible pair consisting of repeated entry.

From the testing table, the testing graph is constructed. The testing graph is constructed in the following way.

- The number of vertices of the testing graph is equal to the number of output compatible pairs taken in the lower half of the testing table. The labels of the vertices are the compatible pairs in the lower half of the testing table.
- A directed arc is drawn from vertex S_iS_j to vertex S_pS_q ($p \neq q$) if S_pS_q is an implied pair for S_iS_j .

The machine is information lossless of finite order if the testing graph is loop-free and its order $\mu = l + 2$, where l is the length of the longest path of the testing graph.

Consider the following examples to clarify the method described previously.

Example 4.18 Test whether the following machine is information lossless or not. If lossless, find its order.

Solution:

Present State	Next State, z	
	X = 0	X = 1
A	C, 1	D, 1
B	A, 1	B, 1
C	B, 0	A, 0
D	D, 0	C, 0

The first step to test whether a machine is lossless or not is to construct a testing table. The testing table is divided into two halves.

Present State	Z = 0	Z = 1
A	–	CD
B	–	AB
C	AB	–
D	CD	–
AB	–	(AC)(BC) (AD)(BD)
CD	(BD)(BC) (AD)(AC)	–
AC	–	–
AD	–	–
BC	–	–
BD	–	–

The testing graph consists of six vertices.

The testing table does not contain any repeated entry. The machine is a lossless machine. The testing graph as shown in Fig. 4.25 does not contain any loop. The order of losslessness is $\mu = 1 + 2 = 3$. The length of the longest path of the graph is 1.

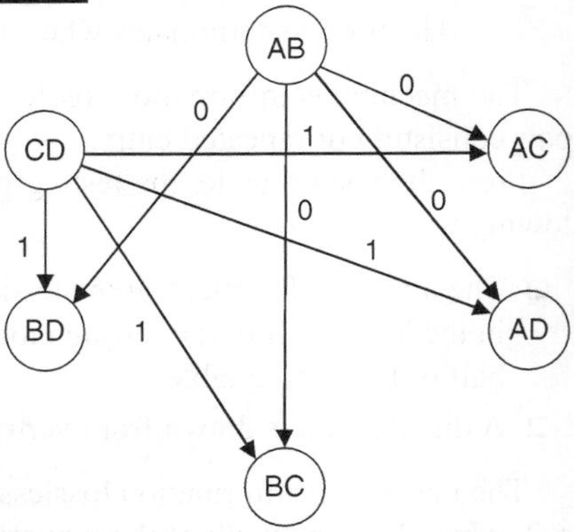

Fig. 4.25 Testing Graph for Information Losslessness

Example 4.19 Find the input string which is applied on state 'A' producing the output string 00011 and the final state 'B' for the following machine.

Solution:

	Next State, z	
Present State	X = 0	X = 1
A	A, 0	B, 0
B	C, 0	D, 0
C	D, 1	C, 1
D	B, 1	A, 1

First, we need to prove whether the machine is information lossless. For this, we need to construct a testing table for information lossless.

Present State	z = 0	z = 1
A	AB	–
B	CD	–
C	–	CD
D	–	AB
AB	(AC)(BC) (AD)(BD)	–
CD	–	(BD)(BC) (AD)(AC)
AC	–	–
AD	–	–
BC	–	–
BD	–	–

The testing table does not contain any repeated entry. So, the machine is information lossless.
Now, we need to construct the output successor table

Present State	z = 0	z = 1
A	A/0, B/1	–
B	C/0, D/1	–
C	–	C/1, D/0
D	–	A/1, B/0

The input string is applied on state A and has produced output 0. From the output successor table, it is clear that the next states are A or B. By this process, the transition is given in Fig. 4.26.

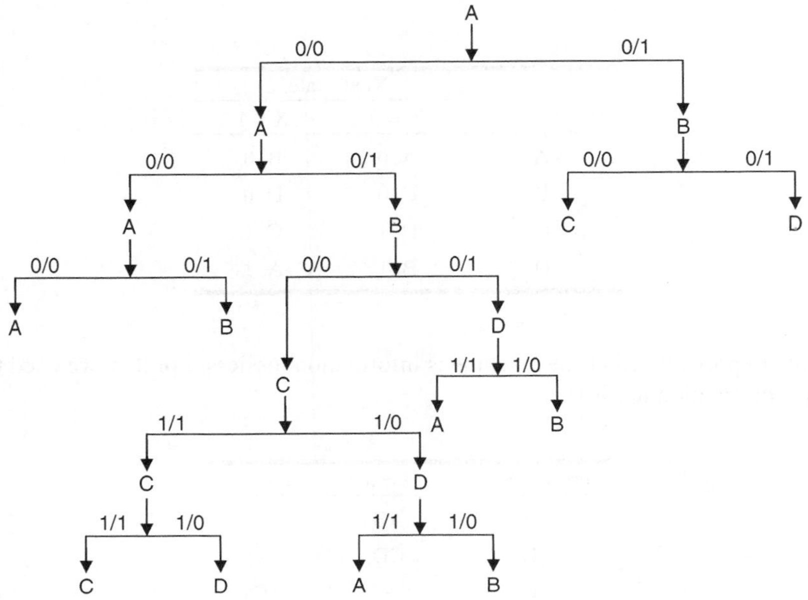

Fig. 4.26

After traversing the total output string by the output successor table, we have got a state B. So, the input string is 01000.

(Input string retrieval is possible only for an information lossless machine. This can be illustrated in the following example.)

Example 4.20 Find the input string which is applied on state 'A' producing the output string 00101 and the final state 'C' for the following machine.

Solution:

Present State	Next State, z	
	X = 0	X = 1
A	B, 0	B, 0
B	C, 0	D, 0
C	D, 0	C, 0
D	A, 0	C, 1

The machine is not information lossless as, in the testing table, for information losslessness, there is a repeated entry BB for the state 'A' and output '0'. Yet, again we try to find the input sequence by constructing the output successor table.

The output successor table for the given machine is

Present State	Next State, Z	
	z = 0	z = 1
A	B/0, B/1	–
B	D/0, D/1	–
C	D/0, C/1	–
D	A/0	C/1

The input string is applied on the state A and has produced output 0. From the output successor table, it is clear that the next states are B with input 0 or B with input 1. By this process, the transition is given in Fig. 4.27.

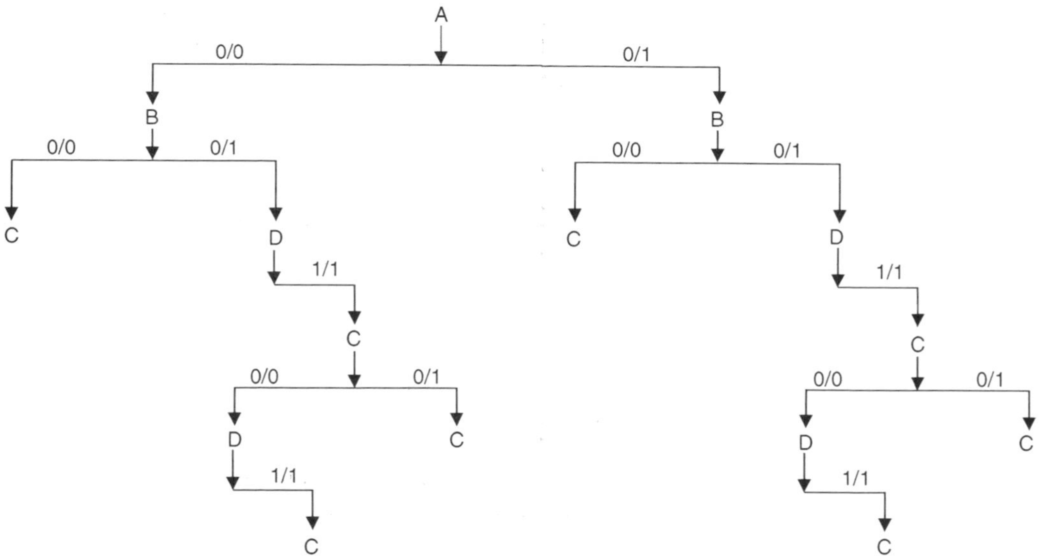

Fig. 4.27

We are getting C as the final state for two input sequences 01101 and 11101. So, we cannot uniquely determine the input string. We can conclude that input string retrieval is not possible for the information lossy machine.

Example 4.21 Test whether the following machine is information lossless or not. If lossless, find its order.

Solution:

Present State	Next State, z	
	X = 0	X = 1
A	B, 0	C, 0
B	D, 0	E, 1
C	A, 1	E, 0
D	E, 0	D, 0
E	A, 1	E, 1

The first step to test whether a machine is lossless or not is to construct a testing table. The testing table is divided into two halves.

Present State	z = 0	z = 1
A	BC	–
B	D	E
C	E	A
D	DE	–
E	–	AE
BC	DE	AE
DE	–	–
AE	–	–

The testing table does not contain any repeated entry. The machine is an information lossless machine. The testing graph for the machine is given in Fig. 4.28.

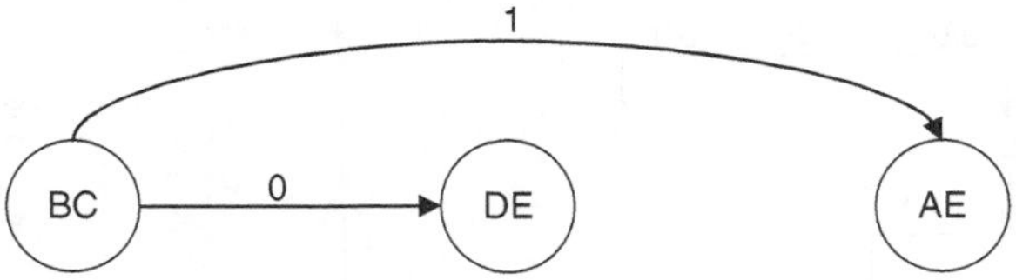

Fig. 4.28 *Testing Graph for Information Losslessness*

The testing graph for information losslessness is loop-free. The order of losslessness is $\mu = 1 + 2 = 3$. The length of the longest path of the graph is 1.

4.11 Inverse Machine

An *inverse machine* M^i is a machine which is developed from the given machine M with its output sequence and produces the input sequence given to machine M, after at most a finite delay.

A deterministic inverse machine can be constructed if and only if the given machine is lossless. The machine can produce the input sequence applied to the original machine after at most a finite delay if and only if M is lossless of a finite order.

Consider the following example.

Example 4.22 Construct an inverse machine of the following given machine:

Solution:

Present State	Next State, z	
	X = 0	X = 1
A	B, 0	C, 1
B	C, 1	D, 0
C	A, 0	D, 1
D	A, 1	C, 0

We have to construct the inverse machine of this given machine.

Before constructing the inverse machine, we need to test whether the machine is information lossless or not.

The testing table for information lossless is

Present State	z = 0	z = 1
A	B	C
B	D	C
C	A	D
D	C	A

This is information lossless of the first order, as here the first input symbol can be determined from the knowledge of the initial state and the first output symbol.

As the machine is information lossless, the inverse machine for it can be constructed.

The inverse machine is

Present State	Next State, X	
	z = 0	z = 1
A	B, 0	C, 1
B	D, 1	C, 0
C	A, 0	D, 1
D	C, 1	A, 0

4.12 Minimal Inverse Machine

An inverse machine is only possible if the machine is lossless. First, we have to check if the machine is lossless or not. If it is lossless, only then a minimal inverse machine construction is possible.

Consider the following machine. We have to construct the minimal inverse machine of it.

Present State	Next State, z	
	X = 0	X = 1
A	A, 0	B, 0
B	C, 1	D, 1
C	B, 0	A, 1
D	A, 0	B, 1

First, we have to check if the machine is lossless or not. Construct the following testing table to check for lossnessness.

Present State	Next State, z	
	z = 0	z = 1
A	AB	_, _
B	_, _	CD
C	B	A
D	A	B
AB	_, _	_, _
CD	AB	AB

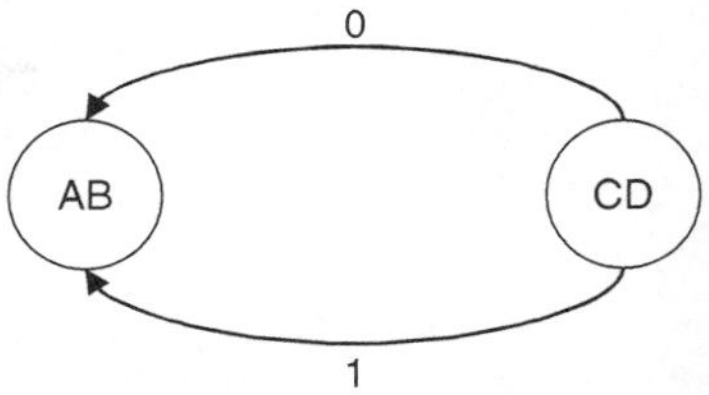

Fig. 4.29 *Testing Graph for Information Losslessness*

The machine is lossless as the testing table does not contain any entry of repeated states.

The testing graph for the information lossless machine is given in Fig. 4.29.

The order of losslessness of the machine is 3.

Let us define a set of triples denoted by (S(t), z(t + 1), and z(t + 2)), where S(t) is the initial state of the machine, z(t + 1) is one of the output symbols that can be produced by a single transition from this state, and z(t + 2) is another output symbol that can follow this initial state and the first output symbol.

Let us take B. The state B has outputs '1' for both input '0' and '1'. So, make the combinations (B, 1) and (B, 1). The next states of B are C and D. C produces outputs '0' and '1' and D produces outputs '0' and '1'. So, the triples are (B, 1, 0) and (B, 1, 1).

Let us take D. The state D has an output '0' for input '0' and an output '1' for input '1'. So, make the combinations (D, 0) and (D, 1). A has the output '0' for all inputs and B has the output '1' for all inputs. So, the triples are (D, 0, 0) and (D, 1, 1).

All the others are constructed in the same process.

For the given machine, the possible triples are

(A, 0, 0)	(B, 1, 0)	(C, 0, 1)	(D, 0, 0)
(A, 0, 1)	(B, 1, 1)	(C, 1, 0)	(D, 1, 1)

The minimal inverse machine is constructed by the following rule.

- The present states are the triples obtained from the given machine.
- The first member is the state to which the machine goes when it is initially in the state, that is, the first member of the present inverse state, and when it is supplied with the first input symbol.

- The second member is the third member of the corresponding present state.

  (Consider (A, 0, 1). The third member is '1'. For the second row, the second member of the next state columns is '1'.

  Consider (C, 1, 0). The third member is '0'. For the sixth row, the second member of the next state columns is '0'.)
- The third member is the present output of the constructing minimal inverse machine.
- (The third member of each entry for z = 0 is '0' and each entry for z = 1 is '1'.)
- The output of the constructing minimal inverse machine is equal to the input at (t –2) to the given information lossless machine.

  (From B, we get D for input 1 and C for input 0. So, for the third row, the outputs are 1 and 0, respectively.

  From D, we get A for input '0'. So, for the seventh row, the outputs are '0' for both z = 0 and z = 1.)

The minimal inverse machine is

Present State	Next State, X	
	z = 0	z = 1
(A, 0, 0)	(A, 0, 0), 0	(A, 0, 1), 0
(A, 0, 1)	(B, 1, 0), 1	(B, 1, 1), 1
(B, 1, 0)	(D, 0, 0), 1	(C, 0, 1), 0
(B, 1, 1)	(C, 1, 0), 0	(D, 1, 1), 1
(C, 0, 1)	(B, 1, 0), 0	(B, 1, 1), 0
(C, 1, 0)	(A, 0, 0), 1	(A, 0, 1), 1
(D, 0, 0)	(A, 0, 0), 0	(A, 0, 1), 0
(D, 1, 1)	(B, 1, 0), 1	(B, 1, 1), 1

In the previous machine, state (A, 0, 1), (D, 1, 1) and state (A, 0, 0), (D, 0, 0) are equivalent as they produce the same next states and same outputs. To minimize the machine, assume

(A, 0, 0) as S_1 (A, 0, 1) as S_2
(B, 1, 0) as S_3 (B, 1, 1) as S_4
(C, 0, 1) as S_5 (C, 1, 0) as S_6

There is no need to name (D, 0, 0) and (D, 1, 1) as they are equivalent with S_1 and S_2, respectively. The minimized machine is

Present State	Next State, X	
	z = 0	z = 1
S_1	S_1, 0	S_2, 0
S_2	S_3, 1	S_4, 1
S_3	S_1, 1	S_5, 0
S_4	S_6, 0	S_2, 1
S_5	S_3, 0	S_4, 0
S_6	S_1, 1	S_2, 1

What We Have Learned So Far

1. Formal language and automata theory covers all the aspects of the theoretical part of computer science.
2. The circuits, whose operations are controlled by clock pulses are called synchronous circuit. The operations in asynchronous circuit are controlled by a number of completion and initialization signals. Here, the completion of one operation is the initialization of the execution of the next consecutive operation.
3. The circuit whose output depends on the present state only is called a combinational circuit.

   i.e., O/P = Func.(present I/P)

   If the output is the function of external input and the present stored information, then the circuit is called a sequential circuit.

   i.e., O/P = Func.(external I/P and present stored information)
4. The machine whose past histories can affect its future behaviour in a finite number of ways is called an FSM.

   An FSM has a finite number of states.
5. No FSM can be designed for an infinite sequence. No FSM can multiply two arbitrary large binary numbers. These are the limitations of FSM.
6. Two states of a machine are called equivalent if they produce the same output sequence for an input sequence applied to the states, separately considering those two states as initial states.
7. The equivalent partition of a machine M can be defined as a set of maximum number subsets of states where states which reside in the same subset are equivalent for any input given to the states.
8. The equivalent partition is unique.
9. The machines where for all states for all inputs, the next state or outputs or both are not mentioned are called incompletely specified machine.
10. A minimal machine is the minimum of the machines obtained by minimizing an incompletely specified machine.
11. A merger graph is an undirected graph where a compatible graph is a directed graph.
12. A merger table can be used as a substitution of the merger graph.
13. An FSM M is called a finite memory machine of order μ if μ is the least integer so that the present state of the machine M can be obtained uniquely from the knowledge of the last μ number of inputs and the corresponding μ number of outputs.
14. A sequential machine M is called definite if there exists a least integer μ, so that the present state of the machine M can be uniquely obtained from the past μ number of inputs to the machine M. μ is called the order of definiteness of the machine.
15. A machine is called information lossless if its initial state, final state, and output string are sufficient to determine uniquely the input string.
16. A machine which is not lossless is called lossy.

Solved Problems

1. Design a two input two output sequence detector which generates an output '1' every time the sequence 1011 is detected. And for all other cases, output '0' is generated. Overlapping sequences are also counted. Draw only the state table and the state diagram.

Solution: The sequence is 1011. We have to start from S_1. If we get input 0, then there is no chance to get 1011, so it is confined in S_1 producing output 0. If we get input 1, then there is a chance to get 1011, and so the control moves to S_2 producing output 0 (as we have not got 1011 as input still). In S_2, if we get 0, then there is a chance to get 1011, and so the control moves to S_3 producing output 0. In S_2, if we get input 1, then there is a chance to get 1011, considering the last 1. So, the control will be confined in S_2, producing output 0.

In S_3, if we get input 0, then there is no chance to get 1011. In this case, we have to start again from the beginning, i.e., from S_1. So, the control moves to S1 producing output 0. In the state S_3, if we get input 0, then there is no chance to get 1011 considering any of the fourth or third and fourth or second, third and fourth or first, second, third, and fourth input combination. In this case, we have to start again from the beginning, i.e., from S_1. So, the control moves to S_1 producing output 0. If we get input 1, then the string 1011 is achieved, and so the output 1 is produced. As the overlapping sequence is also accepted, the control moves to S_2 so that by getting 011, the sequence detector can produce 1.

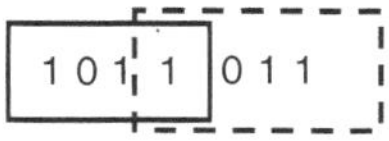

The state diagram is given.

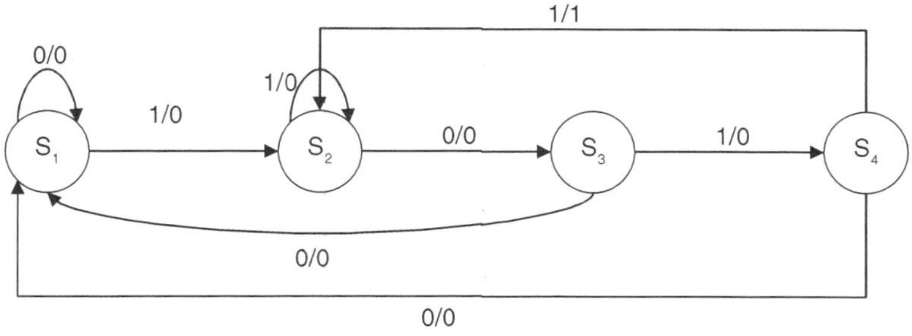

The state table for the sequence detector is

Present State	Next State, O/P X = 0	= 1
S_1	S_1, 0	S_2, 0
S_2	S_3, 0	S_2, 0
S_3	S_1, 0	S_4, 0
S_4	S_1, 0	S_2, 1

2. Design a two input two output sequence detector which generates an output '1' every time the sequence 10101 is detected. And for all other cases, output '0' is generated. Overlapping sequences are also counted. Draw only the state table and the state diagram and make the state assignment.

Solution:

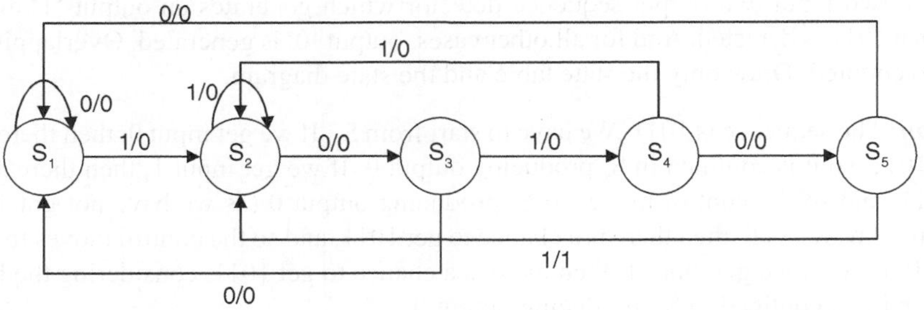

The state diagram is constructed previously.
Here, the overlapping sequence is 101.

The state table for the sequence detector is

Present State	Next State, O/P	
	X = 0	= 1
S_1	S_1, 0	S_2, 0
S_2	S_3, 0	S_2, 0
S_3	$S_{4,}$ 0	S_1, 0
S_4	$S_{5,}$ 0	S_2, 0
S_5	S_1, 0	S_2, 1

For doing the state assignment, we have to assign the states to some binary numbers. Here are five states, so we have to take a binary string of length three because $2^3 = 8 > 5$.

Let us assign 000 as S_1, 001 as S_2, 011 as S_3, 010 as S_4, and 100 as S_5.

After doing the state assignment, the state table becomes

Present State	Next State, (Y_1Y_2)		O/P (z)	
(y_2y_1)	X = 0	= 1	= 0	= 1
000	000	001	0	0
011	011	001	0	0
011	010	000	0	0
010	100	001	0	0
100	000	001	0	1

3. Find the equivalent partition for the following machine.

Present State	Next State, z X = 0	X = 1
A	B, 0	C, 1
B	A, 0	E, 1
C	D, 1	E, 1
D	E, 1	B, 1
E	D, 1	B, 1

Solution: For a string length 0 (i.e., for no input), there is no output. So, all the states are equivalent. It is called 0-equivalent. We can write

$$P_0 = (ABCDE).$$

For string length 1, there are two types of inputs—0 and 1. The states A and B give output 0 for input 0 and states C, D, and E give output 1 for input 0. All of the states give output 1 for input 1.

So, the states in the set P_0 are divided into (AB) and (C, D, E). We can write

$$P_1 = ((AB)\ (CDE)).$$

Here A and E are 1-distinguishable because they produce different outputs for input string length 1.

For input string length 2, check the distinguishability by the next state combination.

The states A and B for input 0 produce next states B and A, respectively, and produce next states C and E for input 1. B and A belong to same set, and C and E also belong to the same set. So, AB cannot be partitioned for the input string length 2.

The states CDE for input 0 produce next states D, E, D, respectively, and produce next states E, B, B for input 1. D, E belong to same set but B and E belong to different sets. So, the set (CDE) is portioned into (C) and (DE). The new partition becomes

$$P_2 = ((AB)\ ((C)\ (DE))).$$

For input string length 3, we have to perform the same as like string length 2. The states A, B for input 0 produce next states B and A, respectively, and produce next states C and E for input 1. B, A belong to same set, but C and E belong to different sets. So, the set (AB) is partitioned into (A) and (B).

C is a single state and cannot be partitioned further.

The states D and E produce next states E and D, respectively, for input 0 and produce next state B for each of the states for input 1. D and E belong to same set. So, D and E cannot be divided. The new partition becomes

$$P_3 = (A)(B)(C)(DE).$$

Next, we have to check for input string length 4. Three subsets contain single state—A, B, and C, and cannot be partitioned further.

The states D and E produce the next states E and D, respectively, for input 0 and the next state B for each of the states for input 1. D and E belong to same set, and so D and E cannot be partitioned further. The partition is the same as P_3. So, the partition P_3 is the equivalent partition of the machine.

Minimization: We know that the equivalent partition is unique. So, P3 = (A)(B)(C)(DE) is the unique combination. Here, every single set represents one state of the minimized machine.

Let us rename these partitions for simplification.
Rename (A) as S_1, (B) as S_2, (C) as S_3 and (DE) as S_4.
The minimized machine becomes

Present State	Next State, z	
	X = 0	X = 1
S_1(A)	S_2, 0	S_3, 1
S_2(B)	S_1, 0	S_4, 1
S_3(C)	S_4, 1	S_4, 1
S_4(DE)	S_4, 1	S_2, 1

4. Find the equivalent partition of the following machine.

Present State	Next State, z	
	X = 0	X = 1
A	B, 0	D, 1
B	D, 1	F, 1
C	F, 1	B, 1
D	F, 0	A, 1
E	C, 0	A, 1
F	C, 1	B, 1

Solution: The partitions are

P_0 = (ABCDEF)
P_1 = (ADE)(BCF) (Depending on o/p for i/p 0)
P_2 = (ADE)(B)(CF) (For i/p 0, the next state of B goes to another set)
P_3 = (A)(DE)(B)(CF) (For i/p 0, the next state of A goes to another set)
P_4 = (A)(DE)(B)(CF)

As P_3 and P_4 are the same, P_3 is the equivalent partition.

Minimization: We know that the equivalent partition is unique. So, P_4 = (A)(DE)(B)(CF) is the unique combination. Here, every single set represents one state of the minimized machine.

Let us rename these partitions for simplification.
Rename (A) as S_1, (B) as S_2, (DE) as S_3, and (CF) as S_4.
The minimized machine becomes

Present State	Next State, z	
	X = 0	X = 1
S_1(A)	S_2, 0	S_3, 1
S_2(B)	S_3, 1	S_4, 1
S_3(DE)	S_4, 0	S_1, 1
S_4(CF)	S_4, 1	S_2, 1

5. Find the equivalent partition of the following machine.

Present State	Next State, z	
	X = 0	X = 1
A	C, 1	D, 0
B	D, 1	E, 0
C	B, 1	E, 1
D	B, 1	A, 0
E	D, 1	B, 1

Solution: The partitions are

P_0 = (ABCDEF)
P_1 = (AD)(BCE) (Depending on o/p for i/p 1)
P_2 = (AD)(BE)(C) (For i/p 0, the next state of C goes to another set)
P_3 = (A)(D)(BE)(C) (For i/p 0, the next state of A and D goes to a different set)
P_4 = (A)(D)(BE)(C)

As P_3 and P_4 are the same, P_3 is the equivalent partition.

Minimization: We know that the equivalent partition is unique. So, P_4 = (A)(D)(BE)(C) is the unique combination. Here, every single set represents one state of the minimized machine.

Let us rename these partitions for simplification.

Rename (A) as S_1, (BE) as S_2, (C) as S_3, and (D) as S_4.

The minimized machine becomes

Present State	Next State, z	
	X = 0	X = 1
S_1(A)	S_3, 1	S_4, 1
S_2(BE)	S_4, 1	S_2, 0
S_3(C)	S_2, 1	S_2, 1
S_4(D)	S_2, 1	S_1, 0

6. Simplify the following incompletely specified machine.

Present State	Next State, Z		
	I_1	I_2	I_3
A	D, 1	E, 1	__
B	B, 0	E, _	C, _
C	C, _	C, 0	B, _
D	B, 0	D, _	E, _
E	__	B, 0	A, _

Solution: Put a temporary state T in the next state place, where the next states are not specified. If the output is not mentioned, there is no need to put any output.

As a temporary state T is considered, T is put in the present state column with the next state T for all inputs with no output.

The simplified machine becomes

Present State	Next State, z		
	I_1	I_2	I_3
A	D, 1	E, 1	T, _
B	B, 0	E, _	C, _
C	C, _	C, 0	B, _
D	B, 0	D, _	E, _
E	T, _	B, 0	A, _
T	T, 0	T, 0	T, 0

7. Minimize the following incompletely specified machine.

Present State	Next State, z		
	I_1	I_2	I_3
A	A, 1	D, _	C, _
B	A, _	D, _	E, _
C	E, 0	A, 1	_, _
D	E, _	A, 1	_, _
E	E, 0	_, _	C, _

Solution: In an incompletely specified machine, all the next states or all the outputs or both are not mentioned. We can minimize an incompletely specified machine by using the merger graph and compatible graph method. From the merger graph, we have to find the compatible pair and from that compatible pair and implied pair we have to construct compatible graph. From the compatible graph, we have to find the closed partition. And, the closed partitions are the minimized states of the machine.

Merger Graph: The machine consists of five states. So, the merger graph consists of five nodes, named A, B, C, D, and E. The outputs of A and B do not differ, and so there is an arc between A and B. For input I_3, the next state conflicts—so the arc is an interrupted arc and in the interrupted portion the conflicting next state pair (CE) is placed.

For states A and C, the output conflicts, and so there is no arc between A and C.

For the states C and D, the outputs as well as next states do not conflict. So, an uninterrupted arc is placed between C and D.

By this process, the merger graph is constructed. The merger graph is as follows

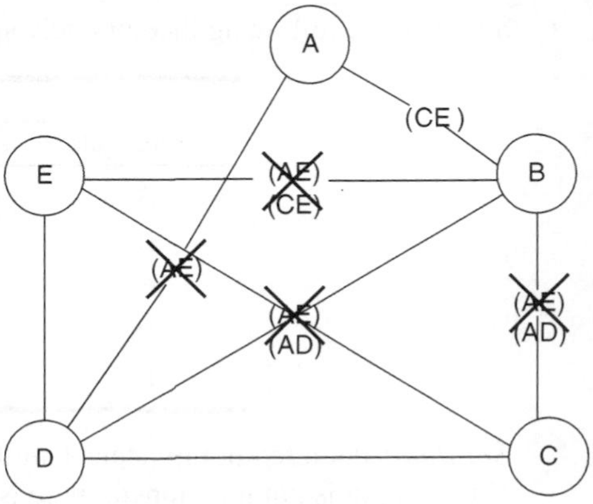

As there is no arc between A and E, the arc between (AD), (BE), (BD), and (BC) are also crossed.

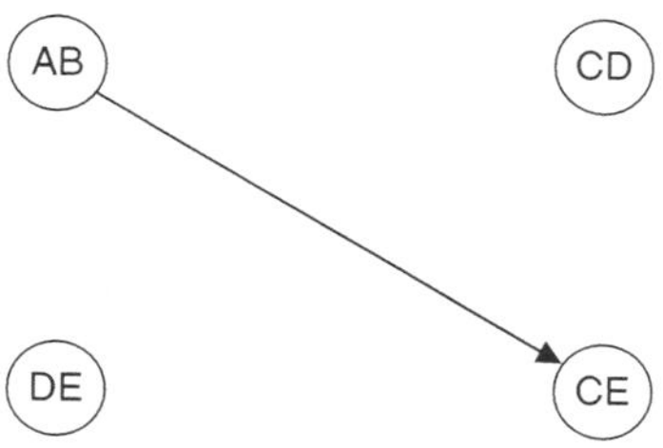

Compatible Graph: (CE) is the implied pair for (AB). A directed arc is drawn from (AB) to (CE). The compatible graph consists of four vertices (AB), (CD), (DE), and (CE).

A subgraph of a compatibility graph is said to be closed cover for the machine, if for every vertex in the subgraph, all outgoing edges and their terminal vertices also belong to the subgraph, and every state of the machine is covered by at least one vertex of the subgraph.

For the constructed compatible graph, (AB), (CD), and (CE) form a closed covering. The states of the minimized machine are (AB) and (CDE). Rename (AB) as S_1 and (CDE) as S_2.

The minimized machine is

Present State	Next State, z		
	I_1	I_2	I_3
S_1	S_1, 1	S_2, –	S_2, –
S_2	S_2, 0	S_1, 1	S_2, –

8. Construct a compatible graph for the following incompletely specified machine.

Present State	Next State, z			
	I_1	I_2	I_3	I_4
A	A, O_1	E, O_2	–, –	A, O_2
B	–, –	C, O_3	B, O_1	D, O_4
C	A, O_1	C, O_3	–, –	–, –
D	A, O_1	–, –	–, –	D, O_4
E	–, –	E, O_2	F, O_1	–, –
F	–, –	G, O_3	F, O_1	G, O_4
G	A, O_1	–, –	–, –	G, O_4

Solution: To construct a compatible graph, we have to first find compatible pairs. To find the compatible pairs, we need to construct a merger graph. But this machine has seven states, and so it is difficult to construct a merger graph for this machine. So, we have to construct a merger table to find compatible pairs.

Merger Table

B	×					
C	×	√				
D	×	√	√			
E	√	×	×	√		
F	×	(CG)(DG)	(CG)	(DG)	×	
G	×	(DG)	√	√	√	√
	A	B	C	D	E	F

Compatible pairs are (AE), (BC). (BD), (CD), (CG), (DE), (DG), (EG), (GF), (BF), (BG), (CF) and (DF). If (CG) and (DG) are compatible, then (BF) is compatible. If (DG) is compatible, then (BG) is compatible. If (CG) is compatible, then (CF) is compatible. If (DG) is compatible, then (DF) is compatible.

Compatible Graph

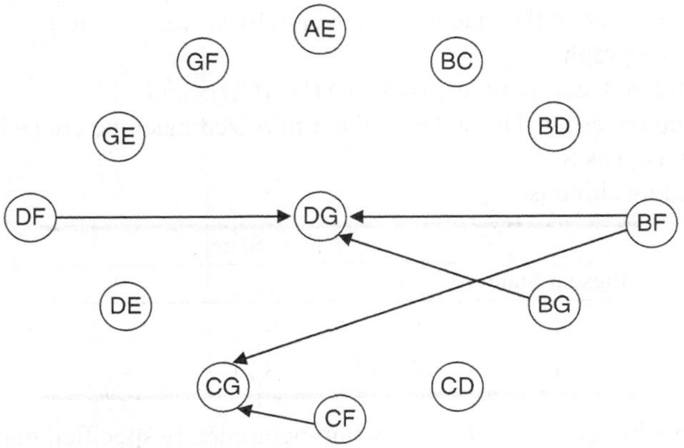

The closed partitions are (AE), (BF), (BG), (CF), (CG), (DF), and (DG). The states of the minimized machine are (AE), (BCDFG). Rename them as S_1, S_2.

The minimized machine is

Present State	Next State, z			
	I_1	I_2	I_3	I_4
S_1	S_1, O_1	S_1, O_2	S_2, O_1	S_1, O_2
S_2	S_1, O_1	S_2, O_3	S_2, O_1	S_2, O_4

9. Consider the following machine M_1

Present State	Next State, z			
	I_1	I_2	I_3	I_4
A	–	C, 1	E, 1	B, 1
B	E, 0	F, 1	–	–
C	F, 0	F, 1	–	–
D	–	–	B, 1	–
E	–	F, 0	A, 0	D, 1
F	C, 0	–	B, 0	C, 1

a) Construct a merger table for M_1.
b) Find the set of compatibles.
c) Draw a compatibility graph for M_1. Describe the procedure used by you.
d) Obtain a closed covering of M_1.
e) Construct a minimized machine M_1^* of M_1. [WBUT 2004 [IT]]

Solution:

a) Merger table for M_1

B	√				
C	(FC)	(EF)			
D	(BE)	√	√		
E	×	√	×	×	
F	×	(CE) ×	√	×	(AB)(CD)
	A	B	C	D	E

(BF) will be crossed as (CE) is crossed.

b) The set of compatible pairs are (AB), (AC), (AD), (BC), (BD), (BE), (CD), (CF), and (EF)

c) (CF) is the implied pair for (AC)
(BE) is the implied pair for (AD)
(EF) is the implied pair for (BC)
(AB) and (CD) are the implied pair for (EF)
The compatible graph is

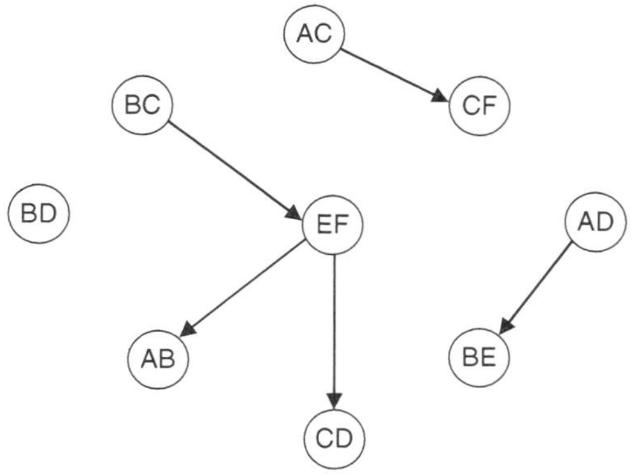

d) In the compatible graph, the closed coverings are [(AB)(CD)(EF)], [(AB)(BC)(CD)(EF)], [(AC)(CF)(AD)(BE)], etc. Among these, the minimum is [(AB)(CD)(EF)].

e) Let us rename the states as S_1 (AB), S_2 (CD), and S_3 (EF).
The minimized machine M_1^* is

	Next State, z			
Present State	I_1	I_2	I_3	I_4
S_1	S_3, 0	S_2, 1	S_3, 1	S_1, 1
S_2	S_3, 0	S_3, 1	S_1, 1	–
S_3	S_2, 0	S_3, 0	S_1, 0	S_2, 1

10. Test whether the following machine is finite or not.

Present State	Next State, O/P	
	X = 0	X = 1
A	A, 0	B, 1
B	C, 0	D, 1
C	B, 1	A, 0
D	D, 1	C, 0

Solution:

Testing Table–Testing Graph Method: A table which is divided into two halves is made. The machine has two inputs and two outputs. There are four input–output combinations namely 0/0, 0/1, 1/0, and 1/1. The upper half of the machine contains single state input–output combination and the lower half contains two state input–output combinations. There are four states, and so six combination pairs are made. The testing table is as follows:

Present State	0/0	0/1	1/0	1/1
A	A	–	–	B
B	C	–	–	D
C	–	B	A	–
D	–	D	C	–
AB	AC	–	–	BD
AC	–	–	–	–
AD	–	–	–	–
BC	–	–	–	–
BD	–	–	–	–
CD	–	BD	AC	–

In the testing table, there are six present states combinations. So, in the testing table, there are six nodes. There is a directed arc with a label of input output combination, from S_iS_j [$i \neq j$] to S_pS_q [$p \neq q$], if S_pS_q is the implied pair of S_iS_j. The testing graph for finite memory is

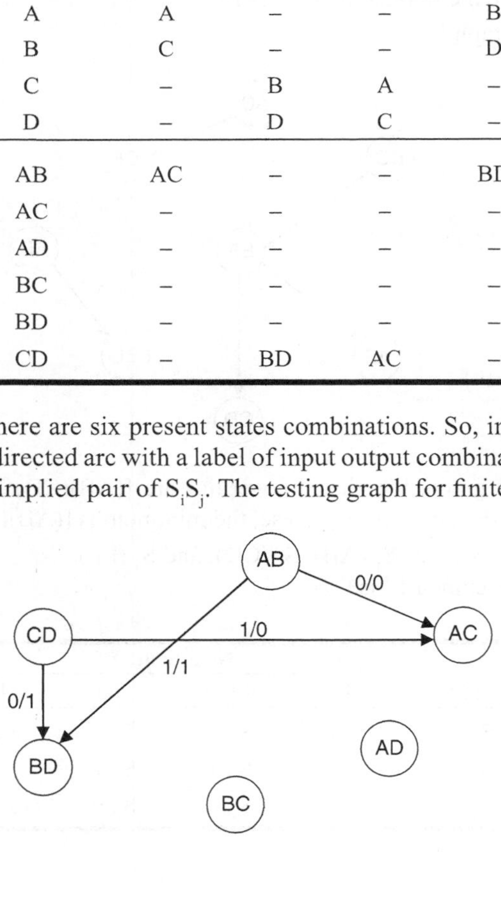

The testing graph is loop-free. So, the machine is finite. The longest path in the testing graph is 1. The order of finiteness of the machine $\mu = 1 + 1 = 2$.

This can be proved by the vanishing connection matrix method.

In the testing table, six present state pairs are formed. Therefore, the connection matrix consists of six rows and six columns. The rows and columns are labelled with the pair of present state combinations.

In the matrix, the (i, j)th entry will be 1 if there is an entry in (S_aS_b) $(a \neq b)$ and (S_pS_q) $(p \neq q)$ combination in the corresponding testing table. Otherwise, the entry will be 0.

The connection matrix of the above machine is as follows.

	AB	AC	AD	BC	BD	CD
AB	0	1	0	0	1	0
~~AC~~	0	0	0	0	0	0
~~AD~~	0	0	0	0	0	0
~~BC~~	0	0	0	0	0	0
~~BD~~	0	0	0	0	0	0
CD	0	1	0	0	1	0

In the above matrix, the rows AC, AD, BC, and BD contain all '0'. So, the rows labelled AC, AD, BC, and BD and the corresponding columns vanish.

The remaining matrix becomes

	AB	CD
~~AB~~	0	0
~~CD~~	0	0

In the above matrix, the rows AB and CD contain all '0'. So, the rows labelled AB and CD and the corresponding columns vanish.

The matrix vanishes. The number of steps required to vanish the matrix is 2. So, the order of finiteness of the machine $\mu = 2$.

11. Test whether the following machine is finite or not.

Present State	Next State, O/P	
	X = 0	X = 1
A	B, 1	C, 0
B	A, 0	D, 1
C	B, 0	A, 0
D	C, 1	A, 1

Solution: There are four states, and so six combination pairs are made. The testing table is as follows.

Present State	0/0	0/1	1/0	1/1
A	–	B	C	–
B	A	–	–	D
C	B	–	A	–
D	–	C	–	A
AB	–	–	–	–
AC	–	–	AC	–
AD	–	BC	–	–
BC	AB	–	–	–
BD	–	–	–	AD
CD	–	–	–	–

In the testing table, there are six present states combinations. So, in the testing table, there are six nodes. There is a directed arc with a label of input–output combination, from S_iS_j $[i \neq j]$ to S_pS_q $[p \neq q]$, if S_pS_q is the implied pair of S_iS_j. The testing graph for finite memory is

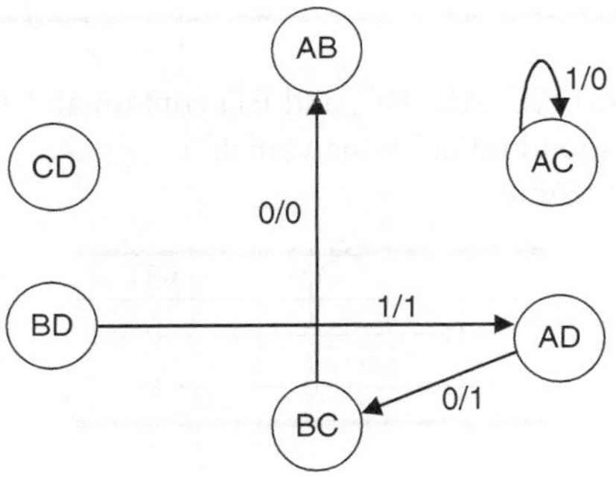

In the testing graph, there is a loop from AC to AC with label 1/0. The testing graph is not loop-free. So, the machine is not of finite memory.

This can be proved by the vanishing connection matrix method.

In the testing table, six present state pairs are formed. Therefore, the connection matrix consists of six rows and six columns. The rows and columns are labelled with the pair of present state combinations.

The connection matrix of the previous machine is as follows.

	AB	AC	AD	BC	BD	CD
AB	0	0	0	0	0	0
AC	0	1	0	0	0	0
AD	0	0	0	1	0	0
BC	1	0	0	0	0	0
BD	0	0	1	0	0	0
CD	0	0	0	0	0	0

In the previous matrix, the rows AB and CD contain all '0'. So, the rows labelled AB and CD and the corresponding column vanish.

The remaining matrix becomes

	AC	AD	BC	BD
AC	1	0	0	0
~~AD~~	~~0~~	~~0~~	~~1~~	~~0~~
BC	0	0	0	0
BD	0	1	0	0

In the previous matrix, the row AD contains all '0'. So the row labelled AD and the corresponding column vanish.

The remaining matrix becomes

	AC	BC	BD
AC	1	0	0
~~BC~~	~~0~~	~~0~~	~~0~~
~~BD~~	~~0~~	~~0~~	~~0~~

In the previous matrix, the rows BC and BD contain all '0'. So, the rows labelled BC and BD and the corresponding column vanish.

The remaining matrix becomes

	AC
AC	1

The matrix does not vanish. So, the machine is not of finite memory.

12. Test whether the following machine is finite or not.

	Next State, O/P	
Present State	X = 0	X = 1
A	B, 0	A, 1
B	D, 1	C, 0
C	C, 0	D, 1
D	A, 1	B, 0

Soution: There are four states, and so six combination pairs are made. The testing table is as follows.

Present State	0/0	0/1	1/0	1/1
A	B	–	–	A
B	–	D	C	–
C	C	–	–	D
D	–	A	B	–
AB	–	–	–	–
AC	(BC)	–	–	(AD)
AD	–	–	–	–
BC	–	–	–	–
BD	–	(AD)	(BC)	–
CD	–	–	–	–

In the testing table, there are six present states combinations. So, in the testing table, there are six nodes. The testing graph for finite memory is

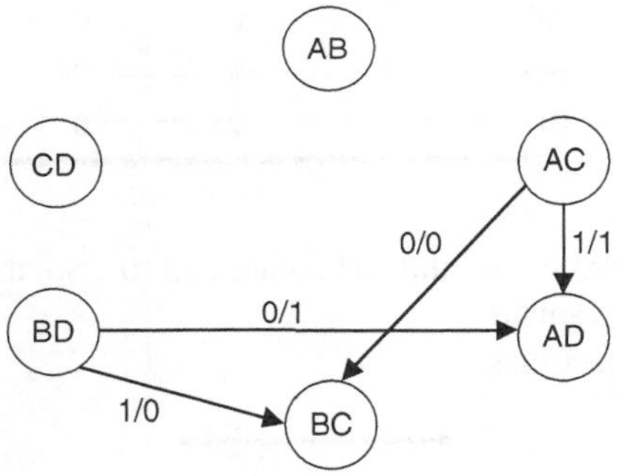

The testing graph is loop-free. So, the machine is finite. The longest path in the testing graph is 1. The order of finiteness of the machine $\mu = 1 + 1 = 2$

Proof by Vanishing Connection Matrix: In the testing table, six present state pairs are formed. Therefore, the connection matrix consists of six rows and six columns. The rows and columns are labelled with the pair of present state combinations.

The connection matrix of the previous machine is as follows.

	AB	AC	AD	BC	BD	CD
AB	0	0	0	0	0	0
AC	0	0	1	1	0	0
AD	0	0	0	0	0	0
BC	0	0	0	0	0	0
BD	0	0	1	1	0	0
CD	0	0	0	0	0	0

In the above matrix, the rows AB, AD, BC, and CD contain all ‘0’. So, the rows labelled AB, AD, BC, and CD and the corresponding column vanish.

The remaining matrix becomes

	AC	BD
~~AC~~	~~0~~	~~0~~
~~BD~~	~~0~~	~~0~~

The matrix vanishes. So, the machine is of finite memory. The connection matrix vanishes in two steps. So, the order of finiteness of the machine $\mu = 2$.

13. Find whether the following machine is definite or not.

	Next State, O/P	
Present State	X = 0	X = 1
A	D	E
B	A	B
C	C	B
D	C	B
E	A	B

Solution: There are three methods to check whether a machine is definite or not:

i) Synchronizing tree method
ii) Contracted table method
iii) Testing table–Testing graph method

(i) *Synchronizing Tree Method:* (ABCDE) is the present state combination of the machine. It has two types of inputs, 0 and 1. (ABCDE) with input 0 produces the next state combination (DACCA). It can be grouped into two distinct states (ACD). (ABCDE) with input 1 produces the next state combination (EBBBB) which can be grouped into (BE).

In the next level, (ACD) with input 0 produces the next state combination (CD). With input 1, (ACD) produces the next state combination (BE). The state combination (BE) with input 0 produces the single next state (A). With input 1, the state combination (BE) produces the single state B. As A and B are single states, it need not be derived again. By this process, (CD) and (BE) are also derived for input 0 and 1. The synchronizing tree is

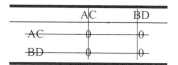

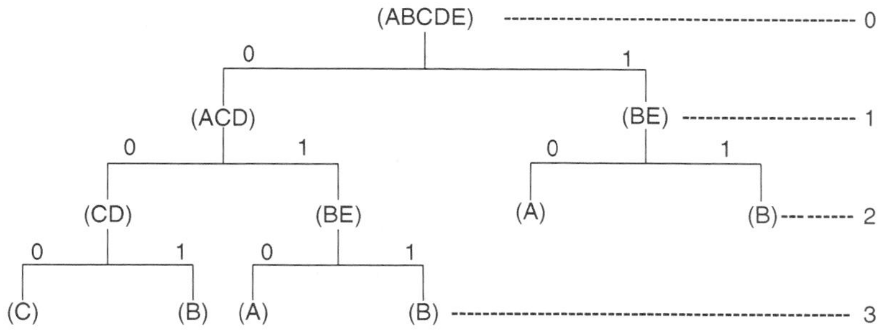

As the leaf nodes are of single state, stop constructing the tree. The order of definiteness is the maximum label of the synchronizing tree. In this case, the order is 3.

(ii) *Contracted Table Method:* In the previous machine, C and D have the same next state combination (both cases C and B). B and E also have the same next state combination (both cases C and B) Therefore, C and D are equivalent states and B and E are also equivalent states. Among C and D, let us take only C, and among B and E let us take B. That is, all the D in the state table are replaced by C and all the E in the state table are replaced by B. The original and contracted table is

Present State	Next State, O/P X = 0	X = 1
A	D	E
B	A	B
C	C	B
D	C	B
E	A	B

↓

Present State	Next State, O/P X = 0	X = 1
A	C	B
B	A	B
C	C	B

In the previous machine, A and C have the same next state combination (both cases C and B). Therefore, A and C are equivalent states. Among A and C, let us take only A. That is, all the C in the state table are replaced by A. The contracted table becomes

Present State	Next State, O/P X = 0	X = 1
A	A	B
B	A	B

In the previous machine, A and B have the same next state combination (both cases A and B). Therefore, A and B are equivalent states. Among A and B, let us take only A. That is, all the B in the state table are replaced by A. The contracted table becomes

Present State	Next State, O/P X = 0	X = 1
A	A	A

The obtained table consists of a single row. So, the machine is definite. Its order is the number of steps required to obtain the single state machine, and here it is 3.

(iii) ***Testing Table–Testing Graph Method:*** A table with two halves is made. The upper half contains the input and next state combination. The lower half of the table contains all the combination of present states taking two into combination. Here, for five present states (A, B, C, D, E), there are $^5C_2$, which is 10, combinations. The table is

Present State	0	1
A	D	E
B	A	B
C	C	B
D	C	B
E	A	B
AB	AD	BE
AC	CD	BE
AD	CD	BE
AE	AD	BE
BC	AC	BB
BD	AC	BB
BE	AA	BB
CD	CC	BB
CE	AC	BB
DE	AC	BB

The testing graph is

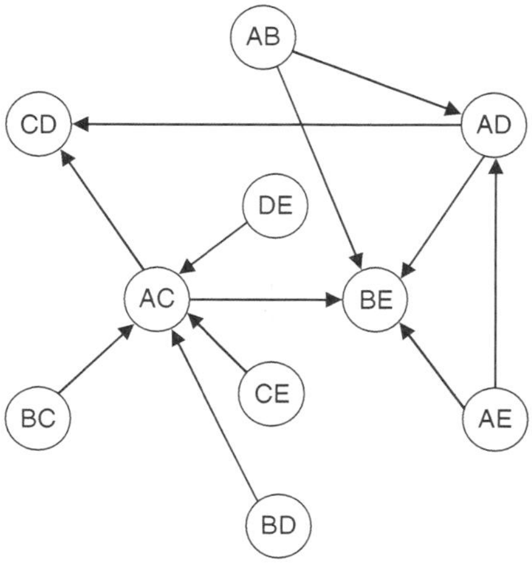

The testing graph is loop-free. So, the machine is of definite memory. The longest path of the testing graph is 2 (AE → AD → BE or AB → AD → BE). So, the order of definiteness of the machine is 2 + 1 = 3.

(We can use any of the methods to check definiteness of the machine. Among them, the contracted table method is the best. The synchronizing tree method is good, but if the machine is not definite the leaf nodes of the synchronizing tree will not become single state. And the construction will be an endless process. If the number of states of the machine is large, then the number of nodes will also be large. It will be difficult to construct the testing table and testing graph.)

14. Find whether the following machine is definite or not.

Present State	Next State, O/P X = 0	Next State, O/P X = 1
A	C	B
B	E	F
C	A	F
D	E	B
E	C	D
F	E	F

Solution: There are three ways to check whether a machine is definite or not. But among them, the contracted table method is the best.

In the previous machine, B and F produce the same next state (both cases E and F). So, B and F are equivalent states. Take any of them—let us take B. That is, all F in the state table is replaced by B. The contracted table from the original one is

Present State	Next State, O/P X = 0	Next State, O/P X = 1
A	C	B
B	E	F
C	A	F
D	E	B
E	C	D
F	E	F

↓

Present State	Next State, O/P X = 0	Next State, O/P X = 1
A	C	B
B	E	B
C	A	B
D	E	B
E	C	D

In the previous state table, B and D produce the same next states (both cases E and B). So, they are equivalent states. Let us take only B and replace all the occurrences of D by B. The contracted table becomes

Present State	Next State, O/P	
	X = 0	X = 1
A	C	B
B	E	B
C	A	B
E	C	B

In the previous state table, A and E are equivalent states as they produce the same next state combination (both cases C and B). Let us take only A and replace all occurrences of E in the table by A. The contracted table is

Present State	Next State, O/P	
	X = 0	X = 1
A	C	B
B	A	B
C	A	B

In the earlier state table, B and C are equivalent states as they produce the same next state combination for input 0 and 1. Let us take only B and replace all the occurrences of C in the table by B. The contracted table becomes

Present State	Next State, O/P	
	X = 0	X = 1
A	B	B
B	A	B

The state table does not become a single row table. So, the machine is not definite.

15. Find whether the following machine is definite or not.

Present State	Next State, O/P	
	X = 0	X = 1
A	B	C
B	D	A
C	B	E
D	D	E
E	B	E
F	B	C

Solution: The contracted table method is used to check whether the given machine is definite or not. In the previous machine, the present states C and E produce the same next state combination for input 0 and 1. In the same way, A and F also produce the same next state combination (both cases B and C). So both C and E, and A and F are equivalent state. Keep C and remove E from the present state column and replace all occurrences of E in the next state column by C. In the same way, keep A and remove F from the present state column and replace all occurrences of F in the next state column by A. The contracted table becomes

	Next State, O/P	
Present State	X = 0	X = 1
A	B	C
B	D	A
C	B	C
D	D	C

In the previous table, A and C are equivalent states as they produce the same next state combination for input 0 and 1. Remove C from the present state column and replace all occurrences of C in the next state portion by A. The contracted table becomes

	Next State, O/P	
Present State	X = 0	X = 1
A	B	A
B	D	A
D	D	A

In the previous state table, B and D are equivalent states as they produce the same next state combination for input 0 and 1. Take B, remove D, and replace all occurrences of D in the next state portion by B. The contracted table becomes

	Next State, O/P	
Present State	X = 0	X = 1
A	B	A
B	B	A

In the previous state table, A and B are equivalent states as they produce the same next state combination for input 0 and 1. Replace B by A and remove B as a present state. The contracted table is

	Next State, O/P	
Present State	X = 0	X = 1
A	A	A

The obtained table consists of a single row. So, the machine is definite. Its order is the number of steps required to obtain the single state machine, and here it is 4.

16. Test whether the following machine is information lossless or not. If lossless, find its order.

Present State	Next State, O/P	
	X = 0	X = 1
A	B, 1	A, 1
B	D, 0	B, 0
C	D, 0	B, 1
D	E, 1	C, 1
E	A, 1	B, 0

Solution: The first step to test whether a machine is lossless or not is to construct a testing table. The testing table is divided into two halves.

Present State	Next State	
	z = 0	z = 1
A	–	(AB)
B	(BD)	–
C	D	B
D	–	(CE)
E	B	A
(AB)	–	–
(BD)	–	–
(CE)	(BD)	(AB)

The testing table does not contain any repeated entry. The machine is an information lossless machine.

The testing graph for the machine is

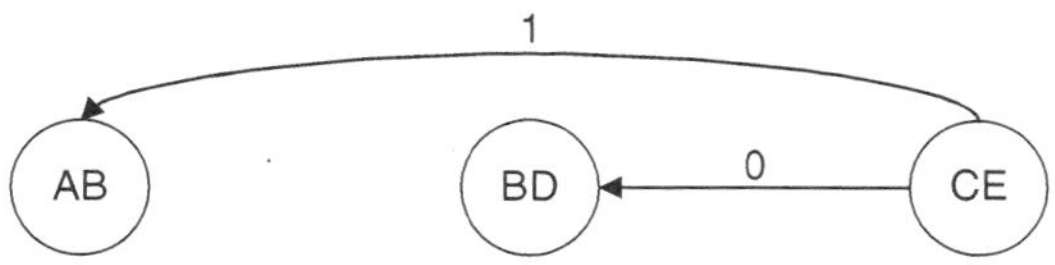

The testing graph for information losslessness is loop-free. The order of losslessness is $\mu = 1 + 2 = 3$. The length of the longest path of the graph is 1.

17. Test whether the following machine is information lossless or not. If lossless, find its order.

Present State	Next State, O/P	
	X = 0	X = 1
A	B, 1	C, 1
B	D, 0	E, 0
C	A, 1	F, 1
D	C, 0	B, 0
E	F, 1	A, 1
F	E, 0	D, 0

Solution: First, we have to construct a testing table for information losslessness for testing whether the machine is information lossless or not. The testing table is divided into two halves.

Present State	Next State	
	z = 0	z = 1
A	–	BC
B	DE	–
C	–	AF
D	BC	–
E	–	AF
F	DE	–
AF	–	–
BC	–	–
DE	–	–

The testing table does not contain any repeated entry. The machine is an information lossless machine.

The testing graph for the machine is

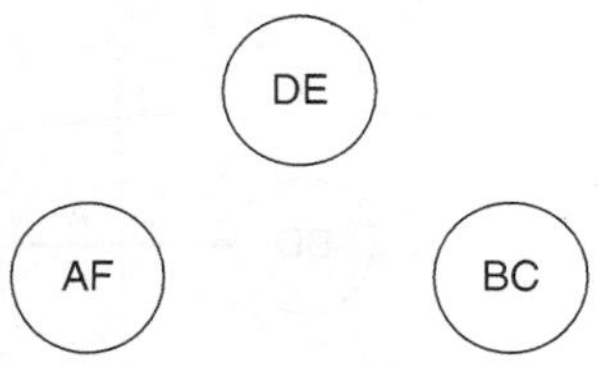

The testing graph for information losslessness is loop-free. The order of losslessness is $\mu = 0 + 2 = 2$. The length of the longest path of the graph is 0.

18. Test whether the following machine is information lossless or not. If lossless, find its order.

Present State	Next State, O/P	
	X = 0	X = 1
A	B, 0	E, 0
B	E, 0	D, 0
C	D, 1	A, 0
D	C, 1	E, 0
E	B, 0	D, 0

Solution: First, we have to construct a testing table for information losslessness for testing whether the machine is information lossless or not. The testing table is divided into two halves.

Present State	Next State	
	z = 0	z = 1
A	BE	–
B	DE	–
C	A	D
D	E	C
E	BD	–
BD	(DE)(EE)	–
BE	(BD)(DE)	–
DE	(BE)(DE)	–

The testing table contains repeated entry (EE). Therefore, the machine is a lossy machine.

19. Test whether the following machine is information lossless or not. If lossless, find its order.

Present State	Next State, O/P	
	X = 0	X = 1
A	B, 1	C, 0
B	B, 1	D, 1
C	E, 1	B, 0
D	A, 0	E, 0
E	F, 0	D, 1
F	A, 1	D, 0

Solution: First, we have to construct a testing table for information losslessness for testing whether the machine is information lossless or not. The testing table is divided into two halves.

Present State	Next State	
	z = 0	z = 1
A	C	B
B	–	(BD)
C	B	E
D	(AE)	–
E	F	D
F	D	A
AE	CF	BD
BD	–	–
CF	BD	AE

The testing table does not contain any repeated entry. The machine is an information lossless machine.

The testing graph for the machine is

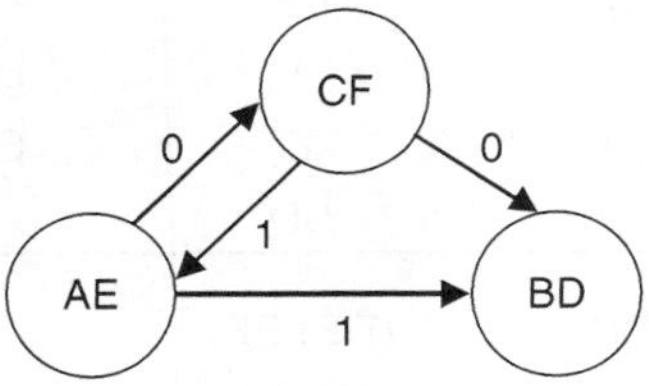

The testing graph contains a loop. So, the machine is not information lossless of finite order. The order of losslessness $\mu = 2 + 2 = 4$. The length of the longest path of the graph is 2.

20. The state table of a finite state machine M is as follows. Check whether the machine M is information lossless. If it is information lossless, then determine the order of losslessness.

[WBUT 2003]

Present State	Next State, O/P	
	X = 0	X = 1
A	A, 0	B, 0
B	C, 0	D, 0
C	D, 1	C, 1
D	B, 1	A, 0

Solution: First, we have to construct a testing table for information losslessness for testing whether the machine is information lossless or not. The testing table is

Present State	Next State	
	z = 0	z = 1
A	(AB)	–
B	(CD)	–
C	–	(CD)
D	–	(AB)
(AB)	(AC)(AD) (BC)(BD)	–
(CD)	–	(AC)(AD) (BC)(BD)
(AC)	–	–
(AD)	–	–
(BC)	–	–
(BD)	–	–

The testing table does not contain any repeated entry. The machine is an information lossless machine.

To find the order of losslessness, a testing graph needs to be constructed.

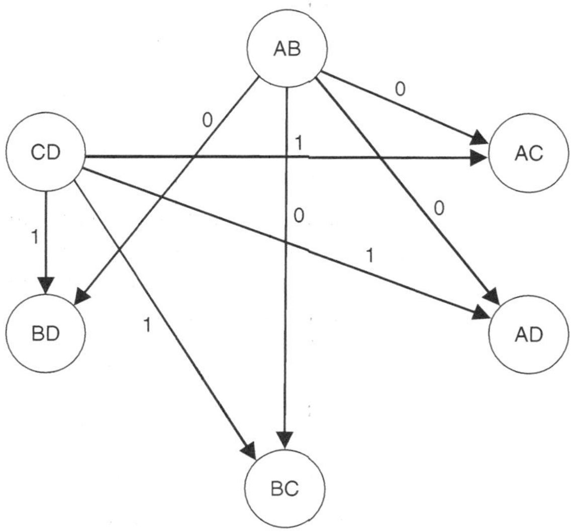

The testing graph does not contain any loop. The order of losslessness is $\mu = 1 + 2 = 3$. The length of the longest path of the graph is 1.

21. Find the input string which is applied on state 'C' producing the output string 1110000010 and the final state 'B' for the following machine.

Present State	Next State X = 0	Next State X = 1
A	E, 1	F, 0
B	D, 0	C, 1
C	F, 1	A, 0
D	C, 0	E, 0
E	B, 0	D, 1
F	D, 1	F, 1

Solution: First, we need to prove that the machine is information lossless. For this, we need to construct a testing table for information lossless. If the machine is information lossless, then only a single input string can be found for a single beginning state and single final state.

The testing table for information lossless is

Present State	Next State z = 0	Next State z = 1
A	F	E
B	D	C
C	A	F
D	(CE)	–
E	B	D
F	–	(DF)
CE	AB	DF
AB	DF	CE
DF	–	–

The testing table does not contain any repeated entry. The machine is an information lossless machine.

The output successor table for the given machine is

Present State	Next State, I/P z = 0	Next State, I/P z = 1
A	F, 1	E, 0
B	D, 0	C, 1
C	A, 1	F, 0
D	(C, 0) (E, 1)	–
E	B, 0	D, 1
F	–	(D, 0) (F, 1)

The input string is applied on state C and has produced output 1. From the output successor table, it is clear that the next states are F with input 0. By this process, the transition is like the following

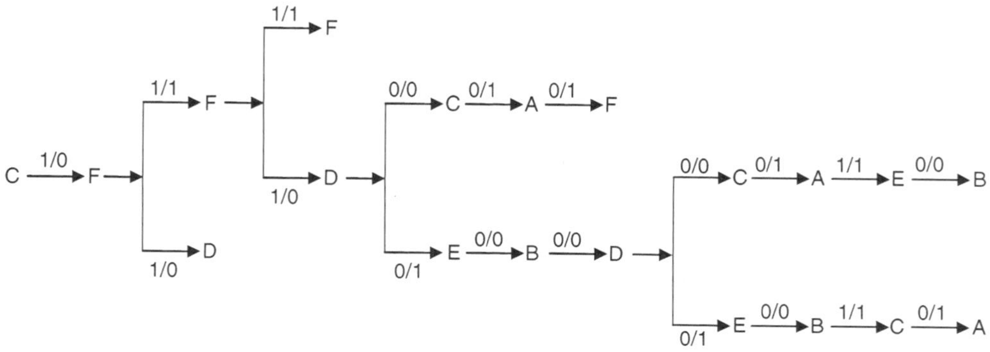

The beginning state C and the final state B are obtained from one path with the input string 0101000110.

22. Find the input string which is applied on state 'D' producing the output string 10011110 and the final state 'D' for the following machine.

Present State	Next State, Z	
	X = 0	X = 1
A	A, 1	C, 1
B	E, 0	B, 1
C	D, 0	A, 0
D	C, 0	B, 0
E	B, 1	A, 0

Solution: First, we need to prove that the machine is information lossless. For this, we need to construct a testing table for information lossless. If the machine is information lossless, then only a single input string can be found for a single beginning state and single final state. The testing table for information lossless is

Present State	Next State	
	z = 0	z = 1
A	–	(AC)
B	E	B
C	(AD)	–
D	(BC)	–
E	A	B

AC	–	–
AD	–	–
BC	(AE)(DE)	–
AE	–	(AB)(BC)
DE	(AB)(AC)	–
AB	–	(AB)(BC)

The testing table does not contain any repeated entry. The machine is an information lossless machine. The output successor table for the given machine is

Present State	Next State, I/P	
	z = 0	z = 1
A	–	(A, 0), (C, 1)
B	E, 0	B, 1
C	(D, 0), (A, 1)	–
D	(C, 0), (B, 1)	–
E	A, 1	B, 0

The transition is like the following

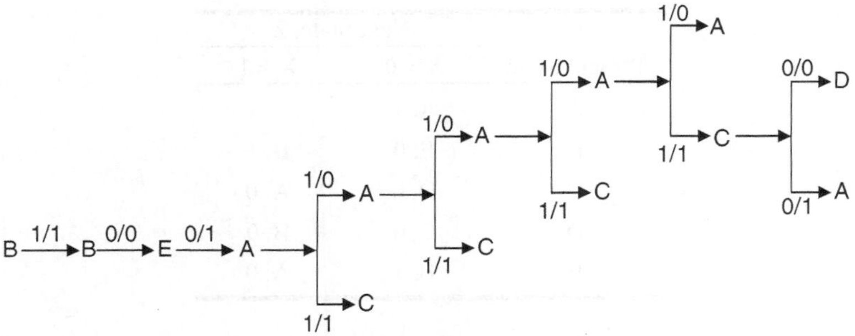

The beginning state B and the final state D are obtained from one path with the input string 10100010.

23. Retrieve the input sequence from the machine when it was initially in state B, has, in response to yet unknown input sequence, produced the output sequence 01110, and terminated in state B. [WBUT 2003]

Present State	Next State	
	X = 0	X = 1
A	A, 0	B, 0
B	C, 0	D, 0
C	D, 1	C, 1
D	B, 1	A, 1

Solution: To retrieve the input sequence, first we need to check whether the machine is information lossless or not. The machine is information lossless as given in Example 4.20.

The output successor table for the given machine is

Present State	Next State, I/P	
	z = 0	z = 1
A	(A, 0), (B, 1)	–
B	(C, 0), (D, 1)	–
C	–	(D, 0), (C, 1)
D	–	(B, 0), (A, 1)

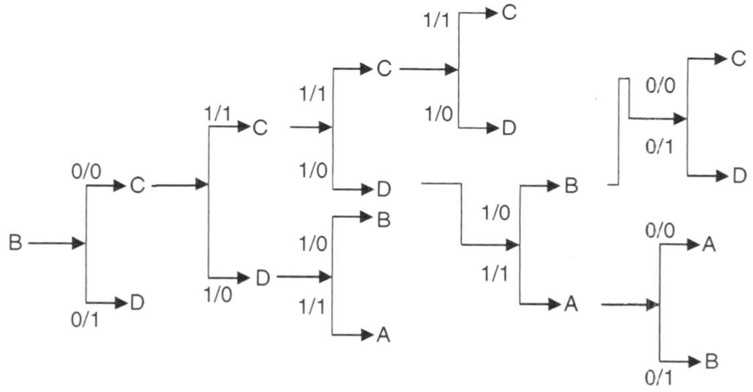

The final state is B, and thus the input is 01011.

24. Design a minimal inverse of the machine shown in the following table.

[WBUT 2007, 2008]

Present State	Next State	
	X = 0	X = 1
A	C, 0	D, 1
B	D, 0	C, 0
C	A, 1	B, 1
D	C, 1	D, 1

Solution: To find the minimal inverse machine, first we need to find the order of losslessness of the machine.

Present State	z = 0	z = 1
A	C	D
B	D	C
C	AB	–
D	–	CD
AB	CD	CD
CD	–	–

The testing graph is

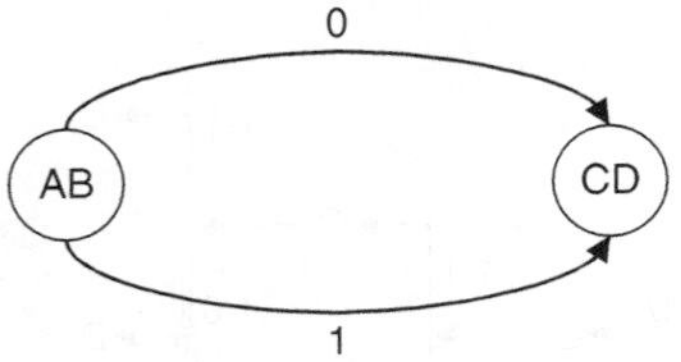

It is loop-free. And the order of finiteness is 3.

Therefore, if we know the initial state and the values of three successive outputs produced by transitions from the initial state, we can determine the first input given to the machine. We need to determine a set of triples, denoted by {S(t), z(t + 1), and z(t + 2)}. There are four states of the machine. The triples are

(A, 0, 0)	(B, 0, 1)	(C, 0, 0)	(D, 1, 0)
(A, 1, 1)	(B, 1, 0)	(C, 0, 1)	(D, 1, 1)

	Next State, X	
Present State	z = 0	z = 1
(A, 0, 0)	(C, 0, 0), 0	(C, 0, 1), 0
(A, 1, 1)	(D, 1, 0), 1	(D, 1, 1), 1
(B, 0, 1)	(D, 1, 0), 0	(D, 1, 1), 0
(B, 1, 0)	(C, 0, 0), 1	(C, 0, 1), 1
(C, 0, 0)	(A, 0, 0), 0	(B, 0, 1), 1
(C, 0, 1)	(B, 1, 0), 1	(A, 1, 1), 0
(D, 1, 0)	(C, 0, 0), 0	(C, 0, 1), 0
(D, 1, 1)	(D, 1, 0), 1	(D, 1, 1), 1

Here, states (A, 0, 0), (D, 1, 0) and (A, 1, 1) and (D, 1, 1) are equivalent states as they produce the same next state and same output. Let us assign (A, 0, 0) as S_1, (A, 1, 1) as S_2, (B, 0, 1) as S_3, (B, 1, 0) as S_4, (C, 0, 0) as S_5 and (C, 0, 1) as S_6.

The minimal inverse machine is

Present State	Next State, X	
	z = 0	z = 1
S_1	S_5, 0	S_6, 0
S_2	S_1, 1	S_2, 1
S_3	S_1, 0	S_2, 0
S_4	S_5, 1	S_6, 1
S_5	S_1, 0	S_3, 1
S_6	S_4, 1	S_2, 0

Multiple Choice Questions

1. A palindrome cannot be recognized by any FSM because
   a) An FSM cannot remember arbitrary, large amount of information
   b) An FSM cannot deterministically fix the mid-point
   c) Even if the mid-point is known, an FSM cannot find whether the second half of the string matches the first half
   d) All of these
2. The basic limitation of an FSM is that
   a) It cannot remember arbitrary, large amount of information
   b) It cannot recognize grammars that are regular
   c) It sometimes recognizes grammars that are not regular
   d) All of these
3. The operation of synchronous circuits are controlled by
   a) Input
   b) State
   c) Output
   d) Clock pulses
4. The output of combinational circuit depends on
   a) Present state
   b) Past input
   c) Present input
   d) Present stored information and present input
5. The output of sequential circuit depends on
   a) Present state
   b) Past input
   c) Present input
   d) Present stored information and present input
6. For which of the following, FSM cannot be designed
   a) Addition of two binary numbers
   b) Subtraction of two binary numbers
   c) Multiplication of two binary numbers
   d) All of these
7. Which is true of the following?
   a) A merger graph is a directed graph
   b) A compatible graph is a directed graph
   c) Both are not directed
   d) None of these
8. A merger table is a substitute of
   a) Merger graph
   b) Compatible graph
   c) Minimized machine
   d) FSM
9. Two states are called 1-equivalent if
   a) Both of the states produce 1
   b) If both of the states produce the same output for string length 1

c) If both of the states produce the same output for same input.
d) If both of the states produce the same output for input 1.

10. Which is true for an incompletely specified machine?
a) All Next states are not mentioned
b) All outputs are not mentioned
c) All inputs are not mentioned
d) Both a and b

11. Compatible pairs are obtained from
a) Merger graph
b) Compatible graph
c) Testing table
d) Testing graph

12. The number of vertices of a merger graph is
a) The number of states of the machine
b) The number of compatible pairs
c) The number of states combinations
d) None of these

13. The number of vertices of a compatible graph is
a) The number of states of the machine
b) The number of compatible pairs
c) The number of states combinations
d) None of these

14. An FSM M is called a finite memory machine of order μ if the present state of the machine M can be obtained from
a) The last μ number of inputs and the corresponding μ number of next states
b) The last μ number of inputs and the corresponding μ number of outputs
c) The last μ number of next states and the corresponding μ number of outputs
d) The last μ number of inputs

15. A sequential machine M is called definite if there exists a least integer μ, so that the present state of the machine M can be uniquely obtained from the
a) Past μ number of inputs
b) Past μ number of outputs
c) Past μ number of next states
d) Past μ number of inputs, outputs, and next states

16. Which is sufficient to find the initial state of an information lossless machine from the input string?
a) Next state and output string
b) Final state and next state
c) Final state and output string
d) Final state, next state, and output string

17. The input sequence of an information lossless machine can be determined from the knowledge of
a) Only the output sequence
b) The output sequence and initial state
c) The output sequence, initial state, and final state
d) The initial state

18. Which is true of the following?
a) All information lossless machines are of finite order
b) Some information lossless machines are of finite order
c) Those machines which are not lossless of finite order are lossy
d) None of the above

19. The following diagram represents an FSM which takes as input a binary number from the least significant bit.

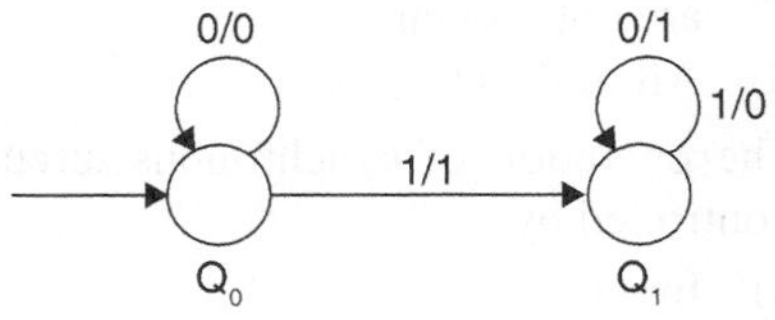

Which one of the following is true?
(a) It computes 1's complement of the input number
(b) It computes 2's complement of the input number
(c) It increments the input number
(d) It decrements the input number

Answers:

1. d	2. a	3. d	4. c	5. d	6. c	7. b	8. a	9. b	10. d
11. a	12. a	13. b	14. b	15. a	16. c	17. c	18. b	19. b #	

Hints:

19 Take 1101 → the output is 0011, which is 2's complement of 1101.

Fill in the Blanks

1. A synchronous circuit is controlled by ___________.
2. The operations in an asynchronous circuit are controlled by a number of ___________.
3. The output of combinational circuits is a function of ___________.
4. The output of sequential circuits is a function of ___________.
5. Binary adder is a ___________.
6. Modulo 3 binary counter has ___________ states.
7. An FSM can be defined as a type of machine whose past histories can affect its ___________ in a ___________ of ways.
8. Two states S_i and S_j of machine M are said to be equivalent if they produce the same ___________ for an ___________ applied to the machine M, considering S_i and S_j as the initial states.
9. If two states are k-distinguishable, then they are ___________ equivalent for k = k to 1.
10. The equivalent partition is ___________.
11. Incompletely specified machines are those machine where for all states for all inputs the ___________ or ___________ or both are not mentioned.
12. Minimal machine is the ___________ of the machines obtained by minimizing an incompletely specified machine.
13. A merger graph of a machine M of 'n' states is an ___________ graph.
14. In a merger graph, there will be an ___________ arc if the next states of the two states (vertices) do not conflict.
15. In a merger graph, there will be no arc between the two vertices if the ___________ of the pair of states conflict.
16. Two states, say S_i and S_j, are said to be compatible, if both of them give the same ___________ strings for all ___________ strings applied to the machine separately, considering S_i and S_j as the initial states.
17. A compatibility graph is a ___________ graph.
18. A subgraph of a compatibility graph is called ___________ if for every vertex in the subgraph, all outgoing arcs and the terminal vertices of the arcs also belong to the subgraph.
19. A merger table is a substitute application of ___________.

20. From a merger graph, we can get the compatible pairs and ________.
21. A vanishing connection matrix is used to find whether a machine is ________ or not.
22. A synchronizing tree method is used to find whether a machine is ________ or not.
23. A contracted table method is used to find whether a machine is ________ or not.
24. The order of definiteness μ = ________, if the length of the longest path in the testing graph is l.
25. The order of definiteness of a machine is the ________ of the synchronizing tree.
26. A machine is called information lossless if its ________ state, ________ state, and ________ string are sufficient to determine uniquely the input string.
27. A machine which is not lossless is called ________.

Answers:

1. clock pulses	2. completion and initialization signals	3. present I/P
4. external I/P and present stored information	5. sequential circuit	6. three
7. future behaviour, finite number	8. output sequences, input string	9. k – 1
10. unique	11. next state, output	12. minimum
13. undirected	14. uninterrupted	15. outputs
16. output, input	17. directed	18. closed
19. merger graph	20. implied pairs	21. finite
22. definite	23. definite	24. 1 + 1
25. maximum label	26. initial, final, output	27. lossy

Exercise

1. Define an FSM. Mention the capabilities and limitations of an FSM.
2. What is the usefulness of a sequence detector? Can it be called a machine? Mention points in support of your answer. What do you mean by overlapping sequence? What is the need of considering overlapping sequence in constructing a sequence detector?
3. i) Design a two input two output sequence detector which generates an output '1' every time the sequence 0101 is detected. And for all other cases, output '0' is generated. Overlapping sequences are also counted.

   ii) Design a two input two output sequence detector which generates an output '1' every time the sequence 1101 is detected. And for all other cases output '0' is generated. Overlapping sequences are also counted.
4. Design a modulo 8 binary counter using JK flip flop.
5. What is the benefit of minimizing an FSM? What do you mean by equivalent partition of a machine? Prove that equivalent partition of an FSM is unique.

   i) Find the equivalent partition for the following machine.

Present State	Next State, O/P X = 0	X = 1
A	E, 0	G, 0
B	G, 0	F, 0
C	H, 0	B, 1
D	G, 0	A, 1
E	A, 0	G, 0
F	A, 0	A, 0
G	F, 0	A, 0
H	C, 1	A, 1

From here, minimize the machine.

ii) Find the equivalent partition for the following machine.

Present State	Next State, O/P X = 0	X = 1
A	E, 0	D, 1
B	F, 0	D, 0
C	E, 0	B, 1
D	F, 0	B, 0
E	C, 0	F, 1
F	B, 0	C, 0

Find the shortest input sequence that distinguishes state A from state E.
From here, minimize the machine.

6. Why are incompletely specified machines called incomplete? What is the necessity of simplifying these types of machines?
   i) Simplify the following incompletely specified machine

Present State	Next State, O/P X = 0	X = 1
A	B, 1	C, 0
B	A, 1	_, 1
C	_, _	B, 0

   ii) Simplify the following incompletely specified machine

Present State	Next State, O/P X = 0	X = 1
A	E, 1	D, 0
B	E, 0	_, _

Present State	Next State, O/P X = 0	X = 1
C	_, 0	B, _
D	A, 0	D, 1
E	A, _	B, 0

7. Define minimal machine. What are differences between minimized machine and minimal machine? Why can the general equivalent partition method not be used to minimize an incompletely specified machine?
   i) Find a minimal machine from the minimum machines for the following machine considering the unspecified outputs as '0' or '1'.

Present State	Next State X = 0	X = 1
A	B, 1	C, _
B	A, _	C, 0
C	A, 1	B, 0

8. Define a merger graph. Can a merger graph be drawn for a completely specified machine? Give reason in favor of your answer.
   i) Develop a merger graph for the following incompletely specified machine. From there, find the compatible pairs.

Present State	Next State, z I_1	I_2	I_3	I_4
A	D, 1	C, _	_, _	D, 1
B	_, _	D, _	A, 0	_, _
C	E, 1	_, _	B, 0	C, 1
D	_, _	D, 1	E, 0	C, 1
E	_, _	A, 0	_, _	_, _

   ii) Define a compatible pair and implied pair. What is the relation between these two? Define a compatible graph. What are differences between a merger graph and a compatible graph? Define closed covering and minimum closed covering.
   iii) For the previously given machine develop a compatible graph and from there develop the minimal machine.
9. Develop a merger graph from the following machine.

Present State	Next State, z		
	I_1	I_2	I_3
A	C, 0	E, 1	_, _
B	C, 0	E, _	_, _
C	B, _	C, 0	A, _
D	B, 0	C, _	E, _
E	_, _	E, 0	A, _

From there, find the compatible pairs. Check whether the same compatible pairs are obtained or not from the merger table constructed from the previous machine.

From the compatible pairs, develop a compatible graph, and hence find the minimal machine.

10. What do you mean by finite memory? What is the usefulness of a machine to be a finite memory?
    i) Find out whether the following machines are of finite memory or not by the testing table – testing graph and vanishing connection matrix method. If finite, find its order.

Present State	Next State, O/P	
	X = 0	X = 1
A	E, 0	G, 0
B	G, 0	F, 0
C	H, 0	B, 1
D	G, 0	A, 1
E	A, 0	G, 0
F	A, 0	A, 0
G	F, 0	A, 0
H	C, 1	A, 1

Present State	Next State, O/P	
	X = 0	X = 1
A	D, 0	C, 0
B	A, 0	E, 0
C	E, 0	B, 0
D	C, 0	D, 0
E	E, 0	B, 1

Present State	Next State, O/P	
	X = 0	X = 1
A	B, 0	D, 0
B	C, 0	C, 0
C	D, 0	A, 0
D	D, 0	A, 1

11. Define a definite memory machine. What are the differences between a finite memory machine and a definite memory machine?
    i) Test whether the following machines are definite or not using the synchronizing tree, contracted table and the testing table–testing graph method. If definite, find its order.

Present State	Next State X = 0	Next State X = 1
A	B	C
B	A	D
C	B	A
D	C	A

Present State	Next State X = 0	Next State X = 1
A	D	A
B	B	B
C	A	A
D	C	B

Present State	Next State X = 0	Next State X = 1
A	B, 1	C, 0
B	D, 1	B, 1
C	E, 1	B, 0
D	A, 0	E, 0
E	F, 0	D, 1
F	D, 0	A, 0

12. What do you mean by information losslessness of a machine? What is the usefulness of it? Find whether the following machines are information lossless or not. If lossless of definite order, find the order of losslessness.

Present State	Next State, O/P X = 0	Next State, O/P X = 1
A	A, 1	C, 1
B	E, 0	B, 1
C	D, 0	A, 0
D	C, 0	B, 0
E	B, 1	A, 0

Present State	Next State, O/P X = 0	Next State, O/P X = 1
A	D, 1	E, 0
B	A, 0	B, 1
C	C, 0	B, 0
D	C, 1	B, 1
E	A, 0	B, 0

13. Find the input string applied to state A of the following machine where the final state is B and the output string is 1110000010.

Present State	Next State, z X = 0	Next State, z X = 1
A	B, 1	C, 0
B	D, 1	B, 1
C	E, 1	B, 0
D	A, 0	E, 0
E	F, 0	D, 1
F	D, 0	A, 1

Regular Expression

5

Introduction

Regular expression, in short RE, is the language part of the type 3 grammar, which is called regular grammar. RE is accepted by the machine called finite automata (FA). RE was first proposed by an American mathematician *Stephen Kleene* as a notion of describing the algebra of a regular set. Later on, *Ken Thompson* used the RE in early computer text editor 'QED' and Unix editor 'ed'. QED uses RE for searching text. RE became useful in the Unix text processing program 'grep' and modern programming languages such as 'PERL'. A Canadian computer scientist *Henry Spencer* wrote 'regex', which is widely used as a software library for RE.

They are called 'regular' because they describe a language with very 'regular' properties. In the previous chapter, we have already discussed about FA. Now, in this chapter, we shall learn about RE and some important properties related to RE.

5.1 Basics of Regular Expression

An RE can be defined as a language or string accepted by an FA.

We know that an FA consists of 5 touples $\{Q, \Sigma, \delta, q_0, F\}$. Among them, an RE is a string on Σ, i.e., it consists of only input alphabets.

In short, a regular expression is written as RE or regex or regexp.

5.1.1 Some Formal Recursive Definitions of RE

1. Any terminals, i.e., the symbols that belong to Σ are RE. The null string(Λ) and the null set($\emptyset$) are also RE.
2. If P and Q are two REs, then the union of the two REs denoted by P+Q is also an RE.
3. If P and Q are two REs, then their concatenation denoted by PQ is also an RE.
4. If P is an RE, then the iteration (repetition or closure) denoted by P* is also an RE.
5. If P is an RE, then (P) is an RE.
6. The expressions got by the repeated application of the rules from (1) to (5) over Σ are also REs.

5.2 Basic Operations on Regular Expression

Three basic operations are union, concatenation, and closure.

- **The operation of union on RE:** If R_1 and R_2 are two REs over Σ, then $L(R_1 \cup R_2)$ is denoted by $L(R_1 + R_2)$.
  $L(R_1 \cup R_2)$ is a string from R_1 or a string from R_2.
  $L(R_1 + R_2) = L(R_1) \cup L(R_2)$.
- **The operation of concatenation on RE:** If R_1 and R_2 are two REs over Σ, then $L(R_1 \cap R_2)$ is denoted by $L(R_1R_2)$.
  $L(R_1 \cap R_2)$ is a string from R_1 followed by a string from R_2.
  $L(R_1R_2) = L(R_1)L(R_2)$
- **The operation of closure on RE:** If R_1 and R_2 are two REs over Σ, then $L(R^*)$ is a string obtained by concatenating n elements for $n \geq 0$.

5.2.1 Kleene's Closure

It is defined as a set of all strings including null (Λ) obtained from the alphabets of the set Σ. It is denoted as Σ*.

If Σ = {0}, then Σ* = {Λ, 0, 00, 000, ...}
If Σ = {0, 1}, then Σ* = {Λ, 0, 1, 01, 00, 11, 010, 011, 100, ...}

5.2.1.1 Difference Between Closure and Kleene's Closure

- Closure is nothing but the iteration of 0 to ∞ times, but Kleene's closure is the set including Λ.
- Closure is applied on RE, but Kleene's closure is applied on Σ.
- If R = 01, then the closure on R denoted by R* are Λ, 01, 0101, 010101......etc., i.e., the iteration of the same string 01.
- If Σ = {0, 1}, then Kleene's closure is denoted by Σ* = {Λ, 0, 1, 01, 00, 11, 010, 011, 100,.......}, i.e., the set of any combinations of 0 and 1 including Λ.

Precedence of Operator: Among the previously discussed operators, Kleene star has the highest priority. Then, concatenation comes into play and finally union has the lowest priority.

REs can be described in English language. Consider the following examples.

Example 5.1 Describe the following REs in English language.

(i) 10*1
(ii) a*b*
(iii) (ab)*
(iv) (a* + b*) c*
(v) (00)*(11)*1

Solution:

(i) The strings that we get from the RE 10*1 are {11, 101, 1001, 10001.....}. From the set, we can conclude that the strings that are generated from the REs are beginning with 1 and ending with 1. In between 1 and 1, there are any numbers of 0.

So, we can describe in English like this:

Set of all strings beginning and ending with 1 with any number of 0 in between the two 1s.

(ii) The strings that we get from the RE a*b* are a* . b*, i.e.,

$$\{\Lambda, a, aa, aaa, \ldots\ldots\}.\{\Lambda, b, bb, bbb, \ldots\ldots\} = \{\Lambda, a, b, aa, bb, ab, abb, aab, \ldots\ldots\}.$$

From the set, we can conclude that the strings that are generated from the RE contain any number of 'a' followed by any number of 'b'.

So, we can describe in English like this:

Set of all strings of 'a' and 'b' containing any number of 'a' followed by any number of 'b'.

(iii) The strings that we get from the RE (ab)* are {Λ, ab, abab, ababab, ...}.

In English, it is described as:

Set of all strings of a and b with equal number of a and b containing 'ab' as repetition.

(iv) (a* + b*) means any combination of a or any combination of b. c* means any combination of c, which is concatenated with (a* + b*).

In English, it can be described as:

Any combination of 'a' or any combination of 'b' followed by any combination of 'c'.

(v) The strings that we get from the RE (00)*(11)*1 are {1, 001, 111, 00111, ...}. In the string, there are always odd number of 1 and even number of 0.

In English, it can be described as:

Set of all strings of '0' and '1' with even number of 0 followed by odd number of 1.

Alternatively, the RE described in English language can be denoted using alphabet and symbols. The following examples describe this.

Example 5.2 Built regular expression of the following

(i) An RE consists of any combination of a and b, beginning with a and ending with b.
(ii) A language of any combination of a and b containing abb as a substring.
(iii) RE of a and b containing at least 2 'a's.
(iv) Write an RE for the language $L = \{a^n b^m \mid \text{where } m + n \text{ is even}\}$.
(v) An RE of a and b, having exactly one 'a'.

Solution:

(i) The RE consists of any combination of a and b, i.e., (a + b)*. But the RE starts with a and ends with b. In between a and b, any combination of a and b occurs. So, the RE is

$$L = a(a + b)^*b.$$

(ii) In the RE, abb occurs as a substring. The RE consists of any combination of a and b. Before the substring, abb may occur at the beginning, at the middle, or at the last. If abb occurs at the beginning, then after abb there is any combination of a and b. If abb occurs at the middle, then before and after abb there are any combination of a and b. If abb occurs at the last, then before abb there is any combination of a and b.

The RE is L = (a + b)*abb(a + b)*.

(iii) In the RE, there are at least two 'a's. The expression consists of a and b. Therefore, any number of 'b' can occur before the first 'a' and before the second 'a', i.e., after the first 'a' and after the second 'a'.

So, the RE will be L = b*ab*ab*.

(iv) The RE consists of n number of 'a' followed by m number of 'b'. But m + n is always even. This is possible if

(a) m and n both are even or (b) m and n both are odd.

If m and n are both even, then the expression is (aa)*(bb)*.

If m and n are both odd, then the expression is (aa)*a(bb)*b.

By combining these two, the final RE becomes

$$L = (aa)^*(bb)^* + (aa)^*a(bb)^*b.$$

(v) In the RE, there is exactly one 'a'. Before and after a, there is any combination of b.

Therefore, the RE is L = b*ab*.

5.3 Identities of Regular Expression

An *identity* is a relation which is tautologically true. In mathematics, an equation which is true for every value of the variable is called an identity equation. As an example, $(a + b)^2 = a^2 + 2ab + b^2$ is an identity equation. Based on these identities, some other problems can be proved. In the RE also, there are some identities which are true for every RE. In this section, we shall discuss those identities related to RE.

1. $\emptyset + R = R + \emptyset = R$

   Proof: LHS: $\emptyset + R$
   $= \emptyset \cup R$
   $= \{\ \} \cup \{\text{Elements of R}\} = R = \text{RHS}$

2. $\emptyset R = R \emptyset = \emptyset$

   Proof: LHS: $\emptyset R$
   $= \emptyset \cap R$
   $= \{\ \} \cap \{\text{Elements of R}\} = \emptyset = \text{RHS}$

   Note: In both the previous cases, Ø denotes a null set, which contains nothing.

3. $\Lambda R = R\Lambda = R$

   Proof: LHS: ΛR
   = Null string concatenated with any symbol of R
   = Same symbol $\in$ R = R = RHS

4. $\Lambda^* = \Lambda$ & $\emptyset^* = \Lambda$

   Proof: LHS: $\Lambda^* = \{\Lambda, \Lambda\Lambda, \Lambda\Lambda\Lambda \ldots\ldots.\}$
   $= \{\Lambda, \Lambda, \Lambda, \ldots..\}$ [according to identity (3)]
   $= \Lambda = \text{RHS}.$

   Same for Ø *.

5. R + R = R

   Proof: LHS: $R + R = R(\Lambda + \Lambda) = R\Lambda$
   $= R$ [according to identity (3)] = RHS

6. R*R* = R*

   Proof: LHS: $R^*R^* = \{\Lambda, R, RR......\}\ \{\Lambda, R, RR......\}$
   $= \{\Lambda\Lambda, \Lambda R, \Lambda RR, ..., R\Lambda, RR, ..., RR\Lambda, RRR, ...\}$
   $= \{\Lambda, R, RR, RRR, ...\}$ [using identity (3)] = R* = RHS

7. R*R = RR*

   Proof: LHS: $R^*R = \{\Lambda, R, RR, RRR, ...\}\ R$
   $= \{\Lambda R, RR, RRR, RRRR, ...\}$
   $= R\{\Lambda, R, RR, RRR, ...\} = RR^* =$ RHS

8. (R*)* = R*

   Proof: LHS: $(R^*)^* = \{\Lambda, R^*R^*, R^*R^*R^*, ...\}$
   $= \{\Lambda, R^*, R^*, ...\}$ [using identity (6)]
   $= \{\Lambda, \{\Lambda, R, RR, RRR, ...\}, \{\Lambda, R, RR, RRR, ...\}, ...\}$
   $= R^* =$ RHS

9. $\Lambda + RR^* = \Lambda + R^*R = R^*$

   Proof: LHS: $\Lambda + RR^* = \Lambda + R\ \{\Lambda, R, RR, RRR, ...\}$
   $= \Lambda + \{R\Lambda, RR, RRR, RRRR, ...\}$
   $= \Lambda + \{R, RR, RRR, RRRR, ...\}$
   $= \{\Lambda, R, RR, RRR, RRRR, ...\} = R^* =$ RHS

10. (PQ)*P = P(QP)*

    Proof: LHS: $(PQ)^*P = \{\Lambda, PQ, PQPQ, PQPQPQ, ...\}P$
    $= \{P, PQP, PQPQP, PQPQPQP, ...\}$
    $= P\{\Lambda, QP, QPQP, QPQPQP, ...\}$
    $= P(QP)^* =$ RHS

11. (P + Q)* = (P*Q*)* = (P* + Q*)* (D'Morgan's theorem)

12. (P + Q)R = PR + QR

    Proof: Let $a \in (P+Q)R$
    $= a \in PR$ or QR = RHS.

Example 5.3 From the identities of RE, prove that

$(1 + 100^*) + (1 + 100^*)(0 + 10^*)(0 + 10^*)^* = 10^*(0 + 10^*)^*$.

Solution: LHS.

$$(1 + 100^*) + (1 + 100^*)(0 + 10^*)(0 + 10^*)^*$$
$$= (1 + 100^*)\ (\Lambda + (0 + 10^*)(0 + 10^*)^*)$$
$$= (1 + 100^*)(0 + 10^*)^* \text{ (according to } \Lambda + RR^* = R^*)$$
$$= 1(\Lambda + 00^*)(0 + 10^*)^*$$
$$= 10^*(0 + 10^*)^* = \text{RHS.}$$

Example 5.4 From the identities of RE, prove that the following three are equivalent

(i) (011((11)* + (01)*)*)*011
(ii) 011(((1 + 0)1)*011)*
(iii) 011(((11)*(01)*)*011)*

Solution: Let P = (11) and Q = (01).
We know (P + Q)* = (P*Q*)* = (P* + Q*)*.
Thus, ((11) + (01))* = ((11)*(01)*)* [in c] = ((11)* + (01)*)*[in a] = ((1 + 0)1)* [in b].
Consider 011 as R and (P + Q)* as S.
We know that (RS)*R = R(SR)*
Thus, (011((11)* + (01)*)*)*011 = 011((11)* + (01)*)*011)*
= 011((11) + (01))*011)*
= 011(((1 + 0)1)*011)*.

Similarly, (011((11)* + (01)*)*)*011 = 011((11)* + (01)*)*011)*
= 011((11) + (01))*011)*
= 011(((11)*(01)*)*011)*.

As (i) ≡ (ii) and (i) ≡ (iii), (i) ≡ (ii) ≡ (iii).

Example 5.5 From the identities of RE, prove that

$$10 + (1010)^*[\wedge + (1010)^*] = 10 + (1010)^*.$$

Solution: LHS 10 + (1010)*[∧ + (1010)*]
⇒ 10 + ∧(1010)* + (1010)* (1010)*
⇒ 10 + (1010)* + (1010)* As ∧R = R and RR = R
⇒ 10 + (1010)* = RHS. As R + R = R.

5.4 The Arden's Theorem

The Arden's theorem is used to construct the RE from a transitional system of FA.

Theorem 5.1: Statement: Let P and Q be two REs over Σ. If P does not contain Λ, then the equation R = Q + RP has a unique (one and only one) solution, R = QP*.

Proof: Now, point out the statements in the Arden's theorem in general form.

- **P** and **Q** are two REs.
- **P** does not contain the **Λ** symbol.
- **R = Q + RP** has a solution, i.e., **R = QP***.
- This solution is the one and only solution of the equation.

If R = QP* is a solution of the equation R = Q + RP, then by putting the value of R in the equation, we shall get the value '0'.

$$R = Q + RP$$
$$R - Q - RP = 0$$
$$\text{LHS } R - Q - RP$$

(Putting the value of R in the LHS, we get)

$$
\begin{aligned}
&QP^* - Q - QP^* P \\
&= QP^* - Q(\wedge + P^* P) \\
&= QP^* - QP^* \text{ [As } (\Lambda + R^*R) = R^*] \\
&= 0.
\end{aligned}
$$

So, from here it is proved that R = QP* is a solution of the equation R = Q + RP.
Now, we have to prove that R = QP* is the one and only solution of the equation R = Q + RP.
As R = Q + RP, put the value of R again and again in the RHS of the equation.

$$
\begin{aligned}
R &= Q + RP \\
&= Q + (Q + RP)P \\
&= Q + QP + RPP \\
&= Q(\Lambda + P) + RPP \\
&= Q(\Lambda + P) + (Q + RP)PP \\
&= Q(\Lambda + P) + QPP + RPPP \\
&= Q(\Lambda + P + PP) + RPPP.
\end{aligned}
$$

After several steps, we shall get

$$R = Q(\Lambda + P + P^2 + P^3 + \cdots P^n) + RP^{n+1}.$$

Now let a string w belong to R. If w is a Λ string, then in which part will the Λ belong?
This string will belong to either the $Q(\Lambda + P + P^2 + P^3 + \cdots P^n)$ part or the RP^{n+1} part. But, according to point number (ii), P does not contain Λ, and so the string w does not belong to RP^{n+1}. So, it will obviously belong to $Q(\Lambda + P + P^2 + P^3 + \cdots P^n)$ which is nothing but QP^*. [$(\Lambda + P + P^2 + P^3 + \cdots P^n)$ is any combination of P including Λ.]

As this string belongs only in one part, R and QP^* represent the same set. That means $R = QP^*$ is the unique solution of the equation R = Q + RP.

5.4.1 Process of Constructing Regular Expression from Finite Automata

There are some assumptions:

- In the transitional graph, there must be no Λ-move.
- In the FA, there is only one initial state.

Now, we have to construct equations for all the states. There are n number of equations if there are n-states.

For any FA, these equations are constructed in the following way.

< State name > = Σ [< State name from which inputs are coming >. < input >]

For the beginning state, there is an arrow at the beginning coming from no state. So, a Λ is added with the equation of the beginning state.

Then, these equations have to be solved by the identities of RE.The expression obtained for the final state and consists of only the input symbol (Σ) is the Regular Expression for the Finite Automata.

The following examples describe the process of the construction of an RE from FA.

Example 5.6 Construct an RE from the given FA in Fig. 5.1 by the algebraic method using Arden's theorem

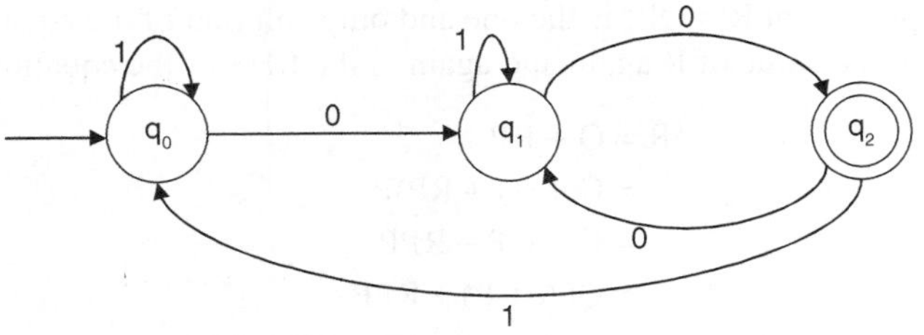

Fig. 5.1

Solution:

For the previously given FA, the equations are

$$q_0 = q_0.1 + q_2.1 + \Lambda \quad (1)$$

$$q_1 = q_0.0 + q_1.1 + q_2.0 \quad (2)$$

$$q_2 = q_1.0 \quad (3)$$

Put the value of q_2 in equation no. (2).

$$q_1 = q_0.0 + q_1.1 + q_1.0.0$$

$$= q_0.0 + q_1(1 + 0.0) \quad (4)$$

This equation is in the form R = Q + RP, where R = q_1, Q = q_0.0, and P = (1 + 0.0).

So, the solution of the equation is R = QP*, i.e., $q_1 = q_0.0.\ (1 + 0.0)^*$.

Putting the value $q_2 = q_1.0$ in equation no. (1), we get

$$q_0 = q_0.1 + q_1.0.1 + \Lambda.$$

Again putting the value of q_1 in the previous equation, we get

$$q_0 = q_0.1 + q_0.0.\ (1 + 0.0)^*.0.1 + \Lambda$$

$$q_0 = q_0(1 + 0.(1 + 0.0)^*.0.1) + \Lambda.$$

This equation is in the form R = Q + RP, where R = q_0, Q = Λ, and P = (1 + 0.(1 + 0.0)*.0.1). So, the solution of the equation is

$$q_0 = \Lambda.(1 + 0.(1 + 0.0)^*.0.1)^*$$

$$= (1 + 0.(1 + 0.0)^*.0.1)^*.$$

Putting the value of q_0 in the solution of q_1, we get

$$q_1 = (1 + 0.(1 + 0.0)^*.0.1)^*.0.\ (1 + 0.0)^*.$$

Replacing the value of q_1 in equation (3), we get

$$q_2 = (1 + 0.(1 + 0.0)^*.0.1)^*.0.\ (1 + 0.0)^*.0.$$

As q_2 is the final state of the FA, all the strings will halt on this state. Therefore, the RE is $(1 + 0.(1 + 0.0)^*.0.1)^*.0.\ (1 + 0.0)^*.0$, thus accepting the FA.

Example 5.7 Construct an RE from the given FA in Fig. 5.2 by the algebraic method using Arden's theorem.

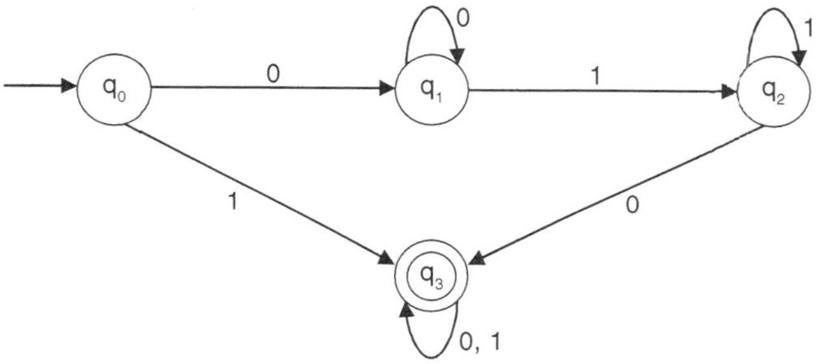

Fig. 5.2

Solution: For the previously given FA, the equations are

$$q_0 = \Lambda \quad (1)$$
$$q_1 = q_0 0 + q_1 0 \quad (2)$$
$$q_2 = q_1 1 + q_2 1 \quad (3)$$
$$q_3 = q_2 0 + q_0 1 + q_3 (0 + 1) \quad (4)$$

Putting the value of q_0 in equation q_1, it becomes $q_1 = 0 + q_1 0$. The equation is in the form R = Q + RP, where R = q_1, Q = 0, and P = 0. By the Arden's theorem, the solution of the equation is R = QP*, i.e.,

$$q_1 = 00^*.$$

Putting the value of q_1 in the equation (3), we get $q_2 = 00^*1 + q_2 1$.

The equation is in the form R = Q + RP, where R = q_2, Q = 00*1, and P = 1. By the Arden's theorem, the solution of the equation is R = QP*, i.e.,

$$q_2 = 00^*11^*.$$

Putting the value of q_2 and q_0 in equation (4), we get

$$q_3 = (00^*11^*0 + 1) + q_3 (0 + 1).$$

The equation is in the form R = Q + RP, where R = q_3, Q = (00*11*0 + 1), and P = (0 + 1). By the Arden's theorem, the solution of the equation is R = QP*, i.e.,

$$q_3 = (00^*11^*0 + 1)\ (0 + 1)^*.$$

As q_3 is the final state, the RE generated from the given FA is (00*11*0 + 1) (0 + 1)*.

Example 5.8 Construct an RE from the given FA in Fig. 5.3 by the algebraic method using Arden's theorem

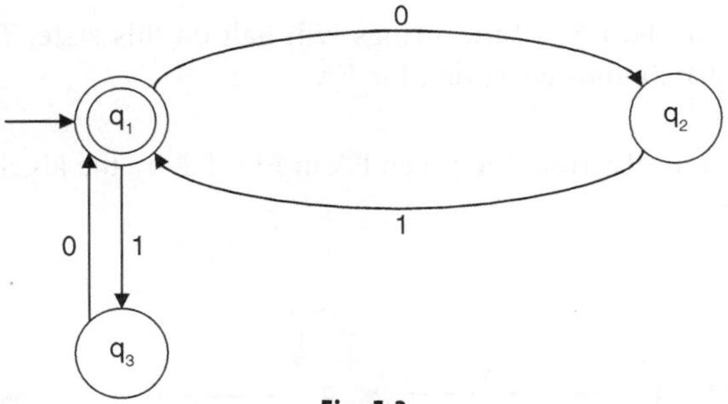

Fig. 5.3

Solution: For the given FA, the equations are

$$q_1 = q_2 1 + q_3 0 + \Lambda \quad (1)$$
$$q_2 = q_1 0 \quad (2)$$
$$q_3 = q_1 1 \quad (3)$$

Substituting the value of q_2 and q_3 in q_1, we get

$$q_1 = q_1 01 + q_1 10 + \Lambda$$
$$q_1 = q_1(01 + 10) + \Lambda.$$

If q_1 is treated as R, Λ as Q, and (01 + 10) as P, then the equation becomes R = Q + RP.
The solution for the equation is R = QP*.

So, $q_1 = \Lambda(01 + 10)^*$, i.e., $q_1 = (01 + 10)^*$ (As Λ.R = R).

As q_1 is the only final state, the string accepted by the FA is (01 + 10)*.

(Note: In the automata the transition from q_2 and q_3 for input 1 is missing. A dead state may be added to complete this FA, but the equation for dead state will be in no need in solving the automata. From here, it is also proved that the dead state is only necessary to prove that a string is not accepted by an FA by totally traversing the string.)

Example 5.9 Construct an RE from the given FA in Fig. 5.4 by the algebraic method using Arden's theorem.

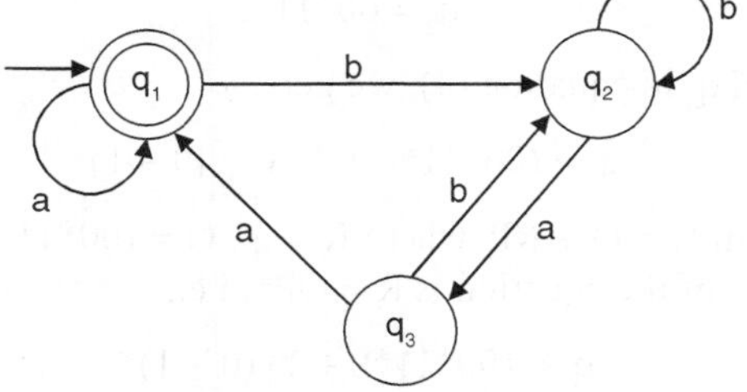

Fig. 5.4

Solution: For the given FA, the equations will be

$$q_1 = q_1.a + q_3.a + \Lambda \quad (1)$$
$$q_2 = q_2.b + q_3.b + q_1.b \quad (2)$$
$$q_3 = q_2.a \quad (3)$$

$q_3 = q_2 a$, replace this in equation (2). The equation becomes

$$q_2 = q_2.b + q_2 a.b + q_1.b$$
$$= q_2(b + a.b) + q_1.b.$$

The equation is in the format of R = Q + RP, q_2 [R]= q_2[R](b + a.b)[P] + q_1.b [Q].
The solution is R = QP*, i.e., $q_2 = q_1 b.(b + ab)^*$.
Replace the value of q_2 in equation (3).

$$q_3 = q_1 b.(b + ab)^*.a.$$

Replace the value of q_3 in equation (1).

$$q_1 = q_1.a + q_1 b.(b + ab)^*.a.a + \Lambda$$
$$= q_1.(a + b.(b + ab)^*.a.a) + \Lambda.$$

The equation is in the format of R = Q + RP, q_1[R] = q_1[R].(a + b.(b + ab)*.a.a)[P] + Λ[Q]. The solution of the equation is R = QP*, i.e., $q_1 = \Lambda.(a + b.(b + ab)^*.a.a)^*$,

i.e., $q_1 = (a + b.(b + ab)^*.a.a)^*$ (according to ΛR = RΛ = R).

q_1 is the final state, and so the RE accepting the FA is

$$(a + b.(b + ab)^*.a.a)^*.$$

5.5 Construction of Finite Automata from a Regular Expression

RE is the language format of the FA. Deterministic finite automata (DFA) can be constructed from a given RE in two ways:

1. Conversion of an RE to non-deterministic finite automata (NFA) with ε transition, and then converting it to DFA by ε closure.
2. Direct conversion of an RE to DFA.

5.5.1 Conversion of an RE to NFA with ε transition

For making an FA from an RE, there are two steps as given in the following.

Step I: From the given RE, an equivalent transitional system (FA) with Λ move has to be constructed.

Step II: From the transitional system with Λ moves, a DFA needs to be constructed by the ε closure method.

For constructing step I, again there are two methods (i) top-down approach and (ii) bottom-up approach. In the top-down approach, the FA construction starts from the given RE and ends by reaching a single element for each transition. In the bottom-up approach, the construction starts from the basic element and ends on reaching the RE.

5.5.1.1 Top-Down Approach

This is divided into several steps as given in the following.

- Take two states, one is the beginning state and another is the final state. Make a transition from the beginning state to the final state and place the RE in between the beginning and final states (Fig. 5.5).

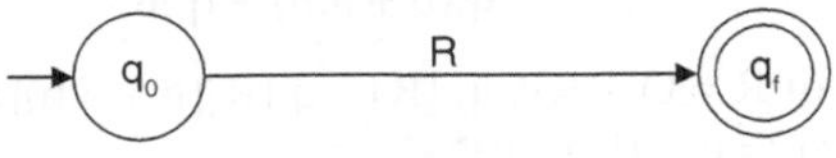

Fig. 5.5

- If in the RE there is a + (union) sign, then there are parallel paths between the two states (Fig. 5.6).

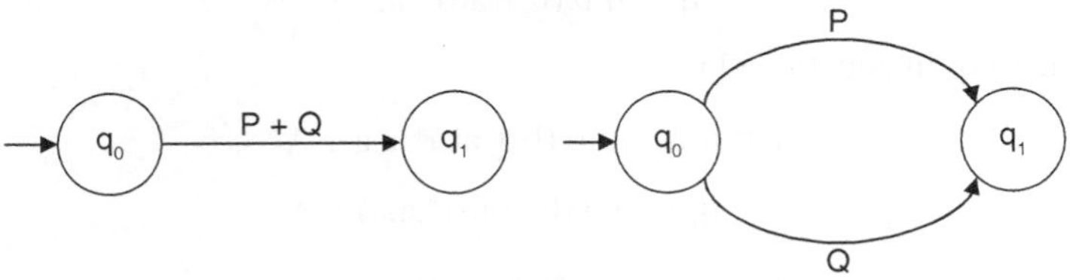

Fig. 5.6

- If in the RE there is a .(dot) sign, then one extra state is added between the two states (Fig. 5.7).

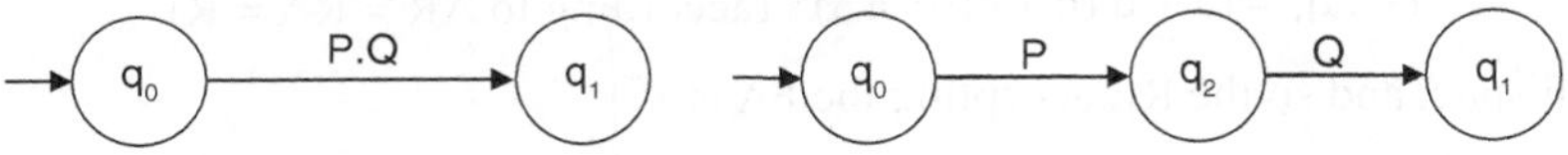

Fig. 5.7

- If in the RE there is a '*' (closure) sign, then a new state is *added in between. A loop is added on the new state and the label Λ is put* between the first to new and new to last (Fig. 5.8)

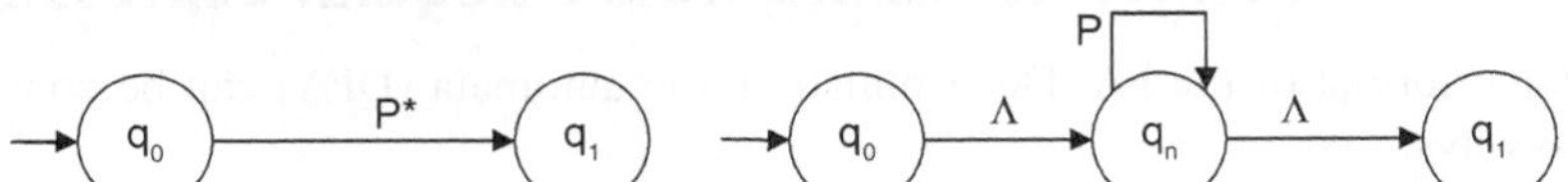

Fig. 5.8

The following examples make the described process clear for understanding.

Example 5.10 Construct an FA equivalent to the RE:

$$L = (a + b)^*(aa + bb)(a + b)^*.$$

Solution:

- **Two states are taken:** one is the beginning state and another is the final state. The RE is placed between the two states with a transition from the beginning state to the final state as given in Fig. 5.9(a)

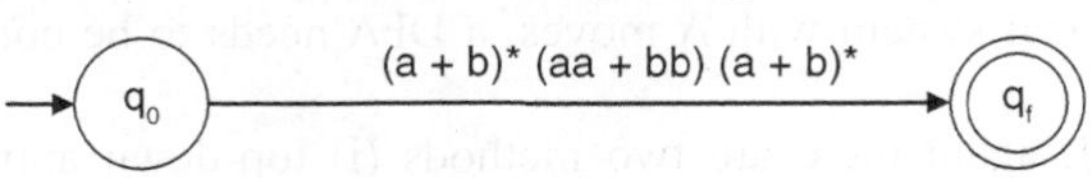

Fig. 5.9(a)

- In the RE, there are two concatenations, between (a + b)* and (aa + bb) and between (aa + bb) and (a + b)*. Two extra states are added between q_0 and q_f as given in Fig. 5.9(b).

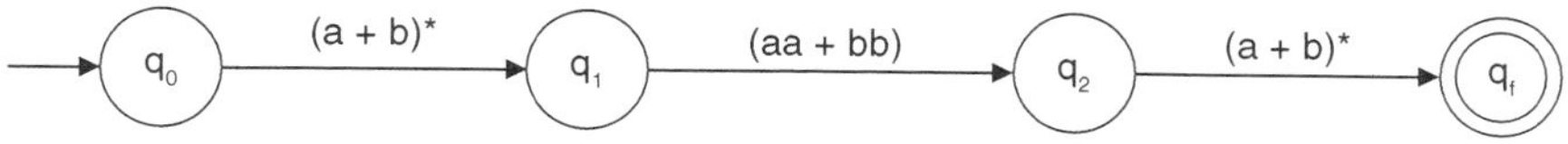

Fig. 5.9(b)

- In aa and bb, there is a + (union) sign. So, between q_1 and q_2, parallel paths are constructed as given in Fig. 5.9(c).

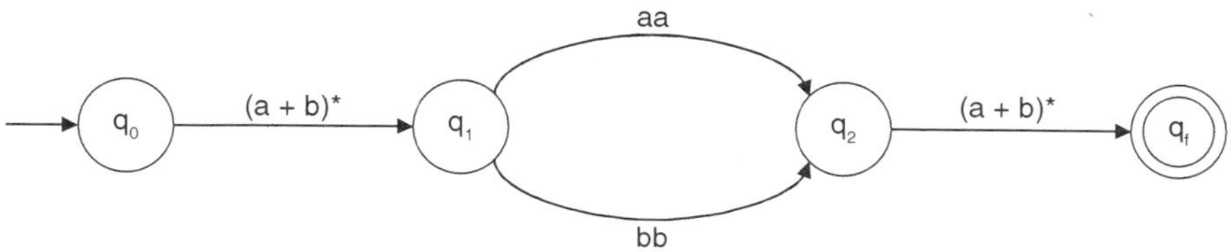

Fig. 5.9(c)

- There are two * (closure) signs between q_0 and q_1, and q_2 and q_f . Two new states q_3 and q_4 are added, respectively, between q_0, q_1 and q_2, q_f. The loops with label a + b are added to q_3 and q_4, making transitions between (q_0, q_3), (q_3, q_1), (q_2, q_4), and (q_4, q_f) with label Λ as in Fig. 5.9(d).

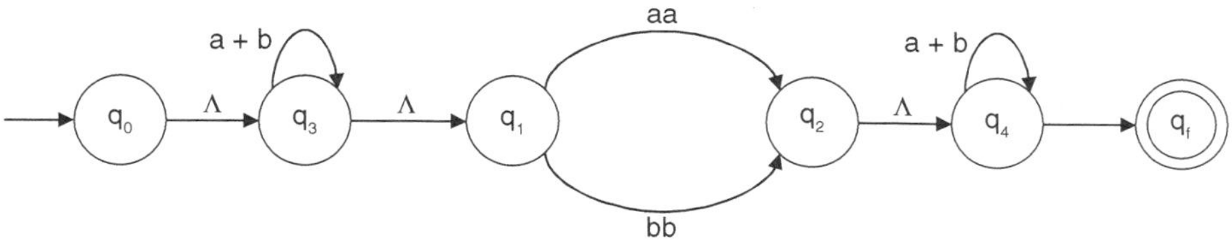

Fig. 5.9(d)

- Removing + (union) and • (concatenation) of the final FA with Λ transitions is given in Fig. 5.9(e).

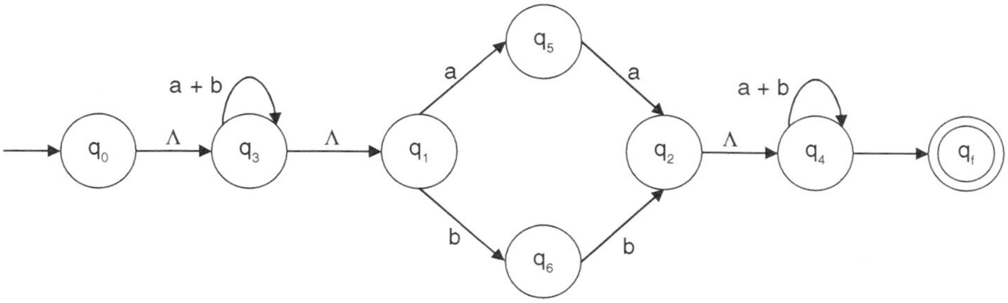

Fig. 5.9(e)

- Between q_0, q_3 and q_4, q_f, there is a label Λ. It means that from q_0 by getting no input it can reach to q_3, and the same thing is for q_4 and q_f. Therefore, q_0 and q_3 may be treated as the same state, and the same for q_4 and q_f. The transitions from q_3 and q_f are added with q_0 and q_4, respectively, with same label.

By removing the Λ transactions, the FA is given in Fig. 5.9(f)

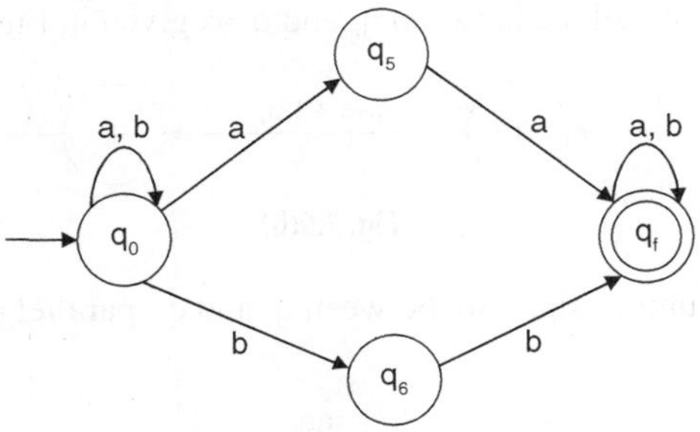

Fig. 5.9(f)

Example 5.11 Construct Finite Automata equivalent to the Regular Expression

$$L = ab(aa + bb)(a + b)^* b.$$

Solution:

Step I: Take a beginning state q_0 and a final state q_f. Between the beginning and final state place the regular expression (Fig. 5.10a).

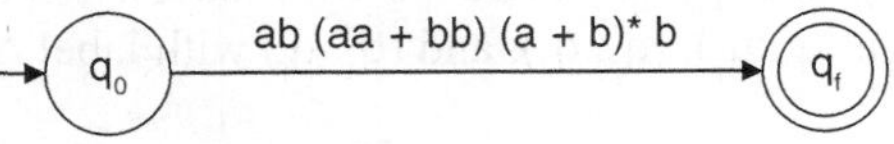

(a)

Step II: There are three dots (.) between ab, (aa + bb), (a + b)*, and b. Three extra states are added between q_0 and q_f (Fig. 5.10b).

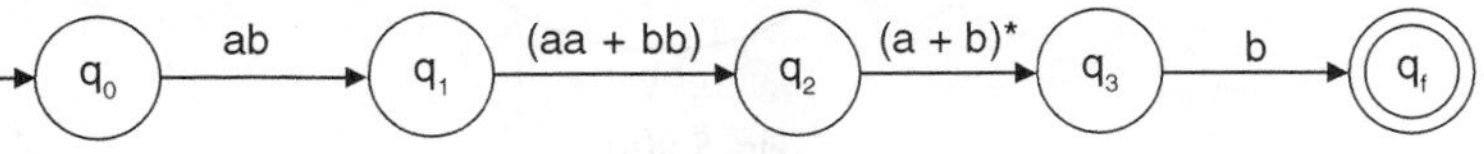

(b)

Step III: Between 'a' and 'b' there is a dot (.), so extra state is added (Fig. 5.10c).

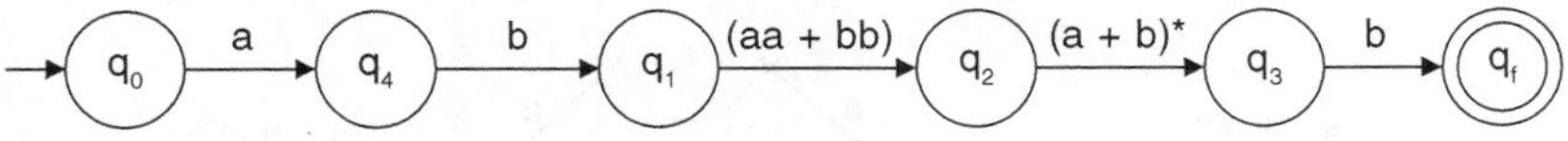

(c)

Step IV: In aa + bb there is a +, therefore there is parallel edges between q_1 and q_2. Between q_2 and q_3 there is (a + b)*. So, extra state q_5 is added between q_2 and q_3. Loop with label a, b is placed on q_5 and Λ transition is made between q_2, q_5 and q_5, q_3 (Fig. 5.10d).

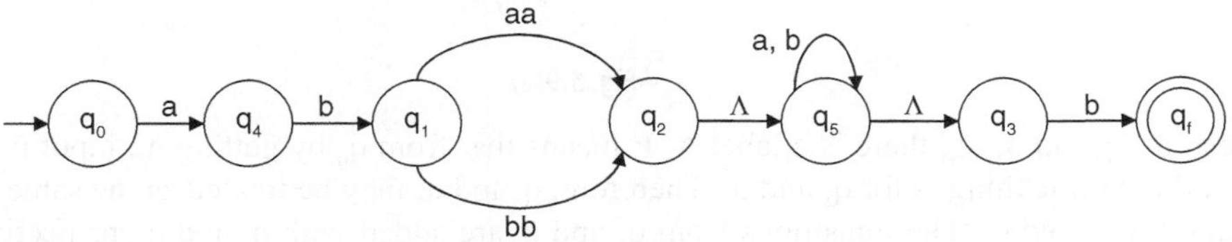

(d)

Step V: In aa and bb there are dots (.). Thus two extra states are added between q_1 and q_2 (one for aa and another bb). The final finite automata for the given regular expression is given in Fig. 5.10e.

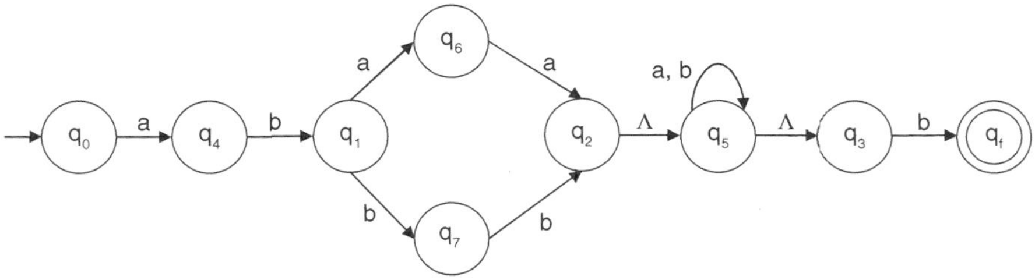

(e)

Figs. 5.10(a) to (e) *Construction of FA for L = ab(aa + bb)(a + b)*b.*

Example 5.12 Construct an FA equivalent to the RE:

$$L = ab + (aa + bb)(a + b)^* b.$$

Solution:

Step I: Take a beginning state q_0 and final state q_f. Between them, place the RE.

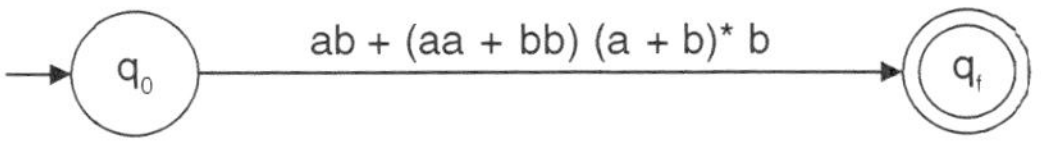

Step II: For ' + ', there will be two parallel edges between q_0 and q_f.

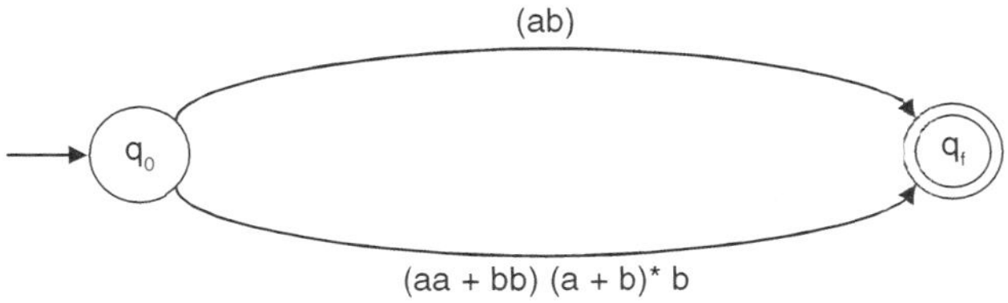

Step III: For '.' dots between a and b, (aa + bb), and (a + b)* and (a + b)* and b, add three new states.

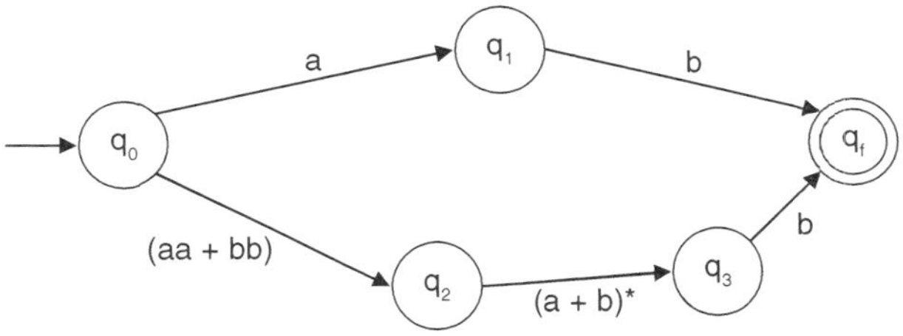

Step IV: For (aa + bb), two parallel edges with two extra states are added.

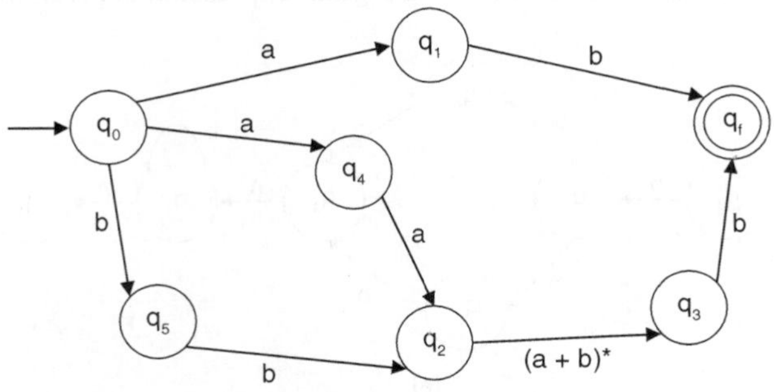

Step V: '*' is removed by putting the loop on another extra state with null transitions between q_2 to the new state and the new state to q_3.

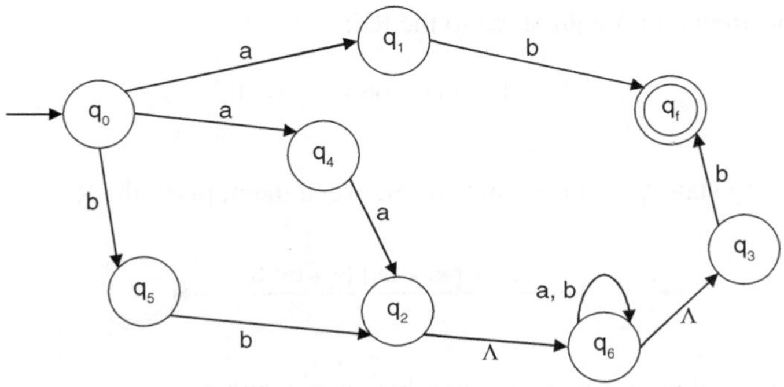

Step VI: Removing the Λ transitions, the final FA is given in Fig. 5.11.

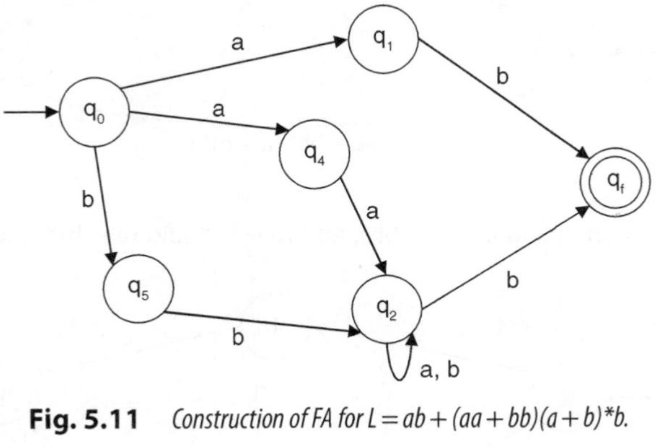

Fig. 5.11 *Construction of FA for $L = ab + (aa + bb)(a + b)^*b$.*

The bottom-up approach to convert an RE to an FA is proposed by Thomson. This method is known as the Thomson construction. The process is described in the following text.

5.5.1.2 Bottom-up Approach (Thomson Construction)

- For input ε, the transition diagram is constructed as in Fig. 5.12(a).
- For input a ∈ ∑, the transition diagram is as shown in Fig. 5.12(b).
- If NFA(r_1) is the NFA for RE r_1 and NFA(r_2) is the NFA for RE r_2, then for the RE $r_1 + r_2$ the transition diagram is as shown in Fig. 5.12(c).
- If NFA(r_1) is the NFA for RE r_1 and NFA(r_2) is the NFA for RE r_2, then for the RE $r_1.r_2$ the transition diagram is as shown in Fig. 5.12(d).
- If NFA(r) is the NFA for RE r, then for the RE r^*, the transition diagram is as shown in Fig. 5.12(e).

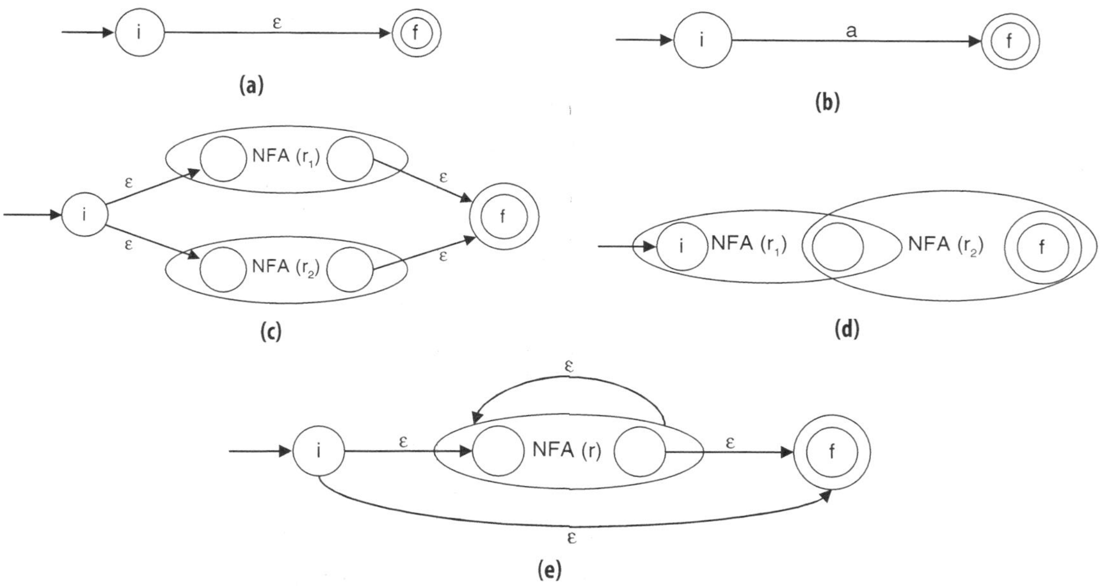

Figs. 5.12(a)–(e) *The Thomson Construction*

Example 5.13 Construct Finite Automata equivalent to the Regular Expression

$$L = ab(aa + bb)(a + b)^*a.$$

Solution:

Step I: The terminal symbols in L are 'a' and 'b'. The transition diagrams for 'a' and 'b' are given in Fig. 5.13(a).

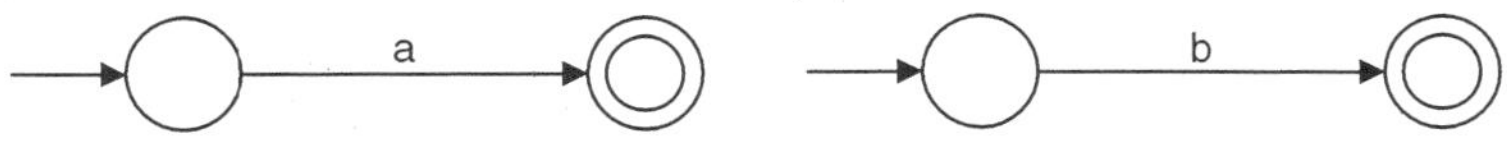

(a)

Step II: The transition diagrams for 'aa', 'ab', 'bb' are given in Fig. 5.13(b).

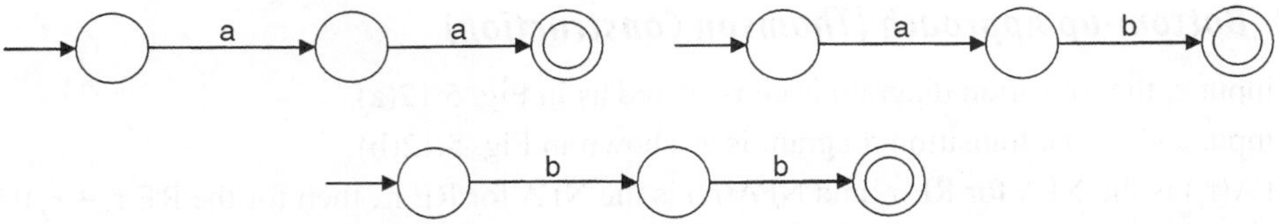

(b)

Step III: The transition diagram for (a+b) is given in Fig. 5.13(c)

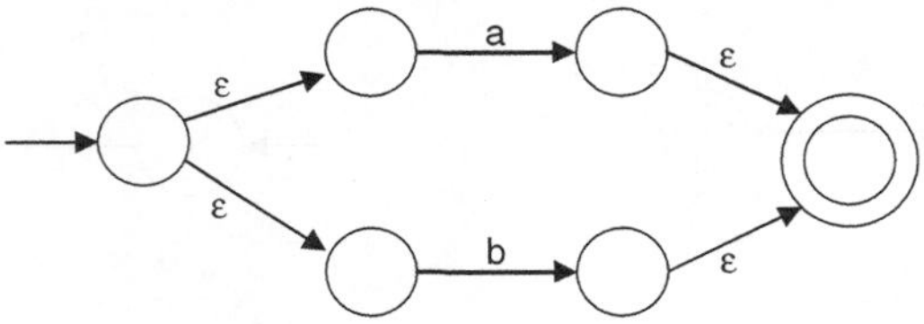

(c)

Step IV: The transition diagram for (a+b)* is given in Fig. 5.13(d).

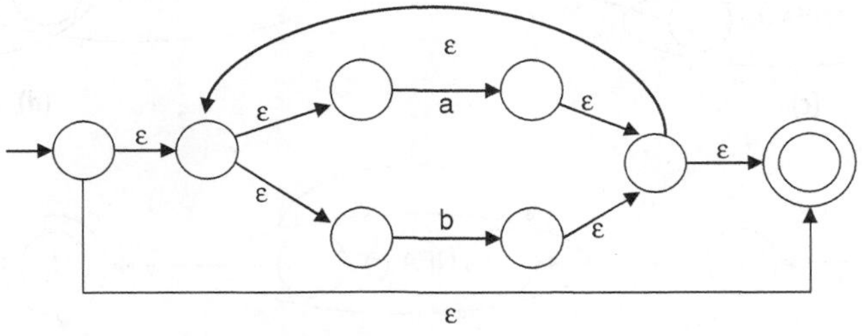

(d)

Step V: For (aa+bb) the transitional diagram is given in Fig. 5.13(e).

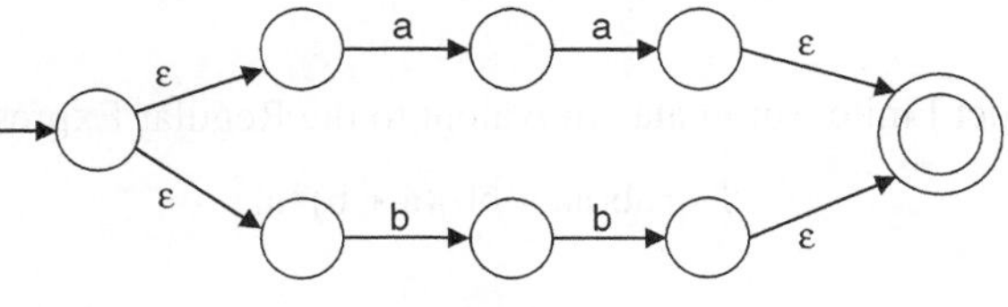

(e)

Step VI: The constructed transitional diagram for ab(aa+bb) is given in Fig. 5.13(f).

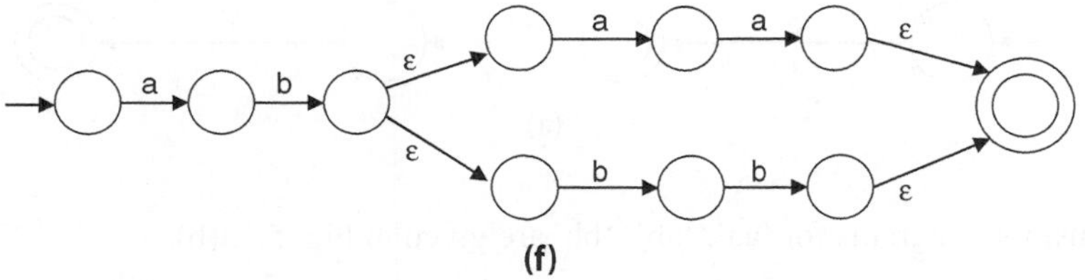

(f)

Step VII: The constructed transitional diagram for ab(aa + bb)(a + b)*a is given in Fig. 5.13(g).

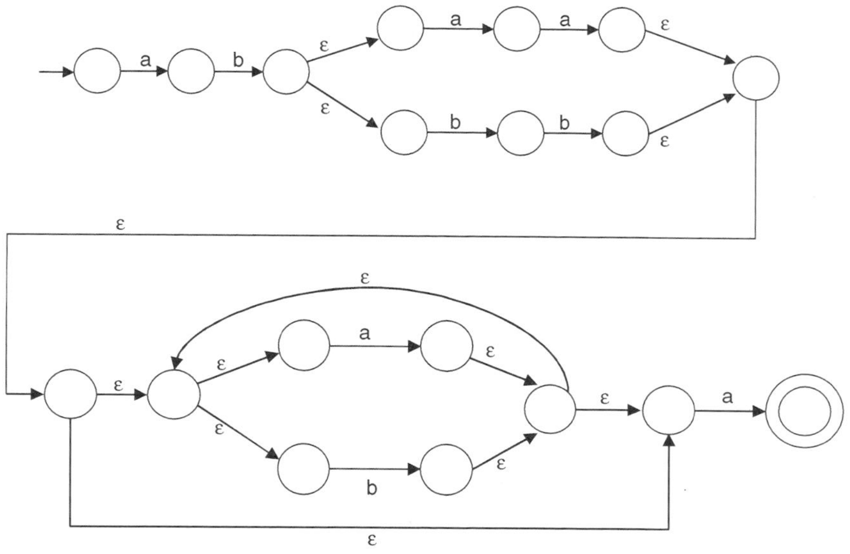

(g)

This can be simplified to Fig. 5.13 (h) by removing ε transitions.

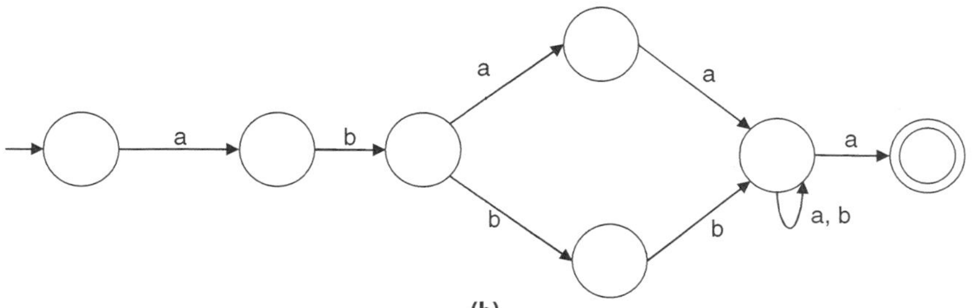

(h)

Figs. 5.13 (a – h) *Construction of FA from RE by Thomson Construction*

Kleene's Theorem: If R is an RE, then there exists an NDFA, M_{NFA} with ε transition, which accepts the language generated by R.

5.5.2 Direct Conversion of RE to DFA

Using this method, a given RE is converted to a DFA without the construction of an NFA. Before describing the process, we need to know a few definitions such as 'firstpos', 'followpos', 'nullable', and the construction processes of those.

firstpos: It is the abbreviation of first position. It is defined as the set of the positions of the first symbols of strings generated by the sub-expression rooted by n.

followpos: It is the abbreviation of follow position. It is defined as the set of the positions of the last symbols of strings generated by the sub-expression rooted by n.

Nullable: It gives two results 'true' or 'false'. It is true if the empty string is a member of strings generated by the sub-expression rooted by n and false otherwise.

The construction of firstpos is made according to the following table.

	nullable(n)	firstpos(n)	lastpos(n)
leaf labelled ε	true	Φ	Φ
leaf labelled with position i	false	{i}	{i}
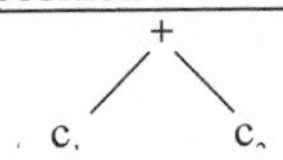	nullable(c_1) or nullable(c_2)	firstpos(c1) ∪ firstpos(c_2)	lastpos(c_1) ∪ lastpos(c_2)
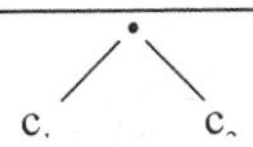	nullable(c_1) and nullable(c_2)	if (nullable(c_1)) firstpos(c_1) ∪ firstpos(c_2) else firstpos(c_1)	if (nullable(c_2)) lastpos(c_1) ∪ lastpos(c_2) else lastpos(c_2)
	true	firstpos(c_1)	lastpos(c_1)

followpos is constructed only for the leaf nodes. It is constructed in the following way.

- If n is a dot (.) node containing the left child c_1 and the right child c_2, and i is a position in lastpos(c_1), then all positions in firstpos(c_2) belong to followpos(i).
- If n is a star node, and i is a position in lastpos(n), then all positions in firstpos(n) belong to followpos(i).

The DFA is constructed by the following steps:

Step I: Make the RE R as augmented by placing an end marker #, and making it M#. Generate a parse tree from M#.

Step II: Calculate the firstpos and lastpos for all the internal and leaf nodes. Calculate the followpos for the leaf nodes.

Step III: Take the firstpos(root) as an unmarked state S of the constructing DFA.

Step IV: while (there exists an unmarked state S in the states of DFA)

do

Mark S and construct a transition from S using the following process for each input symbol 'a' as an alphabet of R

do

let S contain 'a' in position $i_1, i_2, \ldots, i_n$, then

$$S' = \text{followpos}(i_1) \cup \ldots \cup \text{followpos}(i_n)$$
$$\delta(S, a) = S'$$

if (S′ is not empty and have not appeared in the states of the DFA)
put S′ as an unmarked state into the states of the DFA.

Example 5.14 Convert the RE (a + b)*abb directly into an DFA.

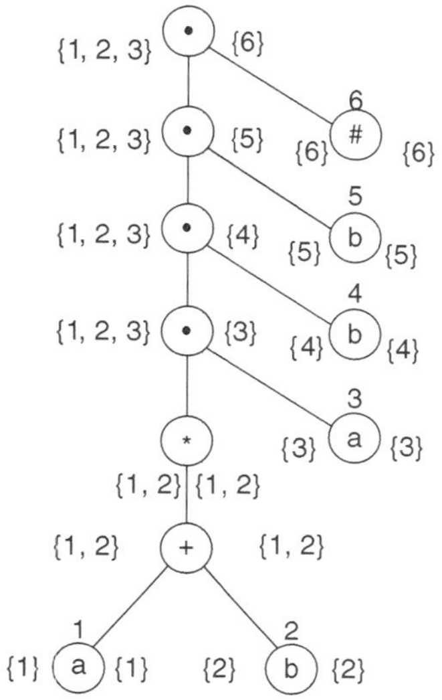

Fig. 5.14

Solution: The augmented RE is (a + b)*abb#

The parse tree constructed from the augmented RE in given in Fig. 5.14.

The LHS of each node (including the leaf) is the firstpost of that node and the RHS of each node is the lastpos of that node. The firstpos and lastpos are constructed from the table given previously.

[Consider the dot (•) node connecting * node and 'a'. The dot node is considered as c_1 and the * node as c_2. The nullable(c_1) is true as c_1 is a star node. So, the firstpos of the dot node is firstpos(c_1) ∪ firstpos(c_2).]

Consider the * node. The lastpos of the * node is {1,2}, and thus the followpos(1) and followpos(2) contain {1, 2}, as {1, 2} is the firstpos of the * node.

followpos(1) = {1, 2}
followpos(2) = {1, 2}

Consider the dot node connecting the * node (as c_1) and 'a' with label '3'(as c_2). The lastpos of c_1 contains {1, 2} and the firstpos of c_2 contains {3}. So, the followpos(1) and followpos(2) contain {3}

followpos(1) = {1, 2, 3}
followpos(2) = {1, 2, 3}

Consider the dot node connecting the dot node (as c_1) and 'b' with label '4'(as c_2). The lastpos of c_1 contains {3} and the firstpos of c_2 contains {4}. So, the followpos(3) contains {4}

followpos (3) = {4}

Consider the dot node connecting the dot node (as c_1) and 'b' with label '5'(as c_2). The lastpos of c_1 contains {4} and the firstpos of c_2 contains {5}. Thus, the followpos(4) contains {5}.

followpos(4) = {5}

Consider the dot node connecting the dot node (as c_1) and '#' with label '6'(as c_2). The lastpos of c_1 contains {5} and the firstpos of c_2 contains {6}. Thus, followpos(5) contains {6}.

followpos(5) = {6}
followpos(6) = { }

The firstpos of the root node is {1,2,3}. Mark it as S_1. The RE contains two symbols 'a' and 'b'. 'a' exists in positions '1' and '3', and 'b' appears in positions '2', '4', and '5'.

(a + b)*a b b #
1   2 3 4 5 6

$\delta(S_1, a)$ = followpos (1) ∪ followpos(3) = {1, 2, 3, 4}

It is other than S_1; therefore, mark it as S_2.

$\delta(S_1, b)$ = followpos(2) = {1, 2, 3} same as S_1.

Take $S_2 = \{1, 2, 3, 4\}$.

$\delta(S_2, a) = \text{followpos}(1) \cup \text{followpos}(3) = \{1, 2, 3, 4\} = S_2$
$\delta(S_2, b) = \text{followpos}(2) \cup \text{followpos}(4) = \{1, 2, 3, 5\}$ New state. Mark it as S_3
$\delta(S_3, a) = \text{followpos}(1) \cup \text{followpos}(3) = \{1, 2, 3, 4\} = S_2$
$\delta(S_3, b) = \text{followpos}(2) \cup \text{followpos}(5) = \{1, 2, 3, 6\}$ New state. Mark it as S_4
$\delta(S_4, a) = \text{followpos}(1) \cup \text{followpos}(3) = \{1, 2, 3, 4\} = S_2$
$\delta(S_4, b) = \text{followpos}(2) = \{1, 2, 3\} = S_1$

All the states are traversed, and no new state appears. Therefore, it is the final DFA. The transitional diagram of the DFA is given is Fig. 5.15.

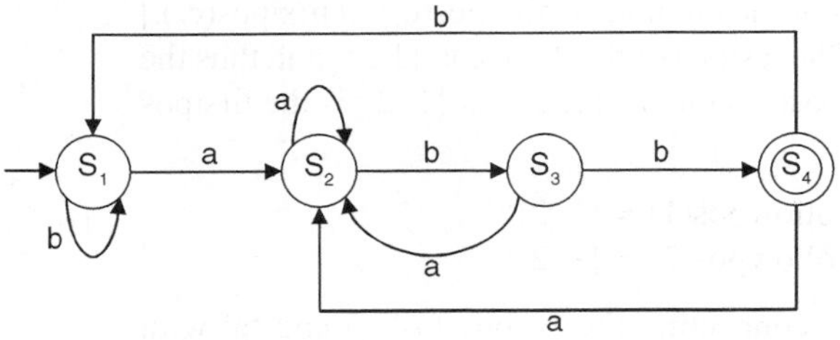

Fig. 5.15

5.6 NFA with ε Move and Conversion to DFA by ε-Closure Method

If any FA contains any ε (null) move or transaction, then that FA is called an NFA with ε moves. An example is shown in Fig. 5.16.

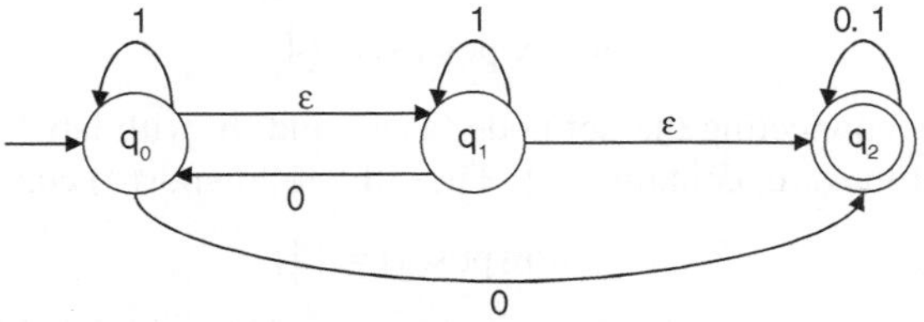

Fig. 5.16

The previous FA in Fig. 5.16 contains two null moves. So, the previous FA is an NFA with null move.

From the definition of an NFA, we know that for an NFA from a single state for a single input the machine can go to more than one state, i.e., $Q \times \Sigma \rightarrow 2^Q$, where 2^Q is the power set of Q.

From a state by getting ε input, the machine is confined into that state. An FA with null move must contain at least one ε move. For this type of FA for input ε, the machine can go to more than one state. (One is that same state and the another is the ε-transaction next state). So, an FA with ε move can be called as an NFA.

For the previous FA,

$$\delta(q_0, \varepsilon) \rightarrow q_0 \text{ and } \delta(q_0, \varepsilon) \rightarrow q_1$$
$$\delta(q_1, \varepsilon) \rightarrow q_1 \text{ and } \delta(q_1, \varepsilon) \rightarrow q_2$$

An NFA with ε move can be defined as

$$M_{\text{NFA null}} = \{Q, \Sigma, \delta, q_0, F\},$$

where Σ: Set of input alphabets including ε and δ is a transitional function mapping $Q \times \Sigma \rightarrow 2^Q$, where 2^Q is the power set of Q.

ε-closure: ε-Closure of a state is defined as the set of all states S, such that it can reach from that state to all the states in S with input ε (i.e., with no input).

Example 5.15 Consider the following NFA in Fig. 5.17 with ε moves.

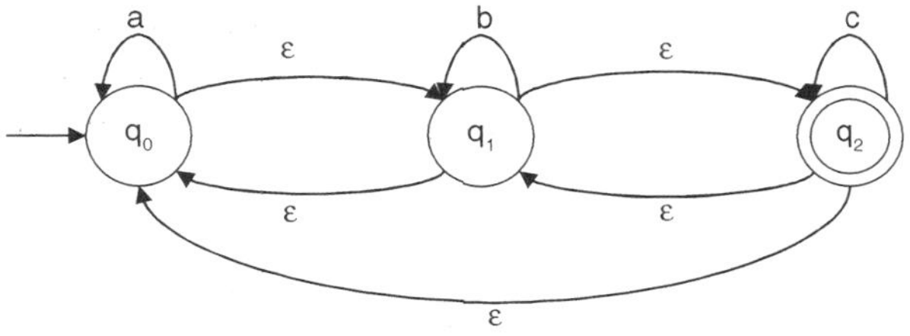

Fig. 5.17

From state q_0 by input ε (i.e., with no input), we can reach states q_1 and q_2. With no input, we can reach the same state, i.e., q_0. So, we can write

$$\varepsilon\text{-closure}(q_0) = \{q_0, q_1, q_2\}$$

Similarly, ε-closure(q_1) = $\{q_0, q_1, q_2\}$.

ε-closure(q_2) = $\{q_0, q_1, q_2\}$ (from q_1 with no input we can reach q_0, from q_2 with no input we can reach q_1, and so from q_2 with no input we can reach q_0).

5.6.1 Conversion of an NFA with ε move to a DFA

In the FA, it is discussed on how to remove ε moves from an FA. The method was the transitional method. In this section, we shall learn the removal of ε moves from an FA using the ε-closure method.

Step I: Find ε-closure of all the states.

Step II: Construct a transitional function δ′ for the beginning state for all inputs. The function δ′ will be calculated similar to the following.

$$\delta'(Q, i/p) = \varepsilon\text{–closure}(\delta(Q, i/p))$$

Let ε–closure(A) = $\{q_1, q_2, q_3\}$ = Q, So, the previous equation will become

$$\delta'(Q, i/p) = \varepsilon\text{-closure}(\delta(\{q_1, q_2, q_3\}, i/p))$$
$$= \varepsilon\text{-closure}(\delta(q_1, i/p) \cup \delta(q_2, i/p) \cup \delta(q_3, i/p)).$$

Step III: First, find the ε-closure of the initial state. Rename the set of states as a new state. Then, find the function δ′ of that state for all inputs. If δ′ of that state for all inputs is constructed, then the state is known to be marked (fully traversed for all the inputs). If any new combinations of state appear in the process of marking other than the marked state, then rename that as a new state.

Step IV: Repeat step III for all new unmarked states. If no new state appears, then stop the construction.

Step V: In this process, the states which contain the final state as entry are the final states of the new NFA without ε move.

The following examples describe the process in detail.

Example 5.16 Convert the following NFA with ε-move in Fig. 5.18 to an equivalent DFA.

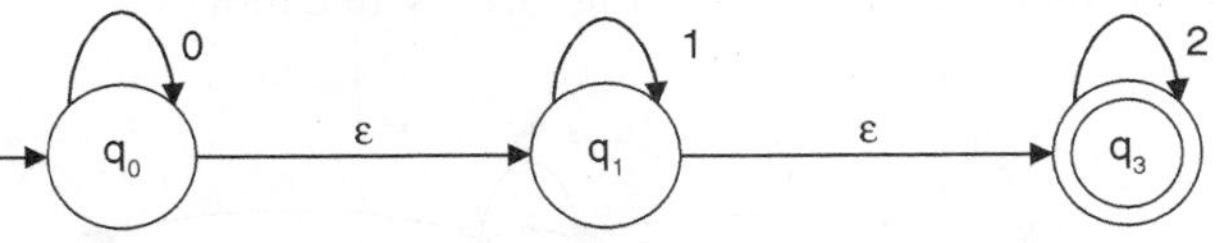

Fig. 5.18

Solution:

$$\varepsilon\text{-closure}(q_0) = \{q_0, q_1, q_2\}$$
$$\varepsilon\text{-closure}(q_1) = \{q_1, q_2\}$$
$$\varepsilon\text{-closure}(q_2) = \{q_2\}$$

ε-closure$(q_0) = \{q_0, q_1, q_2\}$. Let us rename this state as A. Then, construct the δ' function for the new unmarked state A for inputs 0, 1, and 2.

$$\begin{aligned}\delta'(A, 0) &= \varepsilon\text{-closure}(\delta\ (A, 0))\\ &= \varepsilon\text{-closure}\ (\delta((q_0, q_1, q_2), 0))\\ &= \varepsilon\text{-closure}\ (\delta(q_0, 0) \cup \delta(q_1, 0) \cup \delta(q_2, 0))\\ &= \varepsilon\text{-closure}\ (\{q_0\} \cup \{\phi\} \cup \{\phi\})\\ &= \varepsilon\text{-closure}(q_0) = \{q_0, q_1, q_2\} = A\end{aligned}$$

$$\begin{aligned}\delta'(A, 1) &= \varepsilon\text{-closure}(\delta\ (A, 1))\\ &= \varepsilon\text{-closure}\ (\delta((q_0, q_1, q_2), 1))\\ &= \varepsilon\text{-closure}\ (\delta(q_0, 1) \cup \delta(q_1, 1) \cup \delta(q_2, 1))\\ &= \varepsilon\text{-closure}(\{\phi\} \cup \{q_1\} \cup \{\phi\})\\ &= \varepsilon\text{-closure}(q_1) = \{q_1, q_2\}\end{aligned}$$

As it is a new combination of state other than A, rename it as B.

$$\begin{aligned}\delta'(A, 2) &= \varepsilon\text{-closure}(\delta\ (A, 2))\\ &= \varepsilon\text{-closure}\ (\delta((q_0, q_1, q_2)), 2)\\ &= \varepsilon\text{-closure}\ (\delta(q_0, 2) \cup \delta(q_1, 2) \cup \delta(q_2, 2))\\ &= \varepsilon\text{-closure}(\{\phi\} \cup \{\phi\} \cup \{q_2\})\\ &= \varepsilon\text{-closure}(q_2) = \{q_2\}\end{aligned}$$

As it is a new combination of state other than A and B, rename it as C.
As δ' is constructed for all inputs on A, A is marked.

B is unmarked still.
Then, construct the δ′ function for the new unmarked state B for inputs 0, 1, and 2.

$$\begin{aligned}\delta'(B, 0) &= \varepsilon\text{-closure}(\delta\ (B, 0)) \\ &= \varepsilon\text{-closure}\ (\delta(q_1, q_2), 0) \\ &= \varepsilon\text{-closure}\ (\delta(q_1, 0) \cup \delta(q_2, 0)) \\ &= \varepsilon\text{-closure}\ (\{\phi\} \cup \{\phi\}) \\ &= \varepsilon\text{-closure}(\phi) = \phi\end{aligned}$$

$$\begin{aligned}\delta'(B, 1) &= \varepsilon\text{-closure}(\delta\ (B, 1)) \\ &= \varepsilon\text{-closure}\ (\delta(q_1, q_2), 1) \\ &= \varepsilon\text{-closure}\ (\delta(q_1, 1) \cup \delta(q_2, 1)) \\ &= \varepsilon\text{-closure}\ (\{\phi\}) \\ &= \varepsilon\text{-closure}(q_1) = \{q_1, q_2\} = B\end{aligned}$$

$$\begin{aligned}\delta'(B, 2) &= \varepsilon\text{-closure}(\delta\ (B, 2)) \\ &= \varepsilon\text{-closure}\ (\delta(q_1, q_2), 2) \\ &= \varepsilon\text{-closure}\ (\delta(q_1, 2) \cup \delta(q_2, 2)) \\ &= \varepsilon\text{-closure}\ (\{\phi\} \cup \{q_2\}) \\ &= \varepsilon\text{-closure}(q_2) = \{q_2\} = C\end{aligned}$$

As δ′ is constructed for all inputs on B, B is marked.
C is unmarked still.
Then, construct the δ′ function for the new unmarked state C for inputs 0, 1, and 2.

$$\begin{aligned}\delta'(C, 0) &= \varepsilon\text{-closure}(\delta\ (C, 0)) \\ &= \varepsilon\text{-closure}\ (\delta(q_2), 0) \\ &= \varepsilon\text{-closure}\ (\delta(q_2, 0)) \\ &= \varepsilon\text{-closure}\ (\{\phi\}) \\ &= \varepsilon\text{-closure}(\phi) = \phi\end{aligned}$$

$$\begin{aligned}\delta'(C, 1) &= \varepsilon\text{-closure}(\delta\ (C, 1)) \\ &= \varepsilon\text{-closure}\ (\delta(q_2), 1) \\ &= \varepsilon\text{-closure}\ (\delta(q_2, 1)) \\ &= \varepsilon\text{-closure}\ (\{\phi\}) \\ &= \phi\end{aligned}$$

$$\begin{aligned}\delta'(C, 2) &= \varepsilon\text{-closure}(\delta\ (C, 2)) \\ &= \varepsilon\text{-closure}\ (\delta(q_2), 2) \\ &= \varepsilon\text{-closure}\ (\delta(q_2, 2)) \\ &= \varepsilon\text{-closure}\ (\{q_2\}) = \{q_2\} = C\end{aligned}$$

A, B, and C all contain q_2, which is the final state; therefore, A, B, and C are final states.
The equivalent DFA is given in Fig. 5.19.

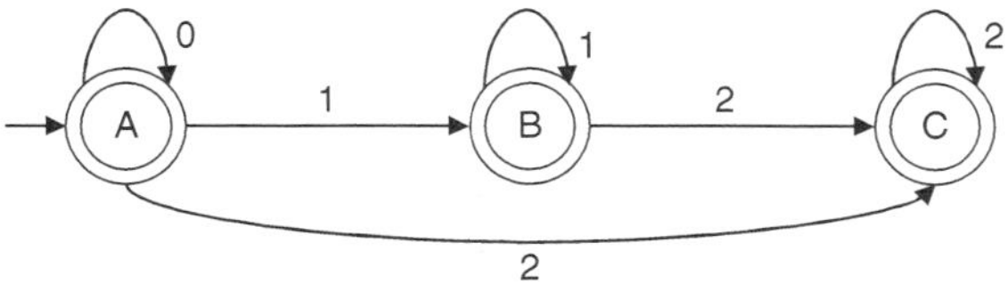

Fig. 5.19

Example 5.17 Convert the following NFA with ε-move in Fig. 5.20 to an equivalent DFA.

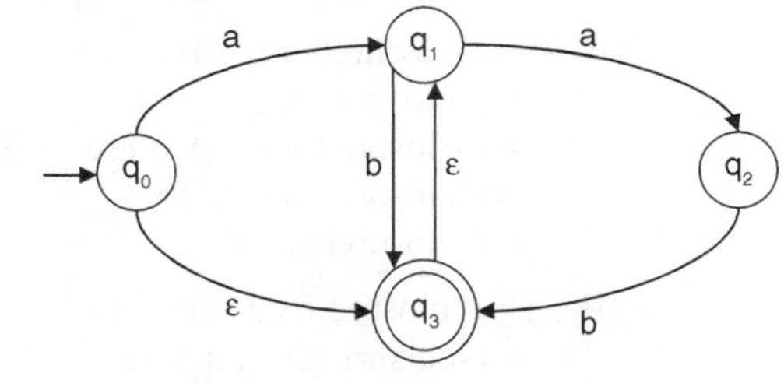

Fig. 5.20

Solution:

$$\varepsilon\text{-closure }(q_0) = \{q_0, q_3, q_1\}$$
$$\varepsilon\text{-closure }(q_1) = \{q_1\}$$
$$\varepsilon\text{-closure }(q_2) = \{q_2\}$$
$$\varepsilon\text{-closure }(q_3) = \{q_3, q_1\}$$

ε-closure(q_0) = {q_0, q_1, q_3}. Let us rename this state as A. Then, construct the δ′ function for the new unmarked state A for inputs 'a' and 'b'.

$$\begin{aligned}\delta'(A, a) &= \varepsilon\text{-closure}(\delta\ (A, a))\\ &= \varepsilon\text{-closure }(\delta((q_0, q_1, q_3), a))\\ &= \varepsilon\text{-closure }(\delta(q_0, a) \cup \delta(q_1, a) \cup \delta(q_3, a))\\ &= \varepsilon\text{-closure }(\{q_1\} \cup \{q_2\} \cup \{\phi\})\\ &= \varepsilon\text{-closure}(q_1, q_2) = \{q_1, q_2\}\end{aligned}$$

As it is a new combination of state other than A, rename it as B.

$$\begin{aligned}\delta'(A, b) &= \varepsilon\text{-closure}(\delta\ (A, b))\\ &= \varepsilon\text{-closure }(\delta((q_0, q_1, q_3), b))\\ &= \varepsilon\text{-closure }(\delta(q_0, b) \cup \delta(q_1, b) \cup \delta(q_3, b))\\ &= \varepsilon\text{-closure}(\{\phi\} \cup \{q_3\} \cup \{\phi\}) = \varepsilon\text{-closure}(q_1) = \{q_3, q_1\}\end{aligned}$$

As it is a new combination of state other than A and B, rename it as C.
As δ′ is constructed for all inputs on A, A is marked.
B is unmarked still.
Then, construct the δ′ function for the new unmarked state B for inputs 'a' and 'b'.

$$\begin{aligned}\delta'(B, a) &= \varepsilon\text{-closure}(\delta\ (B, a))\\ &= \varepsilon\text{-closure }(\delta((q_1, q_2), a))\\ &= \varepsilon\text{-closure }(\delta(q_1, a) \cup \delta(q_2, a)))\\ &= \varepsilon\text{-closure }(\{q_2\} \cup \{\phi\}) = \varepsilon\text{-closure}(q_2) = \{q_2\}\end{aligned}$$

As it is a new combination of state other than A, B, and C, rename it as D.

$$\begin{aligned}\delta'(B, b) &= \varepsilon\text{-closure}(\delta\ (B, b))\\ &= \varepsilon\text{-closure }(\delta((q_1, q_2), b))\\ &= \varepsilon\text{-closure }(\delta(q_1, b) \cup \delta(q_2, b)))\\ &= \varepsilon\text{-closure }(\{q_3\} \cup \{q_3\}) = \varepsilon\text{-closure}(q_3) = \{q_1, q_3\} = C\end{aligned}$$

As δ′ is constructed for all inputs on B, B is marked.
C is unmarked still.
Then, construct the δ′ function for the new unmarked state C for inputs 'a' and 'b'.

$$\begin{aligned}\delta'(C, a) &= \varepsilon\text{-closure}(\delta\,(C, a)) \\ &= \varepsilon\text{-closure}\,(\delta((q_1, q_3), a)) \\ &= \varepsilon\text{-closure}\,(\delta(q_1, a) \cup \delta(q_3, a))) \\ &= \varepsilon\text{-closure}\,(\{q_2\} \cup \{\phi\}) \\ &= \varepsilon\text{-closure}(q_2) = \{q_2\} = D\end{aligned}$$

$$\begin{aligned}\delta'(C, b) &= \varepsilon\text{-closure}(\delta\,(C, b)) \\ &= \varepsilon\text{-closure}\,(\delta((q_1, q_3), b)) \\ &= \varepsilon\text{-closure}\,(\delta(q_1, b) \cup \delta(q_3, b))) \\ &= \varepsilon\text{-closure}\,(\{q_3\} \cup \{\phi\}) \\ &= \varepsilon\text{-closure}(q_3) = \{q_1, q_3\} = C\end{aligned}$$

As δ′ is constructed for all inputs on C, C is marked.
D is unmarked still.
Then, construct the δ′ function for the new unmarked state D for inputs 'a' and 'b'.

$$\begin{aligned}\delta'(D, a) &= \varepsilon\text{-closure}(\delta\,(D, a)) \\ &= \varepsilon\text{-closure}\,(\delta((q_2), a)) \\ &= \varepsilon\text{-closure}\,(\delta(q_2, a)) \\ &= \varepsilon\text{-closure}\,(\{\phi\}) \\ &= \phi\end{aligned}$$

$$\begin{aligned}\delta'(D, b) &= \varepsilon\text{-closure}(\delta\,(D, b)) \\ &= \varepsilon\text{-closure}\,(\delta((q_2), b)) \\ &= \varepsilon\text{-closure}\,(\delta(q_2, b)) \\ &= \varepsilon\text{-closure}\,(\{q_3\}) \\ &= \{q_1, q_3\} = C\end{aligned}$$

States A and C are final states as the states contain q_3.
The equivalent DFA is given in Fig. 5.21

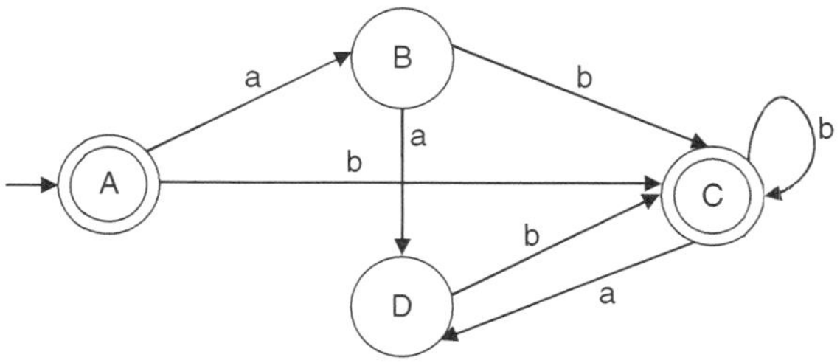

Fig. 5.21

5.7 Equivalence of Two Finite Automata

Two FA are said to be equivalent if they generate the same (equivalent) RE. If both the FA are minimized, they reveal the equivalence by being minimized to the same number of states and the same transitional functions. In this section, we shall discuss the process of proving the equivalence between two FA.

Let there be two FA, M and M′, where the number of input symbols are the same.

- Make a comparison table with n + 1 columns, where n is the number of input symbols.
- In the first column, there will be a pair of vertices (q, q′) where q ∈ M and q′ ∈ M′ . The first pair of vertices will be the initial states of the two machines M and M′. The second column consists of (q_a, q'_a), where q_a is reachable from the initial state of the machine M for the first input, and q'_a is reachable from the initial state of the machine M′ for the first input. The other n – 2 columns consist of a pair of vertices from M and M′ for n – 1 inputs, where n = 2, 3,........... n – 1.
- If any new pair of states appear in any of the n – 1 next state columns, which were not taken in the first column, take that pair in the present state column and construct subsequent column elements like the first row.
- If a pair of states (q,q′) appear in any of the n columns for a pair of states in the present state column, where q is the final state of M and q′ is the non-final state of M′ or vice versa, terminate the construction and conclude that M and M′ are not equivalent.
- If no new pair of states appear, which were not taken in the first column, stop the construction and declare that M and M′ are equivalent.

The following examples describe this in detail.

Example 5.18 Find whether the two DFAs given in Fig. 5.22 are equivalent or not.

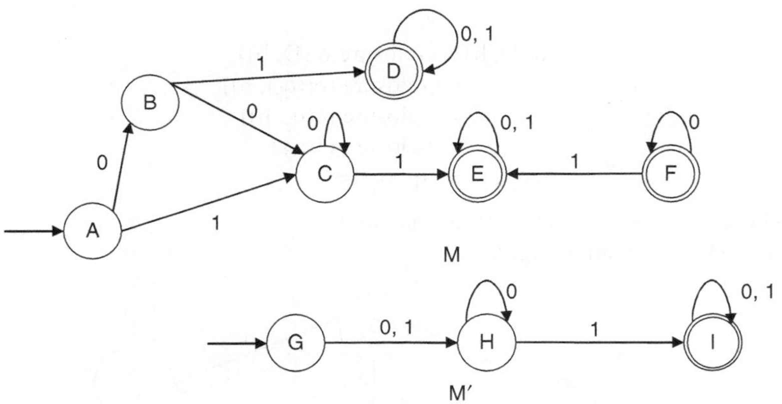

Fig. 5.22

Solution: In the previous two DFAs, the number of input is two 0 and 1. Therefore, the comparison table is of three columns. The beginning state pair is (A, G). From there, for input 0, we are getting (B, H) and for input 1 we are getting (C, H). Both of them are new combination of states. Among them, first take (B, H) in the present state column. The next state pairs for input 0 and 1 are constructed. By this process, the comparison continues.

After the fifth step, further construction is stopped as no new state combinations have appeared. In the whole table, no such type of combination of states appear where one state is the final state of one machine and the other is the non-final state of another machine.

Therefore, the two DFAs are equivalent.

Present State (q, q′)	Next State	
	For I/P 0 (q_0, q'_0)	For I/P 1 (q_1, q'_1)
(A, G)	(B, H)	(C, H)
(B, H)	(C, H)	(D, I)
(C, H)	(C, H)	(E, I)
(E, I)	(E, I)	(E, I)
(D, I)	(D, I)	(D, I)

Example 5.19 Find whether the two DFAs given in Fig. 5.23 are equivalent or not.

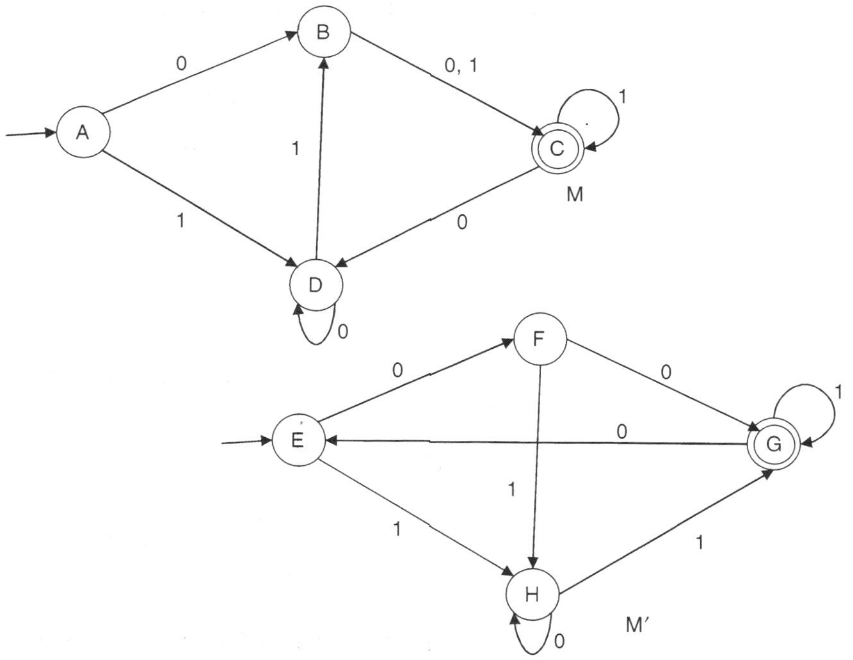

Fig. 5.23

Solution: In the previous two DFAs, the number of input is two 0 and 1. Therefore, the comparison table is of three columns. The beginning state pair is (A, E). From there, for input 0, we get (B, F) and for input 1 we get (D, H). Both of them are new combination of states. Among them, take (B, F) in the present state column. The next state pairs for input 0 is (C, G) and for input 1 is (C, H). (C, H) is a combination of states where C is the final state of M and H is the non-final state of M′. Further construction stops here declaring that the two DFAs are not equivalent.

Present State (q, q′)	Next State	
	For I/P 0 (q_0, q'_0)	For I/P 1 (q_1, q'_1)
(A, E)	(B, F)	(D, H)
(B, F)	(C, G)	(C, H)

5.8 Equivalence of Two Regular Expressions

For every RE, there is an accepting FA. If the FA constructed from both of the REs are the same, then we can say that two REs are equivalent.

Example 5.20 Prove that the following REs are equivalent.

$$L_1 = (a + b)^* \quad L_2 = a^*(b^*a)^*$$

Solution: The FA for L_1 is constructed in Fig. 5.24.

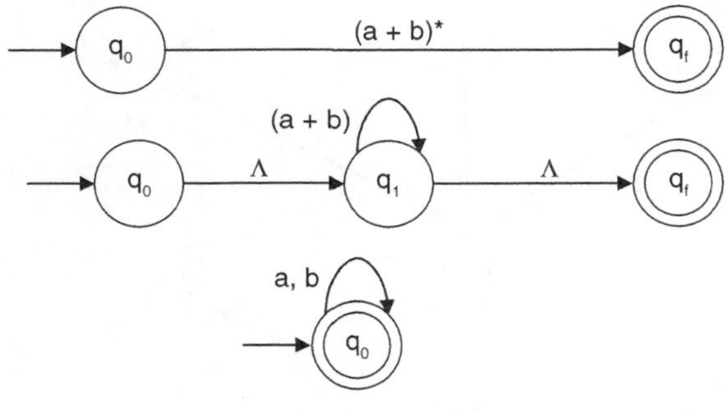

Fig. 5.24

The FA for L_2 is constructed in Fig. 5.25.

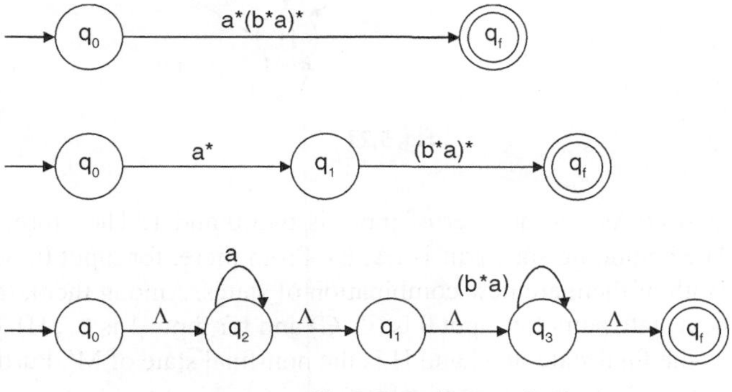

Fig. 5.25 *(Continued)*

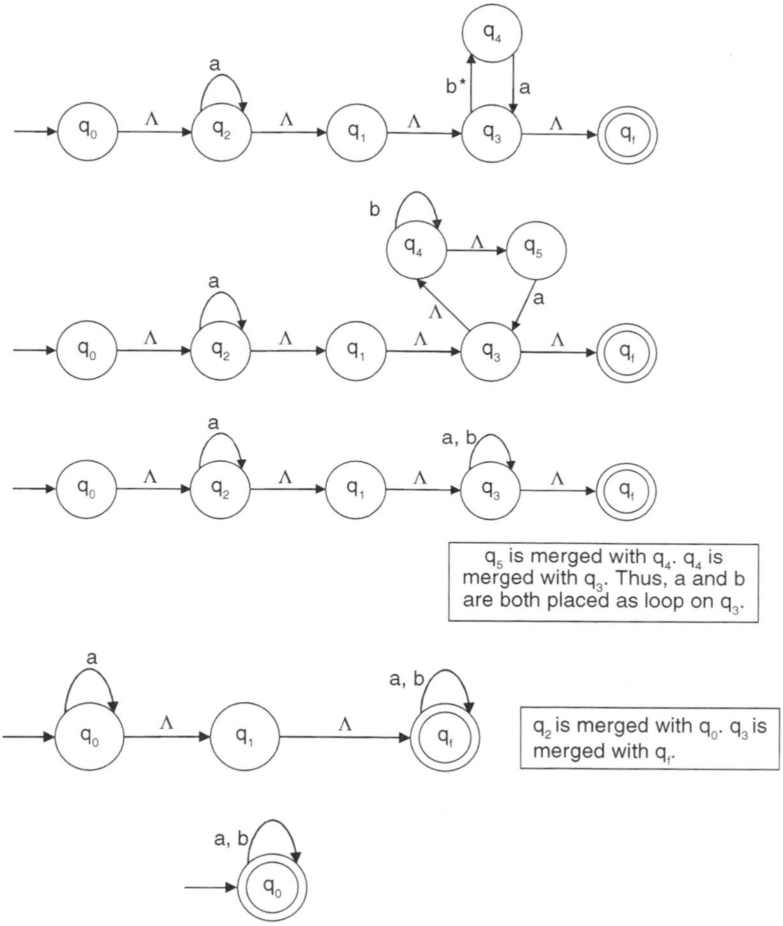

Fig. 5.25

We are seeing that the FA generated by the two REs L_1 and L_2 are the same. So, we can decide that the two FA are equivalent.

Example 5.21 Prove that the following REs are equivalent.

$$L_1 = 1^*(011)^*(1^*(011)^*)^* \quad L_2 = (1 + 011)^*$$

Solution: The FA for L_1 is constructed in Fig. 5.26.

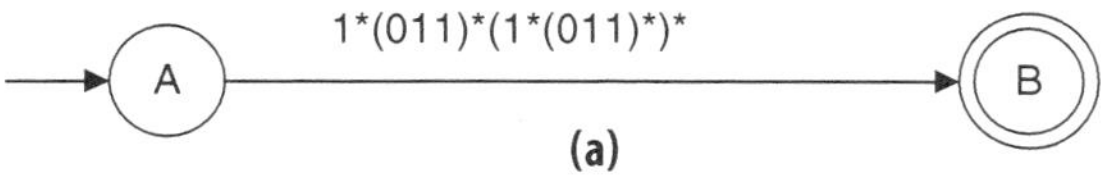

(a)

Fig. 5.26 *(Continued)*

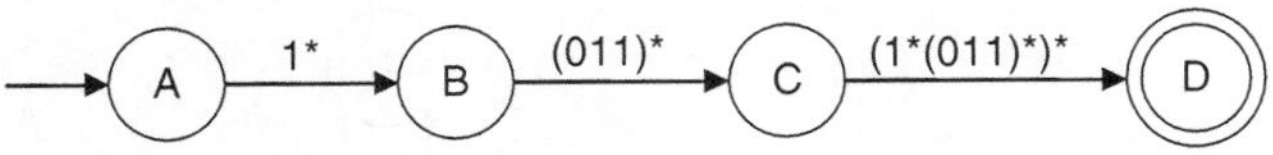

(b)

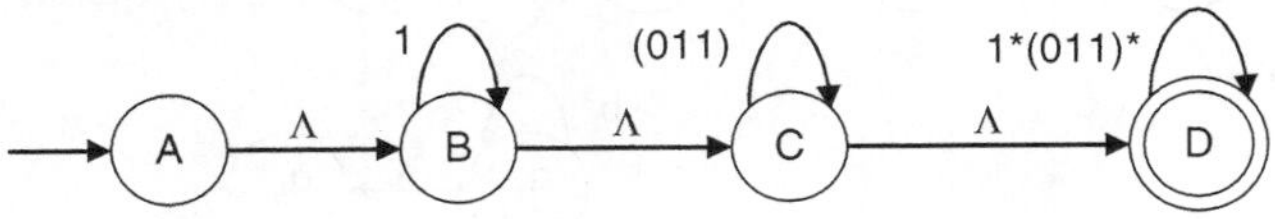

(c)

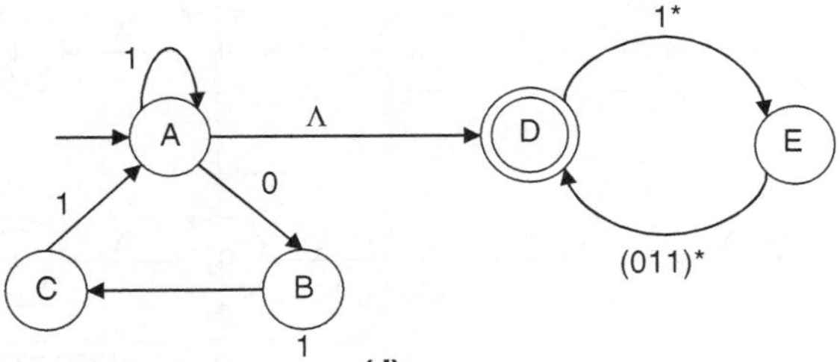

(d)

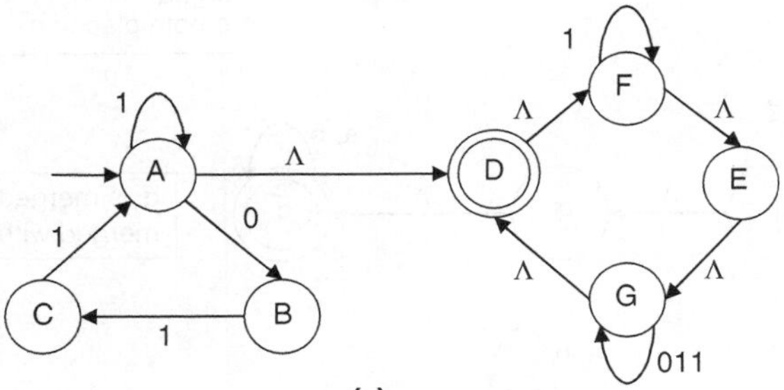

(e)

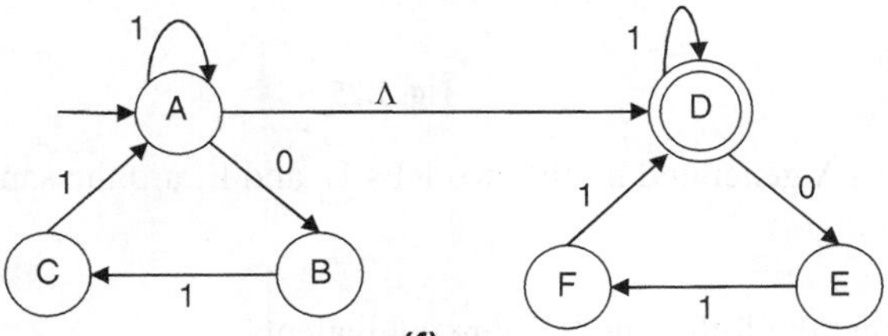

(f)

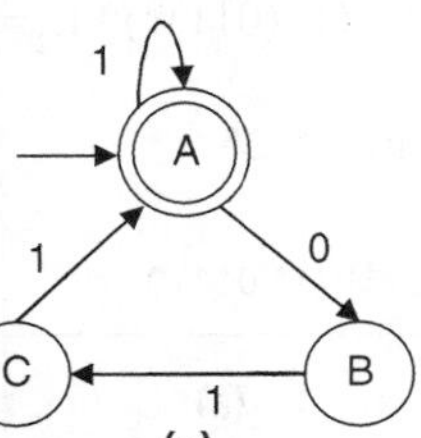

(g)

Fig. 5.26

The FA for L2 is constructed in Fig. 5.27.

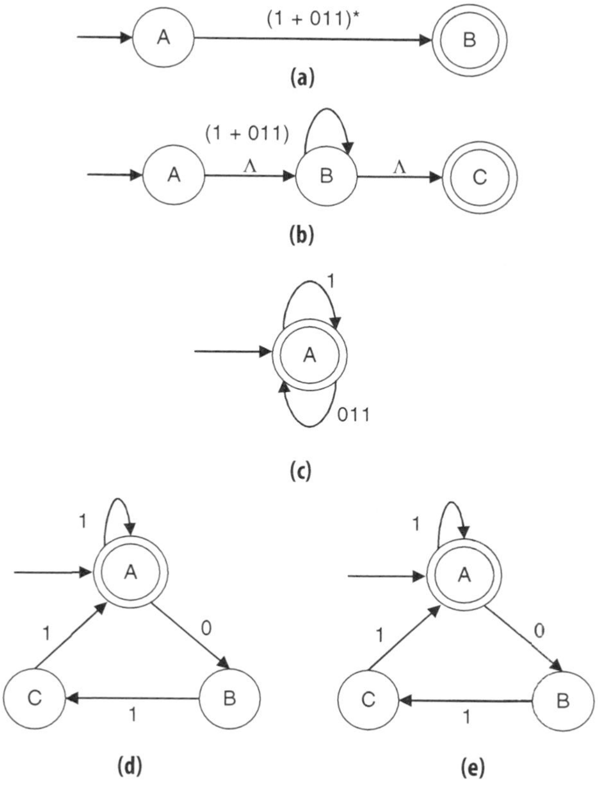

Fig. 5.27

The FA generated by the two REs L_1 and L_2 are the same and, therefore, they are equivalent.

5.9 Construction of Regular Grammar from an RE

We have learnt how to construct a grammar from a language. But in the case of an RE, it is a little difficult to find the exact grammar of it using the conventional methods. The following section describes the process of constructing regular grammar from an RE.

Step I: Construct the equivalent FA for the given RE (eliminate all null moves).

Step II: The number of non-terminals of the grammar will be equal to the number of states of the FA.

Step III: For all transitional functions in the form $\delta(Q_1, a) \rightarrow Q_2$, $[Q_1, Q_2 \in Q$ and $a \in \Sigma]$, the production rule is in the form $A \rightarrow aA_1$. If Q_2 is a final state, then for the transitional function $\delta(Q_1, a) \rightarrow Q_2$ the production rules are $A \rightarrow aA_1$ and $A \rightarrow a$. The start symbol is the symbol representing the initial state of the FA.

The following examples describe the discussed method in detail.

Example 5.22 Construct a regular grammar for the following REs:

(i) a*(a + b)b*

(ii) ab(a + b)*

(iii) 10 + (0 + 11)0*1

Solution:

(i) The FA for the string a*(a + b)b* is constructed in Fig. 5.28.

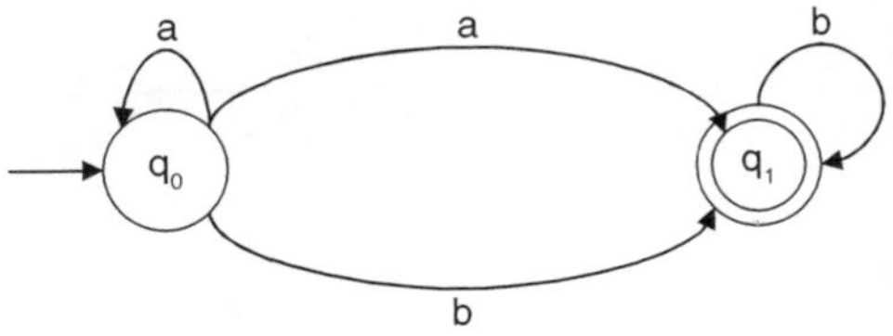

Fig. 5.28

The number of states of the FA is 2, which means that there are two non-terminals in the grammar for the RE. Let us take them as A (for q_0) and B (for q_1).

For the transitional function $\delta(q_0, a) \rightarrow q_0$, the production rule will be $A \rightarrow aA$. For the transitional function $\delta(q_0, a) \rightarrow q_1$, the production rule will be $A \rightarrow aB$. As q_1 is a final state, there will be another production rule $A \rightarrow a$.

$\delta(q_0, b) \rightarrow q_1 : A \rightarrow bB$ and $A \rightarrow b$ (as q_1 is a final state)
$\delta(q_1, b) \rightarrow q_1 : B \rightarrow bB$ and $B \rightarrow b$ (as q_1 is a final state).

The start symbol will be A as q_0 is the beginning state.

The grammar G for the RE a*(a + b)b* is $\{V_N, \Sigma, P, S\}$ where

$VN = \{A, B\}$
$\Sigma = \{a, b\}$
$P : \{A \rightarrow aA/\ bB/a/b,\ B \rightarrow bB/b\}$
$S : \{A\}$.

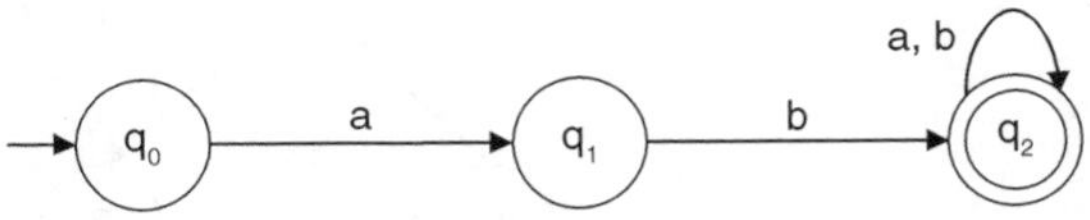

Fig. 5.29

(ii) The DFA for the string ab(a + b)* is given in Fig. 5.29.

There are three states in the DFA. The number of non-terminals for the grammar of the RE will be 3. Let us take them as A (for q_0), B (for q_1), and C (for q_2).

For the transitional function $\delta(q_0, a) \rightarrow q_1$, the production rule will be $A \rightarrow aB$

$\delta(q_1, b) \rightarrow q_2 : B \rightarrow bC$ and $B \rightarrow b$ (as q_2 is a final state)
$\delta(q_2, a) \rightarrow q_2 : C \rightarrow aC$ and $C \rightarrow a$ (as q_2 is a final state)
$\delta(q_2, b) \rightarrow q_2 : C \rightarrow bC$ and $C \rightarrow b$ (as q_2 is a final state)

The start symbol will be A as q_0 is the beginning state.

The grammar G for the RE ab(a + b)* is $\{V_N, \Sigma, P, S\}$ where

$VN = \{A, B, C\}$
$\Sigma = \{a, b\}$
$P : \{A \rightarrow aB,\ B \rightarrow bC/b,\ C \rightarrow aC/bC/a/b\}$
$S: \{A\}$.

(iii) The NFA without null move for the RE 10 + (0 + 11)0*1 is shown in Fig. 5.30.

The equivalent DFA for the previous NFA is given in Fig. 5.31.

There are four states of the DFA for the RE 10 + (0 + 11)0*1. So, for the grammar of the RE, there will be four non-terminals. Let us take them as A (for Q_0), B (for Q_1), C (for Q_2), and D (for Q_3).

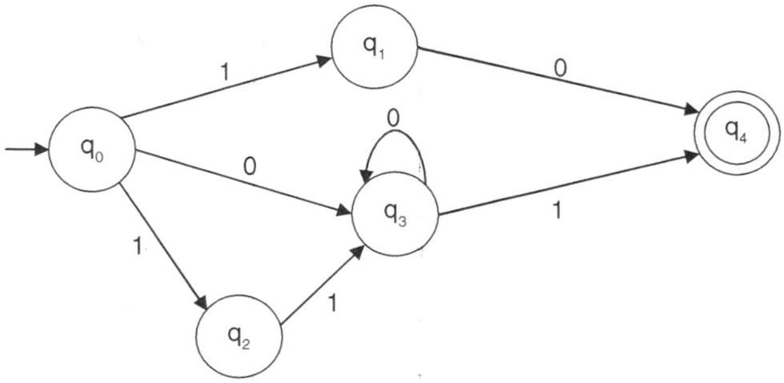

Fig. 5.30

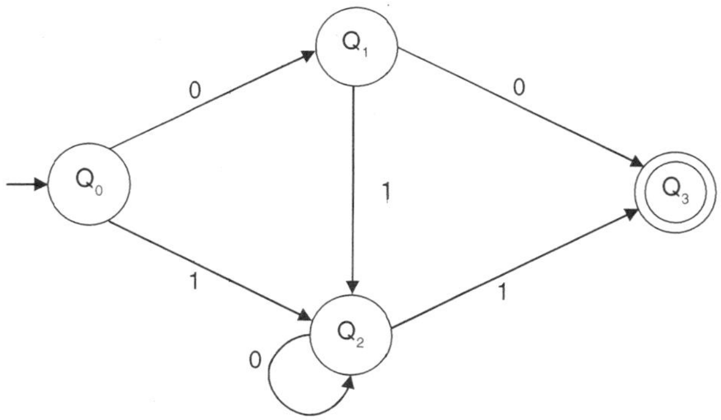

Fig. 5.31

Now, we have to construct the production rules of the grammar.

For the transitional function $\delta(Q_0, 0) \rightarrow Q_1$, the production rule will be A → 0B.

Similar to $\delta(Q_0, 1) \rightarrow Q_2$, the production rule will be A → 1C.

$\delta(Q_1, 0) \rightarrow Q_3$, the production rule will be B → 0D and B → 0 (as Q_3 is the final state)

$\delta(Q_1, 1) \rightarrow Q_2$, the production rule will be B → 1C

$\delta(Q_2, 0) \rightarrow Q_2$, the production rule will be C → 0C

$\delta(Q_2, 1) \rightarrow Q_3$, the production rule will be C → 1D and C → 1 (As Q_3 is the final state).

The start symbol will be A as Q_0 is the beginning state.

The grammar G for the RE 10 + (0 + 11)0*1 is $\{V_N, \Sigma, P, S\}$ where

$$VN = \{A, B, C, D\}$$
$$\Sigma = \{0, 1\}$$

P : {A → 0B/1C, B → 0D/1C/0, C → 0C/1D/1}

S : {A}.

(N.B: From NFA without ∈-move, the regular grammar can be constructed. We can do this by the construction of the equivalent DFA.)

5.10 Constructing FA from Regular Grammar

FA can directly be constructed from regular grammar. This can be considered as a reverse process of constructing the FA from an RE. The following section describes the process of constructing the FA from an RE.

Step I:

- If the grammar does not produce any null string, then the number of states of the FA is equal to the number of non-terminals of the regular grammar +1. Each state of the FA represents each non-terminal and the extra state is the final state of the FA. If it produces a null string, then the number of states is the same as the number of non-terminals.
- The initial state of the FA is the start symbol of the regular grammar.
- If the language generated by the regular grammar contains a null string, then the initial state is also the final state of the constructing FA.

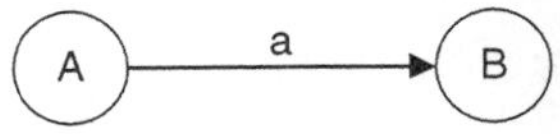

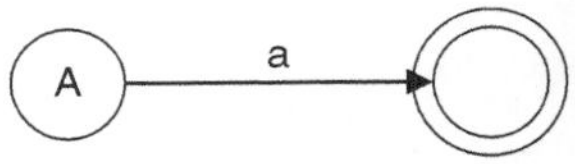

Step II:

- For a production in the form A → aB, make a δ function δ(A, a) → B. There is an arc from state A to state B with label a.
- For a production in the form A → a, make a δ function δ(A, a) → final state.
- For a production A → ∈, make a δ function δ(A, ∈) → A, and A is the final state.

Consider the following examples.

Example 5.23 Convert the following regular grammar into FA:

S → aA/bB/a/b
A → aS/bB/b
B → aA/bS.

Solution: In the grammar, there are three non-terminals, namely S, A, and B. Therefore, the number of states of the FA is four. Let us name the final state as C.

(i) For the production S → aA/bB, the transitional diagram is given in Fig. 5.32(a)

(ii) For the production S → a/b, the transitional diagram including the previous one becomes as Fig. 5.32(b).

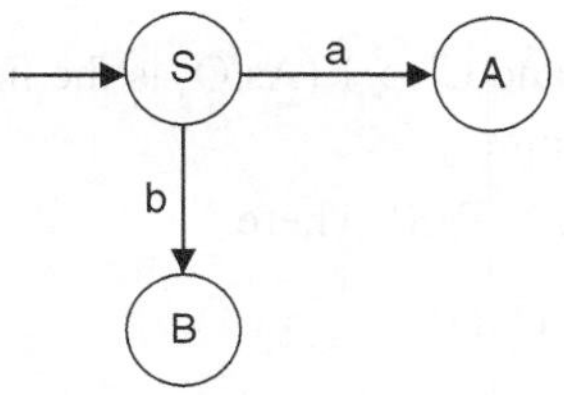

Fig. 5.32(a)

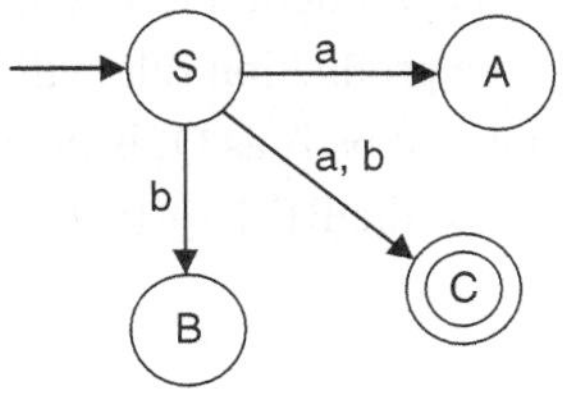

Fig. 5.32(b)

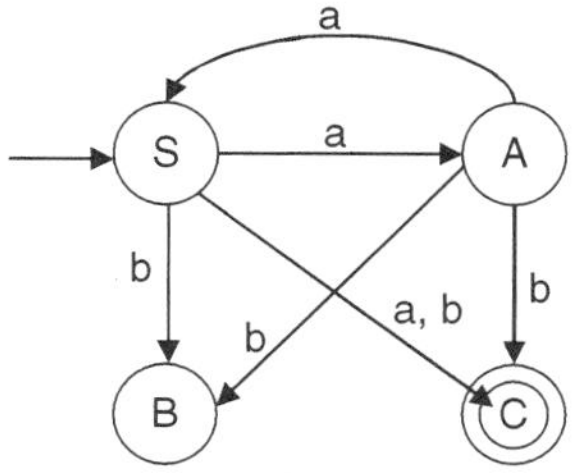

Fig. 5.32(c)

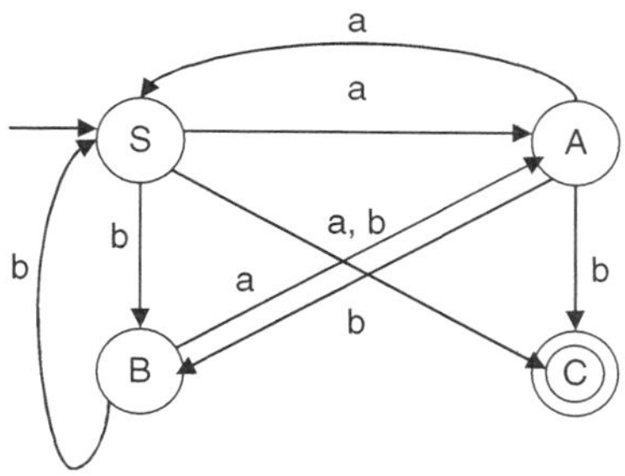

Fig. 5.32(d)

(iii) For the production A → aS/bB/b, the transitional diagram including the previous one looks like Fig. 5.32(c).
For the production B → aA/bS, the transitional diagram including the previous one looks like Fig. 5.32(d)

This is the FA for the given regular grammar.

Example 5.24 Convert the following regular grammar into FA:

S → aA/bS
A → bB/a
B → aS/b.

Solution: In the grammar, there are three non-terminals, namely S, A, and B. So, the number of states of the FA will be four. Let us name the final state as C.

- For the production S → aA/bS, the transitional diagram becomes
- For the production A → bB/a, the transitional diagram including the previous one is
- For the production B → aS/b, the transitional diagram including the previous one looks like

This is the FA for the given regular grammar. The construction process is described in Figs. 5.33(a) to 5.33(c).

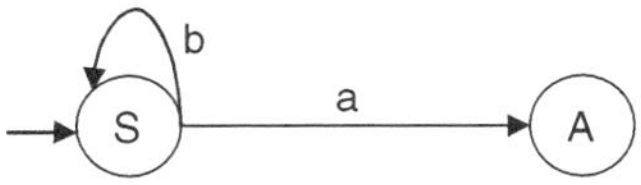

Fig. 5.33(a)

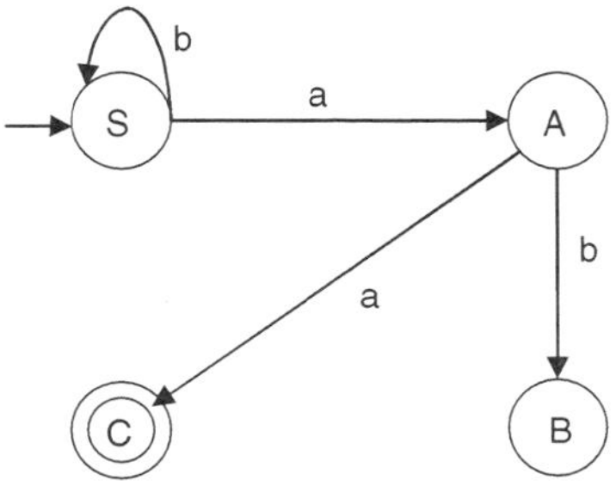

Fig. 5.33(b)

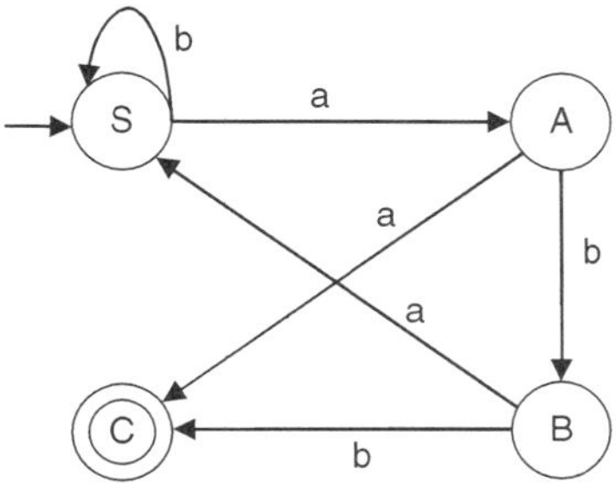

Fig. 5.33(c)

5.11 Pumping Lemma for Regular Expression

There is a necessary condition for an input string to belong to a regular set. This necessary condition is the pumping lemma. Pumping means generating. This lemma is called a pumping lemma because it gives a method of generating many input strings from a given string.

As the pumping lemma is a necessary condition for a string to belong to a regular set, it is used to prove that certain sets are not regular.If any set fulfills all the conditions of Pumping Lemma it can not be said confirm that the set is regular.

But the reverse is true, i.e., if any set breaks the conditions of the pumping lemma, it can be said that the set is not regular. So, the pumping lemma is used to prove that certain sets are not regular.

Theorem 5.2: **Statement of the pumping lemma:** Let $M = \{Q, \Sigma, \delta, q_0, F\}$ be an FA with n number of states. Let L be a regular set accepted by M. Let w be a string that belongs to the set L and $|w| \geq m$. If $m \geq n$, i.e., the length of the string is greater than or equal to the number of states, then there exists x, y, z such that $w = xyz$, where $|xy| \leq n$, $|y| > 0$ and $xy^iz \in L$ for each $i \geq 0$.

(It needs some clarification, after that we shall go to the proof of it.

Let us take the following FA in Fig. 5.34. Here, from q_0, by getting 'a' it goes to q_1 and from q_1 by getting 'b' it goes to q_2, which is the final state. The FA consists of three states, q_0, q_1, and q_2. The string that is accepted by the FA is ba. The length of the string is 2 which is less than the number of states of the FA.

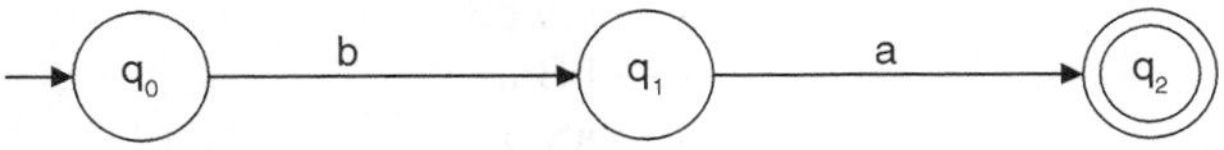

Fig. 5.34

Here, from a state by getting a single input, it goes to a single distinct state. But, if the length of the string is equal or greater than the number of states of the FA, then from a state by getting a single input it is not possible to get all single distinct states. There must be a repetition of at least a single state. This can be described by the following diagram in Fig. 5.35.

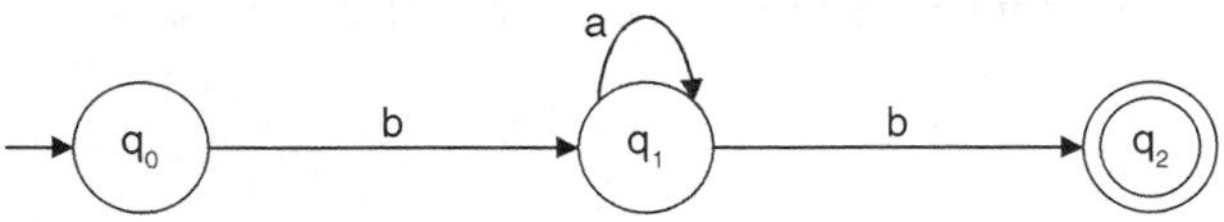

Fig. 5.35

The RE accepted by the automata is ba*b. The expression can be divided into three parts: x, y, z where y is the looping portion, x is the portion before looping, and z is the portion after the looping portion.)

Proof: Let w be a string of length m where m is greater than n, the number of states of the FA accepting the string w as given in Fig. 5.36.

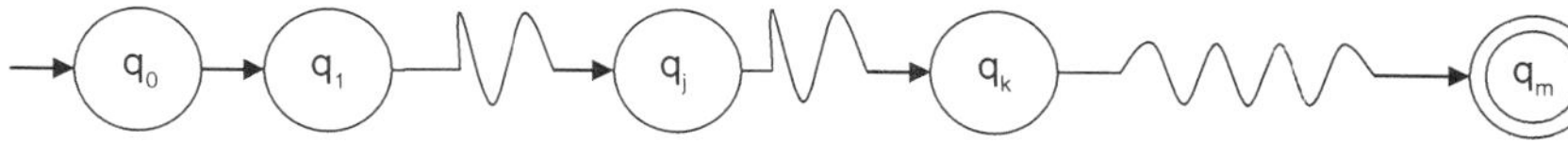

Fig. 5.36

$$w = a_1a_2a_3.......am \mid w \mid = m \geq n$$

Starting from the beginning state q_0, by getting the string w as input, the machine reaches the final state.

In general,

$$\delta(q_0, a_1a2_{a3}.......a_i) = q_i \text{ for } i = 1,2,3, ...m.$$

The set of the states followed by the string w in the transitional path from the beginning state to the final state are $Q_n = \{q_0, q_1, q_2, ... q_m\}$. The number of states in the set is m.

But the number of states of the FA accepting the string w is n. As $m \geq n$, at least two states in the set Q_n must coincide.

Take two integers j and k where $0 \leq j < k \leq n$. Among the various pairs of repeated states, take a pair q_j, q_k. The string w can be decomposed into three substrings: $a_1a_2 ..... a_j$, $a_{j+1} ..... a_k$, and $a_{k+1} ..... a_{m.}$ Let us denote them as x, y, and z, respectively. As $k \leq n$, the length of the string $a_1...a_k$ is $\leq n$, i.e., $| xy | \leq n$ and $w = xyz$.

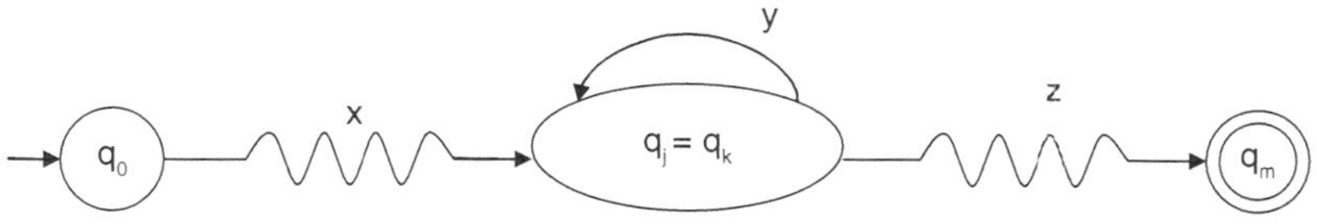

Fig. 5.37

The automaton starts from the initial state q_0. On applying the string x, it reaches the state q_j. On applying the string y, it comes back to q_j as $q_j = q_k$ as denoted in Fig. 5.37. So, on applying the string y, i number of times (where $i \geq 0$) on q_j, it will be in the same state q_j. On applying z on q_j, the automaton will reach the final state. Hence, the string xy^iz with $i \geq 0$ belongs to the language set L., i.e., $xy^iz \in L$.

Application of the pumping lemma: The pumping lemma is used to prove that certain sets are not regular. If an expression satisfies the conditions for the pumping lemma to be good, then it cannot be said that the expression is regular. But the opposite is true. If any expression breaks any condition of the pumping lemma, it can be declared that the expression is not good.

This needs certain steps.

Step I: Assume that the set L is regular. Let n be the number of states of the FA accepting L.

Step II: Choose a string w ($w \in L$) such that $| w | \geq n$. By using the pumping lemma, we can write $w = xyz$, with $| xy | \leq n$ and $| y | > 0$.

Step III: Find a suitable integer i such that $xy^iz \notin L$. This will contradict our assumption. From here, L will be declared as not regular.

The following examples help to clear this.

Example 5.25 Show that $L = \{a^{i^2} \mid i \geq 1\}$ is not regular.

Solution:

Step I: Assume the set L is regular. Let n be the number of states of the FA accepting the set L.

Step II: Let $w = a^{n^2}$. $|w| = n^2$ which is greater than n, the number of states of the FA accepting L. By using the pumping lemma, we can write $w = xyz$ with $|xy| \leq n$ and $|y| > 0$

Step III: Take i = 2. So, the string will become xy^2z.

$$|xy^2z| = |x| + 2|y| + |z| > |x| + |y| + |z| \text{ as } |y| > 0.$$

From step II, we know $|w| = |xyz| = |x| + |y| + |z| = n^2$.
So, $|xy^2z| > n^2$.
Again, $|xy^2z| = |x| + 2|y| + |z| = |x| + |y| + |z| + |y| = n^2 + |y|$.
As $|xy| \leq n$, $|y| \leq n$.
Therefore, $|xy^2z| \leq n^2 + n$.
From the previous derivations, we can write

$$n^2 < |xy^2z| \leq n^2 + n < n^2 + n + n + 1$$
$$n^2 < |xy^2z| < (n + 1)^2.$$

Hence, $|xy^2z|$ lies between n^2 and $(n + 1)^2$. They are the square of two consecutive positive integers. In between the square of two consecutive positive integers, no square of positive integer belongs. But a^{i^2}, where $i \geq 1$ is a perfect square of an integer. So, the string derived from it, i.e., $|xy^2z|$ is also a square of an integer, which lies between the square of two consecutive positive integers. This is not possible.
So, $xy^2z \notin L$. This is a contradiction.
So, $L = L = \{a^{i^2} \mid i \geq 1\}$ is not regular.

Example 5.26 Show that $L = \{a^p \mid p \text{ is prime}\}$ is not regular.

Solution:

Step I: Assume that the set L is regular. Let n be the number of states in the FA accepting L.

Step II: Let p be a prime number which is greater than n. Let the string $w = a^p$, $w \in L$. By using the pumping lemma, we can write $w = xyz$ with $|xy| \leq n$ and $|y| > 0$. As the string w consists of only 'a', x, y, and z are also a string of 'a's. Let us assume that $y = a^m$ for some m with $1 \leq m \leq n$.

Step III: Let us take $i = p + 1$. $|xy^iz|$ will be $|xyz| + |y^{i-1}|$

$$\begin{aligned} |xy^iz| &= |xyz| + |y^{i-1}| \\ &= p + (i - 1)|y| \quad [xyz = a^p] \\ &= p + (i - 1)m \quad [y = a^m] \\ &= p + pm \quad [i = p + 1] \\ &= p(1 + m). \end{aligned}$$

$p(1 + m)$ is not a prime number as it has factors p and $(1 + m)$ including 1 and $p(1 + m)$. So, $xy^iz \notin L$. This is a contradiction.
Therefore, $L = \{a^p \mid p \text{ is prime}\}$ is not regular.

Example 5.27 Show that $L = \{a^{i^3} \mid i \geq 1\}$ is not regular.

Solution:

Step I: Assume that the set L is regular. Let n be the number of states of the FA accepting the set L.

Step II: Let $L = a^{n^3}$. $|w| = n^3$ which is greater than n, the number of states of the FA accepting L. By using the pumping lemma, we can write $w = xyz$ with $|xy| \leq n$ and $|y| > 0$.

Step III: Take $i = 3$. So, the string will become xy^3z.

$$|xy^3z| = |x| + 3|y| + |z| = |x| + |y| + |z| + 2|y| = n^3 + 2|y|$$

As $|xy| \leq n$ so, $|y| \leq n$.
Therefore, $|xy^3z| = n^3 + 2|y| \leq n^3 + n < n^3 + 3n^2 + 3n + 1 < (n + 1)^3$.
As $|y| > 0$, $|xy^3z| = |xyz| + 2|y| > n^3$.
We can write

$$n^3 < |xy^3z| < (n + 1)^3.$$

xy^3z is a perfect cube which lies between n^3 and $(n + 1)^3$, i.e., two consecutive perfect cube numbers. In between the cube of two consecutive positive integers, no cube of positive integer belongs. But a^{i^3}, where $i \geq 1$ is a perfect cube of an integer. So, the string derived from it, i.e., $|xy^3z|$, is also a cube of an integer, which lies between a cube of two consecutive positive integers. This is not possible. So, $xy^3z \notin L$. This is a contradiction. So, $L = \{a^{i^3} \mid i \geq 1\}$ is not regular.

Example 5.28 Show that $L = \{a^nb^n \text{ where } n \geq 1\}$ is not regular.

Solution:

Step I: Let us suppose that the set L is regular. Let n be the number of states of the FA accepting L.

Step II: Let $w = a^nb^n$, $|w| = 2n > n$. By the pumping lemma, we can write $w = xyz$ with $|xy| \leq n$ and $|y| > 0$.

Step III: We want to find a suitable i so that $xy^iz \notin L$.
The string y can be of any of the following:

(i) y is a string of only 'a', i.e., $y = a^k$ for some $k \geq 1$
(ii) y is a string of only 'b', i.e., $y = b^k$ for some $k \geq 1$
(iii) y is a string of both 'a' and 'b', i.e., $y = a^k b^l$ for some $k, l \geq 1$

For case (i), take $i = 0$. As $xyz = a^nb^n$, $xy^0z = xz$ will be $a^{n-k}b^n$.

As $k \geq 1$, $(n-k) \neq n$. So, $xy^0z \notin L$.

For case (ii), take $i = 0$. As $xyz = a^nb^n$, $xy^0z = xz$ will be a^nb^{n-k}.

As $k \geq 1$, $(n-k) \neq n$. So, $xy^0z \notin L$.

For Case (iii), take $i = 2$. As $xyz = a^nb^n$, $xy^2z = xyyz$.

We know $xyz = a^nb^n = a^{n-k}a^k b^l b^{n-l}$. So, xyyz will be

$a^{n-k}a^k b^l a^k b^l b^{n-l} = a^nb^la^kb^n$, which is not in the form a^nb^n. So, $xy^2z \notin L$.
For all the three cases, we are getting contradiction. So, L is not regular.

Example 5.29 Show that L = palindrome over {a, b} is not regular.

Solution:

Step I: Suppose that the set L is regular. Let n be the number of states of the FA accepting L.

Step II: Let w = a^nba^n, | w | = (2n + 1) > n. By the pumping lemma, we can write w = xyz with | xy | ≤ n and | y | > 0.

Step III: We want to find a suitable i so that $xy^iz \notin L$. The string y may consist of only 'a', for example, a^j where j > 0.

Let us take i = 0. xyz = $a^nba^n = a^{n-j}a^j ba^n$. So, $xy^0z = a^{n-j}ba^n$.

As j > 0, n – j ≠ n. So, $a^{n-j}ba^n$ is not a palindrome.

So, $xy^0z \notin L$. Hence, it is proved that L is not regular.

Example 5.30 Show that L = {ww, where w ∈ (a, b)*} is not regular.

Solution:

Step I: Let us consider that the set L is regular. Let n be the number of states of the FA accepting L.

Step II: Let ww = $a^nba^nb \in L$, and so | ww | = 2(n + 1) > n. By applying the pumping lemma, we can write w = xyz with | xy | ≤ n and | y | > 0.

Step III: We want to find a suitable i so that $xy^iz \notin L$. The string consists of 'a' and 'b'. So, y can be any of these.

1. y has no 'b', i.e., y = a^k
2. y has only one 'b' (because the string is a^nba^nb)

Let us take i = 0 in case I. If i = 0, then xy^0z = xz and it is in the form $a^{n-k}ba^nb$ or $a^nba^{n-k}b$.

This is not in the form ww, i.e., xz ∉ L. Hence, L is not regular.

In case II, if we take i = 0, then xy^0z = xz which contains only one 'b', which is not in the form ww, i.e., xz ∉ L. Hence, L is not regular.

Example 5.31 Show that L = {a^nba^n, where n ≥ 1} is not regular.

Solution:

Step I: Assume that L is regular. Let n be the number of states of the FA accepting L.

Step II: Let w = a^nba^n. So, | w | = 2n + 1 > n. According to the pumping lemma, we can write w = xyz, with w = xyz with | xy | ≤ n and | y | > 0.

Step III: We want to find a suitable i so that $xy^iz \notin L$. The string consists of 'a' and 'b'. So, y can be any of these.

1. y has no 'b' i.e. y = a^k
2. y has only one 'b' (because the string is a^nba^n)

For case I, take i = 2.

The string xyz = $a^{n-k} a^kba^n$ or $a^nba^ka^{n-k}$.

So, $xy^2z = a^{n-k}a^{2k}ba^n = a^{n+k}ba^n$ or $a^nba^{2k}a^{n-k} = a^nba^{n+k}$.
As $k \neq 0$, $n+k \neq n$. Therefore, $xy^2z \notin L$.
This is a contradiction, and so L is not regular.
For Case II, let us take i = k. So, $xy^kz = a^nb^ka^n$. This is not in the form a^nba^n. Therefore, $xy^kz \notin L$.
This is a contradiction, and so L is not regular.

Example 5.32 Show that $L = \{a^nb^nab^{n+1}\}$ is not regular.

Solution:

Step I: Assume that L is regular. Let n be the number of states of the FA accepting L.

Step II: Let $w = a^nb^nab^{n+1}$, and so $|w| = (3n + 2) > n$. By using the pumping lemma, we can write w = xyz, with w = xyz with $|xy| \leq n$ and $|y| > 0$.

Step III: We want to find a suitable i so that $xy^iz \notin L$. The string consists of 'a' and 'b'. So, y can be of any of these.

1. y contains only 'a', and let $y = a^k$
2. y contains only 'b', and let $y = b^l$
3. y contains both 'a' and 'b', and let $y = b^ka$

For case I, take i = 0.

$$w = xyz = (a^{n-k})(a^k)(b^nab^{n+1})$$
$$xy^0z = a^{n-k}\, b^nab^{n+1} \text{ which is not in the form } a^nb^nab^n+^1.$$

For case II, take i = 0.

$$w = xyz = (a^nb^{n-l})(b^l)ab^{n+1}$$
$$xy^0z = a^nb^{n-l}\, ab^{n+1} \text{ which is not in the form } a^nb^nab^{n+1}.$$

For case III, take i = 2

$$w = xyz = a^nb^{n-k}(b^ka)^2\, b^{n+1}$$
$$= a^nb^na\, b^kab^{n+1} \text{ which is not in the form } a^nb^nab^{n+1}.$$

In all the three cases, there are contradictions and, hence, the language is not regular.

Example 5.33 Show that the language containing the set of all balanced parenthesis is not regular.

Solution:

Step I: Assume that L is regular. Let n be the number of states of the FA accepting L.

Step II: Let w = (((….()….))) = $(^n)^n$. $|w| = 2n > n$. By using the pumping lemma, we can write w = xyz, with w = xyz with $|xy| \leq n$ and $|y| > 0$.

Step III: We want to find a suitable i so that $xy^iz \notin L$. The string consists of '(' and ')'. So, y can be of any of these

1. y consists of only '(', and let $y = (^k$

2. y consists of only ')', and let y = $)^k$
3. y consists of both '(' and ')', and let y = $(^k)^k$

For case I, take i = 0.

$$w = xyz = (^{n-k}\ (^k)^n$$

$$xy^0z = (^{n-k})^n \text{ which is not a set of all balanced parenthesis.}$$

For case II, take i = 0.

$$w = xyz = (^n)\ ^k)^{n-k}$$

$$xy^0z = (^n)^{n-k} \text{ which is not a set of all balanced parenthesis.}$$

For case III, take i = 2.

$$w = xyz = (^{n-k}(^k)^k)^{n-k}$$

$$xy^2z = (^n)^k(^k)^n \text{ which is not a set of all balanced parenthesis.}$$

In all the three cases, there are contradictions and, hence, the language is not regular.

Example 5.34 Show that L = $\{0^n1^{2n}$, where $n \geq 1\}$ is not regular.

Solution:

Step I: Assume that L is regular. Let n be the number of states of the FA accepting L.

Step II: Let w = 0^n1^{2n}, where $n \geq 1$. | w | = 3n> n. By using the pumping lemma, we can write

$$w = xyz, \text{ with } w = xyz \text{ with } |xy| \leq n \text{ and } |y| > 0.$$

Step III: We want to find a suitable i so that $xy^iz \notin L$. The string consists of '0' and '1'. y can be any of the following forms

1. y consists of only '0', and let y = 0^k
2. y consists of only '1', and let y = 1^k
3. y consists of both '0' and '1', and let y = 0^k1^k

For case I, take i = 0.

$$w = xyz = 0^n1^{2n} = 0^{n-k}0^k1^{2n}$$

$$xy^0z = 0^{n-k}1^{2n}, \text{ as } k \neq 0, (n-k) \neq n.$$

For case II, take i = 0.

$$w = xyz = 0^n1^{2n} = 0^n1^k1^{2n-k}$$

$$xy^0z = 0^n1^{2n-k}, \text{ as } k \neq 0, (2n-k) \neq n.$$

For case III, take i = 2.

$$w = xyz = 0^n1^{2n} = 0^{n-k}0^k1^k1^{2n-k}$$

$$xy^2z = 0^n1^k0^k1^{2n}, \text{ which is not in the form } 0^n1^{2n}$$

For all the three cases, we are getting contradictions and, therefore, L is not regular.

5.12 Closure Properties of Regular Set

A set is closed (under an operation) if and only if the operation on two elements of the set produces another element of the set. If an element outside the set is produced, then the operation is not closed.

Closure is a property which describes when we combine any two elements of the set; the result is also included in the set.

If we multiply two *integer numbers*, we will get another integer number. Since this process is always true, it is said that the integer numbers are 'closed under the operation of multiplication'. There is simply no way to escape the set of integer numbers when multiplying.

Let S = {1,2,3,4,5,6,7,8,9,10....} is a set of integer numbers.

$1 \times 2 = 2$

$2 \times 3 = 6$

$5 \times 2 = 10$

All are included in the set of integer numbers.

We can conclude that integer numbers are closed under the operation of multiplication.

Theorem 5.3: Two REs L_1 and L_2 over Σ are closed under union operation.

Proof: We have to prove that if L_1 and L_2 are regular over Σ, then their union, i.e., $L_1 \cup L_2$ will be also regular.

As L_1 and L_2 are regular over Σ, there must exist FA $M_1 = (Q_1, \Sigma, \delta_1, q_{01}, F_1)$ and $M_2 = (Q_2, \Sigma, \delta_2, q_{02}, F_2)$ such that $L_1 \varepsilon M_1$ and $L_2 \varepsilon M_2$

Assume that there is no common state between Q_1 and Q_2, i.e., $Q_1 \cap Q_2 = \varnothing$.

Define another FA, $M_3 = (Q, \Sigma, \delta, q_0, F)$ where

1. $Q = Q_1 \cup Q_2 \cup \{q_0\}$, where q_0 is a new state $\notin Q_1 \cup Q_2$
2. $F = F_1 \cup F_2$
3. Transitional function δ is defined as $\delta(q_0, \varepsilon) \rightarrow \{q_{01}, q_{02}\}$

   and $\delta(q, \Sigma) \rightarrow \delta_1(q, \Sigma)$ if $q \varepsilon Q_1$

   $\delta(q, \Sigma) \rightarrow \delta_2(q, \Sigma)$ if $q \varepsilon Q_2$.

It is clear from the previous discussion that from q_0 we can reach either the initial state q_1 of M_1 or the initial state q_2 of M_2.

Transitions for the new FA, M, are similar to the transitions of M_1 and M_2.

As $F = F_1 \cup F_2$, any string accepted by M_1 or M_2 will also be accepted by M.

Therefore, $L_1 \cup L_2$ is also regular.

Example:

Let $L_1 = a^*(a + b)b^*$

$L_2 = ab(a + b)^*$

The FA M_1 accepting L_1 is as shown in Fig. 5.38(a).

The FA M_2 accepting L_2 is as shown in Fig. 5.38(b).

The machine M produced by combining M_1 and M_2 is as shown in Fig. 5.38(c).

It accepts $L_1 \cup L_2$.

Theorem 5.4: The complement of an RE is also regular.

If L is regular, we have to prove that L^T is also regular.

As L is regular, there must be an FA, $M = (Q, \Sigma, \delta, q_0, F)$, accepting L. M is an FA, and so M has a transitional system.

Let us construct another transitional system M′ with the state diagram of M but reversing the direction of the directed edges. M′ can be designed as follows:

1. The set of states of M′ is the same as M
2. The set of input symbols of M′ is the same as M

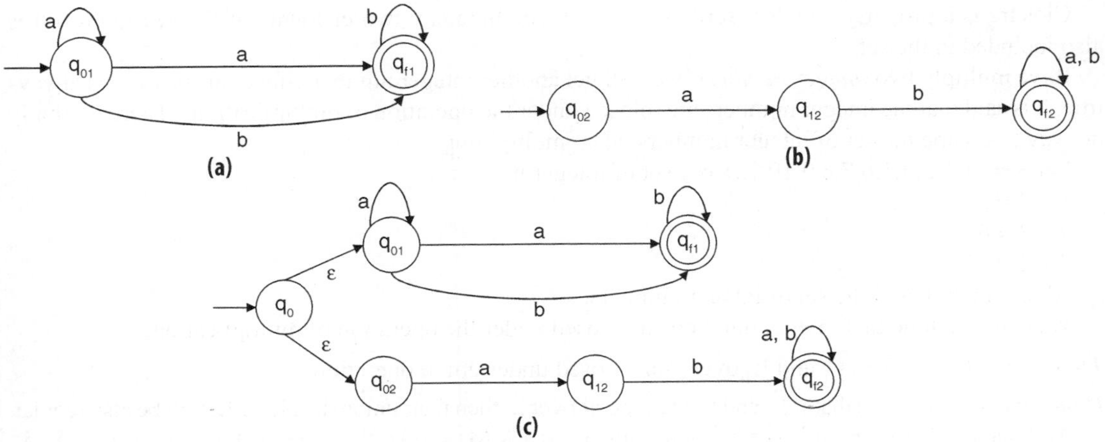

Figs. 5.38 (a)–(c)

3. The initial state of M′ is the same as the final state of M (M′ is the reverse direction of M)
4. The final state of M′ is the same as the initial state of M (M′ is the reverse direction of M)

Let a string w belong to L, i.e., w ∈ M. So, there is a path from q_0 to F with path value w. By reversing the edges, we get a path from F to q_0 (beginning and final state of M′) in M′. The path value is the reverse of w, i.e., w^T. So, w^T ∈ M′.

So, the reverse of the string w is regular.

Example:

Let L = ab(a + b)*.

The FA M accepting L is as shown in Fig. 5.39(a).

The reverse of the FA M′ accepting L^c is shown in Fig. 5.39(b).

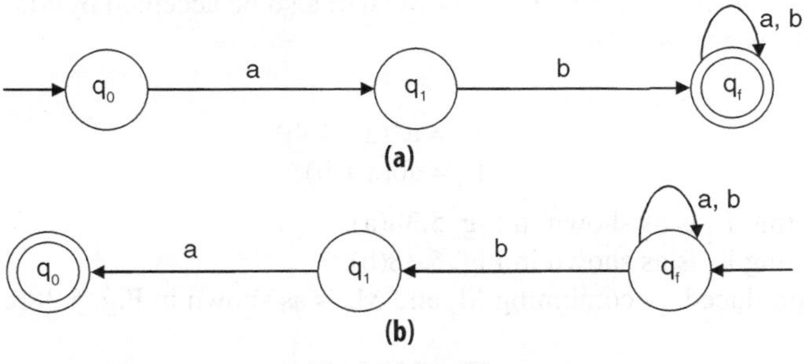

Fig. 5.39 (a)–(b)

M′ accepts (a + b)*ba which is reverse of L.

Theorem 5.5: If L is regular and L is a subset of Σ*, prove that Σ* – L is also a regular set.

As L is regular, there must be an FA, M = (Q, Σ, δ, q_0, F), accepting L. Let us construct another DFA M′ = (Q, Σ, δ, q_0, F′) where F′ = Q–F. So, the two DFA differ only in their final states. A final state of M is a non-final state of M′ and vice versa.

Let us take a string w which is accepted by M′. So, δ(q_0, w) ∈ F′, i.e., δ(q_0, w) ∈ (Q – F).

The string w cannot be accepted by M, because $\delta(q_0, w) \notin F$ [as F does not belong to (Q – F)].
So, $w \notin L$, i.e., $\Sigma^* - L \neq L$. But $\Sigma^* - L$ is accepted by M′ which is an FA.
Therefore, $\Sigma^* - L$ is a regular set.

Theorem 5.6: REs are closed under intersection operation.

Proof: From D'Morgan's theorem, we know

$$L_1 \cap L_2 = (L_1^C \cup L_2^C)^C.$$

We know that if L_1 and L_2 are regular, then L_1^C and L_2^C are also regular.
As L_1^C and L_2^C are regular, $L_1^C \cup L_2^C$ is also regular (RE is closed under union operation).
As $L_1^C \cup L_2^C$ is regular, so complement $(L_1^C \cup L_2^C)C$ is also regular.
So, $L_1 \cap L_2$ is also regular, i.e., the regular sets are closed under intersection.

Theorem 5.7: Two DFA are closed under cross product.

Proof: Let $D_1 = \{Q_1, \Sigma, \delta_1, q_{01}, F_1\}$ and $D_2 = \{Q_2, \Sigma, \delta_2, q_{02}, F_2\}$ are two DFA accepting two RE L_1 and L_2 respectively. Let us construct a new FA, D as follows.

$$D = \{Q, \Sigma, \delta, q_0, F\}$$

where

$$Q = Q_1 \times Q_2$$

$$\delta((S_1, S_2), i/p) = (\delta_1(S_1, i/p), \delta_1(S2, i/p)) \text{ for all } S_1 \in Q_1, S_2 \in Q_2 \text{ and } i/p \in \Sigma.$$

$$q_0 = (q_{01}, q_{02})$$

$$F = F_1 \times F_2$$

Clearly D is a DFA. Thus DFA are closed under cross product.

5.13 Decision Problems of Regular Expression

Decision problems are the problems which can be answered in 'yes' or 'no'. FA are one type of finite state machines which contain a finite number of memory elements. Thus, it can memorize only finite amount of information. FA can answer to those decision problems which require only a finite amount of memory.

The decision problems related to RE are

1. Whether a string x belongs to a regular expression R? ($x \in R$?)
2. Whether the language set of an FA M is empty? [$L(M) = \varnothing$?]
3. Whether the language set of an FA is finite? [L(M) is finite?]
4. Whether there exists any string accepted by two FA, M_1 and M_2?
5. Whether the language set of an FA M_1 is a subset of the language set of another FA M_2? [$L(M_1) \subseteq L(M_2)$?]
6. Whether two FA M_1 and M_2 accept the same language? [$L(M_1) = L(M_2)$]
7. Whether two REs R_1 and R_2 are from the same language set?
8. Whether an FA is the minimum state FA for a language L?

Proof

1. This problem is known as the membership problem. R can be converted to the equivalent FA M (see Section 5.5). x is applied on M. If M reaches its final state upon finishing x; $x \in R$ else $x \notin R$.

2. An FA does not accept any string if it does not have any final state or if the final state is an inaccessible state. Let us calculate the set of state S_k reached from the beginning state q_0 upon applying a string of length k.

$$S_K = \begin{cases} q_0 & \text{if } k = 0 \\ S_{K-1} \cup \{\delta(q, a) \text{ where } q \in S_{K-1} \text{ and } a \in \Sigma\} & \text{if } k > 0 \end{cases}$$

If k = 0, it reaches the beginning state q_0. It reaches S_K if it was in state S_{k-1} and 'a' as input is applied on S_{K-1}.

Compare the set $S_{K.}$

   a. for string length $k \geq 0$ whether a final state appears or
   b. $S_K = S_{K-1}$ (loop) for $k > 0$.

For the case (a), $L(M) \neq \varnothing$, and for case (b) $L(M) = \varnothing$.

3. In the pumping lemma (Section 4.11), it is discussed that for accepting a string of length $\geq n$ (the number of states of an FA), it has to traverse at least one loop. The language accepted by an FA is finite if it has length < n. Formulate an algorithm for testing as follows.

   Give an input to M, the strings of length n or > n in increasing order. If for a string 's' of length in between n and < 2n, it reaches M then L(M) is infinite; else L(M) is finite.

   Infinite means there exists x, y, z such that s = xyz, where $|xy| \leq n$, $|y| > 0$ and $xy^iz \in L$ for each $i \geq 0$ (pumping lemma). Here, y^i is the looping portion, which can generate an infinite number of strings.

4. For this decidable problem, construct an FA M accepting $L(M_1) \cap L(M_2)$. Then apply the decision problem 2 on M.
5. For this decidable problem, construct an FA M accepting $L(M_1) - L(M_2)$. Then, apply decision problem 2 on M. It is true if $L(M_1) - L(M_2) = \varnothing$. (If $L(M_1)$ is a subset of $L(M_2)$, $L(M_1) - L(M_2)$ will produce null.)
6. Two sets A and B are the same if $A \subseteq B$ and $B \subseteq A$. Construct an FA, M, accepting $L(M_1) - L(M_2)$ and M′ accepting $L(M_2) - L(M_1)$ (reducing it to problem v). Problem vi is decidable if $L(M_1) - L(M_2) = \varnothing$ and $L(M_2) - L(M_1) = \varnothing$.
7. There exists an algorithm to convert an RE to FA (see Section 5.5). Reduce this problem to problem (vi).
8. Minimize an FA, M, using 3.13 and generate M′. If the number of states of M and M′ are the same, then it is minimized; else it is not. Hence, decidable.

5.14 'Grep' and Regular Expression

REs have been an integral part of Unix since the beginning. Ken Thompson used the RE in the early computer text editor 'QED' and the Unix editor 'ed' for searching text. 'grep' is a find string command in Unix. It is an acronym for 'globally search a regular expression and print it'. The command searches through files and folders (directories in Unix) and checks which lines in those files match a given RE. 'grep' gives output the filenames and the line numbers or the actual lines that matched the RE.

Let us start with some simple grep commands.

$ grep RE chapter5

This command searches the string 'RE' in the file 'chapter5'.

$ grep 'RE' –i chapter5

This command searches the strings with case insensitive.

$ grep h??a chapter5

This command displays all strings of length 4 starting with 'h' and ending with 'a'.

If we enter into the internal operation of 'grep', we will see that we are searching for a string that belongs to an RE. Let us take the last grep which searches for four length strings starting with 'h' and ending with 'a'. For this case, the RE is $h(ch)^+(ch)^+a$, where 'ch' is any symbol that belongs to a file (Generally symbols available in the keyboard).

5.15 Applications of Regular Expression

RE is mainly used in lexical analysis of compiler design. In the programming language, we need to declare a variable or identifier. That identifier must start with a letter followed by any number of letters or digits. The structure of the identifier is represented by RE. The definition of an identifier in a programming language is

letter → A | B | ... | Z | a | b | ... | z
digit → 0 | 1 | ... | 9
id → letter (letter | digit) *

The definition of an unsigned number in a programming language is

digit → 0 | 1 | ... | 9
digits → digit +
opt-fraction → (. digits) *
opt-exponent → (E (+ | -) * digits) *
unsigned-num → digits opt-fraction opt-exponent

The other application of RE is pattern searching from a given text.

What We Have Learned So Far

1. An RE can be defined as a language or string accepted by an FA.
2. Any terminal symbols, null string (∧), or null set (ϕ) are RE.
3. Union, concatenation, iteration of two REs, or any combination of them are also RE.
4. The Arden's theorem states that if P and Q are two REs over Σ, and if P does not contain Λ, then the equation $R = Q + RP$ has a unique (one and only one) solution $R = QP^*$.
5. The Arden's theorem is used to construct an RE from a given FA by the algebraic method.
6. If any FA contains any ∈ (null) move or transaction, then that FA is called NFA with ∈-moves.
7. The ∈-closure of a state is defined as the set of all states S, such that it can reach from that state to all the states in S with input ∈ (i.e., with no input).
8. The pumping lemma for RE is used to prove that certain sets are not regular.
9. A set is closed (under an operation) if and only if the operation on two elements of the set produces another element of the set.
10. Closure is a property which describes when we combine any two elements of the set, the result is also included in the set.
11. REs are closed under union, complementation, and intersection operation.

Solved Problems

1. Describe the following REs in English language.
   a) a(a+b)*abb b) (a+b)*aba(a+b)* c) (0+1)*1(0+1)*0(0+1)*

   Solution:

   a) The language starts with 'a' and ends with 'abb'. In the middle of 'a' and 'b', there is any combination of 'a' and 'b'.
   Hence, the RE described in English language is
   {Set of any combination of 'a' and 'b' beginning with 'a' and ending with 'abb'}.

   b) The expression is divided into three parts: (a+b)*, aba, and (a+b)*. In each element of the language set, there is aba as a substring. In English language, the RE is described as
   {Set of any combination of 'a' and 'b' containing 'aba' as substring}.

   c) The expression is divided into five parts: (0+1)*, 1, (0+1)*, 0 and (0+1)*. In each element of the language set, there is 1 and 0, where 1 appears first. In English language, the RE is described as

   {Set of any combination of '0' and '1' containing at least one 1 and one 0 where 1 appears first}

2. Find the RE for the following:
   a) The set of language of all strings of 0 and 1 containing exactly two 0's [UPTU 2004]
   b) The set of languages of any combination of 'a' and 'b' beginning with 'a'.
   c) The set of all strings of 0 and 1 that do not end with 11.
   d) The set of languages of any combination of '0' and '1' containing at least one double symbol.
   e) The set of all strings over {a, b} in which the number of occurrences of 'a' is divisible by 3. [UPTU 2004]
   f) The set of all strings where the 10th symbol from the right end is a 1. [JNTU 2007]
   g) The set of languages of any combination of '0' and '1' containing no double symbol.

   Solution:

   a) The language contains exactly two 0's. But the language consists of 0 and 1. 1 may appear at first or in the middle or at the last. Thus, the language is L = 1*01*01*.

   b) The set of any combination of 'a' and 'b' is denoted by $(a+b)^*$. The RE is

   L = a(a+b)*.

   c) The strings may end with 00, 01, or 10. Thus, the language is L = (0+1)*(00 + 01+10).

   d) The language consists of two symbols '0' and '1'. A double (same) symbol means either 00 or 11. The part of the language containing at least one double symbol is denoted by (00 + 11). So, the language of any combination of '0' and '1' containing at least one double symbol is expressed as L = (0+1)* (00+11) (0+1)*.

   e) The number of 'a' is divisible by 3 means that the number of 'a' may be 0, 3, 6, 9, 12, ... The number of 'b' may be 0, 1, 2, 3,.......... The RE is (b*ab*ab*ab*)*.

   f) The 10th symbol from the right hand side should be 'a', whereas the other symbols may be 'a' or 'b'. A string of length n has (n−10) symbols from start and the last 9 symbols are of 'a' or 'b'. The RE is (a + b)*a(a + b)10.

   g) The language consists of two symbols '0' and '1'. A double (same) symbol means either 00 or 11. According to the condition, in the language, 00 or 11 will not appear. The language may start with '0' or start with '1'. The language may start with '0' with no double symbol is (01)*. The language may start with '1' with no double symbol is (10)*.

The expression is L = (01)* + (10)*.

3. Construct an RE from the given FA by the algebraic method using Arden's theorem.

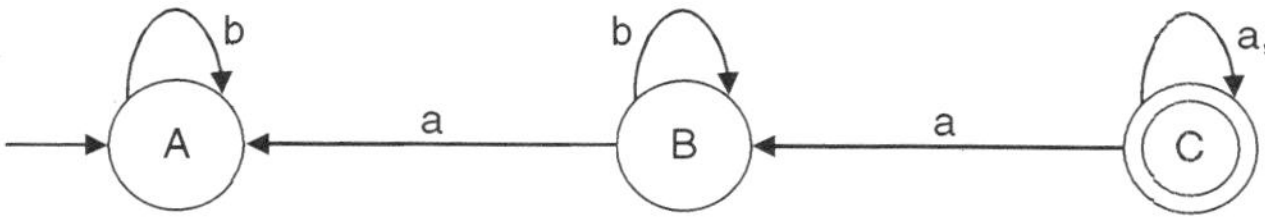

Solution: For the previously given FA, the equations are

$$A = \Lambda + Ab \quad (1)$$
$$B = Aa + Bb \quad (2)$$
$$C = Ba + C(a + b) \quad (3)$$

The equation 1 is in the form R = Q + RP, where R = A, Q = Λ, and P = b. According to the Arden's theorem, the solution for the equation is R = QP*.

So, $A = \Lambda b^* = b^*$ (as $\Lambda R = R\Lambda = R$).

Putting the value of A in equation 2, we get

$$B = b^*a + Bb.$$

The equation is in the format R = Q + RP, where R = B, Q = b*a, and P = b. According to the Arden's theorem, the solution for the equation is R = QP*.

So, $B = b^*ab^*$.

Putting the value of B in equation 3, we get

$$C = b^*ab^*a + C(a+b).$$

The equation is in the format R = Q + RP, where R = C, Q = b*ab*a, and P = (a + b). According to the Arden's theorem, the solution for the equation is R = QP*.

So, $C = b^*ab^*a(a+b)^*$

As C is the final state, the RE accepted by the FA is b*ab*a(a+b)*.

4. Construct an RE from the given FA by the algebraic method using Arden's theorem.

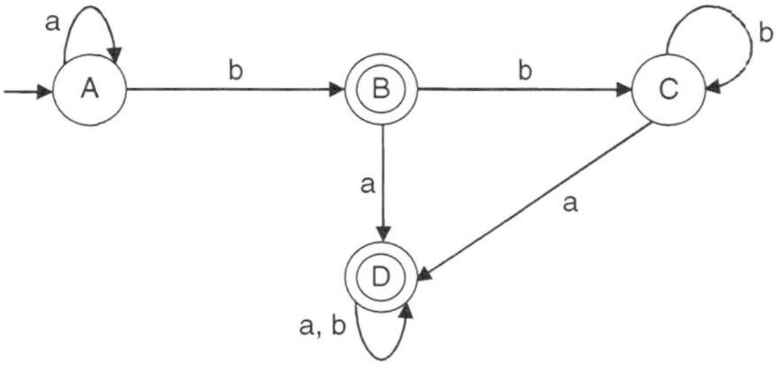

Solution: For the previously given FA, the equations are

$$A = \Lambda + Aa \quad (1)$$
$$B = Ab \quad (2)$$

$$C = Bb + Cb \qquad (3)$$
$$D = Ba + Ca + D\,(a+b) \qquad (4)$$

The equation 1 is in the form R = Q + RP, where R = A, Q = Λ, and P = a. According to the Arden's theorem, the solution for the equation is R = QP*.

So, $A = \Lambda a^* = a^*$ (as $\Lambda R = R\Lambda = R$).

Putting the value of A in equation (2), we get

$$B = a^*b.$$

Putting the value of B in equation (3), we get

$$C = a^*bb + Cb.$$

This is in the format R = Q + RP, where R = C, Q = a*bb, and P = b. According to the Arden's theorem, the solution for the equation is R = QP*.

$$C = a^*bbb^*.$$

Putting the value of B and C in equation (4), we get

$$D = a^*ba + a^*bbb^*a + D(a+b).$$

This is in the format R = Q + RP, where R = D, Q = a*ba + a*bbb*a, and P = a+b. According to the Arden's theorem, the solution for the equation is R = QP*.

$$D = (a^*ba + a^*bbb^*a)(a+b)^*.$$

In the FA, there are two final states B and D. So, the RE accepted by the FA is

$$L = a^*b + (a^*ba + a^*bbb^*a)(a+b)^*.$$

5. Find the RE corresponding to the following FA.

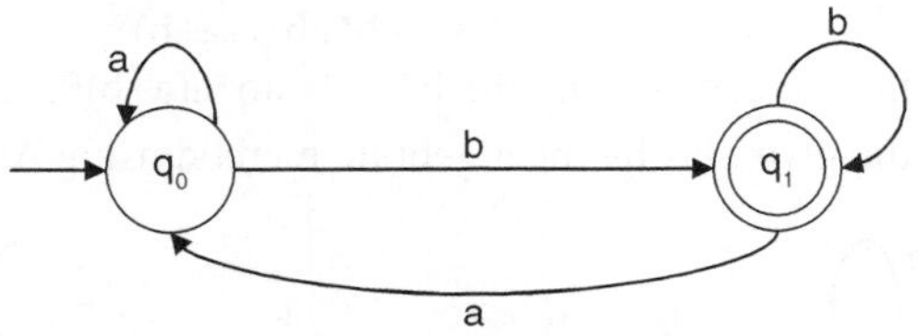

Solution: For the previously given FA, the equations are

$$A = aB + aA + \Lambda \qquad (1)$$
$$B = bA + bB \qquad (2)$$

Take equation number (2)

$$B = bA + bB.$$

The equation is in the form R = Q + RP, where R = B, Q = bA, and P = b.
The solution of the equation is B = bAb*.
Replacing the value of B in equation (1), we get

$A = abAb^* + aA + \Lambda = \Lambda + A(abb^* + a)$. The equation is in the form $R = Q + RP$ where $R = A$, $Q = \Lambda$, and $P = (abb^* + a)$. The solution of the equation is $A = \Lambda(abb^* + a)^* = (abb^* + a)^*$.

Replacing this value in the solution of B, we get $b(abb^* + a)^*b^*$. As B is the final state, the string accepted by the FA is $b(abb^* + a)^*b^*$.

6. Find out the RE of the given diagram. [UPTU 2003]

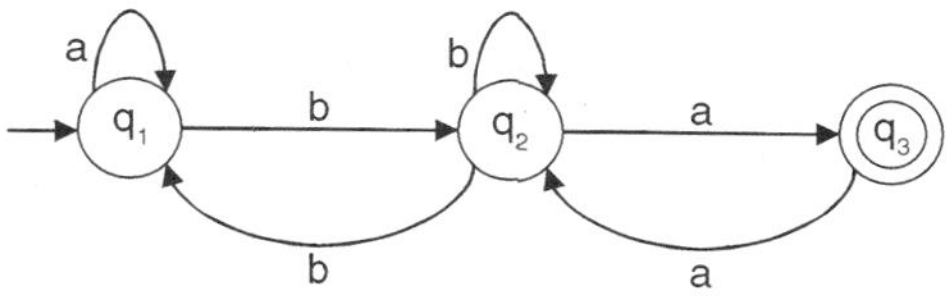

For the previously given FA, the equations are

$$q_1 = q_1a + q_2b + \Lambda \quad (1)$$
$$q_2 = q_1b + q_2b + q_3a \quad (2)$$
$$q_3 = q_2a \quad (3)$$

Replacing the value of q_3 in (2), we get

$$q_2 = q_1b + q_2b + q_2aa$$
$$= q_1b + q_2(b + aa).$$

This is in the form $R = Q + RP$. Hence, the solution is

$$q_2 = q_1b(b + aa)^*.$$

Replacing the value of q_2 in (i), we get

$$q_1 = \Lambda + q_1a + q_2b$$
$$= \Lambda + q_1(a + b(b + aa)^*b).$$

This is in the form $R = Q + RP$. Hence, the solution is

$$q_1 = \Lambda(a + b(b + aa)^*b)^* = (a + b(b + aa)^*b)^*$$
$$q_2 = (a + b(b + aa)^*b)^*\ b(b + aa)^*$$
$$q_3 = (a + b(b + aa)^*b)^*\ b(b + aa)^*\ b.$$

q_3 is the final state and, therefore, the RE accepted by the FA is $(a + b(b + aa)^*b)^*\ b(b + aa)^*\ b$.

7. Find the RE corresponding to the following figure:

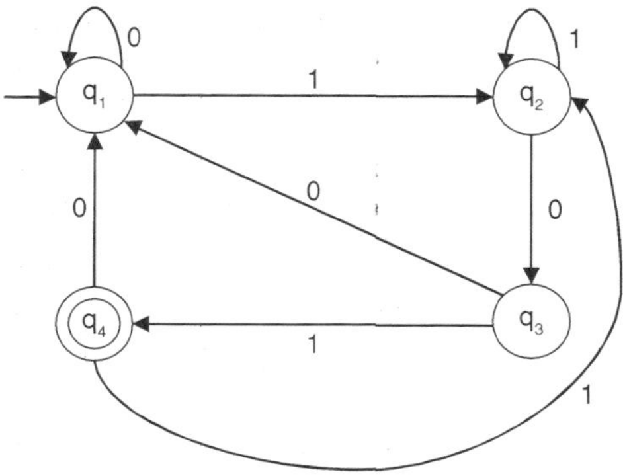

[WBUT 2007]

Solution: There is only one initial state. In the FA, there are no null moves.
The equations for q_1, q_2, q_3, and q_4 are

$$q_1 = q_1.0 + q_3.0 + q_4.0 + \Lambda$$
$$q_2 = q_1.1 + q_2.1 + q_4.1$$
$$q_3 = q_2.0$$
$$q_4 = q_3.1$$

Now, we need to solve the equations by using the Arden's theorem to get the RE.

$$q_4 = q_3.1$$
$$= q_2.0.1$$
$$q_2 = q_1.1 + q_2.1 + q_4.1$$
$$= q_1.1 + q_2.1 + q_2.0.1.1$$
$$= q1.1 + q2(1 + 011)$$

is in the form R = Q + RP, where R = q_2, Q = q_1.1, P = (1 + 011).

The solution of the equation is

$$q_2 = q_1.1\,(1 + 011)^*.$$

From here,

$$q_3 = q_1.1\,(1 + 011)^*0$$
$$q_4 = q_1.1\,(1 + 011)^*01$$
$$q_1 = q_1.0 + q_3.0 + q_4.0 + \Lambda$$
$$= q_1.0 + q_1.1\,(1 + 011)^*00 + q_1.1\,(1 + 011)^*010 + \Lambda$$
$$= \Lambda + q_1\,(0 + 1\,(1 + 011)^*\,(00 + 010)).$$

The equation is in the form R = Q + RP, where R = q_1, Q = Λ, and P = 1 (1 + 011)* (00 + 010).
By applying Arden's theorem, the solution of the equation is R = QP*.

$$q_1 = (0 + 1\,(1 + 011)^*\,(00 + 010))^*$$

The final state of the automaton is q_4 = (0 + 1 (1 + 011)* (00 + 010))*(1 (1 + 011)*01).
The RE corresponding to the given automaton is

$$(0 + 1\,(1 + 011)^*\,(00 + 010))^*(1\,(1 + 011)^*01).$$

8. Construct an FA equivalent to the RE, L = (a + b) + a(a + b)* (ab + ba).

Solution:

Step I: Take a beginning state q_0 and a final state q_f. Between the beginning and final states, place the RE.

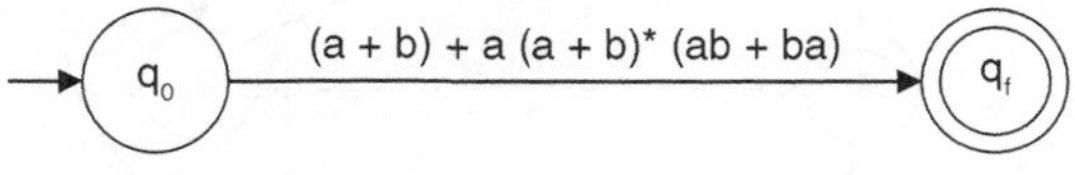

Step II: Between (a+b) and a(a+b)* (ab+ba), there is a + sign, and so there will be parallel edges between q_0 and q_f.

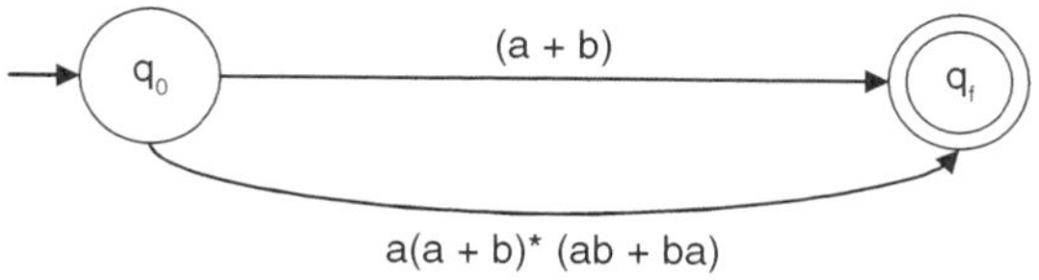

Step III: Between (a + b)* and (ab + ba), there is a .(dot) sign, and so an extra state is added between q_0 and q_f.

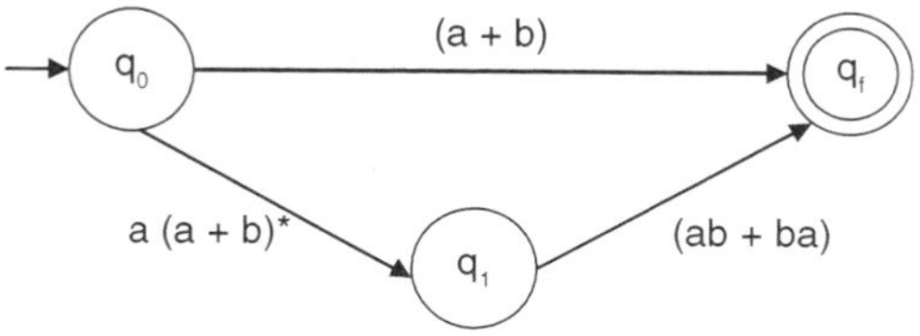

Step IV: Between ab and ba, there is + sign, and so there will be parallel edges between q_1 and q_f.

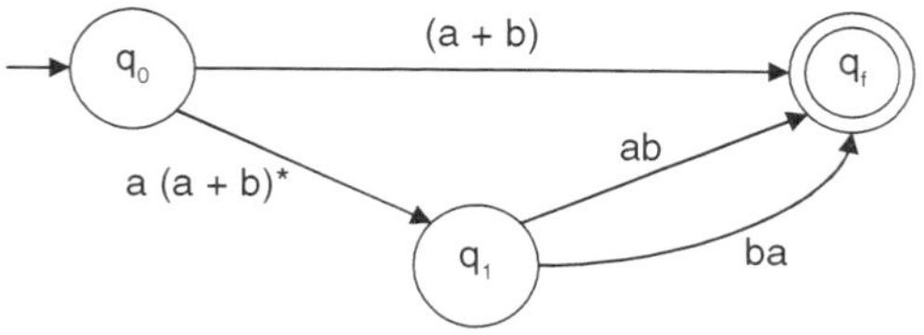

Step V: Between 'a' and b, there is a + sign. So, between q0 and qf there is a parallel edge.

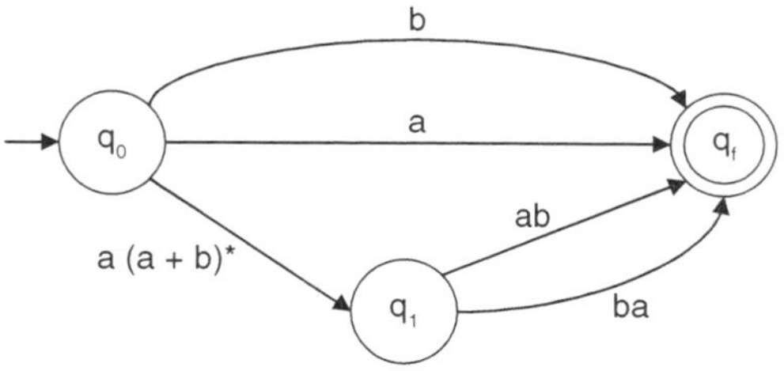

Step VI: Between 'a' and 'b' and between 'b' and 'a', there are .(dots). So, two extra states are added between q_1 and q_f. An extra state is added between q_0 and q_1 for a(a + b)*.

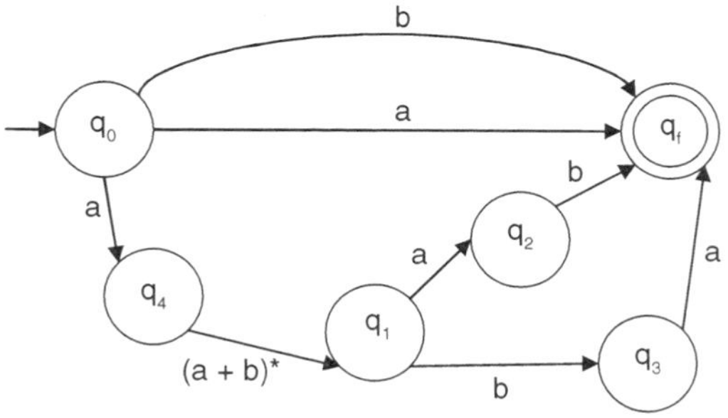

Step VII: The * between q_4 and q_1 is removed by adding an extra state with label a, b, and the ∈ transition from q_4 to that state and from that state to q_1.

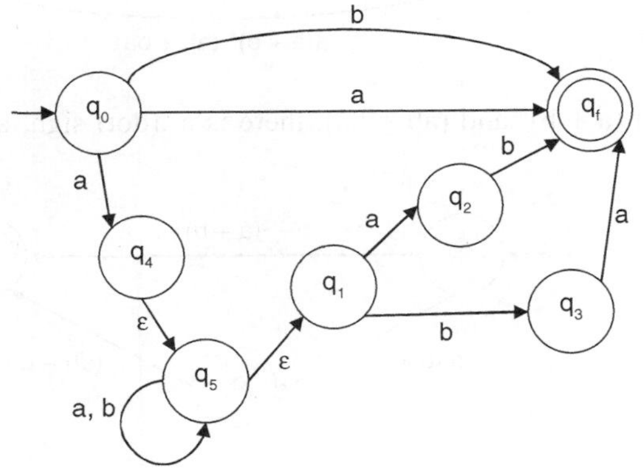

Step VIII: Removing ∈, the automata become

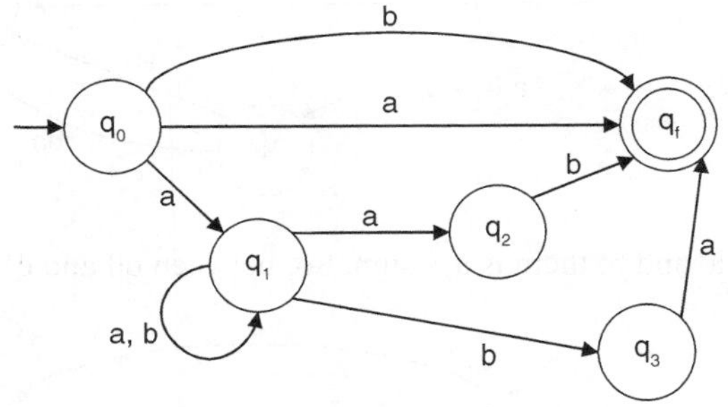

9. Construct an FA equivalent to the RE, L = (00 + 11)* 11 (0 + 1)*0.

Solution:

Step I: Take a beginning state q_0 and a final state q_f. Between the beginning and final state, place the RE.

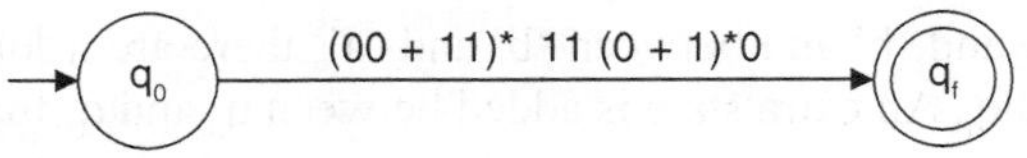

Step II: There are four .(dots) in between $(00 + 11)^*$ and 1, 1 and 1, 1 and $(0 + 1)^*$, and $(0 + 1)^*$ and 0. So, the four extra states are added in between q_0 and q_f.

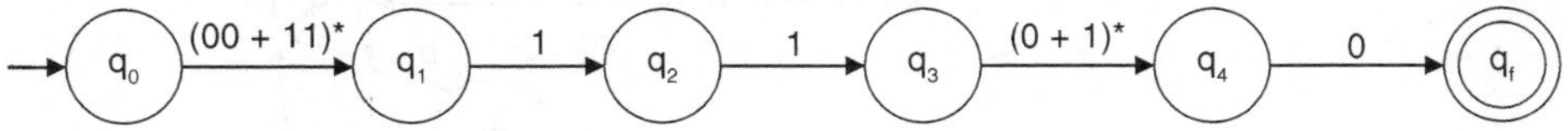

Step III: The * between q_0 and q_1 is removed by adding an extra state with label 00 and 11 as loop and ∈ transition from q_0 to that state and from that state to q_1. The same is applied for the removal of * between q_3 and q_4.

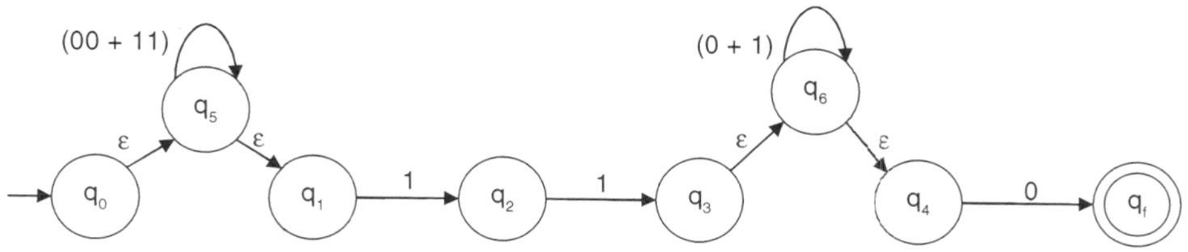

Step IV: Removing the + sign between 00 and 11, parallel edges are added and for two .(dots) signs (between 0, 0 and 1,1), two extra states are added.

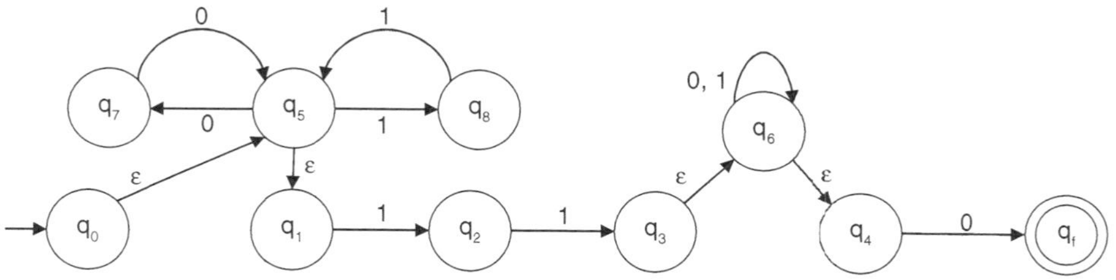

Step V: Use the ∈ removal technique to find the corresponding DFA.

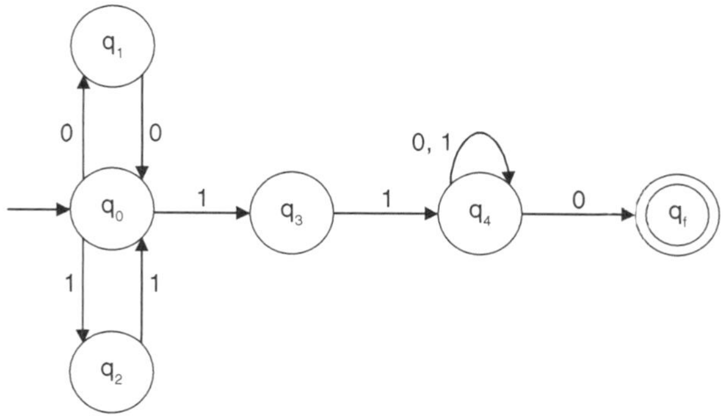

10. Construct an FA for the RE 10 + (0 + 11)0*1. [UPTU 2004]

Solution:

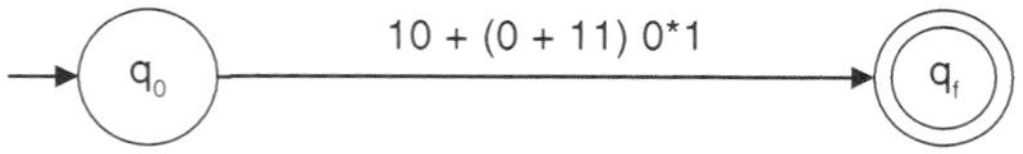

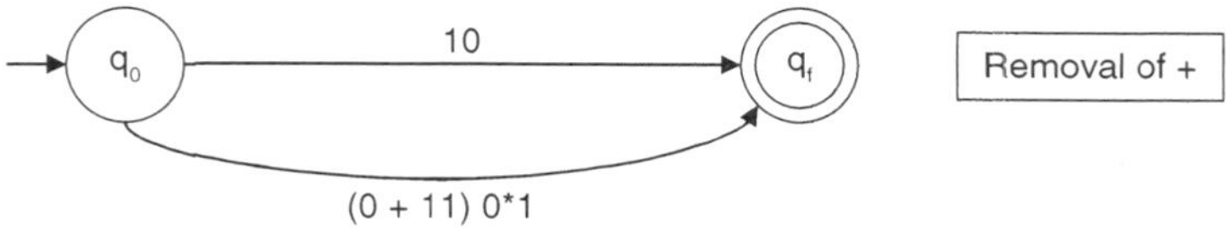

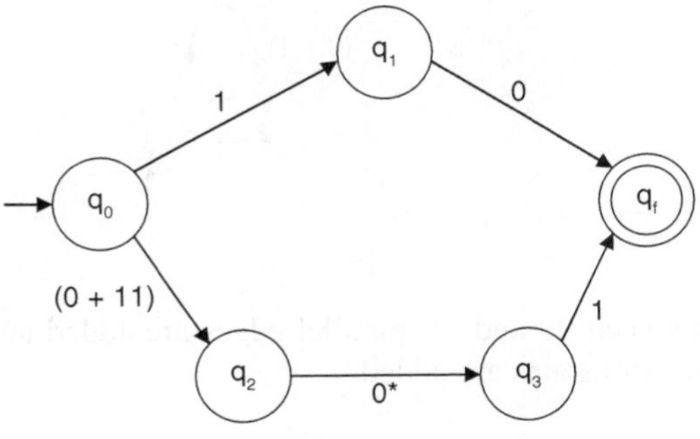

Removal of . (dot)

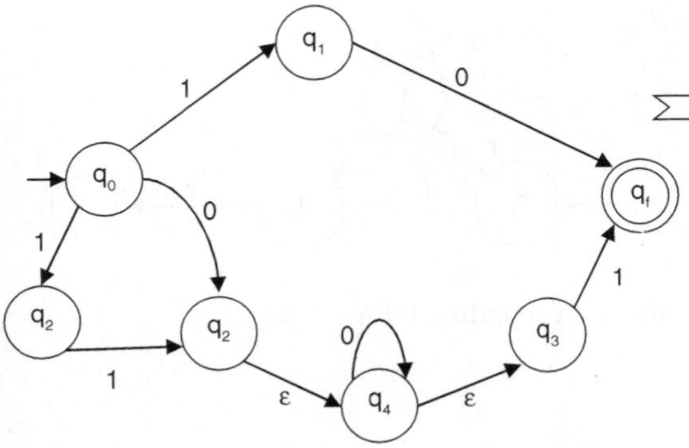

Removal of +, . (dot), and *

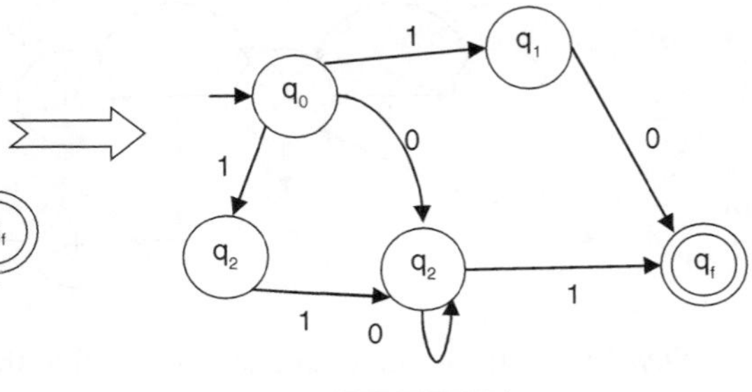

Final FA

11. Convert an RE (0 + 1)*(10) + (00)*(11)* to an NFA with ∈ move.

[Gujrat Technological University 2010]

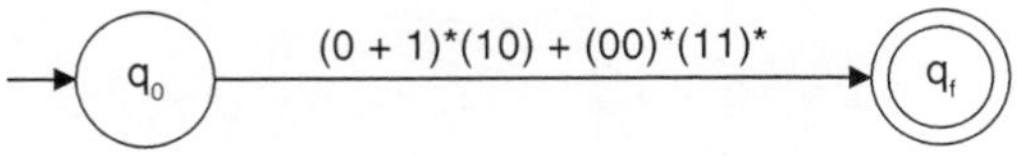

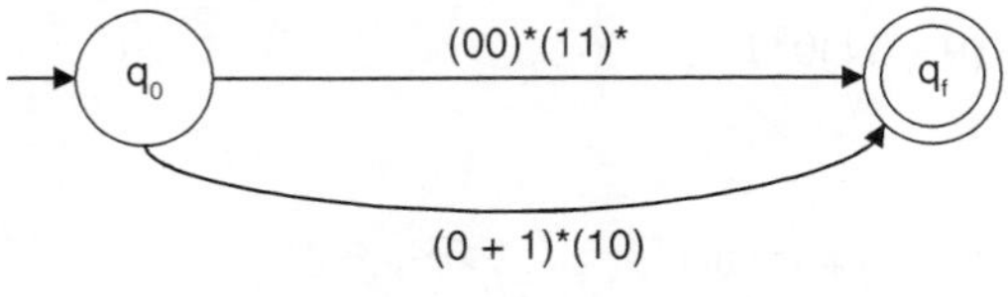

Removal of +

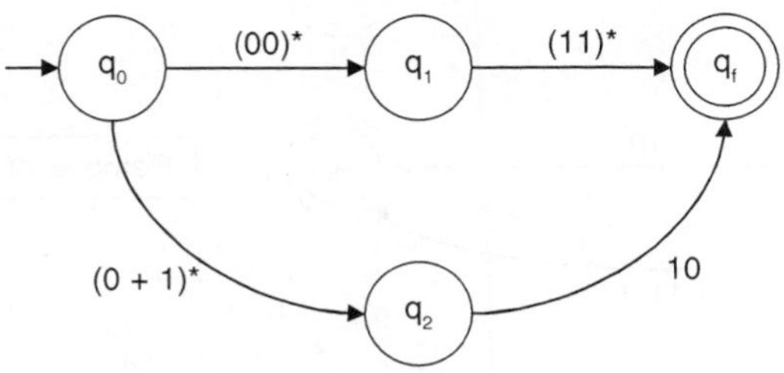

Removal of dot (.)

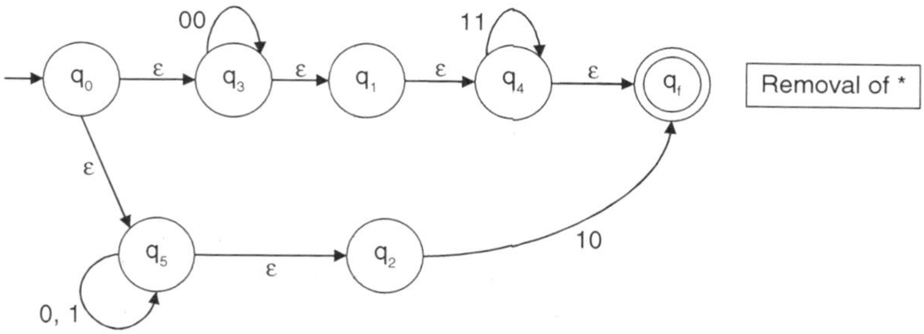

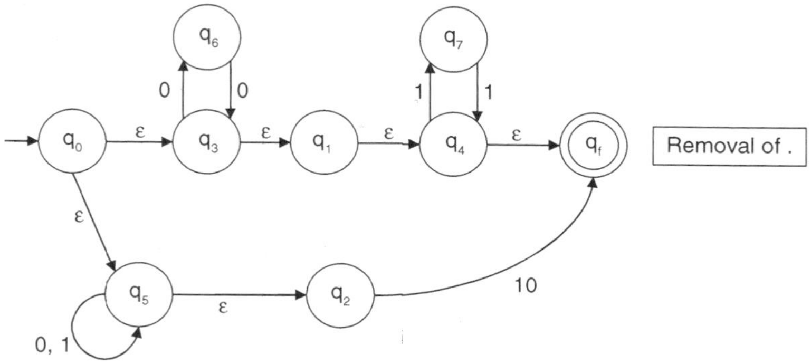

12. Construct an FA equivalent to the regular expression

$$(01(1(10)^* + 111)^* + 0)^*1.$$ [Cochin University 2009]

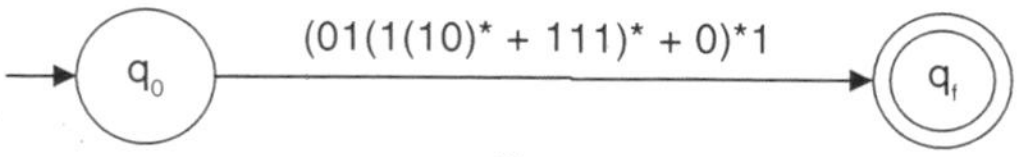

(i)

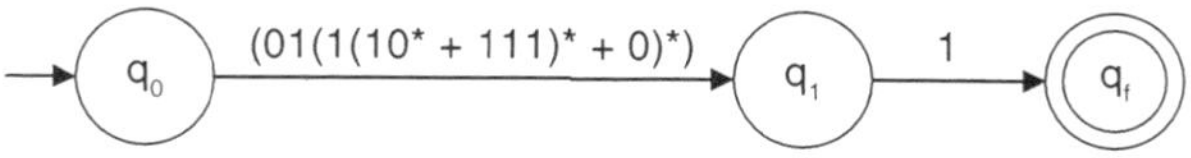

(ii)

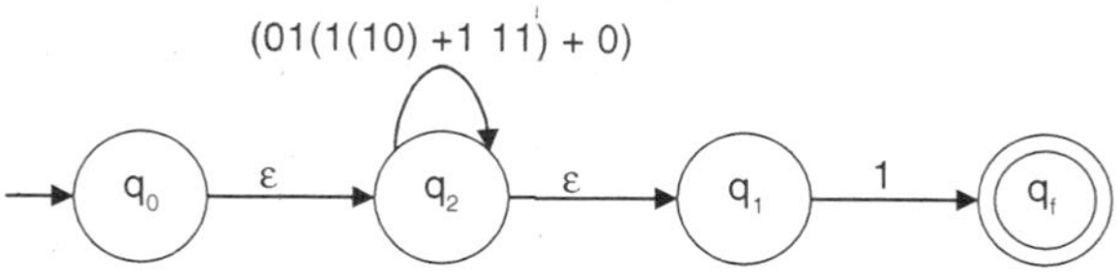

(iii)

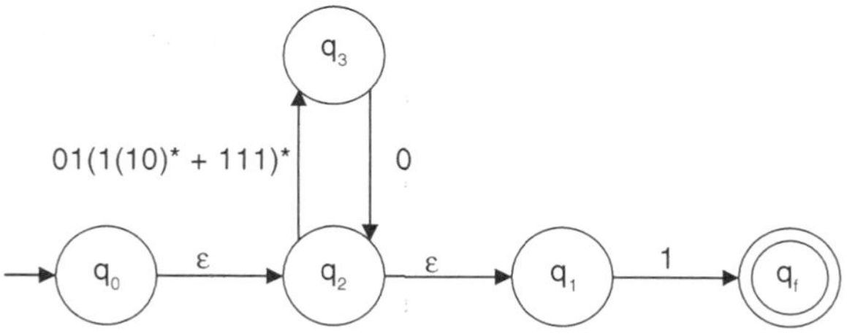

(iv)

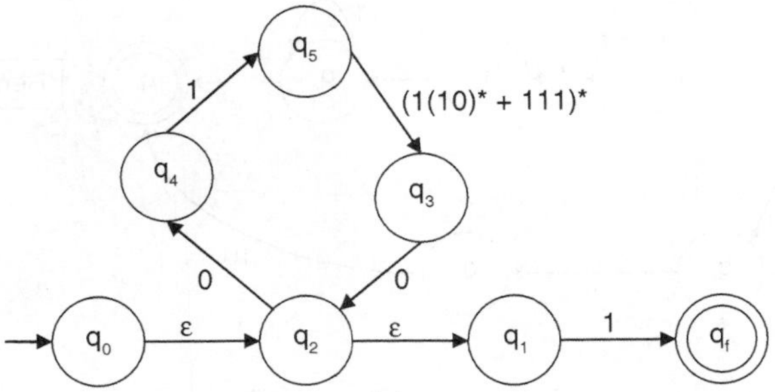
q5
1
(1(10)* + 111)*
q4
q3
0
0
q0
ε
q2
ε
q1
1
qf

(v)

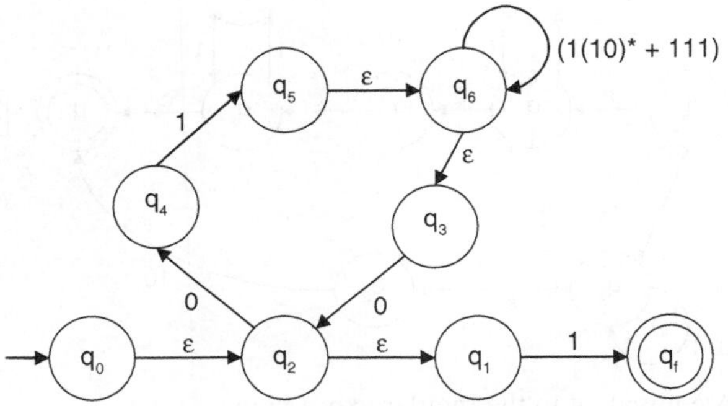
(1(10)* + 111)
q5
ε
q6
1
ε
q4
q3
0
0
q0
ε
q2
ε
q1
1
qf

(vi)

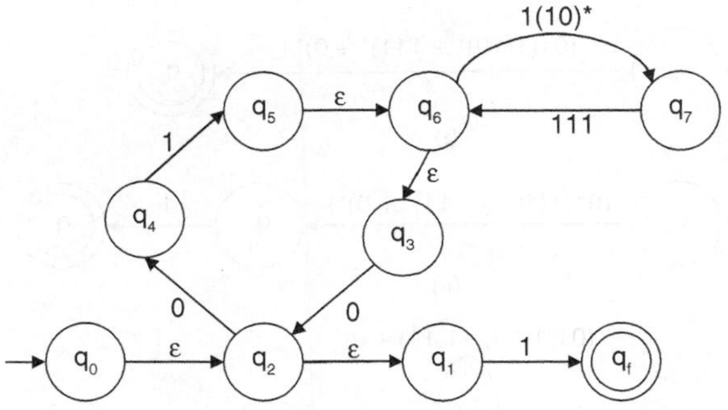
1(10)*
q5
ε
q6
111
q7
1
ε
q4
q3
0
0
q0
ε
q2
ε
q1
1
qf

(vii)

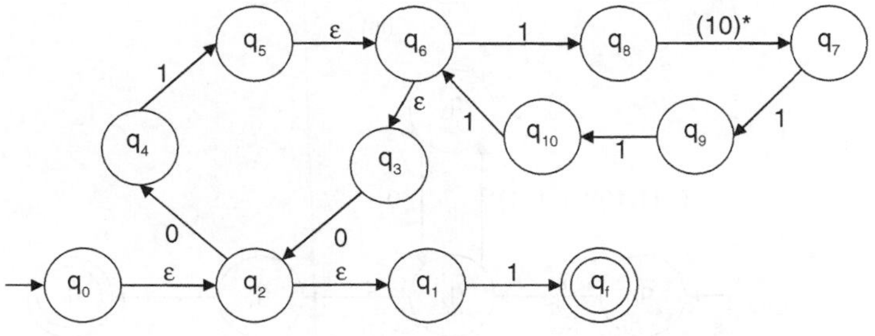
q5
ε
q6
1
q8
(10)*
q7
1
ε
1
q10
1
q9
1
q4
q3
0
0
q0
ε
q2
ε
q1
1
qf

(viii)

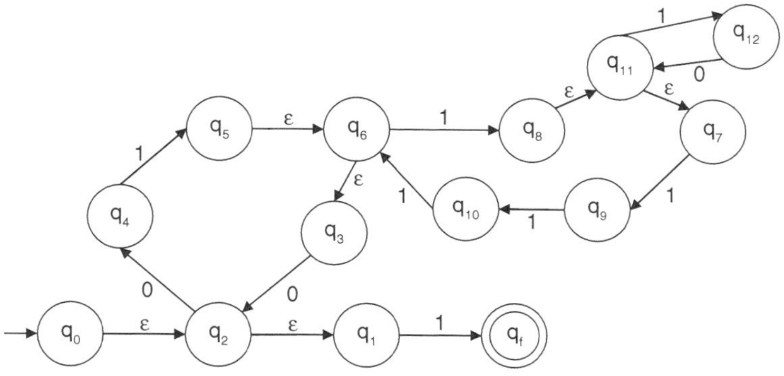

(ix)

13. Convert the following NFA with ∈-move to an equivalent DFA.

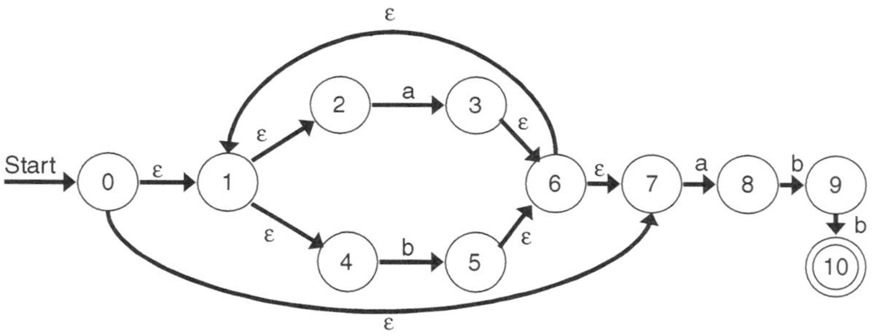

(x)

Solution:

∈-closure(0) = {0, 1, 2, 4, 7} ∈-closure(3) = {1, 2, 3, 4, 6, 7} ∈-closure(6) = {1, 2, 4, 6, 7}
∈-closure(1) = {1, 2, 4} ∈-closure(4) = {4} ∈-closure(7) = {7}
∈-closure(2) = {2} ∈-closure(5) = {1, 2, 4, 5, 6, 7} ∈-closure(8) = {8}
∈-closure(9) = {9} ∈-closure(10) = {10}

As 0 is the beginning state, start with ∈-closure(0) = {0, 1, 2, 4, 7}. Let us rename it as A.
A is still unmarked.
Then, construct a δ′ function for the new unmarked state A for inputs a and b.

$$
\begin{aligned}
\delta'(A, a) &= \in\text{-closure}(\delta\ (A, a)) \\
&= \in\text{-closure}\ (\delta((0, 1, 2, 4, 7), a)) \\
&= \in\text{-closure}\ (3, 8) \\
&= \{1, 2, 3, 4, 6, 7, 8\}.
\end{aligned}
$$

This is a new state, so rename it as B.

$$
\begin{aligned}
\delta'(A, b) &= \in\text{-closure}(\delta\ (A, b)) \\
&= \in\text{-closure}\ (\delta((0, 1, 2, 4, 7), b)) \\
&= \in\text{-closure}\ (5) = \{1, 2, 4, 5, 6, 7\}
\end{aligned}
$$

This is a new state, so rename it as C.

B is still unmarked.

Then, construct a δ' function for the new unmarked state B for inputs a and b.

$$\begin{aligned}\delta'(B, a) &= \in\text{-closure}(\delta\,(B, a))\\ &= \in\text{-closure}\,(\delta((1, 2, 3, 4, 6, 7, 8), a))\\ &= \in\text{-closure}\,(3, 8)\\ &= \{1, 2, 3, 4, 6, 7, 8\} = B\end{aligned}$$

$$\begin{aligned}\delta'(B, b) &= \in\text{-closure}(\delta\,(B, b))\\ &= \in\text{-closure}\,(\delta((1, 2, 3, 4, 6, 7, 8), b))\\ &= \in\text{-closure}\,(5, 9)\\ &= \{1, 2, 4, 5, 6, 7, 9\}\end{aligned}$$

This is a new state, so rename it as D.

C is still unmarked.

Then, construct a δ' function for the new unmarked state C for inputs a and b.

$$\begin{aligned}\delta'(C, a) &= \in\text{-closure}(\delta\,(C, a))\\ &= \in\text{-closure}\,(\delta((1, 2, 4, 5, 6, 7), a))\\ &= \in\text{-closure}\,(3, 8)\\ &= \{1, 2, 3, 4, 6, 7, 8\} = B\end{aligned}$$

$$\begin{aligned}\delta'(C, b) &= \in\text{-closure}(\delta\,(C, b))\\ &= \in\text{-closure}\,(\delta((1, 2, 4, 5, 6, 7), b))\\ &= \in\text{-closure}\,(5) = \{1, 2, 4, 5, 6, 7\} = C\end{aligned}$$

D is still unmarked.

Then, construct a δ' function for the new unmarked state D for inputs a and b.

$$\begin{aligned}\delta'(D, a) &= \in\text{-closure}(\delta\,(D, a))\\ &= \in\text{-closure}\,(\delta((1, 2, 4, 5, 6, 7, 9), a))\\ &= \in\text{-closure}\,(3, 8)\\ &= \{1, 2, 3, 4, 6, 7, 8\} = B\end{aligned}$$

$$\begin{aligned}\delta'(D, b) &= \in\text{-closure}(\delta\,(D, b))\\ &= \in\text{-closure}\,(\delta((1, 2, 4, 5, 6, 7, 9), b))\\ &= \in\text{-closure}\,(5) = \{1, 2, 4, 5, 6, 7, 10\}\end{aligned}$$

This is a new state, so rename it as E.

$$\begin{aligned}\delta'(E, a) &= \in\text{-closure}(\delta\,(E, a))\\ &= \in\text{-closure}\,(\delta((1, 2, 4, 5, 6, 7, 10), a))\\ &= \in\text{-closure}\,(3, 8)\\ &= \{1, 2, 3, 4, 6, 7, 8\} = B\end{aligned}$$

$$\begin{aligned}\delta'(E, b) &= \in\text{-closure}(\delta\,(E, b))\\ &= \in\text{-closure}\,(\delta((1, 2, 4, 5, 6, 7, 10), b))\\ &= \in\text{-closure}\,(5) = \{1, 2, 4, 5, 6, 7\} = C\end{aligned}$$

Here, the final state is E.

The equivalent DFA is

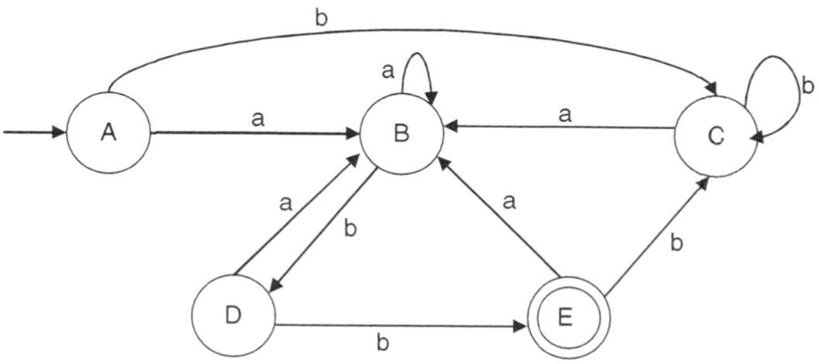

14. Construct a regular grammar for the RE, L = (a + b)*(aa + bb)(a + b)*.
Solution: The NFA for the RE is

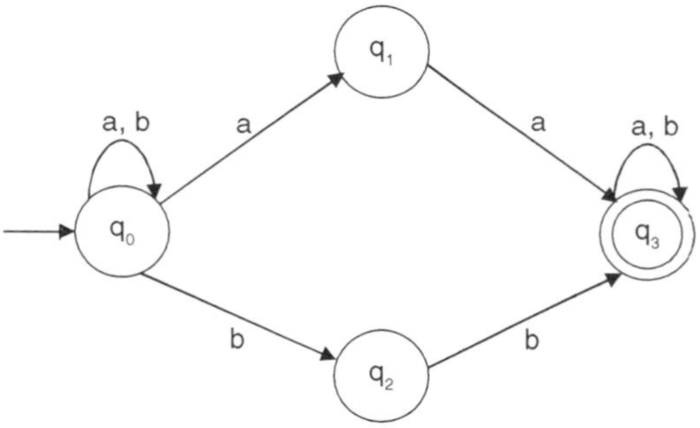

There are four states in the FA. So, in the regular grammar, there are four non-terminals. Let us take them as A (for q_0), B (for q_1), C (for q_2), and D (for q_3).
Now, we have to construct the production rules of the grammar.
For the state q_0, the production rules are

$$A \rightarrow aA, A \rightarrow bA, A \rightarrow aB, A \rightarrow bC.$$

For the state q_1, the production rules are

$$B \rightarrow aD, B \rightarrow a \text{ (as D is the final state).}$$

For the state q_3, the production rules are

$$C \rightarrow b\,D, C \rightarrow b \text{ (as D is the final state).}$$

For the state q_4, the production rules are

$$D \rightarrow aD, D \rightarrow bD, D \rightarrow a/ b.$$

The grammar = {V_N, Σ, P, S}

$$V_N = \{A, B, C, D\} \ \Sigma = \{a, b\}$$
$$P : A \rightarrow aA/bA/aB/bC$$
$$B \rightarrow aD/a$$
$$C \rightarrow bD/b$$
$$D \rightarrow aD/bD/a/b.$$

15. Consider the given FA and construct the smallest DFA which accepts the same language. Draw an RE and the grammar that generates it. [UPTU 2002]

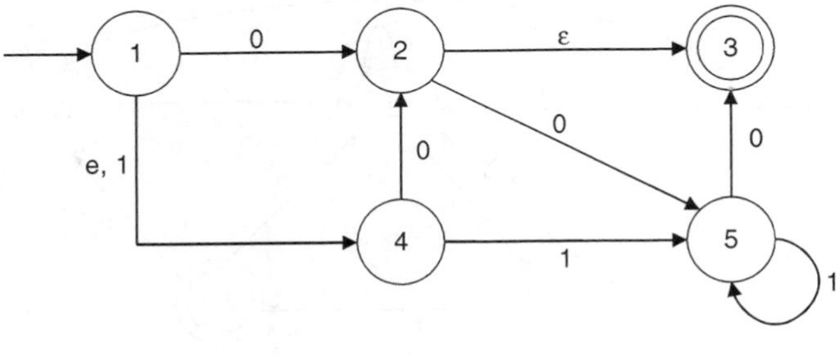

Solution: This problem can be solved by the ∈-closure method.

∈-closure(1) = {1, 4} ∈-closure(3) = {3} ∈-closure(5) = {5}
∈-closure(2) = {2, 3} ∈-closure(4) = {4}

As 1 is the beginning state, start with ∈-closure(1) = {1,4}. Let us rename it as A.
A is still unmarked.
Then, construct a δ′ function for the new unmarked state A for inputs 0 and 1.

δ′(A, 0) = ∈-closure(δ (A, 0))
= ∈-closure (δ ((1,4), 0))
= ∈-closure (2, 2)
= {2, 3}
It is a new state. Mark it as B.

δ′(A, 1) = ∈-closure(δ (A, 1))
= ∈-closure (δ ((1, 4), 1))
= ∈-closure (4, 5)
= ∈-closure (4) ∪ ∈-closure (5) = {4, 5}
It is a new state. Mark it as C.

δ′(B, 0) = ∈-closure(δ (B, 0))
= ∈-closure (δ ((2, 3), 0))
= ∈-closure (5)
= {5}
It is a new state. Mark it as D.

δ′(B, 1) = ∈-closure(δ (B, 1))
= ∈-closure (δ ((2, 3), 1))
= ∈-closure ()
= ∅

δ′(C, 0) = ∈-closure(δ (C, 0))
= ∈-closure (δ ((4, 5), 0))
= ∈-closure (2, 3)
= B

δ′(C, 1) = ∈-closure(δ (C, 1))
= ∈-closure (δ ((4, 5), 1))
= ∈-closure (5)
= {5} = D

δ′(D, 0) = ∈-closure(δ (D, 0))
= ∈-closure (δ ((5), 0))
= ∈-closure (3)
= {3}
It is a new state. Mark it as E.

δ′(D, 1) = ∈-closure(δ (D, 1))
= ∈-closure (δ (5), 1)
= ∈-closure (5)
= {5} = D

δ′(E, 0) = ∈-closure(δ (E, 0))
= ∈-closure (δ (3), 0)
= ∅

δ′(E, 1) = ∈-closure(δ (E, 1))
= ∈-closure (δ (5), 1)
= ∈-closure (5)
= {5} = D

The beginning state is A, and the final state is E.

The transitional diagram of the DFA is

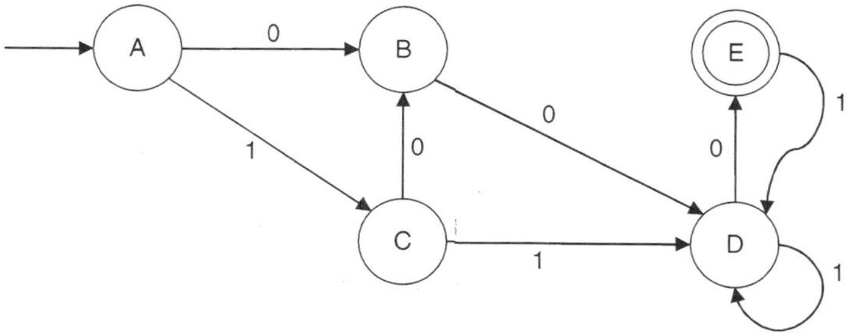

The RE is constructed using the Arden's theorem.

$$A = \wedge$$
$$B = 0A + 0C$$
$$C = 1A$$
$$D = 0B + 1C + 1D + 1E$$
$$E = 0D$$

Replacing A in B and C, we get

$$B = 0 + 0C$$
$$C = 1.$$

Replacing C in B, we get $B = 0 + 01$.
Replacing the new value of B, C, and E in D, we get

$$D = 0(0 + 01) + 11 + 1D + 10D$$
$$= 0(0 + 01) + 11 + D(1 + 10).$$

It is in the format of $R = Q + RP$, where $Q = (0 + (0 + 01) + 11)$, $P = (1 + 10)$.
The solution is $R = QP^*$.

$$D = (0 + (0 + 01) + 11)\ (1 + 10)^*.$$

Replacing the value of D in E, we get $E = 0(0 + (0 + 01) + 11)\ (1 + 10)^*$.
As E is the final state, the RE accepted by the FA is

$$0(0 + (0 + 01) + 11)\ (1 + 10)^*.$$

The regular grammar is constructed as follows taking each state into account.
There are five states in the FA. So, in the regular grammar, there are five non-terminals.
Let us take them as A, B, C, D, and E according to the name of the states.
Now we have to construct the production rules of the grammar.
For the state A, the production rules are

$$A \rightarrow 0B,\ A \rightarrow 1C.$$

For the state B, the production rules are

$$B \rightarrow 0D.$$

For the state C, the production rules are

$$C \rightarrow 0B \quad C \rightarrow 1D.$$

For the state D, the production rules are

$$D \rightarrow 1D \quad D \rightarrow 0E \quad D \rightarrow 0.$$

For the state E, the production rules are

$$E \rightarrow 1D.$$

16. Find the RE recognized by the finite state automaton of the following figure. [GATE 1994]

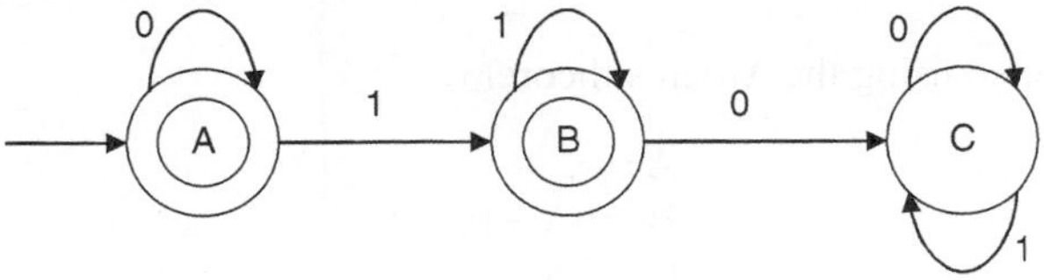

Solution: The equation for the FA is

$$A = 0A + \wedge \quad (1)$$
$$B = 1A + 1B \quad (2)$$
$$C = 0B + 0C + 1C \quad (3)$$

Solving the equation (1) using the Arden's theorem, we get $A = \wedge 0^* = 0^*$.
Putting the value of A in equation (2), we get

$$B = 10^* + 1B.$$

Using the Arden's theorem, we get

$$B = 10^*1^*.$$

Both A and B are final states, and thus the string accepted by the FA is

$$0^* + 10^*1^*$$
$$= 0^* (\wedge + 11^*) = 0^*1^* \text{ (as } \wedge + RR^* = R^*).$$

Multiple Choice Questions

1. The machine format of regular expression is
   a) Finite automata
   b) Push down automata
   c) Turing machine
   d) All of the above
2. The regular expression is accepted by
   a) Finite automata
   b) Push down automata
   c) Turing machine
   d) All of the above
3. The language of all words with at least 2 a's can be described by the regular expression
   a) (ab)*a
   b) (a + b)*ab*a(a + b)*
   c) b*ab*a(a + b)*
   d) All of the above
4. The set of all strings of {0, 1} having exactly two 0's is
   a) 1*01*01* b) {0 + 1)*1
   c) {11 + 0}* d) {00 + 11}*

5. The set of all strings of {0, 1} where 0 is followed by 1 and 1 is followed by 0 is
   a) (01)* b) 0*1*
   c) 0*1 + 1*0 d) 0*1* + 1*0*
6. Which of the strings do not belong to the regular expression (ba + baa)*aaba
   a) baaaba b) babaabaaaba
   c) bababaaba d) baaaaba
7. Which of the following type of language is regular expression?
   a) Type 0 b) Type 1
   c) Type 2 d) Type 3
8. Which of the following RE over {0,1} denotes the set of all strings not containing 100 as substring?
   a) (1 + 0)*0* b) 0*1010*
   c) 0*1*01 d) All of the above
9. $\wedge$ + RR* = ?
   a) R b) R* c) R^+ d) $\wedge$
10. {a/b} denotes the set
   a) {ab} b) {a, b}
   c) {a, b}* d) {ab}*
11. What is the solution for the equation R = Q + RP (If P and Q are RE and P does not contain $\wedge$)?
   a) R = QP* b) R = QP
   c) R = PQ* d) R = P*Q*
12. Which of the languages is accepted by the following FA?

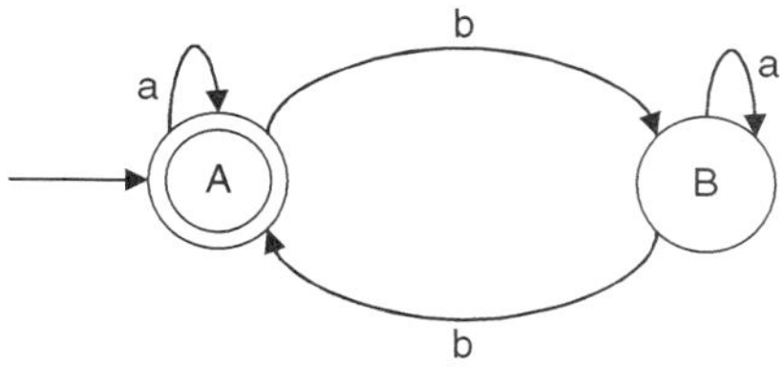

   a) b(a + bba*)*a* b) a(a + bba*)*b*
   c) a(a + bb*a*)*b* d) None of these
13. Which is true for $\in$-closure of a state?
   a) All the states reachable from the state with input null excluding the state
   b) The state only
   c) All the other states
   d) All the states reachable from the state with input null
14. The pumping lemma for regular expression is used to prove that
   a) Certain sets are regular
   b) Certain sets are not regular
   c) Certain regular grammar produce RE
   d) Certain regular grammar does not produce RE
15. Regular sets are closed under
   a) Union b) Concatenation
   c) Kleene closure d) All of the above

Answers:

1. a 2. d 3. b 4. a 5. d 6. c 7. c 8. c 9. b 10. b 11. a 12. a
13. d 14. a 15. d

GATE Questions

1. Which two of the following four regular expressions are equivalent?
   i) (00)*($\in$ + 0) ii) (00)* iii) 0*0* iv) 0(00)*
   a) (i) and (ii) b) (i) and (iv) c) (ii) and (iii) d) (iii) and (iv)
2. Which of the following alternatives is true for the following three regular expressions?
   i) (011((11)* + (01)*)*)*011 ii) 011(((1 + 0)1)*011)* iii) 011(((1 + 0)*1*)*011)*

a) (i) and (iii) are equivalent b) (ii) and (iii) are equivalent
c) (i) and (ii) are equivalent d) No pairs are equivalent

3. Consider the following statements:
i) $Ø^* = \in$ ii) $s(rs + s)^*r = (sr + s)^*sr$ iii) $(r + s)^* = (r^* + s^*) +$
Find out the correct alternatives.
a) All are true b) Only (ii) is false c) only (iii) is false d) Both (ii) and (iii) are false

4. Which regular expression best describes the language accepted by the following non-deterministic automation?

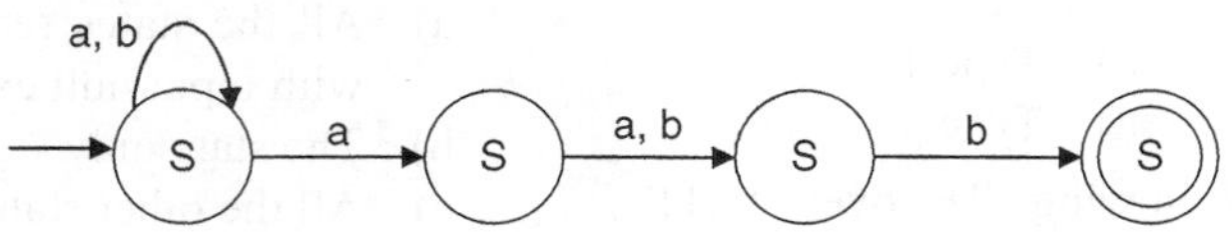

a) $(a + b)^*\, a(a + b)b$ b) $(abb)^*$ c) $(a + b)^*a(a + b)^*b(a + b)^*$ d) $(a + b)^*$

5. Which of the following regular expressions describe the language over {0, 1} consisting of strings that contain exactly two 1's?
a) $(0 + 1)^*11(0 + 1)^*$ b) 0^*110^* c) $0^*10^*10^*$ d) $(0 + 1)^*1(0 + 1)^*1(0 + 1)^*$

6. Find the false statement if S = (a, b)
a) L = $\{a^n b^m, m, n \geq 1\}$ is regular
b) L = {X, where no of a > no of b} is not regular
c) L = $\{a^n b^n, n \geq 1\}$ is regular
d) L = {X, where no of a and no of b are equal} is not regular

7. Which of the following regular expression identified is true?
a) $r^{(*)} = r^*$ b) $(r^*s^*) = (r + s)^*$ c) $(r + s)^* = r^* + s^*$ d) $r^*s^* = r^* + s^*$

8. In some programming languages, an identifier is permitted to be a letter followed by any number of letters or digits. If L and D denote the sets of letters and digits, respectively, which of the following expressions define an identifier?
a) $(L \cup D)^*$ b) $L(L \cup D)^*$ c) $(L.D)^*$ d) $L(L.D)^*$

9. Which two of the following four regular expressions are equivalent?
i) $(00)^*(\in + 0)$ ii) $(00)^*$ iii) 0^* iv) $0(00)^*$
a) (i) and (ii) b) (ii) and (iii) c) (i) and (iii) d) (iii) and (iv)

10. Let $L \subseteq \Sigma^*$, where $\Sigma = \{a, b\}$. Which of the following is true?
a) L = {x | x has equal number of a's and b's} is regular
b) L = $\{a^n b^n \mid n \geq 1\}$ is regular
c) L = {x | x has more a's than b's} is regular
d) L = $\{a^m b^n \mid m \geq 1, n \geq 1\}$ is regular

11. The string 1101 does not belong to the set represented by
a) $110^*(0 + 1)$ b) $1(0 + 1)^*101$ c) $(10)^*(01)^*(00 + 11)^*$ d) $(00 + (11)^*01)^*$

12. Give a regular expression for the set of binary strings where every 0 is immediately followed by exactly k number of 1's and preceded by at least k number of 1's (k is a fixed integer).
    a) $1^*1^k(01^k)^*1^*$ b) $1^*1^k01^k$ c) $1^*(1^k01^k)^*$ d) $1^*(1^k0)^*\,1^*$
13. Let S and T be the languages over $\Sigma = \{a, b\}$ represented by the regular expressions $(a + b^*)^*$ and $(a + b)^*$, respectively. Which of the following is true?
    a) $S \subset T$ b) $T \subset S$ c) $S = T$ d) $S \cap T = \varnothing$
14. The regular expression $0^*(10^*)^*$ denotes the same set as
    a) $(1^*0)^*1^*$ b) $0 + (0 + 10)^*$ c) $(0 + 1)^*10(0 + 1)^*$ d) None of the above
15. Consider the following NFA M

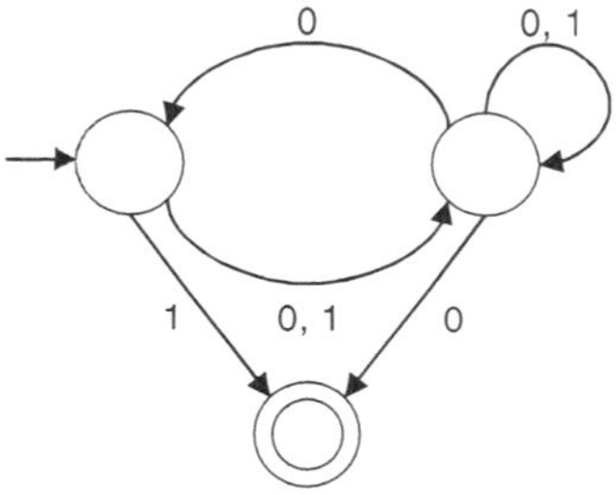

Let the language accepted by M be L. Let L_1 be the language accepted by the NFA M_1 obtained by changing the accepting states of M to a non-accepting state and by changing the non-accepting states of M to accepting states. Which of the following statements is true?
    a) $L_1 = \{0, 1\}^* - L$ b) $L_1 = \{0, 1\}^*$
    c) $L_1 \subseteq L$ d) $L_1 = L$
16. Consider the regular language $L = (111 + 11111)^*$. The minimum number of states in any DFA accepting this language is
    a) 3   b) 5   c) 8   d) 9
17. Which of the following is true?
    (a) Every subset of a regular set is regular.
    (b) Every finite subset of a non-regular set is regular.
    (c) The union of two non-regular sets is not regular.
    (d) Infinite union of finite sets is regular.
18. The language accepted by this automaton is given by the regular expression
    a) $b^*ab^*ab^*ab^*$ b) $(a + b)^*$ c) $b^*a(a + b)^*$ d) $b^*ab^*ab^*$

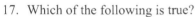

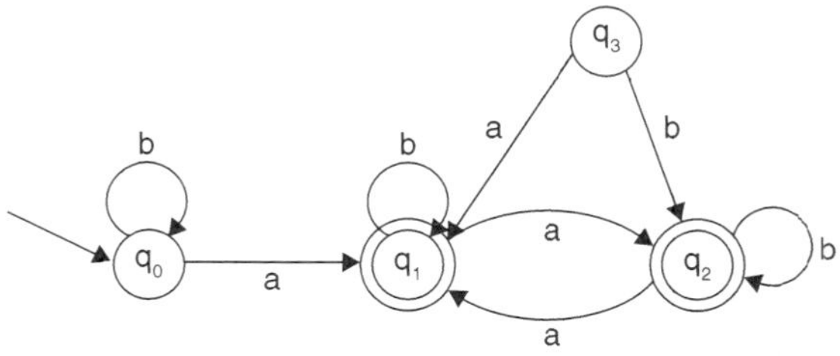

19. The minimum state automation equivalent to the previous FSA has the following number of states.
    a) 1   b) 2   c) 3   d) 4
20. Which of the following is true for the language $\{a^p \mid p \text{ is prime}\}$?
    a) It is not accepted by a Turing machine
    b) It is regular but not context-free
    c) It is context-free but not regular
    d) It is neither regular nor context-free but accepted by a Turing machine.

21. Match the following NFAs with the regular expressions they correspond to.

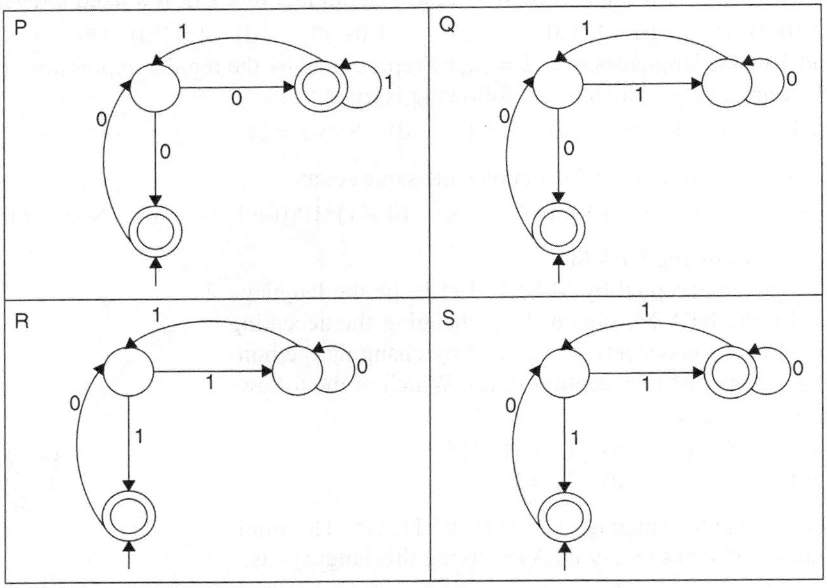

1. $\in + 0(01^*1 + 00)^*01^*$
2. $\in + 0(10^*1 + 00)^*0$
3. $\in + 0(10^*1 + 10)^*1$
4. $\in + 0(10^*1 + 10)^*10^*$

a) P −2, Q −1, R −3, S −4 b) P −1, Q −3, R −2, S −4
c) P −1, Q −2, R −3, S −4 d) P −3, Q −2, R −1, S −4

22. Which of the following are regular sets?

I. $\{a^nb^{2m} \mid n \geq 0, m \geq 0\}$
II. $\{a^nb^m \mid n = 2m\}$
III. $\{a^nb^m \mid n \neq m\}$
IV. $\{xcy \mid x, y \in \{a, b\}^*\}$

a) I and IV only b) I and III only c) I only d) IV only

23. Which one of the following languages over the alphabet {0, 1} is described by the regular expression:

$(0 + 1)^*0(0 + 1)^*0(0 + 1)^*$?

a) The set of all strings containing the substring 00.
b) The set of all strings containing at most two 0's.
c) The set of all strings containing at least two 0's.
d) The set of all strings that begin and end with either 0 or 1

24. Let $L = \{w \in (0 + 1)^* \mid w$ has an even number of 1s$\}$, i.e., L is the set of all bit strings with even number of 1s. Which one of the given regular expressions represent L?

a) $(0^*10^*1)^*$ b) $0^*(10^*10^*)^*$ c) $0^*(10^*1^*)^*0^*$ d) $0^*1(10^*1)^*10^*$

25. What is the complement of the language accepted by the given NFA? Assume that $\Sigma = \{a\}$ and $\in$ is the empty string.

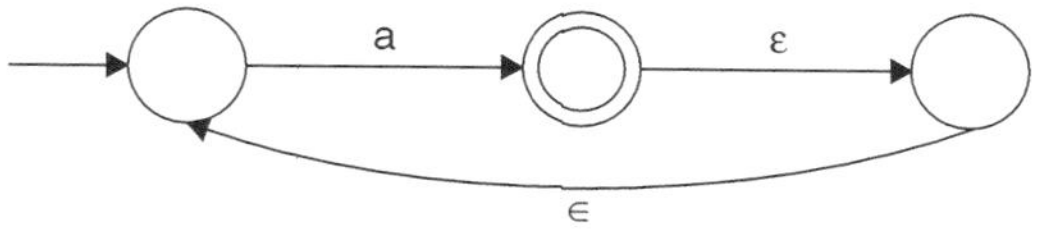

(a) $\in$ (b) $\{\in\}$ (c) a^* (d) $\{a, \in\}$

26. Given the language L = {ab, aa, baa}, which of the following strings are in L*?
    1. abaabaaabaa
    2. aaaabaaaa
    3. baaaaabaaaab
    4. baaaaabaa

    (a) 1, 2, and 3 (b) 2, 3, and 4 (c) 1, 2, and 4 (d) 1, 3, and 4

27. Which one of the following regular expressions is not equivalent to the regular expression $(a + b + c)^*$?

    a) $(a^* + b^* + c^*)^*$ b) $(a^*b^*c^*)^*$ c) $((ab)^* + c^*)^*$ d) $(a^*b^* + c^*)$

28. Let $M = (K, \Sigma, \delta, s, F)$ be a finite state automaton, where $K = \{A, B\}$, $= \{a, b\}$, $s = A$, $F = \{B\}$, $\delta(A, a) = A$, $\delta(A, b) = B$, $\delta(B, a) = B$, and $\delta(B, b) = A$.

    A grammar to generate the language accepted by M can be specified as $G = (V, \Sigma, R, S)$, where $V = K \cup \Sigma$ and $S = A$.

    Which one of the following set of rules will make L (G) = L(M)?

    a) $\{A \rightarrow aB, A \rightarrow bA, B \rightarrow bA, B \rightarrow aA, B \rightarrow \in\}$
    b) $\{A \rightarrow aA, A \rightarrow bB, B \rightarrow bb, B \rightarrow bA, B \rightarrow \in\}$
    c) $\{A \rightarrow bB, A \rightarrow aB, B \rightarrow aA, B \rightarrow bA, B \rightarrow \in\}$
    d) $\{A \rightarrow aA, A \rightarrow bA, B \rightarrow aB, B \rightarrow bA, A \rightarrow \in\}$

Answers:

1. b	2. c	3. d	4. a	5. c	6. c	7. a	8. b	9. c	10. d
11. c	12. a	13. c	14. a	15. b	16. b	17. c	18. c	19. b	20. d
21. c	22. c	23. c	24. a,b,d	25. b	26. c	27. c	28. b		

Hints:

2. See example 5.4.
11. Both generates odd and even length strings
14. $(a + b^*)^* = (a^* (b^*)^*)^* = (a^*b^*)^*$
15. Both of these generate any combination of 0 and 1
16. Construct the transitional diagram.
17.

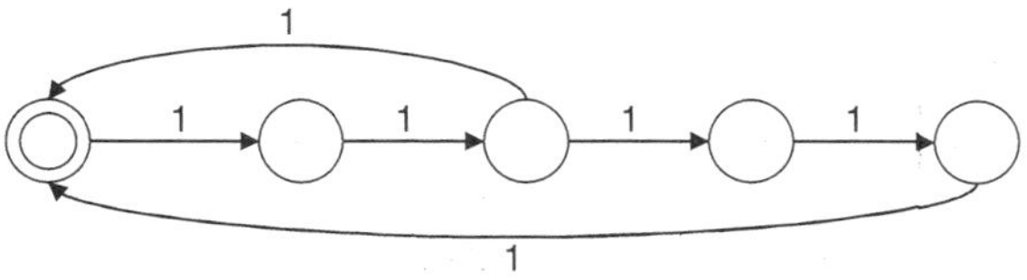

18. a) $(0 + 1)^*$ is regular but 0^n1^n is not regular.
20. The simplified automata is

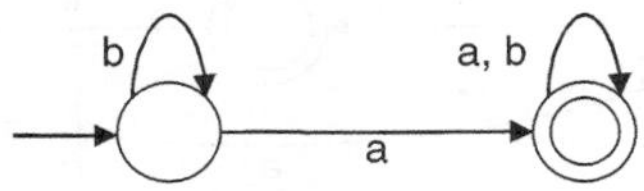

21. It is not regular (proved by the pumping lemma). Not CFG. But accepted by TM as TM accepts unrestricted language.
25. Only c may generate odd number of 1.
26. The language accepted by the FA is a^+. The complement of it is $\Sigma^* - a^+ = \in$.
28. C will not generate ac or bc but all the others will generate.

Fill in the Blanks

1. Regular expression is a string on __________ among $\{Q, \Sigma, \delta, q_0, F\}$
2. If R is a regular expression, then $\Lambda + RR^* =$ __________.
3. According to the Arden's theorem, the solution of the equation $R = Q + RP$ is __________.
4. The set of all states that can be reached from that state to all the states with input $\in$ is called ____.
5. For the following NFA with $\in$-move

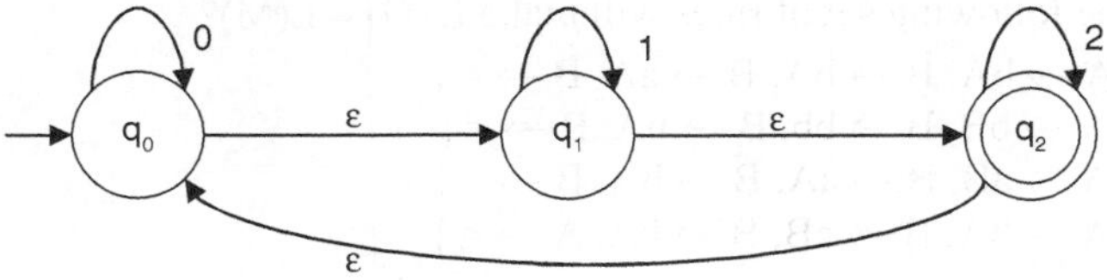

$\in$-closure of q_1 is __________.
6. A grammar where all productions are in the form $A \rightarrow B\alpha$ or $A \rightarrow \alpha$ is called __________.
7. A grammar where all productions are in the form $A \rightarrow \alpha B$ or $A \rightarrow \alpha$ is called __________.
8. The pumping lemma for regular expression is used to prove that certain sets are not __________.
9. The property which describes when we combine any two elements of the set and the result is also included in the set is called __________.
10. If any finite automata contains any $\in$ (null) move or transaction, then that finite automata is called __________.
11. The machine format of regular expression is __________.

Answers:

1. Σ	2. R^*	3. $R = QP^*$
4. $\in$-closure	5. q_0, q_1, q_2	6. left linear grammar
7. right linear grammar	8. regular	9. closure property
10. NFA with Σ moves	11. Finite automata	

Exercise

1. Prove that (0 + 011*) + (0 + 011*)(01 + 0100*)(01 + 0100*)* = 01*(010*)*
2. Find the regular expression for the following.
   a) Set of all strings over (a, b) containing exactly one a
   b) Set of all strings over (a, b) containing at least two a's
   c) Set of all strings over (0, 1) containing exactly two a's
   d) Set of all strings over (a, b) beginning and ending with b
   e) Set of all strings over (0, 1) containing 010 as substring
   f) Set of all strings over (0, 1) not containing substring 00
3. Describe the following regular expressions in English language
   a) b*ab* b) (a + b)* aba(a + b)* c) a(a + b)*ab
4. Find the regular expressions corresponding to the automaton generated by the following finite automata.

   a)

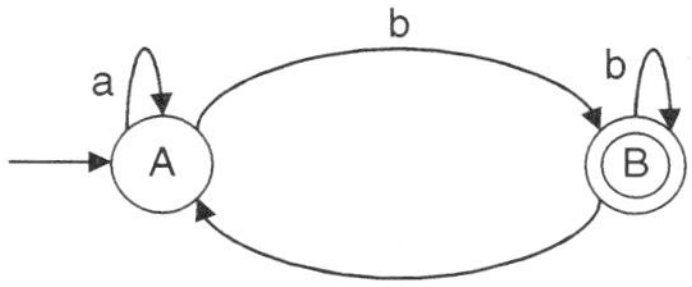

   b)

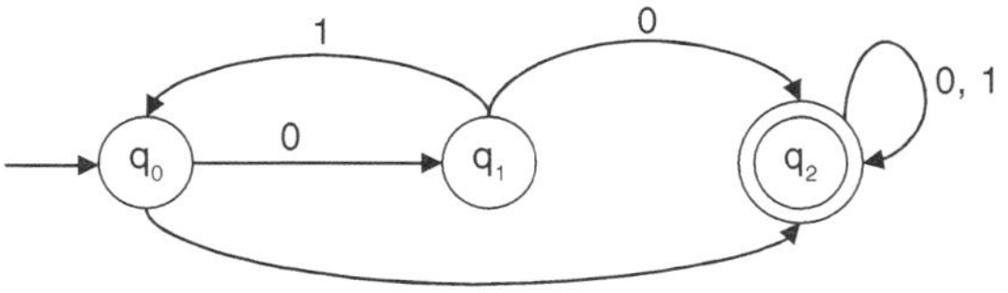

   c)

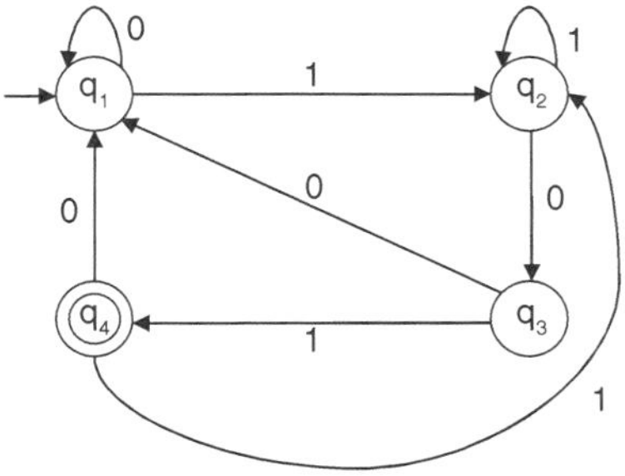

5. Convert the following NFA with null move to an equivalent DFA by the ∈-closure method.

   a)

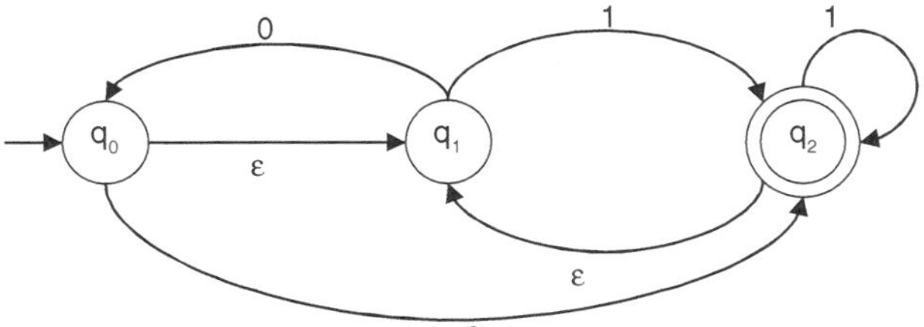

b)

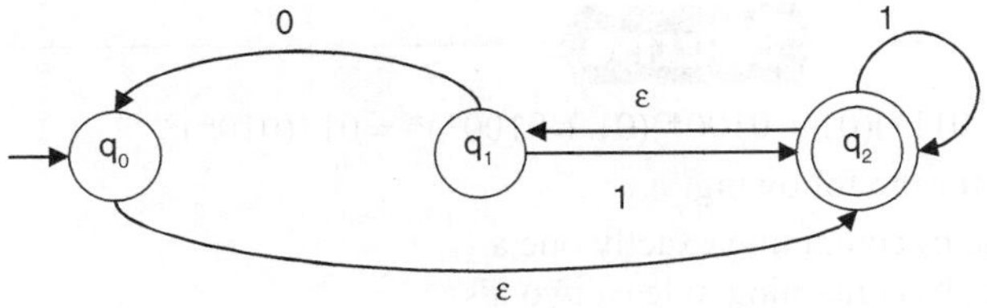

c)

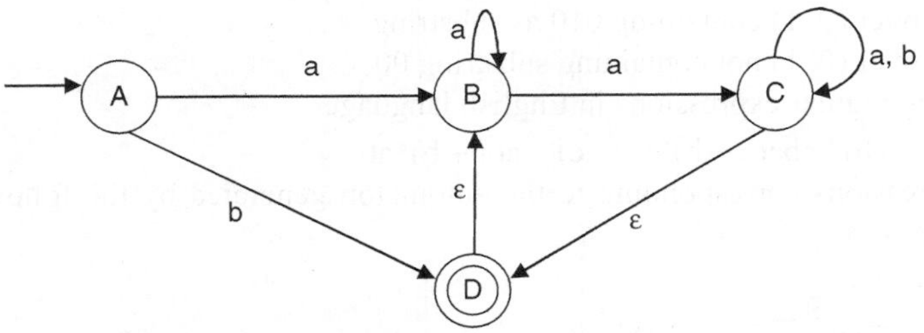

6. Develop the finite automata for the following regular expressions
   a) 10 + (0 + 11)*01
   b) (a*ab + ba)*a
   c) a(ab + a)*(aa + b)
7. Check whether the following two finite automata are equivalent or not.

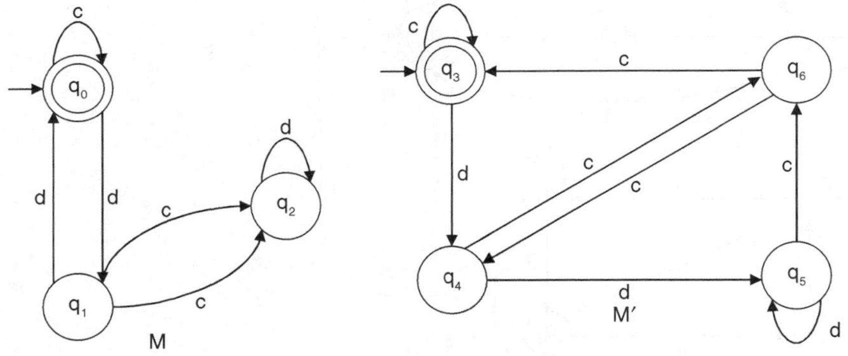

8. Check whether the following two RE are equivalent or not
   L_1 = 10 + (1010)*(1010)* L_2 = 10 + (1010)*
9. Construct the regular grammar for the following finite automata.
   i) ab + (aa + bb)(a + b)*
   ii) 01(00 + 11)*11*
   iii) a*b(a + b)*
   iv) a*(aa + bb)* + b*(ab + ba)*
10. Using the pumping lemma show that the following sets are not regular.
   a) L = $\{a^n b^m$, where $0 < n < m\}$
   b) L = $\{a^n b^{n+1}$, where $n > 0\}$
   c) L = $\{a^i b^j c^k$, where $i, j, k > 0\}$

Context-free Grammar 6

Introduction

Context-free grammar, in short CFG, is type 2 grammar according to the Chomsky hierarchy. The language generated by the CFG is called context-free language (CFL). In the previous chapters, we have discussed about finite automata and regular expression which are related to regular grammar. Regular grammar is a subset of CFG. In this chapter, we shall learn about CFG, how to derive a language from a CFG, simplification of CFG, transformation of CFG into normal form, and the pumping lemma for CFG.

6.1 Definition of Context-free Grammar

According to the Chomsky hierarchy, context-free grammar, or in short CFG, is type 2 grammar.

In mathematical description, we can describe it as

Where all the production are in the form $\alpha \rightarrow \beta$ *where* $\alpha \in V_N$, *i.e., set of non-terminals and* $|\ \alpha\ | = 1$, *i.e., there will be only one non-terminal at the left hand side (LHS) and* $\beta \in V_N \cup \Sigma$, *i.e.,* β *is a combination of non-terminals and terminals.*

Before describing why this type of grammar is called context free, we have to know the definition of context. Non-terminal symbols are the producing symbols because they produce some extra symbols. So, the production rules are in the format of

{A string consists of at least one non-terminal} → {A string of terminals and/or non-terminals}.

If any symbol is present with the producing non-terminal at the LHS of the production rule, then that extra symbol is called context. Context can be of two types: (a) left context and (b) right context.

In CFG, at the LHS of each production rules, there is only one non-terminal. (No context is added with it.) For this reason, this type of grammar is called CFG.

6.1.1 Backus Naur Form (BNF)

This is a formal notation to describe the syntax of a given language. This notation was first introduced by John Backus and Peter Naur. This was first introduced for the description of the ALGOL 60 programming language.

The symbols used in BNF are as follows:

:: = denotes "is defined as"
| denotes "or"
< > used to hold category names

As an example,

<number> :: = <digit> | <number> <digit>
<digit> :: = 0 | 1 | 2 | 3 | 4 | 5 | 6 | 7 | 8 | 9

This can be described as a number is a digit or a number followed by a digit.

A digit is any of the characters from 0 to 9.

From BNF comes the extended BNF which uses some notations of regular expression such as + or *.

As an example, the identifier of a programming language is denoted as

<Identifier> :: = letter(letter | digit)*
<letter> :: = a | ...... | z | A | ... | Z
<digit> :: = 0 | 1 | 2 | 3 | 4 | 5 | 6 | 7 | 8 | 9

BNF is widely used in CFG.

Following are some examples of CFG.

Example 6.1 Construct a CFG for the language $L = \{WCW^R \mid W \in (a, b)^*\}$

Solution: W is a string of any combination of a and b. So, W^R is also a string of any combination of a and b, but it is a string which is reverse of W. C is a terminal symbol like a, b. Therefore, if we take C as a mirror, we will be able to see the reflection of W in the W^R part, for instance, C, abCba, abbaCabba, and so on. That means, there is something that is generating symbols by replacing which it adds the same symbol before C and after C. As $W \in (a, b)^*$, the null symbols are also accepted in the place of a, b. That means, only C is accepted by this language set.

From the previous discussion, the production rules are:

$$S \rightarrow aSb/bSb/C$$

The grammar becomes

$$G = \{V_N, \Sigma, P, S\}$$

where

$V_N : \{S\}$
$\Sigma : \{a, b, C\}$
$P : S \rightarrow aSb/bSb/C$
$S : \{S\}$

Example 6.2 Construct a CFG for the regular expression $(0 + 1)^*\ 0\ 1^*$.

Solution: The regular expression in English language is as follows:

Any combination of 0 and 1 followed by a single 0 and ending with any number of 1.

In this regular expression, a single 0 is in between $(0 + 1)^*$ and 1^*. That is, in the language set, only 0 can exist. For constructing the grammar for this regular expression, keep the single 0 as fixed. Then before and after this 0, consider two non-terminals A and B which can be replaced multiple times.

So, the production rules (P) of the grammar for constructing this regular expression are

$$S \rightarrow ASB$$
$$A \rightarrow 0A/1A/\varepsilon$$
$$B \rightarrow 1B/\varepsilon$$

The grammar is

$$G = \{V_N, \Sigma, P, S\}$$

where

V_N : {S, A, B}
Σ : {0, 1}
S : {S}

Example 6.3 Construct a CFG for the regular expression (011 + 1)* (01)*.

Solution: The regular expression consists of two parts: (011 + 1)* and (01)*. Here, from the start symbol, two non-terminals, each of them producing one part, are taken. (011 + 1) can be written as 011/1. As * represent any combination, null string is also included in the language set.

From the previous discussion, the production rules (P) of the grammar are

$$S \to BC$$
$$B \to AB/\varepsilon$$
$$A \to 011/1$$
$$C \to DC/\varepsilon$$
$$D \to 01$$

The grammar is

$$G = \{V_N, \Sigma, P, S\}$$

where

V_N : {S, A, B, C, D}
Σ : {0, 1}
S : {S}

Example 6.4 Construct a CFG for a string in which there are equal number of binary numbers.

Solution: Binary numbers means 0 and 1. Any string that belongs to the language set consists of an equal number of 0 and 1. These numbers may come in any order, but in the string the number of 0 and 1 are equal.

Hence, the string may start with 0 or may start with 1. 0 can come after 0 or 1 can come after 0. 1 can come after 1 or 0 can come after 1.

The production rules (P) for this grammar are

$$S \to 0S1/1S0/\varepsilon$$

The grammar is

$$G = \{V_N, \Sigma, P, S\}$$

where

V_N : {S}
Σ : {0, 1}
S : {S}

Example 6.5 Construct a CFG for the string $\{abba(baa)^n aab(aaba)^n \mid n \geq 0\}$.

Solution: In English language, the expression can be described as '*abba* followed by any number of *baa* followed by *aab* followed by any number of *aaba*, in which the number of baa and number of *aaba* are same.' The value of n may be 0. So, the part 'baa' and 'aaba' may be omitted—which means the string abbaaab is also accepted by the language set. One thing to be noticed in the string is that the number of (baa) and number of (aaba) are the same.

The production rules (P) for the grammar of this language set are

$$S \rightarrow abbaA$$
$$A \rightarrow (baa)A(aaba)$$
$$A \rightarrow aab$$

The grammar is

$$G = \{V_N, \Sigma, P, S\}$$

where

$V_N : \{S, A\}$
$\Sigma : \{a, b\}$
$S : \{S\}$

Example 6.6 Construct a CFG for $\{a^n b^n c^m d^m \mid n, m \geq 1\}$.

Solution: The string can be described like this: A string of equal number of a and b followed by an equal number of c and d.

In this string, the number of a and the number of b will be the same and the number of c and number of d will be the same. Break the string into two parts A and B. Both are non-terminals. A will generate the string $a^n b^n$ and B will generate the string $c^m d^m$.

Now, the production rules (P) for the grammar of the language will be

$$S \rightarrow AB$$
$$A \rightarrow aAb/ab$$
$$B \rightarrow cBd/cd$$

The grammar will be

$$G = \{V_N, \Sigma, P, S\}$$

where

$V_N : \{S, A, B\}$
$\Sigma : \{a, b, c, d\}$
$S : \{S\}$

Example 6.7 Construct a grammar for the language $\{a^n b^m c^m d^n \mid m, n \geq 1\}$.

Solution: In this string, the number of a and the number of d are the same, and the number of b and the number of c are the same. Here, a comes first followed by b, c, and d. But the problem of breaking the string into two parts is that here b and c come in the middle of a and d. In this case, if we take a non-terminal for generating $b^m c^m$, then the problem will be solved.

Now, the production rules (P) for the grammar of the language will be

$$S \rightarrow aSd/aAd$$
$$A \rightarrow bAc/bc$$

The grammar will be

$$G = \{V_N, \Sigma, P, S\}$$

where

$V_N : \{S, A\}$
$\Sigma : \{a, b, c, d\}$
$S : \{S\}$

6.2 Derivation and Parse Tree

6.2.1 Derivation

In the process of generating a language from the given production rules of a grammar, the non-terminals are replaced by the corresponding strings of the right hand side (RHS) of the production. But if there are more than one non-terminal, then which of the ones will be replaced must be determined. Depending on this selection, the derivation is divided into two parts:

1. **Leftmost derivation:** A derivation is called a leftmost derivation if we replace only the leftmost non-terminal by some production rule at each step of the generating process of the language from the grammar.
2. **Rightmost derivation:** A derivation is called a rightmost derivation if we replace only the rightmost non-terminal by some production rule at each step of the generating process of the language from the grammar.

The following examples discuss the leftmost and rightmost derivations in detail.

Example 6.8 Construct the string 0100110 from the following grammar by using

i) Leftmost derivation
ii) Rightmost derivation

$$S \rightarrow 0S/1AA$$
$$A \rightarrow 0/1A/0B$$
$$B \rightarrow 1/0BB$$

Solution:

i) ***Leftmost Derivation***

$$S \rightarrow 0\underline{S} \rightarrow 01\underline{A}A \rightarrow 010\underline{B}A \rightarrow 0100\underline{B}BA \rightarrow 01001\underline{B}A \rightarrow 010011\underline{A} \rightarrow 0100110$$

(The non-terminals that are replaced are underlined.)

ii) ***Rightmost Derivation***

$$S \rightarrow 0\underline{S} \rightarrow 01A\underline{A} \rightarrow 01\underline{A}0 \rightarrow 010\underline{B}0 \rightarrow 0100B\underline{B}0 \rightarrow 0100\underline{B}10 \rightarrow 0100110$$

(The non-terminals that are replaced are underlined.)

Example 6.9 Construct the string abbbb from the grammar given below by using

i) Leftmost derivation
ii) Rightmost derivation

$$S \rightarrow aAB$$
$$A \rightarrow bBb$$
$$B \rightarrow A/\in$$

Solution:

i) ***Leftmost Derivation***

$$S \rightarrow a\underline{A}B \rightarrow ab\underline{B}bB \rightarrow ab\underline{A}bB \rightarrow abb\underline{B}bbB \rightarrow abbbb\underline{B} \rightarrow abbbb$$

ii) ***Rightmost Derivation***

$$S \rightarrow aA\underline{B} \rightarrow a\underline{A} \rightarrow ab\underline{B}b \rightarrow ab\underline{A}b \rightarrow abb\underline{B}bb \rightarrow abbbb$$

6.2.2 Parse Tree

Parsing a string is finding a derivation for that string from a given grammar.

A parse tree is the tree representation of deriving a CFL from a given context grammar. These types of trees are sometimes called as derivation trees.

A parse tree is an ordered tree in which the LHS of a production represents a parent node and the RHS of a production represents a children node.

There are certain conditions for constructing a parse tree from a given CFG. Those are as follows.

- Each vertex of the tree must have a label. The label is a non-terminal or terminal or null (Λ).
- The root of the tree is the start symbol, i.e., S.
- The label of the internal vertices is a non-terminal symbol $\in V_N$.
- If there is a production $A \rightarrow X_1X_2 \ldots\ldots X_K$, then for a vertex label A, the children of that node will be $X_1X_2 \ldots\ldots X_K$.
- A vertex n is called a leaf of the parse tree if its label is a terminal symbol $\in \Sigma$ or null (Λ).

The parse tree construction is possible only for CFG. This is because the properties of a tree match with the properties of CFG.

Here, we are describing the similar properties of a tree and a CFG.

- For a tree, there must be some root. For every CFG, there is a single start symbol.
- Each node of a tree has a single label. For every CFG at the LHS, there is a single non-terminal.
- A child node is derived from a single parent. For constructing a CFL from a given CFG, a non-terminal is replaced by a suitable string at the RHS (if for a non-terminal, there are multiple productions). Each of the characters of the string is generating a node. That is, for each single node there is a single parent.

For these similarities, a parse tree can be generated only for a CFG.

Note: *A parse tree is not possible with context sensitive grammar because there are production such as $bB \rightarrow aa$ or $AA \rightarrow B$, which means at the LHS there may be more than one symbol.*

Consider the following examples.

Example 6.10 Find the parse tree for generating the string 0100110 from the following grammar.

S → 0S/1AA
A → 0/1A/0B
B → 1/0BB

Solution: For generating the string 0100110 from the given CFG, the leftmost derivation will be

$$S \rightarrow 0\underline{S} \rightarrow 01\underline{A}A \rightarrow 010\underline{B}A \rightarrow 0100\underline{B}BA \rightarrow 01001\underline{B}A \rightarrow 010011\underline{A} \rightarrow 0100110$$

For this derivation, the parse tree is given in Fig. 6.1.

For generating the string 0100110 from the given CFG, the rightmost derivation will be

$$S \rightarrow 0\underline{S} \rightarrow 01A\underline{A} \rightarrow 01\underline{A}0 \rightarrow 010\underline{B}0 \rightarrow 0100B\underline{B}0 \rightarrow 0100\underline{B}10 \rightarrow 0100110$$

For this derivation, the parse tree is shown in Fig. 6.2.

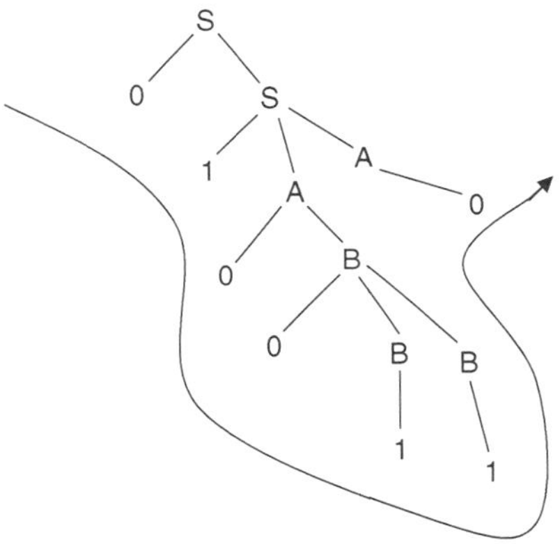

Fig. 6.1 *Parse Tree for Left Most Derivation*

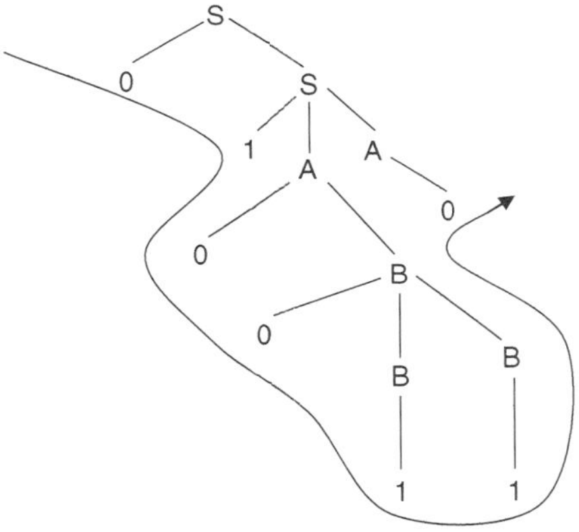

Fig. 6.2 *Parse Tree for Right Most Derivation*

Note: *When you are told to generate a parse tree, you can generate that by the leftmost derivation or the rightmost derivation.*

Example 6.11 Find the parse tree for generating the string 11001010 from the given grammar.

S → 1B/0A
A → 1/1S/0AA
B → 0/0S/1BB

Solution: For generating the string 11001010 from the given CFG the leftmost derivation will be

$$S \rightarrow 1\underline{B} \rightarrow 11\underline{B}B \rightarrow 110\underline{S}B \rightarrow 1100AB \rightarrow 11001B \rightarrow 110010S \rightarrow 1100101B \rightarrow 11001010$$

For this derivation the parse tree is given in Fig. 6.3.

For generating the string 11001010 from the given CFG, the leftmost derivation is shown in Fig. 6.4.

$$S \rightarrow 1\underline{B} \rightarrow 11B\underline{B} \rightarrow 11B0\underline{S} \rightarrow 11B01\underline{B} \rightarrow 11B010\underline{S} \rightarrow 11B0101\underline{B} \rightarrow 11\underline{B}01010 \rightarrow 11001010$$

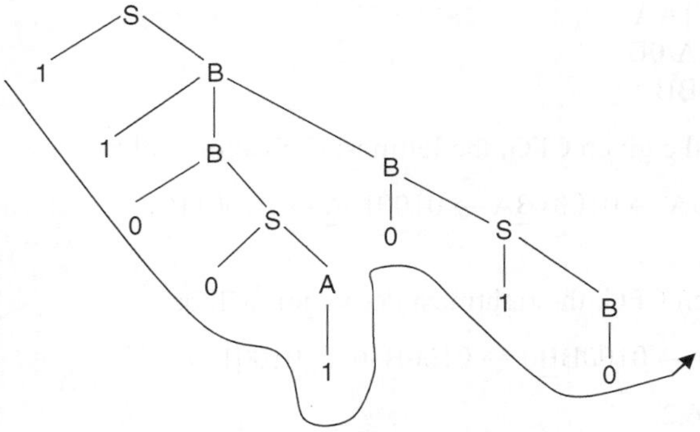

Fig. 6.3 *Parse Tree for Left Most Derivation*

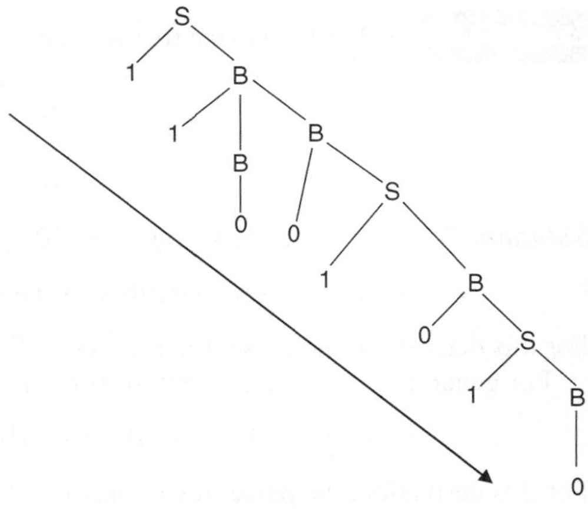

Fig. 6.4 *Parse Tree for Right Most Derivation*

Example 6.12 Construct a parse tree for the string aabbaa from the following grammar.

$$S \rightarrow a/aAS$$
$$A \rightarrow SS/SbA/ba$$

Solution: For generating the string from the given grammar, the derivation will be

$$S \rightarrow a\underline{A}S \rightarrow a\underline{S}bAS \rightarrow aab\underline{A}S \rightarrow aabba\underline{S} \rightarrow aabbaa$$

The derivation tree is given in Fig. 6.5.

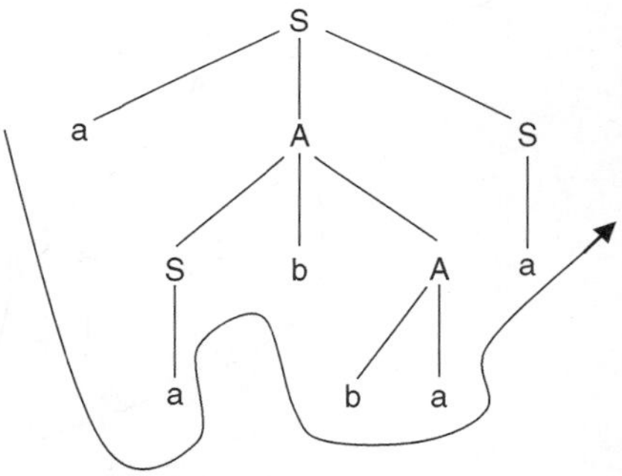

Fig. 6.5 *Parse Tree*

6.3 Ambiguity in Context-free Grammar

For generating a string from a given grammar, we have to derive the string step by step from the production rules of the given grammar. For this derivation, we know two types of derivations: (i) leftmost derivation and (ii) rightmost derivation. Except these two, there is also another approach called the mixed approach. Here, particularly, the leftmost or rightmost is not maintained in each step, whereas any of the non-terminals present in the deriving string is replaced by a suitable production rule.

By this process, different types of derivation can be generated for deriving a particular string from a given grammar. For each of the derivations, a parse tree is generated.

The different parse trees generated from the different derivations may be the same or may be different.

A grammar of a language is called ambiguous if any of the cases for generating a particular string, more than one parse tree can be generated.

The following examples describe the ambiguity in CFG.

(Grammar will be given, and from there one has to consider a string which produces two derivation trees to prove that the grammar is ambiguous.)

Example 6.13 Check whether the grammar is ambiguous or not.

$$S \rightarrow aS/AS/A$$
$$A \rightarrow AS/a$$

Solution: Consider the string ‘aaa’. The string can be generated in many ways. Here, we are giving two ways.

i) $S \rightarrow aS \rightarrow aAS \rightarrow aaS \rightarrow aaA \rightarrow aaa$
ii) $S \rightarrow A \rightarrow AS \rightarrow ASS \rightarrow aAS \rightarrow aaS \rightarrow aaA \rightarrow aaa$

The parse trees for derivation (i) and (ii) are shown in Figs. 6.6 (i) and (ii).

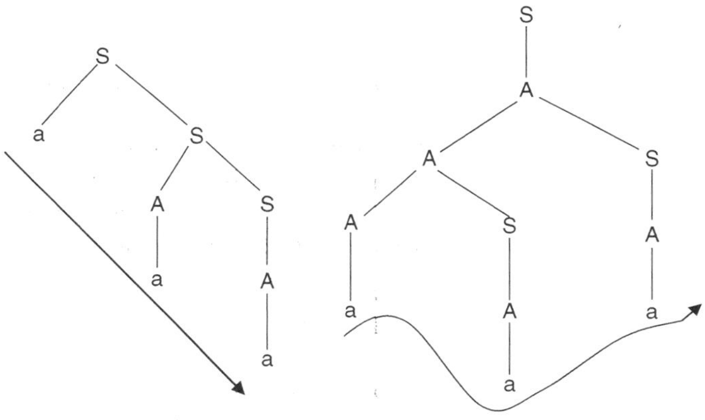

(i) (ii)

Fig. 6.6

Here, for the same string derived from the same grammar we are getting more than one parse tree. So, according to the definition, the grammar is an ambiguous grammar.

Example 6.14 Prove that the following grammar is ambiguous.

V_N: {S}
Σ: {id, +, *}
P: $S \rightarrow S + S/S * S/id$
S: {S}

Solution: Let us take a string id + id*id.

The string can be generated in the following ways.

i) $S \rightarrow S + \underline{S} \rightarrow \underline{S} + S*S \rightarrow id + \underline{S}*S \rightarrow id + id*\underline{S} \rightarrow id + id*id$
ii) $S \rightarrow S*S \rightarrow S + S*S \rightarrow id + S*S \rightarrow id + id*\underline{S} \rightarrow id + id*id$

The parse trees for derivation (i) and (ii) are shown in Figs. 6.7 and 6.8.

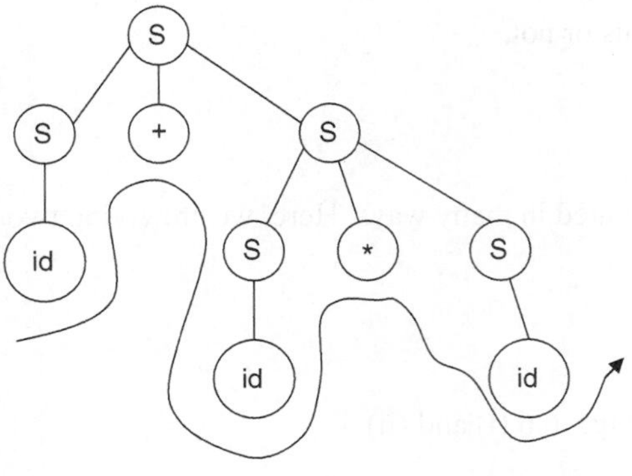

(i)

Fig. 6.7 *Parse Tree for Derivation(i)*

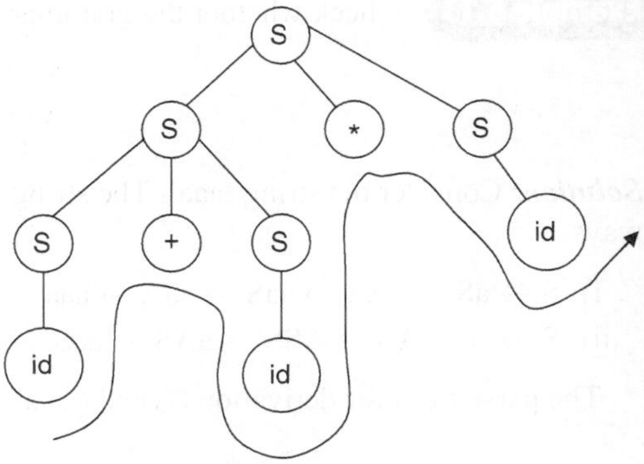

(ii)

Fig. 6.8 *Parse Tree for Derivation(ii)*

As we are getting two parse trees for generating a string from the given grammar, the grammar is ambiguous.

Example 6.15 Prove that the following grammar is ambiguous.

$$S \rightarrow a/abSb/aAb$$
$$A \rightarrow bS/aAAb$$

Solution: Take a string abababb. The string can be generated in the following ways.

i) $S \rightarrow ab\underline{S}b \rightarrow aba\underline{A}bb \rightarrow abab\underline{S}bb \rightarrow abababb$

ii) $S \rightarrow a\underline{A}b \rightarrow ab\underline{S}b \rightarrow aba\underline{A}bb \rightarrow abab\underline{S}bb \rightarrow abababb$

The parse trees for (i) and (ii) are given in Fig. 6.9.

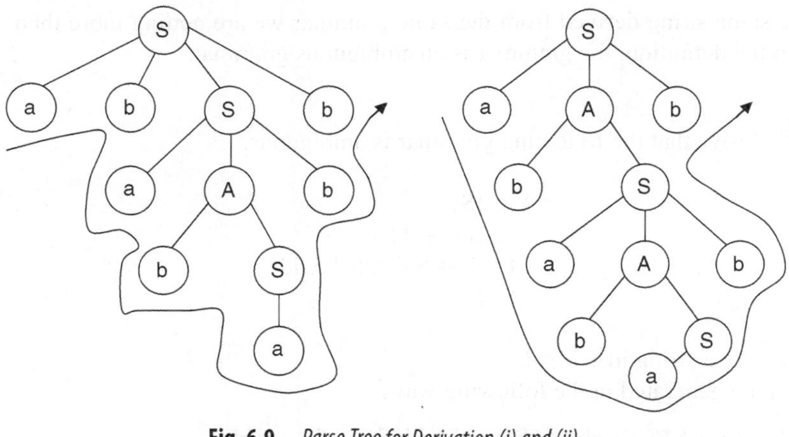

Fig. 6.9 *Parse Tree for Derivation (i) and (ii).*

As we are getting two parse trees for generating a string from the given grammar, the grammar is ambiguous.

Example 6.16 Prove that the following grammar is ambiguous.

$$S \rightarrow 0Y/01$$
$$X \rightarrow 0XY/0$$
$$Y \rightarrow XY1/1$$

Solution: Take a string 000111. The string can be derived in the following ways.

i) $S \rightarrow 0\underline{Y} \rightarrow 0\underline{X}Y1 \rightarrow 00\underline{X}YY1 \rightarrow 000\underline{Y}Y1 \rightarrow 0001\underline{Y}1 \rightarrow 000111$

ii) $S \rightarrow 0\underline{Y} \rightarrow 0X\underline{Y}1 \rightarrow 0\underline{X}XY11 \rightarrow 00\underline{X}Y11 \rightarrow 000\underline{Y}11 \rightarrow 000111$

The parse trees for (i) and (ii) are shown is Fig. 6.10.

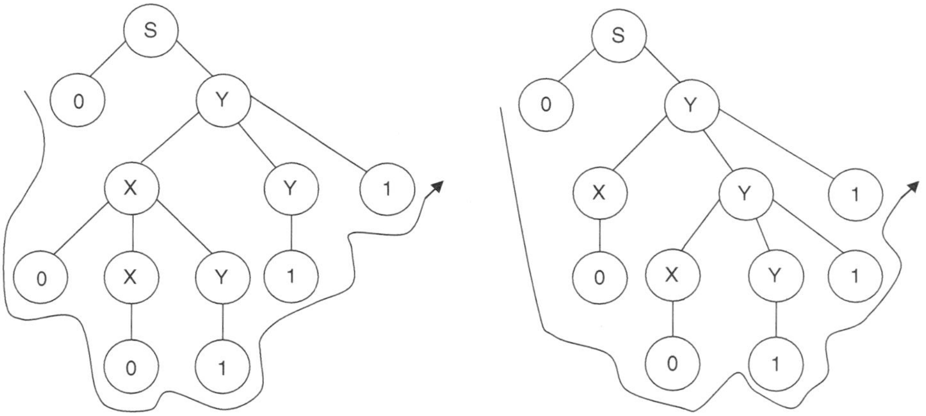

Fig. 6.10 *Parse Tree for Derivation (i) and (ii)*

As we are getting two parse trees for generating a string from the given grammar, the grammar is ambiguous.

In relation to the ambiguity in CFG, there are few more definitions. These are related to a CFL. These are:

1. **Ambiguous CFL:** A CFG G is said to be *ambiguous* if there exists some w ∈ L(G) that has at least two distinct parse trees.
2. **Inherently Ambiguous CFL:** A CFL L is said to be *inherently ambiguous* if all its grammars are ambiguous.
3. **Unambiguous CFL:** If L is a CFL for which there exists an unambiguous grammar, then L is said to be *unambiguous*.

(Even if one grammar for L is unambiguous, then L is an unambiguous language.)

Ambiguous grammar creates problem. Let us take an example.

For a grammar G, the production rule is

$$E \rightarrow E + E/E*E/a.$$

From here, we have to construct a + a*a.

The string can be generated in two different ways

i) E → E + E → E + E*E → a + E*E → a + a*E → a + a*a
ii) E → E*E → E + E*E → a + E*E → a + a*E → a + a*a

So, for these two cases two parse trees are generated as shown in Fig. 6.11.

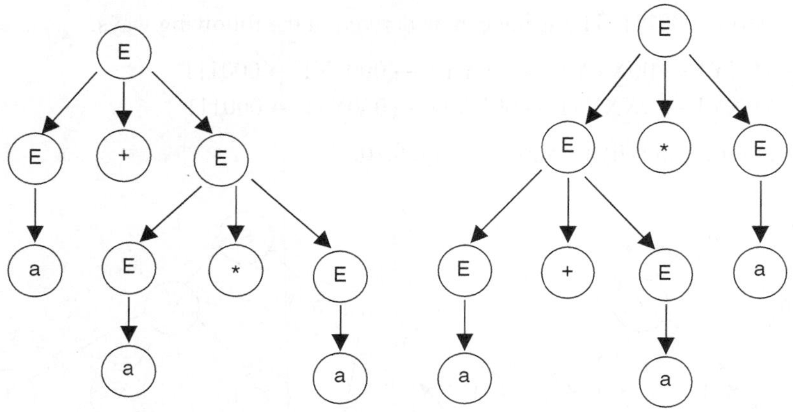

Fig. 6.11 *Parse Tree for Derivation (i) and (ii)*

So, the grammar is ambiguous.
In the place of 'a', put '2'. So, the derivations will be

i) E → E + E → E + E*E → 2 + E*E → 2 + 2*E → 2 + 2*2
ii) E → E*E → E + E*E → 2 + E*E → 2 + 2*E → 2 + 2*2

Up to this step, both of them seem the same. But the real problem is in the parse tree as shown in Fig. 6.12.

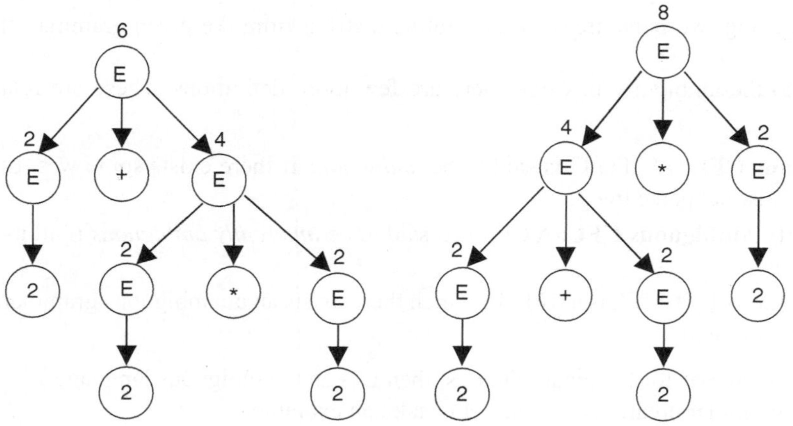

Fig. 6.12 *Annotated Parse Tree for (i) and (ii)*

The correct result is 2 + 2*2 = 2 + 4 = 6 (according to the rules of mathematics * has higher precedence over +).

From here, we can decide that ambiguity is bad for programming language.

There is no particular rule to remove ambiguity from a CFG. Sometimes, ambiguity can be removed by hand. In the previous case, ambiguity can be removed by setting priority to the operators + and *. If * is set a higher priority than +, then ambiguity can be removed.

More bad news is that some CFL have only ambiguous grammar. This means, in no way the ambiguity can be removed. This type of ambiguity is called inherent ambiguity.

6.4 Left Recursion and Left Factoring

6.4.1 Left Recursion

A context-free grammar is called left recursive if a non-terminal 'A' as a leftmost symbol appears alternatively at the time of derivation either immediately (called direct left recursive) or through some other non-terminal definitions (called indirect/hidden left recursive).

In other words, a grammar is left recursive if it has a non-terminal 'A' such that there is a derivation.

$$A \stackrel{+}{\Rightarrow} A\alpha \quad \text{for some string } \alpha$$

6.4.1.1 Direct Left Recursion

Let the grammar be $A \rightarrow A\alpha/\beta$, where α and β consists of terminal and/or non-terminals but β does not start with A.

At the time of derivation, if we start from A and replace the A by the production $A \rightarrow A\alpha$, then for all the time A will be the leftmost non-terminal symbol.

6.4.1.2 Indirect Left Recursion

Take a grammar in the form

$$A \rightarrow B\alpha/\gamma$$
$$B \rightarrow A\beta/\delta$$

where α, β, γ, and δ consist of terminal and/or non-terminal.

At the time of derivation, if we start from A, replace the A by the production $A \rightarrow B\alpha$, and after that replace the B by the production $B \rightarrow A\beta$, then A will appear again as a leftmost non-terminal symbol. This is called the indirect left recursion.

In general, a grammar in the form

$$A_0 \rightarrow A_1\alpha_1/\ldots$$
$$A_1 \rightarrow A_2\alpha_2/\ldots$$
$$A_n \rightarrow A_0\alpha_{(n+1)}/\ldots$$

is the indirect left recursive grammar.

Note: *The left recursive grammar is to be converted into a non-left recursive grammar because the top down parsing technique cannot handle left recursion. The top down parsing is a part of compiler design.*

Immediate left recursion creates problems in top down parsing of syntax analysis in the compiler design. So, immediate left recursion must be removed.

Immediate left recursion can be removed by the following process. This is called the Moore's proposal. For a grammar in the form

$$A \rightarrow A\alpha \mid \beta, \text{ where } \beta \text{ does not start with A,}$$

the equivalent grammar after removing left recursion becomes

$$A \rightarrow \beta A'$$
$$A' \rightarrow \alpha A' \mid \in$$

In general, for a grammar in the form

$$A \rightarrow A\,\alpha_1 \mid \ldots \mid A\,\alpha_m \mid \beta_1 \mid \ldots \mid \beta_n, \text{ where } \beta_1 \ldots \beta_n \text{ do not start with A,}$$

the equivalent grammar after removing left recursion is

$$A \rightarrow \beta_1 A' \mid \ldots \mid \beta_n A'$$
$$A' \rightarrow \alpha_1 A' \mid \ldots \mid \alpha_m A' \mid \in$$

For indirect left recursion removal, there is an algorithm.
Arrange non-terminals in some order: $A_1 \cdots A_n$

```
– for i from 1 to n do {
      – for j from 1 to i – 1 do {
            replace each production
                  Ai → Aj γ
                      by
                  Ai → α1 γ | ... | αk γ
                  where Aj → α1 | ... | αk
      }
      – eliminate immediate left-recursions among Ai productions
}
```

Consider the following examples to make the Moore's proposal clear.

Example 6.17 Remove the left recursion from the following grammar.

$$E \rightarrow E + T \mid T$$
$$T \rightarrow T{*}F \mid F$$
$$F \rightarrow id \mid (E)$$

Solution: In the grammar there are two immediate left recursions $E \rightarrow E + T$ and $T \rightarrow T * F$.
By using Moore's proposal the left recursion $E \rightarrow E + T$ is removed as

$$E \rightarrow TE' \text{ and } E' \rightarrow +TE'/\in$$

The left recursion $T \rightarrow T{*}F$ is removed as

$$T \rightarrow FT'$$
$$T' \rightarrow {*}FT' \mid \in$$

The CFG after removing the left recursion becomes

$$E \rightarrow TE'$$
$$E' \rightarrow +TE' \mid \in$$
$$T \rightarrow FT'$$
$$T' \rightarrow *FT' \mid \in$$
$$F \rightarrow id \mid (E)$$

Example 6.18 Remove the left recursion from the following grammar.

$$S \rightarrow Aa \mid b$$
$$A \rightarrow Sc \mid d$$

Solution: The grammar has indirect left recursion. Rename S as A_1 and A as A_2. The grammar is

$$A_1 \rightarrow A_2a/b$$
$$A_2 \rightarrow A_1c/d.$$

For $i = 1, j = 1$, there is no production in the form $A_1 \rightarrow A_1\alpha$.

For $i = 2, j = 1$, there is a production in the form $A_2 \rightarrow A_1\alpha$. The production is $A_2 \rightarrow A_1c$. According to the algorithm for removal of indirect left recursion, the production becomes

$$A_2 \rightarrow A_2ac/bc$$

This left recursion for the production $A_2 \rightarrow A_2ac/bc/d$ is removed and the production rules are

$$A_2 \rightarrow bcA_2'/dA_2'$$
$$A_2' \rightarrow acA_2'/\in$$

The actual non-left recursive grammar is

$$S \rightarrow Aa/b$$
$$A \rightarrow bcA'/dA'$$
$$A' \rightarrow acA'/\in.$$

Example 6.19 Remove the left recursion from the following grammar.

$$S \rightarrow Aa \mid b$$
$$A \rightarrow Ac \mid Sc \mid f$$

Solution: The grammar has indirect left recursion. Rename S as A_1 and A as A_2. The grammar is

$$A_1 \rightarrow A_2a/b$$
$$A_2 \rightarrow A_2c/A_1d/f$$

For $i = 1, j = 1$, there is no production in the form $A_1 \rightarrow A_1\alpha$.

For $i = 2, j = 1$, there is a production in the form $A_2 \rightarrow A_1\alpha$. The production is $A_2 \rightarrow A_1d$. According to the algorithm for removal of indirect left recursion, the production becomes

$$A_2 \rightarrow A_2ad/bd$$

The A_2 production is $A_2 \rightarrow A_2c/A_2ad/bd/f$.

This left recursion for the production $A_2 \to A_2c/A_2ad/bd/f$ is removed and the production rules are

$$A_2 \to bdA_2'$$
$$A_2 \to fA_2'$$
$$A_2 \to cA_2'/adA_2'$$

The actual non-left recursive grammar is

$$S \to Aa \mid b$$
$$A \to bdA'/fA'$$
$$A' \to cA'/adA'$$

6.4.2 Left Factoring

Let us assume that in a grammar there is a production rule in the form $A \to \alpha\beta_1/\alpha\beta_2/\ldots/\alpha\beta_n$. The parser generated from this kind of grammar is not efficient as it requires backtracking. To avoid this problem, we need to left factor the grammar.

Let the string that we need to construct be $\alpha\beta_2$. The parser is a computer program. It will check only one symbol at a time. So, first it will take 'α'. For getting 'α', there are n number of productions. But which production is to be chosen? Let us assume that it has chosen $A \to \alpha\beta_1$. After traversing 'α', it will take the next symbol 'β_2' in the string. But from the current situation it is not possible to get 'β_2'. So, the parser needs to traverse back again. This is called backtracking. For this reason, we need to left factor a grammar.

To left factor a grammar, we collect all productions that have the same LHS and begin with the same symbols on the RHS. We combine the common strings into a single production and then append a new non-terminal symbol to the end of this new production. Finally, we create a new set of productions using this new non-terminal for each of the suffixes to the common production.

After left factoring, the previous grammar is transformed into:

$$A \to \alpha A_1$$
$$A_1 \to \beta_1/\beta_2/\ldots/\beta_n$$

The following is one example related to the removal of backtracking.

Example 6.20 Left Factor the following grammar.

$$A \to \underline{a}bB \mid \underline{a}B \mid cdg \mid cdeB \mid cdfB$$

Solution: In the previous grammar, the RHS productions abB and aB both start with 'a'. So, they can be left factored. In the same way, cdg, cdeB, and cdfB all start with 'cd'. So, they can also be left factored.

The left factored grammar is

$$A \to aA'$$
$$A' \to bB/B$$
$$A \to cdA''$$
$$A'' \to g/eB/fB$$

6.5 Simplification of Context-free Grammar

CFG may contain different types of useless symbols, unit productions, and null productions. These types of symbols and productions increase the number of steps in generating a language from a CFG. Reduced grammar contains less number of non-terminals and productions, so the time complexity for the language generating process becomes less from the reduced grammar.

CFG can be simplified in the following three processes.

1. Removal of useless symbols
   a) Removal of non-generating symbols
   b) Removal of non-reachable symbols
2. Removal of unit productions
3. Removal of null productions

6.5.1 Removal of Useless Symbols

Useless symbols are of two types:

1. **Non-generating symbols** are those symbols which do not produce any terminal string.
2. **Non-reachable symbols** are those symbols which cannot be reached at any time starting from the start symbol.

These two types of useless symbols can be removed according to the following process.

- Find the non-generating symbols, i.e., the symbols which do not generate any terminal string. If the start symbol is found non-generating, leave that symbol. Because the start symbol cannot be removed, the language generating process starts from that symbol.
- For removing non-generating symbols, remove those productions whose right side and left side contain those symbols.
- Now find the non-reachable symbols, i.e., the symbols which cannot be reached, starting from the start symbol.
- Remove the non-reachable symbols such as rule (b).

[There is no hard and fast rule of which symbol (non-generating or non-reachable) will be removed first but, in most of the cases, the non-generating symbol is removed first, and then the non-reachable symbol is removed. But depending upon the situation, this can be changed.]

Example 6.21 Remove the useless symbols from the given CFG.

$$S \rightarrow AC$$
$$S \rightarrow BA$$
$$C \rightarrow CB$$
$$C \rightarrow AC$$
$$A \rightarrow a$$
$$B \rightarrow aC/b$$

Solution: There are two types of useless symbols: non-generating and non-reachable symbols. First, we are finding non-generating symbols.

Those symbols which do not produce any terminal string are non-generating symbols.

Here, C is a non-generating symbol as it does not produce any terminal string.

So, we have to remove the symbol C. To remove C, all the productions containing C as a symbol (LHS or RHS) must be removed. By removing the productions, the minimized grammar will be

$$S \rightarrow BA$$
$$A \rightarrow a$$
$$B \rightarrow b$$

Now, we have to find non-reachable symbols, the symbols which cannot be reached at any time starting from the start symbol. There is no non-reachable symbol in the grammar. So, the minimized form of the grammar by removing useless symbols is

$$S \rightarrow BA$$
$$A \rightarrow a$$
$$B \rightarrow b$$

Example 6.22 Remove the useless symbols from the given CFG.

$$S \rightarrow aB/bX$$
$$A \rightarrow BAd/bSX/a$$
$$B \rightarrow aSB/bBX$$
$$X \rightarrow SBD/aBx/ad$$

Solution: First, we are going to find non-generating symbols, the non-terminals which do not produce any terminal string.

Here, B is not producing any terminal string, so this is a non-generating symbol. To remove B from the production rules of the given grammar, all the productions containing B as a symbol will be removed. By removing B from the grammar, the modified production rules will be

$$S \rightarrow bX$$
$$A \rightarrow bSX/a$$
$$X \rightarrow ad \qquad (1)$$

Now we have to find non-reachable symbols, the symbols which cannot be reached starting from the start symbol.

Here, in (1), the symbol A cannot be reached by any path at any time starting from the start symbol S. So, A is a non-reachable symbol. To remove A, all the productions containing A as a symbol will be removed. By removing A from the production rules of the grammar, the minimized grammar will be

$$S \rightarrow bX$$
$$X \rightarrow ad$$

Example 6.23 Remove the useless symbol from the given grammar.

$$S \rightarrow aAa$$
$$A \rightarrow bBB$$
$$B \rightarrow ab$$
$$C \rightarrow aB$$

Solution: First, we are going to remove the non-generating symbols, the non-terminals which do not produce any terminal string. Here, all the symbols produce terminal strings. Thus, there is no non-generating symbol.

So, we have to remove the non-reachable symbols.

Here, non-reachable symbols are B and C. But B is a generating symbol, and so B cannot be removed. By removing C, the production rules of the grammar become

$$S \to aAa$$
$$A \to bBB$$
$$B \to ab$$

This is the minimized grammar.

Example 6.24 Reduce the following CFG by removing the useless symbols from the grammar.

$$S \to aC/SB$$
$$A \to bSCa/ad$$
$$B \to aSB/bBC$$
$$C \to aBC/ad$$

Solution: First, we are going to remove the non-generating symbols from the production rules. Here, B is not producing any terminal string, and so it is a non-generating symbol. By removing B, the grammar becomes

$$S \to aC$$
$$A \to bSCa/ad$$
$$C \to ad$$

Now, we have to find the non-reachable symbols, the non-terminals which cannot be reached starting from the start symbol. Here, symbol A cannot be reached by any path starting from the start symbol. So, A will be removed. By removing A, the production becomes

$$S \to aC$$
$$C \to ad$$

This is the reduced format of the grammar by removing the useless symbols.

6.5.2 Removal of Unit Productions

Production in the form *non-terminal → single non-terminal* is called unit production.

Unit production increases the number of steps as well as the complexity at the time of generating language from the grammar. This will be clear if we take an example.

Let there be a grammar

$$S \to AB,\ A \to E,\ B \to C,\ C \to D,\ D \to b,\ E \to a$$

From here, if we are going to generate a language, then it will be generated by the following way

$$S \to \underline{A}B \to \underline{E}B \to a\underline{B} \to a\underline{C} \to a\underline{D} \to ab$$

The number of steps is 6.

The grammar, by removing unit production and as well as minimizing, will be

$$S \to AB,\ A \to a,\ B \to b$$

From this, the language will be generated by the following way

$$S \rightarrow AB \rightarrow aB \rightarrow ab$$

The number of steps is 3.

6.5.2.1 *Procedure to Remove Unit Production*

There is an algorithm to remove unit production from a given CFG.

```
while(there exist a unit production NT → NT)
{
    select a unit production A → B
    for(every non-unit production B → α)
    {
        add production A → α to the grammar.
        eliminate A → B from grammar.
    }
}
```

Example 6.25 Remove the unit production from the following grammar.

$$S \rightarrow AB, A \rightarrow E, B \rightarrow C, C \rightarrow D, D \rightarrow b, E \rightarrow a$$

Solution: In this grammar unit, the productions are A → E, B → C, and C → D. If we want to remove the unit production A → E from the grammar, first we have to find a non-unit production in the form E → {Some string of terminal or non-terminal or both}. There is a production E → a. So, the production A → a will be added to the production rule. And the modified production rule will be

$$S \rightarrow AB, A \rightarrow a, B \rightarrow C, C \rightarrow D, D \rightarrow b, E \rightarrow a$$

For removing another unit production B → C, we have to find a non-unit production C → {Some string of terminal or non-terminal or both}. But there is no such production found.

For removing a unit production C → D we have found a non-unit production D → b. So, the production C → b will be added to the production rule set and C → D will be removed. The modified production rule will be

$$S \rightarrow AB, A \rightarrow a, B \rightarrow C, C \rightarrow b, D \rightarrow b, E \rightarrow a$$

Now the unit production B → C will be removed by B → b as there is a non-unit production C → b. The modified production rule will be

$$S \rightarrow AB, A \rightarrow a, B \rightarrow b, C \rightarrow b, D \rightarrow b, E \rightarrow a$$

[This is the grammar without unit production, but it is not minimized. There are useless symbols (non-reachable symbols) C, D, and E. So, if we want to minimize the grammar, these symbols will be removed. By removing the useless symbols, the grammar will be S → AB, A → a, B → b.]

Example 6.26 Remove the unit production from the following grammar.

$$S \rightarrow AB, A \rightarrow a, B \rightarrow C, C \rightarrow D, D \rightarrow b$$

Solution: The unit productions in the grammar are

$$B \rightarrow C, C \rightarrow D$$

To remove unit production $B \rightarrow C$, we have to search for a non-unit production.

$C \rightarrow$ {Some string of terminal or non-terminal or both}.There is no such production rule found. So, we are trying to remove the unit production $C \rightarrow D$.

There is a non-unit production $D \rightarrow b$. So, the unit production $C \rightarrow D$ will be removed by including a production $C \rightarrow b$.

The modified production rules will be

$$S \rightarrow AB, A \rightarrow a, B \rightarrow C, C \rightarrow b, D \rightarrow b$$

There is a unit production $B \rightarrow C$ that exists. This unit production can be removed by including a production $B \rightarrow b$ to the grammar.

The modified production rules will be

$$S \rightarrow AB, A \rightarrow a, B \rightarrow b, C \rightarrow b, D \rightarrow b$$

In this grammar, no unit production exists.

(This grammar is unit production free, but not minimized. There are useless symbols in the form of non-reachable symbols. Here, the non-reachable symbols are C and D, which cannot be reached starting from the start symbol S. By removing the useless symbols, the grammar will become $S \rightarrow AB$, $A \rightarrow a$, $B \rightarrow b$.)

Example 6.27 Remove the unit production from the following grammar.

$$S \rightarrow aX/Yb/Y$$
$$X \rightarrow S$$
$$Y \rightarrow Yb/b$$

***Solution*:** The unit productions in the grammar are $S \rightarrow Y$ and $X \rightarrow S$. To remove unit production $S \rightarrow Y$, we have to find a non-unit production where $Y \rightarrow$ {Some string of terminal or non-terminal or both}. There is a non-unit production $Y \rightarrow Yb/b$. The unit production $S \rightarrow Y$ will be removed by including production $S \rightarrow Yb/b$ in the production.

The modified production rules will be

$$S \rightarrow aX/Yb/b$$
$$X \rightarrow S$$
$$Y \rightarrow Yb/b$$

In the previous production rules, there is a unit production $X \rightarrow S$. This production can be removed by including the production $X \rightarrow aX/Yb/b$ to the production rules.

The modified production rules will be

$$S \rightarrow aX/Yb/b$$
$$X \rightarrow aX/Yb/b$$
$$Y \rightarrow Yb/b$$

(This is also minimized grammar.)

Example 6.28 Remove the unit production from the following grammar.

$$S \rightarrow AA$$
$$A \rightarrow B/BB$$
$$Y \rightarrow abB/b/bb$$

Solution: In the previous grammar, there is a unit production $A \rightarrow B$. To remove the unit production, we have to find a non-unit production $Y \rightarrow$ {Some string of terminal or non-terminal or both}. There is a non-unit production $B \rightarrow abB/b/bb$. So, the unit production $A \rightarrow B$ can be removed by including the production $A \rightarrow abB/b/bb$ to the production rule.

The modified production rule will be

$$S \rightarrow AA$$
$$A \rightarrow BB/abB/b/bb$$
$$B \rightarrow abB/b/bb.$$

(This is also minimized grammar.)

6.5.3 Removal of Null Productions

A production in the form $NT \rightarrow \in$ is called null production.

If Λ (null string) is in the language set generated from the grammar, then that null production cannot be removed.

That is, if we get, $S \stackrel{*}{\Longrightarrow} \in$, then that null production cannot be removed from the production rules.

6.5.3.1 Procedure to Remove Null Production

Step I: If $A \rightarrow \in$ is a production to be eliminated, then we look for all productions whose right side contains A.

Step II: Replace each occurrence of A in each of these productions to obtain the non-$\in$ production.

Step III: These non-null productions must be added to the grammar to keep the language generating power the same.

Example 6.29 Remove the $\in$ production from the following grammar.

$$S \rightarrow aA$$
$$A \rightarrow b/\in$$

Solution: In the previous production rules, there is a null production $A \rightarrow \in$. The grammar does not produce null string as a language, and so this null production can be removed.

According to first step we have to look for all productions whose right side contains A. There is one $S \rightarrow aA$.

According to the second step, we have to replace each occurrence of ‘A’ in each of the productions. There is only one occurrence of ‘A’ in $S \rightarrow aA$. So, after replacing, it will become $S \rightarrow a$.

According to step three, these productions will be added to the grammar.

So, the grammar will be

$$S \rightarrow aA/a$$
$$A \rightarrow b$$

Now, in the new production rules, there is no null production.

Example 6.30 Remove the $\in$ production from the following grammar.

$$S \rightarrow aX/bX$$
$$X \rightarrow a/b/\in$$

Solution: In the previous grammar, there is a null production $X \rightarrow \in$. In the language set produced by the grammar, there is no null string, and so this null production can be removed.

For removing this null production, we have to look for all the productions whose right side contains X. There are two such productions in the grammar $S \rightarrow aX$ and $S \rightarrow bX$.

We have to replace each occurrence of 'X' in each of the productions to obtain a non-null production. In both the productions, X has occurred only once. So, after replacing, the productions will become $S \rightarrow a$ and $S \rightarrow b$, respectively.

These productions must be added to the grammar to keep the language generating power the same. So, after adding these productions, the grammar will be

$$S \rightarrow aX/bX/a/b$$
$$X \rightarrow a/b$$

Example 6.31 Remove the $\in$ production from the following grammar.

$$S \rightarrow ABaC$$
$$A \rightarrow BC$$
$$B \rightarrow b/\in$$
$$C \rightarrow D/\in$$
$$D \rightarrow d$$

Solution: In the previous grammar, there are two null productions $B \rightarrow \in$ and $C \rightarrow \in$. The grammar does not produce any null string. So, these null productions can be removed to obtain a non-null production.

To remove the null production $B \rightarrow \in$, we have to find the productions whose right side contains B. There are two productions in the grammar $S \rightarrow ABaC$ and $A \rightarrow BC$.

We have to replace each occurrence of B in each of the productions by null to obtain a non-null production. In the previous two cases, B has occurred only once. After replacing B by null, the productions will become $S \rightarrow AaC$ and $A \rightarrow C$. These productions will be added to the grammar to keep the language generating power the same.

To remove the null production $C \rightarrow \in$, we have to find those productions whose right side contains C. There are two productions $S \rightarrow ABaC$ and $A \rightarrow BC$.

We have to replace each occurrence of C in each of the productions by null to obtain a non-null production. In the previous two cases, C has occurred only once. After replacing C by null, the productions will become $S \rightarrow ABa$ and $A \rightarrow B$. These productions will be added to the grammar to keep the language generating power the same.

In the previous two productions $S \rightarrow ABaC$ and $A \rightarrow BC$, both B and C are there. But we have replaced B and C in two steps. If simultaneously B and C are replaced by null, the productions will become $S \rightarrow Aa$ and $A \rightarrow \in$, respectively. These productions will also be added to keep the language generating power the same.

So, after removing the two null productions $B \rightarrow \in$ and $C \rightarrow \in$, the modified grammar will become

$$S \rightarrow ABaC/AaC/ABa/Aa$$
$$A \rightarrow BC/C/B/\in$$
$$B \rightarrow b$$
$$C \rightarrow D$$
$$D \rightarrow d$$

Again, in this grammar, there is a null production A → ∈. By removing the null production according to the previous way, the modified grammar will become

S → ABaC/AaC/ABa/Aa/BaC/aC Ba/a
A → BC/C/B
B → b
C → D
D → d

The grammar is not in minimized format because there are three unit productions A → C, A → B, and C → D. By removing the unit productions from the grammar, the minimized grammar will become

S → ABaC/AaC/ABa/Aa/BaC/aC/Ba/a
A → BC/d/b
B → b
C → d
D → d

Again, in the grammar, there is a non-reachable symbol D. By removing the non-reachable symbol, the minimized grammar will be

S → ABaC/AaC/ABa/Aa/BaC/aC/Ba/a
A → BC/d/b
B → b
C → d

Example 6.32 Remove the ∈ production from the following grammar.

S → ABAC
A → aA/∈
B → bB/∈
C → c

Solution: In the grammar, there are two null productions, A → ∈ and B → ∈. The grammar does not produce any null string. So, these null productions can be removed to obtain a non-null production.

To remove the null production A → ∈, we have to find all the productions whose RHS contains A. There are two productions S → ABAC and A → aA. Among them in the production S → ABAC, there are two A and one B. We have to replace each occurrence of A in each of the productions by null to obtain the non-null production. By replacing, we get S → ABC, S → BAC, S → BC (first rightmost, then leftmost, and then both), and A → a (for A → aA).

To remove the null production B → ∈, we have to find all the productions whose right side contains B. There are two productions S → ABAC and B → bB, each of them having a single occurrence of B. For obtaining non-null productions, we have to replace each occurrence of B in each of the productions by null. By replacing, we get

S → AAC and B → b.

In the production S → ABAC, there are both A and B. But, we have replaced A and B in two separate steps. If A and B are replaced simultaneously by null, the productions will become

S → AC, S → C

So, after removing the two null productions $A \rightarrow \in$ and $B \rightarrow \in$, the modified grammar will become

$$S \rightarrow ABAC/ABC/AAC/BAC/BC/C$$
$$A \rightarrow aA/a$$
$$B \rightarrow bB/b$$
$$C \rightarrow c$$

There is no null production in the grammar.

Example 6.33 Remove the $\in$ production from the following grammar.

$$S \rightarrow aS/A$$
$$A \rightarrow aA/\in$$

Solution: In the previous grammar, there is a null production $A \rightarrow \in$. But in the language set generated by the grammar, there is a null string which can be generated by the following way $S \rightarrow A \rightarrow \in$. As null string is in the language set, and so the null production cannot be removed from the grammar.

6.6 Linear Grammar

A grammar is called linear grammar if it is context free, and the RHS of all productions have at most one non-terminal.

As an example, $S \rightarrow abSc/\in$, the grammar of $(ab)^n c^n$, $n \geq 0$.

There are two types of linear grammar:

1. **Left linear grammar:** A linear grammar is called left linear if the RHS non-terminal in each productions are at the left end. In a linear grammar if all productions are in the form $A \rightarrow B\alpha$ or $A \rightarrow \alpha$, then that grammar is called left linear grammar. Here, A and B are non-terminals and α is a string of terminals.
2. **Right linear grammar:** A linear grammar is called right linear if the RHS non-terminal in each productions are at the right end. In a grammar if all productions are in the form $A \rightarrow \alpha B$ or $A \rightarrow \alpha$, then that grammar is called right linear grammar. Here A and B are non-terminals and α is a string of terminals.

A linear grammar can be converted to regular grammar. Take the following example.

Example 6.34 Convert the following linear grammar into Regular Grammar.

$$S \rightarrow baS/aA$$
$$A \rightarrow bbA/bb$$

Solution: Consider two non-terminals B and C with production $B \rightarrow aS$ and $C \rightarrow bA$. The grammar becomes

$$S \rightarrow bB/aA$$
$$A \rightarrow bC/bb$$
$$B \rightarrow aS$$
$$C \rightarrow bA$$

Still the production A → bb is not a regular grammar. Replace b by a non-terminal D with production D → b. The grammar becomes

S → bB/aA
A → bC/bD
B → aS
C → bA
D → b

Now the grammar is regular.

For a left linear grammar, there exists a right linear grammar and vice versa. Grammar in one form can be converted into another form. The following section describes the process of conversion.

6.6.1 Right Linear to Left Linear

Step I: Generate the regular expression from the given grammar.

Step II: Reverse the regular expression obtained.

Step III: Construct the finite automata of the RE in Step II

Step IV: Generate the regular right linear grammar from the finite automata.

Step V: Reverse the right side of each production of the right linear grammar obtained in Step IV. The resulting grammar is the equivalent left linear grammar.

6.6.2 Left Linear to Right Linear

Step I: Reverse the right side of every production of the left linear grammar.

Step II: Construct the regular expression of the grammar obtained in Step I.

Step III: Reverse the constructed regular expression.

Step IV: Construct the finite automata from the RE obtained in Step III.

Step V: Construct the right linear grammar from the finite automata obtained in Step IV.

Example 6.35 Convert the following right linear grammar to left linear grammar.

S → 10A/1
A → 0A/00

Solution:

i) A = 0A → 00A → .... → 0*00

S → 10A + 1 → (100*00 + 1)

The regular expression generated by the grammar is (100*00 + 1).

ii) Reversing the regular expression, we get (1 + 000*01).

iii) The finite automata obtained from the expression in step II is

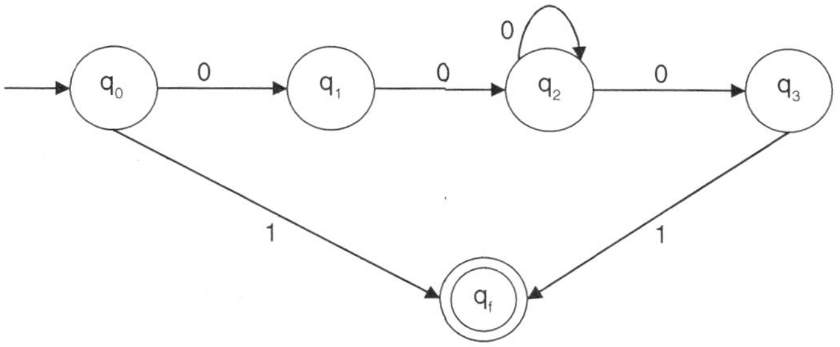

iv) The constructed right linear grammar is

$$A \rightarrow 0B,$$
$$A \rightarrow 1E/1$$
$$B \rightarrow 0C$$
$$C \rightarrow 0D/0C$$
$$D \rightarrow 1E/1$$

Here, E is a useless symbol. Eliminate all productions of E. The new productions are

$$A \rightarrow 0B$$
$$A \rightarrow 1$$
$$B \rightarrow 0C$$
$$C \rightarrow 0D/0C$$
$$D \rightarrow 1$$

v) Reversing the right side of each production of the right linear grammar,

$$A \rightarrow B0/1$$
$$B \rightarrow C0$$
$$C \rightarrow D0/C0$$
$$D \rightarrow 1$$

This is the equivalent left linear grammar to the right linear grammar.

Example 6.36 Convert the following left linear grammar to right linear grammar.

$$S \rightarrow S10/A1$$
$$A \rightarrow A00/00$$

Solution:

i) Reversing the right side of every production, we get

$$S \rightarrow 01S/1A$$
$$A \rightarrow 00A/00$$

ii) A → (00)*00 S → (01)*S/1(00)*00

$$S \rightarrow (01)^*1(00)^*00$$

The regular expression is (01)*1(00)*00.

iii) Reversing the regular expression, we get $00(00)^*1(01)^*$.

iv) The finite automata constructed from the expression in step III is

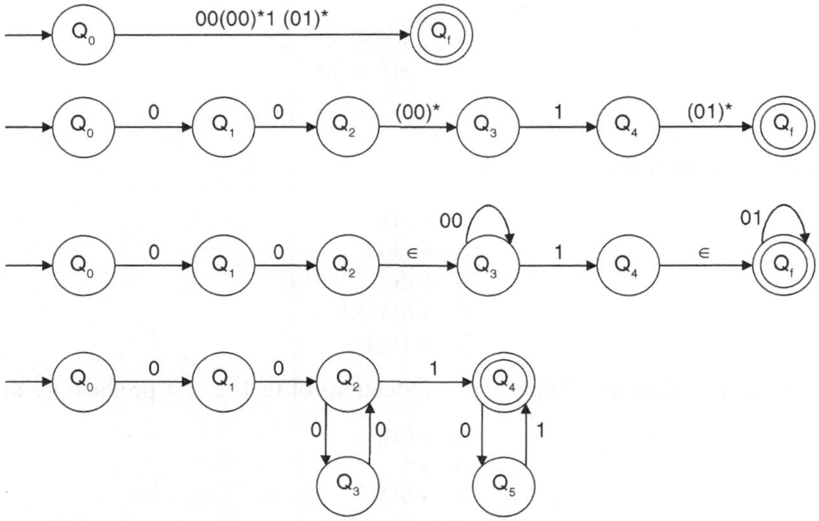

v) The right linear grammar from the finite automata is

A → 0B
B → 0C
C → 0D
D → 0C
C → 0E/1
E → 0F
F → 1E/1

This is the equivalent right linear grammar.

6.7 Normal Form

For a grammar, the RHS of a production can be any string of variables and terminals, i.e., $(V_N \cup \Sigma)^*$. A grammar is said to be in normal form when every production of the grammar has some specific form. That means, instead of allowing any member of $(V_N \cup \Sigma)$ on the RHS of the production, we permit only specific members on the RHS of the production. But these restrictions should not hamper the language generating power of the grammar.

When a grammar is made in normal form, every production of the grammar is converted in some specific form. These help us to design some algorithm to answer certain questions, such as if a CFG is converted into Chomsky normal form (CNF), one can easily answer whether a particular string is generated by the grammar or not. A polynomial time algorithm called CYK algorithm can be constructed to check this. For a CFG converted into CNF, the parse tree generated from the CNF is always a binary tree. Similarly, if a CFG is converted into Greibach normal form (GNF), then a PDA accepting the language generated by the grammar can easily be designed.

In this section we shall learn mainly two types of normal forms: (a) CNF and (b) GNF.

6.7.1 Chomsky Normal Form

A CFG is said to be in CNF if all the productions of the grammar are in the following form.

Non-terminal $\rightarrow$ String of exactly two non-terminals
Non-terminal $\rightarrow$ Single terminal

A CFG can be converted into CNF by the following process.

Step I:

1. Eliminate all the $\in$-production.
2. Eliminate all the unit production.
3. Eliminate all the useless symbols.

Step II:

1. **if** (all the productions are in the form NT $\rightarrow$ string of exactly two NTs or NT $\rightarrow$ Single terminal) Declare the CFG is in CNF. And stop.
2. **else (**follow step III and/or IV and/or V).

Step III: (Elimination of terminals on the RHS of length two or more.)
Consider any production of the form

$$NT \rightarrow T_1T_2 \ldots . T_n$$

where $n \geq 2$ (T means terminal and NT means non-terminal).

For a terminal T_i, introduce a new variable (non-terminal) say NT_1 and a corresponding production $NT_i \rightarrow T_i$. Repeat this for every terminal on the RHS so that every production of the grammar has either a single terminal or two or more variables.

Step IV: (Restriction of the number of variables on the RHS to two.)
Consider any production of the form

$$NT \rightarrow NT_1NT_2 \ldots NT_n$$

where $n \geq 3$.

The production $NT \rightarrow NT_1NT_2 \ldots NT_n$ is replaced by new productions

$$NT \rightarrow NT_1A_1$$
$$A_1 \rightarrow NT_2A_2$$
$$A_2 \rightarrow NT_3A_3$$
$$\ldots\ldots\ldots\ldots$$
$$\ldots\ldots\ldots\ldots$$
$$A_{n-2} \rightarrow NT_{n-1}NT_n$$

Here the A_i's are new variables.

Step V: (Conversion of the string containing terminals and non-terminals to the string of non-terminal on the RHS)

Consider any production of the form

$$NT \rightarrow T_1T_2 \ldots. T_n NT_1 \ldots. NT_{n'},$$

where n ≥ 1.

For the part $T_1T_2\ldots.T_{n'}$, follow Step III.

Check the resultant string by Step II. If not in CNF, follow Step IV.

By this process, the generated new grammar is in CNF.

Example 6.37 Convert the following grammar into CNF.

$S \rightarrow bA/aB$
$A \rightarrow bAA/aS/a$
$B \rightarrow aBB/bS/a$

Solution: The productions $S \rightarrow bA$, $S \rightarrow aB$, $A \rightarrow bAA$, $A \rightarrow aS$, $B \rightarrow aBB$, $B \rightarrow bS$ are not in CNF. So, we have to convert these into CNF. Let us replace terminal 'a' by a non-terminal C_a and terminal 'b' by a non-terminal C_b. Hence, two new productions will be added to the grammar

$$C_a \rightarrow a \text{ and } C_b \rightarrow b$$

By replacing a and b by new non-terminals and including the two productions, the modified grammar will be

$S \rightarrow C_bA/C_aB$
$A \rightarrow C_bAA/C_aS/a$
$B \rightarrow C_aBB/C_bS/a$
$C_a \rightarrow a$
$C_b \rightarrow b$

In the modified grammar, all the productions are not in CNF. The productions $A \rightarrow C_bAA$ and $B \rightarrow C_aBB$ are not in CNF, because they contain more than two non-terminals at the RHS. Let us replace AA by a new non-terminal D and BB by another new non-terminal E. Hence, two new productions will be added to the grammar $D \rightarrow AA$ and $E \rightarrow BB$. So, the new modified grammar will be

$S \rightarrow C_bA/C_aB$
$A \rightarrow C_bD/C_aS/a$
$B \rightarrow C_aE/C_aS/a$
$D \rightarrow AA$
$E \rightarrow BB$
$C_a \rightarrow a$
$C_b \rightarrow b$

Example 6.38 Convert the following grammar into CNF.

$S \rightarrow ABa$
$A \rightarrow aab$
$B \rightarrow Ac$

Solution: All the productions in the grammar are not in CNF. We have to convert them into CNF keeping the language generating power the same.

Let us replace 'a' by D_a, 'b' by D_b, and 'c' by D_c. So, three new productions $D_a \rightarrow a$, $D_b \rightarrow b$, and $D_c \rightarrow c$ are added to the grammar. By replacing 'a' by D_a, ' b' by D_b, and ' c' by D_c and adding three productions, the modified grammar becomes

$$S \rightarrow ABD_a$$
$$A \rightarrow DaD_aD_b$$
$$B \rightarrow AD_c$$
$$D_a \rightarrow a$$
$$D_b \rightarrow b$$
$$D_c \rightarrow c$$

S → ABDa and A → DaDaDb are not in CNF. Take two non-terminals E and F. Replace AB by E and DaDa by F. So, two new productions E → AB and F → DaDa will be added to the grammar. By replacement and adding the productions, the modified grammar will be

$$S \rightarrow ABD_a$$
$$A \rightarrow DaD_aD_b$$
$$B \rightarrow AD_c$$
$$D_a \rightarrow a$$
$$D_b \rightarrow b$$
$$D_c \rightarrow c$$
$$E \rightarrow AB$$
$$F \rightarrow D_aD_a$$

In the previous grammar, all the productions are in the form of Non-terminal → string of exactly two non-terminals or non-terminal → single terminal. So, the grammar is in CNF.

Example 6.39 Convert the following grammar into CNF.

$$E \rightarrow E + E$$
$$E \rightarrow E * E$$
$$E \rightarrow id$$

where $\Sigma = \{+, *, id\}$.

Solution: Except E → id, all the other productions of the grammar are not in CNF. In the grammar E → E + E and E → E * E, there are two terminals + and *. Take two non-terminals C and D for replacing + and *, respectively. Two new productions will be added to the grammar. By replacing + and *, the modified production rules will be

$$E \rightarrow ECE$$
$$E \rightarrow EDE$$
$$E \rightarrow id$$
$$C \rightarrow +$$
$$D \rightarrow *$$

In the production rules E → ECE and E → EDE, there are three non-terminals. Replace EC by another non-terminal F and ED by G. So, two new productions F → EC and G → ED will be added to the grammar. By replacing and adding the new productions, the modified grammar will be

$E \rightarrow FE$
$E \rightarrow GE$
$E \rightarrow id$
$C \rightarrow +$
$D \rightarrow *$
$F \rightarrow EC$
$G \rightarrow ED$

In the previous grammar, all the productions are in the form of non-terminal → string of exactly two non-terminals or non-terminal → single terminal. So, the grammar is in CNF.

Example 6.40 Convert the following grammar into CNF.

$S \rightarrow abAB$
$A \rightarrow bAB/\in$
$B \rightarrow BAa/\in$

Solution: In the grammar, there are two null productions $A \rightarrow \in$ and $B \rightarrow \in$. First, these productions must be removed, and after that the grammar can be converted into CNF. After removing the null production $A \rightarrow \in$, the modified grammar becomes

$S \rightarrow abAB/abB$
$A \rightarrow bAB/bB$
$B \rightarrow Ba$

After removing the null production $B \rightarrow \in$, the modified grammar becomes

$S \rightarrow abAB/abA/abB/ab$
$A \rightarrow bAB/bA/bB/b$
$B \rightarrow Ba/Aa/a$

Now, this grammar can be converted into CNF. In the grammar, except $A \rightarrow b$ and $B \rightarrow a$, all the productions are not in CNF. Let us take two non-terminals C_a and C_b which will replace a and b, respectively. So, two new productions $C_a \rightarrow a$ and $C_b \rightarrow b$ will be added to the grammar. After replacing and adding new productions, the modified grammar is

$S \rightarrow CaC_bAB/C_aC_bA/C_aC_bB/C_aC_b$
$A \rightarrow C_bAB/C_bA/C_bB/b$
$B \rightarrow BC_a/ACa/a$
$C_a \rightarrow a$
$C_b \rightarrow b$

In this modified grammar, $S \rightarrow CaC_bAB$, $S \rightarrow Ca\ CbA$, $S \rightarrow Ca\ C_bB$, and $A \rightarrow C_bAB$ are not in CNF. Let us take two other non-terminals D and E, which will be replaced in the place of CaC_b and AB, respectively. So, two new productions $D \rightarrow CaC_b$ and $E \rightarrow AB$ will be added to the grammar. By replacing and adding two new productions, the modified grammar is

$S \rightarrow DE/DA/DB/C_aC_b$
$A \rightarrow C_bE/C_bA/C_bB/b$
$B \rightarrow BCa/ACa/a$

$$D \rightarrow C_a C_b$$
$$E \rightarrow AB$$
$$C_a \rightarrow a$$
$$C_b \rightarrow b$$

In the previous grammar, all the productions are in the form of non-terminal → string of exactly two non-terminals or non-terminal → single terminal. So, the grammar is in CNF.

6.7.2 Greibach Normal Form

A grammar is said to be in GNF if every production of the grammar is of the form

Non-terminal → (single terminal)(string of non-terminals)
Non-terminal → single terminal

In one line, it can be said that all the productions will be in the form

Non-terminal → (single terminal)(non-terminal)*

(* means any combination of NTs including null)

Application: If a CFG can be converted into GNF, then from that GNF the push down automata (machine format of a CFG) accepting the CFL can easily be designed.

Before going into the details about the process of converting a CFG into GNF, we have to know two lemmas which are useful for the conversion process.

Lemma I: If $G = (V_N, \Sigma, P, S)$ is a CFG, and if $A \rightarrow A\alpha$ and $A \rightarrow \beta_1/\beta_2/\ldots/\beta_n$ belongs to the production rules (P) of G, then a new grammar $G' = (V_N, \Sigma, P', S)$ can be constructed by replacing $A \rightarrow \beta_1/\beta_2/\ldots/\beta_n$ in $A \rightarrow A\alpha$, which will produce

$$A \rightarrow \beta_1\alpha/\beta_2\alpha/\ldots/\beta_n\alpha$$

This production belongs to P′ in G′.

It can be proved that L(G) = L(G′)

Lemma II: Let $G = (V_N, \Sigma, P, S)$ be a CFG and the set of 'A' production which belongs to P be $A \rightarrow A\alpha_1/A\alpha_2/\ldots/A\alpha_m/\beta_1/\beta_2/\ldots/\beta_n$.

Introduce a new non-terminal X. Let $G' = (V'_N, \Sigma, P', S)$, where $(V'_N = (V_N \cup X)$ and P′ can be formed by replacing the A productions by

1. $A \rightarrow \beta_i \qquad 1 \le i \le n$
   $A \rightarrow \beta_i X$
2. $X \rightarrow \alpha_j \qquad 1 \le j \le m$
   $X \rightarrow \alpha_j X$

It can be proved that L(G) = L(G′).

(The main production was $A \rightarrow A\alpha_j/\beta_i$. If we replace $A \rightarrow A\alpha_j$ in the production $A \rightarrow A\alpha_j$, then the production will become $A \rightarrow A\alpha_j\alpha_j$. Then replace A by β_i. The production will become $A \rightarrow \beta_i\ \alpha_j\alpha_j$.

If we replace $A \rightarrow \beta_i$ in the production $A \rightarrow A\alpha_j$, the production will become $A \rightarrow \beta_i\alpha_j$.

Take another non-terminal X, which will be replaced in the place of the string after α_j in both the productions $A \rightarrow \beta_i\ \alpha_j\alpha_j$ and $A \rightarrow \beta_i\alpha_j$. So, the A productions will be $A \rightarrow \beta_i$ and $A \rightarrow \beta_i X$ and X productions will be $X \rightarrow \alpha_j$ or $X \rightarrow \alpha_j\alpha_j\ldots\alpha_j$, i.e., $X \rightarrow \alpha_j X$. So, the lemma II is correct.)

6.7.2.1 *Process for Conversion of a CFG into GNF*

Step I: Eliminate the null productions and the unit productions from the grammar and convert the grammar into CNF.

Step II: Rename the non-terminals of V_N as (A_1, A_2 .... A_n) with the start symbol A_1.

Step III: Using *lemma I* modify the productions such that the LHS variable subscript is less than the RHS starting variable subscript, that is, in mathematical notation, it can be said that all the productions will be in the form $A_i \rightarrow A_jV$ where $i \leq j$.

Step IV: By repeating the applications of *Lemma I* and *Lemma II*, all the productions of the modified grammar will come into GNF.

Example 6.41 Convert the following grammar into GNF.

$$S \rightarrow AA/a$$
$$A \rightarrow SS/b$$

Solution:

Step I: There are no unit productions and no null production in the grammar. The given grammar is in CNF.

Step II: In the grammar, there are two non-terminals S and A. Rename the non-terminals as A_1 and A_2, respectively. The modified grammar will be

$$A_1 \rightarrow A_2A_2/a$$
$$A_2 \rightarrow A_1A_1/b$$

Step III: In the grammar, $A_2 \rightarrow A_1A_1$ is not in the format $A_i \rightarrow A_jV$ where $i \leq j$. Replace the leftmost A_1 at the RHS of the production $A_2 \rightarrow A_1A_1$. After replacing the modified A_2, production will be

$$A_2 \rightarrow A_2A_2A_1/aA_1/b$$

The production $A_2 \rightarrow aA_1/b$ is in the format $A \rightarrow \beta_i$ and the production $A_2 \rightarrow A_2A_2A_1$ is in the format of $A \rightarrow A\alpha_j$ So, we can introduce a new non-terminal B_2 and the modified A_2 production will be (according to Lemma II)

$$A_2 \rightarrow aA_1/b$$
$$A_2 \rightarrow aA_1B_2$$
$$A_2 \rightarrow bB_2$$

And the B_2 productions will be

$$B_2 \rightarrow A_2A_1$$
$$B_2 \rightarrow A_2A_1B_2$$

Step IV: All A_2 productions are in the format of GNF. In the production $A_1 \rightarrow A_2A_2/a$, $A \rightarrow a$ is in the prescribed format. But the production $A_1 \rightarrow A_2A_2$ is not in the format of GNF. Replace the leftmost A_2 at the RHS of the production by the previous A_2 productions. The modified A_1 productions will be

$$A_1 \rightarrow aA_1A_2/bA_2/aA_1B_2A_2/bB_2A_2$$

The B_2 productions are not in GNF. Replace the leftmost A_2 at the RHS of the two productions by the A_2 productions. The modified B_2 productions will be

$$B_2 \rightarrow aA_1A_1/bA_1/aA_1B_2A_1/bB_2A_1$$
$$B_2 \rightarrow aA_1A_1B_2/bA_1B_2/aA_1B_2A_1B_2/bB_2A_1B_2$$

For the given CFG, the GNF will be

$$A1 \rightarrow aA_1A_2/bA_2/aA_1B_2A_2/bB_2A_2/a$$
$$A2 \rightarrow aA_1/b/aA_1B_2/bB_2$$
$$B2 \rightarrow aA_1A_1/bA_1/aA_1B_2A_1/bB_2A_1$$
$$B2 \rightarrow aA_1A_1B_2/bA_1B_2/aA_1B_2A_1B_2/bB_2A_1B_2$$

Example 6.42 Convert the following CFG into GNF.

$$S \rightarrow XY$$
$$X \rightarrow YS/b$$
$$Y \rightarrow SX/a$$

Solution:

Step I: In the grammar, there is no null production and no unit production. The grammar also is in CNF.

Step II: In the grammar, there are three non-terminals S, X, and Y. Rename the non-terminals as A_1, A_2, and A_3, respectively. After renaming, the modified grammar will be

$$A_1 \rightarrow A_2A_3$$
$$A_2 \rightarrow A_3A_1/b$$
$$A_3 \rightarrow A_1A_2/a$$

Step III: In the grammar, the production $A_3 \rightarrow A_1A_2$ is not in the format $A_i \rightarrow A_jV$ where $i \leq j$.

Replace the leftmost A_1 at the RHS of the production $A_3 \rightarrow A_1A_2$ by the production $A_1 \rightarrow A_2A_3$. The production will become $A_3 \rightarrow A_2A_3A_2$, which is again not in the format of $A_i \rightarrow A_jV$ where $i \leq j$. Replace the leftmost A_2 at the RHS of the production $A_3 \rightarrow A_2A_3A_2$ by the production $A_2 \rightarrow A_3A_1/b$. The modified A_3 production will be

$$A_3 \rightarrow A_3A_1A_3A_2/bA_3A_2/a$$

The production $A_3 \rightarrow bA_3A_2/a$ is in the format of $A \rightarrow \beta_i$ and the production $A_3 \rightarrow A_3A_1A_3A_2$ is in the format of $A \rightarrow A\alpha_j$ So, we can introduce a new non-terminal B and the modified A_3 production will be (according to Lemma II)

$$A_3 \rightarrow bA_3A_2$$
$$A_3 \rightarrow a$$
$$A_3 \rightarrow bA_3A_2B$$
$$A_3 \rightarrow aB$$

And B productions will be

$$B \rightarrow A_1A_3A_2$$
$$B \rightarrow A_1A_3A_2B$$

Step IV: All the A_3 productions are in the specified format of GNF.

The A_2 production is not in the specified format of GNF. Replacing A_3 productions in A_2 productions, the modified A_2 production becomes

$$A_2 \rightarrow bA_3A_2A_1/aA_1/bA_3A_2BA_1/aBA_1/b$$

Now, all the A_2 productions are in the prescribed format of GNF.

The A_1 production is not in the prescribed format of GNF. Replacing A_2 productions in A_1, the modified A_1 productions will be

$$A_1 \rightarrow bA_3A_2A_1A_3/aA_1A_3/bA_3A_2BA_1A_3/aBA_1A_3/bA_3$$

All the A_1 productions are in the prescribed format of GNF.

But the B productions are still not in the prescribed format of GNF. By replacing the leftmost A_1 at the RHS of the B productions by A_1 productions, the modified B productions will be

$$B \rightarrow bA_3A_2A_1A_3A_3A_2/aA_1A_3A_3A_2/bA_3A_2BA_1A_3A_3A_2/aBA_1A_3A_3A_2/bA_3A_3A_2$$
$$B \rightarrow bA_3A_2A_1A_3A_3A_2B/aA_1A_3A_3A_2B/bA_3A_2BA_1A_3A_3A_2B/aBA_1A_3A_3A_2B/bA_3A_3A_2B.$$

Now, all the B productions of the grammar are in the prescribed format of GNF.

So, for the given CFG, the GNF will be

$$A_1 \rightarrow bA_3A_2A_1A_3/aA_1A_3/bA_3A_2BA_1A_3/aBA_1A_3/bA_3$$
$$A_2 \rightarrow bA_3A_2A_1/aA_1/bA_3A_2BA_1/aBA_1/b$$
$$B \rightarrow bA_3A_2A_1A_3A_3A_2/aA_1A_3A_3A_2/bA_3A_2BA_1A_3A_3A_2/aBA_1A_3A_3A_2/bA_3A_3A_2$$
$$B \rightarrow bA_3A_2A_1A_3A_3A_2B/aA_1A_3A_3A_2B/bA_3A_2BA_1A_3A_3A_2B/aBA_1A_3A_3A_2B/bA_3A_3A_2B$$

Example 6.43 Convert the following CFG into GNF.

$$S \rightarrow AB/BC$$
$$A \rightarrow aB/bA/a$$
$$B \rightarrow bB/cC/b$$
$$C \rightarrow c$$

Solution:

Step I: In the previous grammar, there is no unit production and no null production. But all productions are not in CNF. Let us take two non-terminals D_a and D_b which will be placed in the place of 'a' and 'b', respectively. So, two new productions $D_a \rightarrow a$ and $D_b \rightarrow b$ will be added to the grammar

$$S \rightarrow AB/BC$$
$$A \rightarrow D_aB/D_bA/a$$
$$B \rightarrow D_bB/CC/b$$
$$C \rightarrow c$$
$$D_a \rightarrow a$$
$$D_b \rightarrow b$$

Now all the productions are in CNF.

Step II: There are six non-terminals in the grammar. Rename the non-terminals as A_1, A_2 ..A_6. After replacing, the modified productions will be

$$A_1 \rightarrow A_2A_3/A_3A_4$$
$$A_2 \rightarrow A_5A_3/A_6A_2/a$$
$$A_3 \rightarrow A_6A_3/A_4A_4/b$$
$$A_4 \rightarrow c$$
$$A_5 \rightarrow a$$
$$A_6 \rightarrow b$$

The productions for A_1, A_2, and A_3 are all in the format $A_i \rightarrow A_jV$ where $i \leq j$. Replace A_6 and A_4 in the productions $A_3 \rightarrow A_6A_3$ and $A_3 \rightarrow A_4A_4$ by $A_6 \rightarrow b$ and $A_4 \rightarrow c$, respectively. The modified A_3 productions will be

$$A_3 \rightarrow bA_3/cA_4/b$$

All the productions are now in the format of GNF.

Replace A_5 and A_6 in the productions $A_2 \rightarrow A_5A_3$ and $A_2 \rightarrow A_6A_2$ by $A_5 \rightarrow a$ and $A_6 \rightarrow b$, respectively. The modified A_2 productions will be

$$A_2 \rightarrow aA_3/bA_2/a$$

All the productions are now in the format of GNF.

The A_1 productions $A_1 \rightarrow A_2A_3/A_3A_4$ are not in the format of GNF. Replace A_2 at the RHS of the production $A_1 \rightarrow A_2A_3$. The modified production will be

$$A_1 \rightarrow aA_3A_3/bA_2A_3/aA_3$$

Replace A_3 at the RHS of the production $A_1 \rightarrow A_3A_4$. The modified production will be

$$A_1 \rightarrow bA_3A_4/cA_4A_4/bA_4$$

So, for the given CFG, the GNF will be

$$A_1 \rightarrow aA_3A_3/bA_2A_3/aA_3/bA_3A_4/cA_4A_4/bA_4$$
$$A_2 \rightarrow aA_3/bA_2/a$$
$$A_3 \rightarrow bA_3/cA_4/b$$
$$A_4 \rightarrow c$$
$$A_5 \rightarrow a$$
$$A_6 \rightarrow b$$

6.8 Closure Properties of Context-free Language

A set is closed (under an operation) if and only if the operation on two elements of the set produces another element of the set. If an element outside the set is produced, then the operation is not closed.

Closure is a property which describes the application of the property on any two elements of the set; the result is also included in the set.

6.8.1 Closed Under Union

Let L_1 be a CFL for the CFG $G_1 = (V_{N1}, \Sigma_1, P_1, S_1)$ and L_2 be a CFL for the CFG $G_2 = (V_{N2}, \Sigma_2, P_2, S_2)$. We have to prove that $L = L_1 \cup L_2$ is also in CFL.

Let us construct a grammar $G = (V_N, \Sigma, P, S)$ using the two grammars G_1 and G_2 as follows.

$$V_N = V_{N1} \cup V_{N2} \cup \{S\}$$
$$\Sigma = \Sigma_1 \cup \Sigma_2$$
$$P = P_1 \cup P_2 \cup \{S \to S_1/S_2\}$$

From the previous discussion, it is clear that the language set generated from the grammar G contains all the strings that are derived from S_1 as well as S_2. So, it is proved that $L = L_1 \cup L_2$.

6.8.2 Closed Under Concatenation

Let L_1 be a CFL for the CFG $G_1 = (V_{N1}, \Sigma_1, P_1, S_1)$ and L_2 be a CFL for the CFG $G_2 = (V_{N2}, \Sigma_2, P_2, S_2)$. We have to prove that $L = L_1L_2$ is also in CFL.

Let us construct a grammar $G = (V_N, \Sigma, P, S)$ using the two grammars G_1 and G_2 as follows.

$$V_N = V_{N1} \cup V_{N2} \cup \{S\}$$
$$\Sigma = \Sigma_1 \cup \Sigma_2$$
$$P = P_1 \cup P_2 \cup \{S \to S_1S_2\}$$

From the previous discussion, it is clear that the language set generated from the grammar G contains all the strings that are derived from S_1 as well as S_2. So, it is proved that $L = L_1L_2$.

6.8.3 Closed Under Star Closure

Let L be a CFL for the CFG $G_1 = (V_{N1}, \Sigma_1, P_1, S_1)$. We have to prove that L* is also in CFL.

Let us construct a grammar $G = (V_N, \Sigma, P, S)$ using the grammar G_1 as follows.

$$V_N = V_{N1} \cup \{S\}$$
$$P = S \to S_1S/\lambda$$

Obviously, G is a context-free grammar. It is easy to see that $L(G) = L^*$.

6.8.4 Closed Under Intersection

Consider two languages $L_1 = \{a^{n+1}b^{n+1}c^n, \text{ where } n, 1 \geq 0\}$
and

$$L_2 = \{a^nb^nc^{n+k}, \text{ where } n, k \geq 0\}.$$

We can easily show that the two languages L_1 and L_2 are context free (by constructing grammar).

Consider $L = L_1 \cap L_2$.

So, $L = a^{n+1}b^{n+1}c^n \cap a^nb^nc^{n+k} = a^nb^nc^n$, where $n \geq 0$.

$a^nb^n c^n$ is a context sensitive language not a context free. As one instance is proved not to be context free then we can decide that context free languages are not closed under intersection.

6.8.5 Not Closed Under Complementation

From the set theory, we can prove $L_1 \cap L_2 = L_1 \cup L_2$. (D' Morgan's Law)

If the union of the complements of L_1 and L_2 are closed, i.e., also context free, then the LHS will also be context free. But we have proved that $L_1 \cap L_2$ is not context free. We are getting a contradiction here. So, CFLs are not closed under complementation.

6.8.6 Every Regular Language is a Context-free Language

From the recursive definition of regular set, we know that ∅ and ∈ are regular expressions. If R is a regular expression, then R + R (union), R.R (concatenation), and R* (Kleene star) are also regular expressions. A regular expression R is a string of terminal symbols. ∅ and ∈ are also CFLs, and we know that CFLs are closed under union, concatenation, and Kleene star. Therefore, we can write that every regular language is a CFL.

6.9 Pumping Lemma for CFL

We have become familiar with the term 'pumping lemma' in the regular expression chapter. The pumping lemma is also related to a CFL. The pumping lemma for CFL is used to prove that certain sets are not context free. Every CFL fulfills some general properties. But if a set or language fulfills all the properties of the pumping lemma for CFL, it cannot be said that the language is context free. But the reverse is true, i.e., if a language breaks the properties it can be said that the language is not context free.

Before discussing the statement of the pumping lemma, we need to prove the following lemma.

'Let G be a CFG in CNF and T be a derivation tree that belongs to G. If the length of the longest path in T is less than or equal to k, then the yield of T is of length less than or equal to 2^{k-1}'.

Proof: We will prove this by the method of induction on k. k is the longest path for all trees with root label A. (called A-Tree). When the longest path of an A-tree is of length 1, the root has only one son and that is the terminal symbol. When the root has two sons, the labels are variable (the CFG is in CNF). So, the yield of T is of length 1. We have got a base for induction.

Let k > 1, and let T be an A-tree with the longest path of length less than or equal to k. The grammar is in CNF, and so the root of T, that is, A, has exactly two sons with labels A_1 and A_2. The length of the longest path of the two subtrees with the two sons as root is less than or equal to k – 1. (Refer to Fig. 6.13.)

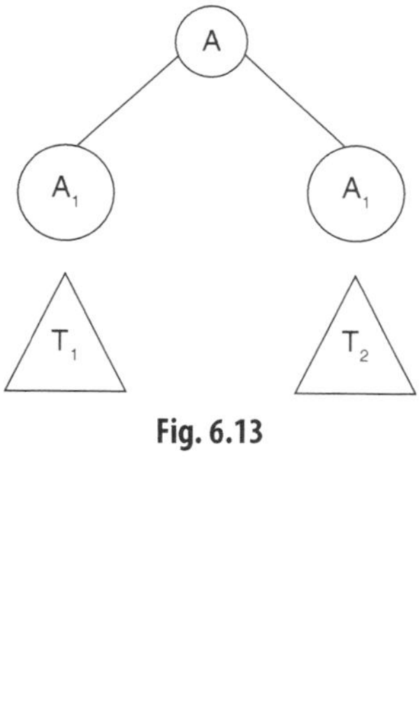

Fig. 6.13

If w_1 and w_2 are the yields of T_1 and T_2, respectively, by induction

$$|w_1| \leq 2^{k-2} \text{ and } |w_2| \leq 2^{k-2}$$

Let w be the yield of T, so $w = w_1w_2$.

$$|w| = |w_1w_2| \leq 2^{k-2} + 2^{k-2} = 2^{k-1}$$

The yield of T is of length less than or equal to 2^{k-1}.

Pumping Lemma for CFL: Let L be a CFL. Then, we can find a natural number n such that

1. Every z ∈ L with z ≥ n can be written as w = uvwxy, for some strings u,v,w,x,y.
2. $|vx| \geq 1$
3. $|vwx| \leq n$
4. $uv^kwx^ky \in L$ for all $k \geq 0$

Proof: We can prove that if G is a CFG, then we can find a CFG G_1 having no null productions such that $L(G_1) = L(G) - \{\Lambda\}$.

If the null string (Λ) belongs to the language set L, we will consider $L - \{\Lambda\}$ and will construct the CNF of the grammar G generating $L - \{\Lambda\}$ (if a null string does not belong to L, we will construct the CNF of G generating L).

Let the number of non-terminals (V_N) of the grammar be m. $| V_N | = m$ and $n = 2^m$. To prove that n is the required number, we start with a string z where $z \in L$, $| z | \geq 2^m$. Construct a derivation tree or parse tree T of the string z. If the length of the longest path in T is almost m, then we can write the length of the yield of T, i.e., $| z | \leq 2^{m-1}$. But $| z | \geq 2^m > 2^{m-1}$. So, T has a path, say Γ of length greater than or equal to m + 1. The path length of greater than or equal to m + 1 needs at least m + 2 vertices, where the last vertex is the leaf. Thus, in Γ, all the labels except the last one are variables. As $| N | = m$, in m + 1 labels of the parse tree T, some labels are repeated.

Choose the repeated label as follows. We start with the leaf of Γ and travel upwards. We stop when we get some label repeated (let us assume that we have got label A as repeated, first).

Let v_1 and v_2 be the vertices with repeated label (say, A) obtained at the time of upward traversal from the leaf of Γ. Between v_1 and v_2, let v_1 be nearer to the root. In Γ, the portion of the path from v_1 to leaf has only one label repeated (say A), and so its length is almost m + 1.

Let T_1 and T_2 be two subtrees taking the root as v_1 and v_2, respectively. Let z_1 and w be the yields of T_1 and T_2, respectively. As Γ is the longest path in T, v_1 to the leaf is the longest path of T_1 and its length is almost m + 1. So, the length of z_1 (yield of T_1), i.e., $| z_1 | \leq 2^m$.

As z is the yield of T and T_1 is a proper subtree of T, the yield of T_1 is z_1. So, the yield of T, i.e., z, can be written as $z = uz_1y$. (Refer to Fig. 6.14.)

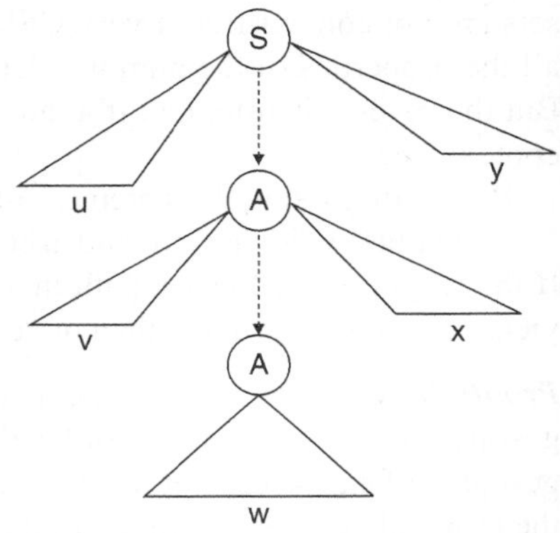

Fig. 6.14 *Derivation for Pumping Lemma for CFL*

As z_1 is the yield of T_1 and T_2 is a proper subtree of T_1, the yield of T_2 is w. So, the yield of T_1, i.e., z_1 can be written as $z_1 = vwx$. T_2 is a proper subtree of T_1, and so $| vwx | > | w |$. So, $| vx | \geq 1$. Thus, we can write z = uvwxy with $| vwx | \leq n$ and $| vx | \geq 1$. This proves (i) to (iii) of the pumping lemma.

As T is a tree with root S (S tree), and T_1 and T_2 are trees with root B (B tree), we can write $S \Rightarrow uAy$, $A \Rightarrow vAx$ and $A \Rightarrow w$.

As $S \Rightarrow uAy \Rightarrow uwy$, $uv^0wx^0y \in L$. For $k \geq 1$, $S \Rightarrow uAy \Rightarrow uv^kAx^ky \Rightarrow uv^k wx^ky \in L$. By this, the theorem is proved.

We use the pumping lemma for CFL to show that certain sets, such as L, are not context free. We assume that the set is context free. By applying the pumping lemma, we get a contradiction.

This proof can be done in the following ways.

Step I: Assume that L is context free. Find a natural number such that $| z | \geq n$.

Step II: So, we can write z = uvwxy for some strings u,v,w,x,y.

Step III: Find a suitable k such that $uv^kwx^ky \notin L$. This is a contradiction, and so L is not context free.

Consider the following examples to get the application of the pumping lemma for CFL.

Example 6.44 Show that $L = \{a^nb^nc^n \text{ where } n \geq 1\}$ is not context free.

Solution:
Step I: Assume that the language set L is a CFL. Let n be a natural number obtained by using the pumping lemma.

Step II: Let $z = a^nb^nc^n$. So, $|z| = 3n > n$. According to the pumping lemma for CFL, we can write $z = uvwxy$, where $|vx| \geq 1$.

Step III: $uvwxy = a^nb^nc^n$. As $1 \leq |vx| \leq n$, ($|vwx| \leq n$, so $|vx| \leq n$) v or x cannot contain all the three symbols a, b, and c. So, v or x will be in any of the following forms.

1. Contain only a and b, i.e., in the form a^ib^j.
2. Or contain only b and c, i.e., in the form b^ic^j.
3. Or contain only the repetition of any of the symbols among a, b, and c.

Let us take the value of k as 2. v^2 or x^2 will be in the form $a^ib^ia^ib^i$ (as v is a string here aba is not equal to a^2b or ba^2) or $b^ic^j\,b^ic^j$. So, uv^2wx^2y cannot be in the form $a^mb^mc^m$. So, $uv^2wx^2y \notin L$.

If v or x contains repetition of any of the symbols among a, b, and c, then v or x will be any of the form of a^i, b^i, or c^i. Let us take the value of k as 0. $uv^0wx^0y = uwy$. In the string, the number of occurrences of one of the other two symbols in uvy is less than n. So, $uv^2wx^2y \notin L$.

Example 6.45 Prove that the language $L = \{a^{i^2}/i \geq 1\}$ is not context free.

Solution:

Step I: Assume that the language set L is a CFL. Let n be a natural number obtained by using the pumping lemma.

Step II: Let $z = a^{i^2}$. So, $|z| = 2^i$. Let $2^i > n$. According to the pumping lemma for CFL, we can write $z = uvwxy$, where $|vx| \geq 1$ and $|vwx| \leq n$.

Step III: The string z contains only 'a', and so v and x will also be a string of only 'a'. Let $v = a^p$ and $x = a^q$, where $(p + q) \geq 1$. Since $n \geq 0$ and $uvwxy = 2^i$, $|uv^nwx^ny| = |uvwxy| + |v^{n-1}x^{n-1}| = 2^i + (p + q)(n - 1)$. As $uv^nwx^ny \in L$, $|uv^nwx^ny|$ is also a power of 2, say 2^j.

$$\begin{aligned} & (p + q)(n - 1) &&= 2^j - 2^i \\ \Rightarrow\ & (p + q)(n - 1) + 2^i &&= 2^j \\ \Rightarrow\ & (p + q)2^{i+1} + 2^i &&= 2^j \\ \Rightarrow\ & 2^i\,(2(p + q) + 1) &&= 2^j \end{aligned}$$

$(p + q)$ may be even or odd, but $2(p + q)$ is always even. However, $2(p + q) + 1$ is odd, which cannot be a power of 2. Thus, L is not context free.

Example 6.46 Prove that $L = \{0^p/\text{where p is prime}\}$ is not a CFL.

Solution:
Step I: Suppose $L = L(G)$ is context free. Let n be a natural number obtained from the pumping lemma for CFL.

Step II: Let p be a number > n, $z = 0^p \in L$. By using the pumping lemma for CFL, we can write $z = uvwxy$, where $|vx| \geq 1$ and $|vwx| \leq n$.

Step III: Let $k = 0$. From the pumping lemma for CFL, we can write uv^0wx^0y, i.e., $uwy \in L$. As uvy is in the form 0^p, where p is prime, $|uwy|$ is also a prime number. Let us take it as q. Let $|vx| = r$. Then, $|uv^qwx^qy| = q + qr = q(1 + r)$. This is not prime as it has factors $q(1 + r)$ including 1 and $q(1 + r)$. So, $uv^qwx^qy \notin L$. This is a contradiction. Therefore, L is not context free.

6.10 Ogden's Lemma for CFL

If L is a context-free language, then there exists some positive integer p such that for any string $w \in L$ with at least p positions marked, then w can be written as

$$w = uvxyz$$

with strings u, v, x, y, and z, such that

1. x has at least one marked position,
2. either u and v both have marked positions or y and z both have marked positions,
3. vxy has at most p marked positions, and
4. uv^ixy^iz is in L for every $i \geq 0$.

Ogden's lemma can be used to show that certain languages are not context free, in cases where the pumping lemma for CFLs is not sufficient. An example is the language, $\{a^ib^jc^kd^l: i = 0 \text{ or } j = k = l\}$. It is also useful to prove the inherent ambiguity of some languages.

6.11 Decision Problems Related to CFG

Decision problems are the problems which can be answered in a 'yes' or 'no'. There are a number of decision problems related to CFG. The following are some of them.

1. Whether a given CFG is empty or not. (Does it generate any terminal string?)
2. Whether a given CFG is finite? (Does it generate up to a certain length word?)
3. Whether a particular string w is generated by a given CFG. (Membership)
4. Whether a given CFG is ambiguous.
5. Whether two given CFG have a common word.
6. Whether two different CFG generate the same language.
7. Whether the complement of a given CFL is also a CFL.

In the following section, we are discussing the first three decision algorithms for CFG.

6.11.1 Emptiness

In a CFG $\{V_N, \Sigma, P, S\}$, if S generates a string of terminal, then it is non-empty; otherwise, the corresponding CFL is empty.

The following algorithm checks whether a CFG is empty or not:

While (there exist a production in the form $NT_i \rightarrow$ T/String of T)

{
Replace each NT at the right hand side of each production by corresponding T or string of T.
}
if (S generates a string of terminal)
return non-empty.
else
empty.

Example 6.47 Check whether the following CFG is empty or not.

S → AB/D
A → aBC
B → bC
C → d
D → CD

Solution: C → d is in the form NT → T. Replacing C → d in A → aBC, B → bC, and D → CD, we get

S → AB/D
A → aBd
B → bd
D → dD

B → bd is in the form NT → string of terminals. Replacing B → bd in S → AB and A → aBd, we get

S → Abd/D
A → abdd
D → dD

A → abdd is in the form NT → String of terminals. Replacing A → abdd in S → Abd, we get

S → abddbd/D
D → dD

As S generates a string of terminals, the CFG is non-empty.

Example 6.48 Check whether the following CFG is empty or not.

S → ABD
A → BC
B → aC
C → b
D → BD

Solution: C → b is in the form NT → T. Replacing C → b in A → BC, B → aC, we get

S → ABD
A → Bb
B → ab
D → BD

B → ab is in the form NT → String of T. Replacing B → ab in S → ABD, A → Bb, D → BD, we get

S → AabD
A → abb
D → abD

A → abb is in the form NT → String of T. Replacing A → abb in S → AabD, we get

S → abbabD
D → abD

'While' part is complete, S does not produce any terminal string. Thus, the CFG is empty.

6.11.2 Finiteness

A language L generated from a given CFG is finite if there are no cycles in the directed graph generated from the production rules of the given CFG. The longest string generated by the grammar is determined by the derivation from the start symbol.

(The number of vertices of the directed graph is the same as the number of non-terminals in the grammar. If there is a production rule S → AB, the directed graph is as shown in Fig. 6.15.)

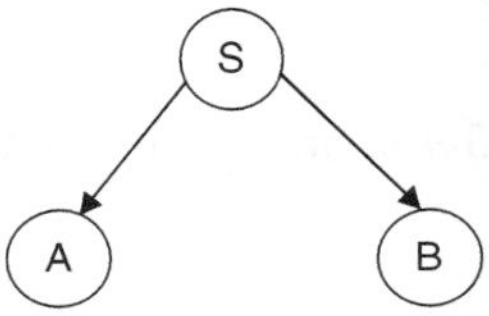

Fig. 6.15

6.11.2.1 Infiniteness

A language L generated from a given CFG is infinite if there is at least one cycle in the directed graph generated from the production rules of the given CFG.

The following examples describe this in detail.

Example 6.49 Verify whether the languages generated by the following grammar are finite or not. If finite, find the longest string generated by the grammar.

a) S → AB
A → BC
B → C
C → a

b) S → AB
A → B
B → SC/a
C → AB/b

Soution:

a) The grammar is not in CNF. Removing the unit productions, the grammar becomes

S → AB
A → BC
B → a
C → a

Now the grammar is in CNF. The directed graph for the grammar is shown in Fig. 6.16.

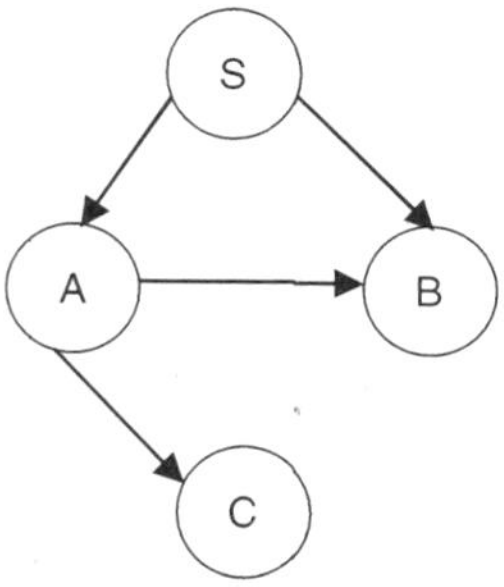

Fig. 6.16

The graph does not contain any loop. So, the language generated by the CFG is finite. The derivation from the grammar is S $\rightarrow$ AB $\rightarrow$ BCB $\rightarrow$ aaa.

Thus, the length of the longest string is 3.

b) In the grammar, there is a unit production A $\rightarrow$ B. By removing the unit production, the grammar becomes

$$S \rightarrow AB$$
$$A \rightarrow SC/a$$
$$B \rightarrow SC/a$$
$$C \rightarrow AB/b$$

The grammar is in CNF. The non-terminal transitional graph for the grammar is shown in Fig. 6.17.

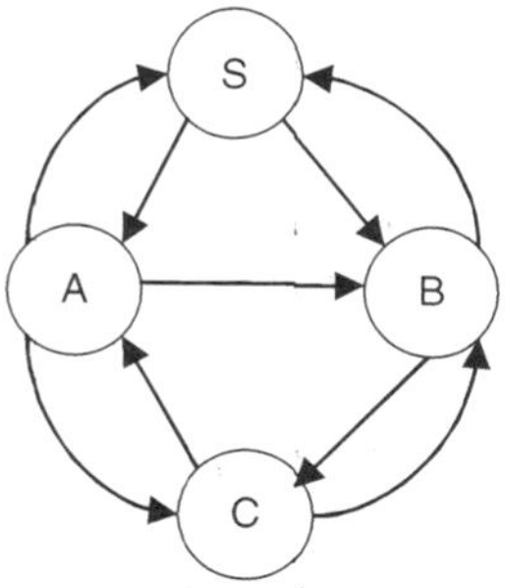

Fig. 6.17

The graph contains loops. So, the language generated by the CFG is infinite.

6.11.3 Membership Problem

Membership problem decides whether a string w is generated by a given CFG.

This is proved by the CYK algorithm. This algorithm was proposed by John Cocke, Daniel Younger, and Tadao Kasami. According to the algorithm, the string of terminals is generated from length 1 to length of w, which is also the string to be checked for membership. If w $\in$ the set of string generated, then w is a member of the strings generated by the CFG. For this, we need to convert the given grammar into CNF. With the help of an example we are showing this.

Example 6.50 Let the grammar converted to CNF is

$$S \to C_bA/C_aB$$
$$A \to C_bD/C_aS/a$$
$$B \to C_aE/C_bS/a$$
$$D \to AA$$
$$E \to BB$$
$$C_a \to a$$
$$C_b \to b$$

Check whether baba is a member of the CFL generated by the CFG.

Solution: Start producing strings of length 1.

String	Producing NT
A	A
A	B
A	C_a
B	C_b

Produce strings of length 2

String	Producing NT
ba	S $(S \to C_bA)$
aa	S $(S \to C_aB)$
aa	D $(D \to AA)$
aa	E $(E \to BB)$

Produce strings of length 3. This may be the first one produced by an NT and the last two produced by another NT or vice versa.

String	Producing NT
baa	A $(A \to C_bD)$
aba	A $(A \to C_aS)$
aaa	A $(A \to C_aS)$
aaa	B $(B \to C_aE)$
bba	B $(B \to C_bS)$
baa	B $(B \to C_bS)$

Produce strings of length 4. This may be the First one produced by an NT and the last three produced by another NT or the first two produced by an NT and the last two produced by another NT or the first three produced by an NT and the last one produced by another NT. Two–two combination is not available for this grammar as there is no production combining two of S, D, and E.

String	Producing NT
abaa	D ($D \rightarrow AA$, $A \rightarrow a$, $A \rightarrow C_bD$)
baaa	D ($D \rightarrow AA$, $A \rightarrow C_bD$, $A \rightarrow a$)
aaba	D ($D \rightarrow AA$, $A \rightarrow a$, $A \rightarrow C_aS$)
abaa	D ($D \rightarrow AA$, $A \rightarrow C_aS$, $A \rightarrow a$)
aaaa	D ($D \rightarrow AA$, $A \rightarrow C_aS$, $A \rightarrow a$)
aaaa	D ($D \rightarrow AA$, $A \rightarrow a$, $A \rightarrow C_aS$)
aaaa	E ($E \rightarrow BB$)
aaaa	E ($E \rightarrow BB$)
abba	E ($E \rightarrow BB$)
bbaa	E ($E \rightarrow BB$)
abaa	E ($E \rightarrow BB$)
baaa	E ($E \rightarrow BB$)
aaaa	S ($S \rightarrow C_aB$)
abba	S ($S \rightarrow C_aB$)
abaa	S ($S \rightarrow C_aB$)
bbaa	S ($S \rightarrow C_bA$)
baba	S ($S \rightarrow C_bA$)
baaa	S ($S \rightarrow C_bA$)

The string in bold is 'baba'. Thus, baba is a member of the CFL generated by the CFG.

6.12 CFG and Regular Language

According to the Chomsky hierarchy, we know that a regular grammar is a subset of CFG. Thus, for every regular language, there exists a CFG. In this section, we shall discuss two theorems related to this

Theorem 1: Every regular language is generated by a CFG.

Proof: Assume that L is regular. Thus, there exists a DFA M = {Q, Σ, δ, q_0, F} accepting L. From the DFA M, we can generate a CFG G = {V_N, Σ, P, S } using the following rules.

1. Every state ∈ Q of M is treated as a non-terminal ∈ V_N. The start state S = q_0 (initial state of DFA).
2. For a transitional function $\delta(q_1, a) \rightarrow q_2$ where $q_1, q_2 \in Q$ and $a \in \Sigma$, add a production $q_1 \rightarrow aq_2$ in P.
3. If q_2 is a final state, add $q_1 \rightarrow aq_2$ and $q_1 \rightarrow a$ in P.

All there productions have a single non-terminal at the LHS. Thus, it is context free.

Theorem 2: The language generated by a regular CFG is a regular language. Or every regular grammar generates a regular language.

Proof: Let G = {V_N, Σ, P, S } be a regular CFG. From this grammar, construct a NDFA M = {Q, Σ, δ, q_0, F} using the following rules.

1. For each production in the form <NT$_1$> → <i/p> <NT$_2$>, add a transitional function

   δ(<NT$_1$>, <i/p>) → <NT$_2$>

2. For each production in the form <NT$_1$ > → <i/p>, add a transitional function

   δ(<NT$_1$>, <i/p>) → final state

We know that a language accepted by an NDFA is regular expression. Thus, it is proved.

Every regular grammar is some right linear grammar. Already it is given that a right linear grammar can be converted to a left linear grammar. Thus, it can be proved that if L is a regular set, then L is generated by some left linear or some by some right linear grammar.

6.13 Applications of Context-free Grammar

The compiler is a program that takes a program written is the source language as input and translates it into an equivalent program in the target language. Syntax analysis in an important phase in the compiler design. In this phase, mainly grammatical errors called syntax errors are checked. The syntax analyzer (parser) checks whether a given source program satisfies the rules implied by a context-free grammar or not. If it satisfies, the parser creates the parse tree of that program. Otherwise, the parser gives the error messages.

In C language, an identifier is described by the following CFG.

The definition of an identifier in a programming language is

letter → A | B |...| Z | a | b | ... | z
digit → 0 | 1 |...| 9
id → letter (letter | digit)*

Let a programmer declare the following variables

```
int capital;
int r_o_i;
int year;
int 1st_year_interest;
```

then the syntax error will be shown in the fourth line as it does not match with the language produced by the grammar.

For the iteration statements (loop) also, there are CFGs.

<iteration statement> → while(<logical expression>) <statement>
/do <statement> while (<logical expression>)
/for(<expression>; <expression>; <expression>) <statement>

The CFG is used to develop extensive markup language (XML). XML is a markup language much like HTML. XML is used as a database and can be applicable to share and store data. It can even be used to construct new languages such as WML.

What We Have Learned So Far

1. According to the Chomsky hierarchy, context-free grammar, in short CFG, is type 2 grammar.
2. In every production of a CFG, at the left hand side there is a single non-terminal.
3. A derivation is called a leftmost derivation if only the leftmost non-terminal is replaced by some production rule at each step of the generating process of the language from the grammar.
4. A derivation is called a rightmost derivation if only the rightmost non-terminal is replaced by some production rule at each step of the generating process of the language from the grammar.
5. Parsing a string is finding a derivation for that string from a given grammar.
6. A parse tree is an ordered tree in which the left hand side of a production represents a parent node and the right hand side of a production represents the children nodes.
7. A grammar of a language is called ambiguous if from any of the cases for generating a particular string, more than one parse tree can be generated.
8. Ambiguity in grammar creates problem at the time of generating languages from the grammar.
9. There is no hard and fast rule for removing ambiguity from a CFG. Some ambiguity can be removed by hand. Some CFGs are always ambiguous. Those are called inherently ambiguous.
10. CFG may contain different types of useless symbols, unit productions, and null productions. These types of symbols and productions increase the number of steps in generating a language from a CFG.
11. Reduced grammar contains less number of non-terminals and productions, and so the time complexity for the language generating process becomes less from reduced grammar.
12. Useless symbols are of two types, non-generating symbols and non-reachable symbols.
13. Non-generating symbols are those symbols which do not produce any terminal string.
14. Non-reachable symbols are those symbols which cannot be reached at any time starting from the start symbol.
15. Production in the form non-terminal $\rightarrow$ single non-terminal is called unit production. Unit production increases the number of steps as well as the complexity at the time of generating the language from the grammar.
16. A production in the form NT $\rightarrow \in$ is called null production. If Λ (null string) is in the language set generated from the grammar, then that null production cannot be removed.
17. When a grammar is made in the normal form, every production of the grammar is converted in some specific form. These help us to design some algorithm to answer certain questions.
18. A CFG is called in CNF if all the productions of the grammar are in the following form.

    Non-terminal $\rightarrow$ string of exactly two non-terminals
    Non-terminal $\rightarrow$ single terminal

19. A grammar is said to be in Greibach normal form (GNF) if every production of the grammar is of the following form.

    Non-terminal $\rightarrow$ (single terminal)(string of non-terminals)
    Non-terminal $\rightarrow$ single terminal

20. Context-free languages are closed under union, concatenation, and star closure.
21. Context-free languages are not closed under intersection and complementation.

22. The pumping lemma and the Ogden's lemma for context-free language are used to prove that certain sets are not context free.
23. A language L generated from a given CFG is finite if there are no cycles in the directed graph generated from the production rules of the given CFG.
24. A language L generated from a given CFG is infinite if there is at least one cycle in the directed graph generated from the production rules of the given CFG.

Solved Problems

1. Construct a CFG for the following
   a) Set of string of 0 and 1 where consecutive 0 can occur but no consecutive 1 can occur.
   b) Set of all (positive and negative) odd integers.
   c) Set of all (positive and negative) even integers.

Solution:

a) There will be two non-terminals, S and A. S will produce production $S \to 0S$. As two consecutive 0 can appear, there will be a production $S \to 0$. In the language set, 01 can appear. So, there will be production $S \to 1$.

The string can start with 1, and so there may be a production $S \to 1S$. But the consecutive 1 will not appear, and so the production will be $S \to 1A$. 'A' can produce 0 but not 1. So, there will be a production $A \to 0$.

If the string starts with 1, then the production rule $S \to 1A$ is applied. As in the language set, there will be no consecutive 1, and so A may produce $A \to 0A$. But if this is the production, then there will be no chance of an occurring 1 in the string except at the last. So, the production will be $A \to 0S$.

So, the grammar is

$$S \to 0S,\ S \to 1A,\ S \to 0,\ S \to 1,\ A \to 0S,\ A \to 0$$

b) An integer is odd if its rightmost entry contains any one of '1', '3', '5', '7', or '9'. If the integer is positive, then its leftmost entry contains a ' + ' sign; if negative, then its leftmost entry contains a '–' sign. The production rules are

S → <sign> <integer>
<integer> → <integer1> <digit1>
<integer1> → <integer1> <digit>/∈
<sign> → +/–
<digit> → 0/1/2/3/4/5/6/7/8/9
<digit1> → 1/3/5/7/9

The set of non-terminals V_N are {S, <sign>, <integer>, <integer1>, <digit>, <digit1>}.

c) An integer is odd if its rightmost entry contains any one of '2', '4', '6', '8', or '0'. If the integer is positive, then its leftmost entry contains a ' + ' sign; if negative, then its leftmost entry contains a '–' sign. The production rules are

S → <sign> <integer>
<integer> → <integer1> <digit1>

<integer1> → <integer1> <digit>/∈
<sign> → +/–
<digit> → 0/1/2/3/4/5/6/7/8/9
<digit1> → 0/2/4/6/8

The set of non-terminals V_N are {S, <sign>, <integer>, <integer1>, <digit>, <digit1>}.

2. Construct the string aaabbabbba from the grammar

S → aB/bA
A → a/aS/bAA
B → b/bS/aBB

By using

a) Leftmost derivation
b) Rightmost derivation.

Solution:

a) ***Leftmost Derivation***

S → aB
→ aaBB [B → aBB]
→ aaaBBB [B → aBB]
→ aaabBB [B → b]
→ aaabbB [B → b]
→ aaabbaBB [B → aBB]
→ aaabbabB [B → b]
→ aaabbabbS [B → bS]
→ aaabbabbbA [S → bA]
→ aaabbabbba [A → a]

b) ***Rightmost Derivation***

S → aB
→ aaBB [B → aBB]
→ aaBbS [B → bS]
→ aaBbbA [S → bA]
→ aaBbba [A → a]
→ aaaBBbba [B → aBB]
→ aaaBbbba [B → b]
→ aaabSbbba [B → bS]
→ aaabbAbbba [S → bA]
→ aaabbabbba [A → a]

3. Show the derivation tree for the string ‘aabbbb’ with the following grammar.

S → AB/∈
A → aB
B → Sb [UPTU 2003]

Solution:

S → AB	
→ aBB	[A → aB]
→ aSbB	[B → Sb]
→ aABbB	[S → AB]
→ aaBBbB	[A → aB]
→ aaSbSbbSb	[B → Sb]
→ aabbbb	[S → ∈]

4. Let G be the grammar S → aB/ba, A → a/aS/bAA, B → b/bS/aBB. For the string aaabbabbba find a
   a) leftmost derivation
   b) rightmost derivation
   c) parse tree. [WBUT 2009]

Solution:

a) ***Leftmost Derivation***

S → aB → aaBB → aaaBBB → aaabBB → aaabbSB → aaabbaBB → aaabbabB → aaabbabbS → aaabbabbba

b) ***Rightmost Derivation***

S → → aB → aaBB → aaBbS → aaBbba → aaaBBbba → aaaBbbba → aaabSbbba → aaabbabbba

c) ***Parse Tree***

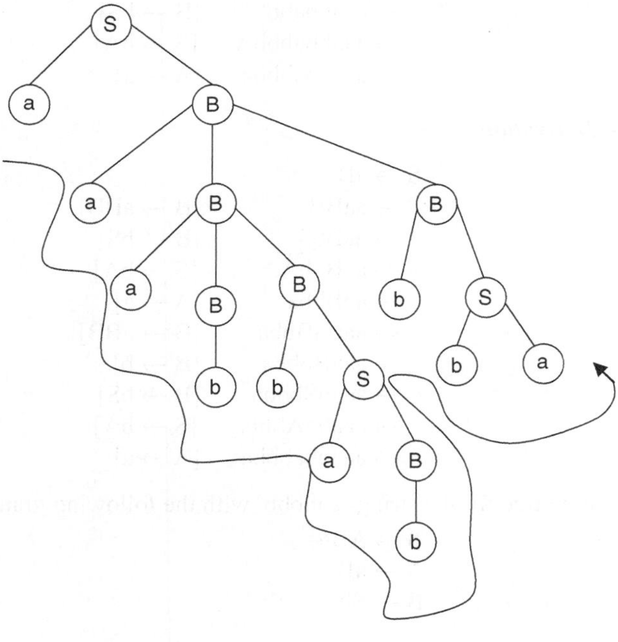

5. Check whether the following grammar is ambiguous or not.

$$S \rightarrow a/Sa/bSS/SSb/SbS$$

Solution: Let us take a string baababaa. Let us try to generate the string by deriving from the given CFG.

Derivation 1: The string is derived by the leftmost derivation, i.e., only the leftmost non-terminal is replaced by a suitable non-terminal.

S → bSS → baS → baSbS → baabS → baabSbS → baababS → baababSa → baababaa

Derivation 2: The string is derived by the rightmost derivation, i.e., only the rightmost non-terminal is replaced by a suitable non-terminal.

S → SbS → SbSa → Sbaa → bSSbaa → bSSbSbaa → bSSbabaa → bSababaa → baababaa

The parse tree for the first derivation is as follows:

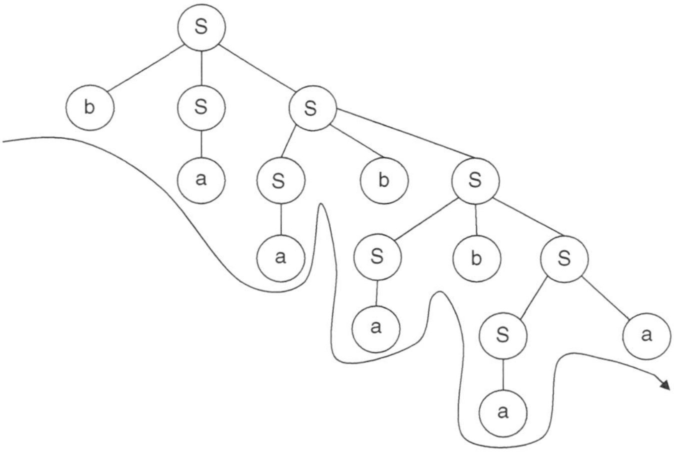

The parse tree for the second derivation is as follows:

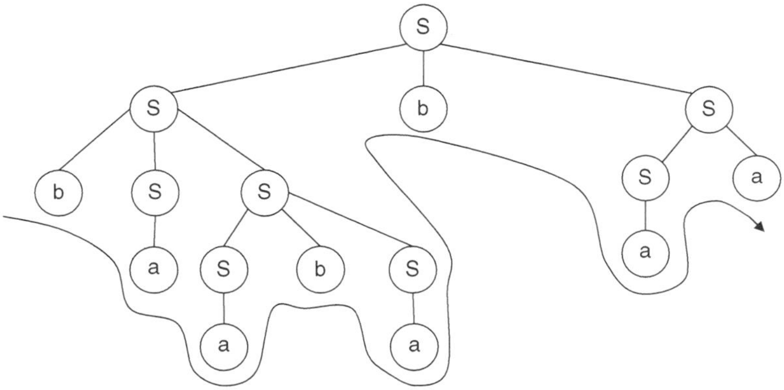

For the string baababaa, two different parse trees are constructed. So, the given CFG is ambiguous.

6. Check whether the following grammar is ambiguous or not.

$$S \rightarrow SS/a/b$$

Solution: Let us take a string abba. Let us try to generate the string by deriving from the given CFG.

Derivation 1: The string is derived by the leftmost derivation, i.e., only the leftmost non-terminal is replaced by a suitable non-terminal.

$$S \rightarrow SS \rightarrow Sa \rightarrow SSa \rightarrow Sba \rightarrow SSba \rightarrow Sbba \rightarrow abba$$

Derivation 2: The string is derived by the leftmost derivation, i.e., only the leftmost non-terminal is replaced by a suitable non-terminal.

$$S \rightarrow SS \rightarrow aS \rightarrow aSS \rightarrow abS \rightarrow abSS \rightarrow abbS \rightarrow abba$$

The parse tree for the first derivation is as follows:

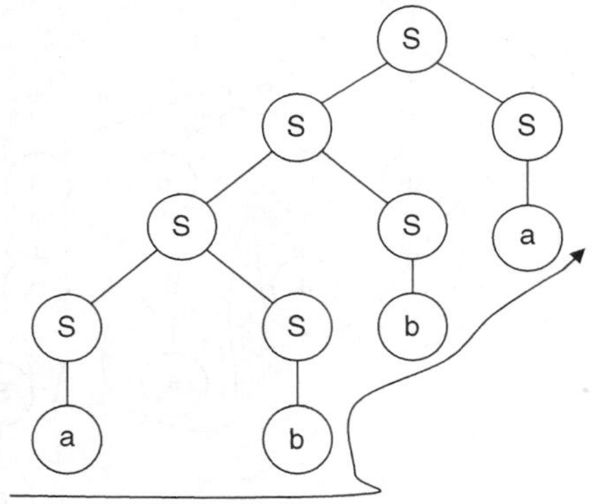

The parse tree for the second derivation is as follows:

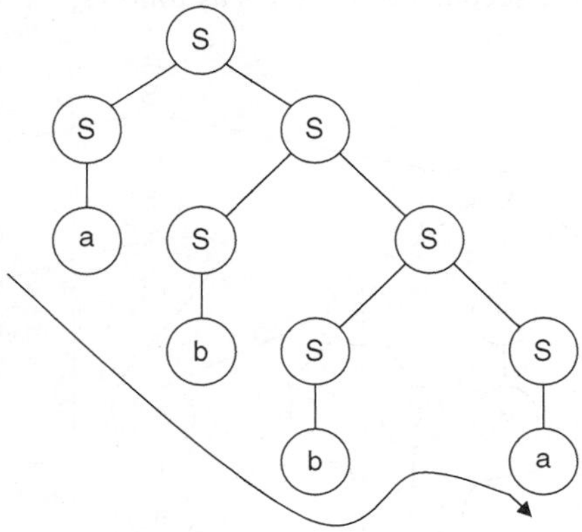

For the string abba, two different parse trees are constructed. So, the given CFG is ambiguous.

7. Consider G whose productions are $S \rightarrow aAS/a$, $A \rightarrow SbA/SS/ba$. Show that $S \rightarrow aabbaa$ by constructing a derivation tree, by rightmost derivation, whose yield is aabbaa. [WBUT 2008]

Solution: S is the start symbol of the grammar

$$
\begin{aligned}
S &\rightarrow aAS \\
&\rightarrow aAa \; [S \rightarrow a] \\
&\rightarrow aSbAa \; [A \rightarrow SbA] \\
&\rightarrow aSbbaa \; [A \rightarrow ba] \\
&\rightarrow aabbaa \; [S \rightarrow a]
\end{aligned}
$$

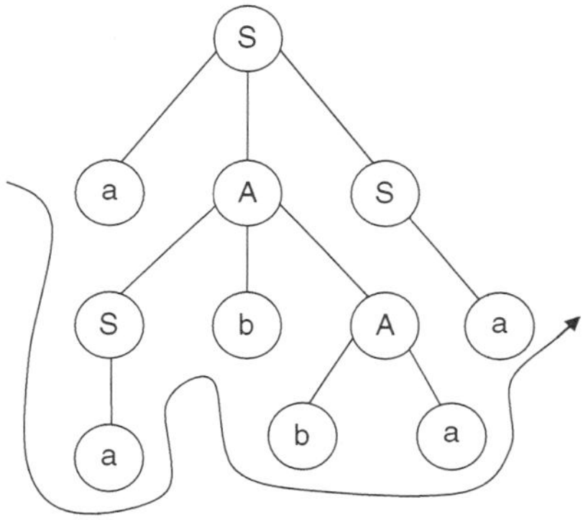

8. Show that the grammar G with production

$$
\begin{aligned}
S &\rightarrow a/aAb/abSb \\
A &\rightarrow aAAb/bS
\end{aligned}
$$

is ambiguous. [JNTU 2007]

Solution: Consider a string abababb. The language can be constructed in two ways.

i) $S \rightarrow abSb \rightarrow abaAbb \rightarrow ababSbb \rightarrow abababb$
ii) $S \rightarrow aAb \rightarrow abSb \rightarrow ababSbb \rightarrow abababb$

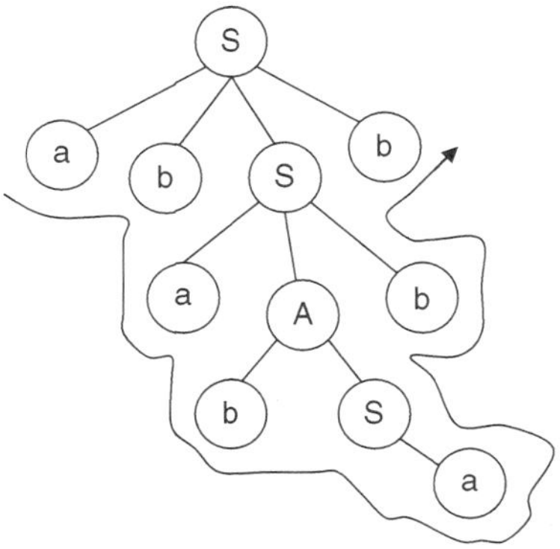

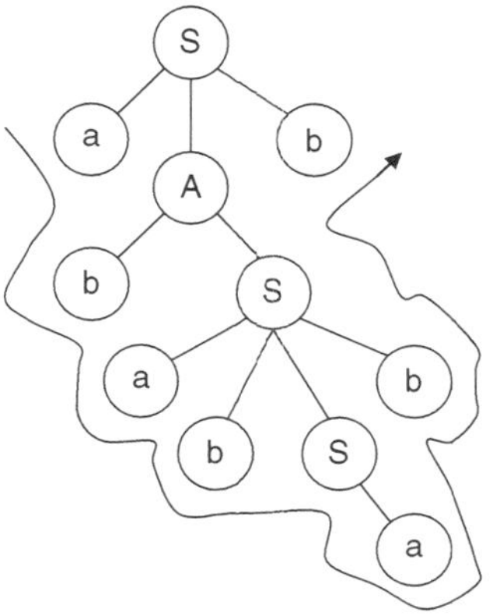

As there are two parse trees for generating a single string, we can say that the grammar is ambiguous.

9. Check whether the following grammar is ambiguous.

S → iCtS/iCtSeS/a
C → b [WBUT 2009]

Solution: Let us take a language ibtibtaea.
The language can be constructed in two ways.

i) S → iCtS → iCtiCtSeS → ibtibtaea (replacing all 'C' by b and all 'S' by a)
ii) S → iCtSeS → iCtiCtSeS → ibtibtaea (replacing all 'C' by b and all 'S' by a)

So, for generating the language, two parse trees can be generated. The parse trees are as follows:

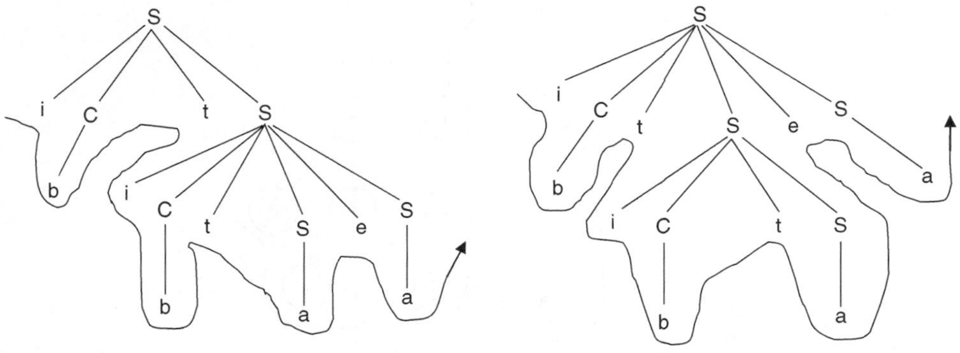

As there are two parse trees for generating a single string, we can say that the grammar is ambiguous.

10. Simplify the following CFG.

S → AaB/aaB
A → D
B → bbA/∈
D → E
E → F
F → aS

Solution: Simplifying a CFG means

a) Removing useless symbols.
b) Removing unit production.
c) Removing null production.

Simplify the CFG by the following steps.

Remove Null Production: In the grammar, there is a null production B → ∈. The grammar does not produce any null string. So, the null production can be removed. To remove the null productions, find all the productions whose right side contains a non-terminal B. There are two productions, S → AbB and S → aaB. To remove the null production, replace each occurrence of B by ∈ and add that new production to the grammar to keep the language generating power the same.

The grammar after removing the null production becomes

$$S \rightarrow AaB/aaB/Aa/aa$$
$$A \rightarrow D$$
$$B \rightarrow bbA$$
$$D \rightarrow E$$
$$E \rightarrow F$$
$$F \rightarrow aS$$

Remove Unit Production: In the grammar, there are three unit productions, A → D, D → E, and E → F. To remove the unit production E → F, we have to find a non-unit production F → {Some string of terminal or non-terminal or both}. There is such a production F → aS. Remove E → by that non-unit production F → aS.

After removing the unit production E → F, the grammar becomes

$$S \rightarrow AaB/aaB/Aa/aa$$
$$A \rightarrow D$$
$$B \rightarrow bbA$$
$$D \rightarrow E$$
$$E \rightarrow aS$$
$$F \rightarrow aS$$

By the same process, after removing all the unit productions, the grammar becomes

$$S \rightarrow AaB/aaB/Aa/aa$$
$$A \rightarrow aS$$
$$B \rightarrow bbA$$
$$D \rightarrow aS$$
$$E \rightarrow aS$$
$$F \rightarrow aS$$

Remove Useless Symbols: In a grammar, there are two types of useless symbols.

1. Non-generating symbol
2. Non-reachable symbol

In the grammar, all are generating symbols, and so we have to find the non-reachable symbol.

In the grammar D, E, and F are non-reachable symbols because they are not reached any way starting from the start symbol. After removing the useless symbols, the grammar becomes

$$S \rightarrow AaB/aaB/Aa/aa$$
$$A \rightarrow aS$$
$$B \rightarrow bbA$$

11. Find a reduced grammar equivalent to the grammar G, whose productions are the following:

$$S \rightarrow AB/CA$$
$$B \rightarrow BC/AB$$
$$A \rightarrow a$$
$$C \rightarrow aB/b$$

[UPTU 2004]

Solution: Reduction of a CFG is done by

a) Removing useless symbols.
b) Removing unit production.
c) Removing null production.

Removing useless symbol: In the production rules, there is no non-reachable symbol. But B is a non-generating symbol as it does not generate any terminal string. Removing 'B', the modified production rules become

$$S \rightarrow CA$$
$$A \rightarrow a$$
$$C \rightarrow b$$

In the modified production rules, there is no unit production or null production. Thus, it is the reduced grammar equivalent to G.

12. S → AB, A → a, B → C/b, C → D, D → E, E → a. Remove the unit productions. [WBUT 2008]

Solution: In the previous grammar, B → C, C → D, and D → E are unit productions.

To remove the unit production D → E, find all the productions whose right hand side contains a terminal symbol. There is a production E → a, and so the D production becomes D → a. The modified production rules become

$$S \rightarrow AB$$
$$A \rightarrow a$$
$$B \rightarrow C/b$$
$$C \rightarrow D$$
$$D \rightarrow a$$
$$E \rightarrow a$$

To remove the unit production C → D, we need to find a production whose right hand side contains a terminal symbol. There is a production D → a, and so the C production become C → a. The modified production rules become

$$S \rightarrow AB$$
$$A \rightarrow a$$
$$B \rightarrow C/b$$
$$C \rightarrow a$$
$$D \rightarrow a$$
$$E \rightarrow a$$

By the same process, the B production becomes B → a. The modified productions with no unit productions become

$$S \rightarrow AB$$
$$A \rightarrow a$$
$$B \rightarrow a/b$$
$$C \rightarrow a$$
$$D \rightarrow a$$
$$E \rightarrow a$$

13. Remove all the unit productions and all the $\in$ productions from the following CFG.

$$S \rightarrow aA/aBB$$
$$A \rightarrow aaA/\in$$
$$B \rightarrow bB/bbC$$
$$C \rightarrow B$$

[Andhra University 2005]

Solution: The production $A \rightarrow \in$ is a $\in$ production. The language set produced by the grammar does not contain any null string. Therefore, this $\in$ production can be removed. For removing the $\in$ production, we have to find all the productions whose right hand side contains A. The productions are $S \rightarrow aA$ and $A \rightarrow aaA$. Replacing A by $\in$, the modified productions become

$$S \rightarrow aA/aBB/a$$
$$A \rightarrow aaA/aa$$
$$B \rightarrow bB/bbC$$
$$C \rightarrow B$$

14. Find a reduced grammar equivalent to the following grammar.

$$S \rightarrow aAa$$
$$A \rightarrow bBB$$
$$B \rightarrow ab$$
$$C \rightarrow aB$$

[WBUT 2009]

Solution: The original grammar is

$$S \rightarrow aAa$$
$$A \rightarrow bBB$$
$$B \rightarrow ab$$
$$C \rightarrow aB$$

First, we have to remove useless symbols. Here, C is non-reachable symbol. So, the production $C \rightarrow aB$ is removed.

After removal of the non-reachable symbol, the modified production rules become

$$S \rightarrow aAa$$
$$A \rightarrow bBB$$
$$B \rightarrow ab$$

Here, the symbol A is not producing any terminal string, and so it is the non-generating symbol. But the symbol cannot be removed as it is attached with the starting symbol production

$$S \rightarrow aAa$$

So, the reduced grammar is

$$S \rightarrow aAa$$
$$A \rightarrow bBB$$
$$B \rightarrow ab$$

15. Convert the following grammar into CNF.

$$S \rightarrow aA/bB$$
$$A \rightarrow aBB/bS/b$$
$$B \rightarrow bAA/aS/a$$

Solution: A CFG is called a CNF if all the productions of the grammar are in the following form:

Non-terminal → string of exactly two non-terminals
Non-terminal → single terminal

The previous grammar does not contain any useless symbol, unit production, or null production. So, the grammar can be directly converted to CNF.

In the previous grammar, S → bA, S → aB, B → bAA, B → aS, A → aBB, and B → bS are not in CNF. So, these productions are needed to be converted into CNF. But the productions A → b and B → a are in CNF.

Consider two extra non-terminals C_a and C_b and two production rules $C_a \rightarrow a$ and $C_b \rightarrow b$. Replace 'a' by C_a and 'b' by C_b in the previous productions. The modified production rule becomes

$$S \rightarrow C_aA/C_bB$$
$$A \rightarrow C_aBB/C_bS/b$$
$$B \rightarrow C_bAA/C_aS/a$$
$$C_a \rightarrow a$$
$$C_b \rightarrow b$$

In the grammar, $A \rightarrow C_aBB$ and $B \rightarrow C_bAA$ are not in CNF as they contain a string of three non-terminals at the right hand side of the production. Introduce two production rules X → BB and Y → AA and replace BB by X and AA by Y. The modified production rule becomes

$$S \rightarrow C_aA/C_bB$$
$$A \rightarrow C_aX/C_bS/b$$
$$B \rightarrow C_bY/C_aS/a$$
$$C_a \rightarrow a$$
$$C_b \rightarrow b$$
$$X \rightarrow BB$$
$$Y \rightarrow AA$$

Here, all the productions are in the specified format of CNF. The CFG is converted to CNF.

16. Convert the following grammar into CNF.

$$S \rightarrow aAbB$$
$$A \rightarrow aA/a$$
$$B \rightarrow bB/b$$

[Cochin University 2006]

Solution: The productions S → aAbB, A → aA, and B → bB are not in CNF. Consider two productions $C_a \rightarrow a$ and $C_b \rightarrow b$. The modified production becomes

$$S \rightarrow C_aAC_bB$$
$$A \rightarrow C_aA/a$$
$$B \rightarrow C_bB/b$$
$$C_a \rightarrow a$$
$$C_b \rightarrow b$$

Replacing C_aA by A and C_bB by B, the modified production becomes

$S \rightarrow AB$
$A \rightarrow C_aA/a$
$B \rightarrow C_bB/b$
$C_a \rightarrow a$
$C_b \rightarrow b$

Now the grammar is in CNF.

17. Convert the following grammar into CNF.

$S \rightarrow aA/B/C/a$
$A \rightarrow aB/E$
$B \rightarrow aA$
$C \rightarrow cCD$
$D \rightarrow abd$

Solution: The CFG is not simplified as it contains useless symbols and unit productions. First, we have to simplify the CFG and then it can be converted into CNF.

In the grammar, E is the useless symbol (non-generating symbol) as it does not produce any terminal symbol. So, the production $A \rightarrow E$ is removed. The modified grammar becomes

$S \rightarrow aA/B/C/a$
$A \rightarrow aB$
$B \rightarrow aA$
$C \rightarrow cCD$
$D \rightarrow abd$

There are two unit productions in the grammar. After removing the unit productions, the grammar becomes

$S \rightarrow aA/cCD/a$
$A \rightarrow aB$
$B \rightarrow aA$
$C \rightarrow cCD$
$D \rightarrow abd$

In the grammar, except $S \rightarrow a$, all other productions are not in CNF.

Consider four extra non-terminals C_a, C_b, C_c, C_d and two production rules $C_a \rightarrow a$ and $C_b \rightarrow b$, $C_c \rightarrow c$ and $C_d \rightarrow d$. Replace 'a' by C_a and 'b' by C_b, c by C_c and d by C_d in the previous productions. The modified production rule becomes

$S \rightarrow C_aA/C_cCD/a$
$A \rightarrow C_aB$
$B \rightarrow C_aA$
$C \rightarrow C_cCD$
$D \rightarrow C_aC_bC_d$
$C_a \rightarrow a$
$C_b \rightarrow b$
$C_c \rightarrow c$
$C_d \rightarrow d$

Here, $S \rightarrow C_cCD$, $C \rightarrow C_cCD$, and $D \rightarrow C_aC_bC_d$ are not in CNF. Introduce two production rules $X \rightarrow CD$ and $Y \rightarrow C_bC_d$ and replace CD by X and C_bC_d by Y. The modified production rule becomes

$$S \rightarrow C_aA/C_cX/a$$
$$A \rightarrow C_aB$$
$$B \rightarrow C_aA$$
$$C \rightarrow C_cX$$
$$D \rightarrow C_aY$$
$$C_a \rightarrow a$$
$$C_b \rightarrow b$$
$$C_c \rightarrow c$$
$$C_d \rightarrow d$$
$$X \rightarrow CD$$
$$Y \rightarrow C_bC_d$$

Here, all the productions are in the specified format of CNF. The CFG is converted to CNF.

18. Convert the following grammar into CNF.

$$E \rightarrow E + T/T$$
$$T \rightarrow (E)/a$$

Solution: The grammar contains two non-terminal symbols E and T and four terminal symbols +, (,), and a. The grammar contains a unit production $E \rightarrow T$. First, the unit production has to be removed. After removing the unit production $E \rightarrow T$, the modified grammar becomes

$$E \rightarrow E + T/(E)/a$$
$$T \rightarrow (E)/a$$

In the previous grammar, except $E \rightarrow a$ and $T \rightarrow a$ all the other productions are not in CNF. Introduce three non-terminals A, B, and C and three production rules $A \rightarrow +$, $B \rightarrow ($ and $C \rightarrow)$ and an appropriate terminal by appropriate non-terminals. The modified production rules become

$$E \rightarrow EAT/BEC/a$$
$$T \rightarrow BEC/a$$
$$A \rightarrow +$$
$$B \rightarrow ($$
$$C \rightarrow)$$

In the previous grammar, $E \rightarrow EAT$, $E \rightarrow BEC$, and $T \rightarrow BEC$ are not in CNF. Consider two non-terminals X and Y and two production rules $X \rightarrow AT$ and $Y \rightarrow EC$. The modified production rules become

$$E \rightarrow EX/BY/a$$
$$T \rightarrow BY/a$$
$$A \rightarrow +$$
$$B \rightarrow ($$
$$C \rightarrow)$$
$$X \rightarrow AT$$
$$Y \rightarrow EC$$

Here, all the productions are in the specified format of CNF. The CFG is converted to CNF.

19. Convert the following grammar into CNF.

$$S \rightarrow ABb/a$$
$$A \rightarrow aaA/B$$
$$B \rightarrow bAb/b$$

Solution: The grammar contains unit production $A \rightarrow B$. After removing the unit production, the grammar is

$$S \rightarrow ABb/a$$
$$A \rightarrow aaA/bAb/b$$
$$B \rightarrow bAb/b$$

Except $S \rightarrow a$ and $B \rightarrow b$, the productions of the grammar are not in CNF.

Consider two non-terminals C_a and C_b and two production rules $C_a \rightarrow a$ and $C_b \rightarrow b$ and replace 'a' by C_a and 'b' by C_b in the appropriate production. The modified production rules become

$$S \rightarrow ABC_b/a$$
$$A \rightarrow C_aC_aA/C_bAC_b/b$$
$$B \rightarrow C_bAC_b/b$$
$$C_a \rightarrow a$$
$$C_b \rightarrow b$$

Consider three non-terminals X, Y, and Z and three production rules $X \rightarrow BC_b$, $Y \rightarrow AC_b$, and $Z \rightarrow C_aA$. Replace the appropriate group of non-terminals in the production by an appropriate new non-terminal. The production rule becomes

$$S \rightarrow AX/a$$
$$A \rightarrow C_aZ/C_bY/b$$
$$B \rightarrow C_bY/b$$
$$C_a \rightarrow a$$
$$C_b \rightarrow b$$
$$X \rightarrow BC_b$$
$$Y \rightarrow AC_b$$
$$Z \rightarrow C_aA$$

20. Convert the following CFG into an equivalent grammar in CNF.

$$S \rightarrow aAbB$$
$$A \rightarrow abAB/aAA/a$$
$$B \rightarrow bBaA/bBB/b$$

[WBUT 2007, 2010]

Solution: In CNF, all the productions will be in the form

Non-terminal $\rightarrow$ String of exactly two non-terminals
Non-terminal $\rightarrow$ single terminal

In the previous grammar, expect $A \rightarrow a$ and $B \rightarrow b$ all the productions are not in CNF.
Consider two productions $C_a \rightarrow a$ and $C_b \rightarrow b$.
The modified grammar becomes

$$S \rightarrow C_aAC_bB$$
$$A \rightarrow C_aC_bAB/C_aAA/a$$

$B \rightarrow C_bBC_aA/C_bBB/b$
$C_a \rightarrow a$
$C_b \rightarrow b$

Replace C_aA by X_1, C_bB by X_2, C_aC_b by X_3, and AB by X_4, The modified grammar becomes

$S \rightarrow X_1X_2$
$A \rightarrow X_3X_4/X_1A/a$
$B \rightarrow X_2X_1/X_2B/b$
$C_a \rightarrow a\ C_b \rightarrow b$
$X_1 \rightarrow C_aA$
$X_2 \rightarrow C_bB$
$X_3 \rightarrow C_aC_b$
$X_4 \rightarrow AB$

The previous grammar is in CNF.

21. Convert the following grammar into CNF.

$S \rightarrow AACD$
$A \rightarrow aAb/\in$
$C \rightarrow aC/a$
$D \rightarrow aDa/bDb/\in$ [Gujrat Technical University 2010]

Solution: The grammar contains ∈ production. First, we have to remove the ∈ productions.
Removing the ∈ productions, the modified grammar becomes

$S \rightarrow AACD/ACD/CD/AAC/C/AC$
$A \rightarrow aAb/ab$
$C \rightarrow aC/a$
$D \rightarrow aDa/aa/bDb/bb.$

The grammar contains the unit production $S \rightarrow C$. Removing the unit production, the grammar becomes

$S \rightarrow AACD/ACD/CD/AAC/AC/aC/a$
$A \rightarrow aAb/ab$
$C \rightarrow aC/a$
$D \rightarrow aDa/bDb/aa/bb$

Now, the grammar can be converted to CNF.
Introduce two new productions $C_a \rightarrow a$ and $C_b \rightarrow b$. Replacing C_a and C_b in appropriate positions in the grammar, the modified grammar becomes

$S \rightarrow AACD/ACD/CD/AAC/AC/C_aC/a$
$A \rightarrow CaAC_b/C_aC_b$
$C \rightarrow C_aC/a$
$D \rightarrow C_aDC_a/C_bDC_b/C_aC_a/C_bC_b$
$C_a \rightarrow a$
$C_b \rightarrow b$

AC can be replaced by S in the 'S' productions for the cases S → ACD/AAC, and AAC can be replaced by S for the production S → AACD. Introduce three non-terminals E, F, and G for C_aA, C_aD, and C_bD, respectively. Including these changes, the modified grammar becomes

$$
\begin{aligned}
S &\rightarrow SD/SD/CD/AS/AC/C_aC/a \\
A &\rightarrow EC_b/C_aC_b \\
C &\rightarrow C_aC/a \\
D &\rightarrow FC_a/GC_b/C_aC_a/C_bC_b \\
C_a &\rightarrow a \\
C_b &\rightarrow b \\
E &\rightarrow C_aA \\
F &\rightarrow C_aD \\
G &\rightarrow C_bD
\end{aligned}
$$

Now, the grammar is in CNF.

22. Convert the following grammar into GNF.

$$
\begin{aligned}
S &\rightarrow A \\
A &\rightarrow aBa/a \\
B &\rightarrow bAb/b
\end{aligned}
$$

Solution: The grammar is not in CNF. So, it has to be converted into CNF. Introduce two non-terminals C_a, C_b and two production rules $C_a \rightarrow a$, $C_b \rightarrow b$.

$$
\begin{aligned}
S &\rightarrow A \\
A &\rightarrow C_aBC_a/a \\
B &\rightarrow C_bAC_b/b \\
C_a &\rightarrow a \\
C_b &\rightarrow b
\end{aligned}
$$

Introduce two non-terminals X and Y and two production rules $X \rightarrow BC_a$ and $Y \rightarrow AC_b$. The production rule becomes

$$
\begin{aligned}
S &\rightarrow A \\
A &\rightarrow C_aX/a \\
B &\rightarrow C_bY/b \\
C_a &\rightarrow a \\
C_b &\rightarrow b \\
X &\rightarrow BC_a \\
Y &\rightarrow AC_b
\end{aligned}
$$

Step I: In the grammar, there is no null production and no unit production. The grammar also is in CNF.

Step II: In the grammar, there are seven non-terminals S, A, B, C_a, C_b, X, and Y. Rename the non-terminals as A_1, A_2, A_3, A_4, A_5, A_6, and A_7, respectively. After renaming the non-terminals, the modified grammar will be

$$
\begin{aligned}
A_1 &\rightarrow A_2 \\
A_2 &\rightarrow A_4A_6/a
\end{aligned}
$$

$$A_3 \rightarrow A_5A_7/b$$
$$A_4 \rightarrow a$$
$$A_5 \rightarrow b$$
$$A_6 \rightarrow A_3A_4$$
$$A_7 \rightarrow A_2A_5$$

Step III: In the previous production, $A_6 \rightarrow A_3A_4$ and $A_7 \rightarrow A_2A_5$ are not in the form $A_i \rightarrow A_jV$, where $i \leq j$.

Using Lemma I, replace $A_3 \rightarrow A_5A_7/b$ in the production $A_6 \rightarrow A_3A_4$. The rule becomes

$$A_6 \rightarrow A_5A_7A_4/bA_4$$

Still $A_6 \rightarrow A_5A_7A_4$ is not in the form $A_i \rightarrow A_jV$, where $i \leq j$. Using Lemma I, replace $A_5 \rightarrow b$ in the production. The modified production is $A_6 \rightarrow bA_7A_4/bA_4$, which is in GNF.

Step IV: Using Lemma I, replace $A_2 \rightarrow A_4A_6/a$ in the production $A_7 \rightarrow A_2A_5$. The production rule becomes $A_7 \rightarrow A_4A_6A_5/aA_5$.

Still $A_7 \rightarrow A_4A_6A_5$ is not in the form $A_i \rightarrow A_jV$, where $i \leq j$. Using Lemma I, replace $A_4 \rightarrow a$ in the production.

The modified production is $A_7 \rightarrow aA_6A_5/aA_5$, which is in GNF.

Lemma II can be applied on the productions $A_2 \rightarrow A_4A_6/a$ and $A_3 \rightarrow A_5A_7/b$.

Applying Lemma II on $A_2 \rightarrow A_4A_6/a$, we get

$$A_2 \rightarrow a/aX$$
$$X \rightarrow A_6/A_6X$$

$X \rightarrow A_6/A_6X$ are not in GNF. Replacing $A_6 \rightarrow bA_7A_4/bA_4$ in the production, the productions $X \rightarrow bA_7A_4/bA_7A_4X/bA_4/bA_4X$ are in GNF.

Applying Lemma II on $A_3 \rightarrow A_5A_7/b$, we get

$$A_3 \rightarrow b/bY$$
$$Y \rightarrow A_7/A_7Y$$

$Y \rightarrow A_7/A_7Y$ are not in GNF. Replacing $A_7 \rightarrow aA_6A_5/aA_5$ in the production, the productions $Y \rightarrow aA_6A_5/aA_6A_5Y/aA_5/aA_5Y$ are in GNF.

$A_1 \rightarrow A_2$ is not in GNF. Replacing $A_2 \rightarrow a/aX$ in the production, we get $A_1 \rightarrow a/aX$, which is in GNF.

The grammar converted into GNF is

$$A_1 \rightarrow aX/a$$
$$A_2 \rightarrow aX/a$$
$$A_3 \rightarrow bY/b$$
$$A_4 \rightarrow a$$
$$A_5 \rightarrow b$$
$$A_6 \rightarrow bA_7A_4/bA_4$$
$$A_7 \rightarrow aA_6A_5/aA_5$$
$$X \rightarrow bA_7A_4/bA_7A_4X/bA_4/bA_4X$$
$$Y \rightarrow aA_6A_5/aA_6A_5Y/aA_5/aA_5Y$$

23. Convert the following grammar to GNF.

i) $S \rightarrow aSa/aSb/\in$

ii) $S \rightarrow aSB/aSbS/\in$ [WBUT 2010]

Solution: *(It can be done by Lemma I and II, but it is done here in a simpler way.)*

i) $S \rightarrow aSa/aSb/\in$

Introduce two productions $C_a \rightarrow a$ and $C_b \rightarrow b$, and the modified productions are

$$S \rightarrow aSC_a/aSC_b/\in$$
$$C_a \rightarrow a$$
$$C_b \rightarrow b$$

ii) $S \rightarrow aSB/aSbS/\in$

Introduce a production $C_b \rightarrow b$, and the modified productions are

$$S \rightarrow aSB/aSC_bS/\in$$
$$C_b \rightarrow b$$

24. Convert the following grammar into GNF.

$$A_1 \rightarrow A_2A_3$$
$$A_2 \rightarrow A_3A_1/b$$
$$A_3 \rightarrow A_1A_2/a$$

[Cochin University 2009]

Solution: In the grammar, the production $A_3 \rightarrow A_1A_2$ is not in the form $A_i \rightarrow A_jV$ where $i \leq j$. Using Lemma 1, replace $A_1 \rightarrow A_2A_3$ in the production $A_3 \rightarrow A_1A_2$. The rule becomes

$$A_3 \rightarrow A_2A_3A_2/a$$

Still it is not in the form $A_i \rightarrow A_jV$, where $i \leq j$. Replacing $A_2 \rightarrow A_3A_1/b$ in $A_3 \rightarrow A_2A_3A_2/a$, we get $A_3 \rightarrow A_3A_1A_3A_2/bA_3A_2/a$.

Now, the modified grammar is

$$A_1 \rightarrow A_2A_3$$
$$A_2 \rightarrow A_3A_1/b$$
$$A_3 \rightarrow A_3A_1A_3A_2/bA_3A_2/a$$

The production $A_3 \rightarrow bA_3A_2/a$ is in the format $A \rightarrow \beta_i$, and the production $A_3 \rightarrow A_3A_1A_3A_2$ is in the format of $A \rightarrow A\alpha_j$. So, we can introduce a new non-terminal B_3, and the modified A_3 production is (according to Lemma II)

$$A_3 \rightarrow bA_3A_2/a$$
$$A_3 \rightarrow bA_3A_2B_3$$
$$A_3 \rightarrow aB_3$$

And the B_3 productions will be

$$B_3 \rightarrow A_1A_3A_2$$
$$B_3 \rightarrow A_1A_3A_2B_3$$

All A_3 productions are in GNF. Replacing A_3 by all A_3 productions in A_2, we get the following productions.

$A_2 \rightarrow bA_3A_2A_1/aA_1/bA_3A_2B_3A_1/aB_3A_1/b$

Now, all A_2 productions are in GNF. Replacing A_2 by all A_2 productions in A_2, we get the following productions.

$A_1 \rightarrow bA_3A_2A_1A_3/aA_1A_3/bA_3A_2B_3A_1A_3/aB_3A_1A_3/bA_3$

Now, all A_1 productions are in GNF. Replacing A_1 by all A_1 productions in two B_3 productions, we get the following productions.

$B_3 \rightarrow bA_3A_2A_1A_3A_3A_2/aA_1A_3A_3A_2/bA_3A_2B_3A_1A_3A_3A_2/aB_3A_1A_3A_3A_2/bA_3A_3A_2$
$B_3 \rightarrow bA_3A_2A_1A_3A_3A_2B_3/aA_1A_3A_3A_2B_3/bA_3A_2B_3A_1A_3A_3A_2B_3/aB_3A_1A_3A_3A_2B_3/bA_3A_3A_2B_3$

Now, the grammar becomes

$A_1 \rightarrow bA_3A_2A_1A_3/aA_1A_3/bA_3A_2B_3A_1A_3/aB_3A_1A_3/bA_3$
$A_2 \rightarrow bA_3A_2A_1/aA_1/bA_3A_2B_3A_1/aB_3A_1/b$
$A_3 \rightarrow bA_3A_2B_3/bA_3A_2/aB_3/a$
$B_3 \rightarrow bA_3A_2A_1A_3A_3A_2/aA_1A_3A_3A_2/bA_3A_2B_3A_1A_3A_3A_2/aB_3A_1A_3A_3A_2/bA_3A_3A_2$
$B_3 \rightarrow bA_3A_2A_1A_3A_3A_2B_3/aA_1A_3A_3A_2B_3/bA_3A_2B_3A_1A_3A_3A_2B_3/aB_3A_1A_3A_3A_2B_3/bA_3A_3A_2B_3$

25. Remove the left recursion from the given grammar.

$$A \rightarrow Ba/b$$
$$B \rightarrow Bc/Ad/e$$

Solution: The grammar has indirect left recursion. To remove the left recursion, rename A as A_1 and B as A_2. The modified production rules become

$$A_1 \rightarrow A_2a/b$$
$$A_2 \rightarrow A_2c/A_1d/e$$

For i = 1 and j = 1, there is no production in the form $A_1 \rightarrow A_1\alpha$.

For i = 2 and j = 1, there is a production in the form $A_2 \rightarrow A_1\alpha$. The production is $A_2 \rightarrow A_1d$. According to the algorithm for removal of indirect left recursion, the production becomes

$$A_2 \rightarrow A_2ad/bd$$

Now, the A_2 production is $A_2 \rightarrow A_2c/A_2ad/bd/e$

$A_2 \rightarrow A_2c$ has immediate left recursion. After removing the left recursion, the production becomes

$$A_2 \rightarrow bdA'_2eA'_2$$
$$A'_2 \rightarrow cA_2/\in$$

$A_2 \rightarrow A_2ad$ has immediate left recursion. After removing the left recursion, the production becomes

$$A_2 \rightarrow bdA''_2/eA''_2$$
$$A''_2 \rightarrow adA''_2/\in$$

The actual non-left recursive grammar is

$$A \rightarrow Ba/b$$
$$B \rightarrow bdB'/eB'/bdB''/eB''$$
$$B' \rightarrow cB'\in$$
$$B'' \rightarrow adB''/\in$$

26. Construct the equivalent finite automata from the following regular grammar.

$$S \rightarrow aS/bA/b$$
$$A \rightarrow aA/bS/a$$

Solution: In the grammar, there are two non-terminals S and A. So, in the finite automata, there are three states. For the production S → aS, the transitional diagram is

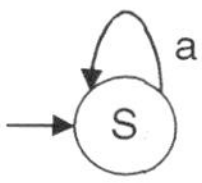

For the production S → bA, the transitional diagram is

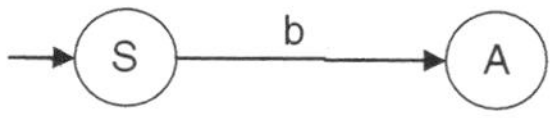

For the production S → b, the transitional diagram is

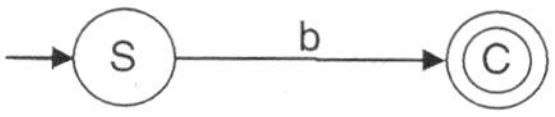

For the 'S' production, the complete transitional diagram is

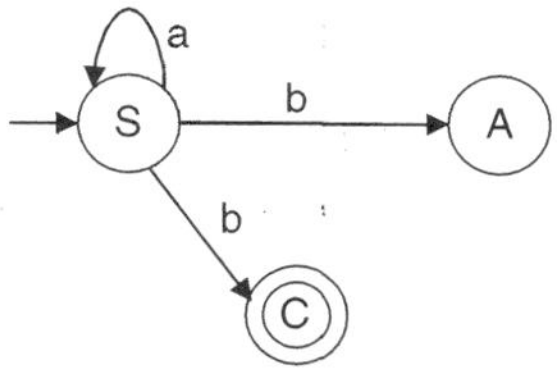

For the 'A' production, the transitional diagram is

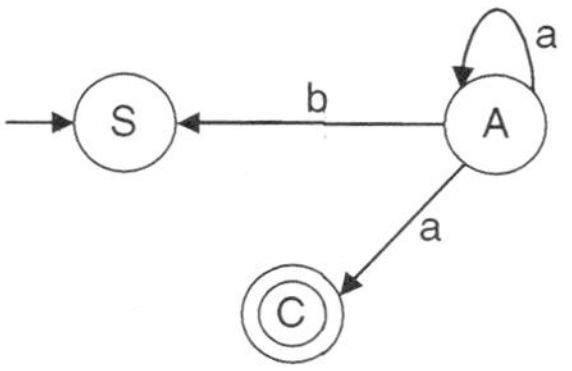

The complete transitional diagram for the previous grammar is

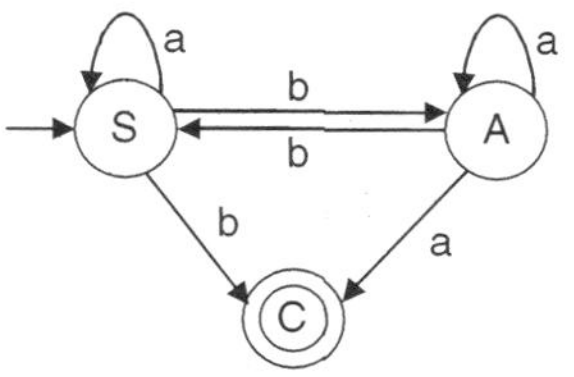

27. Construct the equivalent finite automata from the following regular grammar.

$$S \rightarrow Aa$$
$$A \rightarrow Sb/Ab/\in$$

Solution: The grammar accepts null string. So, A is the final state. For the production rule $A \rightarrow \in$, the transitional diagram is

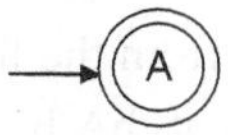

For the production $S \rightarrow Aa$, the transitional diagram is

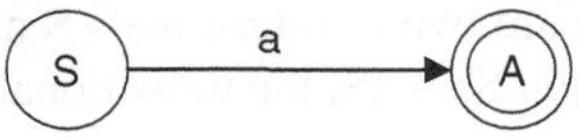

For the production $A \rightarrow Sb/Ab$, the transitional diagram is

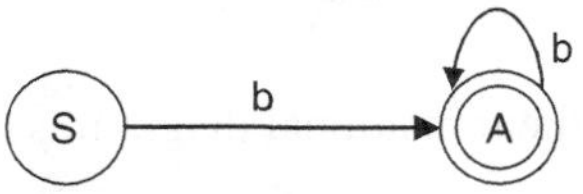

The complete transitional diagram is

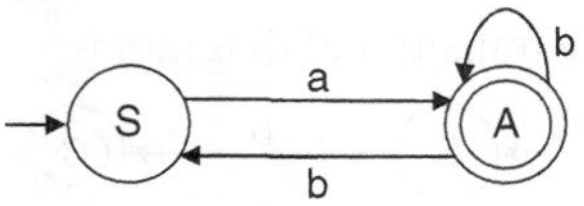

28. Using this lemma, prove that $L = \{a^i b^j \mid j = i^2\}$ is not a CFL. [WBUT 2009 (IT)]

Solution:

Step I: Assume that the language set L is a CFL. Let n be a natural number obtained by using the pumping lemma.
Step II: Let $z = a^i b^{2i}$. So, $|z| = 3i$. Let $3i > n$. According to the pumping lemma for CFL, we can write $z = uvwxy$, where $|vx| \geq 1$ and $|vwx| \leq n$.
Step III: The string z contains 'a' and 'b', and so v and x will be in any of the following forms.

i) Contain only a, i.e., in the form a^x
ii) Contain only b, i.e., in the form b^y
iii) Contain the repetition of 'a' and 'b', i.e., in the form $a^x b^y$

For case (i), if we take k = 2, then uv^2wx^2y cannot be in the form of $a^i b^j \mid j = i^2$,
For case (ii), if we take k = 2, then uv^2wx^2y cannot be in the form of $a^i b^j \mid j = i^2$,
For case (iii), if we take k = 0, then uv^2wx^2y cannot be in the form of $a^i b^j \mid j = i^2$,

For the three cases, we are getting a contradiction, and so $L = \{a^i b^j \mid j = i^2\}$ is not a CFL.

29. Verify whether the languages generated by the following grammar are finite or not. If finite, find the longest string generated by the grammar.

a) $S \rightarrow Ab$
$A \rightarrow aB/a$
$B \rightarrow bS$

b) $S \rightarrow AB$
$A \rightarrow CD$
$B \rightarrow CD$
$C \rightarrow aD$
$D \rightarrow b$

Solution:

a) The grammar is not in CNF. The grammar converted to CNF is

$$S \rightarrow AC$$
$$A \rightarrow AB/a$$
$$B \rightarrow CS$$
$$C \rightarrow b$$

In the grammar, there are four non-terminals. For this reason, in the directed graph for the grammar, there are four nodes. The directed graph is

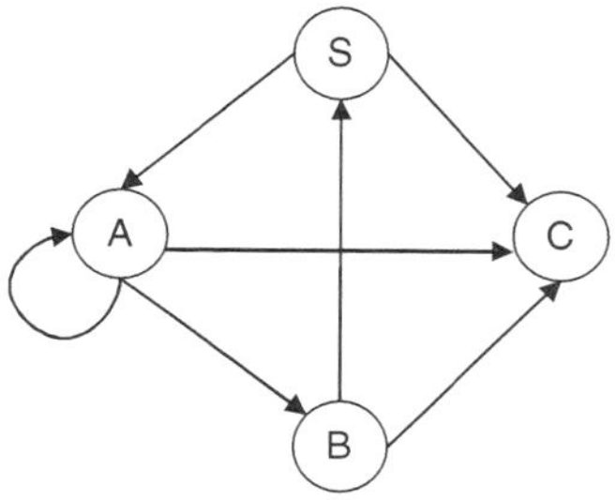

The graph contains loop. So, the language generated by the CFG is infinite.

b) The grammar is not in CNF. To make the grammar in CNF, the productions are as follows

$$S \rightarrow AB$$
$$A \rightarrow CD$$
$$B \rightarrow CD$$
$$C \rightarrow ED$$
$$D \rightarrow b$$
$$E \rightarrow a$$

In the grammar, there are six non-terminals. For this reason, in the directed graph for the grammar, there are six nodes.

The directed graph for the grammar is

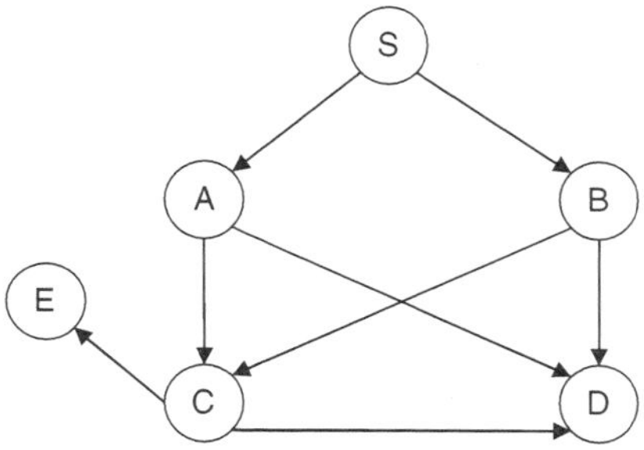

The directed graph does not contain any loop. So, the grammar is finite. The length of the string generated by the grammar is 6.

Multiple Choice Questions

1. Context-free language is ________ language.
   a) Type 0 b) Type 1
   c) Type 2 d) Type 3
2. Parsing a string from a given grammar means
   a) Finding a derivation
   b) Finding a leftmost derivation
   c) Finding a rightmost derivation
   d) Finding a derivation tree.
3. A grammar is called ambiguous if
   a) It generates more than one string
   b) It generates both leftmost and rightmost derivation for a given string
   c) It generates more than one parse tree for a given string
   d) It fulfills both (b) and (c)
4. Which is not true for ambiguous grammar?
   a) Ambiguity creates problem in generating a language from a given grammar
   b) All ambiguity can be removed.
   c) Inherent ambiguity cannot be removed
   d) Some ambiguity can be removed by hand
5. Non-generating symbols are those symbols which
   a) Do not generate any string of non-terminals
   b) Do not generate any null string
   c) Do not generate any string of terminal and non-terminals
   d) Go not generate any string of terminals
6. Useless symbols in CFG are
   a) Non-generating symbols and non-reachable symbols
   b) Null alphabets and null string
   c) Non-terminal symbols
   d) All of these
7. Which of the following is a unit production?
   a) (String of NT) → (String of NT)
   b) (Single NT) → (String of NT)
   c) (Single NT) → (Single NT)
   d) (String of NT) → (Single NT)
8. Which is true for the following CFG?

   $$S \rightarrow aA/\in$$
   $$A \rightarrow bA/a$$

   a) Null production can be removed
   b) Null production cannot be removed
   c) As A does not produce null, null cannot be removed
   d) Both (b) and (c)
9. Which of the following production is in CNF? (more specific)
   a) (NT) → (String of NT)
   b) (NT) → (String of terminal and non-terminal)
   c) (NT) → (String of terminal)
   d) (NT) → (String of exactly two NT)
10. Which of the following production is in GNF? (more specific)
    a) (NT) → (Single T)(String of NT)
    b) (NT) → (Single NT)(String of T)
    c) (NT) → (String of terminal and non-terminal)
    d) (NT) → (String of NT)
11. Which of the following is common in both CNF and GNF?
    a) (NT) → (Single T)(String of NT)
    b) (NT) → (String of exactly two NT)
    c) (NT) → (String of NT)
    d) (NT) → (Single T)
12. Context-free language is not closed under
    a) Union b) Concatenation
    c) Complementation d) Star closure

Answers:

1. c	2. a	3. c	4. b	5. d	6. a
7. c	8. b	9. d	10. a	11. d	12. c

GATE Questions

1. The intersection of CFL and RE is always
   a) CFL b) RE c) CSL d) CFL or CSL
2. Which of the following is in GNF?
   a) $A \rightarrow BC$ b) $A \rightarrow a$ c) $A \rightarrow Ba$ d) $A \rightarrow aaB$
3. Consider an ambiguous grammar G and its disambiguated version D. Let the language recognized by the two grammar be denoted by L(G) and L(D), respectively.
   Which of the following is true?
   a) $L(D) \subset L(G)$ b) $L(D) \supset L(G)$ c) $L(D) = L(G)$ d) L(D) is empty
4. Consider a CFG with the following productions

   $$S \rightarrow AA/B$$
   $$A \rightarrow 0A/A0/1$$
   $$B \rightarrow 0B00/1$$

   S is the start symbol, A and B are non-terminals, and 0 and 1 are the terminals. The language generated by the grammar is
   a) $\{0^n10^{2n} \mid n \geq 1\}$
   b) $\{0^i10^j10^k \mid i\,j, k \geq 0\} \cup \{0^n10^{2n} \mid n \geq 1\}$
   c) $\{0^i10^j \mid i, j \geq 0\} \cup \{0^n10^{2n} \mid n \geq 1\}$
   d) The set of all strings over {0, 1} containing at least two 0's
5. If G is a CFG and w is a string of length l in L(G), how long is a derivation of w in G, if G is a Chomsky normal form?
   a) 2l b) 2l + 1 c) 2l – 1 d) l
6. Context-free languages are
   a) closed under union
   b) closed under complementation
   c) closed under intersection
   d) closed under Kleene closure
7. Which of the following features cannot be captured by context-free grammar?
   a) Syntax of if-then-else statements
   b) Syntax of recursive procedures
   c) Whether a variable has been declared before use
   d) Variable names of arbitrary length
8. Let $\Sigma = \{0, 1\}$, $L = \Sigma^*$ and $R = \{0^n1^n$ such that $n > 0\}$, then the language $L \cup R$ and R are respectively.
   a) regular, regular
   b) not regular, regular
   c) regular, not regular
   d) not regular, not regular
9. The grammar whose productions are

   <Start> → if id then <stmt>
   → if id then <stmt > else <stmt>
   → id: = id

   is ambiguous because

a) the sentence
if a then if b then c : = d
b) the leftmost and rightmost derivations of the sentence
if a then if b then c: = d
give rise to two different parse trees
c) the sentence
if a then if b then c: = d else c: = f
has more than two parse trees
d) the sentence
if a then if b then c: = d else c: = f
has two parse trees

10. If L_1 is a context-free language and L_2 is a regular language, which of the following is/are false?
a) $L_1 - L_2$ is not context free
b) $L_1 \cap L_2$ is context free
c) $\sim L_1$ is context free
d) $\sim L_2$ is regular

11. A grammar that is both left and right recursive for a non-terminal, is
a) Ambiguous
b) Unambiguous
c) Information is not sufficient to decide whether it is ambiguous or unambiguous.
d) None of the above

12. Let L denote the language generated by the grammar $S \rightarrow 0S0/00$. Which of the following is true?
a) $L = 0^+$
b) L is regular but not 0+
c) L is context free but not regular
d) L is not context free

13. Consider the following two statements:

S_1: $\{0^{2n} \mid n \geq 1\}$ is a regular language

S_2: $\{0^m 1^n 0^{m+n} \mid m \geq 1 \text{ and } n \geq 1\}$ is a regular language

Which of the following statements is correct?
a) Only S_1 is correct
b) Only S_2 is correct
c) Both S_1 and S_2 are correct
d) None of S_1 and S_2 is correct

14. Consider the following languages:

$L_1 = \{ww \mid w \in \{a, b\}^*\}$

$L_2 = \{ww^R \mid w \in \{a, b\}^*$ and w^R is the reverse of w$\}$

$L_3 = \{0^{2i} \mid i$ is an integer$\}$

$L_4 = \{0^{i^{\wedge}2} \mid i$ is an integer$\}$ Which of the languages are regular?

a) Only L_1 and L_2
b) only L_2, L_3, and L_4
c) Only L_3 and L_4
d) Only L_3

15. Which of the following grammar rules violate the requirements of an operator grammar? P, Q, and R are non-terminals, and r, s, and t are terminals.

i) $P \rightarrow Q R$ ii) $P \rightarrow Q s R$ iii) $P \rightarrow \in$ iv) $P \rightarrow Q t R r$

a) (i) only
b) (i) and (iii) only
c) (ii) and (iii) only
d) (iii) and (iv) only

16. Consider the languages:

$$L_1 = \{a^n b^n c^m \mid n, m > 0\} \text{ and } L_2 = \{a^n b^m c^m \mid n, m > 0\}$$

Which one of the following statements is false?

a) $L_1 \cap L_2$ is a context-free language
b) $L_1 \cup L_2$ is a context-free language
c) L_1 and L_2 are context-free languages
d) $L_1 \cap L_2$ is a context-sensitive language

17. Consider the following statements about the CFG

$$G = \{S \to SS, S \to ab, S \to ba, S \to \in\}$$

i) G is ambiguous
ii) G produces all strings with equal number of a's and b's
iii) G can be accepted by a deterministic PDA.

Which of the following combination expresses all the true statements about G?

a) (i) only
b) (i) and (iii) only
c) (ii) and (iii) only
d) (i), (ii), and (iii)

18. Consider the grammar

$$S \to aB/bA/bAA$$
$$A \to a/aS$$
$$B \to b/bS/aBB$$

Which of the following strings is generated by the grammar?

a) aaaabb
b) aabbbb
c) aabbab
d) abbbba

19. For the correct answer of the previous question, how many derivation trees are there?

a) 1
b) 2
c) 3
d) 4

20. Which of the following statements are true?

i) Every left-recursive grammar can be converted to a right-recursive grammar and vice versa.
ii) All $\in$ productions can be removed from any context-free grammar by suitable transformations.
iii) The language generated by a CFG, all of whose productions are of the form $X \to w$ or $X \to wY$ (where w is a string of terminals and Y is a non-terminal), is always regular.
iv) The derivation trees of strings generated by a CFG in CNF are always binary trees.

a) (i), (ii), (iii), and (iv)
b) (ii), (iii), and (iv) only
c) (i), (iii) and (iv) only
d) (i), (ii) and (iv) only

21. Match the following:

E	Checking that identifiers are declared before their use	P	$L = \{a^n b^m c^n d^m \mid n \geq 1, m \geq 1\}$
F	The number of formal parameters in the declaration of a function agrees with the number of actual parameters in use of that function.	Q	$X \to XbX \mid XcX \mid dXf \mid g$
G	Arithmetic expression with matched pairs of parentheses	R	$L = \{wcw \mid w \in (a \mid b)^*\}$
H	Palindrome	S	$X \to bXb \mid cXc \mid \in$

a) E–P, F–R, G–Q, H–S
b) E–R, F–P, G–S, H–Q
c) E–R, F–P, G–Q, H–S
d) E–P, F–R, G–S, H–Q

22. Which one of the following statements is false?
    a) There exist context-free languages such that all the CFGs generating them are ambiguous.
    b) An unambiguous CFG always has a unique parse tree for each string of the language generated by it.
    c) Both deterministic and non-deterministic push down automata always accept the same set of languages.
    d) A finite set of string from one alphabet is always a regular language.

Answers:

1. a	2. b	3. a	4. c	5. c	6. a, d	7. b	8. d	9. d	10. c	11. c
12. b	13. a	14. d	15. b	16. a	17. d	18. c	19. b	20. c	21. c	22. a

Hints:

5. See the example of conversion of E → E+E/E*E/id. The derivation for id + id*id is
   E → FE → FGE → ECGE → ECEDE → idCEDE → id + EDE → id + idDE → id + id*E → id + id*id (total 9 steps).
8. ∑* is regular, and so it is a CFL. 0n1n is not regular but context free. The union of two context-free languages is context free.
12. The string is even number of '0'. The grammar can be S → 00S/00 which is regular.
15. Operator grammar is a CFG with no ∈ production where no two consecutive symbols on the right sides productions are variables.
17. Consider L = abab. It can be generated by
    i) S → SS → $\underline{S}\underline{S}$S → abSab → abab [S → ∈]
    ii) S → $\overline{S}$S → abab ---

    Thus, it is ambiguous. II and III are true.
18. S → aB → aaBB → aabB → aabbS → aabbaB → aabbab
19. Another derivation is S → aB → aaBB → aaBb → aabSb → aabbAb → aabbab
20. The language set which contains ∈ may contain a production S → ∈. This cannot be removed. Thus, II is false.
21. H is matched with S. Let there be a pseudocode
    Int i, j; //Declaration
    i = i + j; //used
    ijCij (Before C is declaration and after C is use)
    If the code is
    Int i, j; //Declaration
    k = i + j; //used
    ijCkij (Before C is declaration and after C is use)
    k is used but not declared.

Fill in the Blanks

1. According to the Chomsky hierarchy, CFG is type ____________ grammar.
2. The grammar where production rules are in the format of

   {A string consists of at least one non-terminal} → {A string of terminals and/or non-terminals} is ____________ grammar in particular.
3. The grammar S → aSb/bSb/C produces the string ____________.
4. Finding a derivation for a string from a given grammar is called ____________.
5. The tree representation of deriving a context-free language from a given context grammar is called ____________.
6. A parse tree construction is only possible for ____________ grammar.
7. The root of the parse tree of a given context-free language is represented by the ____________ of the corresponding CFG.
8. A CFG G is said to be ____________ if there exists some w ∈ L(G) that has at least two distinct parse trees.
9. A CFL L is said to be ____________ if all its grammars are ambiguous.
10. The CFG where a non-terminal 'A' as a leftmost symbol appears alternatively at the time of derivation either immediately or through some other non-terminal definitions is called ____________ grammar.
11. To avoid the problem of backtracking, we need to perform ____________.
12. In a CFG, the symbols which do not produce any terminal string is called ____________.
13. In a CFG, the symbols which cannot be reached at any time starting from the start symbol are called ____________.
14. In a CFG, non-generating symbols and non-reachable symbols are both called ____________ symbol.
15. In a CFG, the production in the form non-terminal → single non-terminal is called ____________.
16. In a CFG, a production in the form NT → ∈ is called ____________.
17. Normalizing a CFG should not hamper the ____________ power of the grammar.
18. A CFG where all the productions of the grammar are in the form

    Non-terminal → string of exactly two non-terminals

    Non-terminal → single terminal

    is called ____________ normal form.
19. A CFG where all the productions of the grammar are in the form

    Non-terminal → (single terminal)(string of non-terminals)

    Non-terminal → single terminal

    is called ____________ normal form.
20. Context-free languages are not closed under ____________ and ____________.
21. ____________ is used to prove that certain sets are not context free.

22. $a^nb^nc^n$, where $n \geq 1$, is not ______ language but ______ language.
23. If the length of the longest path of the directed graph generated from a CFG is n, then the longest string generated by the grammar is ______.
24. A language L generated from a given CFG is finite if there are no ______ in the directed graph generated from the production rules of the given CFG.

Answers:

1. Two	2. context-free	3. $WCW^R \mid W \in (a, b)^*$
4. parsing	5. parse tree	6. context-free
7. start symbol	8. ambiguous	9. inherently ambiguous
10. left recursive	11. left factoring	12. non-generating symbols
13. non-reachable symbols	14. useless	15. unit production
16. null production	17. language generating	18. Chomsky
19. Greibach	20. intersection, complementation	21. Pumping lemma for CFL
22. context free, context sensitive	23. 2^n	24. cycles

Exercise

1. Construct a CFG for the following.
   a) a^na^m, where $n > 0$ and $m = n + 1$
   b) a^nba^m, where $m, n > 0$
   c) $a^nb^nc^m$, where $n > 0$ and $m = n + 1$
   d) $L = (011 + 1)^*(01)^*$
   e) L = {Set of all integers}

2. a) Construct the string 0110001 from the grammar

   $S \rightarrow AB$
   $A \rightarrow 0A/1B/0$
   $B \rightarrow 1A/0B/1$

   By using
   i) Leftmost derivation
   ii) Rightmost derivation

   b) Construct the string baaabbba from the grammar

   $S \rightarrow AaB/AbB$
   $A \rightarrow Sa/b$
   $B \rightarrow Sb/a$

   By using
   i) Leftmost derivation
   ii) Rightmost derivation

3. a) Find the parse tree for generating the string abaabaa from the given grammar.

   $S \rightarrow aAS/a$
   $A \rightarrow bS$

b) Find the parse tree for generating the string aabbaa from the given grammar.

$S \rightarrow aAS/a$
$A \rightarrow SbA/SS/ba$

4. Show that the following grammars are ambiguous.

a) $S \rightarrow abSb/aAb/a$
$A \rightarrow bS/aAAb$

b) $E \rightarrow E+E/E*E/id$

c) $S \rightarrow aB/bA$
$A \rightarrow aS/bAA/a$
$B \rightarrow bS/aBB/b$

5. Remove the useless productions from the following grammar

a) $S \rightarrow AB/a$
$A \rightarrow b$

b) $S \rightarrow AB/AC$
$A \rightarrow 0A1/1A0/0$
$B \rightarrow 11A/00B/AB$
$C \rightarrow 01C0/0D1$
$D \rightarrow 1D/0C$

6. Remove the unit production from the following grammar:

a) $S \rightarrow SaA$
$A \rightarrow aB/B/b$
$B \rightarrow bC/C/a$
$C \rightarrow ab$

b) $S \rightarrow Aa/B$
$B \rightarrow A/bb$
$A \rightarrow a/bc/B$

7. Remove the null production from the following grammar

a) $S \rightarrow aAB$
$A \rightarrow Bb$
$B \rightarrow \in$

b) $S \rightarrow aA$
$A \rightarrow bB$
$B \rightarrow b$
$B \rightarrow \in$

8. Simplify the following CFG.

$S \rightarrow AB/aB$
$A \rightarrow BC/B/a$
$B \rightarrow C$
$C \rightarrow b/\in$

9. i) Convert the following left linear grammar into right linear grammar.

$S \rightarrow Sab/Aa$
$A \rightarrow Abb/bb$

ii) Convert the following right linear grammar into left linear grammar.

$S \rightarrow aaB/ab$
$B \rightarrow bB/bb$

10. Convert the following linear grammar into regular grammar.

$S \rightarrow 01B/0$
$B \rightarrow 1B/11$

11. Convert the following grammar into CNF.
    a) S → AB
       A → aA/a
       B → ab/bB/b
    b) S → aSa/SSa/a

12. Convert the following grammar into GNF.
    a) S → Abb/a
       A → aaA/B
       B → bAb
    b) S → aSb/bSa/a/b

13. Construct a DFA equivalent to the regular grammar.
    a) S → aS/bA/b
       A → aA/bS/a
    b) S → bA/aB
       A → bA/aS/a
       B → aB/bS/b

14. Prove that $L = a^n b^n c^{2n}$ is not context free by using the pumping lemma for CFL.

15. Verify whether the languages generated by the following grammar are finite or not.
    a) S → aA
       A → BC
       B → SC/b
       C → B/a
    b) S → AB
       A → C/a
       B → AC/b

16. Remove the left recursion from the following grammar and then perform left factoring.

    E → E + T | T
    T → T*F | F
    F → G^F | G
    G → id | (E)

17. Generate the string id + id*id from the grammar

    E → E + E
    E → E*E
    E → id

    where the precedence of operator is given as follows.

	+	*
+	>	<
*	>	>

Are you getting any ambiguity in the grammar?

Pushdown Automata

7

Introduction

Pushdown automata (in short PDA) is the machine format of the context-free language. It is the same as finite automata with the attachment of an auxiliary amount of storage as stack. Already, we have discussed the limitations of finite automata. Pushdown automata is designed to remove those limitations. Remember the example of a^nb^n, which is not recognizable by the finite automata due to the limitation of memory, but pushdown automata can memorize the number of occurrences of 'a' and match it with the number of occurrences of 'b' with the help of the stack. In this chapter, we shall learn about the mathematical representation of PDA, acceptance by a PDA, deterministic PDA, non-deterministic PDA, conversion of CFG to PDA and PDA to CFG, and the graphical representation of PDA.

7.1 Basics of PDA

7.1.1 Definition

A PDA consists of a 7-tuple

$$M = (Q, \Sigma, \Gamma, \delta, q_0, z_0, F)$$

where

Q denotes a finite set of states.
Σ denotes a finite set of input symbols.
Γ denotes a finite set of pushdown symbols or stack symbols.
δ denotes the transitional functions.
q_0 is the initial state of the PDA ($q_0 \in Q$).
z_0 is the stack bottom symbol.
F is the final state of the PDA.

In PDA, the transitional function δ is in the form

$$Q \times (\Sigma \cup \{\lambda\}) \times \Gamma \rightarrow (Q, \Gamma)$$

(The PDA is in some state. With an input symbol from the input tape and with the stack top symbol, the PDA moves to one right. It may change its state and push some symbol in the stack or pop symbol from the stack top.)

7.1.2 Mechanical Diagram of the PDA (Fig. 7.1)

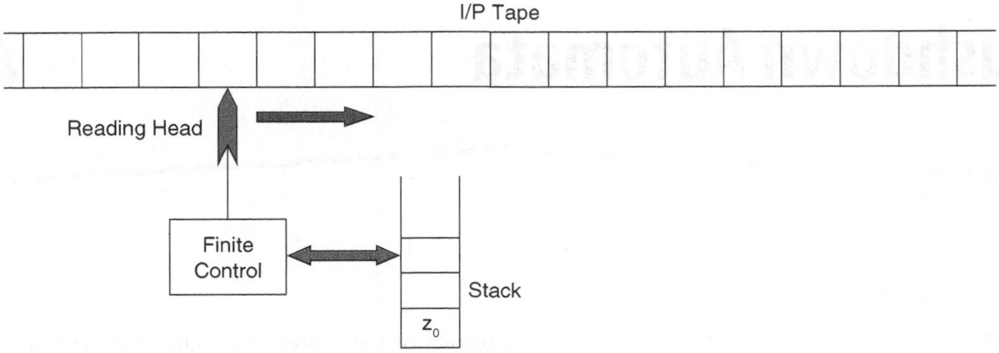

Fig. 7.1

Mechanical diagram of PDA is given in Fig. 7.1. The components of it are described below.

- **Input tape:** The input tape contains the input symbols. The tape is divided into a number of squares. Each square contains a single input character. The string placed in the input tape is traversed from left to right. The two end sides of the input string contain an infinite number of blank symbols.
- **Reading head:** The head scans each square in the input tape and reads the input from the tape. The head moves from left to right. The input scanned by the reading head is sent to the finite control of the PDA.
- **Finite control:** The finite control can be considered as a control unit of a PDA. An automaton always resides in a state. The reading head scans the input from the input tape and sends it to the finite control. A two-way head is also added with the finite control to the stack top. Depending on the input taken from the input tape and the input from the stack top, the finite control decides in which state the PDA will move and which stack symbol it will push to the stack or pop from the stack or do nothing on the stack.
- **Stack:** A stack is a temporary storage of stack symbols. Every move of the PDA indicates one of the following to the stack
  - One stack symbol may be added to the stack (push)
  - One stack symbol may be deleted from the top of the stack (pop)

  The stack is the component of the PDA which differentiates it from the finite automata. In the stack, there is always a symbol z_0 which denotes the bottom of the stack.

7.1.2.1 Why Pushdown?

Push is an operation related to the stack. By this operation, one symbol is added to the stack top.

In finite automata, the states act as a form of primitive memory. The states memorize the non-terminals encountered at the time of derivation of a string. So, only the state is suitable for traversing a regular expression as in the case of finite automata. Now, let us consider a case of L = $\{a^nb^n$, where $n \geq 1\}$.It is not a regular expression but a context-free language. Here, n is any number. In the string, there is an equal number of 'a' and 'b'. In the string, all 'a's appear before 'b'. So, the machine for traversing a^nb^n has to remember n number of a's to traverse an equal number of 'b'. n is any number. Thus, to memorize the number of a's in the string, the machine requires an infinite number of states, i.e., the machine will not be a finite state machine.

To remove this bottleneck, we need to add an auxiliary memory in the form of a stack. For each occurrence of 'a' in the string L, one symbol is pushed into the stack. For each occurrence of b (after finishing 'a'), one symbol will be popped from the stack. By this process, the matching of 'a' and 'b' will be done.

By adding a stack to the finite automata, the PDA is generated.

7.1.2.2 Instantaneous Description (ID)

It describes the configuration of the PDA at a given instance. ID remembers the information of the state and the stack content at a given instance of time.

An ID is a format of triple (q, w, κ), where q ∈ Q (finite set of states), w ∈ Σ (finite set of input alphabets), and κ ∈ Γ (finite set of stack symbols).

7.2 Acceptance by a PDA

There are two ways to declare a string to be accepted by a PDA:

1. Accepted by an empty stack (store)
2. Accepted by the final state

7.2.1 Accepted by an Empty Stack (Store)

We know in each PDA there is a stack attached. In the stack at the bottom, there is a symbol called the stack bottom symbol, say z_0. In each move of the PDA, one symbol called the stack symbol is either pushed in or popped from the stack. But the symbol z_0 still remains in the stack.

A string w may be declared accepted by an empty stack after processing all the symbols of w, if the stack is empty after reading the rightmost input character of the string w. In mathematical notation, we can say that if M = {Q, Σ, Γ, δ, q_0, z_0, F} is a PDA, the string w is declared accepted by empty stack if

$$\{w \in \Sigma^*/(q_0, w, z_0) \xrightarrow{*} (q, \lambda, \lambda) \text{ for some } q \in Q\}$$

In general, we can say that a string is accepted by a PDA by empty stack if both the following conditions are fulfilled:

1. The string is finished (totally traversed)
2. The stack is empty

7.2.2 Accepted by the Final State

A string w may be declared accepted by the final state if after total traversal of the string w the PDA enters into its final state. In mathematical notation, we can say that if M = {Q, Σ, Γ, δ, q_0, z_0, F} is a PDA, the string w is declared accepted by the final state if

$$\{w \in \Sigma^*/(q_0, w, z_0) \xrightarrow{*} (q_f, \lambda, z_0) \text{ for } q_f \in F \text{ and } z_0 \text{ is the stack bottom symbol}\}$$

In general, we can say that a string is accepted by a PDA by the final state if both the following conditions are fulfilled:

1. The string is finished (totally traversed)
2. The PDA enters into its final state

[Although there are two ways to declare a string to be accepted by a PDA, it can be proved that both the ways are equivalent. In general, for both the cases all the steps are the same, except the last step. In the last step, it is differentiated whether the string is accepted by the empty stack or by the final state.

if in the last step $\delta(q_i, \lambda, z_0) \rightarrow (q_i, \lambda)$, then it is accepted by the empty stack
if $\delta(q_i, \lambda, z_0) \rightarrow (q_f, z_0)$, then it is accepted by the final state.]

Theorem 7.1: A language is accepted by a PDA by the empty stack if and only if the language is accepted by a PDA by the final state.

Proof: Let $M_1 = \{Q, \Sigma, \Gamma, \delta, q_0, z_0, F\}$ be a PDA. Let it accept a language L by the final state. Let there exist another PDA M_2, which accepts the same language L by the empty stack.

Let $M_2 = \{Q', \Sigma', \Gamma', \delta', q'_0, z'_0, F'\}$,

where $Q' = Q \cup \{q'_0, q_e\}$, where $q'_0, q_e \notin Q$
$\Sigma' = \Sigma$
$\Gamma' = \Gamma \cup \{z'_0\}$, where $z'_0 \notin \Gamma$

The transitional function δ' is defined as follows:

a) $\delta'(q'_0, \lambda, z'_0)$ contains $(q_0, z_0z'_0)$
b) $\delta'(q_i, a, z)$ is the same as $\delta(q_i, a, z)$ for all $q_i \in Q$, $a \in \Sigma \cup \{\lambda\}$, and $z \in \Gamma$
c) $\delta'(q_i, \lambda, z)$ contains (q_e, λ) for $q_i \in F$ and for all $z \in \Gamma \cup \{z'_0\}$
d) $\delta'(q_e, \lambda, z)$ contains (q_e, λ) for all $z \in \Gamma \cup \{z'_0\}$

[Here (a) describes that M_2 enters in the initial configuration of M_1 with z'_0 as the leftmost push down symbol. (b) describes to make M_2 simulate the behaviour of M_1, thus reading an input word w if M_1 reads w and enters a final state. Once w is read and M_2 enters a final state of M_1, the effect of (c) and (d) are to empty the pushdown store.]

Here, $w \in M_1$ if and only if

$$(q_0, w, z_0) \overset{*}{\Rightarrow} (q_f, \lambda, z_0)$$

where q_f is the final state of the machine M_1.

By rule (a), we can write

$$\begin{aligned}(q'_0, w, z'_0) &\Rightarrow (q_0, w, z'_0z_0)\\ &\overset{*}{\Rightarrow} (q_f, \lambda, z'_0z_0) \text{ — by rule (b) where } q_f \text{ is the final state}\\ &\Rightarrow (q_e, \lambda, \lambda) \text{ — by rules (c) and (d)}\end{aligned}$$

Conversely, let $M_2 = (Q', \Sigma', \Gamma', \delta', q'_0, z'_0, \phi)$ be a PDA accepting L by the empty store.

Then,

$$M_1 = \{Q, \Sigma, \Gamma, \delta, q_0, z_0, F\}$$

where $Q = Q' \cup \{q_0, q_j\}$, where $q_0, q_j \notin Q'$
$\Gamma = \Gamma' \cup \{z_0\}$, where $z_0 \notin \Gamma'$
$F = \{q_j\}$ and δ are defined as follows:

1. $\delta(q_0, \lambda, z_0)$ contains $(q'_0, z_0, z'_{0)}$.
2. $\delta(q_i, a, z)$ is the same as $\delta'(q_i, a, z)$ for all $q_i \in Q'$, $a \in \Sigma \cup \{\lambda\}$, and $z \in \Gamma'$.
3. $\delta(q_i, \lambda, z_0)$ contains (q_j, z_0) for all $q_i \in Q'$.

From these two, it can be proved that a language is accepted by a PDA by the empty stack if and only if the language is accepted by a PDA by the final state.

Example 7.1 Construct a PDA to accept a given language L by both the empty stack and the final state where $L = (a^n b^n$, where $n \geq 1)$.

Solution: The string is $a^n b^n$ where $n \geq 1$. The remarks we can draw from the string $a^n b^n$, where $n \geq 1$, are the following.

- The string consists of two alphabets 'a' and 'b'.
- All 'a' appear before 'b'.
- The number of 'a' is equal to the number of 'b'.
- There is at least one 'a' and one 'b' in the string.

We have to check for equal number of 'a' and 'b'. Let us take a stack symbol z_1, which is pushed into the stack as an 'a' is traversed from the input tape. At the beginning, the PDA is in state q_0 and the stack top is z_0 (as at the beginning there is no other symbol in the stack). In this state, it gets an input 'a' from the input tape (*the string starts with 'a'*). So, a stack symbol z_1 is pushed to the stack. The δ function is

$$\delta(q_0, a, z_0) \rightarrow (q_0, z_1 z_0)$$

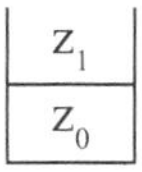

(*At the right hand side, the stack denotes the placement of the symbols. $z_1 z_0$ denotes that there was z_0 in the stack. Now, the symbol z_1 is pushed into the stack. It is given for better understanding of the placement of the stack symbols.*)

From the next input 'a' from the input tape, the PDA is in state q_0 and the stack top is z_1, so another z_1 is pushed into the stack. The δ function is

$$\delta(q_0, a, z_1) \rightarrow (q_0, z_1 z_1)$$

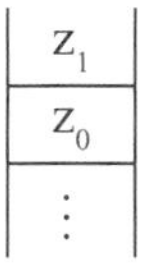

By this δ function, all the remaining 'a's are traversed. After the end of 'a' in the input tape, 'b' occurs. When the first 'b' occurs on the input tape, at that time the PDA is in state q_0 with the stack top z_1 (*Just before it, the 'a's are traversed.*). When 'b' is traversed, the stack top z_1 will be popped and the PDA will change its state to q_1. (*So that there will be no chance to push z_1 in the stack if any 'a' occurs. The PDA is designed only for string $a^n b^n$, but the input string may not be in the form of $a^n b^n$. In that case, the string will not be accepted by the PDA.*) The δ function is

$$\delta(q_0, b, z_1) \rightarrow (q_1, \lambda)$$

When the next 'b' appears, the PDA is in state q_1 with the stack top z_1. That z_1 is popped. The δ function is

$$\delta(q_1, b, z_1) \rightarrow (q_1, \lambda)$$

By this δ function, all the remaining 'b's are traversed.

If the input string is in the form of $a^n b^n$, where $n \geq 1$, then after traversing the last 'b' the PDA will be in state q_1 with the stack top z_0 *(all the other stack symbols are removed)*.

If we have to design the PDA accepted by the empty stack, the δ function will be

$$\delta(q_1, \lambda, z_0) \rightarrow (q_1, \lambda)$$

If we have to design the PDA accepted by the final state, the δ function will be

$$\delta(q_1, \lambda, z_0) \rightarrow (q_f, z_0)$$

So, the PDA for $L = (a^n b^n$, where $n \geq 1)$ is

$$Q = \{q_0, q_1, q_f\}$$
$$\Sigma = \{a, b\}$$
$$\Gamma = \{z_0, z_1\}$$
$$q_0 = \{q_0\}$$
$$z_0 = \{z_0\}$$
$$F = \{q_f\}$$

δ is defined as follows:

$$\delta(q_0, a, z_0) \rightarrow (q_0, z_1 z_0)$$
$$\delta(q_0, a, z_1) \rightarrow (q_0, z_1 z_1)$$
$$\delta(q_0, b, z_1) \rightarrow (q_1, \lambda)$$
$$\delta(q_1, b, z_1) \rightarrow (q_1, \lambda)$$
$\delta(q_1, \lambda, z_0) \rightarrow (q_1, \lambda)$ accepted by the empty stack

$\delta(q_1, \lambda, z_0) \rightarrow (q_f, z_0)$ accepted by the final state

Example 7.2 Construct a PDA to accept $L = (a, b)^*$ with equal number of 'a' and 'b', i.e., $n_a(L) = n_b(L)$ by the empty stack and the final state.

Solution: The string is any combination of 'a' and 'b' including null and the number of 'a' is equal to the number of 'b'. In the language set, a string may start with 'a' or a string may start with 'b'. 'a' can come after 'a' or 'b' can come after 'a', and similarly 'a' can come after 'b' or 'b' can come after 'b'. If a string starts with 'a', then push a stack symbol z_1 in the stack. If a string starts with 'b', then push a stack symbol z_2 in the stack. For these two cases, the PDA is in state q_0. The δ function for these two cases are

$$\delta(q_0, a, z_0) \rightarrow (q_0, z_1 z_0)$$

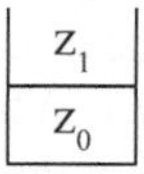

$$\delta(q_0, b, z_0) \rightarrow (q_0, z_2 z_0)$$

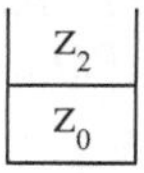

If the string starts with 'a' and 'a' comes after 'a', then the PDA is in state q_0 with the stack top z_1. Another z_1 will be pushed to the stack. The δ function is

$$\delta(q_0, a, z_1) \rightarrow (q_0, z_1z_1)$$

If the string starts with 'a' and 'b' comes after 'a', then one match of 'a' and 'b' is got and the stack top, i.e., z_1 will be popped from the stack. The δ function is

$$\delta(q_0, b, z_1) \rightarrow (q_0, \lambda)$$

If the string starts with 'b' and 'b' comes after 'b', then the PDA is in state q_0 with the stack top z_2. Another z_2 will be pushed to the stack. The δ function is

$$\delta(q_0, b, z_2) \rightarrow (q_0, z_2z_2)$$

If the string starts with 'b' and 'a' comes after 'b', then one match of 'a' and 'b' is got and the stack top, i.e., z_2 will be popped from the stack. The δ function is

$$\delta(q_0, a, z_2) \rightarrow (q_0, \lambda)$$

By these δ functions, all the input symbols of the input tape will be traversed. The PDA will be in state q_0 with the stack top z_0.

If we have to design the PDA accepted by the empty stack, the δ function will be

$$\delta(q_0, \lambda, z_0) \rightarrow (q_0, \lambda)$$

If we have to design the PDA accepted by the final state, the δ function will be

$$\delta(q_0, \lambda, z_0) \rightarrow (q_f, z_0)$$

(In the language set, a null string can occur. If a null string occurs in the language set, then it will also be accepted by the last two δ functions directly.)

So, the PDA for L = (a, b)* with equal number of 'a' and 'b', i.e., $n_a(L) = n_b(L)$ is

$$Q = \{q_0, q_f\}$$
$$\Sigma = \{a, b\}$$
$$\Gamma = \{z_0, z_1, z_2\}$$
$$q_0 = \{q_0\}$$
$$z_0 = \{z_0\}$$
$$F = \{q_f\}$$

δ is defined as follows:

$$\delta(q_0, a, z_0) \rightarrow (q_0, z_1z_0)$$
$$\delta(q_0, b, z_0) \rightarrow (q_0, z_2z_0)$$
$$\delta(q_0, a, z_1) \rightarrow (q_0, z_1z_1)$$
$$\delta(q_0, b, z_1) \rightarrow (q_0, \lambda)$$
$$\delta(q_0, b, z_2) \rightarrow (q_0, z_2z_2)$$
$$\delta(q_0, a, z_2) \rightarrow (q_0, \lambda)$$
$\delta(q_0, \lambda, z_0) \rightarrow (q_0, \lambda)$ accepted by the empty stack
$\delta(q_0, \lambda, z_0) \rightarrow (q_f, z_0)$ accepted by the final state

Example 7.3 Construct a PDA to accept $L = (a, b)^+$ with an equal number of 'a' and 'b', i.e., $n_a(L) = n_b(L)$ by the empty stack and the final state.

Solution: In the string $(a, b)^+$, there will be at least one 'a' and one 'b' with an equal number of 'a, and 'b'. So, the PDA will be made such that the null string will not be accepted. All the conditions are the same like $(a, b)^*$, but to avoid null occurrence the transitions will be like this.

On getting an 'a' at the beginning, change the state from q_0 to q_1 and push z_1 in the stack. From the next input, all the transitions will be on state q_1.

In a similar way, if the string starts with 'b', change the state from q_0 to q_1 and push z_2 in the stack. From the next input, all the transitions will be on state q_1. Changing the state to q_1 removes the possibility of accepting null string. There may come a situation when z_0 will be stacktop but the string is not totally traversed (example: abba). Following two extra δ functions are added to push z_1 or z_2 in the stack.

$$\delta(q_1, a, z_0) \rightarrow (q_1 , z_1z_0)$$
$$\delta(q_1, b, z_0) \rightarrow (q_1 , z_2z_0)$$

Other transitional functions (δ) will be

$$\delta(q_0, a, z_0) \rightarrow (q_1, z_1z_0)$$
$$\delta(q_0, b, z_0) \rightarrow (q_1, z_2z_0)$$
$$\delta(q_1, a, z_1) \rightarrow (q_1, z_1z_1)$$
$$\delta(q_1, b, z_1) \rightarrow (q_1, \lambda)$$
$$\delta(q_1, b, z_2) \rightarrow (q_1, z_2z_2)$$
$$\delta(q_1, a, z_2) \rightarrow (q_1, \lambda)$$
$$\delta(q_1, \lambda, z_0) \rightarrow (q_1, \lambda) \text{ accepted by the empty stack}$$
$$\delta(q_1, \lambda, z_0) \rightarrow (q_f, z_0) \text{ accepted by the final state}$$

[If in the language set there is a null string, there exists no transitional function in the PDA, because the null string will only be traversed if the machine is in state q_1. For coming to q_1, it needs to traverse at least one 'a' or one 'b'. It is clear from the discussion that the PDA will not accept null string, i.e., it is for L = $(a, b)^+$ with an equal number of 'a' and 'b'.]

Example 7.4 Construct a PDA to accept the language L = {WCW^R, where W ∈ $(a, b)^+$ and W^R is the reverse of W} by the empty stack and by the final state.

Solution: W is a string which consists of any combination of 'a' and 'b' not including null. The string W may start with 'a' or may start with 'b'. In the string, 'a' may come after 'a' (for example, aa) or 'b' may come after 'a' (for example, ab). Or, 'a' may come after 'b' (ba) or 'b' may come after 'b' (for example, bb).

W^R is the reverse of W. It means that in W^R if 'a' occurs at the ith place from the beginning, then 'a' will occur at the ith place from the end in W. C is the marker which denotes the end of W and the beginning of W^R.

The PDA will be designed as follows.

The PDA is in state q_0 with the stack top z_0. In the string W, if 'a' is traversed as the start symbol, a stack symbol z_1 will be pushed and the PDA will change its state to q_1 (to avoid accepting the null string). If 'b' is traversed a stack symbol z_2 will be pushed, the PDA will change its state to q_1 (to avoid accepting the null string). The δ function will be

$$\delta(q_0, a, z_0) \rightarrow (q_1, z_1z_0)$$

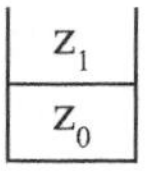

$$\delta(q_0, b, z_0) \rightarrow (q_1, z_2z_0)$$

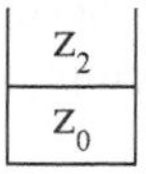

If in W, 'a' occurs after 'a', then the state is q_1 and the stack top is z_1. One z_1 will be pushed to the stack. If 'b' occurs after 'a', then the state is q_1 and the stack top is z_1. One z_2 will be pushed to the stack. The δ function will be

$$\delta(q_1, a, z_1) \rightarrow (q_1, z_1 z_1) \quad \begin{array}{|c|} \hline z_1 \\ \hline z_1 \\ \hline \vdots \\ \end{array}$$

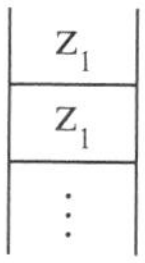

$$\delta(q_1, b, z_1) \rightarrow (q_1, z_2 z_1) \quad \begin{array}{|c|} \hline z_2 \\ \hline z_1 \\ \hline \vdots \\ \end{array}$$

If in W, 'a' occurs after 'b', then the state is q_1 and the stack top is z_2. One z_1 will be pushed to the stack. If 'b' occurs after 'b', then the state is q_1 and the stack top is z_2. One z_2 will be pushed to the stack. The δ function will be

$$\delta(q_1, a, z_2) \rightarrow (q_1, z_1 z_2) \quad \begin{array}{|c|} \hline z_1 \\ \hline z_2 \\ \hline \vdots \\ \end{array}$$

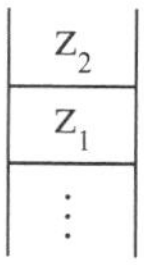

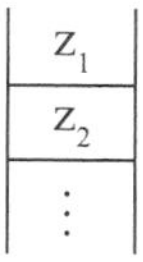

$$\delta(q_1, b, z_2) \rightarrow (q_1, z_2 z_2) \quad \begin{array}{|c|} \hline z_2 \\ \hline z_2 \\ \hline \vdots \\ \end{array}$$

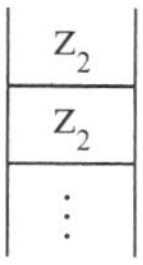

By the already given transitional functions, the string W will be traversed.

At the end of W, the input head will come to the symbol C to traverse it. C indicates that the string W is finished and W^R will start. Before traversing C, either 'a' or 'b' is traversed. When C is going to be traversed, at the stack top there will be either z_1 (if just before 'a' is traversed) or z_2 (if just before 'b' is traversed). Upon traversing C only there will be a state change from q_1 to q_2, but no operation will be done on the stack. The δ function will be

$$\delta(q_1, C, z_1) \rightarrow (q_2, z_1)$$
$$\delta(q_1, C, z_2) \rightarrow (q_2, z_2)$$

After traversing C, the reverse string of W, i.e., W^R will come. If 'a' comes at the ith place from the beginning, then 'a' will come at the ith place from the end in W. If for the string W^R as an input 'a' comes, the stack top will be z_1 and the state will be q_2. If for the string W^R as an input 'b' comes, the stack top will be z_2 and the state will be q_2. This stack top will be popped from the stack. The δ function will be

$$\delta(q_2, a, z_1) \rightarrow (q_2, \lambda)$$
$$\delta(q_2, b, z_2) \rightarrow (q_2, \lambda)$$

By these δ functions, all the input symbols of the input tape will be traversed. The PDA will be in state q_2 with the stack top z_0.

If we have to design the PDA accepted by the empty stack, the δ function will be

$$\delta(q_2, \lambda, z_0) \rightarrow (q_2, \lambda)$$

If we have to design the PDA accepted by final state, the δ function will be

$$\delta(q_2, \lambda, z_0) \rightarrow (q_f, z_0)$$

The PDA for L = {WCW^R, where W ∈ $(a, b)^+$ and W^R is the reverse of W} is

$$Q = \{q_0, q_1, q_2, q_f\}$$
$$\Sigma = \{a, b, C\}$$
$$\Gamma = \{z_0, z_1, z_2\}$$
$$q_0 = \{q_0\}$$
$$z_0 = \{z_0\}$$
$$F = \{q_f\}$$

δ is defined as follows:

$$\delta(q_0, a, z_0) \rightarrow (q_1, z_1z_0)$$
$$\delta(q_0, b, z_0) \rightarrow (q_1, z_2z_0)$$
$$\delta(q_1, a, z_1) \rightarrow (q_1, z_1z_1)$$
$$\delta(q_1, b, z_1) \rightarrow (q_1, z_2z_1)$$
$$\delta(q_1, a, z_2) \rightarrow (q_1, z_1z_2)$$
$$\delta(q_1, b, z_2) \rightarrow (q_1, z_2z_2)$$
$$\delta(q_1, C, z_1) \rightarrow (q_2, z_1)$$
$$\delta(q_1, C, z_2) \rightarrow (q_2, z_2)$$
$$\delta(q_2, a, z_1) \rightarrow (q_2, \lambda)$$
$$\delta(q_2, b, z_2) \rightarrow (q_2, \lambda)$$

$\delta(q_2, \lambda, z_0) \rightarrow (q_2, \lambda)$ accepted by the empty stack

$\delta(q_2, \lambda, z_0) \rightarrow (q_f, z_0)$ accepted by the final state

Example 7.5 Construct a PDA to accept the language L = {WCW^R, where W ∈ (a, b)* and W^R is the reverse of W} by the empty stack and by the final state.

Solution: The string W may contain a null string. If W is null, then W^R is null. It means in the language set, there will be only C if W is null. The PDA will always be in state q_0 at the time of traversing the string W. The δ functions for traversing W will be

$$\delta(q_0, a, z_0) \rightarrow (q_0, z_1z_0)$$
$$\delta(q_0, b, z_0) \rightarrow (q_0, z_2z_0)$$
$$\delta(q_0, a, z_1) \rightarrow (q_0, z_1z_1)$$
$$\delta(q_0, b, z_1) \rightarrow (q_0, z_2z_1)$$
$$\delta(q_0, a, z_2) \rightarrow (q_0, z_1z_2)$$
$$\delta(q_0, b, z_2) \rightarrow (q_0, z_2z_2)$$

C indicates the end of W and the beginning of W^R. When C will be traversed, a state change from q_0 to q_1 will occur. The string W may be null. The set L may contain only C. So, the stack top will be one of the following at the time of traversing C.

May be z_1 (if just before 'a' is traversed)
May be z_2 (if just before 'b' is traversed)
May be z_0 (if no input alphabet is traversed, i.e., W is null).
The δ function for traversing C will be

$$\delta(q_0, C, z_1) \to (q_1, z_1)$$
$$\delta(q_0, C, z_2) \to (q_1, z_2)$$
$$\delta(q_0, C, z_0) \to (q_1, z_0)$$

By the previous given δ functions, all the input symbols of the input tape will be traversed. The PDA will be in state q_1 with the stack top z_0.

If we have to design the PDA accepted by the empty stack, the δ function will be

$$\delta(q_1, \lambda, z_0) \to (q_1, \lambda)$$

If we have to design the PDA accepted by the final state, the δ function will be

$$\delta(q_1, \lambda, z_0) \to (q_f, z_0)$$

Example 7.6 Construct a PDA to accept the language $L = \{a^n b^m c^m d^n$, where $m, n \geq 1\}$ by the empty stack and by the final state.

Solution: The string is $a^n b^m c^m d^n$, where $m, n \geq 1$. The remarks we can draw from the string are as follows:

1. The string consists of a, b, c, and d.
2. 'b' will follow 'a', 'c' will follow 'b', and 'd' will follow 'c'.
3. The number of 'a' is equal to the number of 'd' and the number of 'b' is equal to the number of 'c'.
4. In all the strings of the language set, there will be at least one 'a', one 'b', one 'c', and one 'd'.

We need to check the equal number of 'a' and 'd' and the equal number of 'b' and 'c'. The string starts with 'a', so in the input tape 'a' will come first with state q_0 and the stack top z_0. A stack symbol z_1 will be pushed to the stack. From the next appearance of 'a' in the input tape, the stack top will be z_1. Another z_1 will be pushed to the stack. The δ function for traversing 'a' will be

$$\delta(q_0, a, z_0) \to (q_0, z_1 z_0)$$
$$\delta(q_0, a, z_1) \to (q_0, z_1 z_1)$$

After complete traversal of 'a' in the input tape, 'b' will come to be traversed. At the time of traversal of the first 'b' in the input tape, the PDA will be in state q_0 with the stack top z_1 (before it all 'a's are traversed). A state change from q_0 to q_1 will occur and a stack symbol z_2 will be added to the stack. From the next appearance of 'b' in the input tape, the state will be q_1 and the stack top will be z_2. Another z_2 will be pushed to the stack. The δ function for traversing 'b' will be

$$\delta(q_0, b, z_1) \to (q_1, z_2 z_1)$$
$$\delta(q_1, b, z_2) \to (q_1, z_2 z_2)$$

After complete traversal of 'b' in the input tape, 'c' will come to be traversed. The number of 'b' and the number of 'c' are equal in all strings that belong to the language set L. At the time of traversal of the first ‘c’ in the input tape, the PDA will be in state q_1 with the stack top z_2 (before it, all ‘b’s are traversed). A state change from q_1 to q_2 will occur and the stack top symbol z_2 will be popped from the stack. From the next appearance of ‘b’ in the input tape, the state will be q_2 and the stack top will be z_2.The symbol z_2 will be popped from the stack. The δ function for traversing ‘b’ will be

$$\delta(q_1, c, z_2) \rightarrow (q_2, \lambda)$$
$$\delta(q_2, c, z_2) \rightarrow (q_2, \lambda)$$

When all the ‘c’ will be traversed in the stack, there will be z_1, which was added at the time of traversal of ‘a’ as the stack top, and the PDA will be in state q_2. At the time of traversal of the first ‘d’ in the input tape, the PDA will be in state q_2 with the stack top z_1. A state change from q_2 to q_3 will occur and the stack top symbol z_1 will be popped from the stack. From the next appearance of ‘d’ in the input tape, the state will be q_3 and the stack top will be z_1.The symbol z_1 will be popped from the stack. The δ function for traversing ‘b’ will be

$$\delta(q_2, d, z_1) \rightarrow (q_3, \lambda)$$
$$\delta(q_3, d, z_1) \rightarrow (q_3, \lambda)$$

By the previous given δ functions, all the input symbols of the input tape will be traversed. The PDA will be in state q_3 with stack top z_0.

If we have to design the PDA accepted by the empty stack, the δ function will be

$$\delta(q_3, \lambda, z_0) \rightarrow (q_3, \lambda)$$

If we have to design the PDA accepted by the final state, the δ function will be

$$\delta(q_3, \lambda, z_0) \rightarrow (q_f, z_0)$$

The PDA for $L = \{a^n b^m c^m d^n$, where $m, n \geq 1\}$ is

$$Q = \{q_0, q_1, q_2, q_3, q_f\}$$
$$\Sigma = \{a, b, c, d\}$$
$$\Gamma = \{z_0, z_1, z_2\}$$
$$q_0 = \{q_0\}$$
$$z_0 = \{z_0\}$$
$$F = \{q_f\}$$

δ is defined as follows:

$$\delta(q_0, a, z_0) \rightarrow (q_0, z_1z_0)$$
$$\delta(q_0, a, z_1) \rightarrow (q_0, z_1z_1)$$
$$\delta(q_0, b, z_1) \rightarrow (q_1, z_2z_1)$$
$$\delta(q_1, b, z_2) \rightarrow (q_1, z_2z_2)$$
$$\delta(q_1, c, z_2) \rightarrow (q_2, \lambda)$$
$$\delta(q_2, c, z_2) \rightarrow (q_2, \lambda)$$
$$\delta(q_2, d, z_1) \rightarrow (q_3, \lambda)$$
$$\delta(q_3, d, z_1) \rightarrow (q_3, \lambda)$$
$$\delta(q_3, \lambda, z_0) \rightarrow (q_3, \lambda) \text{ accepted by the empty stack}$$
$$\delta(q_3, \lambda, z_0) \rightarrow (q_f, z_0) \text{ accepted by the final state}$$

Example 7.7 Construct a PDA to accept the language $L = \{a^n b^n c^m d^m$, where $m, n \geq 1\}$ by the empty stack and by the final state.

Solution: Here, the number of 'a' and the number of 'b' are the same and the number of 'c' and the number of 'd' are the same. The state change will occur in transition from 'a' to 'b', 'b' to 'c', and 'c' to 'd'. Upon traversing 'a', the stack symbol z_1 will be pushed to the stack. The number of 'b' is the same as the number of 'a'. The number of z_1 pushed to the stack will be equal to the number of 'a', i.e., the number of 'b'. For traversal of each 'b' in the input tape, one z_1 will be popped from the stack.

When the first 'c' will be traversed, the stack top will be z_0. Upon traversing 'c', the stack symbol z_2 will be pushed to the stack. Upon traversing 'd', those z_2 will be popped from the stack.

By these processes, all the strings will be traversed. The PDA for accepting the string $a^n b^n c^m d^m/m$, $n \geq 1$ will be

$$Q = \{q_0, q_1, q_2, q_3, q_f\}$$
$$\Sigma = \{a, b, c, d\}$$
$$\Gamma = \{z_0, z_1, z_2\}$$
$$q_0 = \{q_0\}$$
$$z_0 = \{z_0\}$$
$$F = \{q_f\}$$

The transition function for constructing PDA for the string $a^n b^n c^m d^m/m$, $n \geq 1$ will be

$$\delta(q_0, a, z_0) \rightarrow (q_0, z_1 z_0)$$
$$\delta(q_0, a, z_1) \rightarrow (q_0, z_1 z_1)$$
$$\delta(q_0, b, z_1) \rightarrow (q_1, \lambda)$$
$$\delta(q_1, b, z_1) \rightarrow (q_1, \lambda)$$
$$\delta(q_1, c, z_0) \rightarrow (q_2, z_2 z_0)$$
$$\delta(q_2, c, z_2) \rightarrow (q_2, z_2 z_2)$$
$$\delta(q_2, d, z_2) \rightarrow (q_3, \lambda)$$
$$\delta(q_3, d, z_2) \rightarrow (q_3, \lambda)$$
$$\delta(q_3, \lambda, z_0) \rightarrow (q_3, \lambda) \text{ accepted by the empty stack}$$
$$\delta(q_3, \lambda, z_0) \rightarrow (q_f, z_0) \text{ accepted by the final state}$$

Example 7.8 Construct a PDA to accept the language $L = \{a^n b^{2n}$, where $n \geq 1\}$ by the empty stack and by the final state.

Solution: In the language set L, each string is in the form $a^n b^{2n}$ where $n \geq 1$. The number of 'b' is double the number of 'a'. At the time of traversing a single 'a' from the input tape, two z_1 as stack symbols will be pushed to the stack.

$$\delta(q_0, a, z_0) \rightarrow (q_0, z_1 z_1 z_0)$$

z_1
z_1
z_0

$$\delta(q_0, a, z_1) \rightarrow (q_0, z_1z_1z_1)$$

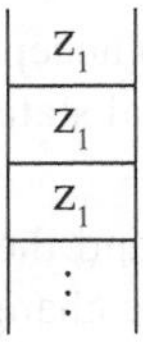

These two z_1 will be popped at the time of traversing the two 'b's. (A single 'a' is equal to two 'b's.) The δ function for traversing 'b' will be

$$\delta(q_0, b, z_1) \rightarrow (q_1, \lambda)$$
$$\delta(q_1, b, z_1) \rightarrow (q_1, \lambda)$$

By the previous given δ functions, all the input symbols of the input tape will be traversed. The PDA will be in state q_1 with the stack top z_0.

If we have to design the PDA accepted by the empty stack, the δ function will be

$$\delta(q_1, \lambda, z_0) \rightarrow (q_1, \lambda)$$

If we have to design the PDA accepted by the final state, the δ function will be

$$\delta(q_1, \lambda, z_0) \rightarrow (q_f, z_0)$$

The PDA will be as follows:

$$Q = \{q_0, q_1, q_f\}$$
$$\Sigma = \{a, b\}$$
$$\Gamma = \{z_0, z_1\}$$
$$q_0 = \{q_0\}$$
$$z_0 = \{z_0\}$$
$$F = \{q_f\}$$

The transitional function will be as follows:

$$\delta(q_0, a, z_0) \rightarrow (q_0, z_1z_1z_0)$$
$$\delta(q_0, a, z_1) \rightarrow (q_0, z_1z_1z_1)$$
$$\delta(q_0, b, z_1) \rightarrow (q_1, \lambda)$$
$$\delta(q_1, b, z_1) \rightarrow (q_1, \lambda)$$
$$\delta(q_1, \lambda, z_0) \rightarrow (q_1, \lambda)$$
$$\delta(q_1, \lambda, z_0) \rightarrow (q_f, z_0)$$

Example 7.9 Construct a PDA for the language $L = \{a^nCb^{2n}, \text{ where } n \geq 1\}$.

Solution: In the language set, the number of 'b' is twice the number of 'a'. 'C' is the middle element, which indicates the end of 'a' and the beginning of 'b'. For each 'a', two z_1 are inserted in the stack, and for each 'b', one z_1 is popped from the stack.

The transitional functions are

$$\delta(q_0, a, z_0) \rightarrow (q_1, z_1z_1z_0)$$
$$\delta(q_1, a, z_1) \rightarrow (q_1, z_1z_1z_1)$$

$\delta(q_1, C, z_1) \rightarrow (q_2, z_1)$
$\delta(q_2, b, z_1) \rightarrow (q_2, \lambda)$
$\delta(q_2, \lambda, z_0) \rightarrow (q_f, z_0)$ accepted by the final state
$\delta(q_2, \lambda, z_0) \rightarrow (q_2, \lambda)$ accepted by the empty stack

Example 7.10 Construct a PDA for the language $L = \{a^m b^n$ where $m \leq n$ and $m, n \geq 1\}$.

Solution: In the language set, the number of 'a' is less than the number of 'b'. At the time of traversing 'a', one z_1 is added to the stack. For m number of 'a', m number of z_1 are added to the stack. At the time of traversing m number of 'b', the stack top is z_1, which is popped. If $n > m$, then at the time of traversing the last $(n - m)$ number of 'b', the stack top is z_0.

The transitional functions are

$\delta(q_0, a, z_0) \rightarrow (q_1, z_1z_0)$
$\delta(q_1, a, z_1) \rightarrow (q_1, z_1z_1)$
$\delta(q_1, b, z_1) \rightarrow (q_2, \lambda)$
$\delta(q_2, b, z_1) \rightarrow (q_2, \lambda)$
$\delta(q_2, b, z_0) \rightarrow (q_2, z_0)$
$\delta(q_2, \lambda, z_0) \rightarrow (q_f, z_0)$ accepted by the final state
$\delta(q_2, \lambda, z_0) \rightarrow (q_2, \lambda)$ accepted by the empty stack.

Example 7.11 Construct a PDA to accept the language $L = \{a^n b^m c^{n+m}$, where $n > 0, m > 0\}$.

Solution: In the language set, the total number of 'c' is equal to the total number of 'a' and 'b'. The PDA can be designed in the following way: at the time of traversing n number of 'a', n number of 'z_1' are added to the stack and for traversing m number of 'b', m number of z_1 are added to the stack. At the time of traversing $(n + m)$ number of c, all the z_1 are popped from the stack.

The transitional functions are

$\delta(q_0, a, z_0) \rightarrow (q_0, z_1z_0)$
$\delta(q_1, a, z_1) \rightarrow (q_1, z_1z_1)$
$\delta(q_1, b, z_1) \rightarrow (q_2, z_1z_1)$
$\delta(q_2, b, z_1) \rightarrow (q_2, z_1z_1)$
$\delta(q_2, c, z_1) \rightarrow (q_3, \lambda)$
$\delta(q_3, c, z_1) \rightarrow (q_3, \lambda)$
$\delta(q_3, \lambda, z_0) \rightarrow (q_f, z_0)$ accepted by the final state
$\delta(q_3, \lambda, z_0) \rightarrow (q_3, \lambda)$ accepted by the empty stack

Example 7.12 Construct a PDA to accept the language $L = \{a^n b^m c^{n+m}$, where $n \geq 0, m > 0\}$.

Solution: The string is the same as the previous one, but in the language set the number of 'a' may be zero. In the case when the number of 'a' is zero, the string starts from 'b'.

The transitional functions are

$$\delta(q_0, a, z_0) \rightarrow (q_0, z_1z_0)$$
$$\delta(q_1, a, z_1) \rightarrow (q_1, z_1z_1)$$
$$\delta(q_1, b, z_1) \rightarrow (q_2, z_1z_1)$$
$$\delta(q_2, b, z_1) \rightarrow (q_2, z_1z_1)$$
$$\delta(q_0, b, z_0) \rightarrow (q_2, z_1z_0) \text{ // If the string starts with b, i.e., there is no 'a'}$$
$$\delta(q_2, c, z_1) \rightarrow (q_3, \lambda)$$
$$\delta(q_3, c, z_1) \rightarrow (q_3, \lambda)$$
$$\delta(q_3, \lambda, z_0) \rightarrow (q_f, z_0) \text{ accepted by the final state}$$
$$\delta(q_3, \lambda, z_0) \rightarrow (q_3, \lambda) \text{ accepted by the empty stack}$$

Example 7.13 Construct a PDA to accept the language $L = \{a^nb^{n+m}c^m, \text{ where } n, m > 0\}$.

Solution: In the language set, the number of 'b' is the total number of 'a' and 'c'. At the time of traversal of n number of 'a', n number of z_1 are pushed in the stack, and at the time of traversal of n number of 'b', n number of z_1 are popped from the stack. At the time of traversal of the rest m number of 'b', m number of z_2 are pushed into the stack, which are popped at the time of traversal of m number of 'c'.

The transitional functions are

$$\delta(q_0, a, z_0) \rightarrow (q_0, z_1z_0)$$
$$\delta(q_1, a, z_1) \rightarrow (q_1, z_1z_1)$$
$$\delta(q_1, b, z_1) \rightarrow (q_2, \lambda)$$
$$\delta(q_2, b, z_1) \rightarrow (q_2, \lambda)$$
$$\delta(q_2, b, z_0) \rightarrow (q_3, z_2z_0)$$
$$\delta(q_3, b, z_2) \rightarrow (q_3, z_2z_2)$$
$$\delta(q_3, c, z_2) \rightarrow (q_4, \lambda)$$
$$\delta(q_4, c, z_2) \rightarrow (q_4, \lambda)$$
$$\delta(q_4, \lambda, z_0) \rightarrow (q_f, z_0) \text{ accepted by the final state}$$
$$\delta(q_4, \lambda, z_0) \rightarrow (q_4, \lambda) \text{ accepted by the empty stack}$$

Example 7.14 Construct a PDA to accept the language $L = \{a^nb^{n+m}c^m, \text{ where } n \geq 0, m > 0\}$.

Solution: The string is the same as the previous one, but in the language set the number of 'a' may be zero. In the case when the number of 'a' is zero, then the string starts from 'b'.

The transitional functions are

$$\delta(q_0, a, z_0) \rightarrow (q_0, z_1z_0)$$
$$\delta(q_1, a, z_1) \rightarrow (q_1, z_1z_1)$$
$$\delta(q_1, b, z_1) \rightarrow (q_2, \lambda)$$
$$\delta(q_2, b, z_1) \rightarrow (q_2, \lambda)$$
$$\delta(q_2, b, z_0) \rightarrow (q_3, z_2z_0)$$
$$\delta(q_0, b, z_0) \rightarrow (q_3, z_2z_0) \text{ // If the string starts with b, i.e., there is no 'a'}$$
$$\delta(q_3, b, z_2) \rightarrow (q_3, z_2z_2)$$

$\delta(q_3, c, z_2) \rightarrow (q_4, \lambda)$
$\delta(q_4, c, z_2) \rightarrow (q_4, \lambda)$
$\delta(q_4, \lambda, z_0) \rightarrow (q_f, z_0)$ accepted by the final state
$\delta(q_4, \lambda, z_0) \rightarrow (q_4, \lambda)$ accepted by the empty stack

Example 7.15 Construct a PDA to accept the following language $L = \{a^{2n}b^n$, where $n > 0\}$.

Solution: The language consists of 2n number of 'a' and n number of 'b' belong to the language set. The PDA can be designed as follows—at the time of traversing the first 'a', two z_1 are added to the stack with a state change from q_0 to q_1. At the time of traversing the second 'a', one z_1 is popped from the stack with a state change from q_1 to q_0. By this process after traversing 2n number of 'a', only n number of z_1 exist in the stack. At the time of traversing the first 'b', the stack top is z_1 and the state is q_0. Those z_1 are popped at the time of traversing n number of 'b'.

The transitional functions are

$\delta(q_0, a, z_0) \rightarrow (q_1, z_1z_1z_0)$
$\delta(q_1, a, z_1) \rightarrow (q_0, \lambda)$
$\delta(q_0, a, z_1) \rightarrow (q_1, z_1z_1z_0)$
$\delta(q_0, b, z_1) \rightarrow (q_2, \lambda)$
$\delta(q_2, b, z_1) \rightarrow (q_2, \lambda)$
$\delta(q_2, \lambda, z_0) \rightarrow (q_f, z_0)$ // Accepted by final state
$\delta(q_2, \lambda, z_0) \rightarrow (q_2, \lambda)$ // Accepted by empty stack

Example 7.16 Construct a PDA to accept the language $L = \{a^nb^nc^m$, where $n, m \geq 1\}$ by the empty stack and by the final state.

Solution: The language is in the form $a^nb^nc^m$, where $n, m \geq 1$. In the language set, the number of 'a' and the number of 'b' are the same, but the number of 'c' is different. All the strings in the language set start with n number of 'a's followed by n number of 'b's and ends with m number of 'c's. Here, m and n are both ≥ 1, and the null string does not belong to the language set.

The PDA for the language can be designed in the following way.

When traversing the 'a's from the input tape, the z_1's are pushed in the stack one by one. At the time of traversing the 'b's, all the z_1's which were pushed into the stack are popped one by one. When the first 'c' will be traversed, at that time the machine is in state q_1 and stack top z_0. At the time of traversal of m number of 'c's, no operation is done on the stack.

The transition function for constructing the PDA for the string $a^nb^nc^m$ where $m, n \geq 1$ are

$\delta(q_0, a, z_0) \rightarrow (q_0, z_1z_0)$
$\delta(q_0, a, z_1) \rightarrow (q_0, z_1z_1)$
$\delta(q_0, b, z_1) \rightarrow (q_1, \lambda)$
$\delta(q_1, b, z_1) \rightarrow (q_1, \lambda)$
$\delta(q_1, c, z_0) \rightarrow (q_1, z_0)$
$\delta(q_1, \lambda, z_0) \rightarrow (q_1, \lambda)$ accepted by the empty stack
$\delta(q_1, \lambda, z_0) \rightarrow (q_f, z_0)$ accepted by the final state

Example 7.17 Construct a PDA to accept the language $L = \{a^n b^m c^n$, where $n, m \geq 1\}$ by the empty stack and by the final state.

Solution: The language is in the form $a^n b^m c^n$, where $n, m \geq 1$. In the language set, the number of 'a's are equal to the number of 'c's. In the language set, 'a's are separated from 'c's by m number of 'b's. As m, $n \geq 1$, the null string does not belong to the language set. In the language set, the number of 'a' is equal to the number of 'c' but not with the number of 'b'.

The PDA can be designed in the following way.

When traversing 'a's from the input tape, z_1's are pushed in the stack one by one. When the first 'b' is read by the input head, the machine is in state q_0 and stack top z_1. One state change has occurred. [If after 'b' again 'a' comes as input, the language is in a wrong format. But that 'a' will also be traversed as there will be a function $\delta(q_0, a, z_1) \rightarrow (q_0, z_1z_1)$ to traverse that 'a'. To avoid this mismatch and to guarantee that the PDA is only for accepting $a^n b^n c^m$, where $n, m \geq 1$, the state change is required.] At the time of traversing 'b', no operation is done on the stack.

When 'c' is going to be traversed, the machine is in state q_1, with the stack top z_1. The stack top is popped and a state change has occurred (to avoid appearance of 'b' after 'c').

The transition function for constructing the PDA for the string $a^n b^n c^m$ where $m, n \geq 1$ will be

$\delta(q_0, a, z_0) \rightarrow (q_0, z_1z_0)$
$\delta(q_0, a, z_1) \rightarrow (q_0, z_1z_1)$
$\delta(q_0, b, z_1) \rightarrow (q_1, z_1)$
$\delta(q_1, b, z_1) \rightarrow (q_1, z_1)$
$\delta(q_1, c, z_1) \rightarrow (q_2, \lambda)$
$\delta(q_2, c, z_1) \rightarrow (q_2, \lambda)$
$\delta(q_2, \lambda, z_0) \rightarrow (q_2, \lambda)$ accepted by the empty stack
$\delta(q_2, \lambda, z_0) \rightarrow (q_f, z_0)$ accepted by the final state

Example 7.18 Construct a PDA to accept the language L = {w, where number of 'a' + number of 'b' = number of 'c'}.

Solution: According to the given specification, any string that belongs to the language set consists of any combination of 'a', 'b', and 'c', and the number of 'c' is equal to the number of 'a' and the number of 'b'. In the string 'a', 'b' and 'c' can occur in any sequence by fulfilling the given condition.

The transitional functions are

$\delta(q_0, a, z_0) \rightarrow (q_0, z_1z_0)$ // 'a' traversed first or when the stack contains only z_0
$\delta(q_0, b, z_0) \rightarrow (q_0, z_1z_0)$ // 'b' traversed first or when the stack contains only z_0
$\delta(q_0, c, z_0) \rightarrow (q_0, z_2z_0)$ // 'c' traversed first or when the stack contains only z_0
$\delta(q_0, c, z_2) \rightarrow (q_0, z_2z_2)$ // 'c' traversed after 'c'
$\delta(q_0, a, z_1) \rightarrow (q_0, z_1z_1)$ // 'a' traversed after 'a' or 'b'
$\delta(q_0, b, z_1) \rightarrow (q_0, z_1z_1)$ // 'b' traversed after 'a' or 'b'
$\delta(q_0, c, z_1) \rightarrow (q_0, \lambda)$ // 'c' traversed after 'a' or 'b'
$\delta(q_0, a, z_2) \rightarrow (q_0, \lambda)$ // 'a' traversed after 'c'
$\delta(q_0, b, z_2) \rightarrow (q_0, \lambda)$ // 'b' traversed after 'c'
$\delta(q_0, \lambda, z_0) \rightarrow (q_f, z_0)$ // accepted by the final state
$\delta(q_0, \lambda, z_0) \rightarrow (q_0, \lambda)$ // accepted by the empty stack

7.3 DPDA and NPDA

There are two types of PDA:

1. Deterministic pushdown automata (DPDA)
2. Non-deterministic pushdown automata (NPDA)

7.3.1 Deterministic Pushdown Automata (DPDA)

A PDA is said to be a DPDA if all derivations in the design give only a single move.

If a PDA being in a state with a single input and a single stack symbol gives a single move, then the PDA is called DPDA.

As an example, for $L = \{a^n b^n \text{ where } n \geq 1\}$, a DPDA can be designed if the transitional functions are as follows:

$$\delta(q_0, a, z_0) \rightarrow (q_0, z_1z_0)$$
$$\delta(q_0, a, z_1) \rightarrow (q_0, z_1z_1)$$
$$\delta(q_0, b, z_1) \rightarrow (q_1, \lambda)$$
$$\delta(q_1, b, z_1) \rightarrow (q_1, \lambda)$$
$$\delta(q_1, \lambda, z_0) \rightarrow (q_1, \lambda) \text{ accepted by the empty stack}$$
$$\delta(q_1, \lambda, z_0) \rightarrow (q_f, z_0) \text{ accepted by the final state}$$

In the previous PDA, in a single state with a single input and a single stack symbol, there is only one move. So, the PDA is deterministic.

A context-free language is called deterministic context-free language if it is accepted by a DPDA.

7.3.2 Non-deterministic Pushdown Automata (NPDA)

A pushdown automata is called non-deterministic if one of the derivations generates more than one move.

If a PDA being in a state with a single input and a single stack symbol gives more than one move for any of its transitional functions, then the PDA is called NPDA.

A context-free language is called non-deterministic context-free language if it is accepted by a NPDA.

The following are some examples of NPDA.

Example 7.19 Design a non-deterministic PDA for accepting the string $\{WW^R \text{ where } W \in (a, b)^+$ and W^R is the reverse of $W\}$ by the empty stack and by the final state.

Solution: W is a string which consists of 'a' and 'b'. W^R is the reverse string of W. Let W = abaa, then W^R will be aaba. WW^R will be abaaaaba. Traversing 'a' in the input tape, z_1 will be pushed into the stack, and traversing 'b' in the input tape z_2 will be pushed into the stack. When W^R will start, for the traversal of the first symbol, the state of the PDA will be changed and the stack top will be popped. From the next input symbols belonging to W^R, the stack top will be popped.

Here, there is a problem. To the PDA, it is not assigned where W is ended and W^R is started. From the beginning state, the PDA will traverse the total string assuming it as W. Sometimes (in some books) attempt has been done to differentiate W and W^R by traversing a λ symbol and changing the state in between W and W^R. But this is also wrong, as all the input symbols are placed on the input tape from left to right in a continuous fashion. There is no gap in between the two input symbols as they are placed one by one in each square from left to right on the input tape.

From the previous discussion, it is clear that a DPDA cannot be designed for WW^R.

But, this is possible for an NPDA. Our problem is to find the middle of the string where W ends and W^R starts. There is a chance to be the middle of the string where two 'a's or two 'b's appear, because the last alphabet of W is the first alphabet of W^R. Make two transitional functions when the stack top is z_1 and 'a' is traversed as input and the stack top is z_2 and 'b' is traversed as input. (All the other δ functions are the same as WCW^R). The PDA for accepting the string $WW^R / W \in (a, b)^+$, where W^R is the reverse of W, is

$$Q = \{q_0, q_1, q_2, q_f\}$$
$$\Sigma = \{a, b\}$$
$$\Gamma = \{z_0, z_1, z_2\}$$
$$q_0 = \{q_0\}$$
$$z_0 = \{z_0\}$$
$$F = \{q_f\}$$

The transitional function will be as follows:

$$\delta(q_0, a, z_0) \rightarrow (q_1, z_1z_0)$$
$$\delta(q_0, b, z_0) \rightarrow (q_1, z_2z_0)$$
$$\delta(q_1, a, z_1) \rightarrow (q_1, z_1z_1), (q_2, \lambda)$$
$$\delta(q_1, b, z_1) \rightarrow (q_1, z_2z_1)$$
$$\delta(q_1, a, z_2) \rightarrow (q_1, z_1z_2)$$
$$\delta(q_1, b, z_2) \rightarrow (q_1, z_2z_2), (q_2, \lambda)$$
$$\delta(q_2, a, z_1) \rightarrow (q_2, \lambda)$$
$$\delta(q_2, b, z_2) \rightarrow (q_2, \lambda)$$
$$\delta(q_2, \lambda, z_0) \rightarrow (q_2, \lambda) \text{ accepted by the empty stack}$$
$$\delta(q_2, \lambda, z_0) \rightarrow (q_f, z_0) \text{ accepted by the final state}$$

Consider a string abbbba. The string is in the form WW^R, where W = abb.

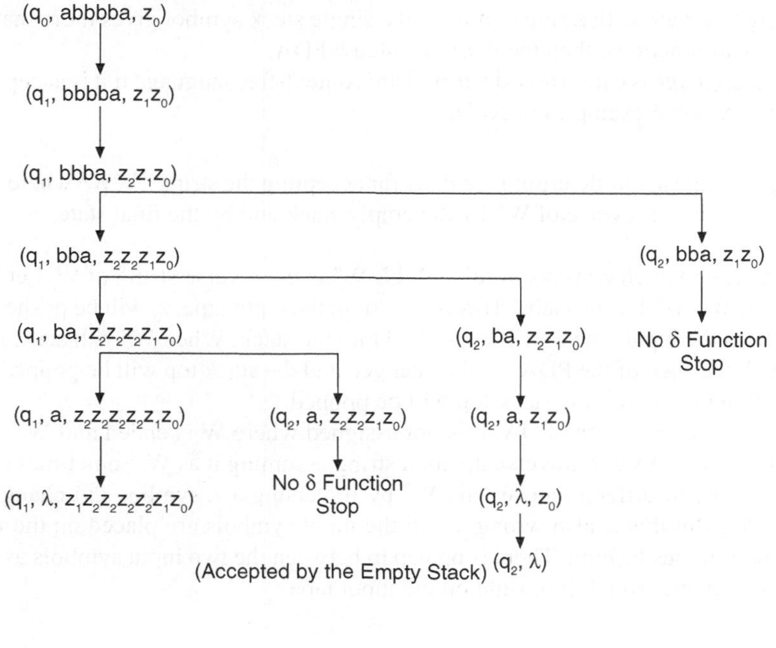

Example 7.20 Design a for accepting the string {WW^R where W ∈ (a, b)* and W^R is the reverse of W} by the empty stack and by the final state.

Solution: The null string may belong to the language set. The PDA will be designed in such a way that the null string can be accepted by the PDA. The transitional functions will be as follows:

$\delta(q_0, a, z_0) \rightarrow (q_0, z_1z_0)$
$\delta(q_0, b, z_0) \rightarrow (q_0, z_2z_0)$
$\delta(q_0, a, z_1) \rightarrow (q_0, z_1z_1), (q_1, \lambda)$
$\delta(q_0, b, z_1) \rightarrow (q_0, z_2z_1)$
$\delta(q_0, a, z_2) \rightarrow (q_0, z_1z_2)$
$\delta(q_0, b, z_2) \rightarrow (q_0, z_2z_2), (q_1, \lambda)$
$\delta(q_1, a, z_1) \rightarrow (q_1, \lambda)$
$\delta(q_1, b, z_2) \rightarrow (q_1, \lambda)$
$\delta(q_0, \lambda, z_0) \rightarrow (q_1, \lambda)$ // This function is for accepting the null string
$\delta(q_1, \lambda, z_0) \rightarrow (q_1, \lambda)$ accepted by the empty stack
$\delta(q_1, \lambda, z_0) \rightarrow (q_f, z_0)$ accepted by the final state

Example 7.21 Design a NPDA for accepting the string L = {(Set of all palindromes over a, b)} by the empty stack and by the final state.

Solution: A string which when read from left to right and right to left gives the same result is called a palindrome. Palindromes can be of two types (a) odd palindrome and (b) even palindrome. Odd palindromes are those in which the number of characters is odd, and even palindrome are those in which the number of characters is even.

The string 'a' is a palindrome, and 'b' is a palindrome. 'aa' is a palindrome, and 'bb' is a palindrome. The string 'aba' is a palindrome, 'aaa' and 'bbb' are also palindromes. The null string is also a palindrome.

The PDA will be

$Q = \{q_0, q_1, q_f,\}$
$\Sigma = \{a, b\}$
$\Gamma = \{z_0, z_1, z_2\}$
$q_0 = \{q_0\}$
$z_0 = \{z_0\}$
$F = \{z_f\}$

The transitional functions are as follows:

$\delta(q_0, a, z_0) \rightarrow (q_0, z_1z_0), (q_1, z_0)$ // for 'a'
$\delta(q_0, b, z_0) \rightarrow (q_0, z_2z_0), (q_1, z_0)$ // for 'b'
$\delta(q_0, a, z_1) \rightarrow (q_0, z_1z_1), (q_1, z_1), (q_1, \lambda)$ // for 'aaa' and 'aa'
$\delta(q_0, b, z_1) \rightarrow (q_0, z_2z_1), (q_1, z_1)$ // for 'aba'
$\delta(q_0, a, z_2) \rightarrow (q_0, z_1z_2), (q_1, z_2)$ // for 'bab'
$\delta(q_0, b, z_2) \rightarrow (q_0, z_2z_2), (q_1, z_2), (q_1, \lambda)$ // for 'bbb' and 'bb'
$\delta(q_0, \lambda, z_0) \rightarrow (q_1, z_0)$ // for null string

$\delta(q_1, a, z_1) \rightarrow (q_1, \lambda)$
$\delta(q_1, b, z_2) \rightarrow (q_1, \lambda)$
$\delta(q_1, \lambda, z_0) \rightarrow (q_1, \lambda)$ accepted by the empty stack
$\delta(q_1, \lambda, z_0) \rightarrow (q_f, z_0)$ accepted by the final state

7.4 Construction of PDA from CFG

PDA is the machine accepting a context-free language. A context-free language is generated from context-free grammar. Now, the question is whether the PDA can be directly generated from the CFG. The answer is 'yes'. The following section describes the process of generating a PDA directly from the CFG.

Step I: Convert the PDA into a Greibach Normal Form (GNF).

Step II: First the start symbol S of the CFG is put to the stack by the transition function

$$\delta(q_0, \in, z_0) \rightarrow (q_1, Sz_0)$$

Step III: For a production in the form $\{NT_1\} \rightarrow \{\text{Single T}\}\{\text{String of NT}\}$, the transitional function will be

$$\delta(q_1, T, NT_1) \rightarrow (q_1, \text{String of NT})$$

Step IV: For a production in the form $\{NT_1\} \rightarrow \{\text{Single T}\}$, the transitional function will be

$$\delta(q_1, T, NT_1) \rightarrow (q_1, \lambda)$$

Step V: For accepting a string, two transitional functions are added, one for being accepted by the empty stack and one for being accepted by the final state.

Follow the examples for constructing a PDA directly from the CFG.

Example 7.22 Construct a PDA that accepts the language generated by the following grammar.

$S \rightarrow aB$
$B \rightarrow bA/b$
$A \rightarrow aB$

Show an ID for the string abab for the PDA generated.

Solution: The context-free grammar is in GNF.

First, the start symbol S is pushed into the stack by the following production

$$\delta(q_0, \in, z_0) \rightarrow (q_1, Sz_0)$$

For the production $S \rightarrow aB$, the transitional function is

$$\delta(q_1, a, S) \rightarrow (q_1, B)$$

For the production $A \rightarrow aB$, the transitional function is

$$\delta(q_1, a, A) \rightarrow (q_1, B)$$

For the production $B \rightarrow bA/b$, the transitional function is

$$\delta(q_1, b, B) \rightarrow (q_1, A), (q_1, \lambda)$$

So, the PDA for the previous CFG is

$$Q = \{q_0, q_1\}$$
$$\Sigma = \{a, b\}$$
$$\Gamma = \{z_0, S, A, B\}$$
$$q_0 = \{q_0\}$$
$$z_0 = \{z_0\}$$
$$F = \{\varnothing\}$$

δ functions are defined as follows:

$$\delta(q_0, \in, z_0) \rightarrow (q_1, Sz_0)$$
$$\delta(q_1, a, S) \rightarrow (q_1, B)$$
$$\delta(q_1, a, A) \rightarrow (q_1, B)$$
$$\delta(q_1, a, A) \rightarrow (q_1, B)$$
$$\delta(q_1, b, B) \rightarrow (q_1, A), (q_1, \lambda)$$
$$\delta(q_1, \lambda, z_0) \rightarrow (q_f, z_0) \quad \text{// accepted by the final state}$$
$$\delta(q_1, \lambda, z_0) \rightarrow (q_1, \lambda) \quad \text{// accepted by the empty stack}$$

ID for the String abab

$$\delta(q_0, \in, z_0) \rightarrow (q_1, Sz_0)$$
$$\delta(q_1, abab, S) \rightarrow (q_1, Bz_0)$$
$$\delta(q_1, bab, B) \rightarrow (q_1, Az_0)$$
$$\delta(q_1, ab, A) \rightarrow (q_1, Bz_0)$$
$$\delta(q_1, b, B) \rightarrow (q_1, z_0)$$
$$\delta(q_1, \lambda, z_0) \rightarrow (q_f, z_0) \quad \text{// accepted by the final state}$$
$$\delta(q_1, \lambda, z_0) \rightarrow (q_1, \lambda) \quad \text{// accepted by the empty stack}$$

[*The process is very easy. First, try to generate the string from the given grammar by leftmost derivation. Then, according to the non-terminal symbols replaced use the particular transitional function to give the ID.*

For the previous case, the string 'abab' is generated like the following derivation (leftmost) $S \rightarrow a\underline{B} \rightarrow ab\underline{A} \rightarrow aba\underline{B} \rightarrow abab$.

The first transitional function is adding S in the stack. After that, the S is replaced by B for the input 'a' using the transitional function $\delta(q_1, a, S) \rightarrow (q_1, B)$

The process continues like this till the end of the string.]

Example 7.23 Construct an equivalent PDA for the following context-free grammar.

$$S \rightarrow aAB/bBA$$
$$A \rightarrow bS/a$$
$$B \rightarrow aS/b$$

Show an ID for the string abbaaabbbbab for the PDA generated with stack description.

Solution: The context-free grammar is in GNF.

First, the start symbol S is pushed into the stack by the following production:

$$\delta(q_0, \in, z_0) \rightarrow (q_1, Sz_0)$$

For the production $S \rightarrow aAB$, the transitional function is

$$\delta(q_1, a, S) \rightarrow (q_1, AB)$$

For the production $S \rightarrow bBA$, the transitional function is

$$\delta(q_1, b, S) \rightarrow (q_1, BA)$$

For the production $A \rightarrow bS$, the transitional function is

$$\delta(q_1, b, A) \rightarrow (q_1, S)$$

For the production $B \rightarrow aS$, the transitional function is

$$\delta(q_1, a, B) \rightarrow (q_1, S)$$

For the production $A \rightarrow a$, the transitional function is

$$\delta(q_1, a, A) \rightarrow (q_1, \lambda)$$

For the production $B \rightarrow b$, the transitional function is

$$\delta(q_1, b, B) \rightarrow (q_1, \lambda)$$

For acceptance, the transitional function is

$\delta(q_1, \lambda, z_0) \rightarrow (q_f, z_0)$ // accepted by the final state
$\delta(q_1, \lambda, z_0) \rightarrow (q_1, \lambda)$ // accepted by the empty stack

ID for the String abbaaabbbbab

$\delta(q_0, \in, z_0) \rightarrow (q_1, Sz_0)$

S	z_0

$\delta(q_1, \text{abbaaabbbbab}, S) \rightarrow (q_1, AB)$

A	B	z_0

$\delta(q_1, \text{bbaaabbbbab}, A) \rightarrow (q_1, S)$

S	B	z_0

$\delta(q_1, \text{baaabbbbab}, S) \rightarrow (q_1, BA)$

B	A	B	z_0

$\delta(q_1, \text{aaabbbbab}, B) \rightarrow (q_1, S)$

S	A	B	z_0

$\delta(q_1, \text{aabbbbab}, S) \rightarrow (q_1, AB)$

A	B	A	B	z_0

$\delta(q_1, \text{abbbbab}, A) \rightarrow (q_1, \lambda)$

B	A	B	z_0

$\delta(q_1, \text{bbbbab}, B) \rightarrow (q_1, \lambda)$

A	B	z_0

$\delta(q_1, \text{bbbab}, A) \rightarrow (q_1, S)$

S	B	z_0

$\delta(q_1, \text{bbab}, S) \rightarrow (q_1, BA)$

B	A	B	z_0

$\delta(q_1, bab, B) \rightarrow (q_1, \lambda)$

A	B	z_0

$\delta(q_1, ab, A) \rightarrow (q_1, \lambda)$

B	z_0

$\delta(q_1, b, B) \rightarrow (q_1, \lambda)$

z_0

$\delta(q_1, \lambda, z_0) \rightarrow (q_f, z_0)$ // accepted by the final state
$\delta(q_1, \lambda, z_0) \rightarrow (q_1, \lambda)$ // accepted by the empty stack

Example 7.24 Construct an equivalent PDA for the following context-free grammar.

$$S \rightarrow 0BB$$
$$B \rightarrow 0S/1S/0$$

Show an ID for the string 010000 for the PDA generated.

Solution: First, the start symbol S is pushed into the stack by the following production

$$\delta(q_0, \in, z_0) \rightarrow (q_1, Sz_0)$$

For the production $S \rightarrow 0BB$, the transitional function is

$$\delta(q_1, 0, S) \rightarrow (q_1, BB)$$

For the production $B \rightarrow 0S$, the transitional function is

$$\delta(q_1, 0, B) \rightarrow (q_1, S)$$

For the production $B \rightarrow 1S$, the transitional function is

$$\delta(q_1, 1, B) \rightarrow (q_1, S)$$

For the production $B \rightarrow 0$, the transitional function is

$$\delta(q_1, 0, B) \rightarrow (q_1, \lambda)$$

For acceptance, the transitional function is

$\delta(q_1, \lambda, z_0) \rightarrow (q_f, z_0)$ // accepted by the final state
$\delta(q_1, \lambda, z_0) \rightarrow (q_1, \lambda)$ // accepted by the empty stack

ID for the String 010000

$(q_0, 010000, z_0) \rightarrow (q_1, 010000, Sz_0) \rightarrow (q_1, 10000, BBz_0) \rightarrow (q_1, 0000, SBz_0) \rightarrow (q_1, 000, BBBz_0) \rightarrow$ $(q_1, 00, BBz_0) \rightarrow (q_1, 0, Bz_0) \rightarrow (q_1, \lambda, z_0) \rightarrow (q_1, \lambda)$ // By empty stack

Example 7.25 Construct an equivalent PDA for the following context-free grammar.

$$S \rightarrow aA$$
$$A \rightarrow aABC/bB/a$$
$$C \rightarrow c$$

Show an ID for the string aabbbc for the PDA generated.

Solution: First, the start symbol S is pushed into the stack by the following production

$$\delta(q_0, \in, z_0) \rightarrow (q_1, Sz_0)$$

For the production $S \rightarrow aA$, the transitional function is

$$\delta(q_1, a, S) \rightarrow (q_1, A)$$

For the production $A \rightarrow aABC$, the transitional function is

$$\delta(q_1, a, A) \rightarrow (q_1, ABC)$$

For the production $A \rightarrow bB$, the transitional function is

$$\delta(q_1, b, A) \rightarrow (q_1, B)$$

For the production $A \rightarrow a$, the transitional function is

$$\delta(q_1, a, A) \rightarrow (q_1, \lambda)$$

For the production $C \rightarrow c$, the transitional function is

$$\delta(q_1, c, C) \rightarrow (q_1, \lambda)$$

For acceptance, the transitional function is

$\delta(q_1, \lambda, z_0) \rightarrow (q_f, z_0)$ // accepted by the final state
$\delta(q_1, \lambda, z_0) \rightarrow (q_1, \lambda)$ // accepted by the empty stack

ID for the String aabbbc

$(q_0, aabbbc, z_0) \rightarrow (q_1, aabbbc, Sz_0) \rightarrow (q_1, abbbc, Az_0) \rightarrow (q_1, bbbc, ABCz_0) \rightarrow (q_1, bbc, BBC) \rightarrow (q_1, bc, BC) \rightarrow (q_1, c, C) \rightarrow (q_1, \lambda, z_0) \rightarrow (q_f, z_0)$ // accepted by the final state.

Example 7.26 Construct an equivalent PDA for the following context-free grammar.

$$S \rightarrow aSbb/aab$$

Show an ID for the string aaabbb for the PDA generated.

Solution: The given grammar is not in GNF. First, we need to convert the grammar into GNF. Replace ‘b’ by B and ‘a’ by A. The converted grammar in GNF is as follows:

$$S \rightarrow aSBB/aAB$$
$$A \rightarrow a$$
$$B \rightarrow b$$

Now, the grammar can be easily converted into GNF.

The transitional functions for the PDA accepted by the language generated by the grammar are as follows:

$$\delta(q_0, \in, z_0) \rightarrow (q_1, Sz_0)$$
$$\delta(q_1, a, S) \rightarrow (q_1, SBB)$$
$$\delta(q_1, a, S) \rightarrow (q_1, AB)$$
$$\delta(q_1, a, A) \rightarrow (q_1, \lambda)$$

$\delta(q_1, a, B) \to (q_1, \lambda)$
$\delta(q_1, \lambda, z_0) \to (q_f, z_0)$ // accepted by the final state
$\delta(q_1, \lambda, z_0) \to (q_1, \lambda)$ // accepted by the empty stack

ID for the String aaabbb

$\delta(q_0, \text{aaabbb}, z_0) \to (q_1, \text{aaabbb}, Sz_0) \to (q_1, \text{aabbb}, SBBz_0) \to (q_1, \text{abbb}, ABBBz_0) \to (q_1, \text{bbb}, BBBz_0)$ $\to (q_1, \text{bb}, BBz_0) \to (q_1, \text{b}, Bz_0) \to (q_1, \lambda, z_0) \to (q_f, z_0)$ // accepted by the final state.

7.5 Construction of CFG Equivalent to PDA

Consider a PDA $M = (Q, \Sigma, \Gamma, \delta, q_0, z_0, F)$ which accepts a language L by empty stack. (Accepted by empty stack and accepted by final state is equivalent). A CFG, $G = (V_N, \Sigma, P, S)$, equivalent to M can be constructed using the following rules.

Let S be the start symbol.

1. For $q_i \in Q$, add a production rule $S \to [q_0, z_0, q_i]$ to P in G. If a PDA contains n number of states, from the start symbol there will be n number of productions.
2. For the δ function $\delta(q, a, Y) \to (r, \varepsilon)$ where $(q, r) \in Q$, $a \in \Sigma$, and $Y \in \Gamma$, add a production rule $[qYr] \to a$ in P.
3. For the transitional function $\delta(q, a, Y) \to (r, Y_1, Y_2 \ldots\ldots Y_k)$ where $(q, r) \in Q$, $a \in \Sigma$, and $Y, Y_1, Y_2 \ldots\ldots Y_k \in \Gamma$, then for each choice of $q_1, q_2, q_k \in Q$, add production $[qYq_k] \to a[rY_1q_1][q_1Y_2q_2] \ldots\ldots [q_{k-1}X_kq_k]$ P of the grammar.

Example 7.27 Convert the following PDA into equivalent CFG.

$\delta(q_0, a, z_0) \to (q_0, z_1z_0)$
$\delta(q_0, a, z_1) \to (q_0, z_1z_1)$
$\delta(q_0, b, z_1) \to (q_1, \lambda)$
$\delta(q_1, b, z_1) \to (q_1, \lambda)$
$\delta(q_1, b, z_0) \to (q_1, z_2z_0)$
$\delta(q_1, b, z_2) \to (q_1, z_2z_2)$
$\delta(q_1, c, z_2) \to (q_2, \lambda)$
$\delta(q_2, c, z_2) \to (q_2, \lambda)$
$\delta(q_2, \lambda, z_0) \to (q_2, \lambda)$ // accepted by the empty stack

Solution: The PDA contains three states: q_0, q_1, and q_2. The following productions are added to the CFG [According to rule (1)]

$$S \to [q_0\ z_0\ q_0]/[q_0\ z_0\ q_1]/[q_0\ z_0\ q_2]$$

The transitional functions (iii), (iv), (vii), (viii), and (ix) are in the form $\delta(q, a, Y) \to (r, \varepsilon)$. Thus, the following five productions are added to the CFG [According to rule (2)]

$(q_0\ z_1\ q_1) \to b$ // From production (iii)
$(q_1\ z_1\ q_1) \to b$ // From production (iv)
$(q_1\ z_2\ q_2) \to c$ // From production (vii)
$(q_2\ z_2\ q_2) \to c$ // From production (viii)
$(q_2\ z_0\ q_2) \to \in$ // From production (ix)

For the remaining transitional functions, the productions are as follows:

$$\delta(q_0, a, z_0) \rightarrow (q_0, z_1z_0)$$

$$(q_0\, z_0\, q_0) \rightarrow a(q_0\, z_1\, q_0)\,(q_0\, z_0\, q_0)$$
$$(q_0\, z_0\, q_0) \rightarrow a(q_0\, z_1\, q_1)\,(q_1\, z_0\, q_0)$$
$$(q_0\, z_0\, q_0) \rightarrow a(q_0\, z_1\, q_2)\,(q_2\, z_0\, q_0)$$
$$(q_0\, z_0\, q_1) \rightarrow a(q_0\, z_1\, q_0)\,(q_0\, z_0\, q_1)$$
$$(q_0\, z_0\, q_1) \rightarrow a(q_0\, z_1\, q_1)\,(q_1\, z_0\, q_1)$$
$$(q_0\, z_0\, q_1) \rightarrow a(q_0\, z_1\, q_2)\,(q_2\, z_0\, q_1)$$
$$(q_0\, z_0\, q_2) \rightarrow a(q_0\, z_1\, q_0)\,(q_0\, z_0\, q_2)$$
$$(q_0\, z_0\, q_2) \rightarrow a(q_0\, z_1\, q_1)\,(q_1\, z_0\, q_2)$$
$$(q_0\, z_0\, q_2) \rightarrow a(q_0\, z_1\, q_2)\,(q_2\, z_0\, q_2)$$

$$\delta(q_0, a, z_1) \rightarrow (q_0, z_1z_1)$$

$$(q_0\, z_1\, q_0) \rightarrow a(q_0\, z_1\, q_0)\,(q_0\, z_1\, q_0)$$
$$(q_0\, z_1\, q_0) \rightarrow a(q_0\, z_1\, q_1)\,(q_1\, z_1\, q_0)$$
$$(q_0\, z_1\, q_0) \rightarrow a(q_0\, z_1\, q_2)\,(q_2\, z_1\, q_0)$$
$$(q_0\, z_1\, q_1) \rightarrow a(q_0\, z_1\, q_0)\,(q_0\, z_1\, q_1)$$
$$(q_0\, z_1\, q_1) \rightarrow a(q_0\, z_1\, q_1)\,(q_1\, z_1\, q_1)$$
$$(q_0\, z_1\, q_1) \rightarrow a(q_0\, z_1\, q_2)\,(q_2\, z_1\, q_1)$$
$$(q_0\, z_1\, q_2) \rightarrow a(q_0\, z_1\, q_0)\,(q_0\, z_1\, q_2)$$
$$(q_0\, z_1\, q_2) \rightarrow a(q_0\, z_1\, q_1)\,(q_1\, z_1\, q_2)$$
$$(q_0\, z_1\, q_2) \rightarrow a(q_0\, z_1\, q_2)\,(q_2\, z_1\, q_2)$$

$$\delta(q_1, b, z_0) \rightarrow (q_1, z_2z_0)$$

$$(q_1\, z_0\, q_0) \rightarrow b(q_1\, z_2\, q_0)\,(q_0\, z_0\, q_0)$$
$$(q_1\, z_0\, q_0) \rightarrow b(q_1\, z_2\, q_1)\,(q_1\, z_0\, q_0)$$
$$(q_1\, z_0\, q_0) \rightarrow b(q_1\, z_2\, q_2)\,(q_2\, z_0\, q_0)$$
$$(q_1\, z_0\, q_1) \rightarrow b(q_1\, z_2\, q_0)\,(q_0\, z_0\, q_1)$$
$$(q_1\, z_0\, q_1) \rightarrow b(q_1\, z_2\, q_1)\,(q_1\, z_0\, q_1)$$
$$(q_1\, z_0\, q_1) \rightarrow b(q_1\, z_2\, q_2)\,(q_2\, z_0\, q_1)$$
$$(q_1\, z_0\, q_2) \rightarrow b(q_1\, z_2\, q_0)\,(q_0\, z_0\, q_2)$$
$$(q_1\, z_0\, q_2) \rightarrow b(q_1\, z_2\, q_1)\,(q_1\, z_0\, q_2)$$
$$(q_1\, z_0\, q_2) \rightarrow b(q_1\, z_2\, q_2)\,(q_2\, z_0\, q_2)$$

$$\delta(q_1, b, z_2) \rightarrow (q_1, z_2z_2)$$

$$(q_1\, z_2\, q_0) \rightarrow b(q_1\, z_2\, q_0)\,(q_0\, z_2\, q_0)$$
$$(q_1\, z_2\, q_0) \rightarrow b(q_1\, z_2\, q_1)\,(q_1\, z_2\, q_0)$$
$$(q_1\, z_2\, q_0) \rightarrow b(q_1\, z_2\, q_2)\,(q_2\, z_2\, q_0)$$
$$(q_1\, z_2\, q_1) \rightarrow b(q_1\, z_2\, q_0)\,(q_0\, z_2\, q_1)$$
$$(q_1\, z_2\, q_1) \rightarrow b(q_1\, z_2\, q_1)\,(q_1\, z_2\, q_1)$$
$$(q_1\, z_2\, q_1) \rightarrow b(q_1\, z_2\, q_2)\,(q_2\, z_2\, q_1)$$
$$(q_1\, z_2\, q_2) \rightarrow b(q_1\, z_2\, q_0)\,(q_0\, z_2\, q_2)$$

$$(q_1\ z_2\ q_2) \to b(q_1\ z_2\ q_1)\ (q_1\ z_2\ q_2)$$
$$(q_1\ z_2\ q_2) \to b(q_1\ z_2\ q_2)\ (q_2\ z_2\ q_2).$$

Example 7.28 Convert the following PDA into an equivalent CFG.

$\delta(q_0, a, z_0) \to (q_1, z_1z_0)$
$\delta(q_0, b, z_0) \to (q_1, z_2z_0)$
$\delta(q_1, a, z_1) \to (q_1, z_1z_1)$
$\delta(q_1, b, z_1) \to (q_1, \lambda)$
$\delta(q_1, b, z_2) \to (q_1, z_2z_2)$
$\delta(q_1, a, z_2) \to (q_1, \lambda)$
$\delta(q_1, \lambda, z_0) \to (q_1, \lambda)$ // accepted by the empty stack

Solution: The PDA contains two states: q_0 and q_1. The following productions are added to the CFG [according to rule (1)].

$$S \to [q_0\ z_0\ q_0]/[q_0\ z_0\ q_1]$$

Transitional functions (iv), (vi), and (vii) are in the form $\delta(q, a, Y) \to (r, \in)$. Thus, the following three productions are added to the CFG [according to rule (2)]

$(q_1\ z_1\ q_1) \to b$ // From production (iii)
$(q_1\ z_2\ q_2) \to a$ // From production (iv)
$(q_1\ z_0\ q_1) \to \varepsilon$ // From production (vii)

For the remaining transitional functions, the productions are as follows:

$\boldsymbol{\delta(q_0, a, z_0) \to (q_1, z_1z_0)}$

$(q_0\ z_0\ q_0) \to a(q_0\ z_1\ q_0)\ (q_0\ z_0\ q_0)$
$(q_0\ z_0\ q_0) \to a(q_0\ z_1\ q_1)\ (q_1\ z_0\ q_0)$
$(q_0\ z_0\ q_1) \to a(q_0\ z_1\ q_0)\ (q_0\ z_0\ q_1)$
$(q_0\ z_0\ q_1) \to a(q_0\ z_1\ q_1)\ (q_1\ z_0\ q_1)$

$\boldsymbol{\delta(q_0, b, z_0) \to (q_1, z_2z_0)}$

$(q_0\ z_0\ q_0) \to b(q_0\ z_2\ q_0)\ (q_0\ z_0\ q_0)$
$(q_0\ z_0\ q_0) \to b(q_0\ z_2\ q_1)\ (q_1\ z_0\ q_0)$
$(q_0\ z_0\ q_1) \to b(q_0\ z_2\ q_0)\ (q_0\ z_0\ q_1)$
$(q_0\ z_0\ q_1) \to b(q_0\ z_2\ q_1)\ (q_1\ z_0\ q_1)$

$\boldsymbol{\delta(q_1, a, z_1) \to (q_1, z_1z_1)}$

$(q_1\ z_1\ q_0) \to a(q_1\ z_1\ q_0)\ (q_0\ z_1\ q_0)$
$(q_1\ z_1\ q_0) \to a(q_1\ z_1\ q_1)\ (q_1\ z_1\ q_0)$
$(q_1\ z_1\ q_1) \to a(q_1\ z_1\ q_0)\ (q_0\ z_1\ q_1)$
$(q_1\ z_1\ q_1) \to a(q_1\ z_1\ q_1)\ (q_1\ z_1\ q_1)$

$$\delta(q_1, b, z_2) \rightarrow (q_1, z_2z_2)$$

$$(q_1\ z_2\ q_0) \rightarrow b(q_1\ z_2\ q_0)\ (q_0\ z_2\ q_0)$$
$$(q_1\ z_2\ q_0) \rightarrow b(q_1\ z_2\ q_1)\ (q_1\ z_2\ q_0)$$
$$(q_1\ z_2\ q_1) \rightarrow b(q_1\ z_2\ q_0)\ (q_0\ z_2\ q_1)$$
$$(q_1\ z_2\ q_1) \rightarrow b(q_1\ z_2\ q_1)\ (q_1\ z_2\ q_1)$$

7.6 Graphical Notation for PDA

The mathematical notation for a PDA is $(Q, \Sigma, \Gamma, \delta, q_0, z_0, F)$. In a PDA, the transitional function δ consists of three touples: first is a present state, second is the present input, and the third is the stack top symbol, which generates one next state and the stack symbol(s) if a symbol is pushed into the stack or $\in$, if the top most symbol is popped from the stack.

In the graphical notation of the PDA, there are states. Among them a circle with an arrow indicates a beginning state and a state with double circle indicates a final state. The state transitions are denoted by arrows. The labels of the state transitions consists of input symbol, previous stack top symbol (at t_{i-1}) and the current stack top symbol (at t_i) which is added after the transitions or null symbol (if a symbol is popped).

Example 7.29 Construct a PDA with a graphical notation to accept L = (a, b)* with equal number of 'a' and 'b', i.e., $n_a(L) = n_b(L)$ by the final state.

Solution: At the beginning of transition, the PDA is in state q_0 with stack z_0. The string may start with 'a' or 'b'. If the string starts with 'a', one z_1 is pushed into the stack. If the string starts with 'b', one z_2 is pushed into the stack. If 'b' is traversed after 'a', and the stack top is z_1, that stack top is popped. If 'a' is traversed after 'b', and the stack top is z_2, that stack top is popped. If 'a' is traversed after 'a', and the stack top is z_1, one 'z_1' is pushed into the stack. If 'b' is traversed after 'b', and the stack top is z_2, one 'z_2' is pushed into the stack.

The PDA in graphical notation is as follows:

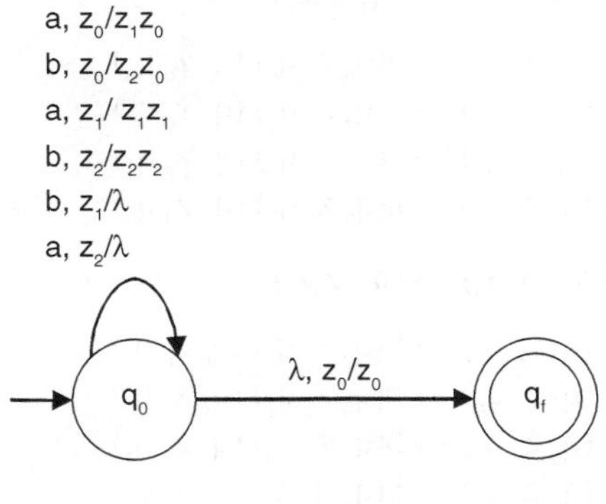

Example 7.30 Construct a PDA with a graphical notation to accept the language L = {WCW^R, where $W \in (a, b)^+$ and W^R is the reverse of W} by the final state.

Solution: At the beginning of the transition, the machine is in state q_0 and the stack top symbol is z_0. If it gets the input symbol 'a', one z_1 is pushed into the stack; if it gets the input symbol 'b', one z_2 is pushed into the stack and the state is changed from q_0 to q_1. In state q_1, if it gets the input 'a' and the stack top

z_1 or z_2, another z_1 or z_2 is pushed into the stack, respectively. In state q_1, if it gets the input 'b' with the stack top z_2 or z_1, another z_2 or z_1 is pushed into the stack, respectively.

C is the symbol which differentiates W with W^R. Before C, W is traversed. So, at the time of traversing C, the stack top symbol may be z_1 or z_2. In state q_1, if the PDA gets C as input and the stack top z_1 or z_2, no operation is done on the stack, but only the state is changed from q_1 to q_2. After C, the string W^R is traversed. If the machine gets 'a' as input, the stack top must be z_1. And that z_1 is popped from the stack top. If the machine gets 'b' as input, the stack top must be z_2. And that z_2 is popped from the stack top. The state is not changed. By this process, the whole string WCW^R is traversed. In state q_2, if the machine gets no input but stack top z_0, the machine goes to its final state q_f.

The graphical notation for the PDA is as follows:

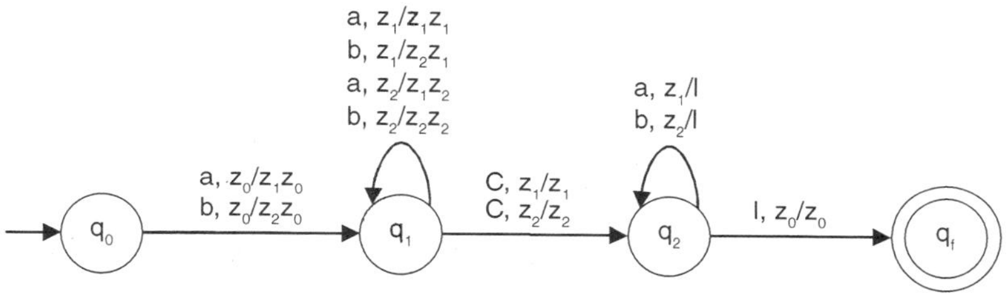

7.7 Two-stack PDA

Finite automata recognize regular languages such as $\{a^n \mid n \geq 0\}$. Adding one stack to a finite automata, it becomes PDA which can recognize context-free language $\{a^n b^n, n \geq 0\}$. In the case of context-sensitive language such as $\{a^n b^n c^n, n \geq 0\}$, the PDA is helpless because of only one auxiliary storage. Now the question arises—if more than one stack is added in the form of auxiliary storage with PDA, does its power increase or not.

From this question, the concept of a two-stack PDA has come. Not only two stacks, but more than two stacks can be added to a PDA.

A PDA can be deterministic or non-deterministic, but two-stack PDA is deterministic and it accepts all context-free languages, which may be deterministic or non-deterministic, with context-sensitive language such as $\{a^n b^n c^n, n \geq 0\}$. In the Turing machine chapter, we shall learn that two-stack PDA is equivalent to the Turing machine. There, we shall also learn a theorem called the Minsky's theorem.

Definition: A two-stack PDA consists of a 9-tuple

$$M = (Q, \Sigma, \Gamma, \Gamma', \delta, q_0, z_1, z_2, F)$$

where

Q denotes a finite set of states.
Σ denotes a finite set of input symbols.
Γ denotes a finite set of first stack symbols.
Γ' denotes a finite set of second stack symbols
δ denotes the transitional functions.
q_0 is the initial state of PDA $[q_0 \in Q]$.
z_1 is the stack bottom symbol of stack 1.
z_2 is the stack bottom symbol of stack 2.
F is the final state of PDA.

In PDA, the transitional function δ is in the form

$$Q \times (\Sigma \cup \{\lambda\}) \times \Gamma \times \Gamma' \rightarrow (Q, \Gamma, \Gamma')$$

Example 7.31 Construct a two-stack PDA for the language $L = \{a^n b^n c^n, n \geq 0\}$.

Solution: While scanning 'a', push X into stack 1. While scanning 'b', push 'Y' into stack 2. While scanning 'c' with stack top X in 1 and stack top Y in 2, pop X and Y from stack 1 and 2, respectively. The transitional functions are as follows:

$$\delta(q_0, \lambda, z_1, z_2) \rightarrow (q_f, z_1, z_2)$$
$$\delta(q_0, a, z_1, z_2) \rightarrow (q_0, Xz_1, z_2)$$
$$\delta(q_0, a, X, z_2) \rightarrow (q_0, XX, z_2)$$
$$\delta(q_0, b, X, z_2) \rightarrow (q_0, X, Yz_2)$$
$$\delta(q_0, b, X, Y) \rightarrow (q_0, X, YY)$$
$$\delta(q_0, c, X, Y) \rightarrow (q_1, \lambda, \lambda)$$
$$\delta(q_1, c, X, Y) \rightarrow (q_1, \lambda, \lambda)$$
$$\delta(q_1, \lambda, z_1, z_2) \rightarrow (q_{f}, z_1, z_2) \text{ // accepted by the final state.}$$

Theorem 7.2: Intersection of RE and CFL is CFL.

Proof: Let L is a CFL accepted by a PDA M1 = $\{Q_1, \Sigma, \Gamma_1, \delta_1, q_{01}, z_0, F_1\}$ and R be a regular expression accepted by a FA $M_2 = \{Q_2, \Sigma, \delta_2, q_{02}, F_2\}$. A new PDA $M_3 = \{Q, \Sigma, \Gamma, \delta, q_0, z_0, F\}$ is designed which performs computation of M_1 and M_2 in parallel and accepts a string accepted by both M_1 and M_2. M_3 is designed as follows.

$$Q = Q_1 \times Q_2 \text{ [Cartesian product]}$$
$$\Sigma = \Sigma$$
$$\Gamma = \Gamma_1$$
$$F = F_1 \times F_2$$

δ: $[(S_1, S_2), i/p, Z) \rightarrow (q_1, q_2, Z')$ if $\delta_1(S_1, i/p, Z) \rightarrow (q_1, Z')$ and $\delta_2(S_2, i/p) \rightarrow q_2$
[Z and $Z' \in \Gamma$]

$$q_0 = q_{01} \cup q_{02}$$

The transitional function of M_3 keeps track of transaction from S_1 to q_1 in PDA M_1 and S_2 to q_2 in FA M_2 for same input alphabet. Thus a string W accepted by M_3 if and only if, it is accepted by both M_1 and M_2. Therefore $W \in L(M_1) \cap L(M_2)$. As W is accepted by a PDA, W is a CFL.

What We Have Learned So Far

1. Pushdown automata (in short PDA) is the machine format of context-free language.
2. The mechanical diagram of a pushdown automata contains the input tape, reading head, finite control, and a stack.
3. A pushdown automata consists of a 7-tuple $M = (Q, \Sigma, \Gamma, \delta, q_0, z_0, F)$ where Q, Σ, q_0, and F have their original meaning, Γ is finite set of stack symbols, and z_0 is the stack bottom symbol.
4. In a PDA, the transitional function δ is in the form $Q \times (\Sigma \cup \{\lambda\}) \times \Gamma \rightarrow (Q, \Gamma)$.

5. Instantaneous description (ID) remembers the information of state and stack content at a given instance of time.
6. There are two ways to declare a string to be accepted by pushdown automata: (a) accepted by the empty Stack (store) and (b) accepted by the final state.
7. A string is declared accepted by a PDA by the empty stack if the string is totally traversed and the stack is empty.
8. A string is declared accepted by a PDA by the final state if the string is totally traversed and the machine reaches its final state.
9. A string is accepted by a PDA by the empty stack if and only if the language is accepted by that PDA by the final state.
10. If a PDA being in a state with a single input and a single stack symbol gives a single move, then the PDA is called deterministic pushdown automata.
11. If a PDA being in a state with a single input and a single stack symbol gives more than one move for any of its transitional functions, then the PDA is called non-deterministic pushdown automata.
12. From context-free grammar, the NPDA can be directly constructed.

Solved Problems

1. Design a PDA to accept the language of nested balanced parentheses (where the number of opening and closing parenthesis is greater than 0).

Solution: A nested balanced parenthesis is in the form ((())). More precisely,

```
(
  (
     (
     )
  )
)
```

This type of nested balanced parenthesis has two types of symbols: open parenthesis '(' and close parenthesis ')'. As it is nested, first the PDA has to traverse the open parenthesis and then the close parenthesis.

The transitional function for traversing the first open parenthesis is [the beginning state is q_0, input symbol is '(' and stack top is z_0.)

$$\delta(q_0, (, z_0) \rightarrow (q_0, z_1z_0)$$

From then on, the state is q_0, the input symbol is '(' [if the number of open parenthesis is >1], and the stack top is z_0. The transitional function for traversing the next open parenthesis is

$$\delta(q_0, (, z_0) \rightarrow (q_0, z_1z_1)$$

By using the previous transitional function, all the remaining open parentheses will be traversed.

After traversing all open parentheses, the PDA will get the first close parenthesis. At that time the state is q_0, the input symbol is ')', and the stack top is z_1. After getting the ')' as an input symbol, the stack top (here z_1) will be popped from the stack and the state will be changed to q_1. The transitional function is

$$\delta(q_0,), z_1) \rightarrow (q_1, \lambda)$$

After that, the state is q_1, the input symbol is ')', and the stack top is z_1 (if the string contains multiple open and close parentheses). The transitional function is

$$\delta(q_1,), z_1) \rightarrow (q_1, \lambda)$$

When all the input symbols will be traversed, the state is q_1, the input symbol is λ, and the stack top is z_0.

If we have to design the PDA accepted by the empty stack, the δ function will be

$$\delta(q_1, \lambda, z_0) \rightarrow (q_1, \lambda)$$

If we have to design the PDA accepted by the final state, the δ function will be

$$\delta(q_1, \lambda, z_0) \rightarrow (q_f, z_0)$$

The transitional functions for the PDA is

$$\delta(q_0, (, z_0) \rightarrow (q_0, z_1z_0)$$
$$\delta(q_0, (, z_0) \rightarrow (q_0, z_1z_1)$$
$$\delta(q_0,), z_1) \rightarrow (q_1, \lambda)$$
$$\delta(q_1,), z_1) \rightarrow (q_1, \lambda)$$
$$\delta(q_1, \lambda, z_0) \rightarrow (q_1, \lambda) \quad \text{// accepted by the empty stack}$$
$$\delta(q_1, \lambda, z_0) \rightarrow (q_f, z_0) \quad \text{// accepted by the final state}$$

2. Design a PDA to accept the language of balanced parentheses (where the number of opening and closing parentheses is greater than 0).

Solution: There are two types of parentheses—open parentheses and close parentheses. Balanced parenthesis means for each open parenthesis there is a close parenthesis. It may be nested like ((( ))) or may be non-nested like (( ) ( )) or ( )( )( ) or any other form. But for each open parenthesis, there is a close parenthesis. (The definition that a balanced parenthesis means equal number of open and close parentheses is wrong.)

As an example,) ( )( ) (is not a balanced parenthesis. The string must start with an open parenthesis and after that, open parenthesis or close parenthesis can appear, but for any position of the string, the number of open parenthesis is greater than or equal to the number of close parenthesis.

The transitional function for traversing the first open parenthesis is [the beginning state is q_0, the input symbol is '(', and the stack top is z_0.]

$$\delta(q_0, (, z_0) \rightarrow (q_1, z_1z_0)$$

(As the null string is not accepted by the PDA)

After the first open parenthesis, there may be an open parenthesis or close parenthesis. If the open parenthesis is to be traversed, then a z_1 is added at the stack top. The transitional function is

$$\delta(q_1, (, z_1) \rightarrow (q_1, z_1z_1)$$

If a close parenthesis is to be traversed, then one z_1 is popped from the stack top. The transitional function is

$$\delta(q_1,), z_1) \rightarrow (q_1, \lambda)$$

It may happen that the state is q_1, the stack top is z_0, and the input symbol is '('. [As an example, for the case (( ))( ).] The transitional function for the case is

$$\delta(q_1, (, z_0) \rightarrow (q_1, z_1z_0)$$

[But the type of case—where the state is q_1, the stack top is z_0, and the input symbol is ')' cannot appear.]

If we have to design the PDA accepted by the empty stack, the δ function is

$$\delta(q_1, \lambda, z_0) \rightarrow (q_1, \lambda)$$

If we have to design the PDA accepted by the final state, the δ function is

$$\delta(q_1, \lambda, z_0) \rightarrow (q_f, z_0)$$

The transitional functions for the entire PDA are

$$\delta(q_0, (, z_0) \rightarrow (q_1, z_1z_0)$$
$$\delta(q_1, (, z_1) \rightarrow (q_1, z_1z_1)$$
$$\delta(q_1,), z_1) \rightarrow (q_1, \lambda)$$
$$\delta(q_1, (, z_0) \rightarrow (q_1, z_1z_0)$$
$$\delta(q_1, \lambda, z_0) \rightarrow (q_1, \lambda) \quad \text{// accepted by the empty stack}$$
$$\delta(q_1, \lambda, z_0) \rightarrow (q_f, z_0) \quad \text{// accepted by the final state}$$

3. Design a PDA to accept the following language $L = 0^n1^n \mid n > 1$. [UPTU 2002]

   Solution: Already solved.

4. Design a PDA to accept the language $L = a^mb^nc^{m-n}$, where $m >= n$ and $m, n > 0$

   Solution: The language consists of 'a', 'b', and 'c'. Here, in every string, all 'a' will appear before 'b' and after 'b', 'c' will appear. The number of 'c' is the subtraction of the number of 'a' and the number of 'b'. At the beginning, the machine is in state q_0 with the stack top z_0 and the input 'a'. After traversing, the machine will add z_1 in the stack. The transitional function is

$$\delta(q_0, a, z_0) \rightarrow (q_0, z_1z_0)$$

If the number of 'a' is greater than one, then the machine has to traverse 'a' with state q_0 and stack top z_1. Another z_1 is added to the stack top. The transitional function is

$$\delta(q_0, a, z_1) \rightarrow (q_0, z_1z_1)$$

After traversal of all 'a', 'b' will appear. When 'b' will come, at that time, the state is q_0, and the stack top is z_1 .The stack top is popped from the top. The transitional function is

$$\delta(q_0, b, z_1) \rightarrow (q_1, \lambda)$$

If the number of ‘b’ is greater than one, then the machine has to traverse ‘b’ with state q_1 and stack top z_1. (The number of ‘b’ greater than one means that the number of ‘a’ was greater than one.) The stack top is popped from the top. The transitional function is

$$\delta(q_1, b, z_1) \rightarrow (q_1, \lambda)$$

If the number of ‘b’ is less than the number of ‘a’, then ‘c’ will appear after full traversal of all the ‘b’s. At that time, the state is q_1, and the stack top is z_1. After traversal of ‘c’, the stack top is popped. The transitional function is

$$\delta(q_1, c, z_1) \rightarrow (q_2, \lambda)$$

If the number of ‘c’ is greater than one, then the machine have to traverse ‘c’ with state q_2 and stack top z_1. (The number of ‘c’ greater than one means that the number of ‘a’ was greater than one.) The stack top is popped. The transitional function is

$$\delta(q_2, c, z_1) \rightarrow (q_2, \lambda)$$

If we have to design the PDA accepted by the empty stack, the δ function is

$$\delta(q_2, \lambda, z_0) \rightarrow (q_2, \lambda)$$

If we have to design the PDA accepted by the final state, the δ function is

$$\delta(q_2, \lambda, z_0) \rightarrow (q_f, \lambda)$$

It may also happen that m = n, i.e., the number of ‘a’ is equal to the number of ‘b’. At that case, there is no ‘c’ in the language. The string ends after full traversal of all ‘b’. After traversal of all ‘b’s, the state of the machine is q_1. If we have to design the PDA accepted by the empty stack, the δ function is

$$\delta(q_1, \lambda, z_0) \rightarrow (q_1, \lambda)$$

If we have to design the PDA accepted by the final state, the δ function is

$$\delta(q_1, \lambda, z_0) \rightarrow (q_f, \lambda)$$

The transitional functions for the entire PDA are

$$\delta(q_0, a, z_0) \rightarrow (q_0, z_1z_0)$$
$$\delta(q_0, a, z_1) \rightarrow (q_0, z_1z_1)$$
$$\delta(q_0, b, z_1) \rightarrow (q_1, \lambda)$$
$$\delta(q_1, b, z_1) \rightarrow (q_1, \lambda)$$
$$\delta(q_1, c, z_1) \rightarrow (q_2, \lambda)$$
$$\delta(q_2, c, z_1) \rightarrow (q_2, \lambda)$$
$\delta(q_2, \lambda, z_0) \rightarrow (q_2, \lambda)$ // accepted by the empty stack if m > n
$\delta(q_2, \lambda, z_0) \rightarrow (q_f, \lambda)$ // accepted by the final state if m > n
$\delta(q_1, \lambda, z_0) \rightarrow (q_1, \lambda)$ // accepted by the empty stack if m = n
$\delta(q_1, \lambda, z_0) \rightarrow (q_f, \lambda)$ // accepted by the final state if m = n

5. Design a PDA to accept the language $L = a^nb^mc^{m-n}$, where m > n and m, n > 0. Show an ID for the string aabbbbcc.

Solution: The language consists of 'a', 'b', and 'c'. Here, in every string all 'a' will appear before 'b', and after 'b', 'c' will appear. The number of 'c' is the subtraction of the number of 'b' and the number of 'a'. In the string, the number of 'b' is the sum of the number of 'a' and the number of 'c'. It can be designed as follows. For n number of 'a', n number of z_1 are added to the stack. For the next n number of b, those z_1 are popped from the stack and z_0 becomes the stack top. For the remaining (m – n) number of 'b', (m – n) number of z_2 are added to the stack. For (m – n) number of 'c', (m – n) number of z_2 are popped from the stack.
The transitional functions are designed as follows.

At the beginning, the machine is in state q_0 with the stack top z_0 and the input 'a'. After traversing, the machine will add z_1 in the stack. The transitional function is

$$\delta(q_0, a, z_0) \rightarrow (q_0, z_1z_0)$$

If the number of 'a' is greater than one, then the machine has to traverse 'a' with state q_0 and the stack top z_1. Another z_1 is added to the stack top. The transitional function is

$$\delta(q_0, a, z_1) \rightarrow (q_0, z_1z_1)$$

After traversal of all 'a', 'b' will appear. When 'b' will come, at that time the state is q_0, and the stack top is z_1 .The stack top is popped from the top. The transitional function is

$$\delta(q_0, b, z_1) \rightarrow (q_1, \lambda)$$

If the number of 'a' is greater than one, then the number of 'b' is also greater than one because m = (m – n) + n. The machine has to traverse 'b' with the state q_1 and the stack top z_1. The stack top is popped from the top. The transitional function is

$$\delta(q_1, b, z_1) \rightarrow (q_1, \lambda)$$

After some time, all z_1 are popped from the stack. The stack top becomes z_0. As m is greater than n, the input symbol is still 'b' with the machine state q_1. In this situation, z_2 is pushed into the stack. The transitional function is

$$\delta(q_1, b, z_0) \rightarrow (q_1, z_2z_0)$$

If m – n is greater than one, i.e., the number of 'c' is more than one, then the machine state is q_1, the input symbol is 'b', and the stack top is z_2. Another z_2 is pushed into the stack. The transitional function is

$$\delta(q_1, b, z_2) \rightarrow (q_1, z_2z_2)$$

After full traversal of all 'b's,'c' appears in the string. When 'c' appears, the machine state is q_1 and the stack top is z_2. The stack top z_2 is popped from the stack. The transitional function is

$$\delta(q_1, c, z_2) \rightarrow (q_2, \lambda)$$

If m – n is greater than one, i.e., the number of 'c' is more than one, the machine has to traverse 'c' with state q_2 and stack top z_2. The stack top z_2 is popped from the stack. The transitional function is

$$\delta(q_2, c, z_2) \rightarrow (q_2, \lambda)$$

If we have to design the PDA accepted by the empty stack, the δ function is

$$\delta(q_2, \lambda, z_0) \rightarrow (q_2, \lambda)$$

If we have to design the PDA accepted by the final state, the δ function is

$$\delta(q_2, \lambda, z_0) \rightarrow (q_f, \lambda)$$

The transitional functions for the entire PDA are

$$\delta(q_0, a, z_0) \rightarrow (q_0, z_1z_0)$$
$$\delta(q_0, a, z_1) \rightarrow (q_0, z_1z_1)$$
$$\delta(q_0, b, z_1) \rightarrow (q_1, \lambda)$$
$$\delta(q_1, b, z_1) \rightarrow (q_1, \lambda)$$
$$\delta(q_1, b, z_0) \rightarrow (q_1, z_2z_0)$$
$$\delta(q_1, b, z_2) \rightarrow (q_1, z_2z_2)$$
$$\delta(q_1, c, z_2) \rightarrow (q_2, \lambda)$$
$$\delta(q_2, c, z_2) \rightarrow (q_2, \lambda)$$
$$\delta(q_2, \lambda, z_0) \rightarrow (q_2, \lambda) \text{ // accepted by the empty stack}$$
$$\delta(q_2, \lambda, z_0) \rightarrow (q_f, \lambda) \text{ // accepted by the final state}$$

ID for the String aabbbc

$\delta(q_0, \text{aabbbbcc}, z_0) \rightarrow (q_0, \text{abbbbcc}, z_1z_0) \rightarrow (q_0, \text{bbbbcc}, z_1z_1z_0) \rightarrow (q_1, \text{bbbcc}, z_1z_0)$
$\rightarrow (q_1, \text{bbcc}, z_0) \rightarrow (q_1, \text{bcc}, z_2z_0) \rightarrow (q_1, \text{cc}, z_2z_2z_0) \rightarrow (q_2, \text{c}, z_2z_0)$
$\rightarrow (q_2, \lambda, z_0) \rightarrow (q_2, \lambda)$ // accepted by empty stack

6. Construct a PDA for the following $L = \{a^ncb^{2n}/n \geq 1\}$ over alphabet {a, b, c}. [UPTU 2003]

   Solution: For each 'a', push two 'X' in the stack. Traverse 'c' with no action. Pop 'X' for each 'b'.

   The transitional functions are

$$\delta(q_0, a, z_0) \rightarrow (q_1, XXz_0)$$
$$\delta(q_1, a, X) \rightarrow (q_1, XXX)$$
$$\delta(q_1, c, X) \rightarrow (q_2, X)$$
$$\delta(q_2, b, X) \rightarrow (q_2, \lambda)$$
$$\delta(q_2, \lambda, z_0) \rightarrow (q_2, \lambda) \text{ // accepted by the empty stack}$$
$$\delta(q_2, \lambda, z_0) \rightarrow (q_f, \lambda) \text{ // accepted by the final state}$$

7. Design a PDA for hypertext markup language (HTML) consisting of all tags having immediate closing tags within the <BODY> </BODY> tag.

   Show an ID for the following language

```
<HTML>
  <HEAD>
  <TITLE>
  My First Web Page
  </TITLE>
  </HEAD>
  <BODY>
  <B> First Web Page </B>
  <P> HTML </B> is an <I>interpreted language. </I> No error is produced in HTML. </P>
  </BODY>
</HTML>
```

Solution: Hypertext markup language or HTML is a language used for webpage designing. This language is an interpreted language and produces no error. The language has a basic structure in the following form.

```
<HTML>
  <HEAD>
    <TITLE>
    --------------
    </TITLE>
  </HEAD>
<BODY>
Some other tags with closing tags
</BODY>
</HTML>
```

The tags can be of capital letter, small letter, or a mixture of both. The tags within < > are called open tags and the tags within < / > are called closing tags. Most of the tags in HTML have closing tags, but some of the tags such as <BR>, <HR> have no closing tags.

In the <BODY> </BODY> tag, all the other tags appear. The tags may appear in any order. But for all the opening tags, there is a closing tag except the tags which do not have any closing tag. For the line HTML </B> is an <I>interpreted language. </I>, the output is HTML is an *interpreted language.*

If the line is in the form HTML is an <I>interpreted </B>language. </I>, the output is HTML is an *interpreted language.*

In this section, we are interested in tags with immediate closing tags like the first example, i.e., the tag within the body which has opened first will close at last.

Every HTML file starts with a <HTML> tag. At state q_0, when the machine gets <HTML> as input, it pushes $Z_{\text{<HTML>}}$ in the stack. The transitional function is

$$\delta(q_0, \text{<HTML>}, Z_0) \rightarrow (q_1, Z_{\text{<HTML>}} Z_0)$$

In every HTML document after <HTML>, the <HEAD> tag comes. The transitional function for traversing <HEAD> is

$$\delta(q_1, \text{<HEAD>}, Z_{\text{<HTML>}}) \rightarrow (q_2, Z_{\text{<HEAD>}} Z_{\text{<HTML>}})$$

In every HTML document after <HEAD>, the <TITLE> tag comes. The transitional function for traversing <TITLE> is

$$\delta(q_2, \text{<TITLE>}, Z_{\text{<HEAD>}}) \rightarrow (q_3, Z_{\text{<TITLE>}} Z_{\text{<HEAD>}})$$

When the machine gets a closing </TITLE> tag, it pops the stack top symbol $Z_{\text{<TITLE>}}$. The transitional function is

$$\delta(q_3, \text{</TITLE>}, Z_{\text{<title>}}) \rightarrow (q_3, \lambda)$$

When the machine gets a closing </HEAD> tag, it pops the stack top symbol $Z_{\text{<HEAD>}}$. The transitional function is

$$\delta(q_3, \text{</HEAD>}, Z_{\text{<HEAD>}}) \rightarrow (q_3, \lambda)$$

Now, the stack top is $Z_{\text{<HTML>}}$. The next symbol is <BODY>. The transitional function for traversing <BODY> tag is

$$\delta(q_3, \text{<BODY>}, Z_{\text{<HTML>}}) \rightarrow (q_4, Z_{\text{<BODY>}} Z_{\text{<HTML>}})$$

Within the <BODY> tag, any tag can come. Let us assume that a tag named <tag j> has come. Let the state of the PDA be q_i, where i is greater than or equal to 4. For traversing that tag, the transition function is

$$\delta(q_i, \text{<tag j>}, Z_{\text{<tag j'>}}) \rightarrow (q_{i+1}, Z_{\text{<tag j>}} Z_{\text{<tag j'>}})$$

If the <tag j> appears with the immediate closing tags, then its closing tag will get the symbol $Z_{\text{<tag j>}}$ as the stack top. That stack top is popped.

$$\delta(q_i, \text{</tag j>}, Z_{\text{<tag j>}}) \rightarrow (q_i, \lambda)$$

The last two tags are </BODY> and </HTML>. When the </BODY> is traversed, the stack top is $Z_{\text{<BODY>}}$. The stack top is popped by the following transitional function.

$$\delta(q_i, \text{</BODY>}, Z_{\text{<BODY>}}) \rightarrow (q_i, \lambda)$$

When </HTML> is traversed, the stack top is $Z_{\text{<HTML>}}$. The stack top is popped by the following transitional function.

$$\delta(q_i, \text{</HTML>}, Z_{\text{<HTML>}}) \rightarrow (q_i, \lambda)$$

If we have to design the PDA accepted by the empty stack, the δ function is

$$\delta(q_i, \lambda, z_0) \rightarrow (q_i, \lambda)$$

If we have to design the PDA accepted by the final state, the δ function is

$$\delta(q_i, \lambda, z_0) \rightarrow (q_f, \lambda)$$

ID for the given HTML document: It is not possible to show the string in each phase. Only the state and the stack condition are shown in the description.

$$\delta(q_0, \text{<HTML>}, Z_0) \rightarrow (q_1, Z_{\text{<HTML>}} Z_0)$$
$$\delta(q_1, \text{<HEAD>}, Z_{\text{<HTML>}}) \rightarrow (q_2, Z_{\text{<HEAD>}} Z_{\text{<HTML>}} Z_0)$$
$$\delta(q_2, \text{<TITLE>}, Z_{\text{<HEAD>}}) \rightarrow (q_3, Z_{\text{<TITLE>}} Z_{\text{<HEAD>}} Z_{\text{<HTML>}} Z_0)$$
$$\delta(q_3, \text{</TITLE>}, Z_{\text{<title>}}) \rightarrow (q_3, Z_{\text{<HEAD>}} Z_{\text{<HTML>}} Z_0)$$
$$\delta(q_3, \text{</HEAD>}, Z_{\text{<HEAD>}}) \rightarrow (q_3, Z_{\text{<HTML>}} Z_0)$$
$$\delta(q_3, \text{<BODY>}, Z_{\text{<HTML>}}) \rightarrow (q_4, Z_{\text{<BODY>}} Z_{\text{<HTML>}} Z_0)$$
$$\delta(q_4, \text{<B>}, Z_{\text{<BODY>}}) \rightarrow (q_5, Z_{\text{<B>}} Z_{\text{<BODY>}} Z_{\text{<HTML>}} Z_0)$$
$$\delta(q_5, \text{</B>}, Z_{\text{<B>}}) \rightarrow (q_5, Z_{\text{<BODY>}} Z_{\text{<HTML>}} Z_0)$$
$$\delta(q_5, \text{<P>}, Z_{\text{<Body>}}) \rightarrow (q_6, Z_{\text{<P>}} Z_{\text{<BODY>}} Z_{\text{<HTML>}} Z_0)$$
$$\delta(q_6, \text{</B>}, Z_{\text{<P>}}) \rightarrow (q_7, Z_{\text{<B>}} Z_{\text{<P>}} Z_{\text{<BODY>}} Z_{\text{<HTML>}} Z_0)$$
$$\delta(q_7, \text{</B>}, Z_{\text{<B>}}) \rightarrow (q_7, Z_{\text{<P>}} Z_{\text{<BODY>}} Z_{\text{<HTML>}} Z_0)$$

$$\delta(q_7, \text{<I>}, Z_{\text{<P>}}) \rightarrow (q_8, Z_{\text{<I>}} Z_{\text{< P>}} Z_{\text{<BODY>}} Z_{\text{<HTML>}} Z_0)$$
$$\delta(q_8, \text{</I>}, Z_{\text{<P>}}) \rightarrow (q_8, Z_{\text{<P>}} Z_{\text{<BODY>}} Z_{\text{<HTML>}} Z_0)$$
$$\delta(q_8, \text{</P>}, Z_{\text{<P>}}) \rightarrow (q_8, Z_{\text{<BODY>}} Z_{\text{<HTML>}} Z_0)$$
$$\delta(q_8, \text{</BODY>}, Z_{\text{<BODY>}}) \rightarrow (q_8, Z_{\text{<HTML>}} Z_0)$$
$$\delta(q_8, \text{</HTML>}, Z_{\text{<HTML>}}) \rightarrow (q_8, Z_0)$$
$$\delta(q_8, \lambda, Z_0) \rightarrow (q_8, \lambda) \text{ // accepted by the empty stack}$$

8. Design a PDA for the language $L = \{a^n b^m a^m b^n, \text{ where } m, n \geq 1\}$.

Solution: The language is a context-free language. The context-free grammar for the language is

$$S \rightarrow aSb/aAb$$
$$A \rightarrow bAa/ba$$

The grammar is not in GNF. So, first we need to convert the grammar into GNF. Take two non-terminals C_a and C_b and two productions $C_a \rightarrow a$ and $C_b \rightarrow b$. Replacing in appropriate positions, the modified context-free grammar is

$$S \rightarrow aSCb$$
$$S \rightarrow aACb$$
$$A \rightarrow bACa$$
$$A \rightarrow bCa$$
$$C_a \rightarrow a$$
$$C_b \rightarrow b$$

The context-free grammar is in GNF.
First, the start symbol S is pushed into the stack by the following production

$$\delta(q_0, \in, z_0) \rightarrow (q_1, Sz_0)$$

For the production $S \rightarrow aSC_b$, the transitional function is

$$\delta(q_1, a, S) \rightarrow (q_1, SC_b)$$

For the production $S \rightarrow aAC_b$, the transitional function is

$$\delta(q_1, a, S) \rightarrow (q_1, AC_b)$$

For the production $A \rightarrow bAC_a$, the transitional function is

$$\delta(q_1, b, A) \rightarrow (q_1, AC_a)$$

For the production $A \rightarrow bC_a$, the transitional function is

$$\delta(q_1, b, A) \rightarrow (q_1, C_a)$$

For the production $C_a \rightarrow a$, the transitional function is

$$\delta(q_1, a, C_a) \rightarrow (q_1, \lambda)$$

For the production $C_b \rightarrow b$, the transitional function is

$$\delta(q_1, b, C_b) \rightarrow (q_1, \lambda)$$

If we have to design the PDA accepted by the empty stack, the δ function is

$$\delta(q_1, \lambda, z_0) \rightarrow (q_1, \lambda)$$

If we have to design the PDA accepted by the final state, the δ function is

$$\delta(q_1, \lambda, z_0) \rightarrow (q_f, \lambda)$$

9. Construct a PDA for the language $L = \{(ab)^n\} \cup \{(ba)^n\}$ $n \geq 1$.

Solution: The given language is the union of two languages, $(ab)^n$ and $(ba)^n$. The grammar for the first language is

$$S_1 \rightarrow abS_1/ab$$

The grammar for the second language is

$$S_2 \rightarrow baS_2/ba$$

Here, we have considered two start symbols S_1 and S_2, respectively. As two languages are connected with the union symbol, consider another start symbol for the entire language as S. Make a production rule

$$S \rightarrow S_1/S_2$$

The production rule for the grammar satisfying the language is

$$S \rightarrow S_1/S_2$$
$$S_1 \rightarrow ab/abS_1$$
$$S_2 \rightarrow ba/baS_2$$

The production rule after removing the unit productions and the useless symbols is

$$S \rightarrow ab/ba/abS_1/baS_2$$
$$S_1 \rightarrow ab/abS_1$$
$$S_2 \rightarrow ba/baS_2$$

The grammar is not in GNF. Consider two non-terminals C_a and C_b and two productions $C_a \rightarrow a$ and $C_b \rightarrow b$. The modified production rule becomes

$$S \rightarrow aC_b/bC_a/aC_bS_1/bC_aS_2$$
$$S_1 \rightarrow aC_b/aC_bS_1$$
$$S_2 \rightarrow bC_a/bC_aS_2$$
$$C_a \rightarrow a$$
$$C_b \rightarrow b$$

The context-free grammar is in GNF.
First, the start symbol S is pushed into the stack by the following production

$$\delta(q_0, \in, z_0) \rightarrow (q_1, Sz_0)$$

For the production $S \rightarrow aC_b$, the transitional function is

$$\delta(q_1, a, S) \rightarrow (q_1, C_b)$$

For the production $S \rightarrow bC_a$, the transitional function is

$$\delta(q_1, b, S) \rightarrow (q_1, C_a)$$

For the production $S \rightarrow aC_bS_1$, the transitional function is

$$\delta(q_1, a, S) \rightarrow (q_1, C_bS_1)$$

For the production $S \rightarrow bC_aS_2$, the transitional function is

$$\delta(q_1, b, S) \rightarrow (q_1, C_aS_2)$$

For the production $S_1 \rightarrow aC_b$, the transitional function is

$$\delta(q_1, a, S_1) \rightarrow (q_1, C_b)$$

For the production $S_1 \rightarrow aC_bS_1$, the transitional function is

$$\delta(q_1, a, S_1) \rightarrow (q_1, C_bS_1)$$

For the production $S_2 \rightarrow bC_a$, the transitional function is

$$\delta(q_1, b, S_2) \rightarrow (q_1, C_a)$$

For the production $S_2 \rightarrow bC_aS_2$, the transitional function is

$$\delta(q_1, b, S_2) \rightarrow (q_1, C_aS_2)$$

For the production $C_a \rightarrow a$, the transitional function is

$$\delta(q_1, a, C_a) \rightarrow (q_1, \lambda)$$

For the production $C_b \rightarrow b$, the transitional function is

$$\delta(q_1, b, C_b) \rightarrow (q_1, \lambda)$$

If we have to design the PDA accepted by the empty stack, the δ function is

$$\delta(q_1, \lambda, z_0) \rightarrow (q_1, \lambda)$$

If we have to design the PDA accepted by the final state, the δ function is

$$\delta(q_1, \lambda, z_0) \rightarrow (q_f, \lambda)$$

10. Construct a PDA for the language $L = \{a^nb^n\} \cup \{a^mb^{2m}\}$ m, n > 0. [WBUT 2009 (IT)]

Solution: The given language is the union of two languages, a^nb^n and a^mb^{2m}. The grammar for the first language is

$$A \rightarrow aAb \text{ and } A \rightarrow ab$$

The grammar for the second language is

$$B \rightarrow aBbb \text{ and } B \rightarrow abb$$

Here, we have considered two start symbols A and B, respectively. As two languages are connected with a union symbol, consider another start symbol for the entire language as S. Make a production rule

$$S \rightarrow A/B$$

The production rule for the grammar satisfying the language is

$$S \rightarrow A/B$$
$$A \rightarrow aAb/ab$$
$$B \rightarrow aBbb/abb$$

The grammar contains two unit productions $S \rightarrow A$ and $S \rightarrow B$. After removing the unit productions, the production rules become

$$S \rightarrow aAb/aBbb/abb/ab$$
$$A \rightarrow aAb/ab$$
$$B \rightarrow aBbb/abb$$

The CFG is not in GNF. Consider two non-terminals, S_a and S_b, and two extra production rules, $S_a \rightarrow a$ and $S_b \rightarrow b$. Replacing a and b by S_a and S_b, respectively, in suitable places, the modified production rules become

$$S \rightarrow aAS_b/aBS_bS_b/aS_bS_b/aS_b$$
$$A \rightarrow aAS_b/aS_b$$
$$B \rightarrow aBS_bS_b/aS_bS_b$$

The context-free grammar is in GNF.
First, the start symbol S is pushed into the stack by the following production

$$\delta(q_0, \in, z_0) \rightarrow (q_1, Sz_0)$$

For the production $S \rightarrow aAS_b$, the transitional function is

$$\delta(q_1, a, S) \rightarrow (q_1, AS_b)$$

For the production $S \rightarrow aBS_bS_b$, the transitional function is

$$\delta(q_1, a, S) \rightarrow (q_1, BS_bS_b)$$

For the production $S \rightarrow aS_bS_b$, the transitional function is

$$\delta(q_1, a, S) \rightarrow (q_1, S_bS_b)$$

For the production $S \rightarrow aS_b$, the transitional function is

$$\delta(q_1, a, S) \rightarrow (q_1, S_b)$$

For the production $A \rightarrow aAS_b$, the transitional function is

$$\delta(q_1, a, A) \rightarrow (q_1, AS_b)$$

For the production $A \rightarrow aS_b$, the transitional function is

$$\delta(q_1, a, A) \rightarrow (q_1, S_b)$$

For the production $B \rightarrow aBS_bS_b$, the transitional function is

$$\delta(q_1, a, B) \rightarrow (q_1, BS_bS_b)$$

For the production $B \rightarrow aS_bS_b$, the transitional function is

$$\delta(q_1, a, B) \rightarrow (q_1, S_bS_b)$$

If we have to design the PDA accepted by the empty stack, the δ function is

$$\delta(q_1, \lambda, z_0) \rightarrow (q_1, \lambda)$$

If we have to design the PDA accepted by the final state, the δ function is

$$\delta(q_1, \lambda, z_0) \rightarrow (q_f, \lambda)$$

11. Construct an NPDA that accepts the language generated by the productions S → aSa/bSb/c. Show an instantaneous description of this string abcba for this problem. [WBUT 2007]

Solution: The production rules are not in GNF. So, we need to first convert it into GNF. The production rules are

$$S \rightarrow aSa/bSb/c$$

Let us introduce two new productions, $C_a \rightarrow a$ and $C_b \rightarrow b$
The new production rules become

$$S \rightarrow aSC_a$$
$$S \rightarrow bSC_b$$
$$S \rightarrow c$$
$$C_a \rightarrow a$$
$$C_b \rightarrow b$$

Now, all the productions are in GNF.
Now, from these productions, a PDA can be easily constructed.
First, the start symbol S is pushed into the stack by the following production

$$\delta(q_0, \in, z_0) \rightarrow (q_1, Sz_0)$$

For the production $S \rightarrow aSC_a$, the transitional function is

$$\delta(q_1, a, S) \rightarrow (q_1, SC_a)$$

For the production $S \rightarrow bSC_b$, the transitional function is

$$\delta(q_1, b, S) \rightarrow (q_1, SC_b)$$

For the production S → c, the transitional function is

$$\delta(q_1, c, S) \rightarrow (q_1, \lambda)$$

For the production $C_a \rightarrow a$, the transitional function is

$$\delta(q_1, a, C_a) \rightarrow (q_1, \lambda)$$

For the production $C_b \rightarrow b$, the transitional function is

$$\delta(q_1, b, C_b) \rightarrow (q_1, \lambda)$$

For acceptance, the transitional function is

$\delta(q_1, \lambda, z_0) \rightarrow (q_f, z_0)$ // accepted by the final state
$\delta(q_1, \lambda, z_0) \rightarrow (q_1, \lambda)$ // accepted by the empty stack

ID for the String 'abcba'

$\delta(q_0, \in abcba, z_0) \rightarrow \delta(q_1, \underline{a}bcba, Sz_0) \rightarrow \delta(q_1, a\underline{b}cba, SC_az_0) \rightarrow \delta(q_1, ab\underline{c}ba, SC_bC_az_0) \rightarrow \delta(q_1, abc\underline{b}a, C_bC_az_0) \rightarrow \delta(q_1, abcb\underline{a}, C_az_0) \rightarrow \delta(q_1, abcbaB, z_0) \rightarrow (q_f, z_0)$ (Acceptance by FS).

12. Construct a PDA, A, equivalent to the following context-free grammar

$$S \rightarrow 0BB,\ B \rightarrow 0S/1S/0$$

Test whether 0104 is in N(A).

Solution: The CFG is $S \rightarrow 0BB$, $B \rightarrow 0S/1S/0$

All the production rules of the grammar are in GNF.

First, the start symbol S is pushed into the stack by the following production

$$\delta(q_0, \in, z_0) \rightarrow (q_1, Sz_0)$$

For the production $S \rightarrow 0BB$, the transitional function is

$$\delta(q_1, 0, S) \rightarrow (q_1, BB)$$

For the production $B \rightarrow 0S$, the transitional function is

$$\delta(q_1, 0, B) \rightarrow (q_1, S)$$

For the production $B \rightarrow 1S$, the transitional function is

$$\delta(q_1, 1, B) \rightarrow (q_1, S)$$

For the production $B \rightarrow 0$, the transitional function is

$$\delta(q_1, 0, B) \rightarrow (q_1, \lambda)$$

For acceptance, the transitional functions are

$$\delta(q_1, \lambda, z_0) \rightarrow (q_f, z_0) \text{ // accepted by the final state}$$
$$\delta(q_1, \lambda, z_0) \rightarrow (q_1, \lambda) \text{ // accepted by the empty stack}$$

The ID for the String 010000

$(q_0, \underline{\in}\ 010000, z_0) \rightarrow (q_1, \underline{0}10000, Sz_0) \rightarrow (q_1, 0\underline{1}0000, BBz_0) \rightarrow (q_1, 01\underline{0}000, SBz_0)$
$\rightarrow (q_1, 010\underline{0}00, BBBz_0) \rightarrow (q_1, 0100\underline{0}0, BBz_0) \rightarrow (q_1, 01000\underline{0}, Bz_0)$
$\rightarrow (q_1, 010000 \in, z_0) \rightarrow (q_f, 010000\in, z_0)$ (Accepted by the final state).

13. Show that the language $L = \{0^n1^n \mid n \geq 1\} \cup \{0^n1^{2n} \mid n \geq 1\}$ is a context-free language that is not accepted by any DPDA. [UPTU 2005]

Solution: The context-free grammar for the language is

$$S \rightarrow S_1/S_2$$
$$S_1 \rightarrow 0S_11/01$$
$$S_2 \rightarrow 0S_211/011$$

The GNF equivalent to the grammar is

$$S \rightarrow 0S_1A/0A/0S_2A/0AA$$
$$A \rightarrow 1$$

The transitional functions of the PDA equivalent to the grammar are

$$\delta(q_0, \in, z_0) \rightarrow (q_1, Sz_0)$$
$$\delta(q_1, 0, S) \rightarrow (q_1, S_1A)$$
$$\delta(q_1, 0, S) \rightarrow (q_1, A)$$
$$\delta(q_1, 0, S) \rightarrow (q_1, S_2A)$$
$$\delta(q_1, 0, S) \rightarrow (q_1, AA)$$
$$\delta(q_1, 1, A) \rightarrow (q_1, \lambda)$$
$\delta(q_1, \lambda, z_0) \rightarrow (q_f, z_0)$ // accepted by the final state
$\delta(q_1, \lambda, z_0) \rightarrow (q_1, \lambda)$ // accepted by the empty stack.

The PDA is an NPDA, as for the combination $(q_1, 0, S)$, there are four transitional functions.

14. Convert the CFG into an equivalent PDA. [Cochin University 2006]

$$S \rightarrow aAA$$
$$A \rightarrow aS/bS/a$$

Solution: The grammar is in GNF.

First, the start symbol S is pushed into the stack by the following production

$$\delta(q_0, \in, z_0) \rightarrow (q_1, Sz_0)$$

For the production $S \rightarrow aAA$, the transitional function is

$$\delta(q_1, a, S) \rightarrow (q_1, AA)$$

For the production $A \rightarrow aS$, the transitional function is

$$\delta(q_1, a, A) \rightarrow (q_1, S)$$

For the production $S \rightarrow bS$, the transitional function is

$$\delta(q_1, b, S) \rightarrow (q_1, S)$$

For the production $A \rightarrow a$ the transitional function is

$$\delta(q_1, a, A) \rightarrow (q_1, \lambda)$$

For acceptance, the transitional functions are

$\delta(q_1, \lambda, z_0) \rightarrow (q_f, z_0)$ // accepted by the final state
$\delta(q_1, \lambda, z_0) \rightarrow (q_1, \lambda)$ // accepted by the empty stack

15. Construct a PDA equivalent to the grammar $S \rightarrow aAA\ A \rightarrow aS/b$. [Andhra University 2007]

Solution: The grammar is in GNF.

First, the start symbol S is pushed into the stack by the following production

$$\delta(q_0, \in, z_0) \rightarrow (q_1, Sz_0)$$

For the production S → aAA, the transitional function is

$$\delta(q_1, a, S) \rightarrow (q_1, AA)$$

For the production A → aS, the transitional function is

$$\delta(q_1, a, A) \rightarrow (q_1, S)$$

For the production A → b, the transitional function is

$$\delta(q_1, b, A) \rightarrow (q_1, \lambda)$$

For acceptance, the transitional functions are

$\delta(q_1, \lambda, z_0) \rightarrow (q_f, z_0)$ // accepted by the final state
$\delta(q_1, \lambda, z_0) \rightarrow (q_1, \lambda)$ // accepted by the empty stack

16. Construct a PDA equivalent to the following grammar. [JNTU 2008]

S → aBc
A → abc
B → aAb
C → AB
C → c

Solution: The grammar is not in GNF. The grammar is converted into GNF by replacing 'c' by C, adding the production D → b, and replacing b by D and replacing A of C → AB by aDC. The final grammar is

S →aBC
A → aDC
B → aAD
C → aDCB
C → c
D → b

(Now convert it to an equivalent PDA.)

17. Convert the PDA P = ({p, q}, {0, 1}, (x, z_0), δ, q, z_0) to a CFG, if δ is given as

$$\delta(q, 1, z_0) \rightarrow (q, xz_0)$$ [UPTU 2005]

Solution: The PDA contains two states, p and q. Thus, the following two production rules are added to the grammar.

$$S \rightarrow (q\ z_0\ q)/(q\ z_0\ p)$$

For the transitional function $\delta(q, 1, z_0) \rightarrow (q, xz_0)$, the production rules are

$\delta(q, 1, z_0) \rightarrow (q, xz_0)$
$(q\ z_0\ q) \rightarrow 1(q\ x\ q)\ (q\ z_0\ q)$
$(q\ z_0\ q) \rightarrow 1(q\ x\ p)\ (p\ z_0\ q)$
$(q\ z_0\ p) \rightarrow 1(q\ x\ q)\ (q\ z_0\ p)$
$(q\ z_0\ p) \rightarrow 1(q\ x\ p)\ (p\ z_0\ p)$

Multiple Choice Questions

1. PDA is the machine format of
   a) Type 0 language
   b) Type 1 language
   c) Type 2 language
   d) Type 3 language
2. Which is not true for mechanical diagram of PDA?
   a) PDA contains a stack
   b) The head reads as well as writes
   c) The head moves from left to right
   d) The input string is surrounded by an infinite number of blanks in both sides
3. The difference between finite automata and pushdown automata is in ______.
   a) Reading head
   b) Input tape
   c) Finite control
   d) Stack
4. In the PDA, transitional function δ is in the form
   a) $Q \times (\Sigma \cup \{\lambda\}) \times \Gamma \rightarrow (Q, \Gamma)$
   b) $Q \times \Sigma \rightarrow Q$
   c) $Q \times \Sigma \times \Gamma \rightarrow Q$
   d) $Q \times \Gamma \rightarrow Q \times \Sigma$

   (Where Q is the finite set of states, Σ is the set of input alphabets, and Γ is the stack symbols.)
5. Instantaneous description remembers
   a) The information of state and input tape content at a given instance of time.
   b) The information of state and stack content at a given instance of time.
   c) The information of input tape and stack content at a given instance of time.
   d) The information of state, input tape, and stack content at a given instance of time.
6. Which of the following is not possible algorithmically?
   a) RE to CFG
   b) NFA to DFA
   c) CFG to PDA
   d) NPDA to DPDA

Answers:

1. c 2. b 3. d 4. a 5. b 6. d

GATE Questions

1. Which of the following is not accepted by a DPDA but accepted by an NPDA?
   a) $L = a^n b^n$, where $n > 0$
   b) $L = WCW^R$, where $W \in (a, b)^*$
   c) $L = WW^R$, where $W \in (a, b)^+$
   d) $L = a^n b^m c^m d^n$, where $m, n > 0$
2. Which of the followings cannot be designed by a PDA?
   a) $a^n b^n c^i$, where $n, i > 0$
   b) $a^n b^n c^n$, where $n > 0$
   c) $a^n c^i b^n$, where $n, i > 0$
   d) $c^i a^n b^n$, where $n, i > 0$
3. The symbols belong to the stack of a PDA
   a) Terminals only
   b) Non-terminals only
   c) States
   d) Both terminals and non-terminals
4. L is accepted by a PDA where no symbol is necessarily removed from the stack. L is in particular
   a) Context-sensitive
   b) Unrestricted
   c) Context-free
   d) Regular

5. Which of the following languages is accepted by a non-deterministic pushdown automaton (PDA) but not by a deterministic PDA?

   a) Always regular
   b) Never regular
   c) Always a deterministic context-free language
   d) Always a context-free language

6. Let L be a context-free language and M a regular language. Then, the language L∩M is

   a) $\{a^nb^nc^n \mid n \geq 0\}$
   b) $\{a^lb^mc^n \mid l \neq m \text{ or } m = n\}$
   c) $\{a^nb^n \mid n \geq 0\}$
   d) $\{a^mb^n \mid m, n \geq 0\}$

7. Which of the following language over {a, b, c} is accepted by a deterministic pushdown automata?

   a) $\{wcw^R \mid w \in (a, b)^*\}$
   b) $\{ww^R \mid w \in (a, b)^*\}$
   c) $\{a^nb^nc^n \mid n \geq 0\}$
   d) {w | w is a palindrome over {a, b, c}}

8. The transitional functions of a pushdown automata are

$$\delta(q_0, 1, z_0) \rightarrow (q_0, Xz_0)$$
$$\delta(q_0, \in, z_0) \rightarrow (q_0, \in)$$
$$\delta(q_0, 1, X) \rightarrow (q_0, XX)$$
$$\delta(q_1, 1, X) \rightarrow (q_1, \in)$$
$$\delta(q_0, 0, X) \rightarrow (q_1, X)$$
$$\delta(q_1, \in, z_0) \rightarrow (q_1, \in)$$

   What is the language accepted by the PDA by the empty store?

   a) 1^n01^n
   b) 1^n1^n
   c) $1^n01^n \cup \in$
   d) $\in$

9 Let L_0 be the set of all languages accepted by a PDA by the final state and L_E the set of all languages accepted by the empty stack. Which of the following is true?

   a) $L_D = L_E$
   b) $L_D \supset L_E$
   c) $L_E = L_D$
   d) None of the above

10. Let G = ({S},{a, b}, R, S) be a context-free grammar where the rule set R is

$$S \rightarrow a\,S\,b \mid S\,S \mid \in$$

   Which of the following statements is true?

   a) G is not ambiguous
   b) There exists x, y ∈ L(G) such that xy ∉ L(G)
   c) There is a deterministic pushdown automaton that accepts L(G)
   d) We can find a deterministic finite state automaton that accepts L(G)

11. Let N_f and N_p denote the classes of languages accepted by the non-deterministic finite automata and the non-deterministic push-down automata, respectively. Let D_f and D_p denote the classes of languages accepted by the deterministic finite automata and the deterministic push-down automata, respectively. Which one of the following is true?

   a) $D_f \subset N_r$ and $D_P \subset N_P$
   b) $D_f \subset N_r$ and $D_P = N_P$
   c) $D_f = N_f$ and $D_P = N_P$
   d) $D_f = N_f$ and $D_P \subset N_P$

12. Consider the languages:

$$L_1 = \{ww^R \mid w \in \{0, 1\}^*\}$$
$$L_2 = \{w\#w^R \mid w \in \{0, 1\}^*\}, \text{ where \# is a special symbol}$$
$$L_3 = \{ww \mid w \in \{0, 1\}^*\}$$

Which one of the following is true?

a) L_1 is a deterministic CFL
b) L_2 is a deterministic CFL
c) L_3 is a CFL, but not a deterministic CFL
d) L_3 is a deterministic CFL

13. Which one of the following is false?
    a) There is a unique minimal DFA for every regular language
    b) Every NFA can be converted to an equivalent PDA
    c) The complement of every context-free language is recursive
    d) Every nondeterministic PDA can be converted to an equivalent deterministic PDA

14. Let M = (K, Σ, Γ, Δ, s, F) be a pushdown automaton, where

    K = {s, f}, F = {f}, Σ = {a, b}, Γ = {a} and

    $\Delta = \{((s, a, \in), (s, a)), ((s, b, \in), (s, a)), ((s, a, \in), (f, \in)), ((f, a, a), (f, \in)), ((f, b, a), (f, \in))\}$

    Which one of the following strings is not a member of L(M)?

    a) aaa b) aabab c) baaba d) bab

Answers:

1. c	2. b	3. d	4. d	5. b	6. d	7. a
8. c	9. a	10. c	11. d	12. b	13. d	14. d

Hints:

10. The language accepted by the grammar is $a^n b^n a^k b^k \ldots\ldots a^l b^l$.
11. An NFA can be converted to an equivalent DFA but a DPDA is a subset of an NPDA.
13. c is discussed in the recursive function chapter.
14. This came in GATE 2004(IT). Unfortunately, no answer is correct.

Fill in the Blanks

1. The full form of PDA is ____________.
2. ____________ is the machine format of context-free language.
3. The mechanical diagram of pushdown automata is similar to finite automata, but the difference is in ____________.
4. ____________ of PDA describes the configuration of PDA at a given instance.
5. There are two ways to declare a string accepted by a pushdown automata: (a) ____________ and (b) ____________.
6. For directly constructing a PDA from a given grammar, we first need to convert the grammar into ____________ normal form.
7. If a PDA being in a state with a single input and a single stack symbol gives a single move, then the PDA is called ____________ pushdown automata.
8. If a PDA being in a state with a single input and a single stack symbol gives more than one move for any of its transitional functions, then the PDA is called ____________ pushdown automata.

9. The PDA for the string L = {(Set of all palindromes over a, b)} can be designed by ________ pushdown automata and not by ________ pushdown automata.

Answers:

1. pushdown automata
2. Pushdown automata
3. stack
4. Instantaneous description (ID)
5. accepted by the empty stack, accepted by the final state
6. Greibach
7. deterministic
8. non-deterministic
9. non-deterministic, deterministic

Exercise

1. Construct a PDA to accept the following languages by the empty store and the final state.
   a) $L = \{a^n, \text{ where } n > 0\}$
   b) $L = \{a^n, \text{ where } n \geq 0\}$
   c) $L = \{a^n b c^n, \text{ where } n > 0\}$
   d) $L = \{a^n b^{2n}, \text{ where } n > 0\}$
   e) $L = \{a^n b^m c^{m+n}, \text{ where } m, n > 0\}$
   f) $L = \{a^n b a^n, \text{ where } n > 0\}$
   g) $L = \{a^n b^n c^i, \text{ where } n, i > 0\}$
   h) $L = \{a^m b^{m+n} c^n, \text{ where } m, n > 0\}$

   (Hints: For m number of 'a', m number of Z_1 are pushed into the stack. For the first m number of 'b', m number of Z_1 are popped from the stack. For the next n number of 'b', n number of Z_2 are pushed into the stack, which will be popped for the next n number of 'c'.)

   i) $L = \{a^m b^{m+n} a^n, \text{ where } m, n > 0\}$
   j) $L = \{WCW, \text{ where } W \in (a, b)^+\}$
   k) $L = \{WCW, \text{ where } W \in (a, b)^*\}$
2. Construct an equivalent PDA for the following context-free grammars.
   a) $S \rightarrow aB/bA$
      $A \rightarrow aS/bAA/a$
      $B \rightarrow bS/aBB/b$

   b) $S \rightarrow bA/aB$
      $A \rightarrow bA/aS/a$
      $B \rightarrow aB/bS/b$

   c) $S \rightarrow Abb/a$
      $A \rightarrow aaA/B$
      $B \rightarrow bAb$

3. Construct a PDA with graphical notation to accept the following languages by the graphical notation.
   a) $L = \{a^n b^n c^m d^m, \text{ where } m, n \geq 1\}$
   b) $L = \{a^n b^n c^m, \text{ where } n, m \geq 1\}$

Turing Machine

8

Introduction

Around 1936, when there were no computers, *Alen Turing* proposed a model of an abstract machine called the Turing machine which could perform any computational process carried out by the present day's computer. The machine was named after Turing, and it is called the Turing machine. The Turing machine is the machine format of unrestricted language, i.e., all types of languages are accepted by the Turing machine.

Based on the Turing machine, a new theory called the 'theory of undecidable problems' is developed. These types of problems cannot be solved by any computer.

8.1 Basics of Turing Machine

The Turing machine, in short TM, is defined by 7 touples

$$(Q, \Sigma, \Gamma, \delta, q_0, B, F)$$

where

Q : Finite set of states
Σ : Finite set of input alphabets
Γ : Finite set of allowable tape symbol
δ : Transitional function
q_0 : Initial state
B : A symbol of Γ called blank
F : Final state

and δ is a mapping from $Q \times \Gamma \rightarrow (Q \times \Gamma \times \{L, R, H\})$.

This denotes that from one state, by getting one input from the input tape, the machine moves to a state. It writes a symbol on the tape and moves to left, right or halts.

8.1.1 Mechanical Diagram

A TM consists of an input tape, a read–write head, and finite control. The input tape contains the input alphabets with an infinite number of blanks at the left and the right hand side of the input symbols. The read–write head reads an input symbol from the input tape and sends it to the finite control. The machine must be in some state. In the finite control, the transitional functions are written. According to

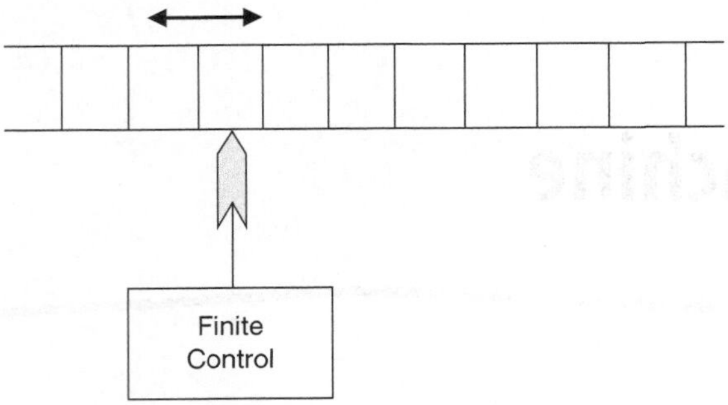

Fig. 8.1 *Mechanical Diagram of Turing Machine*

the present state and the present input, a suitable transitional function is executed. Upon execution of a suitable transitional function, the following operations are performed.

- The machine goes into some state.
- The machine writes a symbol in the cell of the input tape from where the input symbol was scanned.
- The machine moves the reading head to the left or right or halts.

Mechanical diagram of Turing Machine is shown in Fig. 8.1. Two-way finite automata has a movement in both the left and right side like the TM. But, in the TM, both the read and write operations are done from and on the input tape, but for two-way finite automata only the read operation is done, and no write operation is done. A string is accepted by a TM if the string is totally traversed and the machine halts. A string is accepted by a two-way finite automata if the string is totally traversed and the machine reaches to a final state.

8.1.2 Instantaneous Description (ID) in Respect of TM

The ID of the TM remembers the following at a given instance of time.

- The contents of all the cells of the tape, starting from the rightmost cell up to at least the last cell, containing a non-blank symbol and containing all cells up to the cell being scanned.
- The cell currently being scanned by the read–write head and
- The state of the machine.

The following example shows the read–write head movement of the TM.

Example 8.1 Show the movement of the read–write head and the content of the input tape for the string 'abba' traversed by the following TM

$$\delta(q_0, a) \rightarrow (q_0, X, R)$$
$$\delta(q_0, b) \rightarrow (q_0, b, R)$$
$$\delta(q_0, B) \rightarrow (q_1, B, L)$$
$$\delta(q_1, b) \rightarrow (q_1, Y, L)$$
$$\delta(q_1, X) \rightarrow (q_1, X, L)$$
$$\delta(q_1, B) \rightarrow (q_1, B, H)$$

(The left-side diagram is before traversing and the right-side diagram is after traversing.)

Solution:

B	a	b	b	a	B		$\delta(q_0, a) \to (q_0, X, R)$	B	X	b	b	a	B

↑ (pos 1)   ↑ (pos 2)

| B | X | b | b | a | B | $\delta(q_0, b) \to (q_0, b, R)$ | B | X | b | b | a | B |

| B | X | b | b | a | B | $\delta(q_0, b) \to (q_0, b, R)$ | B | X | b | b | a | B |

| B | X | b | b | a | B | $\delta(q_0, a) \to (q_0, X, R)$ | B | X | b | b | X | B |

| B | X | b | b | X | B | $\delta(q_0, B) \to (q_1, B, L)$ | B | X | b | b | X | B |

| B | X | b | b | X | B | $\delta(q_1, X) \to (q_1, X, L)$ | B | X | b | b | X | B |

| B | X | b | b | X | B | $\delta(q_1, b) \to (q_1, Y, L)$ | B | X | b | Y | X | B |

| B | X | b | Y | X | B | $\delta(q_1, b) \to (q_1, Y, L)$ | B | X | Y | Y | X | B |

| B | X | Y | Y | X | B | $\delta(q_1, X) \to (q_1, X, L)$ | B | X | Y | Y | X | B |

| B | X | Y | Y | X | B | $\delta(q_1, B) \to (q_1, B, H)$ | B | X | Y | Y | X | B |

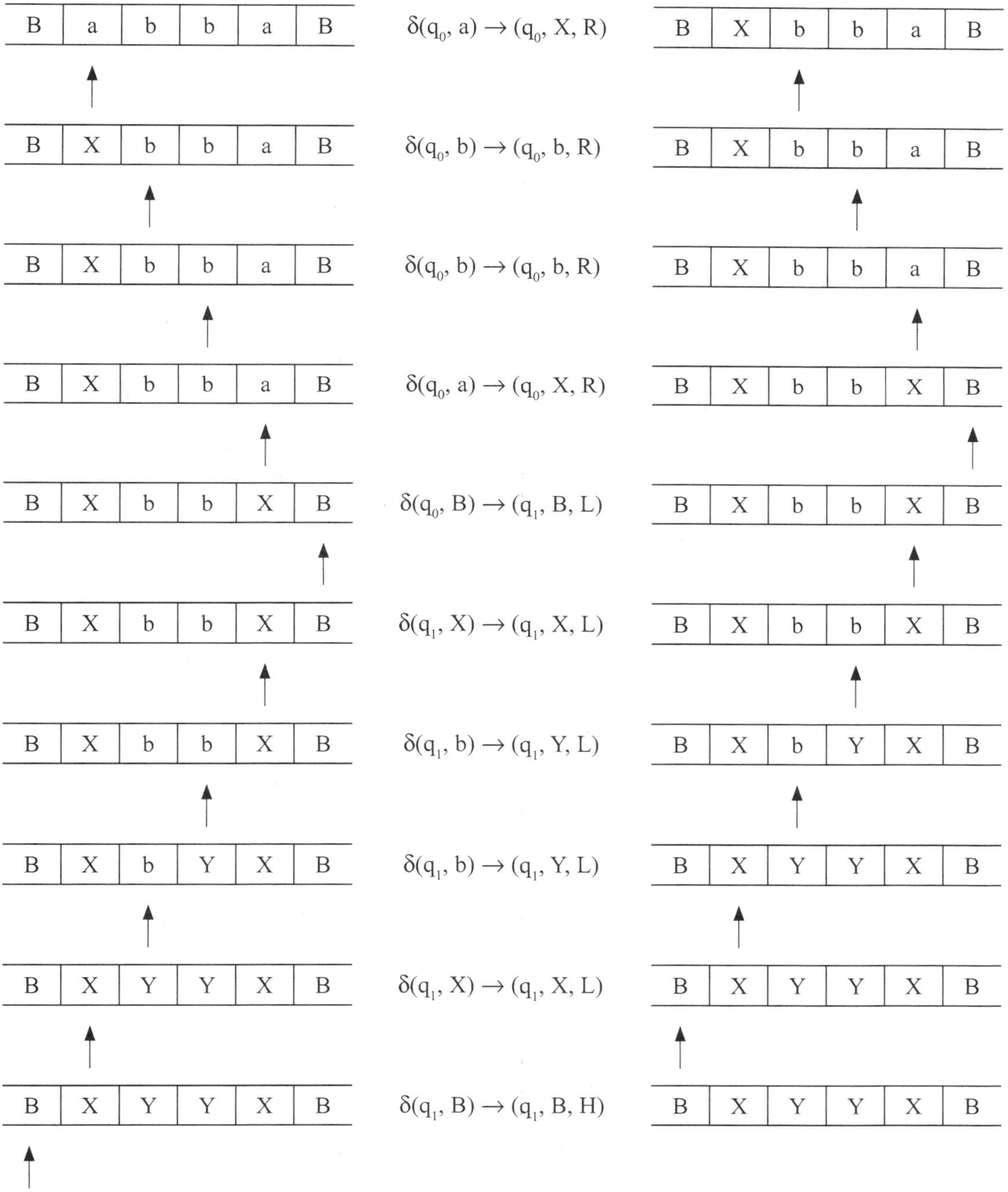

The following are some examples of the TM that accepts some languages.

Example 8.2 Design a TM to accept the language $L = \{a^n b^n, n \geq 1\}$. Show an ID for the string 'aaabbb' with tape symbols.

Solution: The string consists of two types of input alphabets, 'a' and 'b'. The number of 'a' is equal to the number of 'b'. All the 'a's will come before 'b'. In the language set, there is at least one 'a' and one 'b'. The TM can be designed as follows.

When the leftmost 'a' is traversed, that 'a' is replaced by X and the head moves to one right. The transitional function is

$$\delta(q_0, a) \rightarrow (q_1, X, R)$$

Then, the machine needs to search for the leftmost 'b'. Before that 'b', there exist (n – 1) numbers of 'a'. Those 'a' are traversed by

$$\delta(q_1, a) \rightarrow (q_1, a, R)$$

When the leftmost 'b' is traversed, the state is q_1. That 'b' is replaced by Y, the state is changed to q_2, and the head is moved to one left. The transitional functional is

$$\delta(q_1, b) \rightarrow (q_2, Y, L)$$

Then, it needs to search for the second 'a' starting from the left. The first 'a' is replaced by X, which means the second 'a' exists after X. So, it needs to search for the rightmost 'X'. After traversing the leftmost 'b', the head moves to the left to find the rightmost X. Before that, it has to traverse 'a'. The transitional function is

$$\delta(q_2, a) \rightarrow (q_2, a, L)$$

After traversing all the 'a' we get the rightmost 'X'. Traversing the X the machine changes its state from q_2 to q_0 and the head moves to one right. The transitional function is

$$\delta(q_2, X) \rightarrow (q_0, X, R)$$

From the traversal of the second 'b' onwards, the machine has to traverse 'Y'. The transitional function is

$$\delta(q_1, Y) \rightarrow (q_1, Y, R)$$

Similarly, after traversing 'b', the machine has to traverse some Y to get the rightmost 'X'.

$$\delta(q_2, Y) \rightarrow (q_2, Y, L)$$

When all the 'a's are traversed, the state is q_0, because before that state was q_2 and the input was X. The head moves to one right and gets a Y. Getting a Y means that all the 'a's are traversed and the same number of 'b's are traversed. Traversing right, if at last the machine gets no 'b' but a blank 'B', then the machine halts. The transitional functions are

$$\delta(q_0, Y) \rightarrow (q_3, Y, R)$$
$$\delta(q_3, Y) \rightarrow (q_3, Y, R)$$
$$\delta(q_3, B) \rightarrow (q_4, B, H)$$

ID for the String 'aaabbb'

First, the tape symbols are

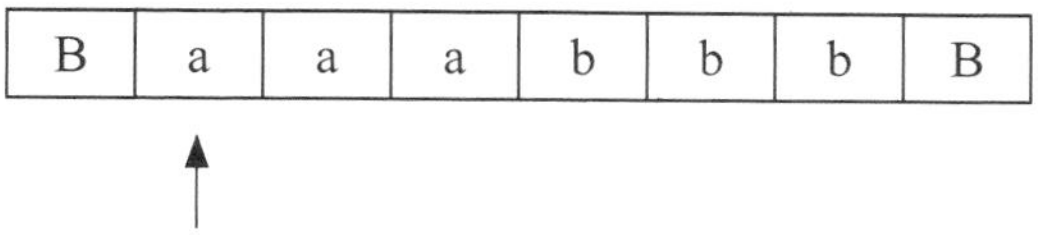

$\delta(q_0, a) \rightarrow (q_1, Xaabbb, R)$

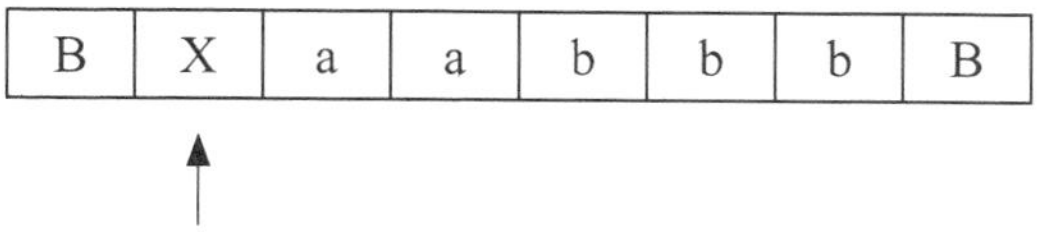

$\delta(q_1, a) \rightarrow (q_1, Xaabbb, R)$

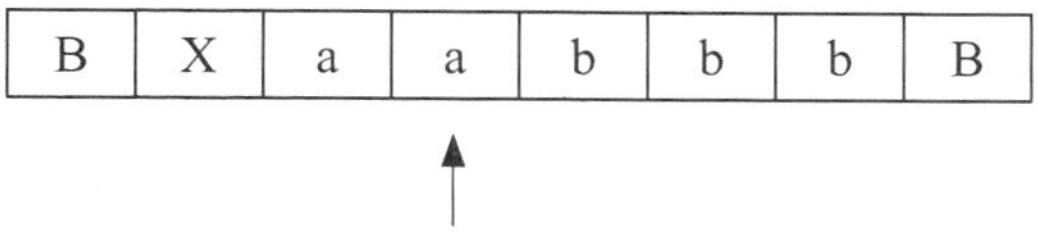

$\delta(q_1, a) \rightarrow (q_1, Xaabbb, R)$

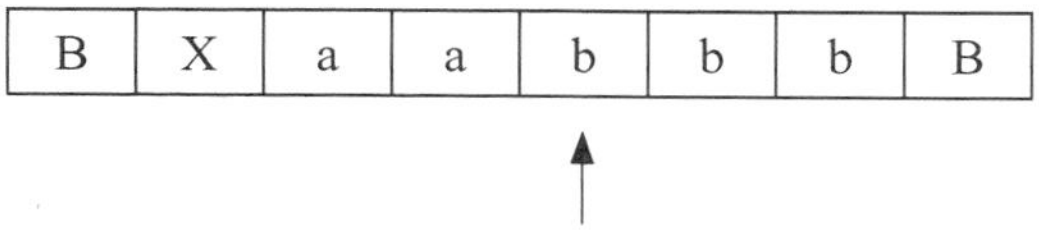

$\delta(q_1, b) \rightarrow (q_2, XaaYbb, L)$

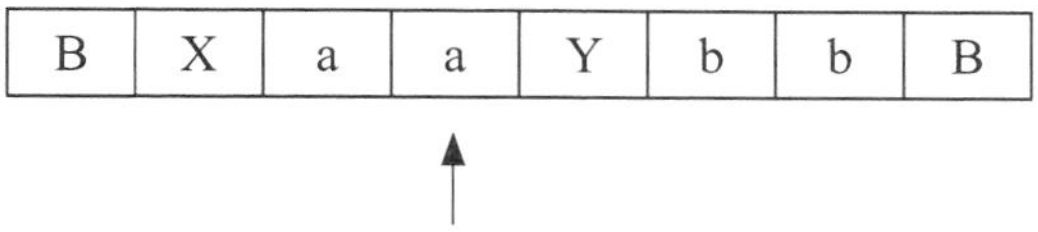

$\delta(q_2, a) \rightarrow (q_2, XaaYbb, L)$

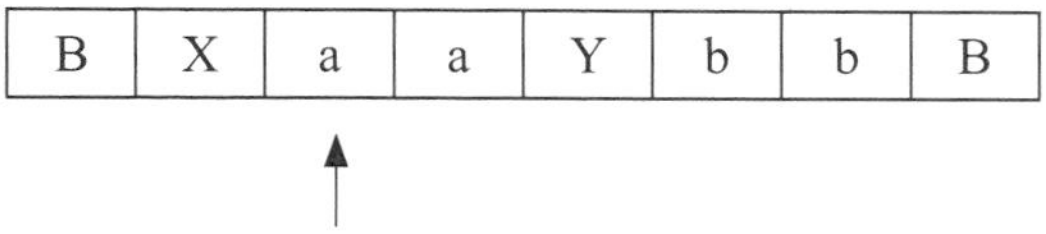

$\delta(q_2, a) \rightarrow (q_2, XaaYbb, L)$

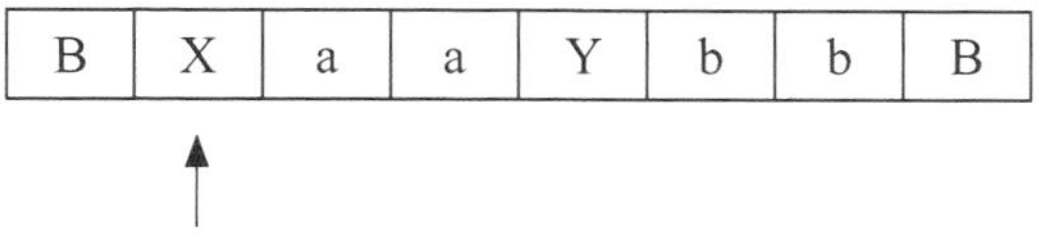

$\delta(q_2, X) \rightarrow (q_0, XaaYbb, R)$

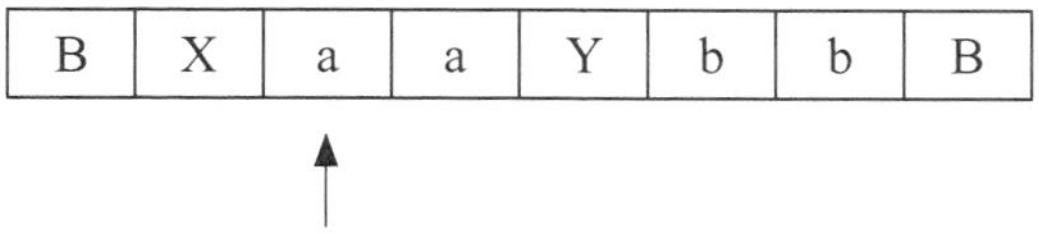

$\delta(q_0, a) \rightarrow (q_1, XXaYbb, R)$

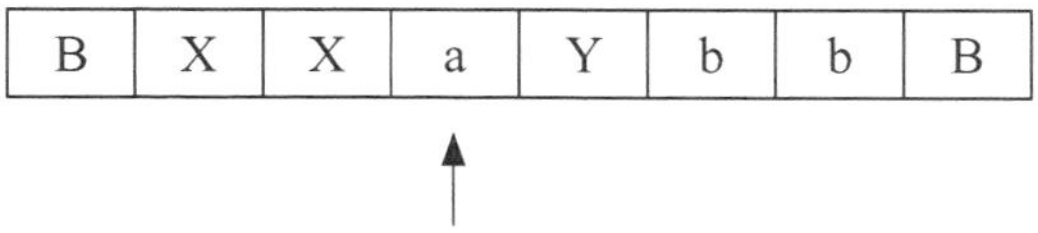

$\delta(q_1, a) \rightarrow (q_1, XXaYbb, R)$

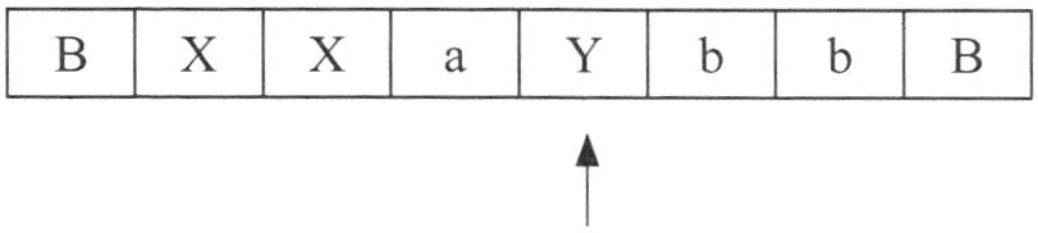

$\delta(q_1, Y) \rightarrow (q_1, XXaYbb, R)$

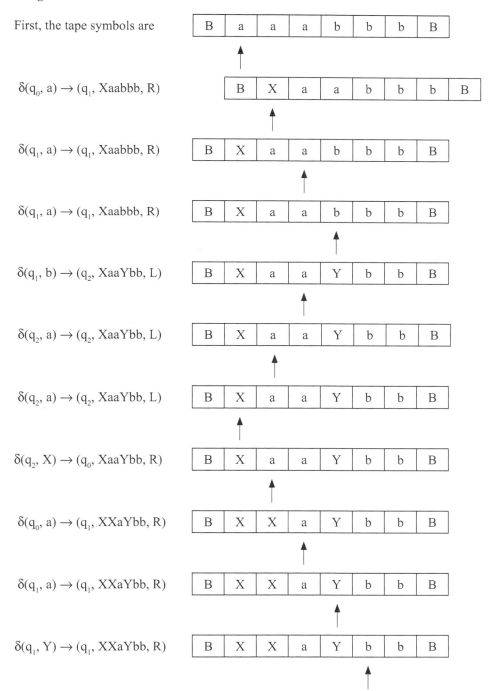

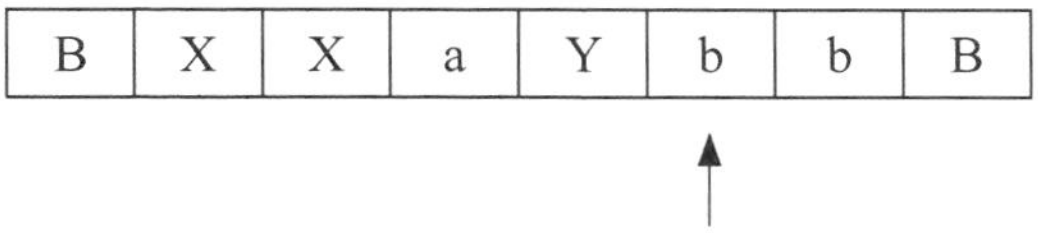

$\delta(q_1, b) \rightarrow (q_2, XXaYYb, L)$

B	X	X	a	Y	Y	b	B
				↑			

$\delta(q_2, Y) \rightarrow (q_2, XXaYYb, L)$

B	X	X	a	Y	Y	b	B
			↑				

$\delta(q_2, a) \rightarrow (q_2, XXaYYb, L)$

B	X	X	a	Y	Y	b	B
		↑					

$\delta(q_2, X) \rightarrow (q_0, XXaYYb, R)$

B	X	X	a	Y	Y	b	B
			↑				

$\delta(q_0, a) \rightarrow (q_1, XXXYYb, R)$

B	X	X	X	Y	Y	b	B
				↑			

$\delta(q_1, Y) \rightarrow (q_1, XXXYYb, R)$

B	X	X	X	Y	Y	b	B
					↑		

$\delta(q_1, Y) \rightarrow (q_1, XXXYYb, R)$

B	X	X	X	Y	Y	b	B
						↑	

$\delta(q_1, b) \rightarrow (q_2, XXXYYY, L)$

B	X	X	X	Y	Y	Y	B
					↑		

$\delta(q_2, Y) \rightarrow (q_2, XXXYYY, L)$

B	X	X	X	Y	Y	Y	B
				↑			

$\delta(q_2, Y) \rightarrow (q_0, XXXYYY, R)$

B	X	X	X	Y	Y	Y	B
			↑				

$\delta(q_0, Y) \rightarrow (q_3, XXXYYY, R)$

B	X	X	X	Y	Y	Y	B
				↑			

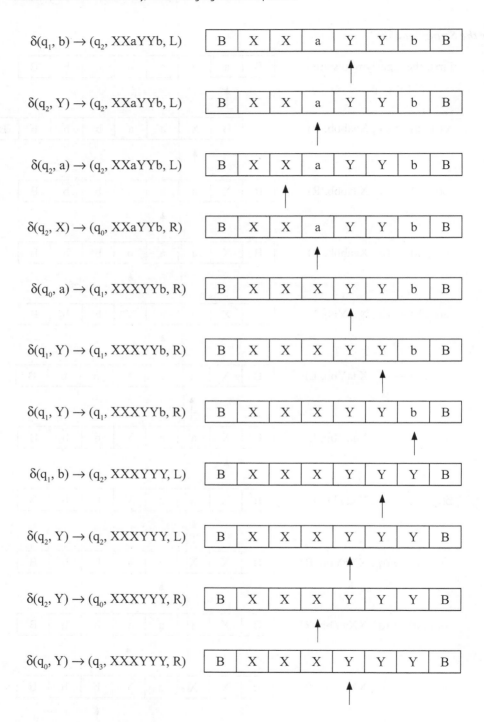

$\delta(q_3, Y) \rightarrow (q_3, XXXYYY, R)$

$\delta(q_3, Y) \rightarrow (q_3, XXXYYY, R)$

$\delta(q_3, Y) \rightarrow (q_3, XXXYYY, R)$

$\delta(q_3, B) \rightarrow (q_3, XXXYYY, H)$

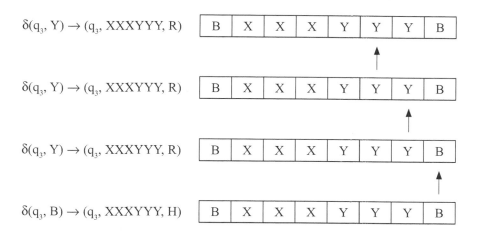

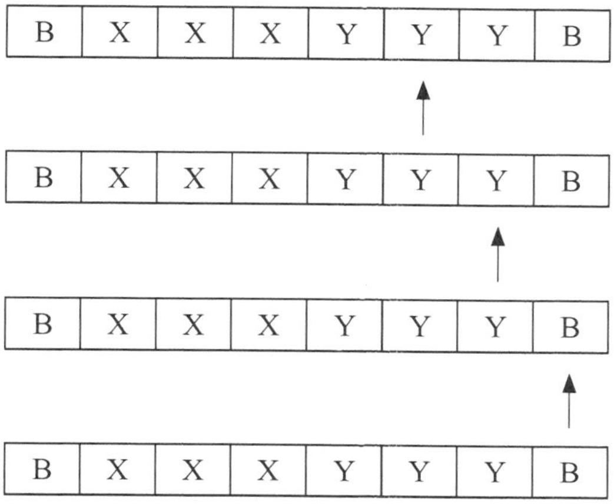

Example 8.3 Design a TM to accept the language

$$L = \{WW^R, \text{ where } W \in (a, b)^+\}$$

Show an ID for the string 'abaaaaba' with tape symbols.

Solution: W^R is the reverse of W. If W starts with 'a' or 'b', W^R ends with 'a' or 'b', respectively. If W ends with 'a' or 'b', W^R starts with 'a' or 'b', respectively. The TM can be designed as follows.

If the string W starts with 'a', upon traversal, that 'a' is replaced by X with a state change from q_0 to q_1, and the head moves to one right. The transitional function is

$$\delta(q_0, a) \rightarrow (q_1, X, R)$$

W^R ends with 'a' if W starts with 'a'. The machine needs to search the end 'a' of W^R. Before that, the machine needs to traverse the end symbols of W and the beginning symbols of W^R. The transitional functions are

$$\delta(q_1, a) \rightarrow (q_1, a, R)$$
$$\delta(q_1, b) \rightarrow (q_1, b, R)$$

After the end symbol of W^R, there exists the blank symbol B. In the traversal process, if the machine gets a B, it traverses back to the left side and gets the end symbol of W^R. The transitional functions are

$$\delta(q_1, B) \rightarrow (q_2, B, L)$$
$$\delta(q_2, a) \rightarrow (q_3, X, L)$$

Now, the machine needs to search for the second symbol of W. Before that, it has to traverse the beginning symbols of W^R and the end symbols of W. The transitional functions are

$$\delta(q_3, a) \rightarrow (q_3, a, L)$$
$$\delta(q_3, b) \rightarrow (q_3, b, L)$$

When the machine gets the rightmost X, it recognizes that the next symbol of W exists after that 'X'. The transitional function is

$$\delta(q_3, X) \rightarrow (q_0, X, R)$$

If the string W starts with 'b', the transitions are the same as the previous one but with some states changed. The transitional functions are

$$\delta(q_0, a) \rightarrow (q_4, Y, R)$$
$$\delta(q_4, a) \rightarrow (q_4, a, R)$$
$$\delta(q_4, b) \rightarrow (q_4, b, R)$$
$$\delta(q_4, B) \rightarrow (q_5, B, L)$$
$$\delta(q_5, b) \rightarrow (q_6, Y, L)$$
$$\delta(q_6, a) \rightarrow (q_6, a, L)$$
$$\delta(q_6, b) \rightarrow (q_6, b, L)$$
$$\delta(q_6, Y) \rightarrow (q_0, Y, R)$$

After the first traversal, i.e., from the second traversal onwards, the machine need not traverse up to the end of W^R. In state q_1 (W starts with 'a') or q_4 (W starts with 'b'), if the machine gets X or Y, it traverses back to the left to point the rightmost 'a' or 'b'. The transitional functions are

$$\delta(q_1, X) \rightarrow (q_2, X, L)$$
$$\delta(q_1, Y) \rightarrow (q_2, Y, L)$$
$$\delta(q_4, X) \rightarrow (q_5, X, L)$$
$$\delta(q_4, Y) \rightarrow (q_5, Y, L)$$

When all the symbols of W and W^R are traversed, the machine gets an X or Y in the state q_0. The machine halts if in state q_0 it gets an X or Y. The transitional functions are

$$\delta(q_0, X) \rightarrow (q_f, X, H)$$
$$\delta(q_0, Y) \rightarrow (q_f, Y, H)$$

The transitional functions can be given in a tabular format as follows.

State	a	B	B	X	Y
q_0	(q_1, X, R)	(q_4, Y, R)	–	(q_f, X, H)	(q_f, Y, H)
q_1	(q_1, a, R)	(q_1, b, R)	(q_2, B, L)	(q_2, X, L)	(q_2, Y, L)
q_2	(q_3, X, L)	–	–	–	–
q_3	(q_3, a, L)	(q_3, b, L)	–	(q_0, X, R)	–
q_4	(q_4, a, R)	(q_4, b, R)	(q_5, B, L)	(q_5, X, L)	(q_5, Y, L)
q_5		(q_6, Y, L)	–	–	–
q_6	(q_6, a, L)	(q_6, b, L)	–	–	(q_0, Y, R)

ID for the String 'abaaaaba': (Here, the symbols in bold represent the read–write head position.)

(q_0, **a**baaaaba) → (q_1, X**b**aaaaba) → (q_1, Xb**a**aaaba) → (q_1, Xba**a**aaba) → (q_1, Xbaa**a**aba)
→ (q_1, Xbaaa**a**ba) → (q_1, Xbaaaa**b**a) → (q_1, Xbaaaab**a**) → (q_1, Xbaaaaba**B**) → (q_2, Xbaaaab**a**B)
→ (q_3, Xbaaaa**b**X) → (q_3, Xbaaaa**b**X) → (q_3, Xbaaa**a**bX) → (q_3, Xbaa**a**abX) → (q_3, Xba**a**aabX)
→ (q_3, Xb**a**aaabX) → (q_3, X**b**aaaabXB) → (q_3, **X**baaaabX) → (q_0, X**b**aaaabX) → (q_4, XY**a**aaabX)
→ (q_4, XYa**a**aabX) → (q_4, XYaa**a**abX) → (q_4, XYaaa**a**bX) → (q_4, XYaaaa**b**X) → (q_4, XYaaaab**X**)
→ (q_5, XYaaaa**b**X) → (q_6, XYaaa**a**YX) → (q_6, XYaa**a**aYX) → (q_6, XYa**a**aaYX) → (q_6, XY**a**aaaYX)
→ (q_0, XY**a**aaaYX) → (q_1, XYX**a**aaYX) → (q_1, XYXa**a**aYX) → (q_1, XYXaa**a**YX) → (q_1, XYXaaa**Y**X)
→ (q_2, XYXaa**a**YX) → (q_3, XYXa**a**XYX) → (q_3, XYX**a**aXYX) → (q_3, XY**X**aaXYX) → (q_0, XYX**a**aXYX)
→ (q_1, XYXX**a**XYX) → (q_1, XYXXa**X**YX) → (q_2, XYXX**a**XYX) → (q_3, XYX**X**XXYX)
→ (q_0, XYXX**X**XYX) → (q_f, XYXXXXYX) [Halt]

Example 8.4 Design a TM to accept the language L = {set of all palindromes over a, b}. Show the IDs for the null string, 'a', 'aba', and 'baab'.

Solution: Palindromes are of two types:

i) Odd palindromes, where the number of characters is odd
ii) Even palindrome, where the number of characters is even. A null string is also a palindrome.

A string which starts with 'a' or 'b' must end with 'a' or 'b', respectively, if the string is a palindrome. If a string starts with 'a', that 'a' is replaced by a blank symbol 'B' upon traversal with a state change from q_1 to q_2 and the right shift of the read–write head. The transitional function is

$$\delta(q_1, a) \rightarrow (q_2, B, R)$$

Then, the machine needs to search for the end of the string. The string must end with 'a', and that 'a' exists before a blank symbol B at the right hand side. Before getting that blank symbol, the machine needs to traverse all the remaining 'a' and 'b' of the string. The transitional functions are

$$\delta(q_2, a) \rightarrow (q_2, a, R)$$
$$\delta(q_2, b) \rightarrow (q_2, b, R)$$
$$\delta(q_2, B) \rightarrow (q_3, B, L)$$
$$\delta(q_3, a) \rightarrow (q_4, B, L)$$

The machine now needs to search for the second symbol (from the starting) of the string. That symbol exists after the blank symbol B, which is replaced at the first. Before that the machine needs to traverse the remaining 'a' and 'b' of the string. The transitional functions are

$$\delta(q_4, a) \rightarrow (q_4, a, L)$$
$$\delta(q_4, b) \rightarrow (q_4, b, L)$$
$$\delta(q_4, B) \rightarrow (q_1, B, R)$$

If the string starts with 'b', the transitional functions are the same as 'a' but some states are changed. The transitional functions are

$$\delta(q_1, b) \rightarrow (q_5, B, R)$$
$$\delta(q_5, a) \rightarrow (q_5, a, R)$$

$$\delta(q_5, b) \rightarrow (q_5, b, R)$$
$$\delta(q_5, B) \rightarrow (q_6, B, L)$$
$$\delta(q_6, b) \rightarrow (q_4, B, L)$$

When all 'a' and 'b' are traversed and replaced by B, the states may be one of q_3 or q_6, if the last symbol traversed is 'a' or 'b', respectively. Transitional functions for acceptance are

$$\delta(q_3, B) \rightarrow (q_7, B, H)$$
$$\delta(q_6, B) \rightarrow (q_7, B, H)$$

A null string is also a palindrome. On the tape, a null symbol means blank B. In state q_1, the machine gets the symbol. The transitional function is

$$\delta(q_1, B) \rightarrow (q_7, B, H)$$

The transitional functions in tabular form are represented as follows.

State	A	b	B
q_1	(q_2, B, R)	(q_5, B, R)	(q_7, B, H)
q_2	(q_2, a, R)	(q_2, b, R)	(q_3, B, L)
q_3	(q_4, B, L)	–	(q_7, B, H)
q_4	(q_4, a, L)	(q_4, b, L)	(q_1, B, R)
q_5	(q_5, a, R)	(q_5, b, R)	(q_6, B, L)
q_6	–	(q_4, B, L)	(q_7, B, H)

ID for Null String

$$(q_1, B) \rightarrow (q_7, B) \text{ [Halt]}$$

ID for the String 'a'

$$(q_1, \mathbf{a}B) \rightarrow (q_2, B\mathbf{B}) \rightarrow (q_3, \mathbf{B}B) \rightarrow (q_7, BB) \text{ [Halt]}$$

ID for the String 'aba'

$(q_1, \mathbf{a}baB) \rightarrow (q_2, B\mathbf{b}aB) \rightarrow (q_2, Bb\mathbf{a}B) \rightarrow (q_2, Bba\mathbf{B}) \rightarrow (q_3, Bb\mathbf{a}B) \rightarrow (q_4, B\mathbf{b}BB) \rightarrow (q_4, \mathbf{B}bBB)$
$\rightarrow (q_1, B\mathbf{b}BB) \rightarrow (q_1, BB\mathbf{B}B) \rightarrow (q_7, BBBB)$ [Halt]

ID for the String 'baab'

$(q_1, \mathbf{b}aabB) \rightarrow (q_5, B\mathbf{a}abB) \rightarrow (q_5, Ba\mathbf{a}bB) \rightarrow (q_5, Baa\mathbf{b}B) \rightarrow (q_5, Baab\mathbf{B}) \rightarrow (q_6, Baa\mathbf{b}B)$
$\rightarrow (q_4, Ba\mathbf{a}BB) \rightarrow (q_4, B\mathbf{a}aBB) \rightarrow (q_4, \mathbf{B}aaBB) \rightarrow (q_1, B\mathbf{a}aBB) \rightarrow (q_2, BB\mathbf{a}BB) \rightarrow (q_2, BBa\mathbf{B}B)$
$\rightarrow (q_3, BB\mathbf{a}BB) \rightarrow (q_3, B\mathbf{B}BBB) \rightarrow (q_7, BBBBB)$ [Halt].

Example 8.5 Design a TM to accept the language $L = a^n b^n c^n$, where $n \geq 1$.

Solution: This is a context-sensitive language. The language consists of a, b, and c. Here, the number of 'a' is equal to the number of 'b' which is equal to the number of 'c'. n number 'c' is followed by n number of 'b' which are followed by n number of 'a'. The TM is designed as follows. For each 'a', search for 'b' and 'c' and again traverse back to the left to search for the leftmost 'a'. 'a' is replaced by 'X', 'b' is replaced by 'Y', and 'c' is replaced by 'Z'.

When all the 'a' are traversed and replaced by 'X', the machine traverses the right side to find if any untraversed 'b' or 'c' is left or not. If not, the machine gets Y and Z and at last a blank. Upon getting the blank symbol, the machine halts.

The transitional functions are

$\delta(q_1, a) \rightarrow (q_2, X, R)$ // 'a' is traversed
$\delta(q_2, a) \rightarrow (q_2, a, R)$ // the remaining 'a's are traversed
$\delta(q_2, b) \rightarrow (q_3, Y, R)$ // 'b' is traversed
$\delta(q_3, b) \rightarrow (q_3, b, R)$ // the remaining 'b's are traversed
$\delta(q_3, c) \rightarrow (q_4, Z, L)$ // 'c' is traversed
$\delta(q_4, b) \rightarrow (q_4, b, L)$ // the remaining 'b's are traversed from right to left
$\delta(q_4, Y) \rightarrow (q_4, Y, L)$ // Y, replacement for 'b' is traversed
$\delta(q_4, a) \rightarrow (q_4, a, L)$ // the remaining 'a's are traversed from right to left
$\delta(q_4, X) \rightarrow (q_1, X, R)$ // rightmost X, replacement of 'a' is traversed
$\delta(q_2, Y) \rightarrow (q_2, Y, R)$ // this is used from the second time onwards
$\delta(q_3, Z) \rightarrow (q_3, Z, R)$ // this is used from the second time onwards
$\delta(q_4, Z) \rightarrow (q_4, Z, L)$ // this is used from the second time onwards
$\delta(q_1, Y) \rightarrow (q_5, Y, R)$ // this is used when all the 'a' are traversed
$\delta(q_5, Y) \rightarrow (q_5, Y, R)$ // this is used when all the 'a' are traversed and the machine is traversing right
$\delta(q_5, Z) \rightarrow (q_6, Z, R)$ // this is used when all the 'a' and 'b' are traversed.
$\delta(q_6, Z) \rightarrow (q_6, Z, R)$ // this is used when all the 'a' and 'b are traversed and the machine is traversing right to search for the remaining 'c'
$\delta(q_6, B) \rightarrow (q_f, B, H)$ // this is used when all the 'a', 'b', and 'c' are traversed

Upon executing the last transitional function, the machine halts.

The transitional functions in tabular form are represented as follows.

State	a	b	c	B	X	Y	Z
q_1	(q_2, X, R)	–	–	–	–	(q_5, Y, R)	–
q_2	(q_2, a, R)	(q_3, Y, R)	–	–	–	(q_2, Y, R)	–
q_3	–	(q_3, b, R)	(q_4, Z, L)	–	–	–	(q_3, Z, R)
q_4	(q_4, a, L)	(q_4, b, L)	–	–	(q_1, X, R)	(q_4, Y, L)	(q_4, Z, L)
q_5	–	–	–	–	–	(q_5, Y, R)	(q_6, Z, R)
q_6	–	–	–	(q_f, B, H)	–	–	(q_6, Z, R)

Example 8.6 Design a TM to perform the concatenation operation on string of '1'. Show an ID for $w_1 = 111$ and $w_2 = 1111$.

Solution: The TM does the concatenation operation on two strings w_1 and w_2. These two strings are placed on the tape of the TM separated by the blank symbol B. After traversal, the blank symbol is removed and the two strings are concatenated.

The transitional functions are

$\delta(q_1, 1) \rightarrow (q_1, 1, R)$ // traversing '1' of the first string
$\delta(q_1, B) \rightarrow (q_2, 1, R)$ // traversing the separating blank symbol and converting it to '1'
$\delta(q_2, 1) \rightarrow (q_2, 1, R)$ // traversing '1' of the second string
$\delta(q_2, B) \rightarrow (q_3, B, L)$ // blank symbol traversal means the end of the second string
$\delta(q_3, 1) \rightarrow (q_4, B, L)$ // the last '1' of the second string is converted to 'B' to keep the number of '1' the same
$\delta(q_4, 1) \rightarrow (q_4, 1, L)$ // traversing the remaining '1' at the left side
$\delta(q_4, B) \rightarrow (q_5, B, H)$ // halts

[This machine is also applicable for performing $Z = X + Y$, where $X = |w_1|$ and $Y = |w_2|$.]

ID for $w_1 = 111$ and $w_2 = 1111$ and the result is 1111111

$(q_1, \mathbf{1}11B1111B) \rightarrow (q_1, 1\mathbf{1}1B1111B) \rightarrow (q_1, 11\mathbf{1}B1111) \rightarrow (q_1, 111\mathbf{B}1111B) \rightarrow (q_2, 1111111B)$
$\rightarrow (q_2, 1\mathbf{1}111111B) \rightarrow (q_2, 11\mathbf{1}11111B) \rightarrow (q_2, 111\mathbf{1}1111B) \rightarrow (q_2, 1111\mathbf{1}111B) \rightarrow$
$(q_3, 11111111\mathbf{B})$
$\rightarrow (q_4, 1111111BB) \rightarrow (q_4, 1111111BB) \rightarrow (q_4, 1111111BB) \rightarrow (q_4, 1111111BB)$
$\rightarrow (q_4, B1111111BB) \rightarrow (q_4, B1111111BB) \rightarrow (q_4, B1111111BB) \rightarrow (q_4, \mathbf{B}1111111BB)$
$\rightarrow (q_5, B1111111BB)$ [Halt]

Example 8.7 Design a TM to perform the following operation

$$f(x, y) = x - y, \text{ where } x > y$$

Show an ID for $4 - 2 = 2$.

Solution: This performs the subtraction operation. Here, x and y both are integer numbers. Let $x = |W_1|$ and $y = |W_2|$, where W_1 and W_2 are the strings of '1'. The two strings are placed on the input tape separated by B.

Starting traversal from the left hand side '1', the machine moves towards the right. It traverses B with a state changed from q_0 to q_1. In state q_1, it again traverses right to find B, that is, the end of the second string. It changes its state and traverses left and changes the rightmost '1' by B (with state change and traversing left). It traverses the separating 'B' symbol and all '1's of the first string and finds the beginning of the first string, which starts after B at the left hand side. It changes the first '1' by B and again traverses right. This process continues till all the '1's of the second string are replaced by 'B'.

The transitional functions are

$\delta(q_0, 1) \rightarrow (q_0, 1, R)$ // all the '1's of the first string are traversed
$\delta(q_0, B) \rightarrow (q_1, B, R)$ // the separating B is traversed
$\delta(q_1, 1) \rightarrow (q_1, 1, R)$ // all the '1's of the second string are traversed
$\delta(q_1, B) \rightarrow (q_2, B, L)$ // to signify the end of the second string

$\delta(q_2, 1) \rightarrow (q_3, B, L)$ // the rightmost '1' of the second string is converted to B
$\delta(q_3, 1) \rightarrow (q_3, 1, L)$ // the remaining '1's of the second string are traversed
$\delta(q_3, B) \rightarrow (q_4, B, L)$ // the separating blank symbol is traversed
$\delta(q_4, 1) \rightarrow (q_4, 1, L)$ // all the '1's of the first string are traversed
$\delta(q_4, B) \rightarrow (q_5, B, R)$ // the blank symbol before the first string is traversed
$\delta(q_5, 1) \rightarrow (q_0, B, R)$ // the leftmost '1' of first string is converted to the blank symbol
$\delta(q_2, B) \rightarrow (q_f, B, H)$ // Halts

ID for 4 − 2 = 2

(q_0, **1**111B11) → (q_0, 1**1**11B11) → (q_0, 11**1**1B11) → (q_0, 111**1**B11) → (q_0, 1111**B**11)
→ (q_1, 1111B**1**1) → (q_1, 1111B1**1**B) → (q_1, 1111B11**B**) → (q_2, 1111B1**1**B)
→ (q_3, 1111B**1**BB) → (q_3, 1111**B**1BB) → (q_4, 111**1**B1BB) → (q_4, 11**1**1B1BB)
→ (q_4, 1**1**11B1BB) → (q_4, B**1**111B1BB) → (q_4, **B**1111B1BB) → (q_5, B**1**111B1BB)
→ (q_5, B**1**111B1BB) → (q_0, BB**1**11B1BB) → (q_0, BB1**1**1B1BB) → (q_0, BB1**1**1B1BB)
→ (q_0, BB11**1**B1BB) → (q_0, BB111**B**1BB) → (q_1, BB111B**1**BB) → (q_1, BB111B1**B**B)
→ (q_2, BB111B**1**BB) → (q_3, BB111**B**BBB) → (q_4, BB11**1**BBBB) → (q_4, BB1**1**1BBBB)
→ (q_4, BB**1**11BBBB) → (q_4, BB111BBBB) → (q_5, BB**1**11BBBB) → (q_0, BBB**1**1BBBB)
→ (q_0, BBB1**1**BBBB) → (q_0, BBB11**B**BBB) → (q_1, BBB11B**B**BB) → (q_2, BBB11**B**BBB)
→ (q_f, BBB11BBBB)

Example 8.8 Design a TM to accept the string L = (a, b)*, where N(a) + N(b) = even.

Solution: The string consists of any combination of 'a' and 'b'. In the string, the total number of characters is even. The TM is designed as follows. If the machine gets 'a' or 'b', it moves right with the state changed from q_0 to q_1. In state q_1, if the machine gets 'a' or 'b', it moves right with state changed from q_1 to q_0. In state q_0, if the machine gets a blank symbol, it halts.

The transitional functions are

$$\delta(q_0, a) \rightarrow (q_1, a, R)$$
$$\delta(q_0, b) \rightarrow (q_1, b, R)$$
$$\delta(q_1, a) \rightarrow (q_0, a, R)$$
$$\delta(q_1, b) \rightarrow (q_0, b, R)$$
$$\delta(q_0, B) \rightarrow (q_f, B, H)$$

Example 8.9 Design a TM for infinite loop.

Solution: An infinite loop means the machine does not halt. It can be designed in the following way.

Let the string consist of only 'a'. If 'a' is traversed in state q_1, the machine moves right. After all the 'a' of the string are traversed, B is traversed and the machine moves left and changes the state to q_2. Getting 'a' in state q_2, the machine moves left again to find the beginning of the string.

When the machine gets 'B', it moves right again with changed state from q_2 to q_1. This process continues and the machine falls in an infinite loop.

The transitional functions are

$$\delta(q_1, a) \rightarrow (q_1, a, R)$$
$$\delta(q_1, B) \rightarrow (q_2, B, L)$$
$$\delta(q_2, a) \rightarrow (q_2, a, L)$$
$$\delta(q_2, B) \rightarrow (q_1, a, R)$$

Example 8.10 Design a TM to perform the function $f(x) = x + 1$. $x > 0$.

Solution: The function adds a '1' with the value of x. If one blank symbol at the right hand side is replaced by '1', then the machine can be easily designed.

The transitional functions are

$$\delta(q_1, 1) \rightarrow (q_1, 1, R)$$
$$\delta(q_1, B) \rightarrow (q_2, 1, L)$$
$$\delta(q_2, 1) \rightarrow (q_2, 1, L)$$
$$\delta(q_2, B) \rightarrow (q_f, B, H)$$

Example 8.11 Design a TM which copies a string of '0' and pastes it just after the string. Show an ID for the string 00.

Solution: We have to design a TM which performs copy paste operation

The machine is designed in the following way.

i) Replace each '0' of the given string by X.
ii) Go to the right and find the blank and traverse left and replace the rightmost X by 0 again.
iii) Then traverse to the right and replace the first 'B' after the rightmost '0' by '0' and traverse left. Repeat the last two steps until there are no more x.

The transitional functions are

$$\delta(q_0, 0) \rightarrow (q_0, X, R)$$
$$\delta(q_0, B) \rightarrow (q_1, B, L)$$
$$\delta(q_1, X) \rightarrow (q_2, 0, R)$$
$$\delta(q_2, 0) \rightarrow (q_2, 0, R)$$
$$\delta(q_2, B) \rightarrow (q_1, 0, L)$$
$$\delta(q_1, 0) \rightarrow (q_1, 0, L)$$
$$\delta(q_1, B) \rightarrow (q_f, B, H)$$

ID for the String 00

$(q_0, \mathbf{0}0BB) \rightarrow (q_0, X\mathbf{0}BB) \rightarrow (q_0, XX\mathbf{B}B) \rightarrow (q_1, X\mathbf{X}BB) \rightarrow (q_2, X0\mathbf{BB}) \rightarrow (q_1, X\mathbf{0}0B) \rightarrow (q_1, \mathbf{X}00B)$
$\rightarrow (q_2, 0\mathbf{0}0B) \rightarrow (q_2, 00\mathbf{0}B) \rightarrow (q_2, 000\mathbf{B}) \rightarrow (q_1, 00\mathbf{0}0) \rightarrow (q_1, 0\mathbf{0}00) \rightarrow (q_1, \mathbf{0}000) \rightarrow (q_1, \mathbf{B}0000)$
$\rightarrow (q_f, B0000)$ [Halt]

Example 8.12 Design a TM which performs 1's complement operation on binary string.

Solution: A complement operation means that '0' is converted to '1' and '1' is converted to '0'.

The transitional functions for the TM are

$$\delta(q_0, 0) \rightarrow (q_0, 1, R)$$
$$\delta(q_0, 1) \rightarrow (q_0, 0, R)$$
$$\delta(q_0, B) \rightarrow (q_f, B, H)$$

Example 8.13 Design a TM to perform 2's complement operation on binary string. Show the IDs for the string 010 and 101.

Solution: To calculate 2's complement, first a binary string is converted into 1's complement, and then with that 1's complement 1 is added to get 2's complement. If in the 1's complement the rightmost bit is '0', then that '0' is converted into '1'. If the rightmost bit is '1', then that '1' is converted into '0' and traverses left to find a '0'. If it gets '1', then that '1' is converted into '0' again. If it gets '0', then that '0' is converted into '1' and traverses left to find the leftmost symbol of the string.

	1	0	0	1	
	0	1	1	0	(1's Complement)
+				1	
	0	1	1	1	(2's Complement)

	0	1	0	0	
	1	0	1	1	(1's Complement)
+				1	
	1	1	0	0	(2's Complement)

The TM is designed in the following way.

Starting from state q_1 if the machine gets '1' or '0', that is converted into '0' or '1', respectively, and the machine traverses right. When the machine gets B, it changes its state from q_1 to q_2 and traverses left. If it gets '0', that is converted into '1' with the state changed from q_2 to q_3 and traverses left to find the leftmost symbol.

$\delta(q_1, 0) \rightarrow (q_1, 1, R)$ // '0' is converted into '1'
$\delta(q_1, 1) \rightarrow (q_1, 0, R)$ // '1' is converted into '0'
$\delta(q_1, B) \rightarrow (q_2, B, L)$ // getting 'B' traverses left
$\delta(q_2, 0) \rightarrow (q_3, 1, L)$ // if the rightmost bit in 1's complement is '0'
$\delta(q_2, 1) \rightarrow (q_2, 0, L)$ // if the rightmost bit in 1's complement is '1' and search for '0'
$\delta(q_3, 0) \rightarrow (q_3, 0, L)$ // left traversal to find the left end
$\delta(q_3, 1) \rightarrow (q_3, 1, L)$ // left traversal to find the left end
$\delta(q_3, B) \rightarrow (q_f, B, H)$ // if 'B', then halts.
$\delta(q_2, B) \rightarrow (q_f, B, H)$ // if in 1's complement all bits are 1's in state q_2, machine gets B

ID for 010

(q_1, B**0**10B) → (q_1, B1**1**0B) → (q_1, B10**0**B) → (q_1, B10**1**B) → (q_2, B10**1**B) → (q_2, B1**0**0B)
→ (q_3, B**1**10B) → (q_f, B110B) [Halt]

ID for 101

(q_1, B**1**01B) → (q_1, B0**0**1B) → (q_1, B01**1**B) → (q_1, B010**B**) → (q_2, B01**0**B) → (q_3, B0**1**1B)
→ (q_3, B**0**11B) → (q_3, **B**011B) → (q_f, B011B) [Halt]

Example 8.14 Design a TM for a set of all strings with equal number of 'a' and 'b'.

Solution: The machine is designed as follows.

The machine first finds an 'a', replaces this by X and moves left. When it gets blank, it again moves right to search for 'b' and replaces it by Y. Again, it traverse left to search for 'a'. The process continues till the right side blank symbol is traversed in state q_0.

The transitional functions are

$\delta(q_0, a) \rightarrow (q_1, X, L)$ // search 'a' and traverse left
$\delta(q_0, b) \rightarrow (q_0, b, R)$
$\delta(q_0, X) \rightarrow (q_0, X, R)$
$\delta(q_0, Y) \rightarrow (q_0, Y, R)$
$\delta(q_1, B) \rightarrow (q_2, B, R)$ // traverse right in search of 'b'
$\delta(q_1, X) \rightarrow (q_1, X, R)$
$\delta(q_1, Y) \rightarrow (q_1, Y, L)$
$\delta(q_1, b) \rightarrow (q_1, b, L)$
$\delta(q_2, a) \rightarrow (q_2, a, R)$
$\delta(q_2, X) \rightarrow (q_2, X, R)$
$\delta(q_2, Y) \rightarrow (q_2, Y, R)$
$\delta(q_2, b) \rightarrow (q_3, Y, L)$
$\delta(q_3, a) \rightarrow (q_3, a, L)$
$\delta(q_3, X) \rightarrow (q_3, X, L)$
$\delta(q_3, Y) \rightarrow (q_3, Y, L)$
$\delta(q_3, B) \rightarrow (q_0, B, R)$ // traverse right in search of 'a'
$\delta(q_0, B) \rightarrow (q_f, B, H)$

Example 8.15 Design a TM to accept the language $L = a^nb^{2n}, n > 0$.

Solution: The number of 'b' is equal to twice the number of 'a'. In the language, all 'a' appear before 'b'. The TM is designed as follows.

Replace one 'a' by X and traverse right to search for the end of the string. After getting a 'B', traverse left and replace two 'b's by Y. In the state q_0, if the machine gets Y as input it halts.

The transitional functions are

$\delta(q_0, a) \rightarrow (q_1, X, R)$
$\delta(q_1, a) \rightarrow (q_1, a, R)$
$\delta(q_1, b) \rightarrow (q_1, b, R)$
$\delta(q_1, B) \rightarrow (q_2, B, L)$
$\delta(q_1, Y) \rightarrow (q_2, Y, L)$ // second traverse onwards
$\delta(q_2, b) \rightarrow (q_3, Y, L)$
$\delta(q_3, b) \rightarrow (q_4, Y, L)$
$\delta(q_4, b) \rightarrow (q_4, b, L)$
$\delta(q_4, a) \rightarrow (q_4, a, L)$
$\delta(q_4, X) \rightarrow (q_0, X, R)$
$\delta(q_0, Y) \rightarrow (q_f, Y, H)$

8.2 Transitional Representation of Turing Machine

The mathematical notation for a TM is (Q, Σ, Γ, δ, q_0, B, F). In a TM, the transitional function δ consists of the form $Q \times \Gamma \to (Q \times \Gamma \times \{L, R, H\})$, where the first is a present state and the second is the present input (single character ∈ Γ). The transition produces a state, a symbol ∈ Γ written on the input tape, and the head moves either left or right or halts.

In the graphical notation of the TM, there are states. Among them, a circle with an arrow indicates a beginning state and a state with double circle indicates a final state, where the machine halts finally. The state transitions are denoted by arrows. The labels of the state transitions consist of the input symbol, the symbol written on the tape after traversal, and the direction of movement of the read–write head (left, right, or halt).

The following are some examples of the TM with the transitional representation.

Example 8.16 Design a TM by the transitional notation for the language $L = a^n b^n$, where $n > 0$.

Solution: When the machine traverses 'a', replace that 'a' by X and traverse right to find the leftmost 'b'. Replace that 'b' by 'Y' and traverse left to find the second 'a'. The second 'a' exists after 'X', which was replaced first for the first 'a'. By this process, when n number of 'a' and n number of 'b' are traversed and replaced by X and Y, respectively, then by getting a blank (B) the machine halts.

The transitional representation of the TM is given in Fig. 8.2.

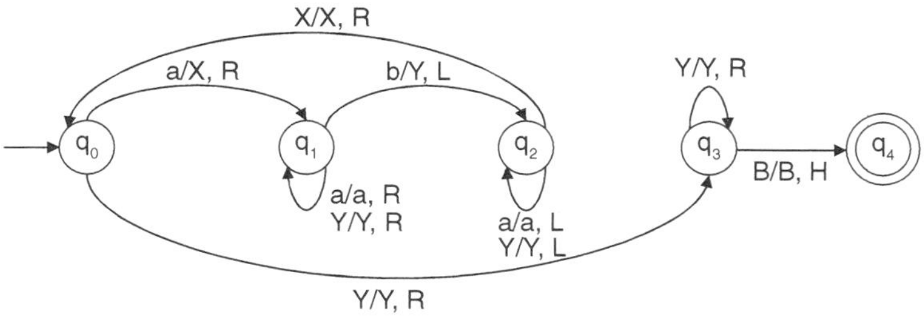

Fig. 8.2 *Transitional Representation of Turing Machine*

Example 8.17 Design a TM by transitional notation for the language L = {set of all palindrome over a, b}.

Solution: A palindrome can be of two types, namely, an odd palindrome and an even palindrome. A null string is also a palindrome. There are three paths to reach to the final state with halt from q_1, the beginning state. This is described in Fig. 8.3.

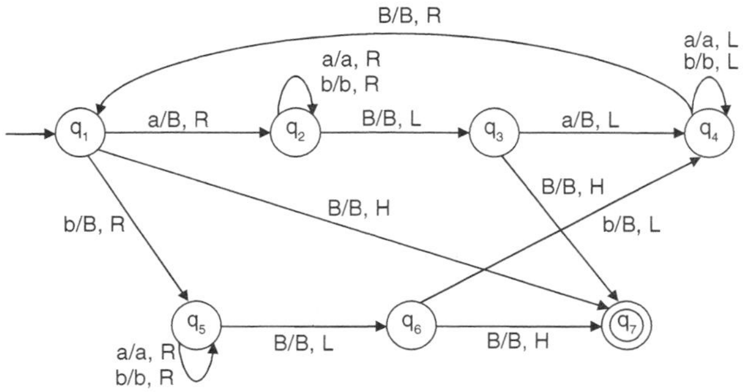

Fig. 8.3 *Turing Machine for Set of All Palindrome Over a, b*

8.3 Non-deterministic Turing Machine

The concept of non-determinism is clear as we have already discussed about NDFA or NDPDA in the earlier chapters. There is also a non-deterministic TM. A non-deterministic TM is defined as (Q, Σ, Γ, δ, q_0, B, F), where δ is $Q \times \Gamma \rightarrow 2^{Q \times \Gamma \times \{L, R\}}$.

Let there exist a transitional function $\delta(q_0, 0) \rightarrow (q_0, 0, R), (q_1, 1, R)$.

For traversing the input symbol, we have to consider two transitions and construct the following transitions accordingly.

The NTM is more powerful than the DTM. An NTM T_1 can be simulated to an equivalent DTM T_2. One of the methods is to construct a computation tree of the NTM and perform a breadth first search of the computation tree from the root until the halt state is reached. At each level of the constructed tree, T_2 uses the transitional functions of T_1 to each configuration at that level and computes its children. These children are stored on a tape for the configuration of the next level. Among the children, if the halting state exists then T_2 halts by accepting the strings. T_2 accepts means that one branch of T_1 accepts it. It can be said that T_2 accepts a string if and only if T_1 accepts it.

Any language accepted by an NTM is also accepted by a DTM. But the time taken by a DTM to accept a language is more than the time taken by an NTM. In most of the cases, it is exponential in length. A famous unsolved problem in computer science, i.e., the P = NP problem, is related to this issue.

Example 8.18 Construct a TM over {a, b} which contains a substring abb.

Solution: In regular expression, it can be written as $(a, b)^*abb(a, b)^*$. In this expression, the substring is important. The string may be abb only. In that case, the machine gets 'a' as input in the beginning state. Before traversing the substring 'abb', there is a chance to traverse 'a' of $\{a, b\}^*$. The transitional functions are

$$\delta(q_1, a) \rightarrow (q_1, a, R), (q_2, a, R)$$
$$\delta(q_1, b) \rightarrow (q_1, b, R)$$
$$\delta(q_2, b) \rightarrow (q_3, b, R)$$
$$\delta(q_3, b) \rightarrow (q_4, b, R)$$
$$\delta(q_4, a) \rightarrow (q_4, a, R)$$
$$\delta(q_4, b) \rightarrow (q_4, b, R)$$
$$\delta(q_4, B) \rightarrow (q_f, B, H)$$

In state q_1 for input 'a', there are two transitional functions. So, it is a non-deterministic TM.

8.4 Conversion of Regular Expression to Turing Machine

Regular expression can be converted to an equivalent TM. The process is as follows.

1. Convert the regular expression to an equivalent automata without the ∈ move.
2. Change both the initial and final states of the automata to an intermediate state.
3. Insert a new initial state with a transition (B, B, R) to the automata's initial state.
4. Convert the transitions with label 'ip' to (ip, ip, R).
5. Insert a new final state with a transition (B, B, R) from the automata's final state to the new final state.

Example 8.19 Construct a TM for the regular expression (a + b)*(aa + bb)(a + b)*.

Solution: The finite automata accepting the regular expression is

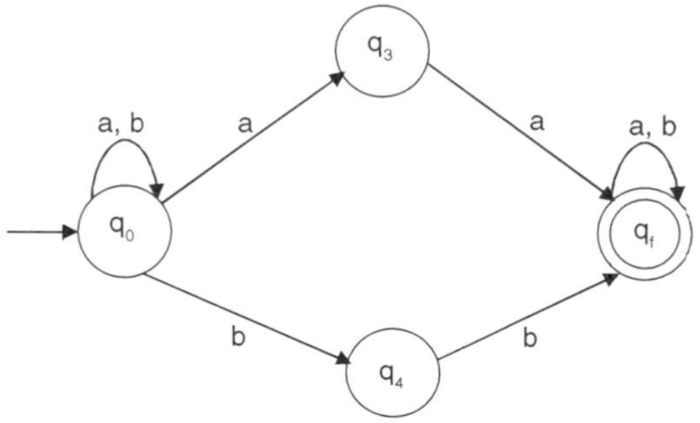

Inserting a new initial state q_i and final state q_H, the finite automata becomes

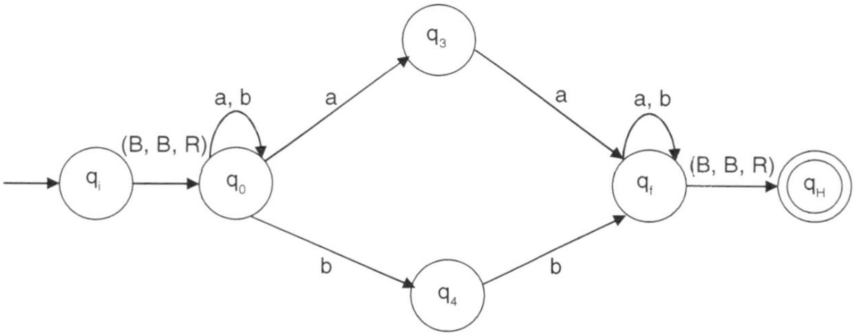

Converting the levels of the inputs, the TM becomes

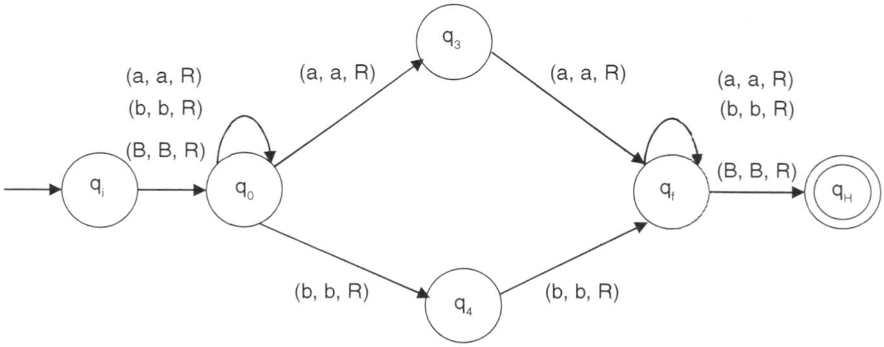

Example 8.20 Construct a TM for the regular expression ab(aa + bb)(a + b)*b.

Solution: The finite automata accepting the regular expression is

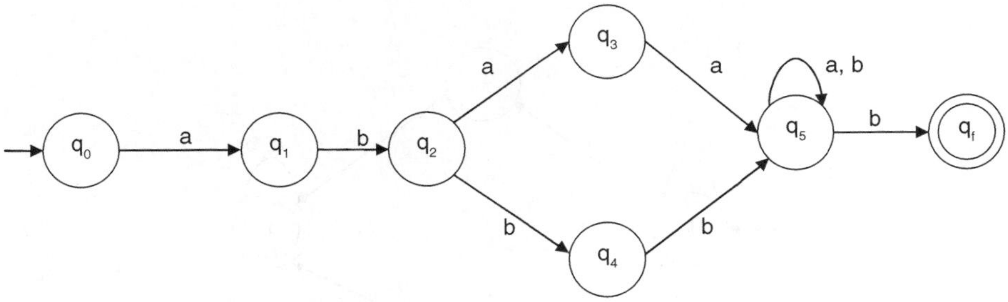

Inserting a new initial state q_i and final state q_H, the finite automata becomes

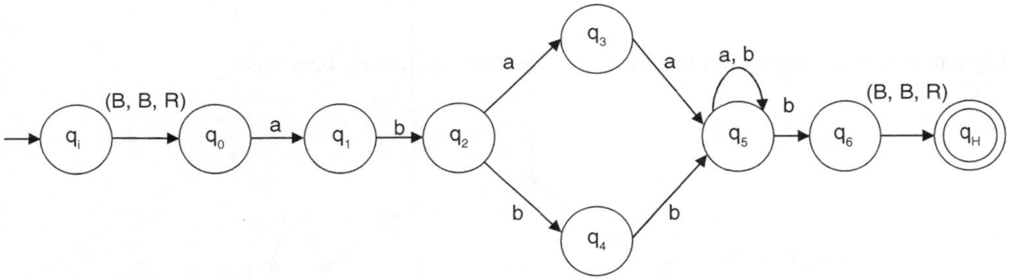

Converting the levels of the inputs, the TM becomes

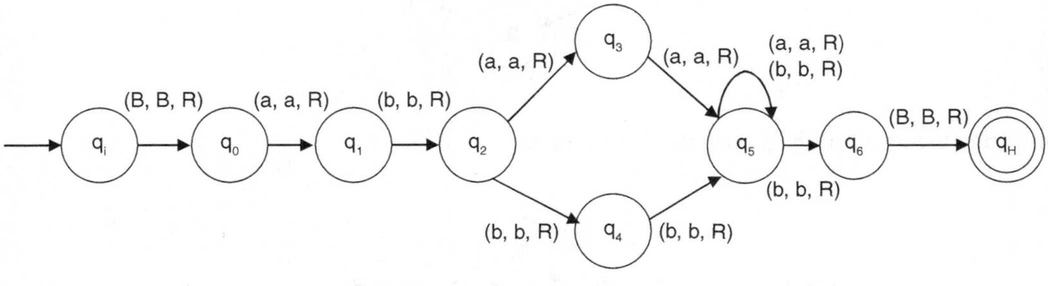

Example 8.21 Design a TM which converts (0, 1)*011(0, 1)* to (0, 1)*100(0, 1)*.

Solution: The problem is to design a TM which searches the substring 011 from a string of 0, 1, and then it replaces 011 by 100. The design is done by the following way.

i) Convert (0, 1)*011(0, 1)* to the accepting DFA.
ii) Convert it to the corresponding TM.
iii) Modify the TM by modifying the transitions of 011 by 100.

The NFA for accepting (0, 1)*011(0, 1)* is

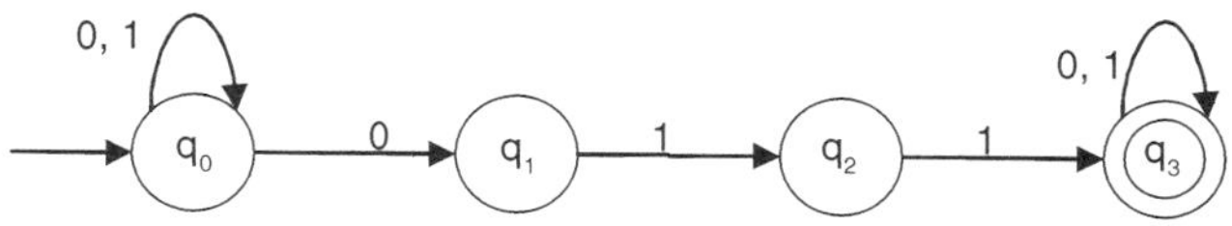

The equivalent DFA is

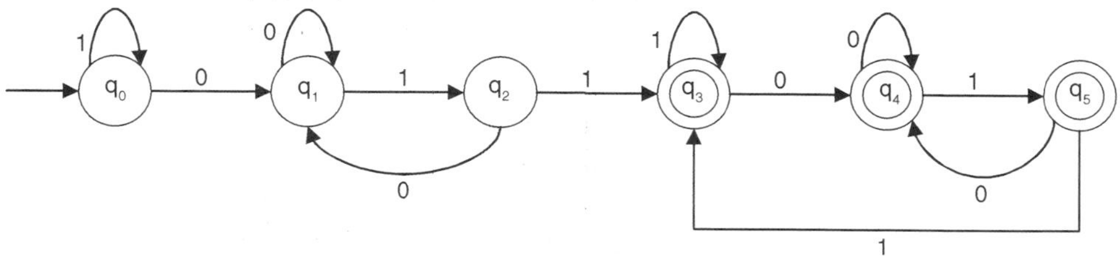

The converted TM is

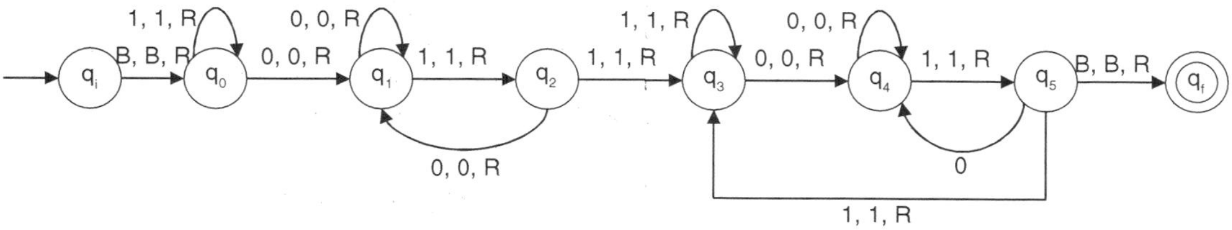

The modified TM is (converting 011 by 100) as follows. The modifications are denoted in bold.

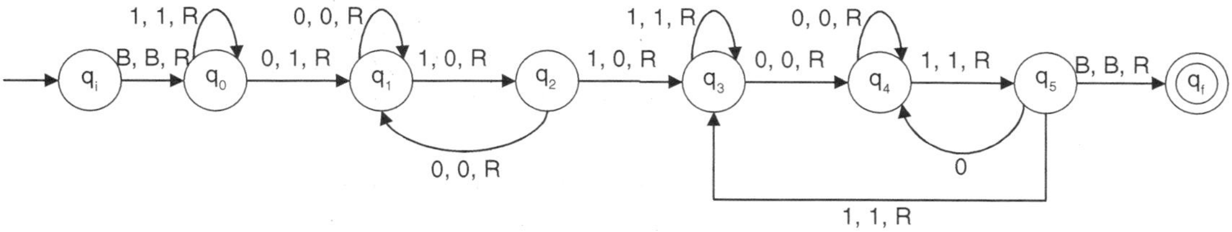

8.5 Two-stack PDA and Turing Machine

Two-stack PDA is already discussed in the pushdown automata chapter. It is also discussed that a PDA cannot be only two stack but may be more than two stacks ($a^nb^nc^nd^n$, $n \geq 0$). In real, the language accepting power of a PDA increases by adding extra stacks. Here, the following question arises 'is two-stack (or more than two stacks) PDA as strong as the TM?'. In this section, we shall learn about a theorem proposed by an American artificial intelligence scientist Marvin Minsky called the Minsky theorem which answers this question.

8.5.1 Minsky Theorem

Any language accepted by a two-stack PDA can also be accepted by some TM and vice versa.

8.5.1.1 General Minsky Model

A general Minsky model is shown in the following figure.

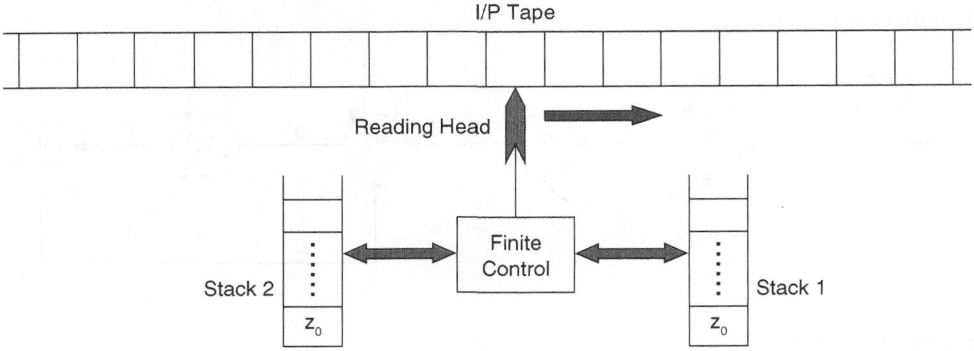

Fig. 8.4 *Two Stack PDA*

The model in Fig. 8.4 is a two-stack PDA containing an input tape, a reading head, and finite control as it was in PDA. The additionals are two stacks S_1 and S_2. Let the language be $a^nb^nc^n$ for $n \geq 1$. Scanning 'a', the symbols (say X) are put into stack S_1. With the stack top 'X' in S_1, 'b' is scanned and the symbols (say Y) are put into stack S_2. At the time of traversing 'c', the stack symbols must be X (for S_1) and Y (For S_2). These are popped for traversing 'c'.

A two-stack PDA is capable of accepting ($a^nb^nc^nd^n$, $n \geq 1$), ($a^nb^nc^nd^ne^n$, $n \geq 1$), and so on, with the additional stack. Accepting these types of languages is not possible for a single-stack PDA but possible for a TM. By this process, the two-stack PDA simulates a TM.

The following example shows the acceptance of ($a^nb^nc^nd^n$, $n \geq 1$) using the two-stack PDA.

Example 8.22 Design a two-stack PDA for ($a^nb^nc^nd^n$, $n \geq 1$).

Solution: Traversing 'a', X are put into stack S_1. Traversing 'b' with the stack top 'X' in S_1, 'X' is popped from S_1, and 'Y' is pushed into S_2. Traversing 'c' with the stack top 'Y' in S_2, 'Y' is popped from S_2, and 'Z' is pushed into S_1. Traversing 'd', 'Z' is popped from S_1. The transitional functions are

$\delta(q_0, \lambda, z_1, z_2) \rightarrow (q_1, z_1, z_2)$ (z_1 and z_2 are stack bottom symbols of S_1 and S_2, respectively.)
$\delta(q_1, a, z_1, z_2) \rightarrow (q_1, Xz_1, z_2)$
$\delta(q_1, a, X, z_2) \rightarrow (q_1, XX, z_2)$
$\delta(q_1, b, X, z_2) \rightarrow (q_2, \lambda, Yz_2)$
$\delta(q_2, b, X, Y) \rightarrow (q_2, \lambda, YY)$
$\delta(q_2, c, z_1, Y) \rightarrow (q_3, Zz_1, \lambda)$
$\delta(q_3, c, Z, Y) \rightarrow (q_3, ZZ, \lambda)$
$\delta(q_3, d, Z, z_2) \rightarrow (q_4, \lambda, z_2)$
$\delta(q_4, d, Z, z_2) \rightarrow (q_4, \lambda, z_2)$
$\delta(q_4, \lambda, z_1, z_2) \rightarrow (q_f, z_1, z_2)$ // accepted by the final state

What We Have Learned So Far

1. The Turing machine was proposed by A.M. Turing in 1936 as a model of any possible combination. Any computational process carried out by the present day's computer can be done on the Turing machine.
2. The Turing machine is the machine format of unrestricted language, i.e., all types of languages are accepted by the Turing machine.
3. Undecidable problems are not solved by the computer as no Turing machine can be developed for these problems.
4. The mathematical description of the Turing machine consists of 7 touples $(Q, \Sigma, \Gamma, \delta, q_0, B, F)$, where Q is the finite set of states, Σ is the finite set of input alphabets, Γ is the finite set of allowable tape symbol, δ is the transitional function, q_0 is the initial state, B is the blank symbol, and F is the final state.
5. The transitional functions of the Turing machine are in the form $Q \times \Gamma \rightarrow (Q \times \Gamma \times \{L, R, H\})$. That is, from one state by getting one input from the input tape the machine moves to a state, writing a symbol on the tape, and moves to left, right, or halts.
6. The mechanical diagram of the Turing machine consists of the input tape, the finite control and the read–write head.
7. Upon execution of a transitional function, the machine goes to some state, writes a symbol in the cell of the input tape from where the input symbol was scanned, and moves the reading head to the left or right or halts.
8. The instantaneous description (ID) of a Turing machine remembers the contents of all cells from the rightmost to at least the leftmost, the cell currently being scanned by the read–write head, and the state of the machine at a given instance of time.

Solved Problems

1. Consider the Turing machine's description in the following table below. Draw the computation sequence of the input string 00. [WBUT 2008]

Present State	Tape Symbol :: b	Tape Symbol :: 0	Tape Symbol :: 1
Q_1	$1Lq_2$	$0Rq_1$	–
Q_2	bRq_3	$0Lq_2$	$1Lq_2$
Q_3	–	bRq_4	bRq_5
Q_4	$0Rq_5$	$0Rq_4$	$1Rq_4$
Q_5	$0Lq_2$	–	–

Solution: $(Q_1, B\underline{0}0B) \rightarrow (Q_1, 0\underline{0}B) \rightarrow (Q_1, 00\underline{B}) \rightarrow (Q_2, 0\underline{0}1) \rightarrow (Q_2, \underline{0}01) \rightarrow (Q_2, \underline{B}001)$ $\rightarrow (Q_3, B\underline{0}01) \rightarrow (Q_4, B\underline{0}1) \rightarrow (Q_4, B0\underline{1}) \rightarrow (Q_4, B01\underline{B}) \rightarrow (Q_5, B010\underline{B}) \rightarrow (Q_2, B01\underline{0}0)$ $\rightarrow (Q_2, B0\underline{1}00) \rightarrow (Q_2, B\underline{0}100) \rightarrow (Q_2, \underline{B}0100) \rightarrow (Q_3, B\underline{0}100) \rightarrow (Q_4, B0\underline{1}00) \rightarrow (Q_4, B01\underline{0}0)$ $\rightarrow (Q_4, B010\underline{0}) \rightarrow (Q_4, B0100\underline{B}) \rightarrow (Q_5, B01000\underline{B}) \rightarrow (Q_2, B0100\underline{0}0)$

(By this process, it will continue as there is no halt in the Turing Machine.)

2. Design a Turing Machine to test a string of balanced parenthesis. Show an ID for ( )(( ))

Solution: There are two types of parentheses—open parentheses and close parentheses. Balanced parenthesis means that for each open parenthesis there is a close parenthesis. It may be nested like ((( ))) or may be non-nested like (( ) ( )) or ( )( )( ) or any other form. But for each open parenthesis, there is a close parenthesis. [The definition that a balanced parenthesis means an equal number of open and close parentheses is wrong.

As an example,) ( )( ) (is not a balanced parenthesis.] The string must start with an open parenthesis and after that open parenthesis or close parenthesis can appear, but for any position of the string, the number of open parenthesis is greater than or equal to the number of close parenthesis.

The Turing Machine is designed as follows. Start from the leftmost symbol and traverse right to find the first close parenthesis. The transitional function is

$$\delta(q_0, \text{‘(’}) \rightarrow (q_0, (, R)$$

Upon getting the first close parenthesis, replace it by ‘X’, change the state to q_1, and move left to find the open parenthesis for the replaced close parenthesis. The transitional function is

$$\delta(q_0, \text{‘)’}) \rightarrow (q_1, X, L)$$

Getting the open parenthesis, replace it by ‘X’ and change the state to q_0. The transitional function is

$$\delta(q_1, \text{‘(’}) \rightarrow (q_0, X, R)$$

Then, traverse towards the right to find the close parenthesis. Here, the machine may have to traverse X, which is the replaced symbol for close parenthesis. The transitional function for traversing this is

$$\delta(q_0, \text{‘X’}) \rightarrow (q_0, X, R)$$

For a nested parenthesis like (( )), the machine has to traverse X, which is the replaced symbol of the open parenthesis at the time of finding the open parenthesis. The transitional function for traversing this is

$$\delta(q_1, \text{‘X’}) \rightarrow (q_1, X, L)$$

Traversing right, if B appears as an input, then there is no parenthesis (open or close) left at the right side. Then, traverse left to find if any parenthesis is left at the left side or not. The transitional function is

$$\delta(q_0, \text{‘B’}) \rightarrow (q_2, B, L)$$

At the time of left traversing, the machine has to traverse X by the following transitional function

$$\delta(q_2, \text{‘X’}) \rightarrow (q_2, B, L)$$

At the left hand side if it gets B, the machine halts.

$$\delta(q_2, \text{‘B’}) \rightarrow (q_3, B, H)$$

ID for ( ) (( ))

(q_0, B( )(( ))B) → (q_0, B( )(( ))B) → (q_1, B(X(( ))B) → (q_0, BXX(( ))B) → (q_0, BXX(( ))B) → (q_0, BXX(( ))B) → (q_0, BXX(( ))B) → (q_0, BXX(( ))B) → (q_1, BXX((X)B) → (q_0, BXX(XX)B) → (q_0, BXX(XX)B) → (q_1, BX X(XXXB) → (q_1, BX X(XXXB) → (q_1, BXX(XXXB) → (q_0, BXXXXXXB) → (q_0, BXXXXXXB) → (q_0, BXXXXXXB) → (q_0, BXXXXXXB) → (q_2, BXXXXXXB) $\overset{*}{\Longrightarrow}$ q_2, BXXXXXXB) → (q_3, BXXXXXXB) [Halt]

3. Design a Turing machine to perform n mod 2. (Or check whether a number is odd or even) From here, by instantaneous description, check whether the number 5 is odd or even.

Solution: n is any number of same input alphabets. Let us take n number of '0's. The machine is in state q_0 and the read–write head is on the beginning input symbol of the input tape. Getting input '0', it moves towards the right replacing '0' by 'B' and changes state to q_1. The transitional function is

$$\delta(q_0, 0) \rightarrow (q_1, B, R)$$

In state q_1, by getting the input '0', it moves towards the right replacing '0' by 'B' and changes state to q_0. The transitional function is

$$\delta(q_1, 0) \rightarrow (q_0, B, R)$$

By the previous two transitional functions, all the '0's are replaced by 'B'. If the number of '0' is even, then the machine will reach to the end of the string with state q_0. At the end, it will get 'B' and traversing that the machine will halt (means the number is even).

The transitional function is

$$\delta(q_0, B) \rightarrow (q_2, B, H)$$

If the number of '0' is even, then the machine will reach to the end of the string with state q_1. With getting input symbol 'B', it will traverse left by changing the state to q_3. The transitional function is

$$\delta(q_1, B) \rightarrow (q_3, B, L)$$

In state q_3, by getting input B, the machine halts by replacing 'B' to '0' (means the number is odd). The transitional function is

$$\delta(q_3, B) \rightarrow (q_2, 0, H).$$

ID for n = 5: The string is 00000.

(q_0, **0**0000B) → (q_1, B**0**000B) → (q_0, BB**0**00B) → (q_1, BBB**0**0B) → (q_0, BBBB**0**B) → (q_1, BBBBB**B**) → (q_3, BBBB**B**B) → (q_2, BBBB0B) [Halt]

4. Construct a Turing machine to compute the function f(w) = w^R, where w ∈ $(a, b)^+$ and w^R is the reverse of w. [UPTU 2004]

Solution: The string consists of 'a' and 'b'. The string may start with 'a' or may start with 'b'. If the string starts with 'a', the reverse of the string will end with 'a'. The same for 'b' also.

If the beginning state is q_1 and if the string starts with 'a', the TM will traverse right replacing 'a' by 'X' and changing the state to q_2. (This q_2 specifies that the symbol was 'a'.) The transitional function is

$$\delta(q_1, a) \rightarrow (q_2, X, R)$$

At the time of traversing right, the machine can get an 'a' or 'b'. The transitional functions for traversing 'a' and 'b' are

$$\delta(q_2, a) \rightarrow (q_2, a, R)$$
$$\delta(q_2, b) \rightarrow (q_2, b, R)$$

If it gets 'B', it means the string ends. The TM changes the state to q_3 and traverses left by the following transitional function

$$\delta(q_2, B) \rightarrow (q_3, B, L)$$

The string may end with 'a' or may end with 'b'. As the beginning symbol was 'a' (which is memorized by the machine by state q_2, then q_3), the end symbol will be replaced by X, whatever it may be. But the end symbol will be the beginning symbol. So, what the end symbol is, it will be memorized by different states. In this case, if the end symbol is 'a', it will be memorized by changing the state q_3 to q_6 and if the end symbol is 'b', it will be memorized by changing q_3 to q_7. The transitional functions are

$$\delta(q_3, a) \rightarrow (q_6, X, L)$$
$$\delta(q_3, b) \rightarrow (q_7, X, L)$$

Traversing left, the TM has to traverse 'a' or 'b'. If the state is q_6, the transitional functions are

$$\delta(q_6, a) \rightarrow (q_6, a, L)$$
$$\delta(q_6, b) \rightarrow (q_6, b, L)$$

At the time of traversing left, the TM may get 'X' or 'Y', the replaced symbols of 'a' or 'b'. The machine is in state q_6, which means the rightmost symbol of the string was 'a'. Upon getting 'X' or 'Y' at the left end, the TM replaces it by 'X' and changes the state to q_1. The transitional functions are

$$\delta(q_6, X) \rightarrow (q_1, X, R)$$
$$\delta(q_6, Y) \rightarrow (q_1, X, R)$$

If the string starts with 'b', the transitions are the same as the string starts with 'a', but only some changes of states. The transitional functions are

$\delta(q_1, b) \rightarrow (q_4, Y, R)$
$\delta(q_4, a) \rightarrow (q_4, a, R)$ // traversing right in search of end of string
$\delta(q_4, b) \rightarrow (q_4, b, R)$
$\delta(q_4, B) \rightarrow (q_5, B, L)$ // end of the string and traversing left
$\delta(q_5, a) \rightarrow (q_6, Y, L)$ // if the end symbol is 'a'
$\delta(q_5, b) \rightarrow (q_7, Y, L)$ // // if the end symbol is 'b'

If the end symbol is 'b' at the time of traversing left, the transitional functions are

$$\delta(q_7, a) \rightarrow (q_7, a, L)$$
$$\delta(q_7, b) \rightarrow (q_7, b, L)$$
$$\delta(q_7, X) \rightarrow (q_1, Y, R)$$
$$\delta(q_7, Y) \rightarrow (q_1, Y, R)$$

By the previous mentioned transitional functions, the total string is traversed and reversed but in the place of 'a' and 'b' there are 'X' and 'Y'.

From the second time of traversing the string, the TM need not find the symbol 'B' to mark the right end of the string. In the state q_2 or q_5, it will get 'X' or 'Y', which are representations of the already traversed symbol. The transitional functions are

$$\delta(q_2, X) \rightarrow (q_3, X, L)$$
$$\delta(q_2, Y) \rightarrow (q_3, Y, L)$$
$$\delta(q_4, X) \rightarrow (q_5, X, L)$$
$$\delta(q_4, Y) \rightarrow (q_5, Y, L)$$

The string may be of even length or odd length. For the odd length string, the middle element may be 'a' or 'b'. If the middle element is 'a', then after replacing the middle element by 'X', the TM moves right but it finds either 'X' or 'Y' (because all the other symbols are already replaced). It backs left by changing the state to q_3 or q_5 using the transitional functions mentioned earlier. At the left, again it gets either 'X' or 'Y'. The TM changes the state to q_8 and moves right by the following transitional functions

$$\delta(q_3, X) \rightarrow (q_8, X, R)$$
$$\delta(q_3, Y) \rightarrow (q_8, Y, R)$$
$$\delta(q_5, X) \rightarrow (q_8, X, R)$$
$$\delta(q_5, Y) \rightarrow (q_8, Y, R)$$

If the string is of even length, then the machine gets 'X' or 'Y' in the state q_1 after replacing all the 'a' and 'b'. The transitional functions are

$$\delta(q_1, X) \rightarrow (q_8, X, R)$$
$$\delta(q_1, Y) \rightarrow (q_8, Y, R)$$

The machine traverses right to find the end of the string consists of 'X' or 'Y'. When it gets 'B', it traverses left by changing the state from q_8 to q_9.

$$\delta(q_8, X) \rightarrow (q_8, X, R)$$
$$\delta(q_8, Y) \rightarrow (q_8, Y, R)$$
$$\delta(q_8, B) \rightarrow (q_9, B, L)$$

In the state q_9 at the time of traversing left, the TM replaces all 'X's by 'a' and all 'Y's by 'b'.

$$\delta(q_9, X) \rightarrow (q_9, a, L)$$
$$\delta(q_9, Y) \rightarrow (q_9, b, L)$$

By this process, when it gets the symbol 'B' at the left it halts.

$$\delta(q_9, B) \rightarrow (q_{10}, B, H)$$

If the string is a blank string (no 'a' and/or 'b'), the machine gets 'B' in the state q_1 and halts.

$$\delta(q_1, B) \rightarrow (q_{10}, B, H)$$

The transitional functions are represented in the tabular format.

State	a	B	X	Y	B
q_1	(q_2, X, R)	(q_4, Y, R)	(q_8, X, R)	(q_8, Y, R)	(q_{10}, B, H)
q_2	(q_2, a, R)	(q_2, b, R)	(q_3, X, L)	(q_3, Y, L)	(q_3, B, L)
q_3	(q_6, X, L)	(q_7, X, L)	(q_8, X, R)	(q_8, Y, R)	–
q_4	(q_4, a, R)	(q_4, b, R)	(q_5, X, L)	(q_5, Y, L)	(q_5, B, L)
q_5	(q_6, Y, L)	(q_7, Y, L)	(q_8, X, R)	(q_8, Y, R)	–
q_6	(q_6, a, L)	(q_6, b, L)	(q_1, X, R)	(q_1, X, R)	–
q_7	(q_7, a, L)	(q_7, b, L)	(q_1, Y, R)	(q_1, Y, R)	–
q_8	–	–	(q_8, X, R)	(q_8, Y, R)	(q_9, B, L)
q_9			(q_9, a, L)	(q_9, b, L)	(q_{10}, B, H)

5. Design a Turing machine for $0^n1^n0^n$ $n \geq 1$. [Cochin University 2000]

Solution: The language consists of 0 and 1. Here, n number of '1' are placed in between n number of '0's and n number of '0'. The Turing machine is designed as follows. For each leftmost '0', search for '1' and the leftmost '0' placed after '1' and again traverse back to the left to search for the leftmost '0'. '0' is replaced by 'X', and '1' is replaced by 'Y'.

When all '0' are traversed and replaced by 'X', the machine traverses right side to find if any untraversed '0' or '1' is left or not. If not, the machine gets Y and X and at last blank. Upon getting the blank symbol, the machine halts.

The transitional functions are

$\delta(q_1, 0) \rightarrow (q_2, X, R)$ // '0' is traversed
$\delta(q_2, 0) \rightarrow (q_2, 0, R)$ // the remaining '0' s are traversed
$\delta(q_2, 1) \rightarrow (q_3, Y, R)$ // '1' is traversed
$\delta(q_3, 1) \rightarrow (q_3, 1, R)$ // the remaining '1's are traversed
$\delta(q_3, 0) \rightarrow (q_4, X, L)$ // '0' placed after '1' is traversed
$\delta(q_4, 1) \rightarrow (q_4, 1, L)$ // the remaining '1's are traversed from right to left
$\delta(q_4, Y) \rightarrow (q_5, Y, L)$ // Y, replacement for '1' is traversed
$\delta(q_5, 0) \rightarrow (q_5, 0, L)$ // the remaining '0's are traversed from right to left
$\delta(q_5, X) \rightarrow (q_1, X, R)$ // rightmost X, replacement of '0' placed before '1' is traversed
$\delta(q_2, Y) \rightarrow (q_2, Y, R)$ // this is used from the second time onwards
$\delta(q_3, X) \rightarrow (q_3, X, R)$ // this is used from the second time onwards
$\delta(q_4, X) \rightarrow (q_4, X, L)$ // this is used from the second time onwards
$\delta(q_1, Y) \rightarrow (q_5, Y, R)$ // this is used when all the '0's placed before '1' are traversed

$\delta(q_5, Y) \rightarrow (q_5, Y, R)$ // this is used when all the '0's placed before '1' are traversed and the machine is traversing right

$\delta(q_5, X) \rightarrow (q_6, X, R)$ // this is used when all '0' (placed before '1') and '1' are traversed.

$\delta(q_6, X) \rightarrow (q_6, X, R)$ // this is used when all '0' (placed before '1') and '1' are traversed and the machine is traversing right to search for the remaining '0' (placed after '1').

$\delta(q_6, B) \rightarrow (q_f, B, H)$ // this is used when all '0' and '1' are traversed.

Upon executing the last transitional function, the machine halts.

6. Design a Turing machine that shifts the input string over alphabet (0, 1) by one position right by inserting '#' as the first character. [JNTU 2008]

Solution: According to the condition, a string B0110B will be B#0110B after traversing through the Turing machine. To design the machine, it has to traverse to the right to the end of the string (determined by the symbol 'B'). Then, it has to traverse left. Getting '0' or '1', it changes to two different states (say, S_0 and S_1), respectively. The states signify the traversed symbol. If in the states the machine gets any symbol ('0' or '1'), it will be replaced by the symbol for which the state is changed and traverse left. This is performed recursively.

$\delta(q_1, 0/1) \rightarrow (q_1, 0/1, R)$ // '0' or '1' is traversed from left to right

$\delta(q_1, B) \rightarrow (q_2, B, L)$ // end of the string

$\delta(q_2, 0) \rightarrow (q_3, 0, R)$ // q_3 signifies that the previous symbol is '0'

$\delta(q_2, 1) \rightarrow (q_4, 1, R)$ // q_4 signifies that the previous symbol is '1'

$\delta(q_3, B) \rightarrow (q_5, 0, L)$ // B after the string is replaced by '0'

$\delta(q_4, B) \rightarrow (q_5, 1, L)$ // B after the string is replaced by '1'

$\delta(q_5, 0/1) \rightarrow (q_2, 0/1, L)$ // whatever the symbol (0 or 1); machine traverse left

$\delta(q_3, 0/1) \rightarrow (q_5, 0, L)$ // if the previous symbol is '0', replace the next as '0'

$\delta(q_4, 0/1) \rightarrow (q_5, 1, L)$ // if the previous symbol is '1', replace the next as '1'

$\delta(q_2, B) \rightarrow (q_6, B, R)$ // reached the left end

$\delta(q_6, 0/1) \rightarrow (q_7, \#, R)$ // Replaced the first symbol by '#'

$\delta(q_7, 0/1) \rightarrow (q_7, 0/1, R)$ // Traverse right

$\delta(q_7, B) \rightarrow (q_f, B, H)$ // Halt

Diagrammatically, it is represented as follows.

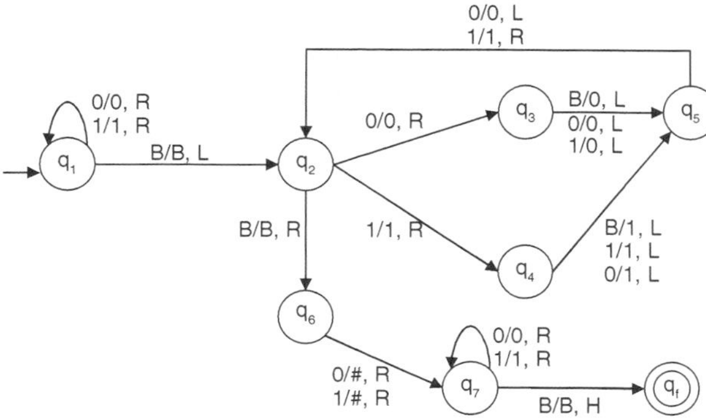

7. Construct a TM over the alphabet {0, 1, #} where '0' indicates blank, which takes a non-null string of '1' and '#' and transfers the rightmost symbol to the left hand end. Thus, 000#1#1#1000 becomes 0001#1#1#000. [Jaipur University 2010]

Solution: The machine has to traverse three symbols '0' as blank, 1, and '#'. To design the machine, traverse right up to '0' (blank). Getting it, traverse left. The symbol may be '1' or '#'. For these two symbols, change to two different states, replace the input by '0' (blank) and traverse left to get '0' (blank). Upon getting it, replace it by '1' or '#' according to the state.

The transitional functions are

$\delta(q_1, 1) \rightarrow (q_1, 1, R)$ // traverse right
$\delta(q_1, \#) \rightarrow (q_1, \#, R)$ // traverse right
$\delta(q_1, 0) \rightarrow (q_2, 0, L)$ // right end, traverse left
$\delta(q_2, 1) \rightarrow (q_3, 0, L)$ // if the rightmost symbol is '1', the state is q_3
$\delta(q_3, 1) \rightarrow (q_3, 1, L)$ // traverse left
$\delta(q_3, \#) \rightarrow (q_3, \#, L)$ // traverse left
$\delta(q_3, 0) \rightarrow (q_5, 1, R)$ // replace the '0' (blank) before the first symbol by '1'
$\delta(q_5, 1/\#) \rightarrow (q_5, 1/\#, R)$ // traverse right
$\delta(q_2, \#) \rightarrow (q_4, 0, L)$ // if the rightmost symbol is '#', the state is q_4
$\delta(q_4, 1) \rightarrow (q_4, 1, L)$
$\delta(q_4, \#) \rightarrow (q_4, \#, L)$
$\delta(q_4, 0) \rightarrow (q_5, \#, R)$ // replace the '0' (blank) before the first symbol by '#'
$\delta(q_5, B) \rightarrow (q_f, B, H)$

8. Design a Turing machine for {a, b}*{aba}{a, b}*. [GTU 2010]

Solution:

Way 1: This is a non-deterministic Turing machine. The important portion of the string is 'aba'. The transitional functions are as follows.

$\delta(q_1, a) \rightarrow (q_1, a, R), (q_2, a, R)$
$\delta(q_1, b) \rightarrow (q_1, b, R)$
$\delta(q_2, b) \rightarrow (q_3, b, R)$
$\delta(q_3, a) \rightarrow (q_4, a, R)$
$\delta(q_4, a) \rightarrow (q_4, a, R)$
$\delta(q_4, b) \rightarrow (q_4, b, R)$
$\delta(q_4, B) \rightarrow (q_f, B, H)$

Way 2: Converting the regular expression to an equivalent finite automata with ∈ moves we get the following.

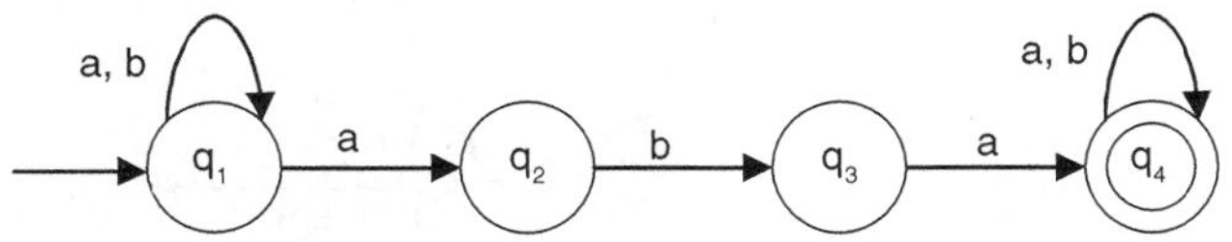

Inserting a new initial state q_i and final state q_H, the finite automata forms as follows.

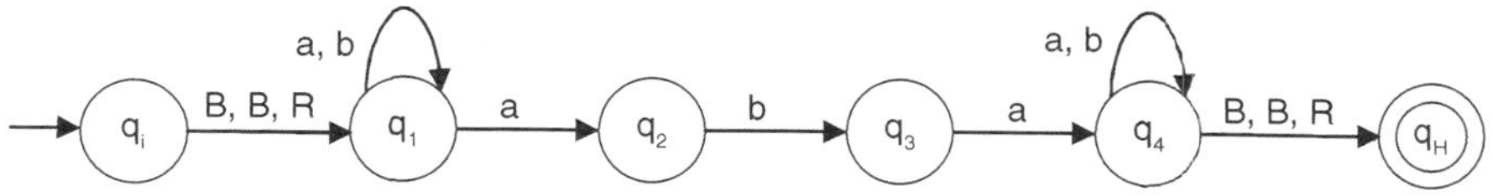

The final Turing machine is formed by modifying all inputs by (ip, ip, R).

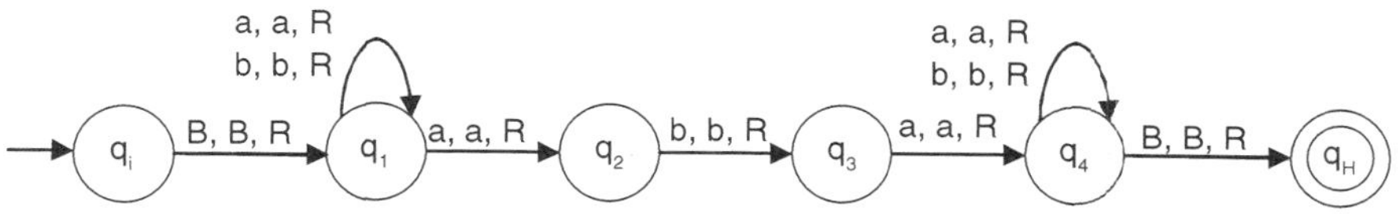

9. Design a Turing machine that accepts $L = \{a^n b^{2n}, n > 0\}$. [UPTU 2005]

Solution: For each 'a', there are two 'b's. Getting 'a' in initial state, replace it by X with the state changed and traverse right. Getting first 'b', replace it by Y with the state changed and traverse right to replace the second 'b' by 'Y'. Then, traverse left. By this process, if it gets Y in initial state, it means no 'a' is left. Traverse right to check whether any 'b' is left or not. If not, halt.

The transitional functions for the Turing machine are

$\delta(q_1, a) \rightarrow (q_2, X, R)$ // replace 'a' by X and traverse right
$\delta(q_2, a) \rightarrow (q_2, a, R)$ // traverse right to find 'b'
$\delta(q_2, b) \rightarrow (q_3, Y, R)$ // getting 'b', replace it by Y
$\delta(q_3, b) \rightarrow (q_3, Y, L)$ // replace the second 'b' by Y and traverse left
$\delta(q_3, Y) \rightarrow (q_3, Y, L)$ // traverse left
$\delta(q_3, a) \rightarrow (q_3, a, L)$ // traverse left in search of 'X'
$\delta(q_3, X) \rightarrow (q_1, X, R)$ // getting 'X', change the state to q_1 and start the phase again
$\delta(q_2, Y) \rightarrow (q_2, Y, R)$ // needed from the second move onwards
$\delta(q_1, Y) \rightarrow (q_4, Y, R)$ // if all 'a' are replaced by 'X'
$\delta(q_4, Y) \rightarrow (q_4, Y, R)$ // traverse right in search of 'b'
$\delta(q_4, B) \rightarrow (q_f, H)$

Multiple Choice Questions

1. The Turing machine is the machine format of____ language.
   a) Type 0 b) Type 1
   c) Type 2 d) Type 3
2. Which is not a part of the mechanical diagram of the Turing machine?
   a) Input tape
   b) Read–write head
   c) Finite control
   d) Stack
3. The difference between the Turing machine and the two-way FA is in the
   a) Input tape
   b) Read–write head
   c) Finite control
   d) All of these

4. The difference between the Turing machine and the pushdown automata is in the
   a) Head movement
   b) Finite control
   c) Stack
   d) All of these
5. Which is not true for the mechanical diagram of the Turing machine
   a) The head moves in both directions
   b) The head reads as well as writes
   c) The input string is surrounded by an infinite number of blank in both sides
   d) Some symbols are pushed into the stack.
6. In the Turing machine, the transitional function δ is in the form
   a) $Q \times \Gamma \rightarrow (Q \times \Gamma \times \{L, R, H\})$
   b) $Q \times \Sigma \rightarrow (Q \times \{L, R, H\})$
   c) $Q \times \Sigma \rightarrow (Q \times \Sigma \{L, R, H\})$
   d) $Q \times \Gamma \rightarrow (Q \times \Sigma \times \{L, R, H\})$

   (where Q is the finite set of states, Σ is the finite set of input alphabets, Γ is the allowable tape symbol, L means left, R means right, and H means halt.)
7. Which of the strings is accepted by the Turing machine?
   a) $L = a^n c^m b^n$, where m, n >0
   b) $L = a^n b^n c^i$, where n, i>0
   c) $L = a^n b^n c^n$, where n >0
   d) All of the above
8. Which is not true for the instantaneous description (ID) of the Turing machine?
   a) It remembers the state of the machine
   b) It remembers the cell currently being scanned by the read–write head
   c) The contents of all the cells of the tape
   d) The content of the cell on which the head was in previously

Answers:

1. a 2. d 3. b 4. d 5. d 6. a 7. d 8. d

GATE Questions

1. Which of the following is not accepted by the PDA but accepted by the two-stack PDA?
   a) $a^n b^n$ b) $a^n b^m c^m d^n$ c) $a^n b^n c^i$ d) $a^n b^n c^n d^n$
2. Which is true of the following?
   A finite automata can be changed to an equivalent Turing machine by
   a) Changing its reading head to read–write head
   b) Adding extra initial and final states with input B, B, R.
   c) Changing its i/p to i/p, i/p, R and adding extra initial and final states with input B, B, R.
   d) None of the above.
3. Consider the following Turing machine

$$\delta(A, a) \rightarrow (B, a, R)$$
$$\delta(B, b) \rightarrow (C, b, R)$$
$$\delta(C, a) \rightarrow (C, a, R)$$
$$\delta(C, b) \rightarrow (D, b, R)$$

   D is the final state. The language accepted by the TM is
   a) aba* b) aba*ab c) aba*b d) a*ba

4. Consider the following Turing machine

$$\delta(A, a) \rightarrow (B, a, R)$$
$$\delta(B, b) \rightarrow (B, b, R)$$
$$\delta(B, a) \rightarrow (C, a, R)$$
$$\delta(C, b) \rightarrow (C, b, R)$$
$$\delta(C, a) \rightarrow (D, a, R)$$

D is the final state. The language accepted by the TM is

a) ab*ab* b) ab*ab*a c) ab + ab + a d) a*(ba)*a

5. Consider the following Turing machine

$$\delta(A, a) \rightarrow (B, a, R)$$
$$\delta(A, b) \rightarrow (B, b, R)$$
$$\delta(B, a) \rightarrow (A, a, L)$$
$$\delta(B, b) \rightarrow (A, b, L)$$
$$\delta(B, \text{Blank}) \rightarrow (C, \text{Blank}, R)$$

C is the final state
On input 'ab', the machine will

a) Halt on accepting state
b) Go into infinite loop
c) Crash
d) Reach to final state but will not halt.

6. When does a Turing machine crash?

a) If the machine traverses all the inputs without traversing some states
b) If it traverses all its states till the input remains
c) If the transitional function is not defined for the present state and the input combination
d) None of these.

7. A single-tape Turing machine M has two states q_0 and q_1, of which q_0 is the starting state. The tape alphabet of M is {0, 1, B} and its input alphabet is {0, 1}. The symbol B is the blank symbol used to indicate the end of an input string. The transition function of M is described in the following table.

	0	1	B
q_0	q_1, 1, R	q_1, 1, R	Halt
q_1	q_1, 1, R	q_0, 1, L	q_0, B, L

The table is interpreted as illustrated in the following.
The entry (q_1, 1, R) in row q_0 and column 1 signifies that if M is in state q_0 and reads 1 on the current tape square, then it writes 1 on the same tape square, moves its tape head one position to the right, and transitions to state q_1.
Which of the following statements is true about M?

a) M does not halt on any string in $(0 + 1)^+$
b) M does not halt on any string in $(00 + 1)^*$
c) M halts on all strings ending in a 0
d) M halts on all strings ending in a 1

8. Consider the languages L_1, L_2, and L_3 as given in the following

$$L_1 = \{0^p1^q \mid p, q \in \}$$
$$L_2 = \{0^p1^q \mid p, q \in N \text{ and } p = q\}$$
$$L_3 = \{0^p1^{qr} \mid p, q, r \in N \text{ and } p = q = r\}$$

Which of the following statements is not true?
a) Pushdown automata (PDA) can be used to recognize L_1 and L_2
b) L_1 is a regular language
c) All the three languages are context free
d) Turing machines can be used to recognize all the languages

Answers:

1. d 2. c 3. c 4. b 5. b 6. c 7. a 8. c

Hints:

5. Aab → aBb → Aab → aBb ..... infinite times

7. For (b), (c), and (d) the machine loops infinitely. It only accepts null string.

8. L_3 is not context free.

Fill in the Blanks

1. All types of languages are accepted by ____________.
2. According to the Chomsky hierarchy, type 0 language is called ____________.
3. The diagram of the Turing machine is like finite automata, but here the head moves ____________.
4. The head of the Turing machine is called ____________.
5. The string $a^nb^nc^n$, $n > 0$ is accepted by ____________.
6. The Turing machine is called ____________.
7. Two-stack PDA is equivalent to ____________.
8. The crash condition occurs if the read-write head of a Turing machine is over the ____________.
9. A Turing machine is said to be in ____________ if it is not able to move further.

Answers:

1. Turing machine
2. unrestricted language
3. in both direction
4. Read–write head
5. Turing machine
6. universal machine
7. Turing machine
8. leftmost cell
9. halt state

Exercise

1. Construct a Turing machine for the following
   a) $L = \{a, b\}^+$
   b) $L = WW$, where $W \in (a, b)^+$
   c) $L = WW^R$, where $W \in (a, b)^+$
   d) L = {all even palindromes over (a, b)}
   e) $L = a^n b^n c^n$, $n > 0$
   f) $L = n - 1$, where $n > 0$
2. Design a Turing machine which acts as an eraser.
   (Hints: Starts from the left hand side, scans each symbol from left to right, and replaces each symbol by blank. Halts if gets blank.)
3. Design a Turing machine which replaces '0' by '1' and '1' by '0' of the string traversed.
4. Design a Turing machine to accept the string $L = \{a^n b^m c^{n+m}$, where $n > 0, m > 0\}$.
   (Hints: Traverse an 'a', replace it by X, and move right to find the first 'c' and replace it by Y. Then, traverse left to find the second 'a'. By this process, replace n number of 'a' by X and n number of 'b' by Y. By the same process, replace m number of 'b' by Z and the last m number of 'c' by Z.)
5. Design a Turing machine to accept the string $L = \{a^m b^{m+n} c^n$, where $m, n > 0\}$.
6. Design a Turing machine by the transitional notation for the following languages
   a) $L = a^n b^n$, $n > 0$
   b) $L = \{a^n b^n c^m d^m$, where $m, n \geq 1\}$
   c) $L = \{a^n b^n c^m$, where $n, m \geq 1\}$
7. Make comments on the following statement that a finite state machine with two stacks is as powerful as a Turing machine.
8. Design a two-stack PDA for adding and subtracting two numbers.
9. Design a two-stack PDA to accept the string $L = \{a^m b^{m+n} c^n$, where $m, n > 0\}$.

Variations of the Turing Machine

9

Introduction

Till now, we have discussed about the Turing machine with a single head and a single tape. It has been clear from the given examples that the Turing machine is a powerful computational machine. But, for some cases, it may seem that the Turing machine takes a lot of time for computation, and it is complex. To understand the computational power of the Turing machine and to reduce the time complexity, researchers have proposed different models of the Turing machine. These models have given us freedom to use a suitable machine wherever applicable to solve a variety of problems by the Turing machine. It can be proved that, computationally, all these Turing machines are equally powerful. That is, what one type of Turing machine can compute, any other type can also compute. However, the efficiency of computation, that is, how fast they can compute, may vary.

9.1 Variations of the Turing Machine

The different types of Turing machine are

- Multi-tape Turing machine
- Multi-head Turing machine
- Two-way infinite tape
- K-dimensional Turing machine
- Non-deterministic Turing machine
- Enumerator

9.1.1 Multi-tape Turing Machine

The name suggests that this type of Turing machine has multiple tapes instead of one. Each of the tapes is connected to the finite control by a read–write head. It means for m tapes there are m read–write heads. The transitional function of the multi-tape Turing machine depends on the present state and the input symbol scanned by each of the read–write head from the respective input tape. Depending on the present state and the input symbols scanned by each of the head, the Turing machine can

- Change its state.
- Write a new symbol on the respective cell of the respective tape from where the inputs were scanned.
- Move the head to one left or one right.

A multi-tape Turing machine having k tapes ($k \geq 1$) is symbolically represented as

$$T_M = (Q, \Sigma, \Gamma, \delta, q_0, B, F)$$

where Q: Finite set of states
Σ: Finite set of input alphabets
Γ: Finite set of allowable tape symbols
q_0: Initial state
B: A symbol of Γ called blank
F: Final states
δ: Transitional function

where δ is a mapping from $Q \times (\Gamma_1, \Gamma_2, \ldots\ldots \Gamma_k) \rightarrow (Q \times (\Gamma_1, \Gamma_2, \ldots\ldots \Gamma_k) \times [(L, R, H), (L, R, H) \ldots$ k times]. In general $Q \times \Gamma^k \rightarrow (Q \times \Gamma^k \times (L, R, H)^k)$.

Diagrammatically, a multi-tape Turing machine can be denoted as shown in Fig. 9.1.

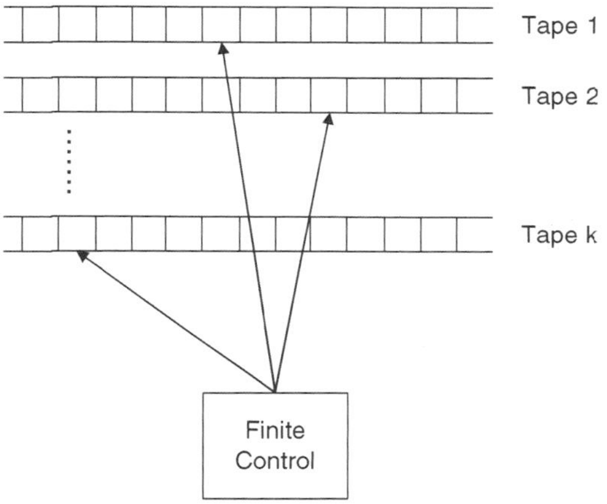

Fig. 9.1 *Multi-tape Turing Machine*

Example 9.1 Design a multi-tape Turing machine for checking whether a binary string is a palindrome or not. Show an ID for 1001.

Solution: We are considering a Turing machine with two tapes. The input is written on the first tape. The machine works by the following process:

i) Copy the input from the first tape to the second tape by traversing the first tape from left to right.
ii) Traverse the first tape again from right to left and point the head to the first symbol of the input on tape 1.
iii) Moves the two heads pointing on the two tapes in opposite directions checking that the two symbols are identical and erasing the copy in tape 2 at the same time.

Here, the moves on two tapes are described in a group.

The string is written on tape 1. Tape 2 is blank. The string may start with '0' or may start with '1'. The machine is in state q_0 initially. Traversing '0' or '1' in tape 1, the blank 'B' in tape 2 is replaced by '0' or '1', respectively, and both the head move to one cell right. The transitional functions are

Tape 1: $\delta(q_0, 0) \rightarrow (q_0, 0, R)$ // for '0'
Tape 2: $\delta(q_0, B) \rightarrow (q_0, 0, R)$

Tape 1: $\delta(q_0, 1) \rightarrow (q_0, 1, R)$ // for '1'
Tape 2: $\delta(q_0, B) \rightarrow (q_0, 1, R)$

When the head of the first tape finds 'B' at the right side, the string is totally traversed. The state is changed to q_1 and head of the first tape starts to traverse to the left. The transitional functions are

Tape 1: $\delta(q_0, B) \rightarrow (q_1, B, L)$
Tape 2: $\delta(q_0, B) \rightarrow (q_1, B, \_)$ (_ means no traversal of the head)

Traversing the first tape from right to left, the head has to traverse '0' and '1' and the head on the second tape has no traversal. The transitional functions are

Tape 1: $\delta(q_1, 0) \rightarrow (q_1, 0, L)$ // for '0'
Tape 2: $\delta(q_1, B) \rightarrow (q_1, B, \_)$

Tape 1: $\delta(q_1, 1) \rightarrow (q_1, 1, L)$ // for '1'
Tape 2: $\delta(q_1, B) \rightarrow (q_1, B, \_)$

When the head of the first tape finds 'B' at the left side, the string is totally traversed from right to left. The state is changed to q_2 and the head of the first tape starts to traverse to the right. The head of the second tape starts to traverse to the left. The transitional functions are

Tape 1: $\delta(q_1, B) \rightarrow (q_2, B, R)$
Tape 2: $\delta(q_1, B) \rightarrow (q_2, B, L)$

Now, both the heads are traversing the same string written on two tapes but in opposite directions. If both the heads traverse '0', or both the heads traverse '1', then it is allowed for them to traverse right or left, respectively. If one head traverses '0' and the other traverses '1' or vice versa, it stops on a state which signifies that the string is not a palindrome. If this situation does not arise and both the heads get blank 'B', then the machine halts on a state which signifies that the string is a palindrome. The transitional functions are

Tape 1: $\delta(q_2, 0) \rightarrow (q_2, 0, R)$ // for '0'
Tape 2: $\delta(q_2, 0) \rightarrow (q_2, B, L)$

Tape 1: $\delta(q_2, 1) \rightarrow (q_2, 1, R)$ // for '1'
Tape 2: $\delta(q_2, 1) \rightarrow (q_2, B, L)$

Tape 1: $\delta(q_2, 0) \rightarrow (q_{no}, 0, \_)$ //q_{no} is the state which signifies not a palindrome
Tape 2: $\delta(q_2, 1) \rightarrow (q_{no}, 1, \_)$ if the first head gets '0' and the second gets '1'.

Tape 1: $\delta(q_2, 1) \rightarrow (q_{no}, 1, \_)$ //q_{no} is the state which signifies not a palindrome
Tape 2: $\delta(q_2, 0) \rightarrow (q_{no}, 0, \_)$ if the first head gets '1' and the second gets '0'.

Tape 1: $\delta(q_2, B) \rightarrow (q_{yes}, B, H)$ // q_{yes} is the state which signifies that it is a palindrome
Tape 2: $\delta(q_2, B) \rightarrow (q_{yes}, B, R)$

The transitional diagram describes it clearly.

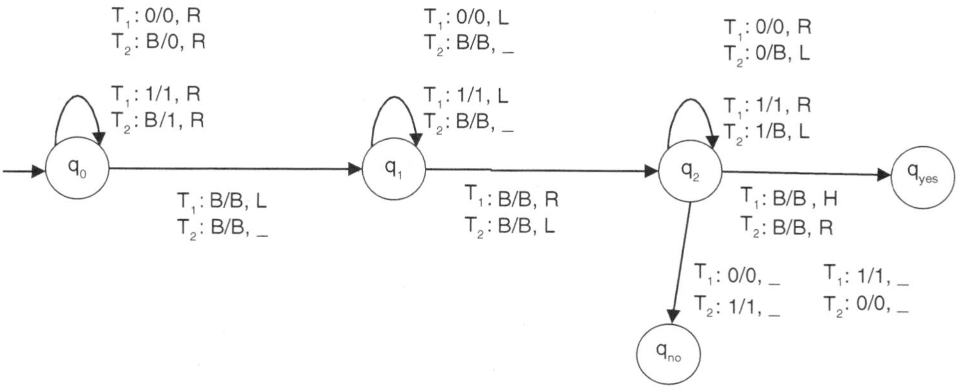

ID for 1001

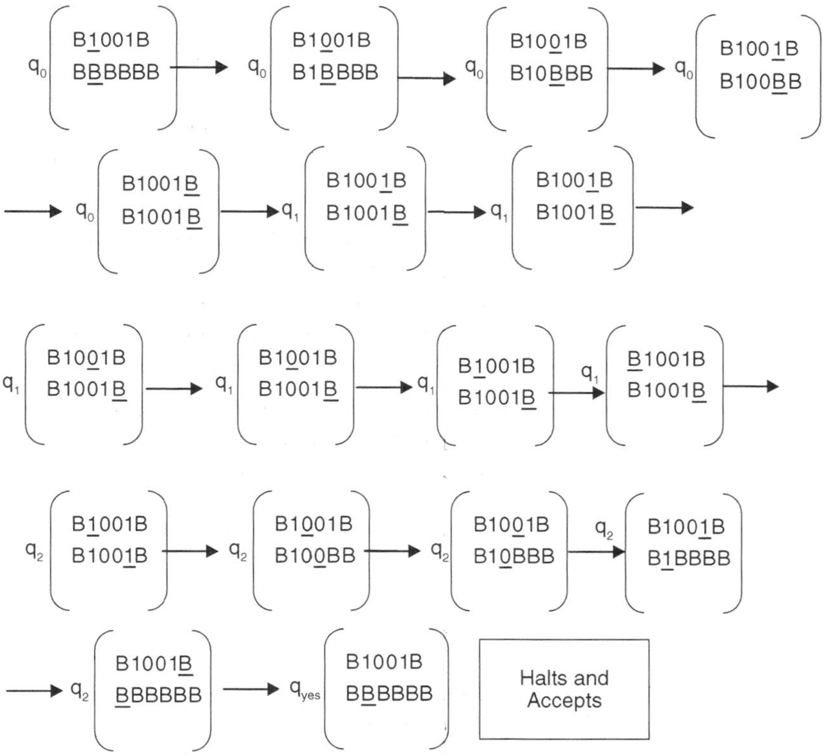

Example 9.2 Design a multi-tape Turing machine for L = $a^nb^nc^n$.

Solution: We are considering a Turing machine with four tapes. The input is written on the first tape. The machine works by the following process

i) Copy all 'a' on tape 2, all 'b' on tape 3, and all 'c' on tape 4.
ii) Traverse tape 2, tape 3, and tape 4 from right to left simultaneously and replace each symbol by a blank.
iii) If all the heads traversing tapes 2, 3, and 4 get B, then declare accept. If not, declare reject.

The transitional representation of the Turing machine is as follows.

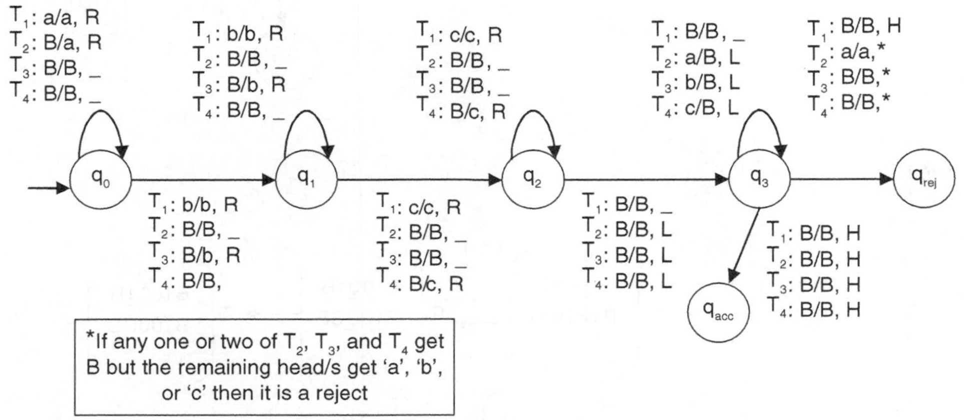

The language acceptance power of a multi-tape Turing machine is the same as the single-tape Turing machine. Only the time complexity may be reduced.

If we take the example of checking whether a string of binary is a palindrome or not, then it can be easily justified.

Consider a single-tape Turing machine for checking an even length string.

i) For checking a string of length n, it needs 2n + 1 steps. *[(Making first symbol to B) + (n right traversal upto B) + (Left traversal with replacing rightmost symbol by B) + ((n − 1) left traversal upto B)]*
ii) From the next traversal, the string of length n is truncated to length (n−2). The number of steps required is (2n − 3). By this process, the last step will be (2n − (2n −1)) = 1.
iii) The total number of steps required is (2n + 1) + (2n − 3) + …… + 1 = [(n + 2)(n + 1)]/2, means $O(n^2)$.

[1 is the (n + 2)/2th element of the series.]

If the string is traversed by a two-tape Turing machine,

i) For copying the string to tape 2, it needs (n + 1) steps.
ii) For left traversal of the string on tape 1, it needs (n + 1) steps.

iii) For right traversal of tape 1 and left traversal of tape 2, it needs (n + 1) steps.
iv) There is a total of 3n + 1 steps, means O(n).

This proves the reduced complexity of the multi-tape Turing machine.

Every multi-tape Turing machine has an equivalent single-tape Turing machine.

Proof: Let M_{MT} be a k tape Turing machine where k > 1. We have to construct a single-tape Turing machine M_{ST} which simulates M.

- M_{ST} simulates the effect of k tapes (k > 1) by storing the information of k tapes on its single tape.
- M_{ST} uses a special symbol # as a mark to separate the content of different tapes.
- M_{ST} tracks the location of the heads of the M_{MT} by marking an _ (underline) on the symbols where there were head positions in M_{MT}.

The process is described in Fig. 9.2

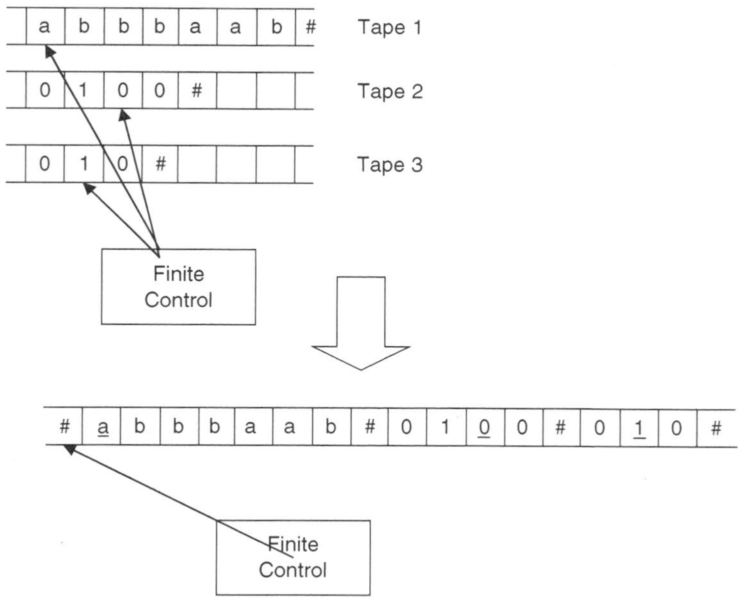

Fig. 9.2 *Conversion of a Multi-tape TM to a Single-tape TM*

9.1.2 Multi-head Turing Machine

The name signifies that this type of Turing machine has multiple heads. All the heads are connected to the finite control. The transitional function for the multi-head Turing machine having n heads can be described as

$$\delta(Q, \Sigma_1, \Sigma_2 \ldots. \Sigma_n) \rightarrow (Q \times (\Gamma_1 \times L, R, N, H) \times (\Gamma_2 \times L, R, N, H) \times \ldots. (\Gamma_n \times L, R, N, H))$$

where Σ_i is the input symbol under head i and Γ_i is the symbol written by replacing Σ_i. N means no movement of the head.

A diagrammatical representation of a multi-head Turing machine is given in Fig. 9.3.

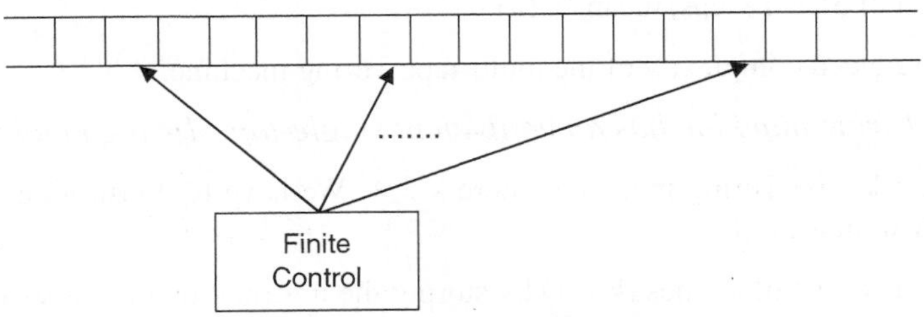

Fig. 9.3 *Multi-head Turing Machine*

We have to be careful about two special cases in designing a multi-head Turing machine.

1. A situation may arise when more than one heads are scanning a particular cell at the same time. But the symbol written by one head is different from the symbol written by the other head.

   In this situation, priority among the heads is defined. The change done by the head with the highest priority will remain.
2. Another situation may arise when a cell under a head is the leftmost cell and the transitional function instructs the head to go left. This condition is called hanging.

Example 9.3 Design a multi-head Turing machine for checking whether a binary string is a palindrome or not. Show the ID for 1001 and 101.

Solution: Let us consider a Turing machine with two heads. The heads are pointing to the two ends of the string on the tape. Both the heads traverse the string in the opposite direction. The head 1 has the priority over head 2.

i) If both of the heads get the same symbols, then it traverses the next input right or left by replacing the present symbol by B.
ii) If both the heads gets B, then halt and declare the string as a palindrome.

The transitional representation of the Turing machine is as follows.

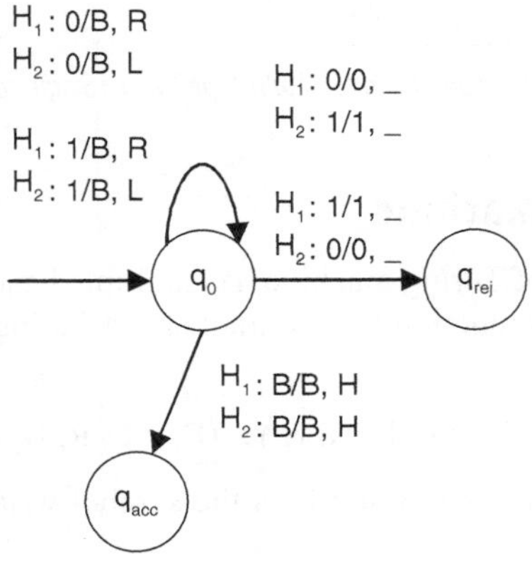

ID for 1001

$$(q_0, \text{B1001B}) \rightarrow (q_0, \text{BB}\mathbf{00}\text{BB}) \rightarrow (q_0, \text{BB}\mathbf{BB}\text{BB}) \rightarrow (q_{acc}, \text{BB}\mathbf{BB}\text{BB})\ [\text{Halt}]$$

ID for 101

$$(q_0, \text{B}\mathbf{101}\text{B}) \rightarrow (q_0, \text{BB}\mathbf{0}\text{BB}) \rightarrow (q_0, \text{B}\mathbf{BBB}\text{B}) \rightarrow (q_{acc}, \text{B}\mathbf{BBB}\text{B})\ [\text{Halt}]$$

Every multi-head Turing machine has an equivalent single-head Turing machine.

Proof: This is proved by converting a multi-head Turing machine to a multi-tape Turing machine and then converting a multi-tape Turing machine to a single-tape and single-head Turing machine.

The conversion of a multi-head Turing machine to a multi-tape Turing machine is done using the following rules.

- Initialize each tape of a multi-tape Turing machine by the input string.
- Whenever head i is moved in a multi-head TM, move the head in the ith tape in the multi-tape TM.
- Whenever any head in a multi-head TM writes a symbol in the input tape, write the same symbol at the same position on every tape in the multi-tape TM.

Convert the multi-tape TM to a single-tape TM.

(The conversion process of a multi-tape TM to a single-tape single-head TM is already discussed.)

This conversion is shown in Fig. 9.4 for the previous example of checking a palindrome.

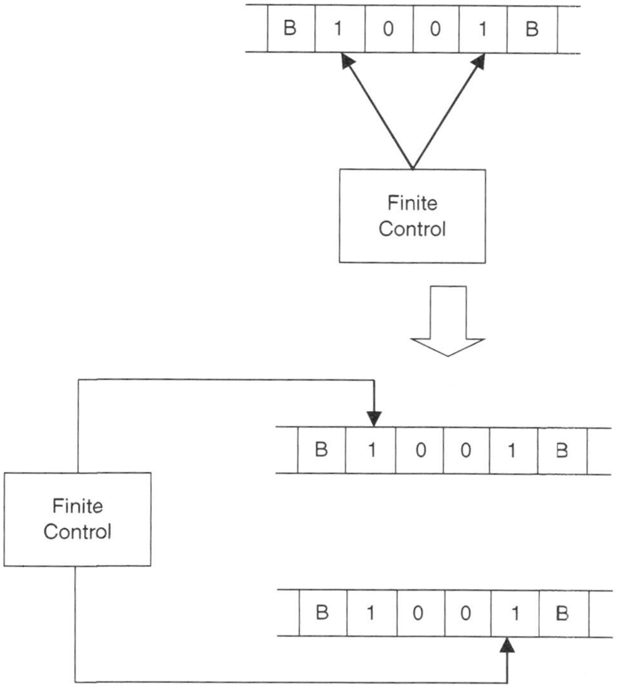

Fig. 9.4 *Conversion of a Multi-head TM to a Multi-tape TM*

9.1.3 Two-way Infinite Tape

In general, in a Turing machine there is a left boundary. If the head crosses that boundary and wants to go left, then the situation is called a crash condition. But the head may traverse the right side up to infinity. In this sense, the input tape of the general Turing machine can be treated as a one-way infinite tape. A typical diagram of the input tape of a general Turing machine is given in Fig. 9.5.

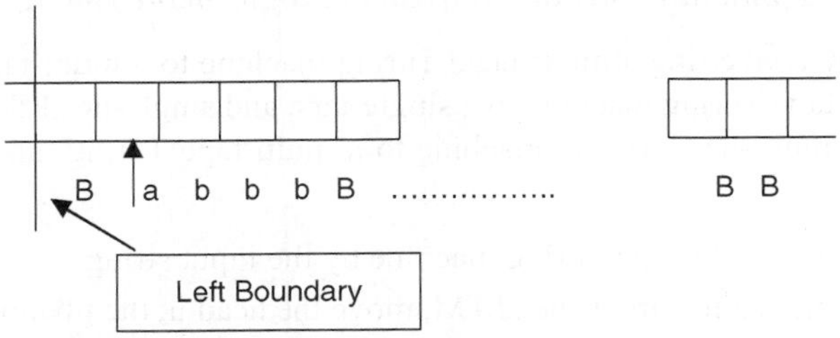

Fig 9.5 *Typical Diagram of an Input Tape of a General Turing Machine*

A Turing machine where there are infinite numbers of sequence of blank on both sides of the tape is called a two-way infinite tape Turing machine. A typical diagram of the input tape of a two-way infinite tape Turing Machine is given in Fig. 9.6.

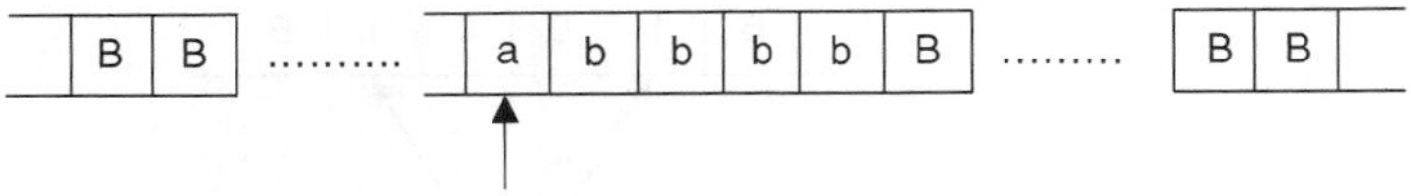

Fig 9.6 *Typical Diagram of a Two-way Infinite Tape*

Every two-way infinite tape Turing machine has an equivalent Turing machine by marking the left-hand side of the input to detect and prevent the head from moving off of that end.

9.1.4 K-dimensional Turing Machine

The input tape of two-dimensional Turing machine is extended to infinity in both sides, but in one direction. If the input tape can extend infinitely in more than one dimension, then the Turing machine is called a multi-dimensional Turing machine. In a general case, just consider k = 2, which means that the input tape is extended to infinity in X and Y direction. For this case, the read–write head can move in the left, right, up, and down directions. The transitional function for a K-dimensional Turing machine is

$$Q \times \Sigma \rightarrow Q \times \Gamma \times (L, R, U, D, H)$$

where L = Left, R = Right, U = Up, D = Down.

The diagram in Fig. 9.7 describes a two-dimensional Turing machine.

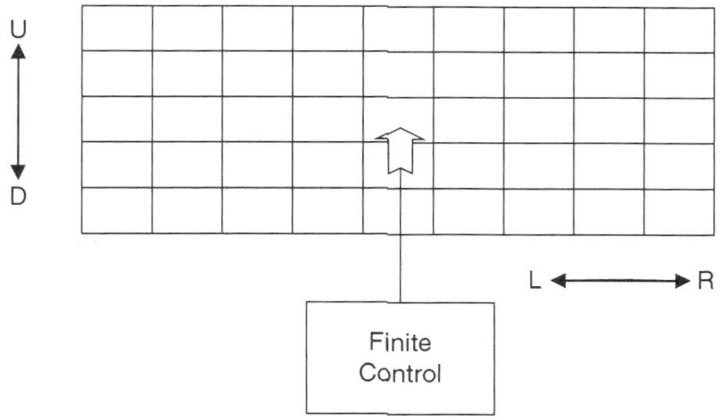

Fig. 9.7 *K-dimensional (k = 2) Turing Machine*

9.1.5 Non-deterministic Turing Machine

We are familiar with the term non-deterministic while discussing finite automata and pushdown automata. If we recall the transitional function of the non-deterministic finite automata, we will see that for a single state and a single input there may be more than one move.

Till now, we have seen different types of Turing machines, but all of them have a unique triplet combination of next state, tape symbol, and move. But in case of a non- deterministic Turing machine, for a pair of present state (q) and input tape symbol (Σ), there may be a finite set of triplets (Q_i, Γ_i, D_i) for i = 1, 2, ... In other words, for a combination of a single present state and a single input symbol, there may be more than one move. The transitional function of a non-deterministic Turing machine is

$$Q \times \Sigma \rightarrow \text{power set of } Q \times \Gamma \times (L, R, H)$$

Example 9.4 Design a Turing machine for 0^n1^m, where m, n > 0 and n may not be equal to m.

Solution: The string contains n number of '0' and m number of '1', but the number of '0' and the number of '1' may not be equal and there is at least one '0' and one '1'. First, the Turing machine traverses (n − 1) number of 0 using the transitional function

$$\delta(q_0, 0) \rightarrow (q_0, 0, R)$$

After the last '0', there are m number of '1'. If the machine starts to traverse '1', it cannot traverse '0' because all '0' will appear before '1' in the string. So, it needs a state change to traverse '1'. But here, there is another problem. From q_0 by getting '1' if the TM changes its state, then the condition that n > 0 is failed. In this case, the necessity of a non-deterministic TM is felt.

If we make another transition from q_0 to a new state q_1 by getting '0', then both the conditions are fulfilled.

The remaining transitional functions of the Turing machine are

$$\delta(q_0, 0) \rightarrow (q_1, 0, R)$$
$$\delta(q_1, 1) \rightarrow (q_1, 1, R)$$
$$\delta(q_1, 1) \rightarrow (q_f, 1, H)$$

The transitional diagram gives it the look of a non-deterministic Turing machine.

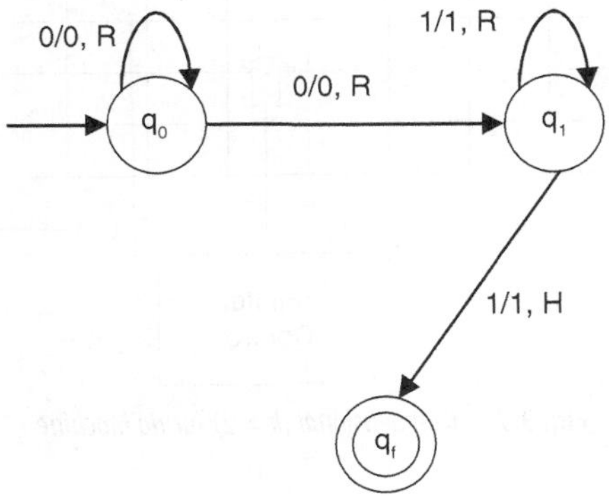

Every non-deterministic Turing machine has an equivalent deterministic Turing machine.

Proof: For accepting a string w by a non-deterministic Turing machine (M_{NT}), it constructs a tree with all possible branches. If the accept state is ever found by the deterministic Turing machine (M_{DT}) on one of these branches, then it accepts. Otherwise simulation will not terminate. The process is as follows.

i) Construct a tree T with all possible branches for accepting a string w.

ii) The root node of T is the start of the configuration and each node N has a **configuration** <current state of M_{NT}, current contents of M_{NT}'s tape, and position of the read head of M_{NT}>

iii) Design a tree D by performing a breadth first search of T.
(This strategy traverses all branches at the same depth before going to traverse any branch at the next depth. Hence, this method guarantees that D will visit each node of T until it encounters an accepting configuration.)

(iv) Use three tapes to simulate D for M_{NT}.
- Tape 1, called input tape, always holds input w
- Tape 2, called simulation tape, is used as M_{NT}'s tape when simulating M on a sequence of non-deterministic choices
- Tape 3, called choice tape, stores the current sequence of non-deterministic choices.

(v) At initial stage, the input tape contains w, and the simulation tape and choice tape are kept blank

(vi) Copy the contents of the input tape on the simulation tape.

(vii) Use tape 2 to simulate M_{NT} with input string w on one branch of its non-deterministic computation.
- Before each step of M_{NT} follow the next symbol on tape 3 to determine the choices to be made among the allowed M_{NT}'s transitional function.
- If tape 3 becomes empty or if the present choice (for M_{NT} 's state and tape 3 input] is invalid, stop traversing the branch and go to step viii.
- If an accepting configuration appears stop traversing; halt and declare accept.

(viii) Replace the string on tape 3 with the lexicographically next string and simulate the next branch of M_{NT}'s computation by going to stage (vi).

9.1.6 Enumerator

Enumerator is a type of Turing machine which is attached to a printer. It has a work tape and an output tape. The work tape is a write only once tape. At each step, the machine chooses a symbol to write from the output alphabet on the output tape. After writing a symbol on the output tape, the head placed on the output tape moves right by one position. The enumerator has a special state, say qp, entering which the output tape is erased and the tape head moves to the leftmost position and finally the string is printed. A string w is printed as output by the enumerator if the output tape contains w at the time the machine enters into qp.

The diagram of an enumerator is shown in Fig. 9.8.

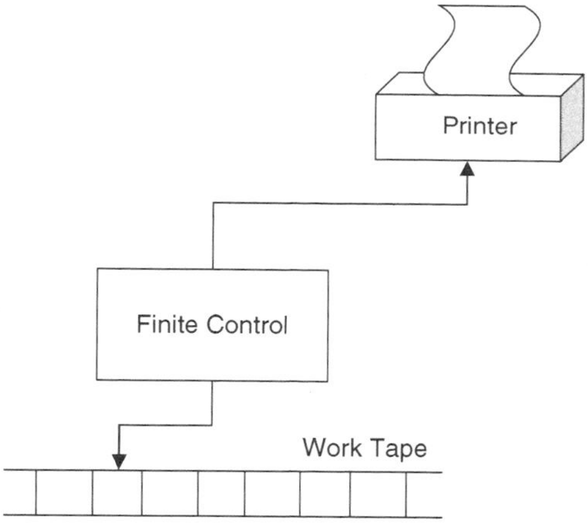

Fig. 9.8 *Enumerator*

An enumerator is defined by 7 touples

$$(Q, \Sigma, \Gamma, \delta, q_0, B, q_p)$$

where Q: Finite set of states
Σ: Finite set of print alphabets
Γ: Finite set of work tape symbols
δ: Transitional function
q_0: Initial state
B: A symbol of Γ called blank
q_p: Final state (print state)

where δ is a mapping from $Q \times \Sigma \times \Gamma \rightarrow (Q \times \Sigma \times \{L, R\} \times \Gamma \times \{L, R\})$.

9.2 Turing Machine as an Integer Function

An integer function is a function which takes its arguments as integers and returns the result in integer. If we consider integer addition, subtraction, multiplication, etc., we see that all these functions take the arguments in integer and produce the result in integer. Let an integer function be $f(i_1, i_2 \ldots\ldots i_k) = M$.

We have to solve this function using the Turing machine. First of all, the input tape must represent the input arguments. Let the number of '0' be used to represent the arguments. But there must be a separator to differentiate the number of '0' for each argument. In the Turing machine, generally, a blank 'B' is used as the separator. In some cases, other symbols may be used as a separator. Initially, the input tape will be $0^{i1}B0^{i2}......B0^{ik}$. After performing the integer function, the result will be M number of 0, which will remain on the tape.

The following are some examples of TM as integer functions.

Example 9.5 Addition of two integers. f(x, y) = x + y.

Solution: The input tape is 0^xB0^y. The calculation process is as follows.

i) Traverse x number of '0' up to 'B'.
ii) On getting 'B', change the state and replace 'B' by 0.
iii) Traverse the second string. 'B' means it is the end of the second string. Change the state and traverse left.
iv) Replace the last '0' of the second string by B and halt.

The transitional diagram of addition function is given in Fig. 9.9.

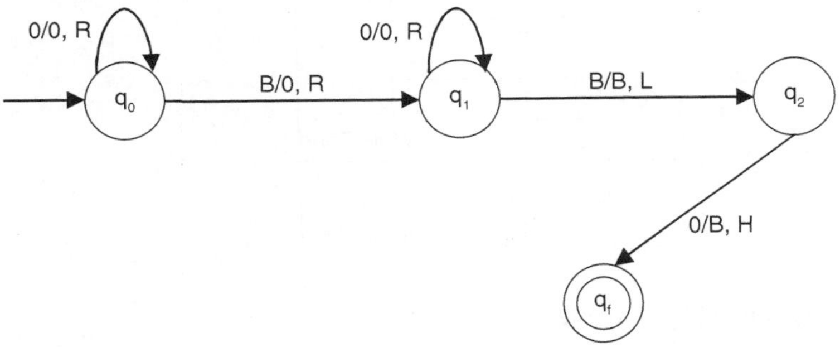

Fig. 9.9

Example 9.6 Compute the function f(x) = x + 2.

Solution: The input tape is 0^x. The calculation process is as follows.

i) Traverse x number of '0' up to 'B'.
ii) Change B by '0' and change the state.
iii) Replace the second B by '0' and halt.

The transitional diagram of the function is given in Fig. 9.10.

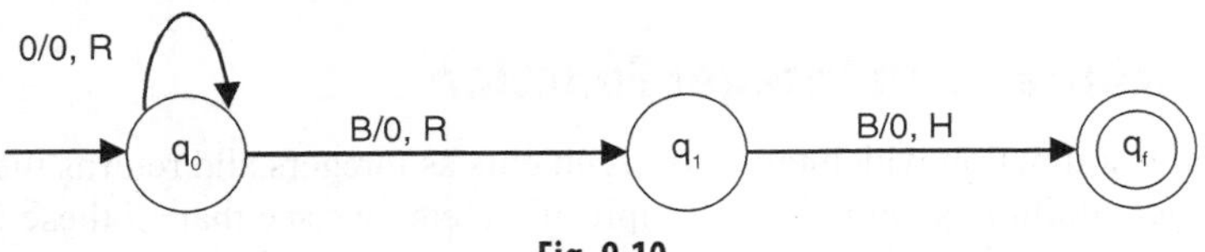

Fig. 9.10

Example 9.7 Subtraction of two integers

$$\begin{aligned} f(x, y) &= x - y && \text{if } x > y \\ &= 0 && \text{if } x \le y. \end{aligned}$$

Solution: The input tape is 0^x10^y. The calculation process is as follows.

i) Replace the first 0 by B, change the state, and traverse right. The transitional function is

$$\delta(q_0, 0) \rightarrow (q_1, B, R)$$

Traverse right for the remaining '0's of the first string. Getting the separator '1', change the state from q_1 to q_2. The transitional functions are

$$\delta(q_1, 0) \rightarrow (q_1, 0, R)$$
$$\delta(q_1, 1) \rightarrow (q_2, 1, R)$$

ii) Replace the leftmost '0' for the second number by '1', change the state, and traverse left to find the replaced B, using transitional functions

$$\delta(q_2, 0) \rightarrow (q_3, 1, L)$$
$$\delta(q_3, 1) \rightarrow (q_3, 1, L)$$
$$\delta(q_3, 0) \rightarrow (q_3, 0, L)$$

Upon getting B, the state is changed to q_0 using the function

$$\delta(q_3, B) \rightarrow (q_0, B, R)$$

From the second iteration onwards, q_2 has to traverse '1' using the transition function

$$\delta(q_2, 1) \rightarrow (q_2, 1, r)$$

If $x > y$, then the state q_2 will get B at the last of the second number representation. x-y number of '0' and y + 1 number of '1' will remain. It changes its state and traverses left using the transitional function

$$\delta(q_2, B) \rightarrow (q_4, B, L)$$

Now, it replaces all the '1' by B and traverses left to find B. The transitional functions are

$$\delta(q_4, 1) \rightarrow (q_4, B, L)$$
$$\delta(q_4, 0) \rightarrow (q_4, 0, L)$$

On getting B, it replaces it by '0' and halts.

$$\delta(q_2, B) \rightarrow (q_6, B, L)$$

If $x < y$, all '0' representing x will finish before y. In the state q_0, the machine gets 1. It changes '1' by B and traverses right with the state change.

$$\delta(q_0, 1) \rightarrow (q_5, B, R)$$

The state q_5 replaces all '0' and '1' by B to make the tape empty, and upon getting B it halts.

$$\delta(q_5, 0) \rightarrow (q_5, B, R)$$
$$\delta(q_5, 1) \rightarrow (q_5, B, R)$$
$$\delta(q_5, B) \rightarrow (q_6, B, H)$$

Example 9.8 Compute the function

$$f(x) = x - 2 \quad \text{if } x > 2$$
$$= 0 \quad \text{if } x \leq 2.$$

Solution: The input tape is 0^x. The calculation process is

i) Traverse x number of '0' up to 'B'.
ii) On getting B, change the state and traverse left.
iii) Replace the first '0' by B with the state change and the second '0' by B with the state change and halt.

The transitional diagram of the function is given in Fig. 9.11.

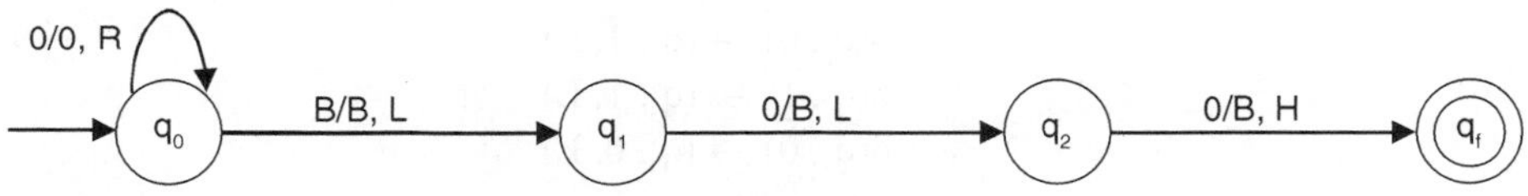

Fig. 9.11

Example 9.9 Multiplication of two integers f(x, y) = x * y.

Solution: Let x = 2 and y = 3. The input tape is in the form

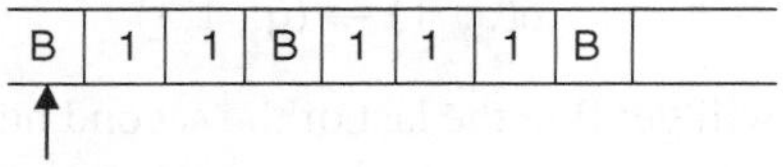

Make the tape in the following form

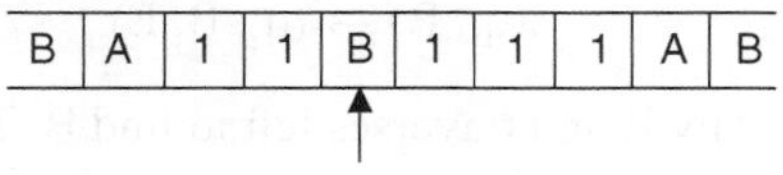

by the transitional functions

$$\delta(q_0, B) \rightarrow (q_1, A, R)$$
$$\delta(q_1, 1) \rightarrow (q_1, 1, R)$$
$$\delta(q_1, B) \rightarrow (q_2, B, R)$$
$$\delta(q_2, 1) \rightarrow (q_2, 1, R)$$
$$\delta(q_2, B) \rightarrow (q_3, A, L)$$
$$\delta(q_3, 1) \rightarrow (q_3, 1, L)$$

Here, A denotes the beginning of the first number and the end of the second number.

The transitional diagram for multiplication operation is given in Fig. 9.12.

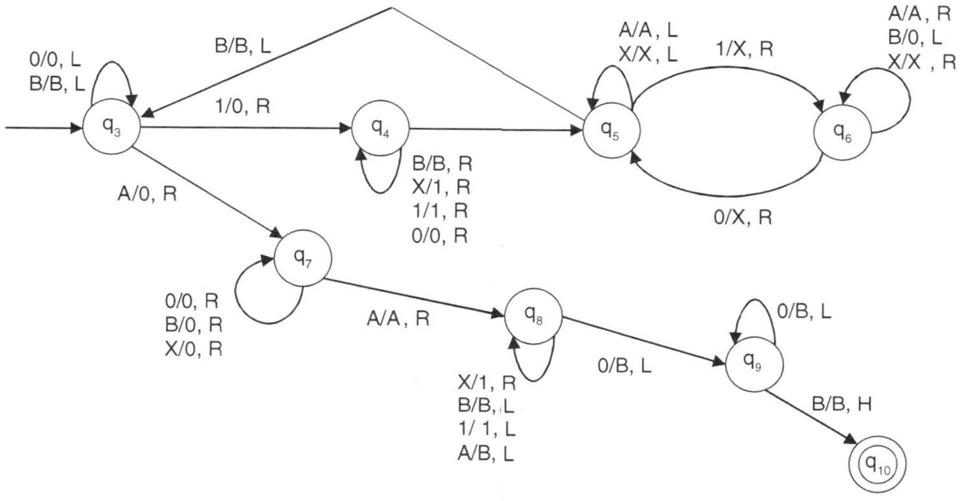

Fig. 9.12

Example 9.10 Th:e remainder after dividing one integer number by 2. f(x, y) = x % 2.

Solution: The remainder of the integer division 2 is either 1 or 0. The input tape is 0^x. The calculation process is as follows.

i) Traverse the string from left to right. Getting 'B', traverse left with a state change.
ii) Replace all the '0' of the tape by 'B' with alternating change of state and traverse left.
iii) Getting 'B', either halt on the final state or traverse the right by replacing one 'B' by '0' and halt depending on the state.

The transitional functions are

$\delta(q_0, B) \rightarrow (q_1, B, R)$
$\delta(q_1, 0) \rightarrow (q_1, 0, R)$ // traverse right
$\delta(q_1, B) \rightarrow (q_2, B, L)$ // end of the string, so traverse left
$\delta(q_2, 0) \rightarrow (q_3, B, L)$] alternating change of state with replacement
$\delta(q_3, 0) \rightarrow (q_2, B, L)$] of 0 by B
$\delta(q_3, B) \rightarrow (q_4, B, R)$ // number is odd, so traverse right
$\delta(q_4, B) \rightarrow (q_5, 0, H)$ // replace the 'B' by '0'
$\delta(q_2, B) \rightarrow (q_5, B, H)$ // number is even, so halt

Example 9.11 Square of an integer. $f(x) = x^2$.

Solution: The square of an integer means the multiplication of the number by the same number. The multiplication function is already described.

The input tape is in the form 1^X. It has to be made in the form $1^X B 1^X$. To make this, the transitional functions are

$\delta(q_0, B) \rightarrow (q_1, X, R)$ // replace the first '1' by X.
$\delta(q_1, 1) \rightarrow (q_1, 1, R)$ // traverse right to find 'B'
$\delta(q_1, B) \rightarrow (q_2, B, R)$
$\delta(q_2, 1) \rightarrow (q_2, 1, R)$ // need from the second traversal
$\delta(q_2, B) \rightarrow (q_3, 1, L)$ // replace one 'B' after the end marker 'B' by '1'
$\delta(q_3, 1) \rightarrow (q_3, 1, L)$ // need from the second traversal
$\delta(q_3, B) \rightarrow (q_4, B, L)$ // traverse left
$\delta(q_4, 1) \rightarrow (q_4, 1, L)$ // traverse left to find the replaced X
$\delta(q_4, X) \rightarrow (q_0, X, R)$
$\delta(q_0, B) \rightarrow (q_5, B, L)$ // if all the 1 are replaced by 'X'
$\delta(q_5, X) \rightarrow (q_5, 1, L)$ // replace all 'X' by '1'
$\delta(q_5, B) \rightarrow$ enter the Turing machine for multiplication

9.3 Universal Turing Machine

From the discussions in the previous sections, it is clear that the Turing machine can perform a large number of tasks. The Turing machine can even perform any computational process carried out by the present day's computer. What is the difference between a Turing machine and real computer? The answer is very simple. The Turing machine is designed to execute only one program but real computers are reprogrammable. A Turing machine is called a universal Turing machine if it simulates the behaviour of a digital computer. A digital computer takes input data from user and produces the output by using an algorithm. A Turing machine can be said to be a universal Turing machine if it can accept (a) the input data and (b) the algorithm for performing the task.

Each task performed by a digital computer can be designed by a Turing machine. So, a universal Turing machine can simulate all the Turing machines designed for each separate task.

How to design a universal Turing machine? The simple answer is to add all the Turing machines designed for each separate task. But, in reality, it is a complex one. We can do this by

- Increasing the number of read–write heads (like multiple head TM)
- Increasing the number of input tapes (like multiple tape TM)
- Increasing the dimension of moving the read–write head (k-dimensional TM)
- Adding special purpose memory like stack.

9.4 Linear-Bounded Automata (LBA)

An LBA is a special type of Turing machine with restricted tape space. The name 'linear bounded' suggests that the machine is linearly bounded. If we compare a LBA with a TM, then we see that the difference is in the operational tape space. For TM, the input tape is virtually infinite in both directions whereas the allowable input tape space of LBA is the linear function of the length of input string. The limitation of tape space makes LBA a real model of computer that exists, than TM in some respect. The diagram of an LBA is shown in Fig. 9.13.

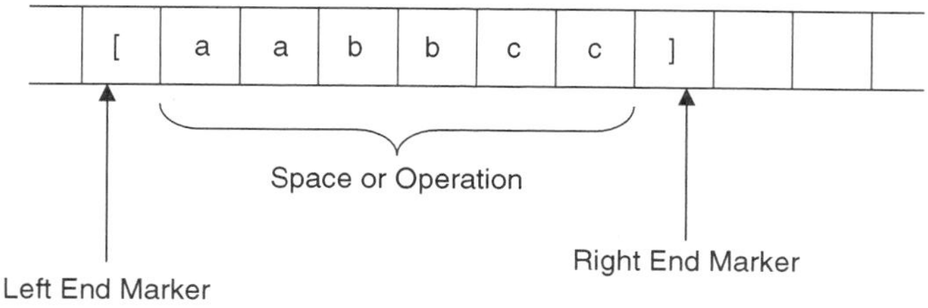

Fig. 9.13 *Linear-bounded Automata*

An TBA is a 7-tuple non-deterministic Turing machine

$$T_M = (Q, \Sigma, \Gamma, \delta, q_0, B, F)$$

where Q: Finite set of states
Σ: Finite set of input alphabets with two end markers '[' and ']'
Γ: Finite set of allowable tape symbols except two end markers '[' and ']'
q_0: Initial state
B: A symbol of Γ called blank
F: Final states
δ: Transitional function

where δ is a mapping from $Q \times \Gamma \rightarrow Q \times 2^Q \times \Gamma \times \{L, R\}$ with two more transitional functions $\delta(q_i, [) \rightarrow (q_j, [, R)$ and $\delta(q_i,]) \rightarrow (q_j,], L)$.

An LBA accepts context-sensitive language. It is powerful than the pushdown automata but less powerful than the Turing machine. The construction of an LBA for a context-sensitive language describes this clearly.

Example 9.12 Construct a linear-bounded automata for the following context-sensitive language.

$$L = \{a^n b^n c^n : n \geq 0\}$$

Solution: The design concept is

i) On each pass, the machine matches the leftmost 'a', 'b', and 'c' and replaces them with 'x'.
ii) If no 'a', 'b', and 'c' are left and it gets '[', then it halts.

The transitional functions are

$\delta(q_0, [) \rightarrow (q_1, [, R)$ // traverse right
$\delta(q_1, a) \rightarrow (q_2, X, R)$ // replace leftmost 'a' by 'X'
$\delta(q_2, a) \rightarrow (q_2, a, R)$ // traverse right
$\delta(q_2, b) \rightarrow (q_3, X, R)$ // replace leftmost 'b' by 'X'
$\delta(q_3, b) \rightarrow (q_3, b, R)$ // traverse right
$\delta(q_3, c) \rightarrow (q_4, X, L)$ // replace leftmost 'c' by 'X' and traverse left
$\delta(q_4, X) \rightarrow (q_4, X, L)$ // traverse left
$\delta(q_4, b) \rightarrow (q_4, b, L)$

$\delta(q_4, a) \rightarrow (q_4, a, L)$
$\delta(q_4, [) \rightarrow (q_1, [, R)$ //this is the left boundary, so traverse right
$\delta(q_1, X) \rightarrow (q_1, X, R)$ // required from the second pass onwards
$\delta(q_2, X) \rightarrow (q_2, X, R)$ // required from the second pass onwards
$\delta(q_3, X) \rightarrow (q_3, X, R)$ // required from the second pass onwards
$\delta(q_1,]) \rightarrow (q_1,], H)$

9.5 Post Machine

The Post machine was proposed by a Polish mathematician Emil Post in 1936. A Post machine is similar to the pushdown automata but it has a queue instead of a stack. A Post machine is always deterministic. According to the property of the queue, any item is added from the rear and is deleted from the front. A Post machine is defined by a 7 tuple

$$M = (Q, \Sigma, \Gamma, \delta, q_0, z_0, A)$$

where Q: Finite set of states
Σ: Finite set of input alphabets
Γ: Finite set of allowable queue symbols
q_0: Initial state
z_0: Queue end symbol (at initial state)
A: set of add state. This concatenates a character with the string from the right end.
δ: Transitional function.

It takes the current state, the symbol currently at the front of the queue, and moves to a state, giving an indication on whether to remove the current symbol from the front of the queue and the symbol to be added at the rear of the queue.

9.6. Church's Thesis

Alonzo Church, an American mathematician, proposed that any machine that can perform a certain list of operations will be able to perform all possible algorithms. Now, the question is 'which machine is this?'. Alen Turing, a Ph.D scholar of Church, proposed that machine called the Turing machine. For writing a program, we need to construct an algorithm first. 'No computational procedure is considered as an algorithm unless it is represented by the Turing Machine.' This is known as the Church thesis or the Church–Turing thesis. This thesis cannot be proved; it is a generally accepted truth. For this reason, it is called a thesis and not a theorem.

What We Have Learned So Far

1. A multi-tape Turing machine has multiple tapes and can be converted to a single-tape Turing machine
2. A multi-tape Turing machine is less complex than a single-tape Turing machine.

3. A multi-head Turing machine has multiple heads attached with a single finite control.
4. Every multi-head Turing machine has an equivalent single-head Turing machine.
5. A Turing machine where there are infinite numbers of sequence of blank on both sides of the tape is called a two-way infinite tape Turing machine.
6. The input tape of a multi-dimensional Turing machine is extended to infinity in more than one dimension.
7. An enumerator is a type of Turing machine attached to a printer.
8. The allowable input tape space of the linear bounded automata is the linear function of the length of input string.
9. A Post machine has a queue in comparison with a PDA which has a stack.

Solved Problems

1. Design a multi-tape Turing machine for accepting the language a^nb^{2n}, $n > 0$.

Solution: Consider a Turing machine of four tapes. Tape 1 contains the input string in the form a^nb^{2n}. Tape 2 is copied with n number of 'a' from a^nb^{2n}. Tape 3 is copied with n number of 'b', and tape 4 is filled with the remaining n number of 'b'.

Traverse tape 2, tape 3, and tape 4 from the right to left simultaneously and replace each symbol by blank.

If all the heads traversing tape 2, 3, and 4 get B at the same time, then declare as accept. If not, declare as reject.

The transitional diagram is shown in the following figure.

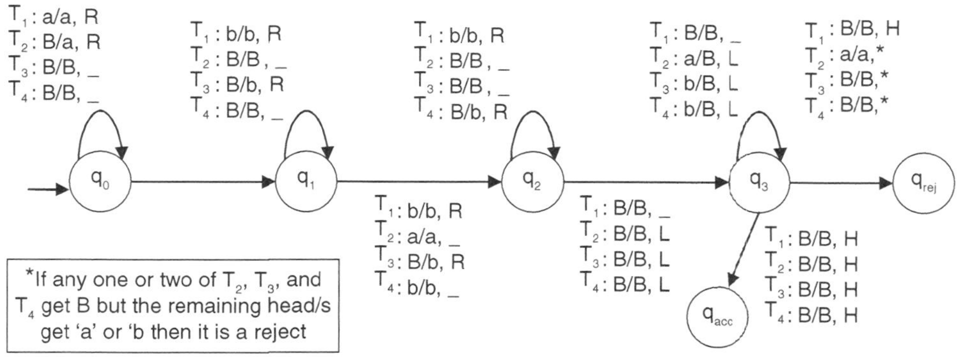

2. Design a multi-head Turing machine for accepting the language $a^nb^nc^n$, $n > 0$.

Solution: Consider a Turing machine with three heads. Initially, all the heads are placed on the first symbol. Place the first head on the first symbol ('a'). Traverse the remaining two heads towards the right. Getting the first 'b', stop traversing the second head. Getting the first 'c', stop traversing the third head.

The transitional representation of the Turing machine is as follows.

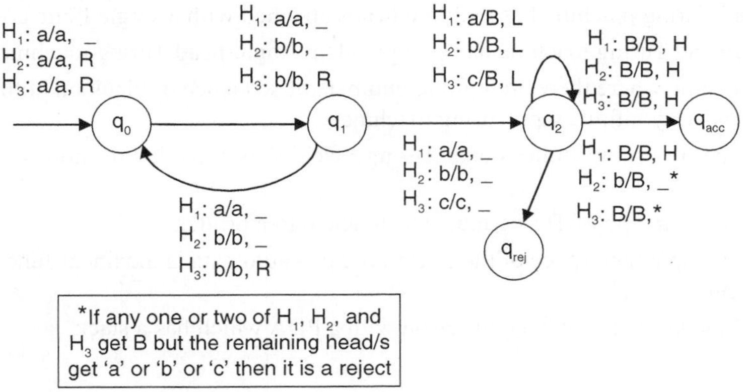

3. Design a multi-tape Turing machine for L = {a, b}*, where N(a) = N(b).

Solution: Consider a three tape Turing machine. The first tape contains the input. The second tape stores 'a' from the first. The third tape stores 'b' from the first. Then traverse the last two tapes to check whether the number of 'a' and 'b' is equal or not.

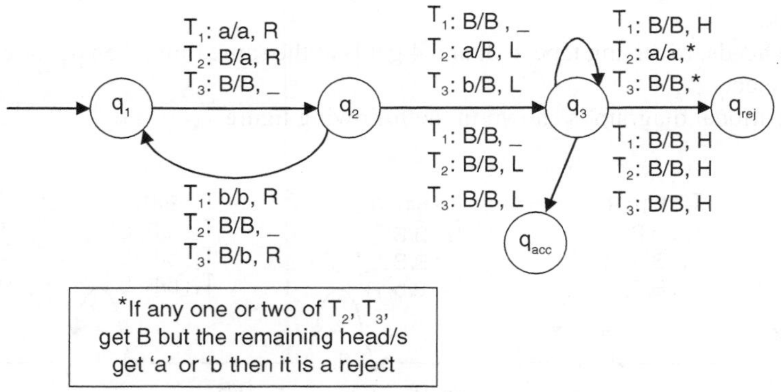

Multiple Choice Questions

1. A multi-tape Turing machine has
   a) Multiple tape
   b) Multiple head
   c) Multiple finite control
   d) Multiple tape and head
2. A multi-head Turing machine has
   a) Multiple tape
   b) Multiple head
   c) Multiple finite control
   d) Multiple tape and head
3. In respect to the k-dimensional Turing machine, a simple Turing machine is
   a) 0 dimensional
   b) 1 dimensional
   c) 2 dimensional
   d) 3 dimensional

4. A simple Turing machine is
   a) 1-way infinite tape
   b) 2-way infinite tape
   c) No way infinite tape
   d) none of these

5. A Post machine has a
   a) Stack
   b) Linear list
   c) Queue
   d) Circular queue

Answers:

1. d 2. b 3. b 4. a 5. c

GATE Questions

1. Which of the following is true (A : Multi-tape TM, B: Multi-head TM, C: Non- deterministic TM, D: K-dimensional TM, E: Single tape TM).
   a) A, B > E b) C > E c) D > A, B > C > E d) A = B = C = D = E
2. Which of the following has a read only tape?
   a) Single-tape TM b) Multi-tape TM
   c) Linear bounded automata d) None of these.
3. Which one of the following conversions is impossible ?
   a) NTM $\rightarrow$ DTM b) Multi-tape TM $\rightarrow$ Single-tape TM
   c) Multi-head TM $\rightarrow$ Single-head TM d) NPDA $\rightarrow$ DPDA
4. The movement of the head is limited to a certain region for which of the following machine
   a) K- dimensional TM b) Linear bounded automata
   c) Pushdown automata d) Two-way infinite tape
5. In which of the cases stated is the following statement true?
   'For every non-deterministic machine M_1, there exists an equivalent deterministic machine M_2 recognizing the same language.'
   a) M_1 is a non-deterministic finite automata
   b) M_1 is a non-deterministic PDA
   c) M_1 is a non-deterministic TM
   d) For no machine M_1 use the above statement true.

Answers:

1. d 2. c 3. d 4. b 5. a

Exercise

1. Construct a multi-tape Turing machine for accepting $L = 0^n1^n0^n$, $n > 0$.
2. Construct a multi-head Turing machine accepting WW^R, where $w \in (a, b)^+$.
3. Design a linear bounded automata accepting WW^R, where $w \in (a, b)^+$.
4. Compute the reverse of a string $w \in (a, b)^+$ using multi-tape Turing machine.

Computability and Undecidability

10

Introduction

According to the Church thesis, the algorithm of any computational procedure can be represented by the Turing machine (TM). Are there some problems for which there are no algorithms to solve the problem? This question divides the set of problems into two parts, namely, decidable and undecidable. If there exists an algorithm to solve a problem, the problem is called decidable; otherwise, it is undecidable. Before discussing undecidability, it is needed to know about the languages accepted by the TM.

10.1 TM Languages

A string w is said to be accepted by a TM if it gets halt(H). What happens if it does not get halt? It may happen that for the current state q and the current input symbol $\in \Sigma$, no transitional function is defined or it can run forever without getting halt. For the first case, the string w does not belong to the defined TM. But, the second case creates an interesting situation.

Let us consider a TM with the transitional functions as follows.

$$\delta(q_0, a) \rightarrow (q_1, X, R)$$
$$\delta(q_1, a) \rightarrow (q_4, X, H)$$
$$\delta(q_1, B) \rightarrow (q_2, B, L)$$
$$\delta(q_2, X) \rightarrow (q_3, a, L)$$
$$\delta(q_3, B) \rightarrow (q_0, B, R)$$
$$\delta(q_0, B) \rightarrow (q_f, B, H)$$

Let us take three input strings, null, 'aa', and 'a'.

1. The null string is accepted because the machine gets 'B', enters state q_f, and halts.
2. For the string 'aa', the first 'a' is traversed replacing 'a' by X. For the second 'a', it enters into q_4 and halts. So, 'aa' is rejected by the defined TM.
3. For the string 'a', the machine runs without halting. The fate of the string 'a' is not decided by the TM.

From the previous discussion, it can be concluded that for a TM, there are two types of languages.

1. A TM acceptable or recursively enumerable
2. Recursive language or Turing decidable.

10.1.1 Turing Acceptable

A language L is Turing acceptable or Turing recognizable or recursively enumerable if there is a TM that halts and accepts L or halts and rejects or loop for an infinite time.

In notation, it can be represented as follows.

For a string w,

$$w \in L, \text{ i.e., the TM halts on a final state}$$

or

$$w \notin L, \text{ i.e., the TM halts on a non-accept state or loops forever}$$

For a Turing acceptable language, there may exist a TM which halts on a large set of input string w, where $w \in L$, but there must exist at least one string $w \notin L$ for which the TM does not halt.

Pictorially, a recursively enumerable language can be represented as in Fig 10.1.

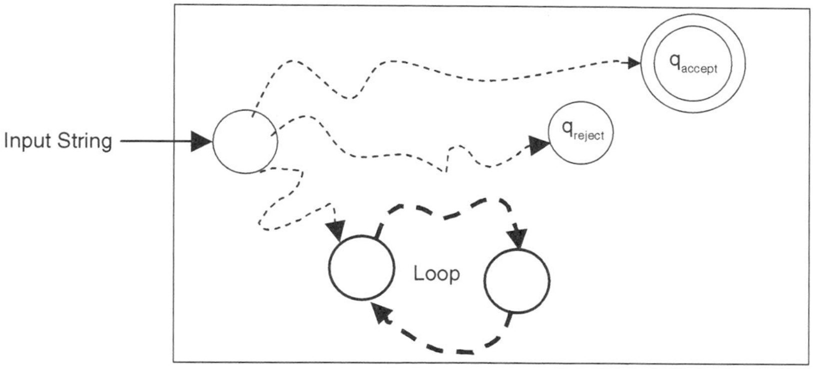

Fig. 10.1 *Turing Acceptable Language*

Sometimes, a situation may appear when for some cases it can be said confirm that the string is accepted, but whether the string is not accepted cannot be decidable.

Consider a TM which finds a symbol X on the tape and halts.

The machine will search for the symbol X by traversing the symbols B. If it finds, it will halt. The transitional functions are

$$\delta(q_0, B) \rightarrow (q_0, B, R)$$
$$\delta(q_0, X) \rightarrow (q_f, H)$$

If a string BBXBBB is taken, the machine can easily detect the symbol and halts. But if the input tape does not contain any X, then the machine will run infinitely. This is called the semi-decidability of the TM.

10.1.2 Turing Decidable

A language L is called Turing decidable (or just decidable), if there exists a TM M such that on input x, M accepts if $x \in L$, and M rejects otherwise. L is called undecidable if it is not decidable.

The decision problems always answer in yes or no. Another important property of decision problems is that, for the problem, the TM must halt.

Let there be a problem of checking whether a number X is prime or not. The algorithm for the problem is:

Step 1: Take the number X as input.

Step 2: Divide the number X by all integer numbers from 2 to $\sqrt{x}$.

Step 3: If any of the numbers divide X, then reject. Otherwise accept.

Step 4: stop

Already it is discussed that if there exists an algorithm for a problem, the corresponding TM can be constructed.

For this case, if an integer X is taken as input it can be said in yes or no whether the integer is prime or not. So, this problem is a Turing decidable problem.

Pictorially, a Turing decidable language or recursive language can be represented in Fig. 10.2.

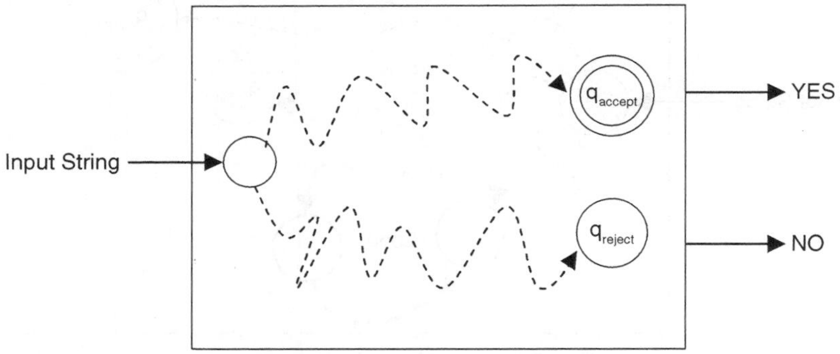

Fig. 10.2 *Turing Decidable Language*

If a language L is recursive, then it is a recursively enumerable language. The TM for the recursive language halts and accepts a string if it belongs to the language set L and halts and rejects if the string does not belong to L. These conditions are satisfied for a recursively enumerable language also. So, all recursive languages are recursively enumerable.

The difference between recursive and recursively enumerable language is that the TM will halt for recursive language but may not halt (loop infinitely) for recursively enumerable language.

This is described in Fig. 10.3.

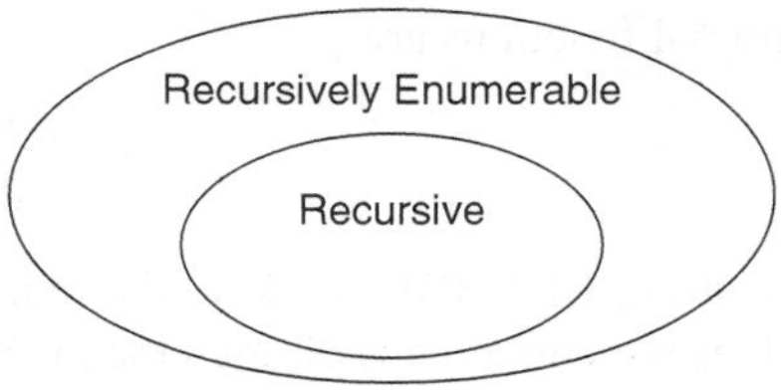

Fig. 10.3 *Relation of Recursive and Recursively Enumerable*

10.2 Unrestricted Grammar

This term was first mentioned in the language and grammar chapter. In the chapter Turing machine, we have learned that unrestricted grammar is accepted by the TM. The term 'unrestricted' signifies that the grammar is not restricted. Here, restriction means the appearance of non-terminal and terminal symbols at both sides.

Definition: *A grammar (V_N, Σ, P, S) is called an unrestricted grammar if all its productions are in the form LS → RS, where LS ∈ $(V_N \cup \Sigma)^+$ and RS ∈ $(V_N \cup \Sigma)^*$.*

In an unrestricted grammar, any combination of terminal and non-terminal can appear at both ends of the production rule. The only restriction is that the null string (λ) cannot appear at the left hand side of the production rule. (Notice for LS, it is + not *.) In the Chomsky hierarchy, type 0 or unrestricted grammar is the superset of all grammars, and thus it accepts more languages than type 1 or other grammars.

Already it is mentioned that unrestricted grammar is accepted by the TM, and the language accepted by a TM is called recursively enumerable language. In the following section, we shall discuss the relation between unrestricted grammar and recursively enumerable language.

Theorem 10.1: Any language set generated by an unrestricted grammar is recursively enumerable.

Proof: Let G = (V_N, Σ, P, S) be an unrestricted grammar. A grammar defines a set of procedures for enumerating all strings ∈ L(G).

- List all w ∈ L(G) such that w is derived from the production rules P in one step. As S is the start symbol, we can write that w is the set of strings derived from S in one step. Symbolically, it can be written as S → w.
- List all w ∈ L(G) such that w is derived from the production rules P in two steps. We can write it as S → x → w.

By the process, the derivation progresses.

If S be the set of all strings, then an enumeration procedure for S is a TM that generates all strings of S one by one in a finite amount of time. If, for a set, there exists an enumeration procedure, the set is countable. The set of strings generated by an unrestricted language is countable. Thus, the derivation procedures can be enumerated on a TM.

Hence it is proved that the language generated by an unrestricted grammar is recursively enumerable.

10.2.1 Turing Machine to Unrestricted Grammar

A TM accepts type 0 languages. Type 0 language is generated by type 0 grammar. Type 0 grammar can be constructed directly from the TM's transitional functions. The rules are discussed in the following.

Let the string to be traversed by a TM M be S which is enclosed between two end markers ψ S $. Note that the IDs are enclosed within brackets. The acceptance of the string S by M means the transformation of ID from [q_0ψ S $] to [$q_f$B]. Here, q_0 is the initial state and q_f is the final or halt state. The length of the ID may change if the read–write head of the TM reaches the left hand side or right hand side brackets. Thus, the productions rules of the grammar equivalent to the transition of ID are divided into two steps: (i) no change in length and (ii) change in length. Assume that the transition table is given as follows.

Input State	a_1	.....	a_j
q_1			
⋮			
q_i			

- **No change in length:**
  - **Right move:** If there is a transitional function

$$\delta(q_i, a_j) \rightarrow (q_k, a_l, R)$$

of the TM, then the equivalent production rule of the grammar is

$$q_i a_j \rightarrow a_l q_k$$

  - **Left move:** If there is a transitional function

$$\delta(q_i, a_j) \rightarrow (q_k, a_l, L)$$

of the TM, then the equivalent production rule of the grammar is in the form

$$a_p q_i a_j \rightarrow q_k a_p a_l \qquad \text{for all} \qquad a_p \in \Gamma$$

(If there are n allowable tape symbols, then for a left movement production n transitional functions are added to the production rule.)

- **Change in length:**
  - **Left bracket '[' at the left end:** If there is a production

$$\delta(q_i, a_j) \rightarrow (q_k, a_l, L)$$

TM going to traverse the left boundary '[': then the production

$$[q_i a_j \rightarrow q_k B a_l$$

is added to the production rule. (Here B represents the blank.)

If B appears next to the left bracket, it can be deleted by the production

$$[B \rightarrow [$$

  - **Right bracket ']' at the right end:** If B appears just before ']', then it can be deleted by the production

$$a_p B] \rightarrow a_p] \qquad \text{for all} \qquad a_p \in \Gamma$$

If TM going to traverse to the right of ']', then the length is increased due to the insertion of B. The productions are

$$q_i] \rightarrow q_i B] \qquad \text{for all} \qquad q_i \in Q$$

- The string is confined by two end markers ψ and \$. To introduce the end markers, the following productions are added.

$$[q_i\, \psi \rightarrow [q_i \qquad \text{for all} \qquad q_i \in Q$$
$$a_p \$ \rightarrow a_p \qquad \text{for all} \qquad a_p\ (\neq B) \in \Gamma$$

To remove brackets '[' and ']' from $[q_f B]$, the following production is added,

$$[q_f B] \rightarrow S$$

Here, S is the start symbol of the grammar.

- Now, reverse the direction of the arrow in the transitional functions. The generated production rules are the production rules of the grammar. The grammar obtained in this process is called generative grammar.

Example 10.1 Consider a TM with the following transitional functions. (This is the TM for 1's complement.) Convert this to an equivalent type 0 grammar.

Input / State	0	1	B
Q_0	Q_0, 1, R	Q_0, 0, R	Q_f, B, R
Q_f			

Solution:

i) For the transitional function $\delta(Q_0, 0) \rightarrow (Q_0, 1, R)$, the production rule is

$$Q_0 0 \rightarrow 1Q_0$$

For transitional function $\delta(Q_0, 1) \rightarrow (Q_0, 0, R)$, the production rule is

$$Q_0 1 \rightarrow 0Q_0$$

For transitional function $\delta(Q_0, B) \rightarrow (Q_f, B, R)$, the production rule is

$$Q_0 B \rightarrow 0Q_f$$

ii) The production rule corresponding to the left end is

$$[B \rightarrow [$$

The production rules corresponding to the right end are

$$0B] \rightarrow 0]\ 1B] \rightarrow 1]\ BB] \rightarrow B]\ \text{[for tape symbols]}$$
$$Q_0] \rightarrow Q_0 B]\ Q_1] \rightarrow Q_1 B]$$

iii) The production rules for introducing end markers are

$$[Q_0\psi \rightarrow [Q_0\ [Q_f\psi \rightarrow [Q_f$$
$$0\$ \rightarrow 0\ 1\$ \rightarrow 1\ \text{[B is excluded]}$$
$$[Q_f B] \rightarrow S$$

By reversing the direction of the arrow, the productions become

$$S \rightarrow [Q_f B],\ 1 \rightarrow 1\$,\ 0 \rightarrow 0\$$$
$$[Q_0 \rightarrow [Q_0\psi,\ [Q_f \rightarrow [Q_f\psi,\ Q_1 B] \rightarrow Q_1]$$
$$Q_0 B] \rightarrow Q_0],\ B] \rightarrow BB],\ 1] \rightarrow 1B]$$
$$0] \rightarrow 0B],\ [\rightarrow [B,\ 0Q_f \rightarrow Q_0 B$$
$$0Q_0 \rightarrow Q_0 1\ 1Q_0 \rightarrow Q_0 0$$

Example 10.2 Consider a TM with the following transitional functions. (This is the TM which accepts an even string of 'a'.) Convert this to an equivalent type 0 grammar. Check it for the string 'aa'.

Input / State	a	B
Q_0	Q_1, B, R	
Q_1	Q_0, B, R	

where Q_0 is the initial and final state.

Solution:

i) For transitional function $\delta(Q_0, a) \rightarrow (Q_1, B, R)$, the production rule is

$$Q_0\, a \rightarrow BQ_1$$

For the second transitional function $\delta(Q_1, a) \rightarrow (Q_0, B, R)$, the production rule is

$$Q_1\, a \rightarrow BQ_0$$

ii) The production rule corresponding to the left end is

$$[B \rightarrow [$$

The production rules corresponding to the right end are

$$aB] \rightarrow a]\quad BB] \rightarrow B]\ \text{[for tape symbols]}$$
$$Q_0] \rightarrow Q_0B]\quad Q_1] \rightarrow Q_1B]$$

iii) The production rules for introducing end markers are

$$[Q_0\psi \rightarrow [Q_0\quad [Q_1\psi \rightarrow [Q_1$$
$$a\$ \rightarrow a\ \text{[B is excluded]}$$
$$[Q_0B] \rightarrow S$$

By reversing the direction of the arrow, the productions become

$$BQ_1 \rightarrow Q_0a\quad BQ_0 \rightarrow Q_1a$$
$$[\rightarrow [Ba]\quad \rightarrow aB]$$
$$B] \rightarrow BB]\quad Q_0B] \rightarrow Q_0]$$
$$Q_1B] \rightarrow Q_1]\quad [Q_0 \rightarrow [Q_0\psi$$
$$[Q_1 \rightarrow [Q_1\psi\quad a \rightarrow a\$$$
$$S \rightarrow [Q_0B]$$

Checking: The string is 'aa'. It means from S we have to produce $[Q_0\psi aa\$]$.

$$S \rightarrow [Q_0B]$$
$$\rightarrow [Q_0]\quad (\text{As } Q_0B] \rightarrow Q_0])$$
$$\rightarrow [BQ_0]\quad (\text{As } [\ \rightarrow [B)$$

$$\begin{array}{ll} \rightarrow [Q_1a] & (\text{As } BQ_0 \rightarrow Q_1a) \\ \rightarrow [BQ_1a] & (\text{As } [\ \ \rightarrow [B) \\ \rightarrow [Q_0aa] & (\text{As } BQ_1 \rightarrow Q_0a) \\ \rightarrow [Q_0\psi aa] & (\text{As } [Q_0 \rightarrow [Q_0\psi) \\ \rightarrow [Q_0\psi aa\$] & (\text{As } a \rightarrow a\$) \end{array}$$

10.2.2 Kuroda Normal Form

The Kuroda normal form is related to unrestricted and context-sensitive grammar. Sige-Yuki Kuroda, a Japanese linguist, proposed this normal form related to context-sensitive grammar. Since then, this is named after her.

Definition: A grammar is in Kuroda normal form if and only if all production rules of it are in the form

$$\begin{aligned} AB &\rightarrow CD \\ A &\rightarrow BC \\ A &\rightarrow B \\ A &\rightarrow a \end{aligned}$$

where A, B, and C are non-terminals and 'a' is terminal.

10.2.3 Conversion of a Grammar into the Kuroda Normal Form

Consider $G_1 = \{V_{N1}, \Sigma, P_1, S\}$ as a context-sensitive grammar which does not produce $\in$. An equivalent grammar $G_2 = \{V_{N2}, \Sigma, P_2, S\}$ in the Kuroda normal form can be constructed from G_1 using the following rules.

1. Replace all terminals $T \in G_1$ by a new production $A_T \rightarrow T$ and add them to G_2.
2. For each production in the form $A \rightarrow A_1A_2 \ldots A_k$ in G_1, add a new production $A \rightarrow A_1B_1$ in G_2 where $B_i \rightarrow A_{i+1}B_{i+1}$ for $i = 1, 2 \ldots k-3$ and the terminating condition is

   $$B_{k-2} \rightarrow A_{k-1}A_k$$

3. For each production in the form $A_1A_2 \ldots A_m \rightarrow B_1B_2 \ldots B_k$, where $2 \le m \le k \ge 3$, the following new productions

   $$\begin{aligned} A_1A_2 &\rightarrow B_1C_1 \\ C_iA_{i+2} &\rightarrow B_{i+1}C_{i+1} \quad \text{for } i = 1 \ldots\ldots 1-2 \\ C_i &\rightarrow B_{i+1}C_{i+1} \quad \text{for } i = 1-1, \ldots\ldots m-3 \\ C_{m-2} &\rightarrow B_{m-1}B_m \end{aligned}$$

   are added to G_2.

10.3 Modified Chomsky Hierarchy

In the grammar and language chapter, we learned about the Chomsky hierarchy. There, the type 0 language was mentioned as an unrestricted language which is accepted by the TM. While discussing the Turing language, we have learned that this language is called the recursively enumerable language.

A recursive language is a subset of the recursively enumerable language. Thus, type 0 language is divided into two parts, namely recursively enumerable language and recursive language. The modified Chomsky hierarchy is given in Fig. 10.4.

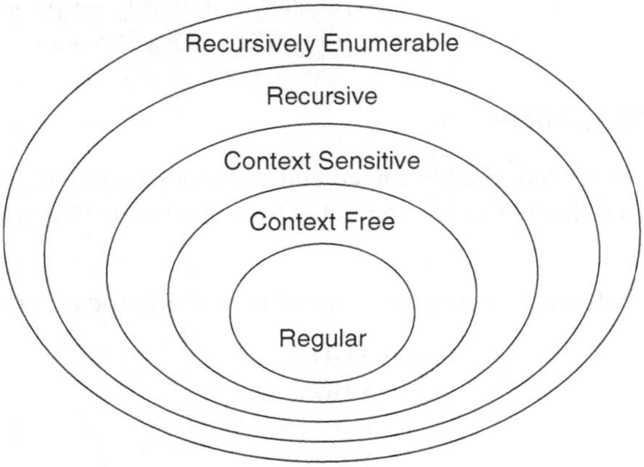

Fig. 10.4 *Modified Chomsky Hierarchy*

10.4 Properties of Recursive and Recursively Enumerable Language

1. **The union of two recursive languages is recursive.**

   Proof: Let L_1 and L_2 be two recursive languages, so there exist two TMs, say T_1 and T_2 accepting L_1 and L_2, respectively. Let us construct a new TM M by combining T_1 and T_2 as given in Fig. 10.5. A string w is given as input to T_1; if it accepts, then M accepts. If it rejects, then the same input is given as input to T_2; if it accepts, then M accepts. From the discussion, it is clear that M accepts $L_1 \cup L_2$.

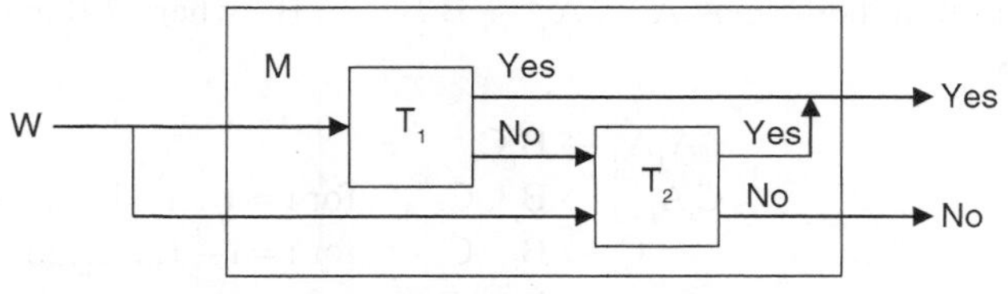

Fig. 10.5 *Union of Two Recursive Languages*

   As there exists a TM for $L_1 \cup L_2$ which guarantees to halt, it is thereby proved that $L_1 \cup L_2$ is recursive.

2. **The union of two recursively enumerable languages is recursively enumerable.**

   Proof: Let L_1 and L_2 be two recursively enumerable languages, so there exist two TMs, say T_1 and T_2 accepting L_1 and L_2, respectively. Let us construct a new TM M by combining T_1 and T_2 as denoted in Fig. 10.6.

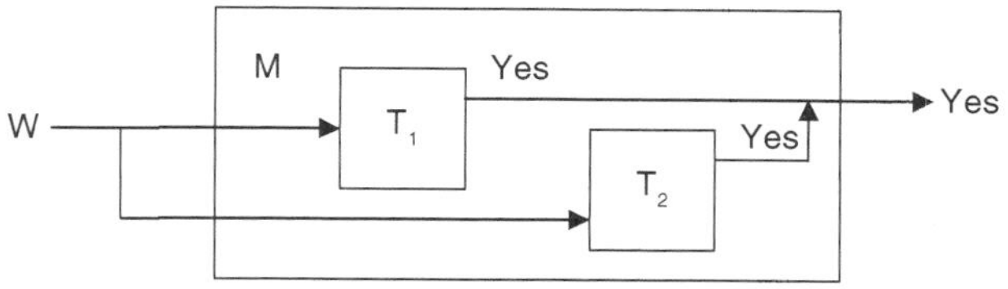

Fig. 10.6 *Union of Two Recursive Enumerable Languages*

A string w is accepted by M if it is accepted by either T_1 or T_2. Acceptance by either T_1 or T_2 means accepting $L_1 \cup L_2$. If w is not accepted by both T_1 and T_2, then the machine M has no guarantee to halt. It proves that $L_1 \cup L_2$ is recursively enumerable.

3. **The intersection of two recursive languages** is **recursive.**

   Proof: Let L_1 and L_2 be two recursive languages, so there exist two TMs, say T_1 and T_2 accepting L_1 and L_2, respectively. Let us construct a new TM M by combining T_1 and T_2 as given in Fig. 10.7.

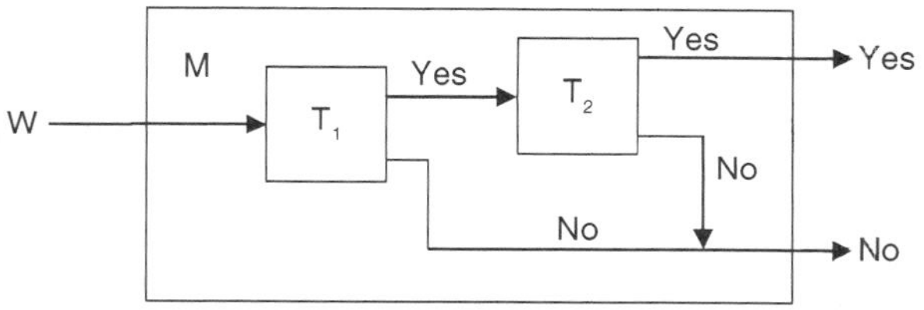

Fig. 10.7 *Intersection of Two Recursive Languages*

A string w is accepted by M if it is accepted by both T_1 and T_2, and rejected if any of them reject. From the discussion it is clear that M accepts $L_1 \cap L_2$. As there exists a TM for $L_1 \cap L_2$ which guarantee to halt; thereby proving that $L_1 \cap L_2$ is recursive.

Algorithmically, it can be written as

```
On input w
    Run T1 on w
if T1 accepts w then
    Run T2 on w
    if M2 accepts w then
    accept else reject
else reject w
```

4. **The intersection of two recursively enumerable languages is recursively enumerable.**

   It can be proved similar to the previous property, excluding the rejection part as given in Fig. 10.8.

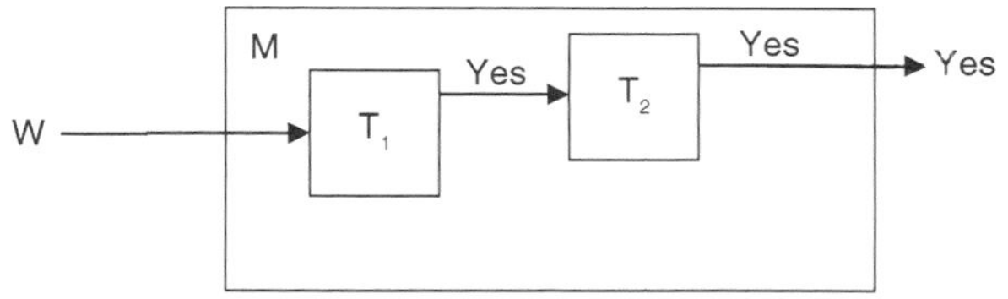

Fig. 10.8 *Intersection of Two Recursively Enumerable Languages*

5. **The complement of a recursive language is recursive.**

Proof: Let L be a recursive language, so there exists a TM T_m accepting L. T_m guarantees to accept and halt for any string w ∈ L and reject and halt for any string w ∉ L.

Let us construct another TM T'_m which simulates M on w; if T_m accepts w, then T'_m rejects it and if T'_m rejects w, then T'_m accepts it as shown in Fig. 10.9.

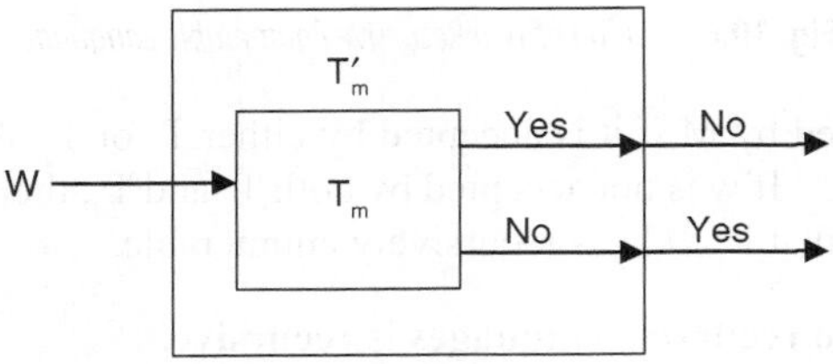

Fig. 10.9 *Complement of Two Recursive Languages*

Clearly, T'_m is the complement of T_m. As it halts on all possible inputs and accepts the complement of L, it is proved that the complement of a recursive language is recursive.

6. **If a language and its complement both are recursively enumerable, then both the language and its complement are recursive.**

Proof: Let L and its complement L′ be recursively enumerable languages, so there exist two TMs, say T_1 and T_2 accepting L and L′, respectively. Let us construct a new TM M which simulates T_1 and T_2 simultaneously, accepting a string w if T_1 accepts it and rejecting a string w if T_2 accepts it, as denoted in Fig. 10.10.

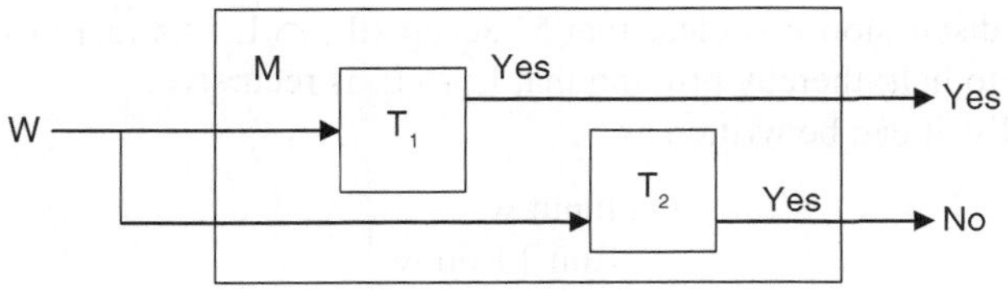

Fig. 10.10

Thus, M will always halt and answer in 'yes' or 'no'. As M guarantees to halt for all possible inputs, L is recursive.

Already it is proved that the complement of a recursive language is recursive, thus proving L′ is recursive.

(Note that the recursively enumerable languages are not closed under complementation. If an RE language is closed under complementation, it is recursive.)

7. **The concatenation of two recursively enumerable languages is recursively enumerable.**

Proof: Let L_1 and L_2 be two recursively enumerable languages, so there exist two TMs, say, T_1 and T_2 accepting L_1 and L_2, respectively. Design a machine M to accept L_1L_2 whose operation is performed by the following algorithm.

Let W∈ L_1L_2. Guess a number n between 0 and | W |. Break W into W_1 and W_2 such that $W = W_1W_2$ and $| W_1 | = n$ and $| W_2 | = | W | - n$.

Set counter i = 0 with one increment up to | W |.

{

Run W_1 on T_1 for n steps

Run W_2 on T_2 for next | W | – n steps.

}

If (T_1 accept W_1 and T_2 accept W_2)

Declare accept

M accepts if the substrings L_1 and L_2 are accepted by T_1 and T_2, respectively. Hence L_1L_2 is recursively enumerable. This is described in Fig. 10.11.

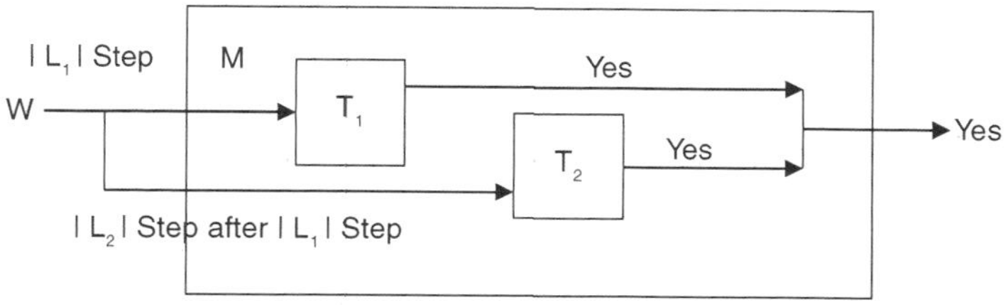

Fig. 10.11 *Concatenation of Two Recursively Enumerable Languages*

8. **The concatenation of two recursive languages is recursive.**

   Proof: Let L_1 and L_2 be two recursive languages, so there exist two TMs, say T_1 and T_2 accepting L_1 and L_2, respectively. Design a machine M to accept L_1L_2 whose operation is performed by the following algorithm.

   Let W ∈ L_1L_2. Guess a number n between 0 and | W |. Break W into W_1 and W_2 such that $W = W_1W_2$ and $|W_1| = n$ and $|W_2| = |W_1W_2| - n$.

   Set counter i = 0 with one increment upto | W |

   Run W_1 on T_1 for n steps

   if (T_1 halts and accept W_1)

   Run W_2 on T_2 for next | W | steps

   if (T_2 halts and accept W_2)

   Declare accept

   Else reject

   Else reject

   M accepts if the substrings W_1 and W_2 are accepted by T_1 and T_2, respectively, and rejects if any of T_1 and T_2 rejects it. Hence, L_1L_2 is recursive.

9. **The Kleene-star operation on a recursively enumerable language is recursively enumerable.**

   Proof: Let L be a recursively enumerable language accepted by the TM T. We have to prove that L* is also recursively enumerable. Design another TM M to accept L* whose operation is performed by the following algorithm.

   Let W ∈ L*. On input W to M.

   If W = λ (null) accept

   If W ≠ λ

   Consider a number n between 1 and | W |

Consider n different numbers NUM_k (k = 0 to n) between 0 and | W | to break the string W into substrings $w_1w_2 \ldots w_n$.

```
Set another counter i = 0 and start it.
Set a counter j = 0 with 1 increment up to NUM_k
{
    Run T on w_i for j steps
    If (T halts and accept w_i)
        Increment i by 1
}
  If(j = = n)
[T accepts each substring w_i (i = 1 to n)]
  declare accept
```

M halts and accepts L* if each of $w_1w_2 \ldots w_n$ are accepted by T. Hence, L* is recursively enumerable.

10. **The Kleene-star operation on a recursive language is recursive.**

    Proof: It can be proved the same way as the previous case with a modification in algorithm that if any of w_i is not accepted by T, then reject.

11. **The intersection of recursive and recursively enumerable language is recursively enumerable.**

    Proof: Let L_1 be recursive and L_2 be a recursively enumerable language. There exist two TMs, say T_1 and T_2 accepting L_1 and L_2, respectively. Design a machine M to accept $L_1 \cap L_2$ whose operation is performed by the following logic.

```
On input w ∈ L_1 ∩ L_2
Simulate w on T_1
  If (T_1 halts and accepts w)
    Erase the tape content of M and copy w on the tape again.
Simulate w on T_2
  If (T_2 halts and accepts w)
    w is accepted by M.
```

Thus, the intersection of recursive and recursively enumerable language is recursively enumerable.

12. **If L is a recursive language, then $\Sigma^* - L$ is also recursive.**

    Proof: As L is recursive, there exists a TM T_m accepting L. Σ^* − **L** is the set of all languages except L which consists of the terminal symbols by which L is constructed.

    A new TM T'_m is constructed as shown in the following diagram (Fig. 10.12) which halts and rejects a string $w \in L$ but halts and accepts a string $w \notin L$.

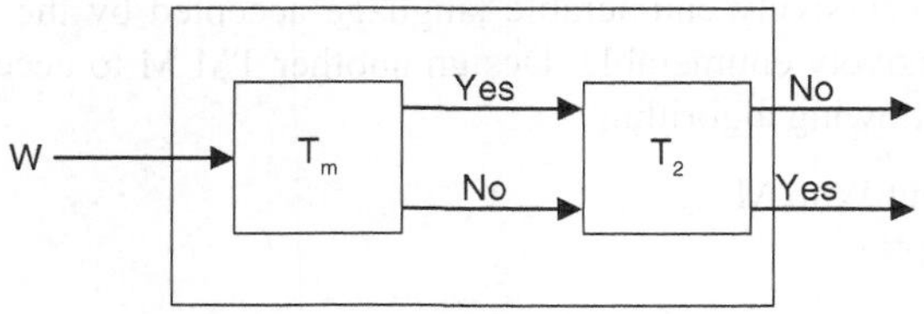

Fig. 10.12 *TM for $\Sigma^* - L$*

The new machine consists of two TMs, T_m and T_2. The design is such that if a string is accepted by T_m, it will be rejected by T_2 and vice versa.

The machine functions as follows: let a string $w \in \Sigma^*$ be given as input to the machine; it passes it to T_m. If $w \in L$, then T_m accepts it and the input is passed to T_2 which rejects it. On the reverse, if $w \notin L$, T_m rejects but T_2 accepts it.

It is found that there exists a TM which halts and answers in 'yes' or 'no' for $\Sigma^* - L$, thus proving it is recursive.

The following table is important for memorizing the closure property of different operations on different languages.

Types / Operations	Regular	Context Free	Recursively Enumerable	Recursive
Union	Yes	Yes	Yes	Yes
Intersection	Yes	No	Yes	Yes
Concatenation	Yes	Yes	Yes	Yes
Complement	Yes	No	No	Yes
Closure	Yes	Yes	Yes	Yes

10.5 Undecidability

In the previous section, it is already discussed that if there exists an algorithm to solve a problem, then the problem is called decidable. The Church–Turing thesis says that the 'no computational procedure is considered as an algorithm unless it is represented by the Turing machine.' It means that for a decidable problem there must exist a TM. Decidable problems are called solvable problems which can be solved by the computer. As solvable problems exist, there must also exist unsolvable problems which are not solved by the computer. Before going into the details of undecidability, the types of problems need to be discussed.

Decidable or Solvable: A language L is called decidable if there exists a TM M such that M accepts and halts on every string in L and rejects and halts on every string in L′.

A computational problem is known as decidable or solvable if the corresponding language is decidable.

Recognizable: A language L is called recognizable if there exists a TM M such that M accepts every string in L and either rejects or fails to halt on every string in L′.

A computational problem is known as recognizable or Turing acceptable if the corresponding language is recognizable.

Undecidable: Undecidable means not decidable. An undecidable language has no decider. There does not exist any TM which accepts the language and makes a decision by halting for every input string (may halt for some strings but not for all).

A computational problem is known as undecidable or unsolvable if the corresponding language is undecidable.

In the following section, we shall check the decidability of some languages related to DFA, NFA, and CFL.

Example 10.3 Prove that the problem that a string w is accepted by a DFA M is decidable.

Solution: According to the Church–Turing thesis, the algorithm for solving a problem must be represented by a TM. There exists an algorithm to check whether a string w is accepted by a DFA or not.

[Recall the algorithm in the finite automata chapter.

On getting input w to the beginning state of the DFA M, it traverses the string according to its transitional functions δ.

- The string is declared accepted if the string is totally traversed and the DFA reaches its final state.
- The string is not accepted if it is totally traversed but the DFA fails to reach to its final state.]

This can be represented by a TM which works as follows:

On getting a string w as an input to a DFA M

i) The TM T simulates the DFA M on input w.
ii) T performs the simulation in a direct way keeping track of M's current state and current position in the input string w.
iii) When T finishes the simulation by traversing the last symbol of w by the DFA M and reaches an accept state, it accepts. Reaching a non-accept state, T rejects.

As for every string w, the DFA halts declaring whether it accepts or rejects; so, the problem is decidable.

In another way, it can be proved.

```
On input w to a DFA M
Run M on the string w.
    If (w is accepted by M)
        Halt and declare accept
    Else
        Halt and declare reject.
```

Example 10.4 Prove that the problem that a string w is accepted by an NFA M is decidable.

Solution: This example can be proved by using the proof of the previous example with a modification.

On getting a string w as an input to a NFA M

Convert the NFA M to an equivalent DFA M′.

(Algorithm exists, so an equivalent TM can be designed.)

Then, the TM T simulates the DFA M′ on input w.

Example 10.5 Prove that the problem that a set of null string is accepted by a DFA M is decidable.

Solution: This can be represented by a TM which works as follows:

i) On getting the input to the DFA M, mark the beginning state of the DFA.
ii) while(new states get marked = true)

```
do
  {
    Mark the state that has an incoming transition from a state already marked.
  }
```

iii) If no accept state is marked, *accept*; otherwise *reject*. Hence, decidable.

Example 10.6 Prove that the problem that a string $L \in \Sigma^*$ is accepted by a DFA M is decidable.

Solution: Construct a DFA M′ by changing the non-final states of M to the final state and final states of M to non-final states. Thus, the new DFA will accept $\Sigma^* - L = \Phi$, i.e., null set.

Construct a TM T which simulates the DFA M′.

Using example 3, if $L(M') = \Phi$ then accept, otherwise reject. Hence, decidable.

Example 10.7 Prove that the problem that the two DFA, M_1 and M_2, which satisfy the same language is decidable.

Solution: Let us assume that this is not true. Consider two languages L_1 and L_2 accepted by the two DFA M_1 and M_2. A new DFA, M, is constructed which accepts only those strings that are accepted by either M_1 or M_2 but not both. From this, it is clear that if M_1 and M_2 recognize the same language, M will accept the null set.

Thus, the language accepted by M is

$$L = (L_1 \cap \overline{L}_2) \cup (\overline{L}_1 \cap L_2)$$

If L is null, then M_1 and M_2 recognize the same language.

If $L = \Phi$, then $(L_1 \cap \overline{L}_2) = \Phi$ and $(\overline{L}_1 \cap L_2) = \Phi$.

If $(L_1 \cap \overline{L}_2) = \Phi$. then $L_1 \subseteq L_2$

(See the diagram.)

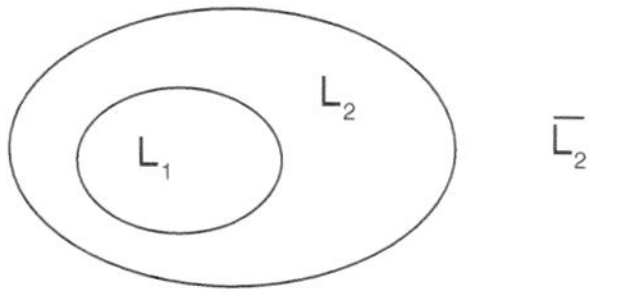

If $(L_2 \cap \overline{L}_1) = \Phi$, then $L_2 \subseteq L_1$

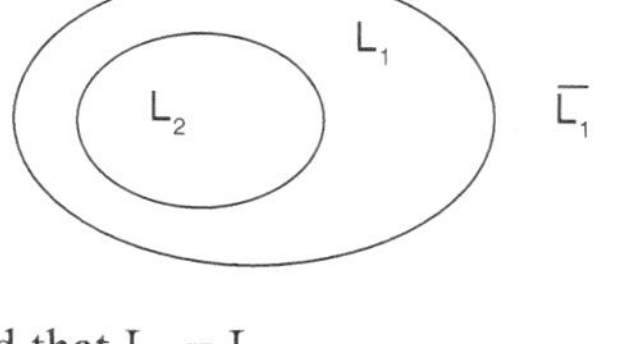

As $L_1 \subseteq L_2$ and $L_2 \subseteq L_1$, it is proved that $L_1 = L_2$.

Now let us come to the original proof.

i) Construct a TM T which simulates on getting the input to M.

ii) Using Example 10.5, if $L(M) = \Phi$ then accept, otherwise reject. Hence, decidable.

Example 10.8 Prove that the problem that a string w is generated by a CFG G is decidable.

Solution: This can be proved by designing a TM which searches w from the set of the languages generated by G. But the set like (a,b)* may contain an infinite number of elements. So, this technique will not work for all cases. A different technique is used to prove this.

Construct a TM T which simulates the construction of w by G.
On getting input G, w, where G is the grammar and w is the string,

i) Convert the grammar G to an equivalent grammar G′, where G′ is the Chomsky normal form of G. (Algorithm exists.)
ii) Derive up to 2n – 1 steps where | w | = n and list all the generated strings in a set S. If n = 0, then make derivations up to 1 step.
iii) If w ∈ S then accept, otherwise reject. Hence, decidable.

Example 10.9 Prove that the problem that a string w is accepted by a PDA P is decidable.

Solution: Construct a TM T which simulates this.
On getting input w, P, where w is the string and P is the PDA,

i) Convert the PDA P to an equivalent CFG G. (Algorithm exists.)
ii) Convert the grammar G to an equivalent grammar G′, where G′ is the Chomsky normal form of G. (Algorithm exists.)
iii) Derive up to 2n – 1 steps where | w | = n and list all the generated strings in a set S. If n = 0, then make derivations up to 1 step.
iv) If w ∈ S then accept, otherwise reject. Hence, decidable.

Example 10.10 Prove if a language L is decidable, then its complement $\overline{L}$ is also decidable.

Solution: As L is decidable, there exists a TM which accepts and halts for every string of L. Let us build another TM T′ using T where the 'accept' states of T are the 'reject' states of T′ and the rejected states of T are the accepted states of T′. T′ works as follows.

For each input string w,

i) Run T on w.
ii) If T halts and accepts, then reject.
If T halts and rejects, then accept.

As T′ halts for every input string, $\overline{L}$ is decidable.

Example 10.11 There is a language which is Turing acceptable and undecidable.

Solution: Let a language L be Turing acceptable. So, the language $\overline{L}$ is not accepted by any TM. It is already proved that the complement of a decidable language is also decidable. As $\overline{L}$ is not Turing acceptable, L is undecidable too. This proves that L is Turing acceptable but undecidable. This is described in Fig. 10.13.

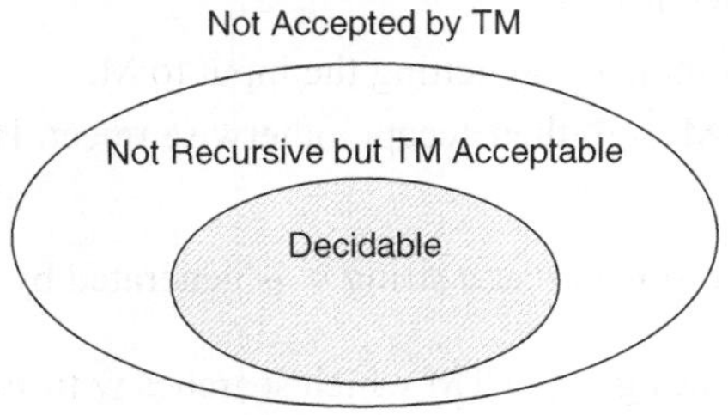

Fig. 10.13 *Turing Acceptable and Undecidable*

10.6 Reducibility

A reduction is a process of converting one problem into another solved problem in such a way that the solution of the second problem can be used to solve the first problem.

This is coming from our daily life experience.

Let us assume that one is travelling in a new city and not aware of the transportation route. The person takes the help of a travel guide book (which contains the bus number, route number, destination, etc.). The book may be thought of as a solved problem. Finding the bus to reach to the destination is the problem. To solve the problem, we take the help of an already solved problem, i.e., the travel guide book.

To find the regular expression accepted by a finite automata is a problem. The Arden theorem is the solved problem. We reduce the first problem to the second problem (Arden theorem) to solve the first.

Reducibility involves two problems where one is a solved problem and another is the problem to be solved. If the problem to be solved is reduced to the solved problem, then the solution of the second problem is used to solve the first. Reducibility has some properties as follows.

Let the two problems be A and B, where B is a solved problem.

1. When A is reduced to B, solving A cannot be harder than solving B.
2. If A is reducible to B and B is a decidable problem, then A is also a decidable problem.
3. If A is an undecidable problem and reducible to B, then B is also an undecidable problem (the important property used in proving that a problem is undecidable).

Theorem 10.2: A language is Turing recognizable if and only if some enumerator enumerates it.

Proof

(IF Part): If an enumerator enumerates a language, there exists a TM which recognizes the language.

Let the language L be enumerated by an enumerator E. Let a TM M recognize L, where M is designed as follows.

On input w,

- ❑ Run E. Every time when E outputs a string, say x, compare it with w.
- ❑ If x = w (i.e., the output of E ever matches with w), accept.

If an algorithm exists, it means that a TM exists. Thus, the TM M can be designed. This proves the 'IF' part.

(Only IF): Suppose the TM M recognizes L. Let $s_1, s_2, s_3, \ldots\ldots s_i$ be a list of all possible strings in Σ^*. Construct an enumerator E_s to generate this sequence. The enumerator works as follows.

- ❑ Repeat the following for input i = 1, 2, 3, ...
- ❑ Run M for i steps on each input $s_1, s_2, s_3, \ldots\ldots s_i$.
- ❑ If any computation is accepted, print out the corresponding S_i.

Example 10.12 Prove that the problem 'an arbitrary string w is accepted by an arbitrary TM M' is undecidable.

Solution: This is known as the membership problem. The basic idea behind the problem is to find a contradiction. First, it is assumed that the problem is decidable. With this assumption, we will prove that every Turing-acceptable language is also decidable. This is a contradiction to our assumption.

Let us assume that the problem is decidable. So, there exists a TM A_{TM} accepting this. The machine A_{TM} takes the input <M, w> where M is the TM and w is the string. If M accepts w, then A_{TM} halts and accepts, if M rejects w then A_{TM} halts and rejects.

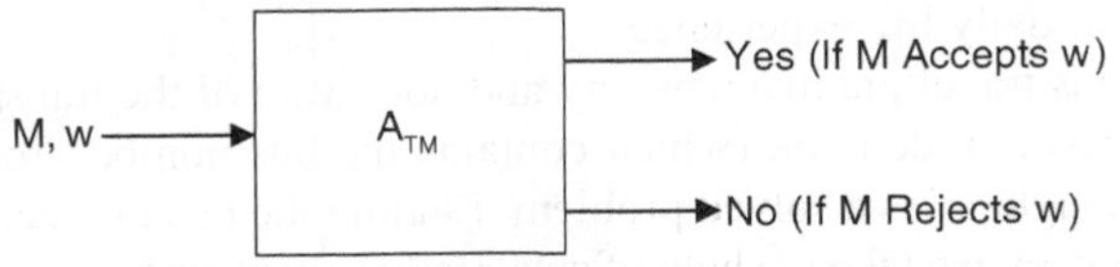

Let L be a recursively enumerable language and M′ be the TM accepting L. Let us build a decider for L reducing it to A_{TM}.

The pair <M′, s> is given as input to A_{TM}, where s is a string. As A_{TM} is the general TM for the arbitrary string w given input to the arbitrary TM M, it will halt for every input. Thus, L is decidable.

From here, it is proved that every Turing-acceptable language is decidable. But already it is proved that there is a Turing-acceptable language which is undecidable. Hence, we get a contradiction. So, our assumption is wrong.

It is proved that the membership problem is undecidable.

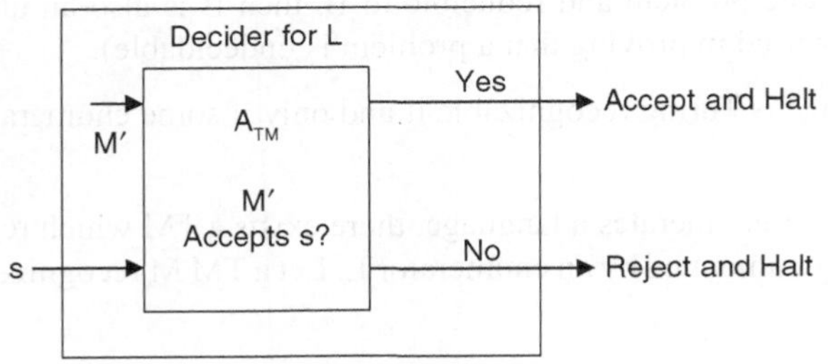

(Is it possible to know that a string w is accepted by a given TM M before traversing the string using the machine? This problem is known as the membership problem. From our general knowledge, we know that the answer is 'no'. No algorithm exists to decide this. This is proved in the previous example.)

Example 10.13 Prove that the problem "a string w halts on a Turing machine M" is undecidable.

Solution: This is known as the halting problem.

Let us assume that the problem is decidable. So, there must exist a TM say A_{TM} to accept this. The machine A_{TM} takes the input <M, w> where M is the TM and w is the string. If M halts on w, then A_{TM} halts and accepts, if M does not halt on w then A_{TM} halts and rejects.

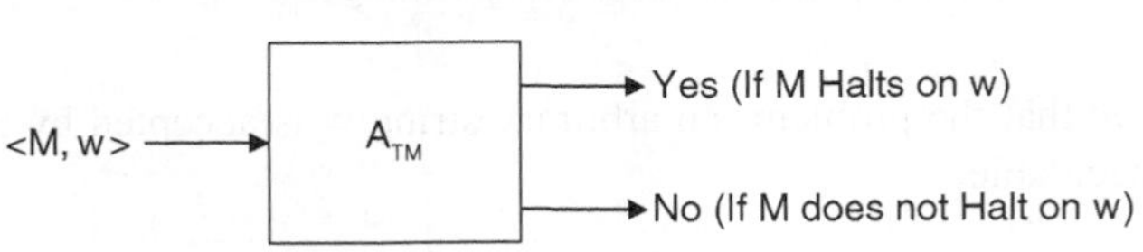

Let L be a recursively enumerable language and M_L be the TM that accepts L. Let us build a decider for L reducing it to A_{TM}.

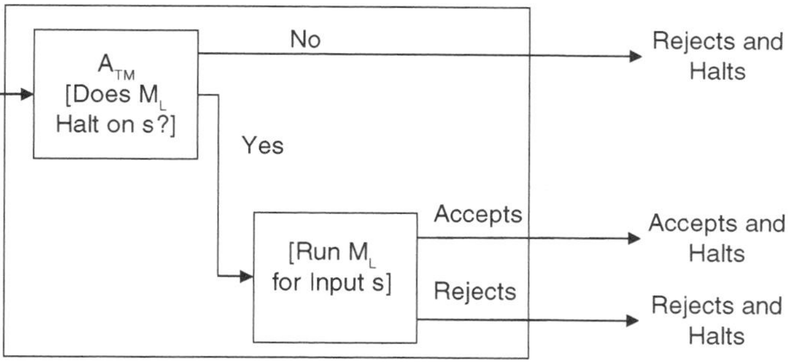

The pair $<M_L, s>$ is given the input to A_{TM}. s is an arbitrary string given as input to M_L. M_L will either halt or not. If M_L halts, then two conditions arise—whether s is accepted by M_L or it is rejected by M_L. In all the three cases, the decider halts on string s. Therefore, L is decidable.

A_{TM} is the general TM for the arbitrary string w given as input to the arbitrary TM M, and thus every Turing-acceptable language is also decidable.

But, it is already proved that there is a Turing-acceptable language which is undecidable. Hence, we get a contradiction. So, our assumption is wrong.

It is proved that the halting problem is undecidable.

(In general, the halting problem can be described that from the description of a computer program, decide whether the program finishes running or continues to run forever.)

Example 10.14 Given a TM *M*, prove that the problem whether or not *M* halts if started with a blank tape is undecidable.

Solution: This problem is called blank tape halting problem. This problem can be proved by reducing the halting problem to it.

Suppose that the blank tape halting problem is decidable. Therefore, it must have a decider to accept it. The decider takes the input of the TM M, halts and accepts if an M halt started with a blank tape; otherwise, it halts and rejects if M does not halt start with a blank tape.

The following is the decider for it.

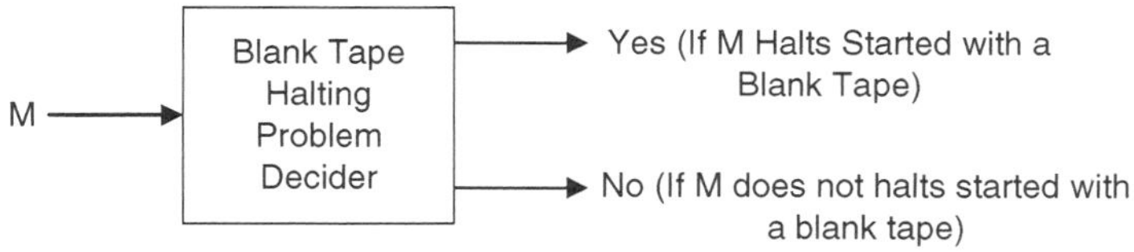

Let there be a decider for deciding whether an arbitrary string w halts on a TM M. The decider takes two inputs—the machine M and the string w. This is the halting problem.

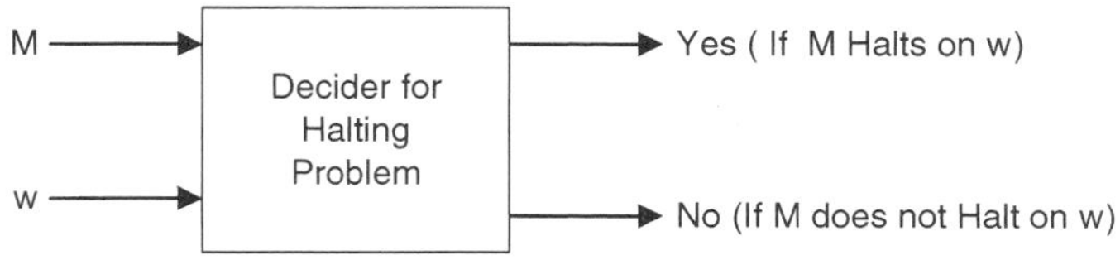

The halting problem can be reduced to the input of the blank tape halting problem decider using the following algorithm.

i) Take the input from the machine M and the string w.
ii) Construct from M a new machine M_w which starts with a blank tape, writes *w* on the tape, and then positions itself in a configuration (q_0, w)

The final decider is given in the following diagram.

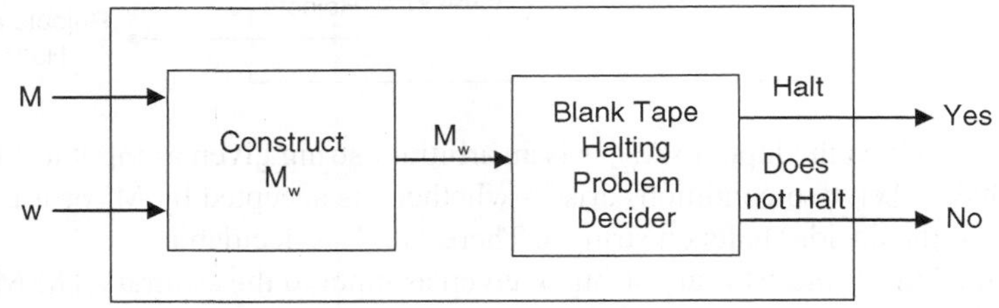

From the discussion, it is clear that M_w halts on a blank tape if and only if *M* halts on *w*. Since this can be done for the arbitrary M and w, an algorithm for the blank-tape halting problem can be converted into an algorithm to solve the halting problem. But already it is proved that halting problem is undecidable. Thus, it is proved that the blank tape halting problem is undecidable.

Example 10.15 Prove that the problem 'does a TM accept a regular language?' is undecidable.

Solution: Suppose that the problem is decidable. Therefore, it must have a decider to accept it. The decider takes the input <M, w> where M is the TM and w is the string. It halts and accepts if w is a regular language $\in \Sigma$. Otherwise it halts and rejects if w is not regular.

The following is the decider for it.

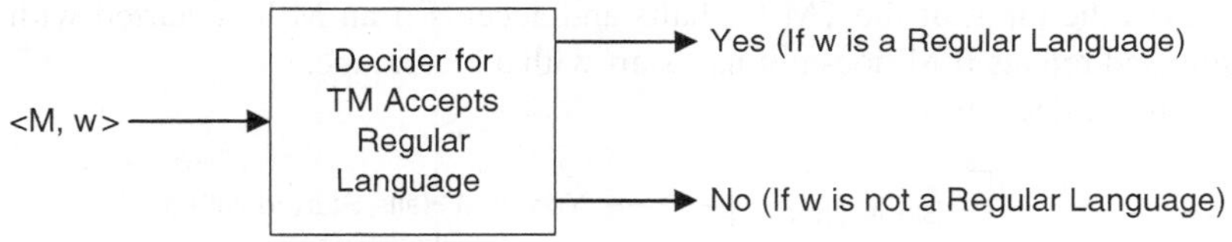

Let there be a decider for deciding whether an arbitrary string w halts on a TM M. The decider takes two inputs—the machine M and the string w. This is the halting problem.

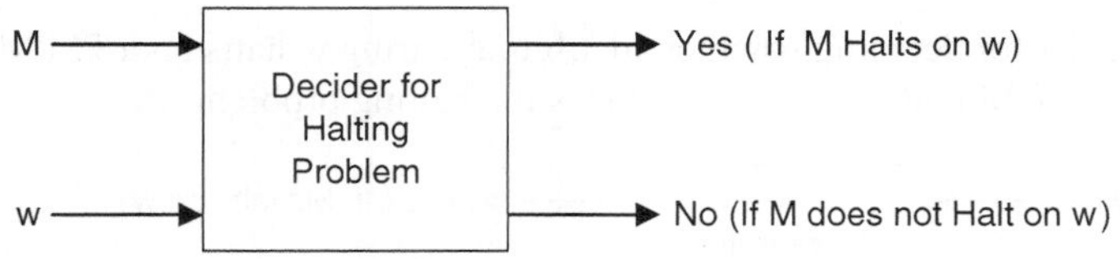

The halting problem can be reduced to the input of the regular language accepted by the TM problem using the following algorithm.

From M and w, design <M, w> such that:

i) if *M* does not accept *w*, then the decider for the regular language accepted by the TM problem accepts the non-regular language.

ii) if *M* accepts *w*, then it accepts the regular language Σ^*.

The final decider is given in the following diagram.

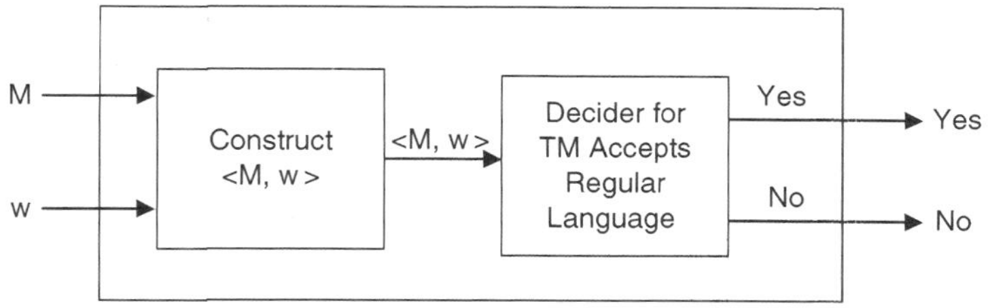

If this problem is decidable, then it can be used to solve the halting problem. But it is already proved that the halting problem is undecidable. Thus, it is proved that the regular language accepted by the TM problem is undecidable.

Example 10.16 Prove that the problem 'language generated by a TM is empty' is undecidable.

Solution: The problem can be denoted symbolically as

$$E_{TM} = \{<M, w> \mid M \text{ is a TM and } L(M) = \varnothing\}$$

We have to prove that E_{TM} is undecidable.

Assume that E_{TM} is decidable; hence, a decider exists. Let R be the TM that decides E_{TM}. Construct a new TM S using R that decides whether the string w halts on the machine M or not (The halting problem).

Upon feeding the input <M, w> to the machine S, it is called R with input M. If R accepts, then S rejects as L(M) is empty. But the problem occurs when R rejects, as we do not know whether M accepts w or not. In this case, the input given to R is needed to be modified according to the following algorithm.

From M and w, design M_w such that:
On any arbitrary string x as input

i) If $x \neq w$, rejects.

ii) If $x = w$, run M on w and accept if M accepts.

(Here, M_w either accepts w or accepts nothing.)
The final decider is given in the following diagram.

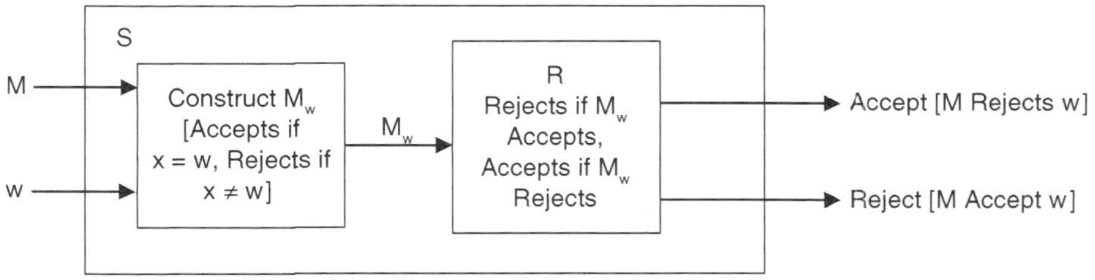

Already, it is proved that S (the halting problem) is undecidable. So, E_{TM} is also undecidable.

Example 10.17 Prove that the problem 'a given TM enters into an arbitrary state $q \in Q$ while traversing a string $w \in \Sigma^+$' is undecidable.

Solution: This problem is called entry state problem. The undecidability of this problem can be proved by reducing the halting problem to it.

Assume that there exists an algorithm A that solves the entry state problem. This can be used to solve the halting problem.

The algorithm takes M and w. Modify M to get M_1 in such a way that M_1 halts in state q if and only if M halts.

M halts on a state q_i for input 'i', because $\delta(q_i, i)$ is undefined.

i) Change every such undefined δ to $\delta(q_i, i) = (q_f, i, R)$ to get M_1 where q_f is a final state.
ii) Apply the entry state algorithm *A* to (M_1, q_i, w).

If A produces the answer yes (i.e., state q_i is entered while traversing w), then (M, w) halts.

If A answers no, then (M, w) does not halt.

From the previous discussion, we are getting an algorithm for designing the halting problem decider. But it is already proved that the halting problem is undecidable. Thus, the entry state problem is also undecidable.

Example 10.18 Prove that the problem whether an arbitrary RE language is empty is undecidable.

Solution: This problem can be proved undecidable by reducing the membership problem to it.

Let us consider that this problem is decidable. Therefore, there is a decider which halts and accepts if the grammar for the RE language produces an empty string; otherwise, it halts and rejects if the grammar generates a non-empty string w.

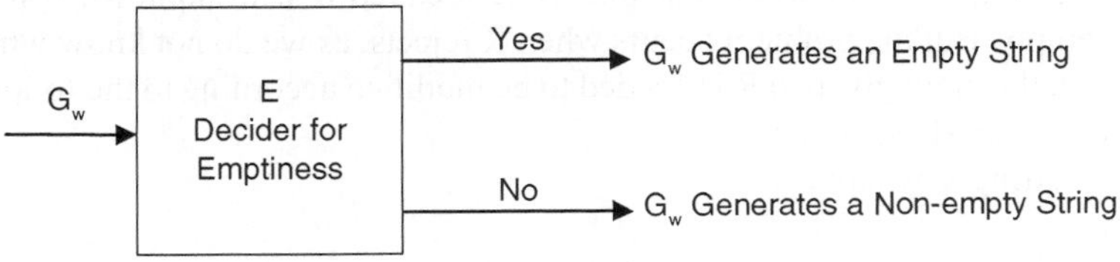

Suppose that the membership problem is decidable. So, the decider A_{TM} halts and accepts if w is accepted by M, otherwise it halts and rejects if w is not accepted by A_{TM}.

Modify the machine M such that for input M and w,

i) M saves its input on some specified location of its input tape.
ii) When M enters its final state upon traversing the input, it checks the stored input. It accepts if and only if it is same as stored in the input tape.

Construct a machine M_w for each w such that

$$L(M_w) = L(M) \cap \{w\}$$

Let T be the algorithm by which a grammar G_w can be constructed from it. If $w \in L(M)$, then the language produced by the grammar G_w, i.e., $L(G_w)$, is non-empty.

Put the output of T as the input to E. But, we know that the halting problem is undecidable, so the problem 'whether an arbitrary RE language is empty' is also undecidable.

Example 10.19 Prove that the problem 'whether a recursively enumerable language L(M) is finite' is undecidable.

Solution: This can be proved by reducing the halting problem to it.

Let M be the TM accepting L. Suppose that M_f is a decider for the finite language problem which takes the TM M as input.

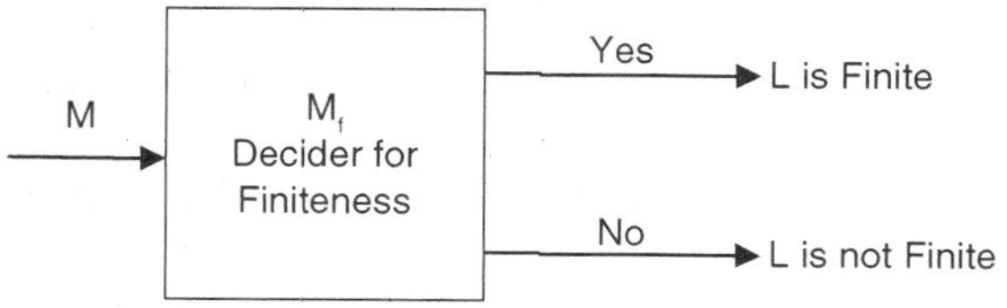

Suppose that there is a decider for the halting problem which takes a TM M and a string w as input.

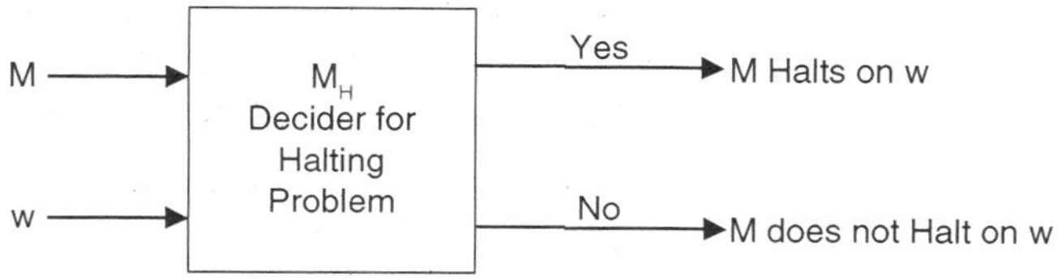

Now, reduce the halting problem to the finite language problem which constructs the machine M_w using the following process.

On input w to M,

i) Design the machine M_w such that when M enters into one of its halting states, M_w enters into its final state. M_w copies w on some special location of its input tape and then performs the same computation as M using the input w written on its input tape.
ii) If M enters into any one of its halting states, M_w halts reaching its final state.
iii) If M does not enter into its halting state, M_w does not halt and thus accepts nothing.

In other words, M_w accepts either the infinite language Σ^+ or the finite language Φ.

The final decider is given in the following diagram.

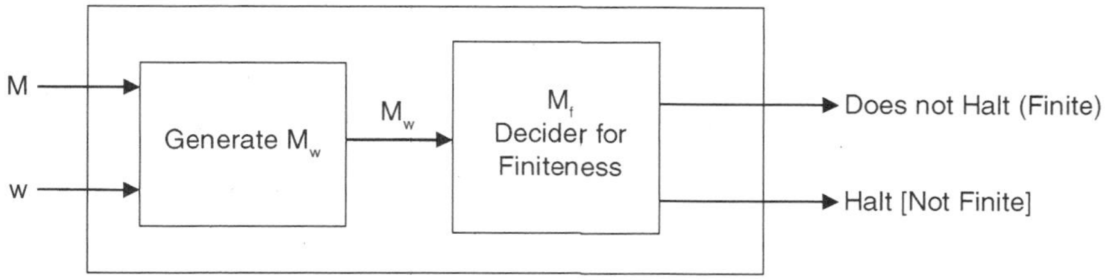

Already, it is proved that the halting problem is undecidable. So, M_f is also undecidable.

Example 10.20 Prove that the problem 'whether L(M) contains two strings of the same length' is undecidable.

Solution: This can be proved by reducing the halting problem to it. Reduce the halting problem to this problem by constructing a machine M_w using the following process.

On input w to M (the halting problem decider)

i) M_w copies w on some special location of its input tape and then performs the same computation as M using the input w written on its input tape.

ii) If w = a or w = b, M enters into one of its halting states and M_w reaches its final state. Thus, $L(M_w) = \{a, b\}$

iii) Otherwise, reject. Thus $L(M_w) = \{\Phi\}$.

The final decider is given in the following diagram.

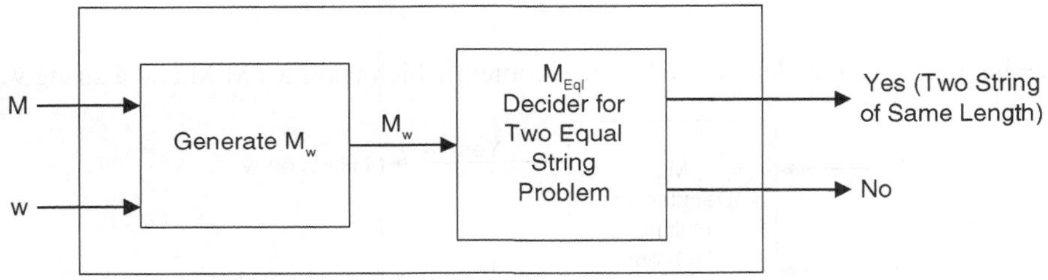

Already, it is proved that the halting problem is undecidable. So, M_{Eql} is also undecidable.

Example 10.21 Prove that the virus detection is undecidable.

Solution: A computer virus is an executable program that can replicate itself and spread from one computer to another. Every file or program that becomes infected can act as a virus itself, allowing it to spread to other files and computers. But the virus detection problem is undecidable! This can be proved using the undecidability of halting problem.

Let us define a virus by an algorithm called the 'is-virus?'

For a program P,

```
define (halts for input P)
        if (is-virus?)            //Execute P assuming P is not a virus.
        {
          // If true
          Virus-code is evaluated. If it is a virus, and P must halt.
        }
          // If false
          Virus-code never executed. Hence, P must not halt.
```

Already, we have proved that the halting problem is undecidable. Hence, the virus detection problem is undecidable.

(As the virus detection problem is undecidable, there exists no anti-virus program! Amazing? Yes, it is true. The various anti-virus software just keep a record of the known viruses. If a program P matches with one of them, then P is declared as a virus. For this reason, we need to update our anti-virus software regularly.)

10.7 Post's Correspondence Problem (PCP)

The Post correspondence problem (PCP) was proposed by an American mathematician Emil Leon Post in 1946. This is a well-known undecidable problem in the theory of computer science. PCP is very useful for showing the undecidability of many other problems by means of reducibility.

Before discussing the theory of PCP, let us play a game. Let there be N cards where each of them are divided into two parts. Each card contains a top string and a bottom string.

Example: Let N = 5 and the cards be the following.

0	011	0	011	1	
10	00	11	0	11	N = 5

Aim of the game: Is to arrange the cards in such a manner that the top and the bottom strings become same.

Solution for the example: Solution is possible and the sequence is 4 3 2 5 1

011	0	011	1	0
0	11	00	11	10
4	3	2	5	1

Let us take another kind of example and try to find the solution.

Example: There are four types of cards, and each are five in number, which means there is a total 20 cards. There are four players. At the beginning of the game, each player will get five random cards. The game starts with the single card shifting, starting from one of the players. Getting a card from the left player, the player shifts one card from his/her collection to the immediate right player. Player able to show 5 cards arranged in a particular sequence, which constructs same top and bottom strings; then the player will win.

The four different cards are

01	1	10	010
0	101	0	1

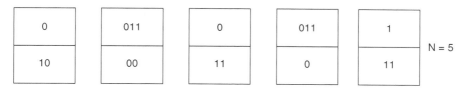

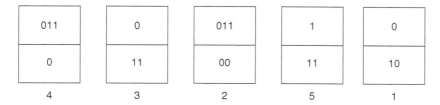

Solution: Yes, it is possible. The sequence is 2 1 4 3 2.

1	01	010	10	1
101	0	1	0	101
2	1	4	3	2

But, for all types of cards, the solution is not possible. The game will continue for an infinite time. There is a condition which must be fulfilled to make the game halt. The condition is

1. At least one card must contain the same starting character of the top and bottom strings. Though if the condition is fulfilled, it cannot be said confirm that the game will halt.

The Post Correspondence Problem: Given two sequences of strings $w_1, w_2, \ldots, w_n$ and $x_1, x_2, \ldots, x_n$ over Σ. The solution of the problem is to find a non-empty sequence of integers $i_1, i_2, \ldots, i_k$ such that $w_{i1}w_{i2} \ldots w_{ik} = x_{i1}x_{i2} \ldots x_{ik}$.

The solution of PCP hides in the condition $w_{i1}w_{i2} \ldots w_{ik} = x_{i1}x_{i2} \ldots x_{ik}$.

If there exists a solution for a given instance of PCP, there may exist multiple solutions for that instance of PCP.

As an example, consider the previous example.

$\Sigma = \{0, 1\}$ and the sequence of string w_i, x_i are as follows.

i	w_i	x_i
1	01	0
2	1	101
3	10	0
4	010	1

The problem has a solution $i_1 = 2, i_2 = 1, i_3 = 4, i_4 = 3, i_5 = 2$ and the string is

1 0 1 0 1 0 1 0 1

Example 10.22 Find whether there is a solution for the following PCP problem.

i	w_i	x_i
1	01	1
2	1	01
3	10	0
4	010	1

Solution: To get a solution, the top and bottom strings of at least one card must start with the same character. But here, the top and bottom strings of no cards start with same character. Thus, this PCP problem does not have any solution.

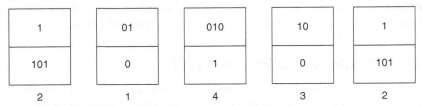

10.8 Modified Post Correspondence Problem

If the first substrings used in a PCP are w_1 and x_1 for all instances, then the PCP is called modified Post correspondence problem (MPCP). Therefore, the MPCP is

Given two sequences of strings $w_1, w_2, \ldots, w_n$ and $x_1, x_2, \ldots, x_n$ over Σ. The solution of the problem is to find a non-empty sequence of integers $i_2, i_3, \ldots, i_k$ such that $w_{i1}w_{i2}w_{i3} \ldots w_{ik} = x_{i1}x_{i2}x_{i3} \ldots x_{ik}$.

The derivation process of a string accepted by a given grammar can be reduced to the MPCP. We know that the MPCP needs two sequence of strings, say, denoted by two sets A and B. The derivation process can be reduced to MPCP with the help of the following table.

Top A	Bottom B	Grammar G
FS →	F	S: Start symbol F: Special symbol
a	a	For every symbol ∈ Σ
V	V	For every non-terminal symbol
E	→ wE	String w E: Special symbol
y	x	For every production x → y
→	→	

Example 10.23 Let the grammar be

$$S \rightarrow aBAb/Aac$$
$$Ab \rightarrow c$$
$$BC \rightarrow acb$$

The string is w = aacb. Convert the derivation to generate the string to MPCP.

Solution: The grammar with the string is mapped in the following table.

w	A	v	B
1	FS →	1	F
2	a	2	a
3	b	3	b
4	c	4	c
5	A	5	A
6	B	6	B
7	C	7	C
8	S	8	S
9	E	9	→ aacbE

w	A	v	B
10	aBAb	10	S
11	Aac	11	Aac
12	C	12	Ab
13	acb	13	BC
14	→	14	→

The string is derived as

$$S \rightarrow aBAb \rightarrow aBC \rightarrow aacb$$

The conversion of the derivation to MPCP is shown as follows.

i) S → aBAb

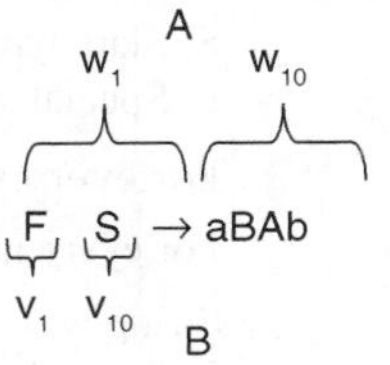

ii) S → aBAb → aBC

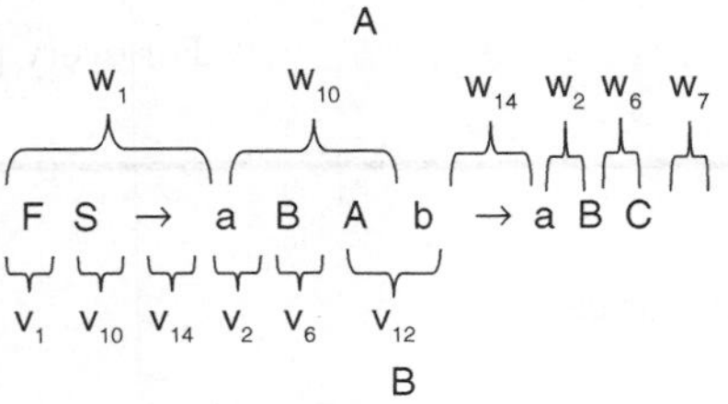

iii) S → aBAb → aBC → aacb

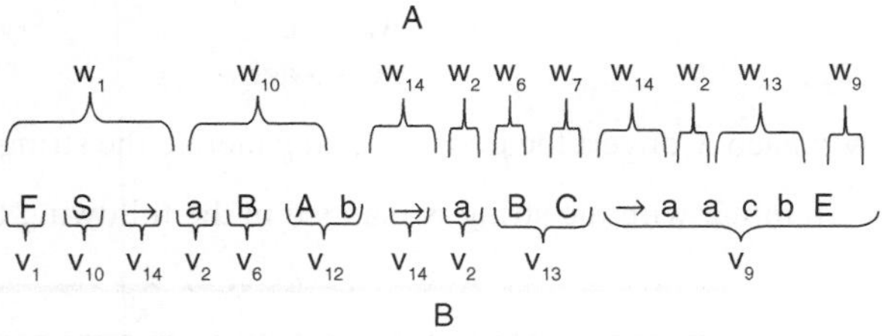

Theorem 10.3: The modified Post correspondence problem is undecidable.

Proof: This can be proved by reducing the membership problem to it.

Let MPC problem be decidable. Thus, there is a decider for the MPC problem which takes two sets A and B and halts and accepts if they satisfy the MPC condition or halts and rejects if they fail to satisfy the MPC condition. The following is the decider for the MPC problem.

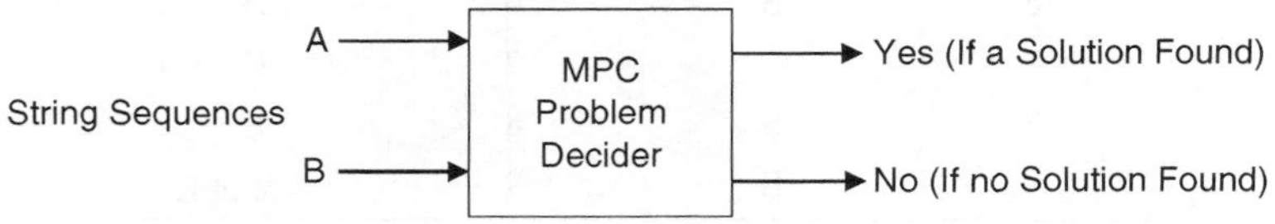

Let there be a decider for deciding whether an arbitrary string w is accepted by a grammar G. The decider takes two inputs, w and G, and halts and accepts if w is accepted or halts and rejects if w is not accepted by G. This is the membership problem. The following is the decider for this.

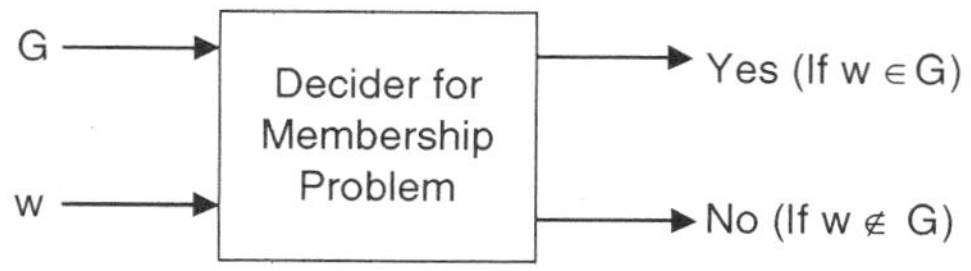

The membership problem with input (G, w) can be reduced to the input of the MPC problem decider (A, B) as described earlier.

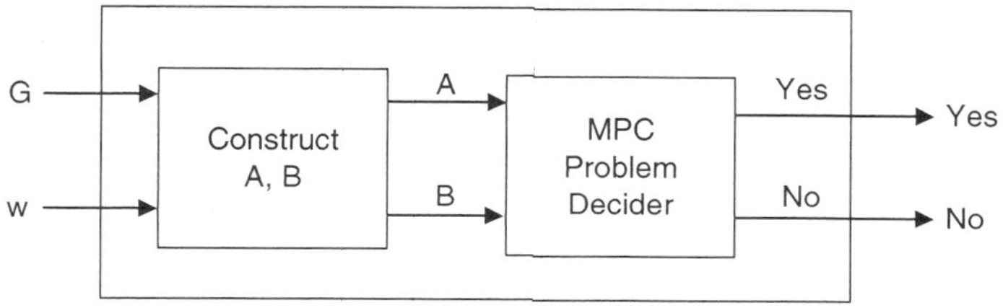

The two set of strings A and B has a solution if and only if w∈ G. Since the membership problem is undecidable, the MPC problem is also undecidable.

The modified PCP can be reduced to a PCP. This reduction can be used to prove that the PCP is undecidable.

10.8.1 Reduction of MPCP to PCP

Given two sequences of strings $w_1, w_2, \ldots, w_n$ and $x_1, x_2, \ldots, x_n$ over Σ. A sequence of integers $i_2, i_3, \ldots, i_k$ exist such that $w_1 w_{i2} w_{i3} \ldots w_{ik} = x_1 x_{i2} x_{i3} \ldots x_{ik}$.

1. Introduce a new symbol *.
2. In the first list w_i (i = 1, 2, ... n), the new symbol * appears after every symbol. In the second list x_i (i = 1, 2, ... n), the new symbol * appears before every symbol.

   (***Example:*** If the w and x part of a MPCP is 01 and 101, respectively, then in PCP this becomes 1*0* and *1*0*1, respectively.)
3. Take the first pair w_1, x_1 from the given MPCP, and add to the PCP instance another pair in which the *s remain as modified in (ii) but an extra * is added to w_1 at the beginning.

   (***Example:*** If in the previous example w and x are w_1 and x_1 part of a MPCP, then in PCP this becomes *1*0* and *1*0*1, respectively.)
4. Since the first list has an extra * at the end, add another pair $ and *$ to the PCP instance. This is referred to as the final pair.

Example 10.24 Convert the following MPCP to an equivalent PCP.

i	w_i	x_i
1	01	011
2	10	0
3	01	11
4	1	0

Solution: This is in MPCP as there is a sequence 1, 4, 3, 2 which generates the string 0110110 in both halves.

Adding * after every symbol in w, before every symbol in x, and adding an extra * at the beginning of w_1, the modified sequence becomes

i	w_i	x_i
1	*0*1*	*0*1*1
2	1*0*	*0
3	0*1*	*1*1
4	1*	*0

Adding extra sequence $ and *$ to the existing sequence as the final pair, the PCP becomes

i	w_i	x_i
1	*0*1*	*0*1*1
2	1*0*	*0
3	0*1*	*1*1
4	1*	*0
5	$	*$

This is a PCP as there exists a sequence 1, 4, 3, 2, 5 which generates the string *0*1*1*0*1*1*0*$.

Theorem 10.4: The post correspondence problem is undecidable.

Proof: Let PCP be decidable. Thus, there is a decider for PCP which takes two sets w and x and halts and accepts if they satisfy the PCP condition or halts and rejects if they fail to satisfy the PCP condition. The following is the decider for PCP.

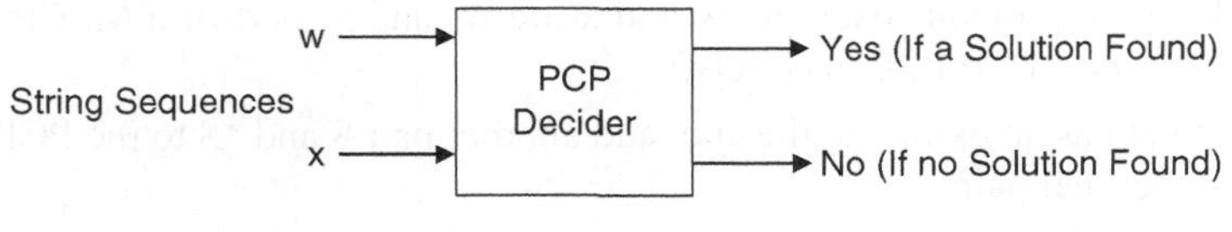

Let there be a decider for MPCP. The decider takes two inputs A and B and halts and accepts if a solution is found or halts and rejects if a solution is not found. The following is the decider for this.

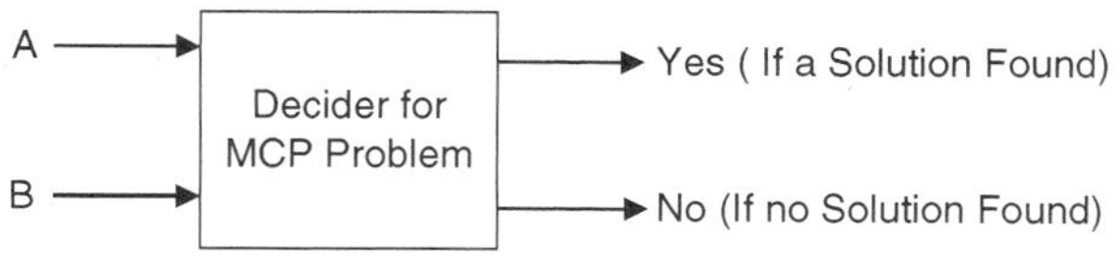

The MPC problem with input (A, B) can be reduced to the input of a PCP decider (w, x) as described earlier.

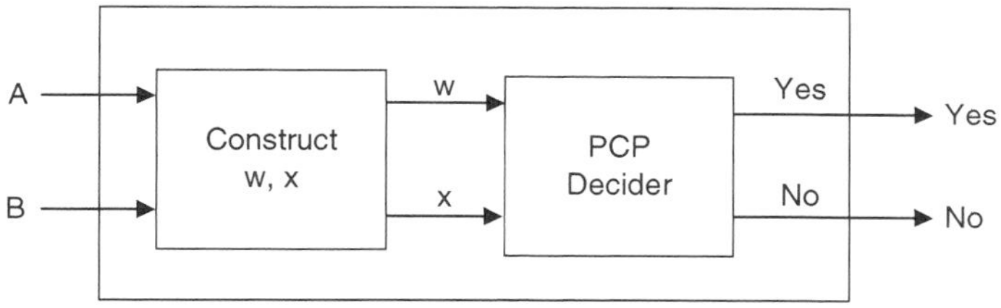

The two set of strings w and x have a solution if and only if the two set of strings A and B have a solution. Since the MPC problem is undecidable, PCP is also undecidable.

Example 10.25 Prove that the problem 'whether an arbitrary CFG G is ambiguous' is undecidable.

Solution: To prove this, we reduce the PCP problem to the A_{Amb} problem, where A_{Amb} is considered as a decider for the ambiguity of the CFG problem. The PCP problem decider takes two lists A and B as input and generates a CFG G. If the PCP decider has a solution for the two lists A and B, then the CFG G is ambiguous.

This reduction is done by the following process.

Let $A = \{w_1, w_2, \ldots\ldots w_n\}$ and $B = \{x_1, x_2, \ldots\ldots x_n\}$.

Let $a_1, a_2 \ldots\ldots a_n$ be a list of new symbols generated by the languages.

$$L_A = \{w_{i1}\ w_{i2} \ldots\ldots w_{im}\ a_{im}\ a_{im-1} \ldots\ldots a_{i2}\ a_{i1},\ \text{where } 1 \le i \le n\ ;\ m \ge 1\}$$

and

$$L_B = \{x_{i1}\ x_{i2} \ldots\ldots x_{im}\ a_{im}\ a_{im-1} \ldots\ldots a_{i2}\ a_{i1},\quad \text{where } 1 \le i \le n\ ;\ m \ge 1\}$$

Both L_A and L_B can be constructed from the following grammar.

$$S_A \to w_i S_A a_i / w_i a_i$$
$$S_B \to x_i S_B a_i / x_i a_i$$

Both of them are CFG.

We can easily construct a new CFG by the union operation of the two CFG since CFG are closed under the union operation.

$$S \rightarrow S_A/S_B$$
$$S_A \rightarrow w_iS_Aa_i/w_ia_i$$
$$S_B \rightarrow x_iS_Ba_i/x_ia_i$$

If the given instance of the PCP has a solution for $i_1, i_2, \ldots.. i_m$, then the string generated in the top and bottom halves are $w_{i1}\ w_{i2} \ldots.. w_{im} = x_{i1}\ x_{i2} \ldots.. x_{im}$. Thus, $L_A = L_B$. For the grammar $L_A \cup L_B$, there exist two derivations of a same string, which means two parse trees. This makes $L_A \cup L_B$ ambiguous.

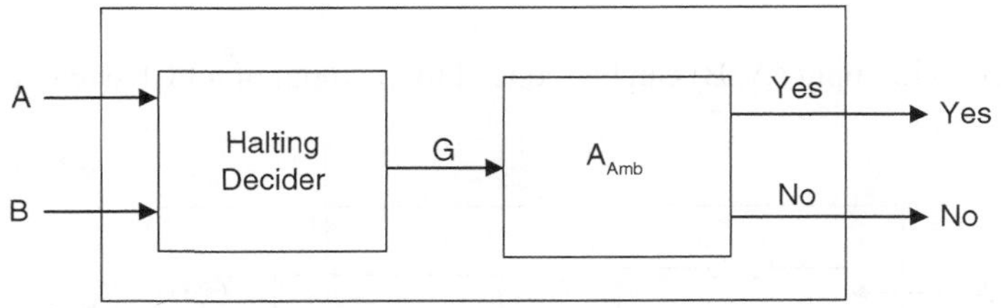

But it is already proved that the halting problem is undecidable. So, the problem A_{Amb} is also undecidable.

What We Have Learned So Far

1. A Turing-acceptable language is called Turing-recognizable or recursively enumerable.
2. A language is recursively enumerable if there is a Turing machine that halts and accepts the language or halts and rejects or loops for an infinite time.
3. A language is called Turing-decidable if there exists a Turing machine which halts and accepts the language. Otherwise, it is undecidable.
4. A recursive language is a subset of a recursively enumerable language.
5. A reduction is a process of converting one problem into another solved problem in such a way that the solution of the second problem can be used to solve the first problem.
6. The complement of a decidable language is also decidable.
7. The problem 'an arbitrary string w is accepted by an arbitrary Turing machine M' is known as the membership problem, which is an undecidable problem.
8. The problem 'a string w halts on a Turing machine M' is known as the halting problem, which is an undecidable problem.
9. Given two sequences of strings $w_1, w_2, \ldots, w_n$ and $x_1, x_2, \ldots, x_n$ over Σ. The solution of the problem is to find a non-empty sequence of integers $i_1, i_2, \ldots, i_k$ such that $w_{i1}w_{i2} \ldots w_{ik} = x_{i1}x_{i2} \ldots x_{ik}$ is known as the Post correspondence problem (PCP).
10. Given two sequences of strings $w_1, w_2, \ldots, w_n$ and $x_1, x_2, \ldots, x_n$ over Σ. The solution of the problem is to find a non-empty sequence of integers $i_2, i_3, \ldots, i_k$ such that $w_1w_{i2}w_{i3} \ldots w_{ik} = x_1x_{i2}x_{i3} \ldots x_{ik}$ is known as the modified Post correspondence problem (MPCP).
11. The PCP and MPCP are both undecidable problems.

Solved Problems

1. Find at least three solutions to the PCP defined by the dominoes.

	1	2	3
X:	1	10	10111
Y:	111	0	10

[UPTU 2002]

Solution: The solution is possible as X1, Y1 and X3, and Y3 have the same initial substring. The solution is 3112.

	3	1	1	2
X	10111	1	1	10
Y	10	111	111	0

Repeating the sequence 3112, we can get multiple solutions.

2. Does the PCP with two lists X = {b, bab^3, ba} and Y = {b^3, ba, a} have a solution.

[UPTU 2003]

Solution: The solution is possible as X1, Y1 and X2, and Y2 have the same initial substring. The solution is 2113.

	2	1	1	3
X	babbb	b	b	ba
Y	ba	bbb	bbb	a

3. Does the PCP with two list X = {(a, ba), (b^{2}a, a^3), (a^{2}b, ba)} has a solution?

Solution: The solution is not possible as none of the top and bottom parts of the list have the same starting character.

4. Show that the following PCP has a solution and give the solution.

	A	B
1	11	111
2	100	001
3	111	11

[JNTU 2008]

Solution: The solution is possible as the A and B parts of the first and third lists have the same starting characters. The solution is 123.

	1	2	3
X	11	100	111
Y	111	001	11

5. The lists A and B given in the following are an instance of PCP.

	A	B
1	0	01
2	0101	1
3	100	0010

Find the solution for the given PCP.

Solution: The solution is 1332.

	1	3	3	2
A:	0	100	100	0101
B:	01	0010	0010	1

6. Prove that if L_1 is regular and L_2 is context free, then $L_1 \cup L_2^c$ is recursive.

Solution: L_1 is regular, so it is recursive. The complement of the context-free language is not context free. But, the context-free language is recursive and the complement of the recursive language is recursive. It is proved that the union of two recursive languages is recursive. Hence, $L_1 \cup L^c_2$ is recursive.

Multiple Choice Questions

1. The Turing machine accepts
   a) Regular language
   b) Context-free language
   c) Context-sensitive language
   d) All of these
2. A language is recursively enumerable if a Turing machine
   a) Halts and accepts
   b) Halts and rejects
   c) loops forever
   d) performs (a), (b), or (c)
3. A language L is called decidable (or just decidable), if there exists a Turing machine M which
   a) Accepts L
   b) Rejects L
   c) Loops forever on L
   d) performs (a), (b), or (c)
4. Find the true statement
   a) A recursively enumerable language is a subset of a recursive language
   b) A recursive language is a subset of a recursively enumerable language
   c) Both are equivalent
   d) Both may loop forever on the input to a Turing machine
5. Which is false for recursive language?
   a) Union of two recursive languages is recursive

b) Complement of a recursive language is recursive
c) Recursive language may loop forever on Turing machine
d) String belongs to a recursive language either accepts or rejects on Turing machine.

6. Find the decidable problem regarding the DFA.
   a) The problem that a set of null strings is accepted by a DFA M
   b) The problem that a string w is accepted by a DFA M
   c) The problem that two DFA, M_1 and M_2, satisfy the same language
   d) All of these

7. Which is true for reducibility?
   a) Converting one problem to another problem.
   b) Converting one solved problem to another unsolved problem.
   c) Converting one unsolved problem into another solved problem to solve the first problem.
   d) Converting one unsolved problem to another unsolved problem.

Answers:

1. d 2. d 3. a 4. b 5. c 6. d 7. c

GATE Questions

1. Which of the following statements is false?
   a) The halting problem of a Turing machine is undecidable.
   b) Determining whether a context-free grammar is ambiguous is undecidable.
   c) Given two arbitrary context-free grammars G_1 and G_2, it is undecidable whether $L(G_1) = L(G_2)$.
   d) Given two regular grammars G_1 and G_2, it is undecidable whether $L(G_1) = L(G_2)$.

2. Which of the following is not decidable?
   a) Given a Turing machine M, a string s, and an integer k, M accepts s within k steps
   b) The equivalence of two given Turing machines
   c) The language accepted by a given finite state machine is not empty.
   d) The language generated by a context-free grammar is non-empty.

3. Consider the following decision problems:

   P_1: Does a given finite state machine accept a given string

   P_2: Does a given context-free grammar generate an infinite number of strings.

   Which of the following statements is true?
   a) Both P_1 and P_2 are decidable
   b) Neither P_1 nor P_2 are decidable
   c) Only P_1 is decidable
   d) Only P_2 is decidable

4. Consider the following problem X.

   Given a Turing machine M over the input alphabet Σ, any state q of M and a word $w \in \Sigma^*$, does the computation of M on w visit the state q?

a) X is decidable
b) X is undecidable but partially decidable
c) X is undecidable and not even partially decidable
d) X is not a decision problem

5. Which of the following is true?
   a) The complement of a recursive language is recursive
   b) The complement of a recursively enumerable language is recursively enumerable
   c) The complement of a recursive language is either recursive or recursively enumerable
   d) The complement of a context-free language is context free

6. L_1 is a recursively enumerable language over Σ. An algorithm A effectively enumerates its words as $w_1, w_2, w_3, \ldots$ Define another language L_2 over $\Sigma \cup \{\#\}$ as $\{w_i \# w_j : w_j \in L_1, i < j\}$. Here, # is a new symbol.

   Consider the following assertions.

   S_1: L_1 is recursive implies that L_2 is recursive

   S_2: L_2 is recursive implies that L_1 is recursive

   Which of the following statements is true?
   a) Both S_1 and S_2 are true
   b) S_1 is true but S_2 is not necessarily true
   c) S_2 is true but S_1 is not necessarily true
   d) Neither is necessarily true

7. Consider three decision problems P_1,P_2, and P_3. It is known that P_1 is decidable and P_2 is undecidable. Which one of the following is true?
   a) P_3 is decidable if P_1 is reducible to P_3
   b) P_3 is undecidable if P_3 is reducible to P_2
   c) P_3 is undecidable if P_2 is reducible to P_3
   d) P_3 is decidable if P_3 is reducible to P_2's complement

8. Let L_1 be a recursive language, and let L_2 be a recursively enumerable but not a recursive language. Which one of the following is true?
   a) $\overline{L_1}$ is recursive and $\overline{L_2}$ is recursively enumerable.
   b) $\overline{L_1}$ is recursive and $\overline{L_2}$ is not recursively enumerable.
   c) $\overline{L_1}$ and $\overline{L_2}$ are recursively enumerable.
   d) $\overline{L_1}$ is recursively enumerable and $\overline{L_2}$ is recursive

9. For $S \in (0 + 1)^*$, let d(s) denote the decimal value of s [e.g., d(101) = 5]. Let $L = \{s \in (0 + 1)^* \mid d(s) \bmod 5 = 2 \text{ and } d(s) \bmod 7 \neq 4\}$

   Which one of the following statements is true?
   a) L is recursively enumerable, but not recursive
   b) L is recursive, but not context free
   c) L is context free, but not regular
   d) L is regular

10. Let L_1 be a regular language, L_2 be a deterministic context-free language, and L_3 a recursively enumerable, but not recursive language. Which one of the following statements is false?
    a) $L_1 \cap L_2$ is a deterministic CFL
    b) $L_3 \cap L_1$ is recursive
    c) $L_1 \cup L_2$ is context free
    d) $L_1 \cap L_2 \cap L_3$ is recursively enumerable

11. Which of the following problems is undecidable?
    a) Membership problem for CFGs.
    b) Ambiguity problem for CFGs.
    c) Finiteness problem for FSAs
    d) Equivalence problem for FSAs.

12. The language $L = \{0^i21^i \mid i \geq 0\}$ over alphabet $\{0, 1, 2\}$ is
    a) not recursive
    b) recursive and is a deterministic CFL.
    c) a regular language.
    d) not a deterministic CFL but a CFL.

13. Which of the following are decidable?
    i) Whether the intersection of two regular languages is infinite
    ii) Whether a given context-free language is regular
    iii) Whether two pushdown automata accept the same language
    iv) Whether a given grammar is context free

    a) (i) and (ii) b) (i) and (iv) c) (ii) and (iii) d) (ii) and (iv)

14. If L and $\overline{L}$ are recursively enumerable, then L is
    a) Regular b) Context free c) Context sensitive d) Recursive

15. Which of the following statements is false?
    a) Every NFA can be converted to an equivalent DFA
    b) Every non-deterministic Turing machine can be converted to an equivalent deterministic Turing machine.
    c) Every regular language is also a context-free language
    d) Every subset of a recursively enumerable set is recursive.

16. Let L1 be a recursive language. Let L_2 and L_3 be the languages that are recursively enumerable but not recursive. Which of the following statements is not necessarily true?
    a) $L_2 - L_1$ is recursively enumerable
    b) $L_1 - L_3$ is recursively enumerable
    c) $L_2 \cap L_1$ is recursively enumerable
    d) $L_2 \cup L_1$ is recursively enumerable

17. Let $L = L_1 \cap L_2$, where L_1 and L_2 are languages as defined in the following:

    $$L_1 = \{a^m b^m c a^n b^n,\ m, n \geq 0),\ L_2 = \{a^i b^j c^k,\ i, j, k \geq 0)$$

    Then, L is
    a) not recursive
    b) regular
    c) context-free but not regular
    d) recursively enumerable but not context free

18. Which of the following problems are decidable?
    i) Does a given program ever produce an output?
    ii) If L is a context-free language, then, is $\overline{L}$ also context free?
    iii) If L is a regular language, then is $\overline{L}$ also regular?
    iv) If L is a recursive language, then is $\overline{L}$ also recursive?

    a) (i), (ii), (iii), (iv) b) (i), (ii) c) (ii), (iii), (iv) d) (iii), (iv)

Answers:

1. d	2. b	3. a	4. b	5. a	6. b	7. b	8. b	9. b
10. b	11. b	12. b	13. b	14. d	15. d	16. b	17. a	18. d

Hints:

1. From a regular grammar, FA can be designed. It can be tested whether two FA are equivalent or not.

3. A given CFG is infinite if there is at least one cycle in the directed graph generated from the production rules of the given CFG in CNF.

4. The problem is undecidable. But if the state is the beginning state, it must be traversed. Thus, it is partially decidable.

8. The complement of a recursively enumerable language is not recursively enumerable.

9. L = 5n + 2 but L ≠ 7n + 4. Hence, we can design a machine which halts and accepts if L = 5n + 2 and halts and rejects if L ≠ 7n + 4. So, it is decidable.

10. L_1 and L_2 are recursive. The intersection of two recursive languages is recursive. But the intersection of the recursive and the recursively enumerable languages are recursively enumerable and not recursive.

12. A DPDA can be designed which PUSH X for each appearance of '0'. No stack operation for traversing 2 and POP X for '1'. A Turing machine can be designed for it where for each '0' it traverses the rightmost '1' by replacing them by X and Y, respectively. If after X, 2 appears and after 2, Y appears, then it halts. Thus, it is recursive.

13. The intersection of two regular language is regular, i.e., CFL. Using Q3, we can find infiniteness.

16. The size of the L_1 set is less than the size of the L_3 set.

17. Two languages are CFL. The intersection of the two CFL is not a CFL, so not regular. (b and c are false). The answer will be 'a' or 'd'.

Exercise

1. Prove that any decision cannot be taken for $A \cup B^C$, if A is recursive and B is recursively enumerable.

2. Prove that $A \cup B^C$ is recursively enumerable if A is recursive and B is recursively enumerable.

3. Prove that the problem that a regular language L accepted by a pushdown automata P is decidable.
4. Prove that the problem that a regular string w is generated by a CFG G is decidable.
5. Prove that the problem that the same string w accepted by two DFA, M_1 and M_2, is decidable.
6. Prove that the problem whether a DFA and a regular expression is equivalent is decidable.
7. Prove that the problem whether two regular expressions R_1 and R_2 are equivalent is decidable.
8. Prove that the problem that an arbitrary Turing machine M ever prints a string 101 on its tape is undecidable.
9. Prove that the problem whether a string accepted by an arbitrary Turing machine M is regular is undecidable.
10. Prove that the problem whether an arbitrary Turing machine accepts a single string is undecidable.
11. Find a solution for the following Post correspondence problem.

$$\text{List A} = \{01, 10, 01, 1\}$$
$$\text{List B} = \{011, 0, 11, 0\}$$

Recursive Function 11

Introduction

In the 'Turing machine as integer function' section of the chapter 'Extension of the Turing Machine', different integer functions such as addition, subtraction, multiplication, remainder finding, square, etc. are constructed using the Turing machine. These are the basic functions. By combining these basic functions, complex functions are constructed. As the basic functions are computable, the complex functions are also computable.

Like the theory of the Turing machines, the recursive function theory is a way to make formal and precise the intuitive, informal, and imprecise notion of an effective method. Like the theory of the Turing machines, the recursive function theory also obeys the converse of the Church's thesis. The base of the recursive function is some very elementary functions that are naturally effective. Then, it provides a few methods for constructing more complex functions from those simpler functions. The initial functions and the steps for building complex functions are very few in number, and very simple in character.

11.1 Function

A function is a relation that associates each element of a set with an element of another set. Let a function f be defined between two sets A and B. There must be a relation between A and B such that for an element $a \in A$, there must be another element $b \in B$, such that (a, b) is in the relation.

The relation given by f between a and b is represented by the ordered pair <a, b> and denoted as f(a) = b, and b is called the image of a under f.

11.1.1 Different Types of Functions

11.1.1.1 One to One or Injective

A function from set A to set B is said to be one-to-one if no two elements of set A have exactly the same elements mapped in set B. In other words, it can be said that f(x) = f(y) if and only if x = y.

Example: f(x) = x + 4. Let x = 1, 2, 3, ... (set of all positive integer numbers). This is an injective function because f(1) = 5, f(2) = 6, f(3) = 7 ......, for no two elements of set A there is exactly the same value in set B.

Diagrammatically, it can be shown as follows.

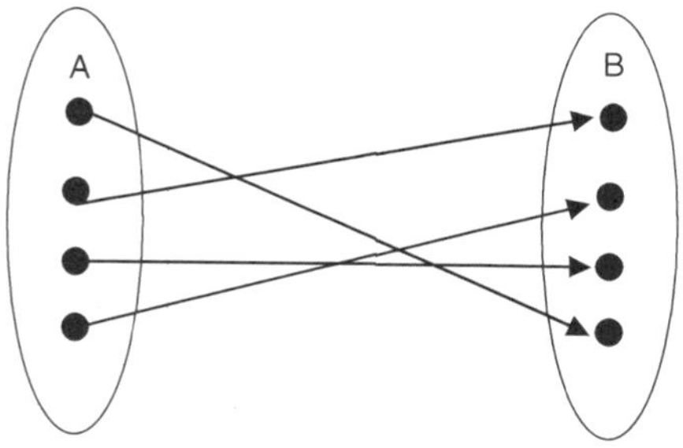

But if there is a function $f(x) = x^2$, where x is the set of real numbers then f(x) is not injective because for x = 2 and –2; in both cases, f(x) results in 4 which violates the condition of the injective function.

11.1.1.2 Onto or Surjective

A function f from set A to a set B is said to be surjective (or onto), if for every $y \in Y$, there is an $x \in X$ resulting in f(x) = y. In other words, each element of the set B can be obtained by applying the function f to some element of A.

Example: f(x) = 2x from $A \rightarrow B$, where A is the set of natural numbers and B is the set of non-negative even numbers. Here, for every element in B, there is an element in A.

Diagrammatically, it can be shown as follows.

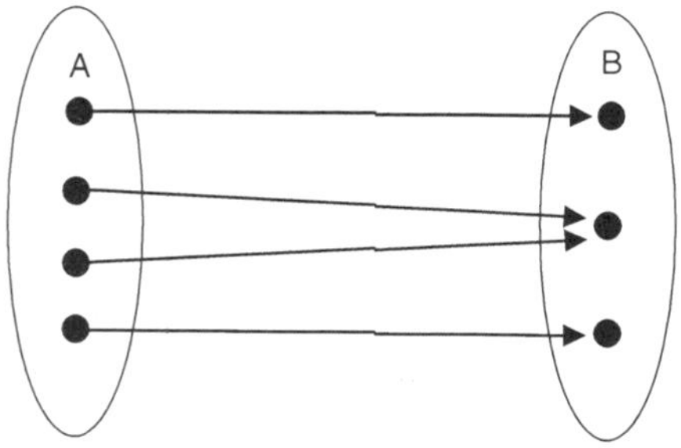

11.1.1.3 Bijective

A function f is said to be bijective if it is both injective and surjective.

Example: The example f(x) = 2x of surjective is also bijective.

Diagrammatically, it can be shown as follows.

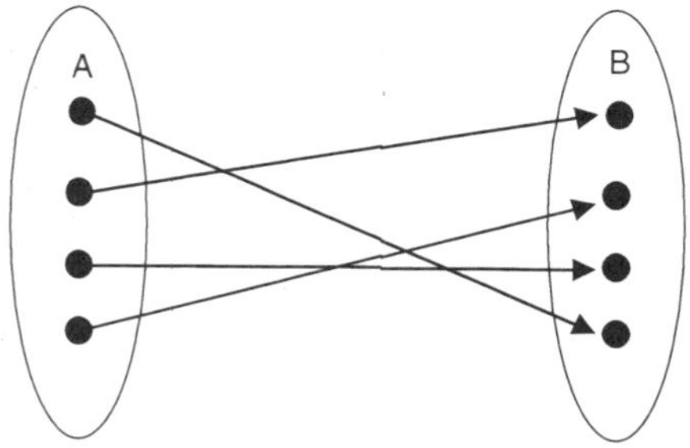

11.1.1.4 Inverse Function

Let us define a function f to be a bijection from a set A to set B. Suppose another function g is defined from B to A such that for every element y of B, g(y) = x, where f(x) = y. Then, the function g is called the inverse function of f, and it is denoted by f^{-1}.

Example: If f(x) = 5x from the set of natural numbers A to the set of non-negative even numbers *B*, then $g(x) = f^{-1}(x) = 1/5\ x$.

11.1.1.5 Composite Function

Let f(x) be a function from a set A to set B, and let g be another function from set B to a set C. Then, the composition of the two functions f and g, denoted by fg, is the function from set A to set C that satisfies fg(x) = f(g(x)) for all x in A.

Example: $f(x) = x^2$ and g(x) = (x + 2). Then, $fg(x) = f(g(x)) = (x + 2)^2$

11.2 Initial Functions

For a set of natural numbers N, three types of functions can be defined.

1. **Zero function**: A zero function returns 0 for all values. It is denoted as

   $$Z(x) = 0, \text{ for all } x \in N, \text{ where N is the set of natural numbers}$$

   The zero function is also called constant function.

2. **Successor function**: A successor function takes one number as argument and returns the succeeding number. It can be denoted as

   $$S(x) = x + 1, \text{ for all } x \in N, \text{ where N is the set of natural numbers}$$

3. **Projection function:** A projection function is an n-ary projection function denoted by P_i^n, where $n \geq 1$ and $1 \leq i \leq n$, takes n arguments, and returns the ith element.
   It can be denoted as $P_i^n(x_1, x_2, x_3, \ldots.. x_n) = x_i$.

11.3 Recursive Function

We are familiar with the term 'recursion' at the time of doing computer programming using some middle level or high level programming languages such as C, C++, Java, etc. A function which calls itself and terminates after a finite number of steps is called a recursive function.

For every recursive, there must be a base condition using which the function terminates.

Example: fact(x) = x × fact(x – 1), where x is a positive integer number and fact(0) = 1.

If x = 4, then

$$\begin{aligned} \text{fact}(4) &= 4 \times \text{fact}(3) \\ &= 4 \times (3 \times \text{fact}(2)) \\ &= 4 \times (3 \times (2 \times \text{fact}(1))) \\ &= 4 \times (3 \times (2 \times (1 \times \text{fact}(0)))) \\ &= 4 \times 3 \times 2 \times 1 \times 1 \\ &= 24 \end{aligned}$$

In this context, we need to know the total recursive, partial recursive, and primitive recursive function.

11.3.1.1 Partial and Total Function

A partial function f from a set A to set B is defined as a function from A′ to B where A′ is a subset of A. For all $x \in A$, there may exist $f(x) = y \in B$ or f(x) is undefined.

Example: Let A and B be two sets of positive integer numbers. A function f(x) is defined as $\sqrt{x}$. The relation $A \rightarrow B$ only exists if $x \in A$ is a perfect square such as 4, 9, 16, 25, … etc. f(16) is defined but f(20) is undefined.

If A′ = A, then the function f(x) is called a total function.

Example: $f(x) = x + 1$, where $x \in$ the set of integer numbers is a total function as f(x) is defined for all values of x.

11.3.1.2 Partial Recursive Function

A function computed by the Turing machine that need not halt for all input is called a partial recursive function. A partial recursive function is allowed to have an infinite loop for some input values.

11.3.1.3 Total Recursive Function

Partial recursive functions for which the Turing machine always halts are called the total recursive function. The total recursive function always returns a value for all possible input values.

From the discussion, it is clear that the total recursive function is a subset of the partial recursive function.

Example: Prove that the addition of two positive integers is a total recursive function.

Solution: $f(x, y) = x + y$, where $x, y \in$ set of positive integer numbers.

$f(x, 0) = x + 0 = x$ is the base condition.

$$f(x, y + 1) = x + y + 1 = f(x, y) + 1$$

Here, recursion occurs.

Thus, the function of the addition of two positive integers is recursive. The function is defined (returns a value) for all value of x, y, which proves it a total recursive.

11.3.1.4 Primitive Recursive Function

Primitive recursive functions are a subset of the total recursive function which can be obtained by a finite number of operations of composition and recursion from the initial functions (zero and successor function).

Primitive recursion is defined for $f(x_1, x_2, \ldots, x_n)$ as

$$f(\,) = g(x_1, x_2, \ldots, x_{n-1}) \quad \text{if } x_n = 0$$
$$= h(x_1, x_2, \ldots, x_n, f(x_1, x_6, \ldots, x_{n-1})) \text{ if } x_n > 0$$

where g and h are primitive recursive functions.

Example: Every primitive recursive function is Turing computable.

Solution: The primitive recursive function consists of the initial function (zero function, successor function, and projection function), composition, and recursion. If there exists a Turing machine for all these, then there exists a Turing machine for the primitive recursive function.

1. **Turing machine for zero function**

$$Z(x) = 0, \text{ for all } x \in N, \text{ where N is the set of natural numbers}$$

It is nothing but an eraser. There exists a Turing machine for an eraser.

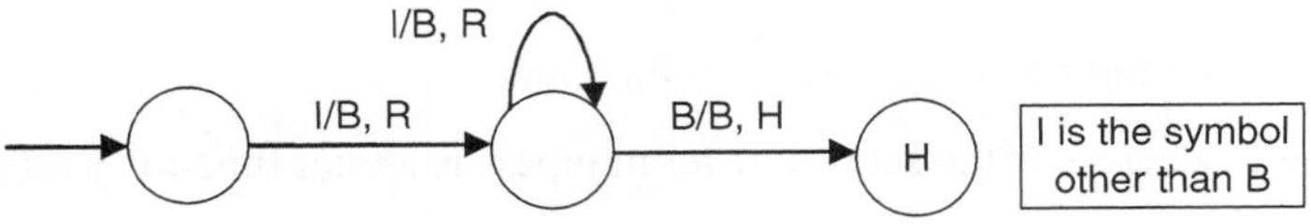

2. **Turing machine for successor function**

$$S(x) = x + 1, \text{ for all } x \in N \text{ where N is the set of natural numbers}$$

The function adds a '1' with the value of x. If one blank symbol at the right hand side is replaced by '1', then the machine can be easily designed.

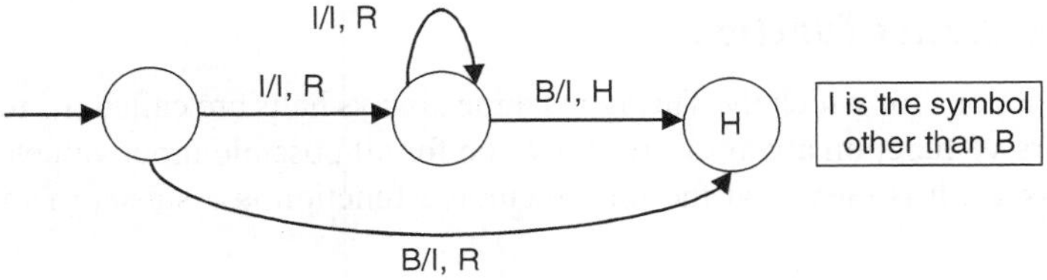

3. A projection function is denoted as $P_i^n (x_1, x_2, x_3, \ldots.. x_n) = x_i$.
   A Turing machine can be designed which takes input $Bx_1Bx_2Bx_3B \ldots\ldots Bx_nB$ and produces the output Bx_iB.
4. A Turing Machine can be designed for *composition* of different functions by combining the respective Turing machines for each of the function in the order in which the functions appear.
5. Turing machine for *recursive functions* can be designed by combining the Turing Machine designed for the simple function with multiple call and the Turing Machine designed for the Base function as termination point.
6. A final TM can be designed using the TMs for steps (4) and (5).

Hence, it is proved that every primitive recursive function is Turing computable.

Example 11.1 Prove that the function f(x, y) = x + y where x, y are positive integers is primitive recursive.

Solution:

$$f(x, y) = x + y$$
$$\Rightarrow f(x,0) = x \ = Z(x)$$
$$f(x, y + 1) = x + y + 1 = f(x, y) + 1 = S(x, f(x, y))$$

The function can be obtained by composition and recursion from the zero and successor functions in a finite number of steps. Thus, it is primitive recursive.

Let f(4, 3) be given.

$$\begin{aligned} f(4, 3) &= S(f(4, 2) \\ &= f(4, 2) + 1 \\ &= S\ (f(4, 1)) + 1 \\ &= f(4, 1) + 1 + 1 \\ &= S(f(4, 0) + 1 + 1 \\ &= f(4, 0) + 1 + 1 + 1 \\ &= 4 + 1 + 1 + 1 \\ &= 7 \end{aligned}$$

Example 11.2 Prove that the function f(x, y) = x − y, where x, y are positive integers, is primitive recursive.

Solution:

$$f(x, y) = x - y$$

$$\Rightarrow f(x,0) = x = Z(x)$$

$$f(x, y + 1) = x - (y + 1) = x - y - 1 = f(x, y) - 1 = S\ (x, f(x, y))$$

The function can be obtained by composition and recursion from the zero and successor functions in a finite number of steps. Thus, it is primitive recursive.

Let f(4, 2) be given.

$$\begin{aligned} f(4, 2) &= S(f(4, 1)) \\ &= f(4, 1) - 1 \\ &= S(f(4, 0)) - 1 \\ &= f(4, 0) - 1 - 1 \\ &= 4 - 1 - 1 \\ &= 2 \end{aligned}$$

Example 11.3 Prove that the function f(x, y) = x * y, where x, y are positive integers, is primitive recursive.

Solution:

$$f(x, y) = x * y$$

$$\Rightarrow f(x,0) = 0 = Z(x)$$

$$f(x, y + 1) = x * (y + 1) = x * y + x = f(x, y) + x = Add(x, f(x, y))$$

We have already proved that addition is primitive recursive, thus multiplication is primitive recursive.

Let f(3, 2) be given.

$$\begin{aligned} f(3, 2) &= S(f(3, 1)) \\ &= f(3, 1) + 3\ [\text{As } S(f(x, y)) = f(x, y) + x] \\ &= S(f(3, 0)) + 3 \\ &= f(3, 0) + 3 + 3 \\ &= 0 + 3 + 3 \\ &= 6 \end{aligned}$$

Example 11.4 Prove that the function $f(x, y) = x^y$, where x, y are positive integers, is primitive recursive.

Solution:

$$f(x, y) = x^y$$
$$\Rightarrow f(x, 0) = x^0 = 1 = Z(x)$$
$$f(x, y + 1) = x^{y+1} = x^y * x = f(x, y) * x = \text{Mult}(x, f(x, y))$$

We have already proved that multiplication is primitive recursive, thus exponential is primitive recursive.

Let f(4, 2) be given.

$$\begin{aligned} f(4, 2) &= S(f(4, 1)) \\ &= f(4, 1) * 4 \\ &= S(f(4, 0)) * 4 \\ &= f(4, 0) * 4 * 4 \\ &= 1 * 4 * 4 \\ &= 16 \end{aligned}$$

Example 11.5 Prove that the function $f(x) = x/2$ if x is even
$= (x - 2)/2$ if x is odd
where x is a positive integer which is primitive recursive.

Solution:

$$f(x) = x/2, \quad \text{if } x \text{ is even}$$
$$f(0) = 0/2 = 0$$

If x is even, then x + 1 is odd.

$$\begin{aligned} f(x + 1) &= (x + 1 - 2)/2 \\ &= x/2 - 1/2 \\ &= f(x) - 1/2 = S(x, f(x)) \end{aligned}$$

If x is odd, then x + 1 is even.

$$\begin{aligned} f(x + 1) &= (x + 1)/2 \\ &= (x-2)/2 + 3/2 \\ &= f(x) + 3/2 \\ &= S(f(x)) \end{aligned}$$

The given function can be rewritten as

$$\begin{aligned} S(f(x)) &= f(x) - 1/2, && \text{if } x \text{ is even} \\ &= f(x) + 3/2, && \text{if } x \text{ is odd} \end{aligned}$$

Hence, it is primitive recursive.

Example 11.6 Prove that the function fact(x) is primitive recursive.

Solution:

$$\text{fact}(0) = 0! = 1 = s(0)$$
$$\text{fact}(x + 1) = x \times \text{fact}(x) = \text{mult}(s(x), \text{fact}(x))$$

As the multiplication function is primitive recursive, fact is also primitive recursive.

Example 11.7 Prove that the bounded addition of the primitive recursive function is also primitive.

Solution: If f(m, n) is a primitive recursive function, then we have to prove that $g(m, n) = \sum_{i=0}^{n} f(m,n)$ is primitive recursive.

$$\sum_{i=0}^{n} f(m,n) = f(m, 0) + f(m, 1) + f(m, 2) + \ldots\ldots f(m, n)$$

$$g(m, 0) = \sum_{i=0}^{0} f(m,0) = 0 = Z(0)$$

$$g(m, n + 1) = \sum_{i=0}^{n+1} f(m,n+1) = \sum_{i=0}^{n} f(m,n) + f(m, n + 1) = g(m, n) + f(m, n + 1)$$

Thus, it is also primitive recursive.

(It can also be proved in the same way that the bounded product of primitive recursive functions is also primitive recursive.)

11.4 Gödel Number

In the late 19th century, two properties of a formal system were the cause of brainstorming for the logicians and mathematicians. The properties were completeness and consistency. Completeness is defined as the possibility to prove or disprove any proposition that can be expressed in the system. On the other hand, consistency is the impossibility to both prove and disprove a proposition in a system. Using the axioms of set theory, many mathematicians try to prove these properties of a formal system.

A British philosopher and logician Bertrand Arthur William Russell in 1901 discovered a paradox known as the Russell's paradox.

11.4.1 Russell's Paradox

S is defined as the set of all sets that are not members of themselves. Is S a member of itself?

From this paradox, two conditions appear for S.

1. If S is an element of S, then S is a member of itself and should not be in S.
2. If S is not an element of S, then S is not a member of itself, and should be in S.

In this context, Kurt Gödel, an Australian (later American), proposed a theorem in 1931 known as the Gödel Incompleteness Theorem. The theorem states that 'if a formal theorem is consistent, then it is impossible to prove within the system that it is inconsistent.'

The concept of the Gödel number was used by Gödel in proving his famous incompleteness theorem. The Gödel numbering is a function that assigns each symbol and a well-formed formula of some formal language to a unique natural number. It can also be used as an encoding scheme that assigns a unique number to each symbol of a mathematical notation. The Gödel numbering encodes a sequence of numbers in a single value. The scheme of converting a sequence of natural numbers uses the factorization of a natural number into a product of prime numbers.

The Gödel number for a sequence $x_0, x_1, \ldots\ldots x_{n-1}$ for n natural numbers is defined as

$$pn(0)^{x_0} \times pn(1)^{x_1} \times pn(2)^{x_2} \times \ldots\ldots pn(n))^{x_n} = 2^{x_0} \times 3^{x_1} \times 5^{x_2} \times \ \ldots\ldots \ pn(n)^{x_n}$$

where pn(n) is the nth prime number.

As an example, the sequence 2, 1, 0, 3 is encoded as $2^2 \cdot 3^1 \cdot 5^0 \cdot 7^3 = 4116$. The sequence 1, 0, 2, 1 is encoded as $2^1 \cdot 3^0 \cdot 5^2 \cdot 7^1 = 350$.

Decoding a Gödel number G to obtain the sequence is the reverse process. The steps are

- Prime factorize the given number.
- Arrange the obtained primes in sequence, starting from the least and raise appropriate power with the number of times they appear. If any prime number in between two does not appear, raise the power as '0'.

As an example, take a number 10780. The prime factorization of the number results in $2 \times 2 \times 5 \times 7 \times 7 \times 11 = 2^2 \times 5^1 \times 7^2 \times 11^1$.

In the sequence '3' is missing, though 3 is a prime number in between 2 and 5. Thus, the modified sequence with power raised to each prime is $2^2 \times 3^0 \times 5^1 \times 7^2 \times 11^1$.

Hence, the sequence is (2, 0, 1, 2, 1).

The function for constructing a Gödel number is not one to one. Two different sequences may have the same Gödel number. These two sequences are different with the number of '0' at last. As an example, the sequences (2, 0, 1) and (2, 0, 1, 0, 0) have the same Gödel number.

11.5 Ackermann's Function

The German mathematician Wilhelm Friedrich Ackermann discovered a function which proves that all total computable functions are not primitive recursive. This function is named after him and known as the Ackermann's function.

For two non-negative integers x and y, the Ackermann's function is defined as

$$A(x, y) = \begin{cases} y+1 & \text{if } x = 0 \\ A((x-1), 1) & \text{if } x > 0 \text{ and } y = 0 \\ A((x-1), A(x, (y-1)) & \text{if } x > 0 \text{ and } y > 0 \end{cases}$$

For every non-negative integer x and y, the function A(x, y) can be computed. So, it is a total function.

$$A(x, 0) = A((x-1), 1) = A((x-2), A((x-1), (1-1))) = A((x-2), A((x-1), 0))$$

(this will continue recursively).

Here, we are not getting a zero function, which proves that it is not primitive recursive.

Example: Compute A(1, 3).

$$A(1, 3) = A(0, A(1, 2)) \text{ As } x, y > 0$$
$$= A(0, A(0, A(1, 1)))$$

$$
\begin{aligned}
&= A(0, A(0, A(0, A(0, 1)))) \\
&= A(0, A(0, A(0, 2))) \\
&= A(0, A(0, 3)) \\
&= A(0, 4) \\
&= 5
\end{aligned}
$$

Example: Compute A(2, 2).

$$
\begin{aligned}
A(2, 2) &= A(1, A(2, 1)) \\
&= A(1, A(1, A(2, 0))) \\
&= A(1, A(1, A(1, 1))) \\
&= A(1, A(1, A(0, A(1, 0)))) \\
&= A(1, A(1, A(0, A(0, 1)))) \\
&= A(1, A(1, A(0, 2))) \\
&= A(1, A(1, 3)) \\
&= A(1, A(0, A(1, 2))) \\
&= A(1, A(0, A(0, A(1, 1)))) \\
&= A(1, A(0, A(0, A(0, A(1, 0))))) \\
&= A(1, A(0, A(0, A(0, A(0, 1))))) \\
&= A(1, A(0, A(0, A(0, 2)))) \\
&= A(1, A(0, A(0, 3))))) \\
&= A(1, A(0, 4)) \\
&= A(1, 5) \\
&= A(0, A(1, 4)) \\
&= A(0, A(0, A(1, 3))) \\
&= A(0, A(0, A(0, A(1, 2)))) \\
&= A(0, A(0, A(0, A(0, A(1, 1))))) \\
&= A(0, A(0, A(0, A(0, A(0, A(1, 0)))))) \\
&\ldots\ldots\ldots\ldots\ldots\ldots\ldots\ldots\ldots\ldots \\
&= A(0, 6) \\
&= 7
\end{aligned}
$$

11.6 Minimalization

Suppose there is a function f(x) which has a least value of x which makes f(x) = 0. Now, it is told to find that least value. We shall consider x = 0, 1, 2, ... and calculate f(x) correspondingly. Where f(x) = 0 is achieved, that value of x is the least value. The process will stop within a finite amount of time after a finite number of steps. Let g(x) be a function which computes the process of finding the least value of x such that f(x) = 0. So, g is computable. It is said that g is produced from f by minimization.

As an example, let f(x, y) = 2x + y – 5, where x and y are both positive integers (I).

This is a total function as for every positive integer value of x and y, there is a value of f(x, y).

We have to find the range of x for which $y \in I$.

Make f(x, y) = 0, i.e., 2x + y – 5 = 0

$\rightarrow y = 5 - 2x$

If x > 2, then there is no value of y such that y ∈ I.
Consider this function as g(x, y). We can write

$$g(x, y) = 5 - 2x \quad \text{for } x \leq 2$$
$$= \text{Undefined} \quad \text{for } x > 2.$$

Let g be a k + 1 argument function. The unbounded minimalization of g is a k argument function f such that for every arguments $n_1, n_2, \ldots.. n_k$

$$f(n_1, n_2, \ldots.. n_k) = \text{minimum value i such that } g(n_1, n_2, \ldots.. n_k, i) \text{ exist.}$$
$$= \text{Undefined otherwise.}$$

But if such an f(x) is given for which it is not known whether there exists an x for which f(x) = 0, then the process of testing may never terminate. This is called unbounded minimalization. If g is produced by unbounded minimization, then it is not effectively computable as there is no surety of termination.

If the searching is set to an upper limit, then the function g can be made computable. The search process will find a value which makes f(x) = 0 and the function g will return that value if such a value is found. Otherwise, g will return 0. This is called bounded minimization as a boundary of search process is specified.

Bounded minimalization is defined by the search operator over the natural numbers $\leq y$

$$f(x_1, \ldots, x_n, y) = \mu^y z[p(x_1, \ldots, x_n, z)]$$

It defines a function f which returns the least value of z satisfying $p(x_1, \ldots, x_n, z)$ or returns the bound. More precisely, it can be written as

$$f(x_1, \ldots, x_n, y) = \begin{cases} \min z \text{ subject to } p(x_1, \ldots, x_n, z) = 1 \text{ and } 0 \leq z \leq y \\ y + 1 \text{ otherwise} \end{cases}$$

11.7 μ Recursive

A function is said to be μ recursive if it can be constructed from the initial functions and by a finite operation of:

- Composition
- Primitive recursion
- Unbounded minimalization (μ operation)

This type of function was first proposed by Gödel and Stephen Kleene.

It can be proved that 'a function is computable by a Turing machine if it is μ recursive'.

11.7.1 Properties of a μ Recursive Function

- The zero function, successor function, and projection function are μ recursive.
- If f is a m variable μ recursive function and g is a n variable μ recursive function, then $f(g_1, g_2, \ldots g_n)$ is also μ recursive.
- If f and g are n and n + 2 variable μ recursive functions, then a new function, say h constructed from f and g by primitive recursion, is μ recursive.

11.8 λ Calculus

An interesting story of proof correction for printing is hidden in λ calculus. We are coming to the point but will first discuss the history of λ calculus. In 1900, a German mathematician David Hilbert gave a speech at the International Congress of Mathematicians in Paris. He pointed out 23 mathematical problems, and among them the 10th was to find an algorithm that tests whether a polynomial has an integral root and which will terminate within a finite number of steps. Church and Turing later on separately proved that the problem is undecidable.

Principia Mathematica, a book on the foundations of mathematics written by Alfred North Whitehead and Bertrand Russell in 1910–1913, depicted the function $f(x) = 2x + 1$ as $\hat{x} \cdot 2x + 1$. At the time of proof correction, this became ^x · 2x + 1 and at the time of the final proof correction it became $\lambda \cdot 2x + 1$!

Should a function have a name? The λ calculus started from this point. Let there be a total function named 'cube' which calculates the cube of an integer. This can be defined as a cube: integer → integer, where $cube(x) = x^3$ such that $x \in$ integer.

There is no significance of the name 'cube' unless it is defined. Now the question comes, whether a function can be defined without a name?

Church used the notation λ (Originally ^, then modified to λ at the time of proof correction by others) to represent a function without a name. He used it as $\lambda x \cdot x^3$, which maps each $x \in$ integer set to $x^3 \in$ another integer set.

Let an expression be given as $x^2 + 2y$. For the value of (2, 3), it gives two different results 10 and 13, because the order is not mentioned. Someone can take $x = 2$ and $y = 3$ whereas someone may take $x = 3$, $y = 2$. The λ notation resolves this ambiguity by specifying the order as $\lambda x \cdot \lambda y \cdot x^2 + 2y$ or $\lambda y \cdot \lambda x \cdot x^2 + 2y$ if $x = 2$ and $y = 3$ or $x = 3$, $y = 2$, respectively.

Let a function be defined as $\lambda x \cdot \lambda y \cdot x^2 + 2y$. It can be redesigned to an equivalent function which accepts a single value and returns another function as output. The output function is the function of a single variable. This is called currying. As an example, the previous function can be redesigned as

$$\lambda x \cdot (\lambda y \cdot x^2 + 2y)$$

where the value of y is given the input first. Let $x = 5$ and $y = 2$.

$$\begin{aligned} &\lambda x \cdot \lambda y \cdot x^2 + 2y \\ &\Rightarrow 5^2 + 2.2 \\ &\Rightarrow 25 + 4 \\ &\Rightarrow 29 \end{aligned}$$

If currying is applied, the value is calculated as

$$\begin{aligned} &(\lambda x.\ (\lambda y.\ x^2 + 2y)(5))\ (2) \\ &\Rightarrow (\lambda y.\ 25 + 2y)(2) \\ &\Rightarrow 25 + 4 \\ &\Rightarrow 29 \end{aligned}$$

To calculate the result of a function using λ calculus, we need to know the β reduction.

A **β reduction** is the process of calculating a result from the application of a function to an expression.

Let a function be given as $\lambda x \cdot x^2 + 2xy$. Suppose we apply the function to value 7. For calculating the result, the value 7 is substituted for every occurrence of x. The resultant function becomes $49 + 14y$.

It can be proved that a function N → N, where N is a set of natural numbers, is a computable function if and only if there exists a λ expression *f* such that for every pair of x, y in N, F(x) = *y* if and only if fx′ = β is reducible to y′, where x′ and y′ are the Church numerals corresponding to x and y, respectively.

11.9 Cantor Diagonal Method

Rational numbers are infinite. Let us represent all rational numbers by a set S1 containing infinite numbers. The decimal numbers are also infinite. Can the decimal number be represented by a set as rational number? The Cantor diagonal method answers this question. Rational numbers are said to be countable infinite, whereas decimal numbers are uncountable. (Consider all real numbers between 0 and 1 which can be represented as the terminating decimal (such as 0.345) or infinitely repeating decimal such as .345345345 …….)

The diagonal principle uses the cardinality principle with contradiction.

Suppose the number of decimal numbers can be represented by a set as rational number and represented by a set S_2. As the two sets (S_1 and S_2) have the same cardinality, there must exist a bijective relation between the elements of the two sets. Consider the decimal numbers between 0 and 1. Make a one to one correspondence with the rational numbers in S_1 as shown in the following.

Rational number	Decimal number
1	$\rightarrow d_1 = 0.\ \mathbf{d_{11}}\ d_{12}\ d_{13}\ d_{14} \ldots\ldots$
2	$\rightarrow d_2 = 0.\ d_{21}\ \mathbf{d_{22}}\ d_{23}\ d_{24} \ldots\ldots$
3	$\rightarrow d_3 = 0.\ d_{31}\ d_{32}\ \mathbf{d_{33}}\ d_{34} \ldots\ldots$
4	$\rightarrow d_4 = 0.\ d_{41}\ d_{42}\ d_{43}\ \mathbf{d_{44}} \ldots\ldots$
	⋮
n	$\rightarrow d_n = 0.\ d_{n1}\ d_{n2}\ d_{n3}\ d_{n4} \ldots\ldots$

Now, consider a decimal number $x = 0 \cdot x_1x_2x_3x_4 \ldots\ldots$, where x_i is any digit other than d_{ii} (the diagonal elements). X is a decimal number and located between 0 and 1. Thus, x must be in the previous list. X is not the first element, as its first position after the decimal point is not d_{11}. x is not the second element, as its second position after the decimal point is not d_{22}. By this process, it is not the nth element as its nth digit is not $d_{nn.}$ Thus, x is not in the list. A contradiction appears. It means that some decimal elements are left to be listed. This proves that 'listing' the decimal numbers is impossible, and so the infinity of decimal numbers is greater than the infinity of counting numbers.

11.10 The Rice Theorem

The Rice theorem states the following.

Let P be a non-trivial property of Turing recognizable languages. The problem 'does L(M) has the property P?' is undecidable.

In other words, it can be said that 'any non-trivial property about the language recognized by a Turing machine is undecidable.'

Before discussing the theorem, we need to discuss two important definitions, namely 'property' and 'non-trivial property'.

Property: Let P be an arbitrary class of Turing recognizable language. P is called a property of L if L ∈ P.

Non-Trivial Property: Let P be a property of Turing recognizable language. The property P is said to be non-trivial if there exists L_1 and L_2, two Turing recognizable languages such that $L_1 \in P$ but $L_2 \notin P$.

Trivial Property: A property P is called a trivial property if P is ∅ or P = L, where L is a Turing recognizable language.

Proof of the Rice Theorem: The Rice theorem is proved by reducing this to halting problem. P is a non-trivial property of the Turing recognizable language. Thus, there exists some language L ∈ P. As L is Turing recognizable, there must exist a Turing machine M_L accepting L.

Assume that P is decidable. So, there must exist a Turing machine M_P which decides whether L(M′) ∈ P or not for any machine M′.

Design a pre-processor M_{Prep} which takes <M, w> as input and generates a new machine M′ such that if M accepts w, then and only L(M′) ∈ L. This is represented by the following figure.

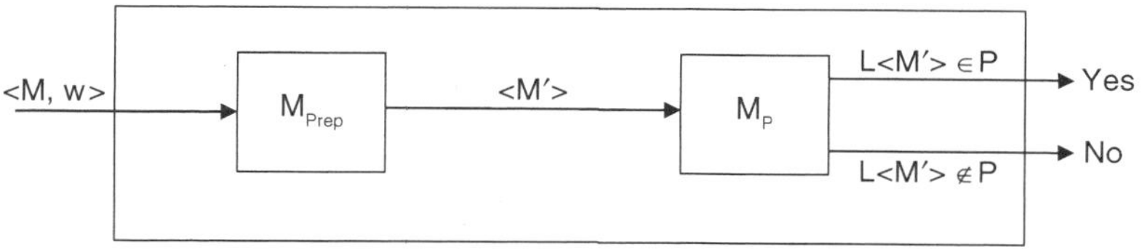

In this section, the internal operation of M_{Prep} is discussed. M_{Prep} takes <M, w> as input and generates M′ as output as discussed previously. We need to know what kind of machine M′ is. M′ simulates M using a fixed input w ignoring the actual input x. If M does not halt on w, M′ will surely not halt on x irrespective of what x is. But if M halts on w, then x is given input on M_L and accepts if M_L halts.

This means that on input w if M halts, then M′ accepts the same string as M_L. This means that L(M′) = L as L ∈ P. On the other hand, if M does not halt on w, then M′ does not accept any input. This means L(M′) = ∅.

From the previous discussion, we are getting a solution to the halting problem. Already it is proved that the halting problem is undecidable. So, our assumption is wrong, which proves that M_P cannot exist.

What We Have Learned So Far

1. In an injective function, there exists a one to one relationship between the elements of two sets.
2. A function is called surjective between two sets if each element of a set can be obtained by applying the function to some element of another set.
3. A function is bijective if it is both injective and surjective.
4. The initial functions are zero function, successor function, and projection function.
5. A zero function returns a 0 for all values.
6. A successor function returns the succeeding number of the argument taken as the input.
7. A projection function P_i^n takes n number of argument as the input and returns the ith element.
8. A partial recursive function is allowed to have an infinite loop for some input values.

9. The total recursive function always returns a value for all possible input values.
10. Primitive recursive functions can be obtained by a finite number of operations of composition and recursion from the initial functions.
11. Every primitive recursive function is Turing computable.
12. The Gödel numbering is a function that assigns each symbol and a well-formed formula of some formal language to a unique natural number.
13. A μ recursive function is constructed from the initial functions and by a finite operation of: composition, primitive recursion, and unbounded minimalization.
14. The Cantor diagonal method proves that real numbers are uncountable.
15. The Rice theorem says that 'any non-trivial property about the language recognized by a Turing machine is undecidable.'

Solved Problems

1. The Fibonacci numbers are defined as follows

$$\text{FIB}(0) = 0$$
$$\text{FIB}\ (1) = 1$$
$$\text{FIB}(n + 1) = \text{FIB}(n) + \text{FIB}(n - 1) \text{ for } n \geq 1$$

Show that FIB: N → N is primitive recursive. [UPTU 2003]

Solution: The Fibonacci series is 0, 1, 1, 2, 3, 5, 8, 13, 21, 34, ……
Let us compute FIB(n).

$$\rightarrow \text{FIB}(n - 1 + 1)$$
$$\rightarrow \text{FIB}(n - 1) + \text{FIB}(n - 2)$$
$$\rightarrow \text{FIB}(n - 2) + \text{FIB}(n - 3) + \text{FIB}(n - 2)$$
$$\rightarrow 2\ \text{FIB}(n - 2) + \text{FIB}(n - 3)$$
$$\rightarrow 2\ [\text{FIB}(n - 3) + \text{FIB}(n - 4)] + \text{FIB}(n - 3)$$
$$\rightarrow 3\ \text{FIB}(n - 3) + 2\ \text{FIB}(n - 4)$$
$$\rightarrow 3\ [\text{FIB}(n - 4) + \text{FIB}(n - 5)] + 2\ \text{FIB}(n - 4)$$
$$\rightarrow 5\ \text{FIB}(n - 4) + 3\text{FIB}(n - 5)$$

Look into the lines in bold. The numbers associated with FIB are (1, 1), (2, 1), (3, 2), (5, 3), …… Introduce a helper function h such that h(y) = (FIB(n – 1), FIB(n)). From the previous sequence, it is clear that h(0) = (1, 1).

$$h(y + 1) = ((\text{left}(h(y) + \text{right}(h(y)), \text{left}(h(y)))$$

To calculate FIB(n), we need to construct (n – 1) sequences [take FIB(6) or FIB(5)] to get the terminating conditions [FIB(0) and FIB (1)] · h(y) returns the sequence of coefficient of FIB(n – k) and FIB(n – k + 1), respectively, which ultimately ends on FIB(1) and FIB(0). Thus, FIB(n + 1) produce results after n number of steps with a finite number of operations of composition and recursion from the initial functions. This proves that FIB(n) is primitive recursive.

2. Find the Gödel equivalent number of the sequence (3, 0, 2, 0, 1, 0, 1).

Solution: There are seven numbers in the sequence. The first five prime numbers are 2, 3, 5, 7, 11, 13, and 17. So, the Gödel equivalent number of the sequence is

$$2^3 \times 3^0 \times 5^2 \times 7^0 \times 11^1 \times 13^0 \times 17^1 = 8 \times 1 \times 25 \times 1 \times 11 \times 1 \times 17$$
$$= 37400.$$

3. Find the sequence for the number 56208 using the Gödel principle.

Solution: The prime factorization of the number is $2 \times 2 \times 2 \times 2 \times 3 \times 11 \times 11$.

$$\rightarrow 2^4 \times 3^1 \times 11^2 = 2^4 \times 3^1 \times 5^0 \times 7^0 \times 11^2$$

Thus, the sequence is (4, 1, 0, 0, 2).

4. Compute A(2, 1) using the Ackermann's function.

Solution: The Ackerman's function is defined as

$$A(x,y) = \begin{cases} y+1 & \text{if } x=0 \\ A((x-1),1) & \text{if } x>0 \text{ and } y=0 \\ A((x-1), A(x,(y-1)) & \text{if } x>0 \text{ and } y>0 \end{cases}$$

So,

$$\begin{aligned} A(2, 1) &= A(1, A(2, 0)) \text{ As } x, y > 0 \\ &= A(1, A(1, 1)) \text{ As } x > 0 \text{ and } y = 0 \\ &= A(1, A(0, A(1, 0))) \text{ As } x, y > 0 \\ &= A(1, A(0, A(0, 1))) \text{ As } x > 0 \text{ and } y = 0 \\ &= A(1, A(0, 2)) \text{ As } x = 0 \\ &= A(1, 2) \text{ As } x = 0 \\ &= A(0, A(1, 1)) \text{ As } x, y > 0 \\ &= A(0, A(0, A(1, 0))) \\ &= A(0, A(0, A(0, 1))) \\ &= A(0, A(0, 2)) \\ &= A(0, 3) \\ &= 4. \end{aligned}$$

Multiple Choice Questions

1. The function where no two elements of one set have exactly the same elements mapped in another set is called
   a) Injective b) Bijective
   c) Surjective d) Composite

2. The function where for every element of one set there is an element in another set is called
   a) Injective b) Bijective
   c) Surjective d) Composite

3. Zero function always returns
   a) One b) Zero
   c) Null d) Element^0
4. Find the true statement
   a) The partial recursive function always returns a value.
   b) The total recursive function may be in an infinite loop.
   c) In both of these cases, the Turing machine must halt.
   d) The partial recursive function may be in an infinite loop.
5. Which is true for the Gödel number
   a) One number can have more than one Gödel sequence
   b) Two different Gödel sequences may produce the same number
   c) (a) and (b) are correct if two sequences are different with the number of '0' at last.
   d) The Gödel sequence of a number is not unique.
6. According to the Ackermann's function, if $x > 0$ and $y = 0$, then A(x, y) is
   a) $y + 1$
   b) $A((x - 1), 1)$
   c) $A((x - 1), A(x, (y - 1))$
   d) $x + 1$
7. A function is said to be μ recursive if it can be constructed from the initial functions and by a finite operation of
   a) Composition
   b) Primitive recursion
   c) Unbounded minimalization (μ operation)
   d) All of these
8. For (3, 2), the function $\lambda y \cdot \lambda x \cdot x^2 + 2y$ returns
   a) 10 b) 13 c) 12 d) 11

Answers:

1. a 2. c 3. b 4. d 5. c 6. b 7. d 8. a

Exercise

1. Prove that proper subtraction of two numbers defined as

$$\text{Psub}(x, y) = \begin{cases} x - y & \text{if } x > y \\ 0 & \text{otherwise} \end{cases}$$

   is primitive recursive.
2. Prove that the modulus operation is primitive recursive.
3. A function f(n) is defined as

$$f(n) = \begin{cases} n/3 & \text{if n is even} \\ (n-2)/3 & \text{if n is odd} \end{cases}$$

   Check whether f(n) is primitive recursive or not.
4. Prove that addition and multiplication of two positive integers is a total function.
5. Show that a function Max defined as

$$\text{Max}(x, y) = \begin{cases} 1 & \text{if } x > y \\ 0 & \text{if } x < y \end{cases}$$

   is primitive recursive.
6. Evaluate the Ackermann's function for A (2, 2), A(3, 1).
7. Calculate $\lambda y \cdot \lambda x \cdot x^3 + 2xy + y^2$ for (3, 1) and (3, 2)

Computational Complexity 12

Introduction

It is already discussed that decidable problems have algorithms. For solving a decidable problem, there may be more than one algorithm. The algorithms may differ in time taken and/or memory required to execute and produce the result, for the same input. This is known as computational complexity. It concerns with the question 'which is the efficient algorithm for solving a decision problem.' While discussing computational complexity, some underlying particulars such as hardware, software, and data structure are generally ignored. Computer science deals with the theoretical foundations of information, computation, and practical techniques for implementation and application. In this chapter, we shall discuss the different types of complexities of an algorithm, polynomial time algorithm, classes of P and NP, and different examples related with these.

12.1 Types of Computational Complexity

Mainly, two types of complexities are measured for an algorithm. These are

1. Time complexity
2. Space complexity

The measurement of complexity is done for the worst cases of an algorithm.

12.1.1 Time Complexity

The time complexity of a program for a given input is the number of elementary instructions (the number of basic steps required) that this program executes. This number is computed based on the length (n) of the input. If T(n) is the time complexity for an algorithm 'A' for a problem 'P', then the number of steps required to produce the output from an input of length n is less than or equal to n. (Here it is assumed that each elementary instruction takes unit time and one processor instruction.)

12.1.2 Space Complexity

The space complexity of a program for a given input is the number of elementary objects that this program needs to store (i.e., memory consumed) during its execution. This number is computed based on the length (n) of the input. If S(n) is the space complexity for an algorithm 'A' for a problem 'P', then the number of memory locations required to produce the output from an input of length n is less than or equal to n.

12.2 Different Notations for Time Complexity

Take four functions:

1. $f_1(n) = 2n$
2. $f_2(n) = n^2$
3. $f_3(n) = 2n^2$
4. $f_4(n) = 2^n + 3$

Now, in the following table, we are calculating the values of the four functions for different values of n.

Function	n = 0	= 2	= 10	= 50	= 100
$f_1(n)$	0	4	20	100	200
$f_2(n)$	0	4	100	2500	10000
$f_3(n)$	0	8	200	5000	20000
$f_4(n)$	4	7	1027	1125899906842627	1267650600228229401496703205379

Among these, consider $f_2(n)$ and $f_3(n)$. Both of them are directly proportional to one another. For $n = 0$, both of them give the same results but for $n > 0$, $f_2(n) < f_3(n)$. If we consider $f_1(n)$ and $f_2(n)$, we shall notice that for $n = 0$ and $n = 2$ both the functions give the same results but for $n > 2$, $f_1(n) < f_2(n)$.

It can be proved that the growth rate of any exponential function is greater than that of any polynomial.

From this discussion, we can draw different growth rate notations. There are mainly five well-known notations for growth rates of functions. These are

1. Big oh notation, denoted as 'O'
2. Big omega notation, denoted as Ω
3. Theta notation, denoted as Θ
4. Little-oh notation, denoted as 'o'
5. Little omega notation, denoted as 'ω'

12.2.1 Big Oh Notation

Let f(n) and g(n) be two functions of positive natural or real numbers. f(n) is said to be O(g(n)) if there exists a positive constant C and a value k such that $f(n) \leq Cg(n)$ for all $n \geq k$. Big oh represents the worst case time complexity of an algorithm.

12.2.1.1 Rules Related to Big Oh

Let there be two functions of n, namely $F_1(n)$ and F_2, of order O(f(n)) and O(g(n)), respectively.

1. $F_1(n) + F_2(n) = \max(O(f(n)), O(g(n)))$.
2. $F_1(n) * F_2(n) = O(f(n)* g(n))$.
3. $C* F_1(n) = O(f(n))$, where C is a constant.

Example 12.1 Find the order of the following functions in big oh notation.

i) $f(n) = K$, where K is a constant
ii) $f(n) = 3n - 2$
iii) $f(n) = 6n^3 + 5n + 1$

Solution:

i) Let $K = 10$. So, $f(n) = 10 \le 11*1$
   Consider $C = 11$ and $g(n) = 1$. Here, $k = 0$.
   Thus, $f(n) = O(1)$.

ii) $f(n) = 3n - 2 \le 4n \quad \text{for } n \ge 0$
   Consider $C = 4$ $k = 0$ and $g(n) = n$.
   Thus, $f(n) = O(n)$.

iii) $f(n) = 6n^3 + 5n + 1 \le 6n^3 + 5n^3 + 1n^3$ for $n \ge 1$ (For $n = 0$ LHS > RHS)
   $\Rightarrow 6n^3 + 5n + 1 \le 12\, n^3$ for $n \ge 1$
   Consider $g(n)$ as n^3. We can write the following here: $f(n) \le 12n^3$ for $n \ge 1$.
   Thus, $f(n) = O(g(n))$, i.e., $O(n^3)$ for $C = 12$ and $k = 1$.

Example 12.2 Check whether $f(n) = 6n^3 + 5n + 1$ be $O(n^4)$

Solution: Consider $C = 2$ and $g(n) = n^4$. The following table is constructed to find the value k.

	f(n)	Cg(n)
n = 0	1	0
n = 1	12	2
n = 2	59	32
n = 3	192	162
n = 4	405	512
n = 5	776	1250

The table shows that f(n) is becoming less after $n = 4$. Thus, $f(n) = O(n^4)$ for $C = 2$ and $k = 4$.

Example 12.3 Can n^3 be $O(f(n))$ where $f(n) = 6n^3 + 5n + 1$?

Solution: If we take $C = 1$ and $k = 1$, then $n^3 \le 6n^3 + 5n + 1$.
   Thus, n^3 is $O(f(n))$ for $C = 1$ and $k = 1$.

12.2.2 Big Omega (Ω) Notation

Let f(n) and g(n) be two functions of positive natural numbers. f(n) is said to be $\Omega(g(n))$ if there exists a positive constant C and a value k such that $f(n) \ge Cg(n)$ for all $n \ge k$. **Big omega** represents the asymptotic lower bound of a given function. It is opposite to the big oh notation.

Example 12.4 Find the order of the following functions in the big omega notation.

i) f(n) = K, where K is a constant
ii) $f(n) = 3n - 2$
iii) $f(n) = 6n^3 + 5n + 1$

Solution:

i) Let K = 10. So, $f(n) = 10 \geq 9 * 1$
Consider C = 9, g(n) = 1 Here, k = 0.
Thus, $f(n) = \Omega(1)$.

ii) $f(n) = 3n - 2 \geq 2n$ for $n \geq 2$
Consider C = 2 k = 2 and g(n) = n.
Thus, $f(n) = \Omega(n)$.

iii) $f(n) = 6n^3 + 5n + 1 \geq 6n^3$ for $n \geq 1$
Consider C= 6 k = 1 and $g(n) = n^3$.
Thus, $f(n) = \Omega(n^3)$.

Example 12.5 Check whether $f(n) = 6n^3 - 3n^2 - 5n + 1$ be $\Omega(n^2)$.

Solution: f(n) is $\Omega(n^2)$ if there exists a positive number C and k ($\leq$ n) such that

$$6n^3 - 3n^2 - 5n + 1 \geq Cn^2$$
$$\Rightarrow 6n^3 - (3 + C)n^2 - 5n + 1 \geq 0$$

Rename $6n^3 - (3 + C)n^2 - 5n + 1$ as T(n).

Lesser the value of (3 + C) makes better the probability of the previous inequality becoming true. Let us take C = 1.

The following table is constructed to find the value k.

	T(n)
n = 0	1
n = 1	–2
n = 2	23
n = 3	112

The previous table shows that the inequalities are true from n = 2 onwards. Thus, $f(n) \geq C\, g(n)$ for C = 1 and k = 2 ($n \geq k$), where $g(n) = n^2$. So $f(n) = \Omega(n^2)$.

12.2.3 Theta Notation (Θ)

Let f(n) and g(n) be two functions of positive natural numbers (both real and integer).

f(n) is called $\Theta(g(n))$ if there exist two positive constants C_1 and C_2 and a value k such that $C_2 g(n) \leq f(n) \leq C_1 g(n)$ for all $n \geq k$.

If we break the notation for a function to be in Θ, we get $f(n) \le C_1g(n)$ and $f(n) \ge C_2g(n)$. In other words, it can be said that for two functions f(n) and g(n), $f(n) = \Theta(g(n))$ if $f(n) = O(g(n))$ and $f(n) = \Omega(g(n))$.

12.2.4 Little-oh Notation (o)

Let f(n) and g(n) be two functions of positive natural numbers (both real and integer). f(n) is called o(g(n)) if there exist a natural number k and a constant $C > 0$ such that $f(n) < Cg(n)$ for all $n \ge k \ge 1$.

12.2.5 Little Omega Notation (ω)

Let f(n) and g(n) be two functions of positive natural numbers (both real and integer). f(n) is called ω(g(n)) if there exist a natural number k and a constant $C > 0$ such that $f(n) > Cg(n)$ for all $n \ge k$.

12.3 Problems and Its Classification

A problem is a question to be answered. A problem consists of a parameter. What is the value of X + Y, is the element X available in list[ ], how can an array be sorted in minimum time, etc. Problems are classified into different types such as the following.

1. Decision problem
2. Optimization problem
3. Search problem
4. Functional problem

Among these, we shall discuss here mainly two classes of problems—decision and optimization problem.

Decision Problem: Decision problem answers in 'yes' or 'no'. Depending on the situation, the answer may be 1 or 0, true or false, success or failure.

- Given two numbers X and Y, 'whether Y is a factor of X?' is a decision problem.
- Given a number N, 'is N a prime number?' is a decision problem.
- Given a list of N items, to sort the list of items in ascending order is not a decision problem.

Optimization Problem: Optimization problem is described as a problem of finding the best solution from all available feasible solutions. In other words, it can be said to find the optimal solution from all available feasible solutions. Optimal solution means the use of resources in the most effective and efficient manner, and yielding the highest possible returns under the circumstances. Optimization problems can be either a maximization or minimization problem.

1. **Shortest path algorithm:** This is the problem of finding a path between two vertices (or nodes) in a graph in such a way that the sum of the weights (cost) of its constituent edges is minimized.
2. **Minimum spanning tree:** Given the connected graph G with positive edge weights (cost), the minimum spanning tree is to find a minimum weight (cost) set of edges that connects all the vertices of the graph.

12.4 Different Types of Time Complexity

Already we have learnt about the different notations of time complexity and different types of problems. For solving these problems, we need to construct algorithms. For those algorithms, times complexities need to be calculated which are needed to analyse the algorithms and find the suitable one. So, there exists a close relation between design analysis of algorithm and computational complexity theory. The former is related to the analysis of the resources (time and/or space) utilized by a particular algorithm to solve a problem and the later is related to a more general question about all possible algorithms that could be used to solve the same problem. There are different types of time complexity for different algorithms. In this section, we shall discuss the different types of time complexities.

12.4.1 Constant Time Complexity

An algorithm whose operation does not depend on the size of the input is called constant time complexity algorithm. Regardless of the size of the input, the algorithm takes the same amount of time for all cases. The complexity is denoted as O(1).

Example 12.6 Find the time complexity for accessing the ith element of an array.

Solution: The operation of the algorithm does not depend on the size of the algorithm. Let the algorithm be

Algorithm:

```
Procedure: Accessing [A, location]
Input: A[ ], i // An array of size n and location
  Local b as integer
  Begin:
    b = A[i]
    return b
End
```

If the size of the array is m, then also the time remains the same.

The algorithm for accessing the highest or lowest element of a sorted list is also in constant time complexity.

12.4.2 Logarithmic Time Complexity

An algorithm which takes logarithmic time based on the input to perform its operation is called a logarithmic time complexity algorithm. For logarithmic time complexity algorithm, the base is usually 2 because computers use the binary number system. For this reason, $\log_2 n$ is usually written as log n. Logarithmic time complexity is denoted as O(log n).

Example 12.7 Find the time complexity of binary search.

Solution: *Binary search* is performed on sorted array. The algorithm compares the key element input by the user with the middle element of the array. If the key is matched with the middle element, return the index of the middle element. If the key is less than the middle element, apply the same algorithm for the left half of the array. If the key is greater than the middle element, apply the same algorithm for the right half of the array. The algorithm for the binary search is

```
int binary_search(int A[ ], int key, int min, int max)
{
  int mid = [(min + max)/2];

  if (A[mid] > key)
    return binary_search(A, key, min, mid-1);
  else if (A[mid] < key)
    return binary_search(A, key, mid + 1, max);
  else
    // key has been found
    return mid;
}
```

Calculation of the mid-element takes a unit time. If the key does not match with the mid-element, the search is limited to n/2 space. Hence, the time taken by the algorithm is

$$T(n) = T\left(\frac{n}{2}\right) + 1$$

$$T\left(\frac{n}{2}\right) = T\left(\frac{n}{2^2}\right) + 1$$

$$T\left(\frac{n}{2^2}\right) = T\left(\frac{n}{2^3}\right) + 1$$

$$\cdots\cdots$$

$$\cdots\cdots$$

$$T\left(\frac{n}{2^{i-1}}\right) = T\left(\frac{n}{2^i}\right) + 1$$

Adding all these equations, we get

$$T(n) = T\left(\frac{n}{2^i}\right) + i$$

Let $2^i = n$. $i = \log_2 n$

$$T(n) = T(1) + \log_2 n.$$

If there is only one element in the array, it takes no time to search. Thus, $T(n) = \log_2 n$.
So, the complexity of the algorithm is $O(\log_2 n)$.

12.4.3 Linear Time Complexity

An algorithm is said to be linear if its running time increases linearly with the size of the input. Linear time complexity is denoted as $O(n)$.

Example 12.8 Find the time complexity of linear search.

Solution: *Linear search* is the simplest method of the searching technique. In this technique, the item to be searched is searched sequentially from the first item of the list until it is found. If the item is found, the location of the item in the list is returned. Following is the algorithm for linear search.

```
Procedure: LinearSearch(List[ ], target)
  Inputs: List[ ] - A list of numbers
  Local: i integers
  Begin:
  int location = -1;
  For i = 0 to List.Size;
    If (List[i] == target)
    location = i;
    return location;
    End If
  Next i
  Return -1
End
```

Start calculating the complexity of the algorithm. The assignment of the value of 'i' to 'location' is of constant time, i.e., O(1). Returning the location also takes O(1). These two statements are executed once if the condition within the 'if' statement holds well. 'If' condition checking takes O(1). The statement 'if' takes maximum time if the condition is true [till the time is O(1) as O(1) + O(1) + O(1) = O(1)]. This 'if' statement is surrounded by a 'for' loop. The 'if' statement is executed as many times as the 'For' loop iterates. For the worst case (if the target element is the last element of list[ ] or if it does not exist), the loop iterates N times, where N is the size of list[ ].

Thus, the time taken by this algorithm is (worst case):

$$\sum_{i=0}^{N-1}\left(\underset{\text{IF}}{1}\right) = N$$

So, the complexity is O(N). Here N is a polynomial of N of degree 1.

12.4.4 Quasilinear Time Complexity

Before discussing quasilinear time complexity, we need the knowledge of linearithmic algorithm. The running time of the algorithms with linearithmic time complexity increases faster than the linear algorithm and slower than the logarithmic algorithm. Linearithmic time complexity is denoted as O(n log n).

Quasilinear time complexity was first proposed by Schnorr and Gurevich and Shelah in their research paper 'Satisfiability is quasilinear complete in NQL' in 1978. An algorithm is said to be in quasilinear time complexity if its running time is of $O(n (\log n)^k)$ where $k \geq 1$. Quasilinear algorithm runs faster than the polynomial algorithms of degree more than 1. Linearithmic algorithms are one type of quasilinear algorithm with k = 1.

Example 12.9 Find the time complexity of quicksort.

Solution: *Quicksort* is one type of divide and conquer type sorting algorithm proposed by T. Hoare. In this algorithm, a pivot element is chosen and the remaining elements of the list are divided into two halves based on the condition whether the elements are greater or smaller than the pivot element. The same quicksort algorithm is run again separately for two halves.

Process:

i) Take an element, called pivot element, from the list.

ii) Rearrange the list in such a way that all lesser elements than the pivot come before the pivot, while all greater elements than the pivot come after it (equal values can go in either direction). This partitioning makes the pivot to be placed in its final position. This is called the **partition** operation.

iii) Recursively, sort the sub-list of lesser elements and the sub-list of greater elements.

A list of size zero or one never needs to be sorted. This is the base case of the recursion.

```
Procedure: QuickSort (A[ ], left, right)
  {
    if (left < right)
      select PIVOT s.t left ≤ PIVOT ≤ right
      PIVOT = PARTITION(A[ ], left, right, PIVOT)
      QuickSort(A[ ], left, PIVOT-1)
      QuickSort(A[ ], PIVOT + 1, right)
  }

Function: PARTITION(A[ ], left, right, PIVOT)
  {
    Value = A[PIVOT]
    swap A[PIVOT] and A[right]      // Move pivot to end
    Index = left
    for i from left to right - 1 // left ≤ i < right
      if A[i] < pivot
        swap A[i] and A[Index]
        Index = Index + 1
    swap A[Index] and A[right]      // Move pivot to its final place
    return Index
  }
```

12.4.5 Average Case Time Complexity

Let the pivot element be the kth smallest element of the array. Then, there are (k – 1) elements to the left of the position of k and (n – k) elements are at the right of the position of k. Thus, for sorting the left half of the array needs T(k – 1), and for sorting the right half of the array needs T(n – k). To place the pivot element, the number of comparisons is (n + 1). For an array of n element, the time complexity for average case quicksort is

$$T(n) = (n+1) + \frac{1}{n}\sum_{1\le k\le n} T(k-1) + T(n-k) \tag{1}$$

$$nT(n) = n(n+1) + \sum_{1\le k\le n} T(k-1) + T(n-k) \tag{2}$$

Replacing n by (n − 1), we get

$$(n-1)T(n-1) = n(n-1) + \sum_{1\le k\le (n-1)} T(k-1) + T(n-k-1) \tag{3}$$

Subtracting equation (3) from (2), we get

$$nT(n) - (n-1)T(n-1) = 2n + 2T(n-1) \tag{4}$$

Note: $\sum_{1\le k\le n} T(k-1) + T(n-k) - \sum_{1\le k\le (n-1)} T(k-1) + T(n-k-1)$

$$= \begin{aligned}&[T(0)+T(n-1)+T(1)+T(n-2)+ \cdots\cdots + T(n-1)+T(0)] - [T(0)+T(n-2)\\ &+T(1)+T(n-3)+ \cdots\cdots + T(n-2)+T(0)]\end{aligned}$$

$$= 2T(n-1)$$

Rearranging equation (4), we get

$$\Rightarrow nT(n) = 2n + (n+1)T(n-1)$$

$$\Rightarrow T(n) = 2 + \frac{(n+1)}{n}T(n-1)$$

$$\Rightarrow \frac{T(n)}{(n+1)} = \frac{1}{n}T(n-1) + \frac{2}{(n+1)}$$

$$= \frac{1}{n-1}T(n-2) + \frac{2}{n} + \frac{2}{(n+1)}$$

$$= \frac{1}{n-2}T(n-3) + \frac{2}{n-1} + \frac{2}{n} + \frac{2}{(n+1)}$$

$$\vdots$$

$$= \frac{T(1)}{2} + 2\left[\sum_{3\le k\le n+1} \frac{1}{k}\right] \le \int_2^{n+1} \frac{1}{k}$$

$$\le [\log x]_2^{n+1} = \log(n+1) - \log_2 2 \le \log(n+1)$$

$$T(n) \le (n+1)\log(n+1)$$

Hence, the time complexity is O(n log n).

12.4.6 Polynomial Time Complexity

In mathematics, a *polynomial* is an expression of finite length constructed by the operations of addition, subtraction, multiplication, and non-negative integer exponents from variables and constants. As an example, $f(x) = x^4 + 2x^3 - 3x + 4$ is a polynomial of degree 4.

An algorithm is said to be in polynomial time complexity if its execution time is upper bounded by a polynomial expression in the size of the input for the algorithm. The complexity is denoted as $O(n^k)$, where k is a constant.

Let us discuss some algorithms to make the idea clear.

Example 12.10 Find the time complexity of bubble sort.

Solution: Consider the example of bubble sort to sort a list of n items.

Here, an array called list is taken as input. The array is sorted by bubble sort() using the following algorithm.

```
Procedure: BubbleSort(List[ ])
  Inputs: List[ ] - A list of numbers
  Locals: i, j - integers
  Begin:
    For i = 0 to List.Size-1
      For j = i + 1 to List.Size-1
        If List[i] > List[j], Then
          Swap List[i] and List[j]
        End If
      Next j
    Next i
End
```

Start calculating the complexity of the algorithm from the 'swap' operation of two elements. Regardless of size of the list, 'swap' always takes the same amount of time, i.e., O(1). The 'swap' operation depends on the 'if statement' condition. If the condition is false, it does not execute swap. In the worst case (if condition holds true), swap executes. Again, the 'if' statement is of O(1). Thus, if the condition is true, it takes maximum time which is still O(1).

The 'if' statement is surrounded by two 'for' loops. The 'if' statement is executed as many times as the inner 'for' loop iterates, and the inner 'for' loop executes as many times as the outer 'for' loop iterates. For each iteration of the outer loop, the inner loop executes $N-1$ times, where N is the size of list[ ]. The outer loop iterates $N - 1$ times. So, the complexity for this algorithm is:

$$\left(\sum_{i=0}^{N-1}\sum_{j=i+1}^{N-1}(1)\right) = \sum_{i=0}^{N-1}(N-1-(i+1)+1)$$

$$\rightarrow \sum_{i=0}^{N-1}(N-i-1) \rightarrow N\sum_{i=0}^{N-1}1-\sum_{i=0}^{N-1}(i+1) \rightarrow N^2-(1, 2, 3 \ldots . N) \rightarrow N^2 - \frac{N(N+1)}{2} \rightarrow \frac{N^2-N}{2}$$

So, the complexity is $O(N^2)$. Here, $\frac{N^2-N}{2}$ is a polynomial of N of degree 2.

Example 12.11 Find the time complexity of matrix multiplication.

Solution: *Matrix multiplication* is a two-dimensional array. Matrix multiplication can be done if the number of rows of the first matrix is equal to the number of columns of the second matrix. Matrix multiplication is done by the following algorithm.

```
Procedure: MatrixMul(MAT1[][], MAT2[][])
Inputs: MAT1[][], MAT2[][]
  Locals: i, j, k - integers
  Mult[][] = 0 -integers
  If No. of row of MAT1 = No. of column of MAT2
    For i = 0 to No. of row of MAT1
      For j = 0 to No. of column of MAT2
        Mult[i][j] = 0
        For k = 0 to No. of row of MAT1
          Mult[i][j]= Mult[i][j] + MAT1[i][k] * MAT2[k][j]
        Next k
          Return Mult[i][j]
      Next j
    Next i
End If
```

Start calculating the complexity of the algorithm from 'mult'. 'Mult' consists of addition and multiplication. Let the time taken for addition and multiplication be t_1 and t_2, respectively. The total time for 'mult' operation is t_1+t_2, and thus constant, i.e., O(1). This operation is surrounded by three 'for' loops of i, j, and k. So, the complexity for this algorithm is:

$$\left(\sum_{i=0}^{N-1}\left(\sum_{j=0}^{N-1}\left(\sum_{k=0}^{N-1}(1)\right)\right)\right) = \sum_{i=0}^{N-1}\left(\sum_{j=0}^{N-1} N\right) = \sum_{i=0}^{N-1} N \sum_{j=0}^{N-1} 1 = \sum_{i=0}^{N-1} N^2 = N^2 \sum_{i=0}^{N-1} 1 = N^3$$

So, the complexity is $O(N^3)$. N^3 is a polynomial of N of degree 3.

12.4.7 Super Polynomial Time Complexity

This is sometimes called exponential time complexity. An algorithm is called super polynomial time complexity if it runs in 2^n steps for an input of size n. The execution time of such algorithms increases rapidly as n grows.

A funny story on exponential growth is the suitable place to mention in this context. An Indian Brahmin Sissa wanted to teach his king a good lesson. The king told this Brahmin proudly that he can pay any amount of paddy grain as he has a large grain container. The Brahmin challenged the king to play chess. The king was defeated. In return, the Brahmin requested the king to pay him such amount of grains so that he can place 1 packet of grain (about 20 kg) on the first box, 2 on the second, 4 on the third, 8 on the fourth, and so on. So, the last box would contain 2^{63} packets of grains. (A chess board contains 64 boxes.) The total amount of grains payable to the Brahmin is $2^{64} - 1$ = 1, 84, 467, 44, 07, 370, 95, 51, 616 packets. India requires several years to produce this amount of paddy grains!

This is called exponential grows. The problems with exponential grows are hard to compute by a computer.

Consider the following problems.

There are 15 different-sized tennis balls. The job of the computer is to arrange the balls such that no two sets of arrangement are equal.

The numbers of such different arrangements are 15! = 1307674368000.

Let there be a computer which can print 10^6 such arrangements in 1 second. This computer will take about 2.5 years to print all such arrangements!

Example 12.12 Find the time complexity of the tower of the Hanoi problem.

Solution: *Tower of Hanoi* is a mathematical puzzle invented by E. Lucas in 1883. This consists of three rods and a number of different-sized disks with a hole in the middle so that the disks can be arranged by sliding them on the rods. At the initial stage, the disks are placed in ascending order with the smallest radius on the top on the leftmost rod.

The objective of the puzzle is to move the entire stack of disks from the leftmost to the rightmost rod, obeying the following conditions:

- At a time, only one disk can be moved.
- In each move, the upper disk from a rod is slided out and placed on the top of the other disks that may already be present on the another rod.
- A larger radius disk is not allowed to be placed on the top of a smaller disk.

To solve this problem, three rods are named as 'source', 'auxiliary', and 'destination'. There are n disks to be shifted from one rod and placed on another rod, which abide by the rules given previously. If the number of disks is 1, it can be easily moved from one rod to another. It can be considered as the base case. For n disks, this can be implemented by the following process.

i) Move the top n – 1 disks from source to auxiliary.
ii) Move the bottom disk from source to destination.
iii) Move n – 1 disks from auxiliary to destination.

The algorithm to implement this is given as follows.

```
Tower of Hanoi (n, source, auxiliary, destination)
  {
    If n = 1 move disk from source to destination; (base case)
    Else,
      {
        Tower of Hanoi(top n - 1, source, destination, auxiliary);
        Move the nth disk from source to destination;
        Tower of Hanoi(n - 1, auxiliary, source, destination);
      }
  }
```

Moving operation takes a unit time. And the recursive functions take T(n – 1) time each. Thus, the time required for running the else part of the algorithm one time is $T(n) = 2T(n-1)+1$. The subsequent operation will continue up to T(2). The total time for the algorithm is

$$T(n) = 2T(n-1) + 1 \quad (1)$$
$$T(n-1) = 2T(n-2) + 1 \quad (2)$$
$$T(n-2) = 2T(n-3) + 1 \quad (3)$$
$$\vdots$$
$$T(2) = 2T(1) + 1 \quad (n-1)\text{th}$$

Multiplying equation (2), (3), (4), ….. (n – 1)th by $2^1, 2^2, \ldots\ldots 2^{n-1}$ on both sides, we get

$$T(n) = 2T(n-1) + 1 \quad (1)$$
$$2[T(n-1) = 2T(n-2) + 1] \quad (2)$$
$$2^2[T(n-2) = 2T(n-3) + 1] \quad (3)$$
$$\vdots$$
$$2^{n-2}[T(2) = 2T(1) + 1] \quad (n-1)^{th}$$

Adding the equations, we get

$$T(n) = (2^0 + 2 + 2^2 + \ldots\ldots + 2^{n-2} + 2^{n-1}) = \frac{1(2^n-1)}{2-1} = 2^n - 1$$

Thus, the complexity of the algorithm is $O(2^n)$

12.5 The Classes P

An algorithm is known as polynomial time complexity problem if the number of steps (or time) required to complete the algorithm is a polynomial function of n (where n is some non-negative integer) for a problem of size n. In other words, for a problem of size n, the time complexity of the algorithm to solve the problem is a function of the polynomial of n.

In general, polynomial class contains all of the problems which are solved using computers easily. The problems those we solve using computers do not require too many computations, but hardly $O(n^3)$ or $O(n^4)$ time. Truly speaking, most of the important algorithms we have learnt are somewhere in the range of $O(\log n)$ to $O(n^3)$. Thus, we shall state that the time complexity of practical computation resides within polynomial time bounds. These classes of problems are written as the P class problem.

Definition1: The class of *polynomial solvable problems*, P, contains all sets in which the membership may be decided by an algorithm whose running time is bounded by a polynomial.

Definition 2: P is defined as the set of *all decision problems* for which an algorithm exists which can be carried out by a deterministic Turing machine in polynomial time.

12.6 Non-polynomial Time Complexity

Non-deterministic polynomial time is abbreviated as NP time. For these types of problems if we are able to certify a solution anyhow then it becomes easy to prove its correctness using the polynomial time algorithm. For this reason, NP class problems are called polynomial time verifiable. In NP problem, it is allowed to make a guess about the answer, and then to verify the correctness of the answer done in polynomial time. Let us consider the case of the vertex cover problem. A vertex cover of a graph G is the set of vertices C such that each edge of G is incident to at least one vertex of C. If a set of vertices is chosen, it can be easily checked in polynomial time whether the set of vertices is a vertex cover of G or not. To find the minimum, such vertex cover is called minimum vertex cover problem. This problem is in class NP.

NP is the set of all decision problems that can be solved in polynomial time on a non-deterministic Turing machine. Already, we have discussed about deterministic and non-deterministic Turing machines. The difference between a deterministic and non-deterministic Turing machine is that the deterministic one operates like a conventional computer, performing each instruction in sequence, forming a computational path, whereas a non-deterministic Turing machine can form a 'branch off' where each branch can execute a different statement in parallel, forming a computational tree. The time required to solve the NP problem using any currently known algorithm increases very quickly as the size of the problem grows.

Definition: NP is defined as the set of all decision problems for which an algorithm exists which can be carried out by a non-deterministic Turing machine in polynomial time.

Consider some examples.

Example 12.13 **Prime Factorization:** A number N can be decomposed into prime numbers where $N \geq 2$ such that $N = p_1 \times p_2 \times \ldots\ldots p_k$, where p_i are k prime numbers. But there are some numbers which have only two prime factors. 'Prime factorization' is finding which prime numbers multiply together to make the original number.

Consider a number N, which has to be broken into two factors. The process is

i) Guess two numbers A and B.

ii) Multiply B times A, and check whether the result matches with N.

The checking can be performed in polynomial time.

The algorithm seems to be easy, but not really. In 2009, several researchers factored a 232-digit number, utilizing several hundreds of machines. It took more than 2 years. These types of large numbers left for prime factorization are called RSA number. There are huge amounts of cash prizes for finding the factors of these large numbers! Prime factorization of such large numbers is used for RSA public key cryptography.

Example 12.14 **Traveling Salesman Problem:** The problem is to find the minimum distance covered by a salesman who wishes to travel around a given set of cities, travelling each city exactly once and returning to the beginning.

Let us consider that an area of travelling consists of 4 cities as given in the following. The cost (distance) to travel from one city to another is given in the following matrix. We are assuming that the cost of ith node to jth node is the same as the cost of jth to ith. [They may be different (if the road is one way).]

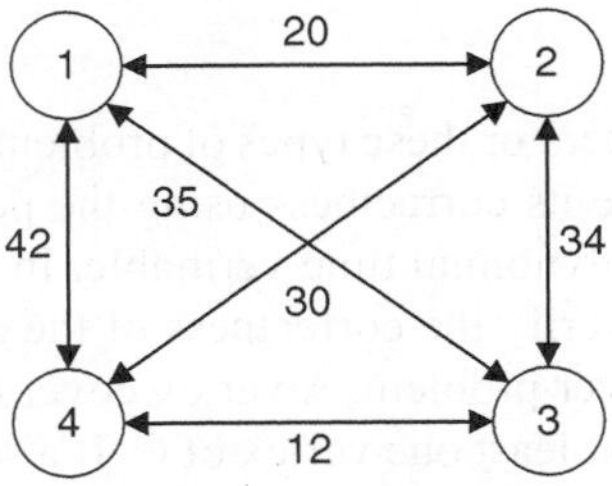

	1	2	3	4
1	0	20	35	42
2	20	0	34	30
3	30	34	0	12
4	41	30	12	0

Consider a tour starts and ends on node 1. From node 1, the person has to follow the path going to any of 2, 3, or 4 and returning to 1. In general, we can say that the tour consists of an edge $\langle 1, k \rangle, k \in V - \{1\}$ (V is the set of all vertices) and a path from k to 1. According to the condition of the problem, the salesman can travel each city exactly once. Thus, the path from k to 1 must go through all the vertices in $V - \{1, k\}$, and this path must be the shortest. Let the length of the shortest path starting from node i and covering all the nodes exactly once and ending on node I be l(i, S). If i = 1, then the function becomes

$$l(1, V-\{1\}) = \min_{2 \le k \le n} \{c_{1k} + l(k, V-\{1,k\})\}$$

c_{1k} is the cost of node 1 to node k.

In general, it can be written as

$$l(i, S) = \min_{j \in S} \{c_{ij} + l(j, S-\{j\})\}$$

where $i \notin S$.

If n = 4, i = 1, then S = {2, 3, 4}.

So, $l(1, \{2, 3, 4\}) = \min\{[c_{12} + l(2, \{3, 4\})], [c_{13} + l(3, \{2, 4\})], [c_{14} + l(4, \{2, 3\})]\}$

$$l(2, \{3, 4\}) = \min\{[c_{23} + l(3, \{4\})], [c_{24} + l(4, \{3\})]\}$$
$$l(3, \{2, 4\}) = \min\{[c_{32} + l(2, \{4\})], [c_{34} + l(4, \{2\})]\}$$
$$l(4, \{2, 3\}) = \min\{[c_{42} + l(2, \{3\})], [c_{43} + l(3, \{2\})]\}$$

$$l(3, \{4\}) = \min\{c_{34} + l(4, \varnothing)\} = 12 + 41 = 53 \; [l(4, \varnothing) = c_{41}]$$
$$l(4, \{3\}) = \min\{c_{43} + l(3, \varnothing)\} = 12 + 30 = 42 \; [l(3, \varnothing) = c_{31}]$$
$$l(2, \{4\}) = \min\{c_{24} + l(4, \varnothing)\} = 30 + 41 = 71 \; [l(4, \varnothing) = c_{41}]$$
$$l(4, \{2\}) = \min\{c_{42} + l(2, \varnothing)\} = 30 + 20 = 50 \; [l(2, \varnothing) = c_{21}]$$
$$l(2, \{3\}) = \min\{c_{23} + l(3, \varnothing)\} = 34 + 30 = 64 \; [l(3, \varnothing) = c_{31}]$$
$$l(3, \{2\}) = \min\{c_{32} + l(2, \varnothing)\} = 34 + 20 = 54 \; [l(2, \varnothing) = c_{21}]$$

$$l(2, \{3, 4\}) = \min\{[34 + 53], [30 + 42]\} = \min\{87, 72\} = 72$$
$$l(3, \{2, 4\}) = \min\{[34 + 71], [12 + 50]\} = \min\{105, 62\} = 62$$
$$l(4, \{2, 3\}) = \min\{[30 + 64], [12 + 54]\} = \min\{94, 66\} = 66$$

$$l(1, \{2, 3, 4\}) = \min\{[20 + 72], [35 + 62], [42 + 66]\} = \min\{92, 97, 108\} = 92$$

If the element of the set S is 3, there are 3 distinct choices for i. If | S | is 2, there are 2 distinct choices for i. In general, for each value of | S | there are n − 1 distinct choices for i.

Not including 1 and I, the number of distinct sets of size k that can be constructed is ${}^{n-2}C_k$. The number of steps required for TSP of node n is

$$\sum_{k=0}^{n-2}(n-1)({}^{n-2}C_k) = (n-1)\sum_{k=0}^{n-2}{}^{n-2}C_k = (n-1)\left[{}^{n-2}C_0 + {}^{n-2}C_1 + \cdots\cdots + {}^{n-2}C_{n-2}\right]$$

We know that $(1+x)^n = {}^nC_0x^0 + {}^nC_1x^1 + {}^nC_2x^2 + \cdots\cdots + {}^nC_nx^n$. If we put x = 1, it becomes 2^n. Therefore, the simplification of the previous formula is $(n - 1)2^{n-2}$.

The complexity of the problem is $O(n2^n)$, and it is an NP problem.

Example 12.15 **Vertex Cover Problem:** It is one type of optimization problem in computer science. Given a graph G, the aim is to find the minimum vertex cover $C \subseteq G$.

Given a graph G = {V, E} and an integer $k \leq V$, the problem is to find V′, subset of at most k vertices, such that each edge of G is incident to at least one vertex in V′.

The algorithm can be constructed by assuming a set of k vertices C for a given graph G. We have to check whether there exists a vertex cover of $\leq$ k vertices.

The checking can be done by the following steps.

i) Check whether C contains at most k number of vertices.

ii) Check whether C is a vertex cover of G.

```
Input G = {V, E}, C= {V'}, where | V' |≤k
for i = 1 to | E |
  check the edges ends on V'i
  remove them from {E}.
End
```

Thus, this checking operation is performed in O(| V′ |+| E |) i.e. in polynomial time.

A graph with n number of vertices can have 2^n number of such vertex combination. For each of the combinations, the same checking algorithm is performed. Among these, the set of vertices with minimum number is the minimum vertex cover of G. So, a minimum vertex cover algorithm is carried out by a non-deterministic Turing machine in polynomial time.

12.7 Polynomial Time Reducibility

In a deterministic Turing machine for a single state and single input, there is only one move. Thus, a problem fed to a deterministic Turing machine is either solved or not. *Polynomial-time reduction* is a reduction which is computable by a deterministic Turing machine in polynomial time. This is called polynomial time many-to-one reduction. A reduction which converts instances of one decision problem into instances of a second decision problem is called many to one reduction proposed by Emil Post (1944). Already we have learnt about Turing reduction where a problem A is reduced to a problem B, to solve A, assuming B is already known. Many-to-one reductions are stronger than the Turing reduction.

The polynomial time reducibility concept is used in the standard definitions of NP complete.

If P_1 and P_2 are two decision problems, then a problem P_1 is called polynomial time reducible to P_2 if there is a polynomial time algorithm which transforms an instance of P_1 to an instance of P_2.

Theorem 1: If there is a polynomial time reduction from problem P_1 to problem P_2 and if problem P_2 is in P, then P_1 also is in P.

Proof: A Problem P_1 is polynomial time reducible to P_2 if there is a polynomial time algorithm which converts an instance of P_1 to an instance of P_2. Let the size of the input of P_1 be n. According to the condition of polynomial time reducibility, the instance of P_2 from the instance of P_1 can be obtained in polynomial time. Let the time complexity of this conversion be $O(n^i)$. According to the big oh notation, f(n) is said to be O(g(n)) if there exists a positive constant C and a value k such that $f(n) \leq Cg(n)$ for all $n \geq k$. Thus, the size of the converted instance of P_2 is less than or equal to Cn^i, where C is a constant. The problem P_2 is in P. Let the complexity of solving a problem in P_2 be $O(m^k)$ where m is the size of the problem instance of P_2. Thus, the instance of P_2 of size Cn^i can be solved in complexity $O((Cn^i)^k)$. The total conversion time from the input in P_1 to the solution by P_2 (in other words the solution of problem P_1) takes $O(n^i + (Cn^i)^k) = O(n^i + C^k n^{ik}) = O(n^{ik})$. It is in polynomial time. It is proved that P_1 is in P.

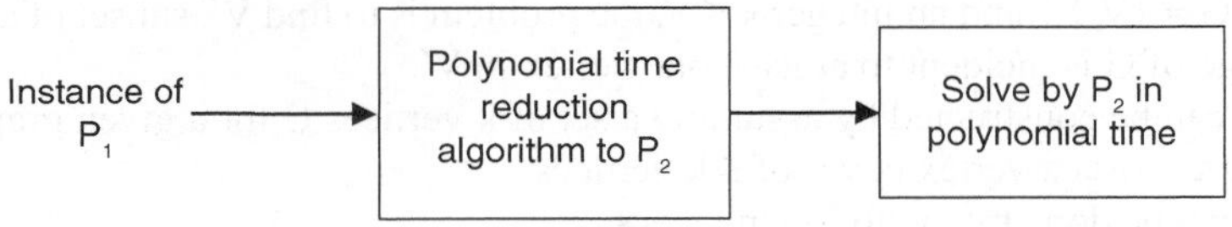

12.8 Deterministic and Non-deterministic Algorithm

An algorithm is called a deterministic algorithm if, for a function, the output generated for a given input is the same for all trials. The behaviour of this type of algorithm is predictable as the underlying machine follows the same sequence of states for a given input to produce output. A mathematical function is deterministic. Hence, the state is known at every step of the algorithm.

An algorithm is called a non-deterministic algorithm if, for a function, the output generated for a given input is different for different trails. An algorithm that solves a problem in non-deterministic polynomial time can run in polynomial time or exponential time depending on the choices it makes during execution. An algorithm is non-deterministic if the underlying machine follows a different sequence of states for a given input. Some examples of non-deterministic algorithm are spanning tree construction algorithm, searching algorithm, merge sort algorithm, vertex set covering problems, etc. To know more about these problems, follow some books on algorithm.

12.8.1 Tractable and Intractable Problem

A problem is called a *tractable problem* if there is a polynomial time algorithm to solve it for all instances. There exists a deterministic Turing machine to solve these problems. In other words, it can be said that a problem is called a tractable problem, when there exists a deterministic Turing machine M such that

- M runs in polynomial time on all input
- for $x \in L$, M gives output 1
- for $x \notin L$, M gives output 0.

A tractable problem may be of constant [O(1)], logarithmic [O(log n) or O(n log n)], linear [O(n)], or polynomial [$O(n^k)$, where k is a finite number]. The upper bound of a tractable problem is O(polynomial).

A problem is called an intractable problem if the optimal algorithm for the problem cannot solve all instances of the problem in polynomial time. These types of problems can be solved in theory, but in practice they take too long for their solutions to be useful. Intractable problems are sometimes called computationally intractable, computationally complex, or computationally hard. An intractable problem may be exponential [$O(k^n)$], factorial [$O(n!)$], or super exponential [$O(n^n)$].

constant	$O(1)$
logarithmic	$O(\log n)$
linear	$O(n)$
n-log-n	$O(n \times \log n)$
quadratic	$O(n^2)$
cubic	$O(n^3)$
exponential	$O(k^n)$, e.g. $O(2^n)$
factorail	$O(n!)$
super-exponential	e.g. $O(n^n)$

12.9 P = NP?—The Million Dollar Question

It is discussed that if a problem can be solved by a deterministic Turing machine in polynomial time, it is called P class problem. If a problem can be solved by a deterministic Turing machine in exponential time, it is called E class problem. The time complexity of E class problems are of $O(2^n)$ or $O(n^n)$. If $n = 1, 2, 3 \ldots$, then E class problem becomes P class problem. Thus, it can be said that $P \subset E$. On the other hand, if a problem can be solved by a non-deterministic Turing machine in polynomial time, it is called NP class problem. In the chapter 'Variation of Turing machine', the conversion process of a non-deterministic Turing machine to a deterministic Turing machine is discussed. There, we have to construct a computational tree consisting of the branches for accepting a string by a non-deterministic Turing machine. Then the tree is traversed by BFS. Traversing this BFS tree means visiting all possible computations of the NTM in the order of increasing length. It stops when it finds an accepting one. Therefore, the length of an accepting computation of the DTM is, in general, exponential in the length of the shortest accepting computation of the NTM. This means that all problems in class NP belong to class E, and they can be solved by a DTM in an exponential time.

It is mentioned that $P \subset E$. Now, NP also belongs to class E. Now, can it be told that P = NP or P ≠ NP? It is one of the unsolved problems in computer science. NP class problems are polynomial time verifiable. Whether S is a solution or not for an NP problem PROB can be verified easily (in polynomial time). But whether the problem PROB can be solved as easy as verification is the basis for P = NP? As an example, consider the subset sum problem. The problem is as follows: is there a non-empty subset from a set of integer numbers whose sum is zero? Let a set of integer numbers be $\{1, 2, -3, 4, -5, 7, 1\}$. Does a subset of this set produce 0? The answer is affirmative as a subset $\{1, -3, -5, 7\}$ produces 0. Thus, the answer is easily verified but does there exist an algorithm to find the answer (construction of the subset) in polynomial time? Till now no such algorithm is found, though there exists an algorithm which can find the answer in $O(2^n)$ complexity, and thus NP.

The P = NP problem was first introduced by S. Cook in his research paper 'The complexity of theorem proving procedures' in 1971 at the third annual ACM symposium on theory of computing. The Clay mathematical institute has declared a one million dollar prize for solving this question.

Scott Aaronson of the MIT believes that 'There is value for creative leaps in the world because P ≠ NP. If P = NP, then everyone who could appreciate a symphony would be Mozart; everyone who could follow a step-by-step argument would be Gauss.'

Still, research is on to find whether P = NP or P ≠ NP. You can also try. If you succeed, then one million dollar is yours! No need for further study to get a job!

12.10 SAT and CSAT Problem

12.10.1 Satisfiability Problem (SAT)

This problem plays an important role in NP problem. So, the discussion of this problem is very important before discussing topics such as NP complete, Cook's theorem, etc.

In mathematics a formula is called satisfiable if it is possible to find an interpretation that makes the formula to return true.

12.10.2 Circuit Satisfiability Problem (CSAT)

A circuit consisting of AND, OR, and NOT GATE (because these three are called basic GATE) can be represented by a Boolean function. Each element of the circuit has a constant number of Boolean input and output. Boolean values are defined on the set of {0, 1} where '0' means FALSE and '1' means TRUE.

Symbolically, OR is represented as $\vee$, AND is represented as $\wedge$, and NOT is represented as $\neg$. The truth tables representing these three gates can be found in Chapter 1.

Given a Boolean function consisting of AND, OR, and NOT gate with single output. Is there an assignment of values to the circuit's inputs in such a way as to make the function evaluate to TRUE? The C-SAT problem is sometimes called as Boolean satisfiability problem.

Before discussing the details, we have to know some basic terms.

Literal: A literal is either a variable or negative of a variable. X_1 and $\neg X_1$ are literal. $X_1 \vee \neg X_1$ always evaluate to true.

Disjunction: Disjunction is defined as a logical formula having one or more literals separated only by ORs. $(X_1 \vee \neg X_2 \vee X_3)$ is a disjunction but $(X_1 \vee \neg X_2 \wedge X_3)$ is not. Disjunction is sometimes referred to as clause.

Conjunction: A logical conjunction is defined as an AND operation on two or more logical values. $(X_1 \vee \neg X_2) \wedge (X_2 \vee X_3)$ is a logical conjunction.

Conjunctive Normal Form (CNF): A Boolean formula is known to be in conjunctive normal form if it is a conjunction of disjunction (clauses). A formula in CNF can contain only AND, OR, and NOT operations. A NOT operation can only be used as a part of literal. Like $\neg X_2$, but not $\neg(X_2 \vee X_3)$.

The examples in (a), (b), and (c) are in CNF

a) $(X_1 \vee \neg X_2) \wedge (X_2 \vee X_3)$
b) $(X_1 \vee \neg X_2 \vee X_3) \wedge (X_2 \vee \neg X_1)$
c) $X_1 \wedge (\neg X_2 \vee X_3) \wedge X_3$

The following three examples in (d), (e), and (f) are not in CNF

d) $\neg(X_1 \vee \neg X_2)$ [NOT problem]
e) $\neg X_2 \vee (X_1 \wedge X_3)$ (AND problem)
f) $X_1 \wedge (X_2 \vee (X_1 \wedge X_3))$ (OR problem)

These three formulas can be converted to equivalent CNF as in (g), (h), and (i).

g) $\neg X_1 \wedge X_2$
h) $(\neg X_2 \vee X_1) \wedge (\neg X_2 \vee X_3)$
i) $X_1 \wedge (X_2 \vee X_1) \wedge (X_2 \vee X_3)$

CNF is used in SAT problems for detecting conflict and to remember partial assignments that do not work.

(In the context-free grammar chapter, we learnt about the Chomsky normal form. Do not mess this CNF with the conjunctive normal form.)

2 SAT: A SAT problem is called 2 SAT if the number of literals in each disjunction (clause) is exactly 2.

3 SAT: A SAT problem is called 3 SAT if the number of literals in each disjunction (clause) is exactly 3.

SAT is in NP because any assignment of Boolean values to Boolean variables that are claimed to satisfy the given logical expression can be verified in polynomial time by a deterministic Turing machine. (Can be proved)

Prove that 2 SAT problem is in P

Proof: Let φ be an instance of 2 SAT. Generate a graph $G(\varphi)$ by the following rules.

1. Vertices of the graph G are the variables and their negations. If the number of variables is n, then the number of nodes is 2n.
2. If there is a clause $(\neg\alpha \vee \beta)$ or $(\beta \vee \neg\alpha)$ in φ, then an edge $\alpha \rightarrow \beta$ is added in G.
3. If there is an edge from $\alpha \rightarrow \beta$ in G, then an edge $\neg\beta \rightarrow \neg\alpha$ is added in G.
4. If there is a clause $(\alpha \vee \beta)$ in φ, then two edges $\neg\alpha \rightarrow \beta$ and $\neg\beta \rightarrow \alpha$ are added in G.

Let us take an example $\varphi = (x_1 \vee x_2) \wedge (x_1 \vee \neg x_3) \wedge (\neg x_1 \vee x_2) \wedge (x_2 \vee x_3)$.

1. There are three variables x_1, x_2, and x_3 in φ. Thus, the graph contains 6 nodes namely x_1, $\neg x_1$, x_2, $\neg x_2$, x_3, and $\neg x_3$.
2. $(x_1 \vee \neg x_3)$ is in the form $(\beta \vee \neg\alpha)$. Thus, an edge $x_3 \rightarrow x_1$ is added to G. According to the rule (c), an edge $(\neg x_1 \rightarrow \neg x_3)$ is also added to G.
3. By the same rule, $x_1 \rightarrow x_2$ and $(\neg x_2 \rightarrow \neg x_1)$ are also added to G.
4. $(x_1 \vee x_2)$ is in the form $(\alpha \vee \beta)$. Thus, two edges $(\neg x_1 \rightarrow x_2)$ and $(\neg x_3 \rightarrow x_1)$ are added to the graph.
5. By the same rule, $(\neg x_2 \rightarrow x_3)$ and $(\neg x_3 \rightarrow x_2)$ are added to the graph for $(x_2 \vee x_3)$.

φ is unsatisfiable if and only if there is a variable x such that there are edges from x to $\neg$x and $\neg$x to x in G.

Suppose such x exists. x may have any value between 0 and 1. Consider x = 1 and an edge x to $\neg$x. (Consider x as α and $\neg$x as β.) It means there is a clause $(\neg x \vee \neg x)$ or $(\neg x \vee \neg x)$ (both are the same) in φ. The clause $(\neg x \vee \neg x)$ returns 0 if x = 1. This makes the total φ as 0 [because the sub-clauses are attached with $\wedge$] which means unsatisfiable.

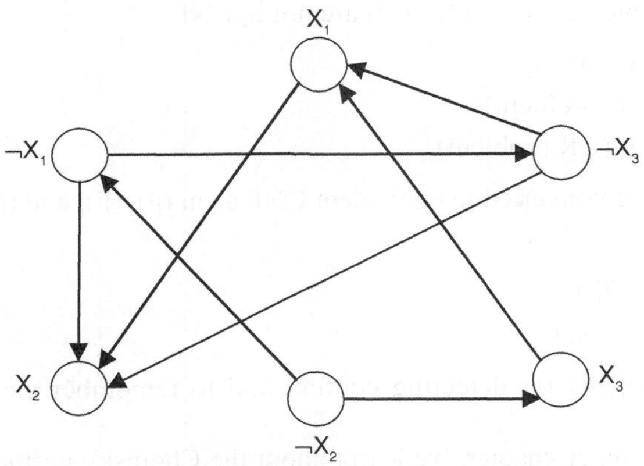

The same thing is true for an edge $\neg x$ to x in G.

For proving the 'only if' part, we have to take the help of contradiction. Consider φ is unsatisfiable.

1. If there is a path from α to ¬α for a node α, then α must be assigned to false.
2. If there is no path from α to ¬α for a node α, then those nodes which are reachable from α must be assigned to false. [According to the rule that if there is an edge α → β, then there are clauses (¬α ∨ β) or (β ∨ ¬α) in φ.]
3. Repeat this for all the nodes.

Now consider the previous graph. There is no edge from x_3 to $\neg x_3$.

If x_3 is true, then x_1 must be false, because this was added for the sub-clause $(x_1 \vee \neg x_3)$ in φ. There is an edge $x_1 \rightarrow x_2$ for the clause $(\neg x_1 \vee x_2)$ in φ. If x_1 is true, then x_2 is false, which means x_3 must be false and $\neg x_3$ must be true.

There is an edge $(\neg x_2 \rightarrow x_3)$ for the clause $(x_2 \vee x_3)$ in φ. If x_2 is false, then x_3 must be false.

Here, we are getting the contradiction for x_3 and $\neg x_3$.

It proves that step (2) cannot exist if φ is unsatisfiable, which justifies that φ is unsatisfiable if and only if there is a variable x such that there are edges from x to ¬x and ¬x to x in G.

For a 2 SAT problem, if there are n variables, then the existence of an edge from x to ¬x and ¬x to x can be found within 2n steps. This signifies that 2 SAT is in P.

Prove that 3 SAT problem is in NP

Proof: A SAT problem φ is called 3 SAT if the number of literals in each clause is exactly 3. Let F be a CNF SAT problem. We have to convert F to a 3 CNF SAT problem F′ such that if F is satisfiable, then F′ is also satisfiable. Let the problem F contain clauses $C_1, C_2 \ldots C_k$. Here, three cases of C may occur.

1. C_i contains exactly 3 literals.
2. C_i contains less than 3 literals.
3. C_i contains more than 3 literals.

We have to concentrate on cases (2) and (3).

Solution for Case (2): If C_i contains less than 3 literals, then the number of literals may be 2 or 1. If it contains 1 literal, say L_1, then replace this by $(L_1 \vee L_1 \vee L_1)$.

If it contains 2 literals, say $(L_1 \vee L_2)$, then replace it by $(L_1 \vee L_2 \vee L_1)$.

Solution for Case (3): If C_i contains more than 3 literals, then consider new literals Y_j, where j = 1 to k − 3. If C_i is in the form $(x_1 \vee x_2 \vee \ldots\ldots \vee x_k)$, where k ≥ 4, replace it by

$$(x_1 \vee x_2 \vee y_1) \wedge (x_3 \vee \neg y_1 \vee y_2) \wedge (x_4 \vee \neg y_2 \vee y_3) \wedge \ldots\ldots \wedge (x_j \vee \neg y_{j-2} \vee y_{i-1}) \wedge \ldots\ldots \wedge (x_{k-1} \vee x_k \vee \neg y_{k-3})$$

The previous equation is true for any value of x_k and y_j.

Let C_i contain 4 literals $(x_1 \vee x_2 \vee x_3 \vee x_4)$. According to the previous equation, it is converted as

$$(x_1 \vee x_2 \vee y_1) \wedge (x_3 \vee x_4 \vee \neg y_1)$$

As $(x_1 \vee x_2 \vee x_3 \vee x_4) = 1$, any of the literals is 1 to make the clause 1. Let x_1 and x_2 be 1 but x_3 and x_4 be 0. If y_1 is 0, then the total clause will be 1. If x_3 and x_4 are 1 but x_1 and x_2 are 0, then y_1 is 1 to make the clause evaluate to 1. From the discussion, it is proved that there exists assignment of literals to make the clause evaluate to 1. Using the rules, F′ is converted to 3 CNF form.

It is also proved that if there is an assignment of literals to make F = 1, then there also exist literals to make F′ = 1.

Now, calculate the time complexity of 3 SAT problem. Let a SAT problem F contain 2 clauses where each clause contains 4 literals. F′ contains $2 \times 2 = 4$ clauses. If the number of clauses is 3 but the number of literals in each clause is 4, then F′ contains $2 \times 2 \times 2 = 2^3$ clauses. If the number of literals in each clause is 5, then F′ contains $3 \times 3 \times 3 = 3^3$ clauses. For l literals with k clauses each, F′ contains $(l - 2)^k$ clauses, where l ≥ 4 and k ≥ 1. It means the time increases exponentially to convert a SAT problem to 3 SAT. But the time to check whether a SAT is in 3-SAT or not can be checked in polynomial time. This means 3-SAT is in NP.

12.11 NP Complete

The concept of NP complete was first introduced by Stephen Cook by the circuit satisfiability problem. Then, Hopcroft and Ullman make it familiar in their book of automata theory. Already we have gained knowledge about the NP problem. These types of problems can be verified in polynomial time (easily) but cannot be solved easily by any algorithm known till date.

The NP complete problem ρ, written as NPC, is the problem which belongs to NP (i.e., can be verified easily), and any NP problem can be converted to ρ by a transformation of the inputs in polynomial time.

NP Complete: A decision problem ρ is in NP complete if

- It is in a class of NP.
- A known NP complete problem can be reducible to ρ in polynomial time.

To prove that a problem is NP complete, first prove that it is in NP, and then reduce some known NP complete problem to it.

The Boolean satisfiability problem (CSAT), traveling salesman problem, subset sum problem, Hamiltonian path problem, vertex cover problem, graph colouring problem, knapsack problem, and N-puzzle problem are some examples of NP complete problem. All these problems can be reducible to the CSAT problem, and thus CSAT is NP complete. TSP can be reducible to the Hamiltonian path problem which is reducible to the vertex cover problem.

Theorem 2: Let P_1 and P_2 be two decision problems in NP, and P_1 is NP-complete. Problem P_2 is NP complete if there is a polynomial time reduction for P_1 to P_2.

Proof: Let L be a language which belongs to an NP complete problem. P_1 is an NP complete problem. According to the definition of the NP complete problem, every problem can be reducible to P_1 in polynomial time. So, L can be reduced to P_1 in polynomial time. Let w be a string of L of size n. The string w can be converted to an instance of P_1, say w′, in polynomial time, say ρ(n), where ρ is a polynomial.

Let P_1 be reducible to P_2 in polynomial time. Suppose this conversion takes ρ′(n) time. Hence, the time taken to transform w to an instance of P_2 (via an instance of P_1) is ρ(n) + ρ′(ρ(n)). The addition of two polynomials is a polynomial. Thus, L is reducible to P_2 in polynomial time.

- P_2 is in NP
- An NP problem L is reducible to P_2 in polynomial time.

So, we can decide that P_2 is NP complete.

Theorem 3: If any NP complete problem is solvable in P, P = NP proves.

Proof: Let ρ be an NP complete problem which is solvable in polynomial time. Let ρ′ be another NP complete problem. According to the definition of NP completeness, ρ′can be reduced to ρ in polynomial time. According to theorem 1, as ρ is in P, ρ′ is also in P. A problem is NP complete means it is in NP. Thus, it is proved that NP = P.

12.11.1 Cook–Levin Theorem

In complexity theory, the Cook–Levin theory is sometimes known as the Cook theorem. The theorem states that *SAT is an NP complete problem.*

Proof: To prove any problem as NP complete, we need to follow two steps.

- The problem is in NP
- Every other language in NP is polynomial time reducible to that problem.

For SAT, the first one is proved in the case of the 3 SAT problem. This section contains the proof of the section (b).

Idea: Consider N to be a non-deterministic TM that decides A ∈ NP. The reduction process takes a string w ∈ A, and generates a Boolean formula F such that N accepts w ⇔ F is satisfiable (the symbol ⇔ signifies denotes). In particular, we choose the Boolean formula F such that its satisfying assignment corresponds to the accepting computation for N to accept string w ∈ A.

The following section describes the process to make the idea come true.

Let N be the non-deterministic Turing machine which decides an NP problem A. Let n^k be the running time of the Turing machine N on an input of length n. Here, k is some constant.

Define a table of dimension $n^k \times n^k$ for N on input w. Each row of the table stores a configuration in the branch of computation of N on w. The table is as follows.

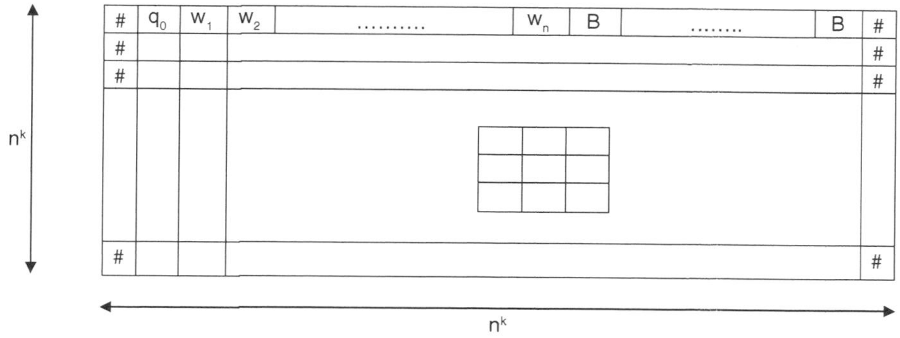

All configuration start and end with an end marker #. Configurations are represented row wise, where the first row represents the starting configuration and the subsequent rows follow its previous row. The table is declared in accepting configuration or, in other words, an accepting computation is achieved if any row of the table is in accepting configuration.

It means deciding whether the Turing machine N accepts w is the same as to decide whether an accepting table for N on w exists. So, our task is now simplified to find a formula F that can check if such an accepting table exists.

The formula F is a combination of some variables. Consider a variable $X_{i,j,s}$ for a cell label (i, j) in the table (i is the number of rows and j is the number of columns) and s is a symbol $\in Q \cup \Gamma \cup \#$. Let us represent this notation as C. The variable $X_{i,j,s} = 1$ if the cell label (i, j) contains symbol s.

For making the formula F to be satisfiable, it must be assured that

- Each cell of the table contains only one symbol.
- The table has accepting configuration.
- Each row of the table is correct.

For the previously mentioned three cases, consider three sub-formula f_{cell}, f_{accept}, and $f_{correct}$, respectively. The formula F is satisfiable if the three formulas are satisfiable. Thus, we can write

$$F = f_{cell} \wedge f_{accept} \wedge f_{correct}$$

A cell contains only one symbol if it contains at least one symbol and at most one symbol. Let f_{cell_1} and f_{cell_2} be two sub-functions for the two conditions.

1. $f_{cell_1} = \vee_{x \in C} X_{ij}$

2. $f_{cell_2} = \wedge_{s,t \in C, s \neq t}((-x_{i,j,s}) \vee (-x_{i,j,t}))$

Thus, $f_{cell} = f_{cell_1} \wedge f_{cell_2}$

For f_{accept}, i.e., to ensure that the table has accepting configuration, the notation is

$$f_{accept} = \vee_{i,j} X_{i,j,q_{accept}}$$

Ensuring each row of the table is correct means to ensure that the first row of the table is correct and the subsequent rows are correct. The following sub-formula is used to ensure that the starting row is correct.

$$f_{Start} = x_{1,1,\#} \wedge x_{1,2,q_0} \wedge x_{1,3,w_1} \wedge .... \wedge x_{1,n+1,w_{n-1}} \wedge x_{1,n+2,w_n} \wedge x_{1,n+3,B} \wedge ... \wedge x_{1,n^{k-1},B} \wedge x_{1,n^k,\#}$$

A window of size 2×3 known as legal window is used to ensure that the remaining rows are correct.

(i, j)	(i, j + 1)	(i, j + 2)
(i + 1, j)	(i + 1, j + 1)	(i + 1, j + 2)

A window of cell (i, j) refers to a group of 2×3 cells having label as mentioned previously. A legal window is a window which does not violate the actions of the transitional functions of the Turing machine N, assuming the configuration of each row follows legally from the configuration of its upper row.

Let us concentrate more on the legal window.

1. The following window is legal for the transitional function $\delta(q_1, b) \rightarrow (q_2, a, L)$.

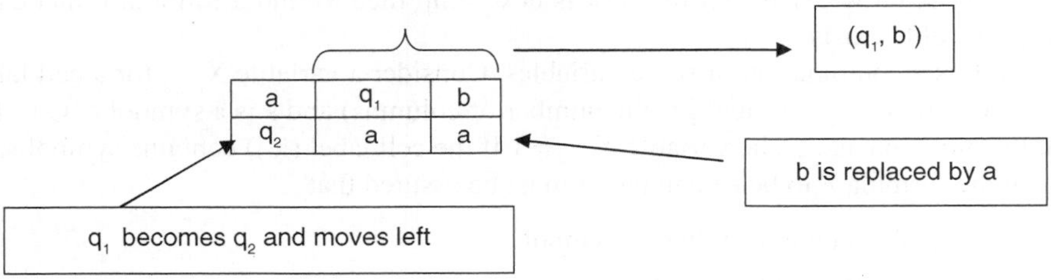

2. The following window is legal for the transitional function $\delta(q_1, b) \rightarrow (q_2, a, R)$.

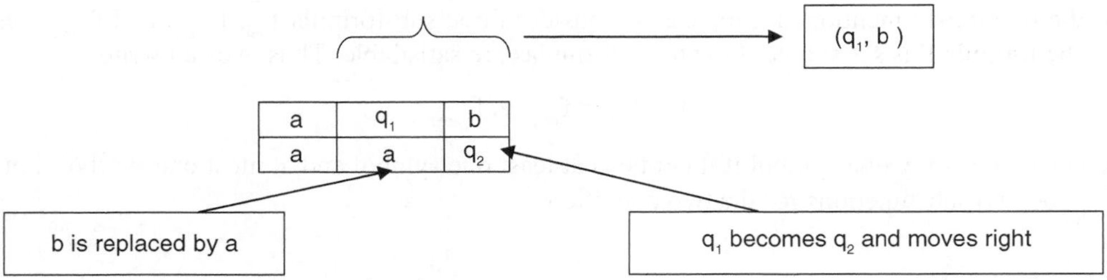

3. The following window is legal for the transitional function $\delta(q_1, a) \rightarrow (q_2, b, R)$.

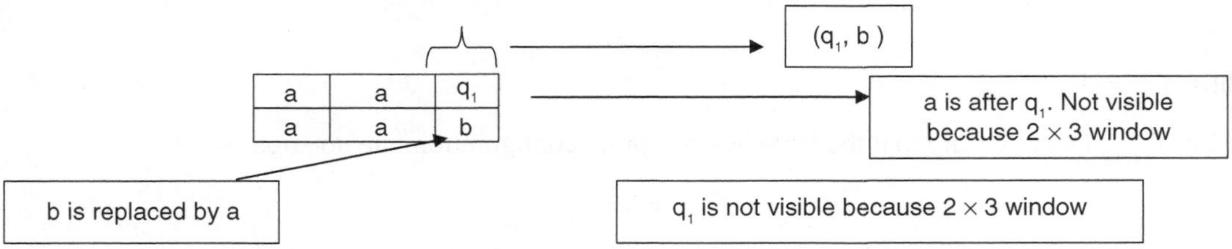

4. The following window is legal for the transitional function $\delta(q_1, b) \rightarrow (q_2, a, L)$.

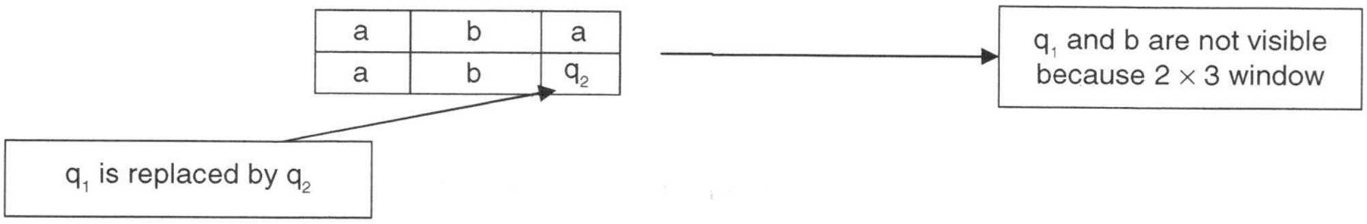

5. The following window is legal for the transitional function $\delta(q_1, b) \rightarrow (q_2, a, L)$.

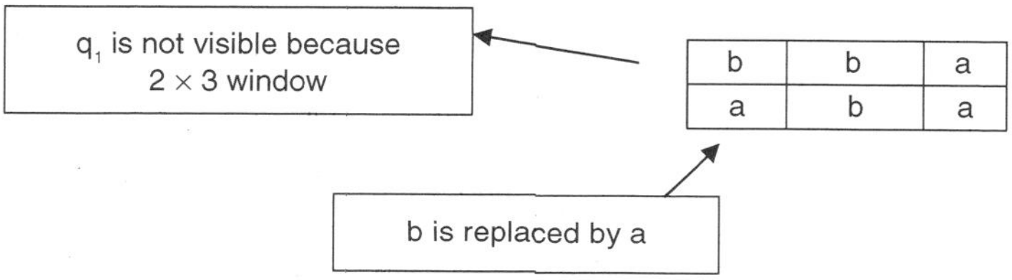

6. The following window is also legal.

#	b	a
#	b	a

Based on the concept of the legal window, the sub-formula f_{move} is constructed as

$$f_{move} = \wedge_i \geq 1, j \leq n^k - 2(\text{window}_{i,j}\text{correct})$$

where the notation 'windows$_{i,j}$ correct' expressed as $\vee_{a_1, a_2, \ldots a_6}$ is a legal window.

$$\left(x_{i,j,a_1} \wedge x_{i,j+1,a_2} \wedge x_{i,j+2,a_3} \wedge x_{i+1,j,a_4} \wedge x_{i+1,j+1,a_5} \wedge x_{i+1,j+2,a_6}\right)$$

From the previous discussions, we can now construct F as

$$F = f_{cell} \wedge f_{accept} \wedge f_{correct}$$
$$= (\wedge_{i,j}\,(f_{cell_1} \wedge f_{cell_2})) \wedge f_{accept} \wedge (f_{start} \wedge f_{move})$$

Here, F is satisfiable means its satisfying assignment represents an n^k X n^k accepting table. This implies that N has an accepting computation on input w, i.e., N accepts w.

In the reverse, if N accepts w, then there must be an accepting computation, and F has a satisfying assignment, i.e., F is satisfiable.

This means for any w, we can construct a Boolean formula F such that N accepts w, which implies F is satisfiable. The construction of the Boolean formula F gives a reduction from deciding a language in NP to deciding whether a formula is in SAT.

Let w be of length n. So, the sub-formula f_{start} can be constructed in $O(n^k)$ time.

The sub-formula $(\wedge_{i,j}\,(f_{cell_1} \wedge f_{cell_2}))$, f_{accept}, $f_{move})$ can be construct in $O(n^{2k})$ time.

So, F can be constructed in polynomial time. Thus, any language in NP is polynomial time reducible to SAT and SAT is in NP, which implies SAT is NP complete.

Example 12.16 Prove that 3 SAT is NP complete.

Solution:

i) 3 SAT is clearly in NP, given an instance of a circuit C and a 'satisfying' setting of Boolean variables. A simulation of the circuit can easily check in polynomial time whether the output is 1 or not.

ii) A SAT problem φ is called 3 SAT if the number of literals in each clause is exactly 3. It is already shown how a SAT formula with clauses less than or more than 3 literals can be converted to a formula with clauses each having exactly 3 literals. This conversion can be made in polynomial time. Hence, SAT can be reduced to 3 SAT in polynomial time. So 3 SAT is NP complete.

Example 12.17 Prove that the vertex cover problem is NP complete.

Solution:

i) That the vertex cover problem is in NP is already proved.

ii) Reduce 3-SAT to vertex cover.

Reduction:

i) For each variable of the Boolean expression in 3 SAT, create two vertices v_i and $\neg v_i$, and connect them by edges. For N such variables, there are 2N vertices and N edges. Let the graph be G_1

ii) To cover G_1, at least n vertices must be taken, one for each pair.

iii) For each clause of the Boolean expression in 3 SAT, create 3 vertices and connect them by edges. If a Boolean expression contains C clauses, then the number of vertices will be 3C. Let the graph be G_2.

iv) To cover G_2, at least 2C vertices must be taken. (To cover each triangle, at least 2 vertices are required.)

v) Finally, the vertices of G_1 and G_2 which share the same literal are connected to produce a graph G with N + 3C vertices.

Example: Let the 3 SAT Boolean formula be

$$F = (U \vee \neg W \vee \neg X) \wedge (\neg U \vee V \vee \neg X)$$

There are 4 variables U, V, W, and X. So, there are 8 vertices and 4 edges.

Construction of G_1

Construction of G_2

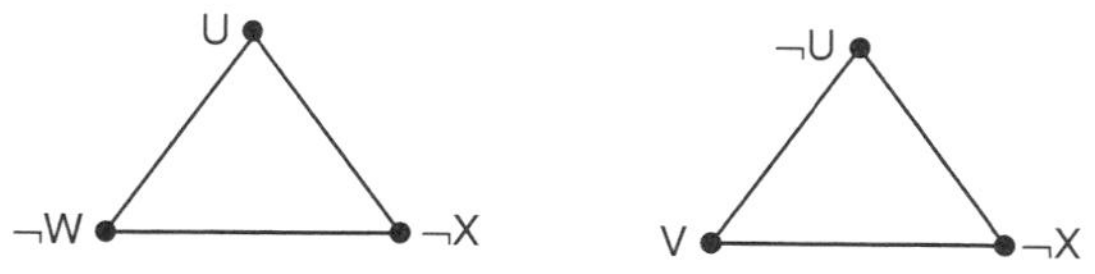

Construction of G

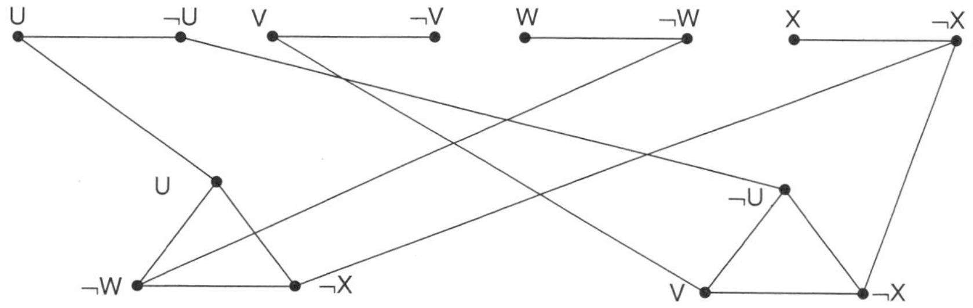

If is it proved that the graph G has a vertex cover of size N + 2C, if and only if the expression is satisfiable, then the reduction is correct.

Proof: Already it is shown that, any cover of G must have at least N + 2C vertices. To prove that the reduction is correct, it must be shown that:

- ***Every satisfying truth assignment gives a cover.***

  Take a Boolean formula in 3 SAT with N literals. Taking N vertices, at least one of the three edges constructing a clause is covered. Taking the other two vertices (total 2C) will cover the other two edges.

- ***Every vertex cover gives a satisfying truth assignment.***

  Every vertex cover must contain N vertices from G_1 and 2C vertices from G_2. Let the vertices from G_1 define the truth assignment. To give the cover, at least one of the three edges of the triangle constructed from a clause is incident on the vertex. By adding the other two vertices (total 2C) to the cover, all the edges of G_2 are associated with it, which forms a cover of N + 2C vertices.

  From here, it is proved that every 3 SAT defines a cover and every cover gives the TRUE value for a Boolean expression in 3 SAT. This conversion is possible in polynomial time.

  As the reduction is possible in polynomial time, we can conclude that the vertex cover problem is NP complete.

Co-NP: The complement of every NP language is Co-NP.

The complement of language in P is also in P. All P are also in NP. The complement of the NP-complete problem is called the Co-NP complete problem, which is a subset of the Co-NP problem.

The relation of NP, P, and Co-NP is given in the following.

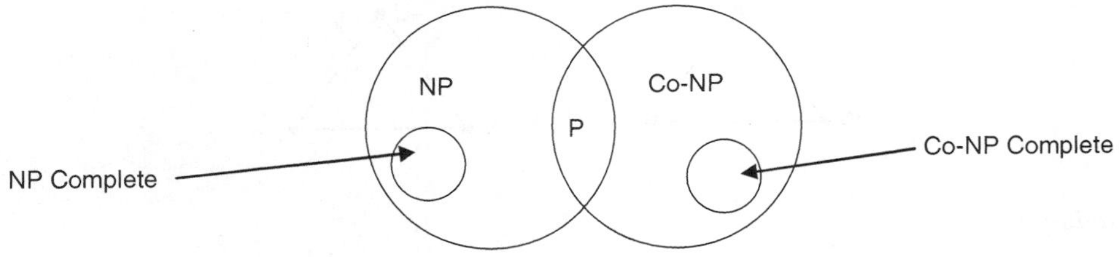

If it is proved that P = NP, then P = NP = Co-NP. NP = Co-NP if there exists some NP complete problem whose complement is in NP. NP = Co-NP? is another unsolvable problem in computer science.

12.12 NP Hard

In complexity theory, the NP hard problems are known informally 'at least as hard as the hardest problems in NP'. A problem P is called NP hard if there is an NP complete problem P_1, which is Turing reducible to P in polynomial time. In this context, we need to know about another machine called the Oracle machine. The Oracle machine is an abstract machine which can be thought of as a Turing machine with a black box. This abstract machine is used to decide certain decision problems in single operation. An Oracle machine can perform all operations performed by a Turing machine, as well as it can answer some questions in the form 'A is a set of natural numbers. If $x \in A$?'. We can say in other words that a problem P is called NP hard if there is an NP complete problem P_1 where P_1 can be solved in polynomial time by an Oracle machine with an oracle for P.

12.12.1 Properties of NP Hard Problems

- P is used to solve P_1, so P is as hard as P_1.
- P does not have to be in NP; for this reason, it does not have to be a decision problem.
- If for any NP-hard problem there is a polynomial time algorithm, then all problems in NP can be solved in polynomial time. From here, we can prove P = NP.
- P = NP does not resolve that NP-hard problems can be solved in polynomial time.

From the discussions of NP-hard problems, we can construct a diagram showing the relations between P, NP, NP complete, and NP hard.

Example:

- Minimum vertex cover problem.
- Given a program and its input, will it run forever?—Halting problem.
- Maximum clique problem.

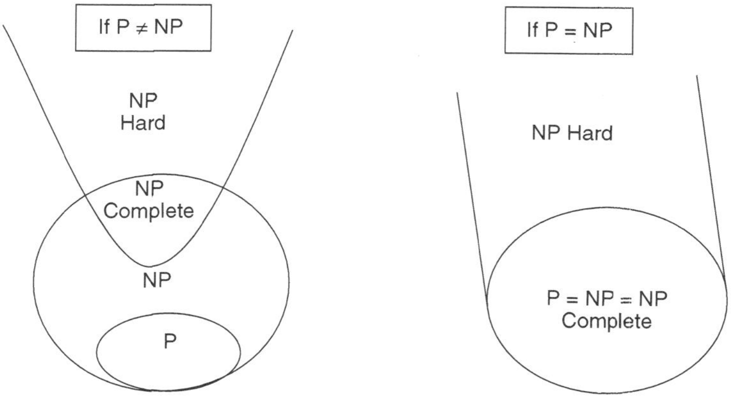

The following are the set of NP complete problems. Here, the arrow direction from A → B signifies that problem A is reducible to problem B.

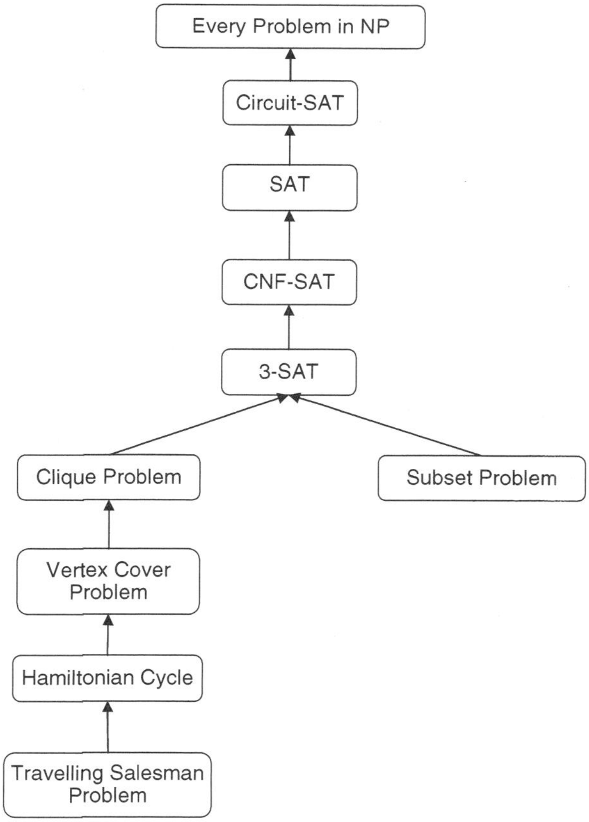

Following are the descriptions of the given problems.

- **Clique problem:** Let a graph G = (V, E), where V is the set of vertices and E is the set of edges. A clique is described as a complete sub-graph of G, that is, any two vertices in the sub-graph are joined by an edge. The size of a clique is denoted by the number of vertices that exist in the sub-graph. The clique problem is 'given a graph G, is there a clique of size n?'
- **Vertex cover problem:** Let a graph G = (V, E), where V is the set of vertices and E is the set of edges. A vertex cover of the graph is a set of vertices such that each edge of the graph is incident to at least one vertex of the set. The vertex cover problem is 'given a graph G, does G has a vertex cover of size n?'
- **Hamiltonian cycle problem:** Let G be an undirected graph of V vertices. The Hamiltonian path is a path which visits every vertices of G exactly once. The Hamiltonian cycle is a Hamiltonian path which contains a cycle. The Hamiltonian cycle problem is 'given a graph G, does G have a Hamiltonian cycle?'
- **Subset sum problem:** Given a set of integer Z and another integer n, the problem is whether there is a non-empty subset S, sum of whose elements is n?

12.13 Space Complexity

Space complexity was mentioned in Section 12.1. To execute an algorithm, memory space is necessary. Although elementary, the Turing machine input tape cell is one kind of a resource. Space complexity of a program is the number of elementary objects that it requires to store during execution. This is computed with respect to the size of the input data. Space complexity shares many features of time complexity. In this section, the Turing machine is considered as a computing model and space means the tape space consumed during the execution of the algorithm.

Definition: Let M be a deterministic Turing machine designed for algorithm A and that halts on all inputs. The space complexity of M is defined as a function f: N → N, where f(n) is the maximum number of tape cells scanned by M on any input of length n.

Let M be a non-deterministic Turing machine designed for algorithm A and where all branches halt on all inputs. Here, space complexity f(n) is the maximum number of tape cells that M scans on any branch during the computation phase on any input of length n.

For an algorithm A and input I, space complexity SPACE(A, I) is denoted as the number of cells (maximum for NTM) used during the computation of A(I) (input I applied on A), when A(I) must halt. SPACE(A, I) = O(f(n)) if | I | = n.

Polynomial Space or PSPACE: The complexity of the decision problems which can be solved by a deterministic Turing machine and a polynomial amount of space are called polynomial space complexity problem or PSPACE problem.

The class PSPACE contains all classes of SPACE(f(n)), where f(n) is a polynomial of finite degree.

Non-polynomial Space or NSPACE: The space complexity for an algorithm performed by a non-deterministic Turing machine is denoted as NSPACE. NSPACE(f(n)) is the set of decision problems that can be solved by a non-deterministic Turing machine with the use of O(f(n)) amount of space. Here, f(n) is the maximum number of tape cells that NTM scans on any branch during the computation phase on any input of length n and must halt.

From the previous discussions, it is clear that PSPACE ⊆ NSPACE.

Relation Between Time and Space

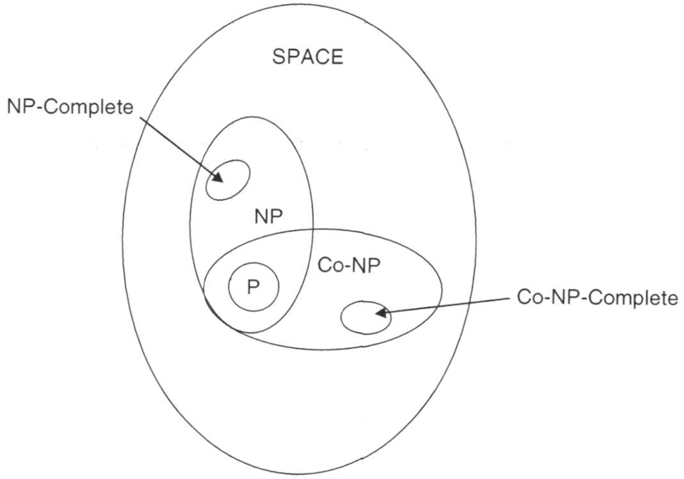

What We Have Learned So Far

1. Algorithms may differ in time taken and/or memory required to execute and produce result, for the same input.
2. Mainly, two types of complexities are measured for an algorithm:- time complexity and space complexity.
3. The time complexity of a program for a given input is the number of elementary instructions (the number of basic steps required) that this program executes.
4. The space complexity of a program for a given input is the number of elementary objects that this program needs to store (i.e., memory consumed) during its execution.
5. The growth rate of any exponential function is greater than that of any polynomial.
6. The well-known notations for the growth rate of functions are big oh notation (O), big omega notation (Ω), theta notation (Θ), little-oh notation (o), and little omega (ω).
7. The problems are mainly classified as decision problem, optimization problem, search problem, and functional problem.
8. An algorithm is known as a polynomial time complexity problem if the number of steps (or time) required to complete the algorithm is a polynomial function of n (where n is some non-negative integer) for a problem of size n.
9. NP is defined as the set of all decision problems for which an algorithm exists which can be carried out by a non-deterministic Turing machine in polynomial time.
10. Polynomial-time reduction is a reduction which is computable by a deterministic Turing machine in polynomial time.
11. If there is a polynomial time reduction from problem P_1 to problem P_2 and if problem P_2 is in P, then P_1 is also in P.

12. A decision problem ρ is NP complete if
    a. it is in a class of NP.
    b. Every NP problem can be reducible to ρ in polynomial time.
13. If any NP complete problem is solvable in P, P = NP proves.

Solved Problems

1. Find the complexity of the following algorithm.

```
int powerA(int x, int n)
  {
    if (n==0)
      return 1;
    if (n==1)
      return x;
    else
      return x * powerA(x, n - 1);
  }
```

Solution: If n = 1, then the result is x. Returning x takes unit time. If n > 1, then the algorithm enters into the else part. This part contains a multiplication and a same algorithm repetition up to n – 1 time. So, we can write T(n) = T(n – 1) + 2 if n > 1.

So, we can write

$$T(n) = T(n-1) + 2 \quad \text{if} \quad n > 1$$
$$T(n) = 1 \quad \text{if} \quad n = 1$$

$$\vdots$$

$$T(n) = T(n-1) + 2$$
$$T(n-1) = T(n-2) + 2$$
$$T(n-2) = T(n-3) + 2$$
$$\vdots$$
$$T(2) = T(1) + 2$$
$$T(1) = 1$$

Adding all these, we get

$$T(n) = (n-1) * 2 + 1 = 2n - 1$$

So, the complexity is O(n). Here, 2n – 1 is a polynomial of n of degree 1.

2. Write the divide and conquer algorithm for merge sort and find its complexity.

[Sri Venkateswara University 2008]

Solution: In merge sort, the n element array is divided into two halves each with n/2 elements (for n = even), or one with n/2 and the other with n/2 + 1 elements. The two halves are sorted and merged. The algorithm for performing merge sort is as follows:

```
mergeSort(int min, int max)
{
  if(min < max)
   {
     m = (min + max)/2
     mergeSort(min, m)
     mergeSort(m + 1, max)
     Merge(min, m, max)
   }
}
```

```
Merge(min, m, max)
{
  i = min, j = min + 1, k = 0
  while((i ≤ min) && j ≤ max)
   {
     if(A[i] < A[j])
      {
        temp[k++] = A[i]
        i++
      }
    elseif(A[i] > A[j])
      {
        temp[k++] = A[j]
        j++
      }
        else
        {
          temp[k++] = A[i]
          temp[k++] = A[j]
        }
   }
  while(i ≤ m)
  {
    temp[k++] = A[i++]
  }
  while(j ≤ q)
  {
    temp[k++] = A[j++]
  }
  for(i = 0, i < n, i++)
  {
  A[i] = temp[i]
  }
}
```

The algorithm `Merge(min, m, max)` is performed in O(n). The algorithm `mergeSort(int min, int max)` of n element becomes half when it is called recursively. Thus, the whole algorithm takes

$$T(n) \le 2T\left(\frac{n}{2}\right) + cn$$

$$\leq 2\left[2T\left(\frac{n}{2^2}\right)+c\frac{n}{2}\right]+cn$$

$$T(n) \leq 2^2 T\left(\frac{n}{2^2}\right)+2cn$$

. . . .

. . . .

$$T(n) \leq 2^i T\left(\frac{n}{2^i}\right)+icn$$

$$\text{If } 2^i = n,\ i = \log_2 n$$

$$T(n) \leq 2^i T(1) + cn \log_2 n$$

T(1) = 1 (time required to sort a list of one element).
Thus, the complexity of merge sort is O(n $\log_2$ n).

3. Prove that the k clique problem is NP complete.

Solution:

Definition: Given a graph G = {V, E} and an integer k, check whether there exists a sub-graph C of G which contains k number of vertices.

(The max clique problem is finding a clique with largest number of vertices. It is an NP hard problem.)

To prove that clique is NP complete, we need to prove:

A. Clique is in NP

- Check whether C contains k number of vertices. This can be done in O(| C |).
- Check whether C is a complete sub-graph. This can be done in ${}^kC_2$ steps.

Thus, checking whether C is a clique of G or not can be performed in polynomial time.

A graph with n number of vertices can have 2^n number of such vertex combination. For each of the combinations, the same checking algorithm is performed. Among these, the clique with k vertices is the k-clique of G. So, a k-clique algorithm is carried out by a non-deterministic Turing machine in polynomial time. Hence, it is in NP.

B. Reduce 3-SAT to k-clique

Reduction:

- For each clause of a Boolean assignment, create distinct vertices for each literal. If a Boolean assignment contains C clauses, then the number of vertices is 3C.
- Join the vertices of the clauses in such a way that
  - Two vertices of the same clause cannot be joined.
  - Two vertices whose literals are the negation of the other cannot be joined.

Example: Consider $F = (X_1 \vee X_2 \vee Y_1) \wedge (X_3 \vee \neg Y_1 \vee \neg Y_2) \wedge (X_4 \vee X_5 \vee \neg Y_2)$

The graph is

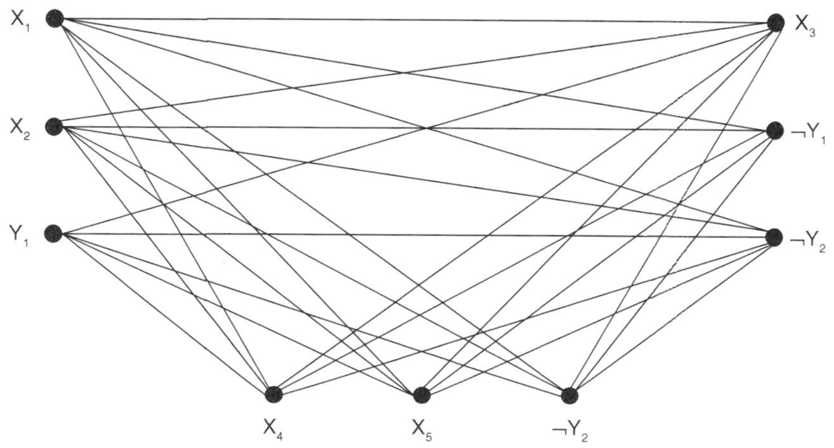

If it can be proved that G has a k-clique if and only if F is satisfiable, then the reduction is correct.

- **If F is satisfiable, G has k-clique:** If the 3-SAT instance F is TRUE, then each clause is TRUE, which means every clause has a TRUE literal. By selecting a corresponding vertex to a TRUE literal in each clause, a clique in G of size k is constructed, where k is the number of clauses of the 3-SAT instance. Because, if there is a missing edge, that would mean that our truth assignment effectively set something to be true and false at the same time (Y_1 and $\neg Y_1$)!
- **If G-has k-clique, F is satisfiable:** Assume that there is a clique of size k in G where k is the number of clauses in F. It is to be proved that there must be a truth assignment that satisfies the given Boolean formula. Clique is a complete graph. Let us assume that the truth assignment induced by the labels of the vertices satisfies F. It signifies that every pair of vertices in the clique has an edge. But the vertices labelling the literals may be set to both true and false. Already it is mentioned that every trio of vertices corresponding to a clause of F have no edges between those vertices, which signifies that there must be a vertex from every clause of F in the clique. This shows that the clauses of F are satisfied as well as F is satisfied.

This reduction is possible in polynomial time. Hence, clique is NP complete.

Multiple Choice Questions

1. Worst case time complexity is denoted by the notation.
   a) Big oh notation
   b) Big omega notation
   c) Theta notation
   d) Little omega notation
2. Best case time complexity is denoted by the notation.
   a) Big oh notation
   b) Big omega notation
   c) Theta notation
   d) Little omega notation

3. If f(n) = O(g(n)) and f(n) = Ω(g(n)), then which is true?
   a) f(n) = ω(g(n))
   b) f(n) = o(g(n))
   c) f(n) = Θ(g(n))
   d) f(n) = θ(g(n))
4. The problem which results in 'yes' or 'no' is
   a) Decision problem
   b) Optimization problem
   c) Search problem
   d) Functional problem
5. Which type of problem is the shortest path algorithm?
   a) Decision problem
   b) Optimization problem
   c) Search problem
   d) Functional problem
6. Which is true for the P class problem?
   a) The number of steps (or time) required to complete the algorithm is a polynomial function of n.
   b) It is computable by a deterministic Turing machine in polynomial time.
   c) It contains all sets in which the membership may be decided by an algorithm whose running time is bounded by a polynomial.
   d) All of these

Answers:

1. a 2. b 3. c 4. a 5. b 6. d

GATE Questions

1. For problems X and Y, Y is NP complete and X reduces to Y in polynomial time. Which of the following is true?
   a) If X can be solved in polynomial time, then so can Y.
   b) X is NP complete
   c) X is NP hard
   d) X is in NP, but not necessarily NP complete.
2. Which of the problems is not NP hard?
   a) Hamiltonian circuit problem
   b) The 0/1 knapsack problem
   c) Finding bi-connected components of a graph
   d) The graph colouring problem
3. Ram and Shyam have been asked to show a certain problem ∏ is NP complete. Ram shows a polynomial time reduction from 3-SAT problem to ∏, and Shyam shows a polynomial time reduction from ∏ to 3 SAT. Which of the following can be inferred from these reductions?
   a) ∏ is NP hard but not NP complete
   b) ∏ is NP, but is not NP complete
   c) ∏ is NP complete
   d) ∏ is neither NP hard nor NP
4. No body knows yet if P = NP. Consider the language L defined as follows.

$$L = \begin{cases} (0+1)^* & \text{if} \quad P = NP \\ \varphi & \text{Otherwise} \end{cases}$$

   Which of the following statements is true?
   a) L is recursive
   b) L is recursively enumerable but not recursive

c) L is not recursively enumerable
d) Whether L is recursive or not will be known after we find out if P = NP

5. Consider two languages L_1 and L_2, each on the alphabet Σ. Let f: $\Sigma \rightarrow \Sigma$ be a polynomial time computable bijection such that $(\forall x)[x \in L_1$ iff $f(x) \in L_2]$. Further, let f^{-1} be also polynomial time computable. Which of the following cannot be true?
   a) $L_1 \in P$ and L_2 is finite
   b) $L_1 \in NP$ and $L_2 \in P$
   c) L_1 is undecidable and L_2 is decidable
   d) L_1 is recursively enumerable and L_2 is recursive

6. The problems 3-SAT and 2-SAT are
   a) Both in P
   b) Both NP complete
   c) NP complete and P
   d) Undecidable and NP complete, respectively

7. Consider the following two problems on undirected graphs:
   i) Given G(V, E), does G have an independent set of size V – 4?
   ii) Given G(V, E), does G have an independent set of size 5?

   Which one of the following is TRUE?
   a) (i) is in P and (ii) is NP complete
   b) (i) is NP complete and (ii) is in P
   c) Both (i) and (ii) are NP complete
   d) Both (i) and (ii) are in P

8. Let S be an NP-complete problem and Q and R be two other problems not known to be in NP. Q is polynomial time reducible to S and S is polynomial time reducible to R. Which one of the following statements is true?
   a) R is NP complete
   b) R is NP hard
   c) Q is NP complete
   d) Q is NP hard

9. Let $SHAM_3$ be the problem of finding a Hamiltonian cycle in a graph G = (V, E) with V divisible by 3 and $DHAM_3$ be the problem of determining if a Hamiltonian cycle exists in such graphs. Which one of the following is true?
   a) Both $DHAM_3$ and $SHAM_3$ are NP hard
   b) $SHAM_3$ is NP hard, but $DHAM_3$ is not
   c) $DHAM_3$ is NP hard, but $SHAM_3$ is not
   d) Neither $DHAM_3$ nor $SHAM_3$ is NP hard

10. Let Π_A be a problem that belongs to the class NP. Then which one of the following is true ?
    a) There is no polynomial time algorithm for Π_A
    b) If Π_A can be solved deterministically in polynomial time, then P = NP
    c) If Π_A is NP hard, then it is NP complete.
    d) Π_A may be undecidable.

11. Assuming $P \neq NP$, which of the following is true?
    a) NP complete = NP
    b) NP complete $\cap$ P = $\varnothing$
    c) NP hard = NP
    d) P = NP complete

12. The recurrence relation capturing the optimal execution time of the tower of Hanoi problem with n discs is
    a) $T(n) = 2T(n-2) + 2$
    b) $T(n) = 2T(n-1) + n$
    c) $T(n) = 2T(n/2) + 1$
    d) $T(n) = 2T(n-1) + 1$

13. A list of n strings, each of length n, is sorted into lexicographic order using the merge-sort algorithm. The worst case running time of this computation is
    a) O(n log n) b) $O(n^2 \log n)$ c) $(n^2 + \log n)$ d) $O(n^2)$

Answers:

1. b	2. c	3. c	4. a	5. b	6. c	7. c
8. c	9. a	10. a	11. b	12. d	13. a	

Hints:

4. There are two options $(0 + 1)^*$ or φ. Both are regular hence both are recursive.

Exercise

1. Determine O(n), Ω(n), and Θ(n) of the following
    a) $f(n) = 3n^2 + 2\log n + 1$
    b) $f(n) = 5n^3 - 3n^2 + 5$
    c) $f(n) = 5n^3 - 3n^2 - 5$
    d) $f(n) = 3n - 3\log n + 2$
2. Find the O or the following recurrence relations
    a) $T(n) = 8T\left(\frac{n}{2}\right) + cn^2$
    b) $T(n) = \begin{cases} c & \text{for } n = 1 \\ aT\left(\frac{n}{b}\right) + cn & \text{for } n > 1 \end{cases}$
    where n is a power of b
    c) $T(n) = 4T\left(\frac{n}{2}\right) + cn^2$
    d) $T(n) = \begin{cases} 0 & \text{for } n = 1 \\ 3T\left(\frac{n}{2}\right) + n - 1 & \text{for } n > 1 \end{cases}$
3. Find the time complexity of the following algorithms
    a)
```
bool IsFirstElementNull(String[ ] strings)
{
  if(strings[0] == null)
  {
    return true;
  }
  return false;
}
```

b)
```
for (int i = 0; i < N; i++)
  for (int j = i + 1; j < N; j++)
    if (A[i] > A[j])
      swap(A[i], A[j])
```

3. Given a sorted array A, determine whether it contains two elements with the difference D. Find the time complexity of the algorithm.
4. Find the time complexity of heap sort operation.
5. Find the time complexity of the Lagrange interpolation polynomial.
6. Prove that the following problem is in NP.
   Given integers n and k, is there a factor f with $1 < f < k$ and f dividing n?
7. Prove that the subset sum problem is NP complete.
8. A set of vertices inside a graph G is an independent set if there are no edges between any two of these vertices. Prove that for a given graph G and an integer k, 'is $G' \subseteq G$ an independent set of size k?' an NP complete problem. (Reduce it to 3 CNF or vertex cover.)
9. Prove that the vertex cover is NP complete by reducing the independent set problem to it.
10. Prove that the graph colouring problem is NP complete.

Basics of Compiler Design

13

Introduction

When we have learnt different programming languages like C, C++ or Java we got familiar with the term compiler. As an example, for C and C++ we have found Turbo C compiler, for Java we have seen Java compiler and so on. Most of the students have a wrong idea that compilation is required only to check the errors. But this is not the only job for a compiler, its job is much wider than error checking. Compiler is basically built on the knowledge of Automata. To learn different phases of compiler we need the knowledge of Finite Automata, Regular Expression, Context Free Grammar etc. Compiler needs a vast discussion, but in this section we shall discuss only those parts which are directly related to Automata.

13.1 Definition

Compiler is a program that takes a program written in a source language and translates it into an equivalent program in a target language. In relation to compiler the source language is human readable language called High Level Language and target language is machine readable code called machine level language. In the operational process of a compiler it reports the programmer the presence of errors in the source program. The block diagram of compiler is shown in Fig. 13.1.

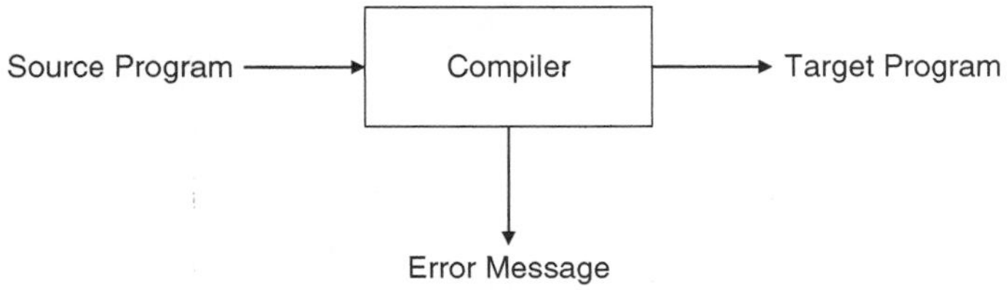

Fig. 13.1 *Block Diagram of Compiler*

13.2 Types of Compiler

Compilers are classified mainly into three types according to their operational process.

1. **Single pass:** In computer programming, a single pass or one-pass compiler is a compiler that scans through the source code of each compilation unit only once. In other words, a single-pass compiler does not 'look back' the code it has previously processed. Sometimes single pass

compiler is called narrow compiler, which points to the limited scope of a one-pass compiler. Some programming languages have been designed specifically to be compiled with one-pass compilers. As an example of such a construct is the forward declaration in Pascal. Normally Pascal requires that procedures to be fully defined before use. This helps a single-pass compiler with its type checking—calling a procedure that has not been defined is a clear error. However, this requirement makes mutually recursive procedures impossible to implement.

2. **Multi-pass:** A multi-pass compiler is a type of compiler that processes the source code or abstract syntax tree constructed from the program several times. This is in contrast to a one-pass compiler, which traverses the program only once. In multipass compiler each pass takes the result of the previous pass as the input, and creates an intermediate output. In this way, the intermediate code is improved pass by pass, and in the final pass the final code is produced.

   Referring to the greater scope of the passes multi-pass compilers are sometimes called wide compilers as they can 'see' the entire program being compiled, instead of just a small portion of it. This type of compilers allows better code generation which is smaller in size, and faster to execute by the computer hardware, compared to the output of one-pass compilers. But the main disadvantage is the higher compilation time and memory consumption. In addition, some languages cannot be compiled in a single pass, as a result of their design.

3. **Load and go:** A load and go compiler generates machine code and then immediately executes it.

   Compilers usually produce either absolute code that is executed immediately upon conclusion of the compilation or object code that is transformed by a linking loader into absolute code. These compiler organizations will be called Load & Go and Link/Load. Both Load & Go and Link/Load compilers use a number of passes to translate the source program into absolute code. A pass reads some form of the source program, transforms it into another form, and normally outputs this form to an intermediate file which may be the input of a later pass.

13.2.1 Difference Between Single Pass and Multi-Pass Compiler

- A one-pass compiler (narrow compiler) is a compiler that passes through the source code of each compilation unit only once whereas a multi-pass compiler (wide compiler) is a type of compiler that processes the source code several times.
- A one-pass compiler is faster than multi-pass compilers as it traverse the source code only once.
- Multi-pass compiler produces smaller targets and faster to execute target code than single-pass compiler.
- Many programming languages like Pascal can only be represented with a single pass compiler, whereas some languages like Java require a multi-pass compiler.

13.3 Major Parts of Compiler

The working process of a compiler is divided into two parts—analysis and synthesis. Analysis phase scan the program and breaks it into constituent pieces. From the pieces an intermediate representation of the source program is created. Synthesis phase takes the intermediate representation and construct the target code for the source program. The phases of a compiler are shown in Fig. 13.2.

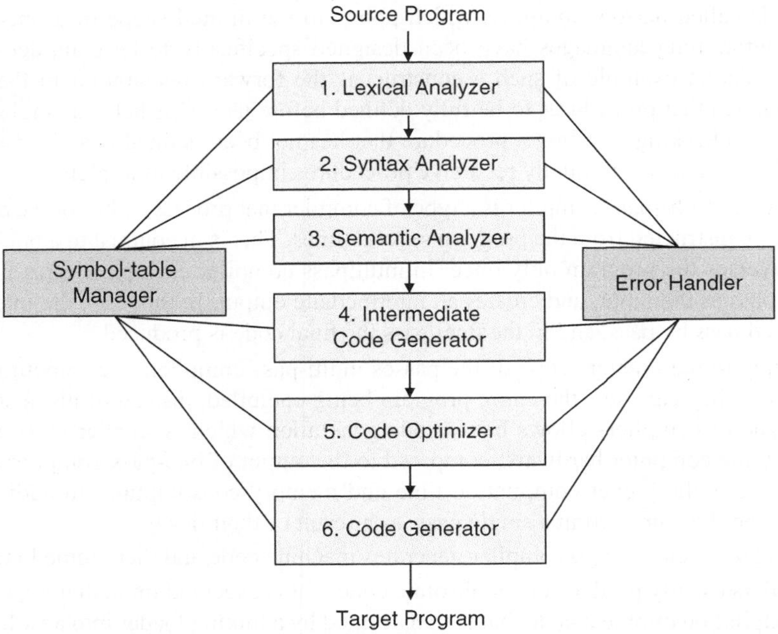

Fig. 13.2 *Different Phases of Compiler*

Analysis part can be divided into three parts:

1. Linear or lexical analysis
2. Hierarchical analysis or syntax analysis
3. Semantic analysis

Synthesis part is also divided into three parts:

1. Intermediate code generator
2. Code optimizer
3. Code generator

Lexical Analysis: This is the initial phase of compiler. The dictionary meaning of lexical is pertaining into words. In this phase the source program is read from left to right and divided into tokens. Tokens are the sequence of characters having a collective meaning. In relation to programming languages tokens are objects like variable name, number, keywords and so on. This phase also performs some secondary tasks like, removing source program's comments, white spaces in the form of blank, tab and newline character and so on.

Syntax Analysis: Syntax analyzer sometimes called parser. This phase takes the list of tokens produced by lexical analysis and arranges these in a tree-structure. Parser checks the input with a generated Context Free Grammar (CFG). If the parser determines the input to be valid one, it represents the input in the form of a parse tree. But if it is grammatically incorrect, the parser declares the presence of syntax error in the input. In this case no parse tree can be produced.

Semantic Analysis: Once a program is found to be syntactically correct—that is grammatically—the next task is to check for semantic correctness. A major job of semantic analysis is type checking where the match between operator and operand is checked. As an example, let an array is declared as integer. So all the elements within the array must be integer. If a float number is used to index the array, error will be reported.

Intermediate Code Generation: This phase is the first phase of synthesis part. Here the tree generated and modifies in syntax or semantic analysis phase is converted into a low level or machine like intermediate representation. This intermediate representation makes the target code easy to produce.

Code Optimization: This phase takes the intermediate representation as input and optimizes it to reduce its time and space complexity. By this the intermediate code is improved and better target code is produced.

Code Generation: Produces the target language in a specific architecture. The target program is normally is a object file containing the machine codes.

Symbol Table: In different phases of compiler the variables are needed. The values of the variables are changed from time to time in different operational phases of the compiler. To keep record of the variables or identifiers memory storage is needed. Symbol table is that storage. A symbol table is a data structure which keeps the record for each identifier, with fields for the attribute of the identifier. This data structure allows us to find the record for each identifier quickly and to store or retrieve data from that record quickly.

Upon detecting the identifiers in the source program by the lexical analyzer, those are entered into the symbol table. The remaining phases enter information about identifiers into the symbol table and then use this information in various ways.

Error Detection and Reporting: Errors can be produced in each phase of the compiler. After detecting an error, a phase may deal with those errors, so that the compilation can proceed and further errors in the source program are detected or simply reports the errors.

The syntax and semantic analysis phase usually handle a large fraction of the errors detectable by the compiler.

Among these phases, lexical analysis and syntax analysis have direct connection with automata theory. Lexical analysis phase deals with regular expression and finite automata whereas syntax analysis needs context free grammar. For this reason these two phases need some of discussion with automata. In this section we shall discuss the parts of lexical and syntax analysis which has connection with automata.

13.3.1 Lexical Analysis

Lexical analysis is the first phase of compiler design. In this phase the source program is read from left to right, character by character and generates token. Lexical analysis does not work single and not return a list of tokens at one shot, it returns a token when the parser asks a token from it. Getting a *get next token* command from parser, lexical analysis scans the source program and upon finding a token it returns it to the parser. The working principle of lexical analysis is shown in Fig. 13.3.

Though token generation process is not so simple, it uses a technique called double buffering which makes the scanning process faster. But that is beyond our discussion.

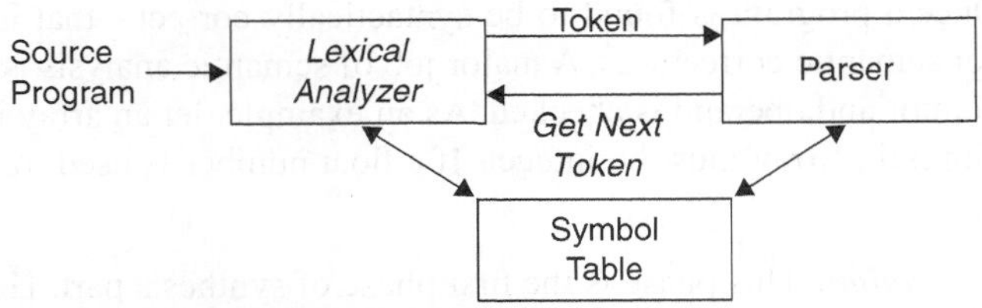

Fig. 13.3 *Working Principle of Lexical Analysis*

13.3.1.1 Major Terms of Lexical Analysis

It is already discussed that lexical analyzer returns tokens. But how a lexical analyzer decides a string of characters and/or numbers as token? For this we have to know the definitions of token, pattern and lexemes.

Token: Sequence of characters having a collective meaning. Token can be differentiated as type token (integer number, real number), punctuation token (void, return) and alphabetic token (keywords).

Comment, macros, whitespace characters (blank, tab, new line) are treated as non tokens.

Pattern: Pattern is the set of rules which characterizes a string as a token. If a scanned string is matched with the set of rules; it is declared as token. As an example the pattern of identifier is id → letter (letter | digit)*. According to this int A2 is a valid identifier but 2A is not.

Lexeme: It is the actual string that fulfills the rules of pattern as classified by a token.

Example: **int principal1** is characterized by the pattern that id → letter (letter | digit)*. Thus it is a valid identifier and it is a type token. The actual string is principal1, thus it is a lexeme.

13.3.2 Syntax Analysis

Syntax analyzer creates the syntactic structure of the given source program. The mostly used syntactic structure is parse tree. For this reason syntax analysis is known as parser. What is the syntax? The syntax of a programming language is the rules described by a context free grammar associated with it. In syntax analysis phase it is checked whether a given source program satisfies the rules of the underlying context free grammar or not. If a valid parse tree is created for the given source program then the program is syntactically correct, otherwise it gives the error message known as syntax error. As an example a = b + c is a valid C statement but b + c = a is not. Because for the second one a valid parse tree can not be generated. Compiler generates an error L value required for this type of C statement.

The rules or grammar related to syntax analysis is context free grammar. Therefore, the grammar may me ambiguous. A very common case of ambiguity occurs in if-else statement in C or Java. Let us take the following example.

Statement → if (condition) statement
Statement → if (condition) statement else statement

Now consider the following C statement

```
if (a > b)
 if(c > d)
 x = y
 else
 x = z
```

This can be derived by the following ways.

statement → if (condition) statement
→ if (condition) if (condition) statement else statement
→ if(a > b) if(c > d) x = y else x = z

or

statement → if (condition) statement else statement
→ if (condition) if (condition) statement else statement
→ if(a > b) if(c > d) x = y else x = z

The parse trees for the previous two derivations are given in Fig. 13.4.

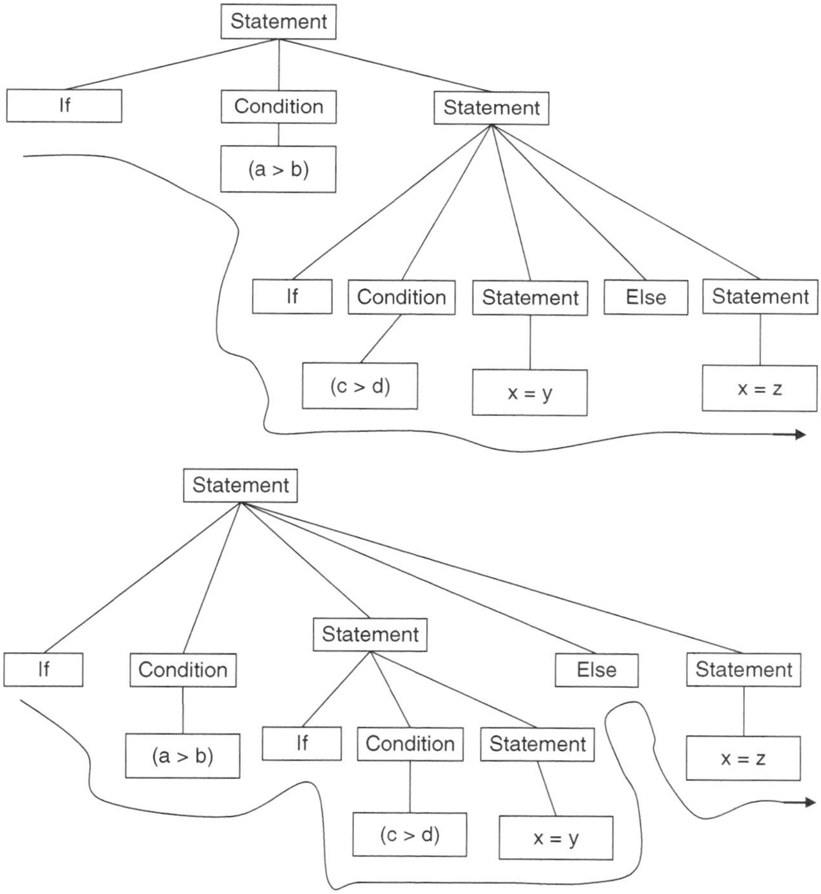

Fig. 13.4 *Nested If-Else Problem*

Now the 'else' part is for which if?—the first or the second. Attaching 'else' with each of two results different. So, ambiguity must be removed. Here the ambiguity is removed by the following rule—'Match each else with the closest previous unmatched if'.

In context free grammar section another common case of ambiguity has already been described for the grammar.

$$E \rightarrow E + E/E - E/E * E/id$$

In this case the ambiguity is removed by putting precedence of operators.

13.3.3 Parser

The syntax analyzer checks whether a given source program satisfies the rules implied by a context-free grammar or not by constructing a parse tree for a given statement. For this reason syntax analyzer is sometimes known as parser. Parser is of mainly two types—top down parser and bottom up parser. In this chapter we shall discuss LL(1) parsing as an example of top down parsing and LR parsing, as an example of bottom up parsing.

Top down parsing technique can not handle left recursive grammar. A grammar is called left recursive if it has a non-terminal 'A' such that there is a derivation.

$$A \overset{+}{\Rightarrow} A\alpha \text{ for some string } \alpha$$

The process of the removal of left recursion is described in context free grammar chapter.

Backtracking is also a problem related to parsing.

Let a grammar is in the form.

$$A \rightarrow aBC/aB$$
$$B \rightarrow bB/bC$$
$$C \rightarrow c$$

And the string to be generated is abbc. Compiler can scan one symbol at a time. As aBC and aB both are started with the symbol 'a'; compiler can take any of the productions to generate abbc. Let it take A → aBC and the derivations are as follows.

$$A \rightarrow a\underline{B}C \rightarrow ab\underline{B}C \rightarrow abbCC \rightarrow abbcC$$

The string abbc is already generated but a non-terminal 'C' still remains. So, the parser has to backtrack to choose the alternative production A → aB, and the following derivations are

$$aB \rightarrow ab\underline{B} \rightarrow abbC \rightarrow abbc$$

It succeeds in generating the string. For backtracking the parser wastes a lot of lime. Hence the backtracking must be removed. Left factoting of the grammar is a process to remove backtracking. Actually top-down parsing technique insists that the grammar must be left factored. The process of left factoring a grammar is described in Context free grammar chapter.

13.3.3.1 Top-down Parsing

In top down parsing a parse is created from root to leaf. Top down parsing corresponds to a preorder traversal of the parse tree. In this parsing a left most derivation is applied at each derivation step. There are mainly two types of top-down parsing—recursive decent parsing and predictive parsing.

Recursive Decent Parsing: Recursive decent parsing is most straight forward form of parsing. In this type of top down parsing the parser attempts to verify the input string read from left to right with the grammar underlying. A basic operation of recursive decent parsing involves reading characters from the input stream one at a time, and matching it with terminals from the production rule of the grammar that describes the syntax of the input. In recursive decent parsing backtracking occurs. This backtracking is caused because in recursive decent parsing left recursion is not removed. Due to this it is not efficient and not widely used.

Predictive Parsing: This is a type of top down parsing which never backtracks. To prevent the occurrence of backtracking from the grammar left recursion is removed and the grammar is left factored. Non recursive predictive parsing is a mixture of recursive decent parsing and predictive parsing, which never backtracks.

13.3.3.2 LL(1) Parsing

LL(1) parsing is an example of predictive parsing. The first 'L' in LL(1) denotes that input string is scanned from left to right. Second 'L' denotes that by left most derivation the input string is constructed. '1' denotes that only one symbol of the input string is used as a look-head symbol to determine parser action. LL(1) is a table driven parser as the parser consults with a parsing table for its actions. A LL(1) parser consists of a input tape, a stack and a parsing table. The block diagram of a LL(1) parser is given in Fig. 13.5.

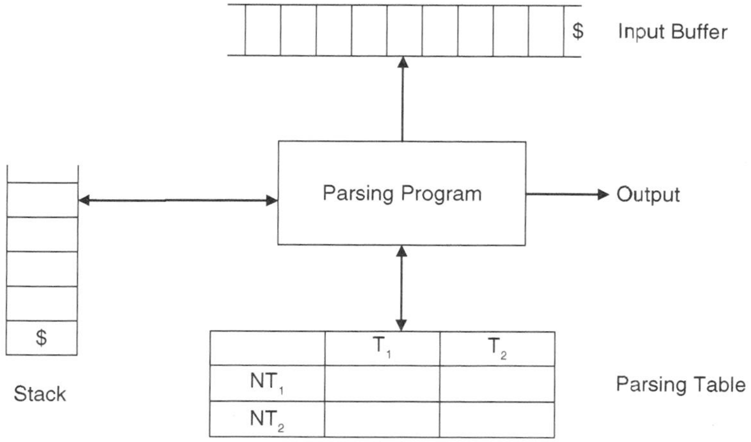

Fig. 13.5 *LL(1) Parser*

Different Components of LL(1) Parser

- **Input buffer:** Input buffer contains the input string to be parsed. It contains a \$ as the end marker of the input string.
- **Stack:** Stack contains the terminal and non-terminal symbols. In brief it can be said that Stack contains grammar symbols. Bottom of the stack contains a special symbol $\$ \notin (V_N \cup \Sigma)$. At the beginning of parsing the stack contains \$S, where S is the start symbol.
- **Parsing table:** It is most important part of a LL(1) parser. It is a two dimensional array whose left most column contains the non-terminal symbols $\in V_N$ and top most row contains the terminal symbol $\in \Sigma \cup \$$. Some entries of the parsing table contain the production rules of the grammar and remaining contains error.

How LL(1) Parser Works? As described in earlier section, the input buffer contains the input string to be parsed ended by a $ and stack contains the grammar symbols including $. The symbols in the input buffer are terminal symbols. As it is LL(1) parser, therefore one symbol of the input buffer is scanned at a time. The parser action is determined by the top of the stack symbol let T_{Stack} and the current scanned input symbol from the input buffer let $I_{i/p}$. Four types actions are possible for a LL(1) parser.

1. **T_{Stack} and $I_{i/p}$ are same terminal symbol:** Parser pops T_{Stack} from the top of the stack and points next symbol in the input buffer.
2. **T_{Stack} is a non-terminal:** Parser consults with the corresponding entries for T_{Stack} and $I_{i/p}$ in the parsing table. The top of the stack is replaced by the right hand side of the production in reverse. [Parsing table entries is a production rule of the grammar]
3. **T_{Stack} is $ and $I_{i/p}$ is $:** parsing successful. Declare accept.
4. **None of the above three:** Error.

Let take an example to describe it in detail.

Example 13.1 Consider the following parsing table.

	int	*	+	(	)	$
E	E → TX			E → TX		
X			X → +E		X → ε	X → ε
T	T → int Y			T → (E)		
Y		Y → *T	Y → ε		Y → ε	Y → ε

Parse the string **(int * int) + int** using the following parsing table.

Solution:

Stack	Input	Output
$E	(int * int) + int $	E → TX
$XT	(int * int) + int $	T → (E)
$X(E)	(int * int) + int $	
$X(E	int * int) + int $	E→ TX
$X(XT	int * int) + int $	T→ int Y
$X(XYint	int * int) + int $	
$X(XY	*int) + int $	Y → *T
$X(XT*	*int) + int $	
$X(XT	int) + int $	T→ int Y
$X(XY int	int) + int $	
$X(XY	) + int $	Y → ε

Stack	Input	Output
$X (X	)+ int $	X → ε
$X (	)+ int $	
$X	+ int $	X → +E
$E +	+ int $	
$E	int $	E → TX
$XT	int $	T → int Y
$XY int	int $	
$XY	$	Y → ε
$X	$	X → ε
$	$	Accept

Hence the string **(int *int) + int** is acepted by the given LL(1) parser.

Example 13.2 Consider the following parsing table.

	id	+	(	(	)	$
E	E → TE′			E → TE′		
E′		E′ → +TE′			E′ → ε	E′ → ε
T	T → FT′			T → FT′		
T′		T′ → (	T′ → −FT′		T′ → ε	T′ → ε
F	F → id			F → (E)		

Check whether the string (id–id) + id is accepted by the parser or not.

Solution:

Stack	Input	Output
$E	(id – id) + id $	E → TE′
$E′T	(id – id) + id $	T → FT′
$E′T′F	(id – id) + id $	F → (E)
$E′T′)E(	(id – id) + id $	
$E′T′)E	id – id) + id $	E → TE′
$E′T′)E′T	id – id) + id $	T → FT′
$E′T′)E′T′F	id – id) + id $	F → id
$E′T′)E′T′id	id – id) + id $	
$E′T′)E′T′	– id) + id $	T′ → –FT′

Stack	Input	Output
\$E′T′)E′T′F–	– id) + id \$	
\$E′T′)E′T′F	id) + id \$	F → id
\$E′T′)E′T′id	id) + id \$	
\$E′T′)E′T′	) + id \$	T′ → ε
\$E′T′)E′	) + id \$	E′ → ε
\$E′T′)	) + id \$	
\$E′T′	+ id \$	T′ → ε
\$E′	+ id \$	E′ → + TE′
\$E′T+	+ id \$	
\$E′T	id \$	T → FT′
\$E′T′F	id \$	F → id
\$E′T′id	id \$	
\$E′T′	\$	T′ → ε
\$E′	\$	E′ → ε
\$	\$	accept

Hence the string is accepted by the given LL(1) parser.

Construction of LL(1) Parsing Table: LL(1) parsing is a table driven parsing. The parsing table is an important part of the parser. In this section we shall learn how to construct the LL(1) parsing table. Before going into the construction process of LL(1) parsing table, two important topics are needed to be discussed—First and Follow.

1. **FIRST:** For a given string $\alpha \in \{V_N \cup \Sigma\}$, First($\alpha$) is the set of terminal symbols that begin the strings derivable from ε. If ε is derived from α then ε is also included in First(α). First can be calculated for terminal, non-terminal and string of terminal and/or non-terminal.
2. **FOLLOW:** Follow is calculated for non-terminal symbols. Follow (NT) is the set of the terminals which occur immediately after (follow) the non-terminal NT in the strings derived from the starting symbol. A terminal t is in Follow(NT) $S \overset{+}{\Rightarrow} \ldots t \ldots$ In the blank space there is non-terminal or terminal symbols.

Rules of Constructing FIRST

- First of ε is ε.
- First of a terminal symbol is that terminal symbol, i.e. if a is a terminal symbol then FIRST (a) = {a}
- First of a string XY is
  - i) FIRST(X) ∪ FIRST(Y) if FIRST(X) contains ε
  - ii) FIRST(X) if FIRST(X) does not contains ε
- If there are productions $NT \rightarrow \alpha_1/\alpha_2/ \ldots\ldots /\alpha_n$
  then $FIRST(NT) = FIRST(\alpha_1) \cup FIRST(\alpha_2) \cup \ldots\ldots \cup FIRST(\alpha_n)$.

Example 13.3 Construct FIRST of the non-terminal symbols of the given grammar.

$$S \rightarrow iEtSS'/a$$
$$S' \rightarrow eS/\varepsilon$$
$$E \rightarrow b$$

[This is the grammar Here i is for 'if', E is for 'Expression', t is for 'then' , S is for 'statement' and e is for 'Else'.]

Solution: The terminal symbols in the grammar are 'i', 't', 'a', 'e' and 'b'.

FIRST(i) = {i}
FIRST(t) = {t}
FIRST(a) = {a}
FIRST(e) = {e}
FIRST(b) = {b}
FIRST(ε) = {ε}

Now calculate the FIRST of the Non-Terminal symbols.

FIRST(S) = FIRST (iEtSS′) ∪ FIRST(a)

FIRST (iEtSS′) =

i) FIRST(i) ∪ FIRST(EtSS′) if FIRST(i) contains ε.
ii) FIRST(i) if FIRST(i) does not contain ε.

Hence FIRST(S) = {i, a}.

FIRST(S′) = FIRST(eS) ∪ FIRST(ε)
FIRST(eS) = FIRST(e) ∪ FIRST(S) if FIRST(e) contains ε.
= FIRST(e) if FIRST(e) does not contain (.

Hence FIRST(S′) = {e,(}.

FIRST(E) =FIRST(b) = {b}

Example 13.4 Construct FIRST of the non-terminal symbols of the given grammar.

$$E \rightarrow TE'$$
$$E' \rightarrow +TE' \mid \varepsilon$$
$$T \rightarrow FT'$$
$$T' \rightarrow -FT' \mid \varepsilon$$
$$F \rightarrow (E) \mid id$$

Solution: The terminal symbols in the grammar are +, –, (,), id.

FIRST(+) = {+}
FIRST(–) = {–}
FIRST('(') = {(}
FIRST(')') = {)}
FIRST(id) = {id}
FIRST(ε) = {ε}

Now calculate the FIRST of the non-terminal symbols.

FIRST(E) = FIRST(TE′)

i) FIRST(T) ∪ FIRST(E′) if FIRST(T) contains ε.
ii) FIRST(T) if FIRST(T) does not contain ε.

FIRST(E′) = FIRST(+TE′) ∪ FIRST(ε)
= FIRST(+) ∪ FIRST(ε) [FIRST(+) does not contain ε]
= {+, ε}

FIRST(T) = FIRST(FT′)

i) FIRST(F) ∪ FIRST(T′) if FIRST(F) contains ε.
ii) FIRST(F) if FIRST(F) does not contain ε.

FIRST(T′) = FIRST(–FT′) ∪ FIRST(ε)
= FIRST(–) ∪ FIRST(ε) [FIRST(–) does not contain ε]
= {–, ε}

FIRST(F) = FIRST((E)) ∪ FIRST(id)
= FIRST('(') ∪ {id} [FIRST('(') does not contain ε]
= {(, id}

As FIRST(F) does not contain ε, therefore

FIRST(T) = FIRST(F) = {(, id}

As FIRST(T) does not contain ε, therefore

FIRST(E) = FIRST(T) = {(, id}

Example 13.5 Construct FIRST of the non-terminal symbols of the given grammar.

E → TX
T → (E) | int Y
X → +E | ε
Y → * T | ε

Solution: The terminal symbols in the grammar are '(', ')',int, +, *.

FIRST('(') = {(}
FIRST(')') = {)}
FIRST(int) = {int}
FIRST (+) = {+}
FIRST (*) = {*}
FIRST (ε) = {ε}

The non-terminal of the grammar are E, T, X and Y.

FIRST(E) = FIRST(TX)

i) FIRST(T) ∪ FIRST(X) if FIRST(T) contains ε.
ii) FIRST(T) if FIRST(T) does not contain ε.

$$\text{FIRST}(T) = \text{FIRST}((E)) \cup \text{FIRST}(\text{int } Y) = \{(, \text{int}\}$$

$$\text{FIRST}(X) = \text{FIRST}(+E) \cup \text{FIRST}(\varepsilon) = \{+, \varepsilon\}$$

$$\text{FIRST}(Y) = \text{FIRST}(*T) \cup \text{FIRST}(\varepsilon) = \{*, \varepsilon\}$$

$$\text{FIRST}(E) = \text{FIRST}(T) = \{(, \text{int}\}$$

Rules of Constructing FOLLOW

- FOLLOW of a start symbol contains $.
- If there is a production rule in the form $A \rightarrow \alpha B\beta$ then everything in FIRST(β) is included in FOLLOW(B) except ε.
- If (there is a production rule in the form $A \rightarrow \alpha B$) OR (a production rule in the form $A \rightarrow \alpha B\beta$ and ε is in FIRST(β)) then everything in FOLLOW(A) is included in FOLLOW(B).

Example 13.6 Construct FOLLOW of the non-terminal symbols of the given grammar.

$$S \rightarrow iEtSS'/a$$
$$S' \rightarrow eS/\varepsilon$$
$$E \rightarrow b$$

Solution: S is the start symbol of the grammar. So, FOLLOW(S) contains $.

$S \rightarrow iEtSS'$ is in the form $A \rightarrow \alpha B\beta$, where

i) **Case 1:** α is i, B is E and β is tSS′. FOLLOW(E) = FIRST(tSS′)–ε.

$$\text{FIRST}(tSS') = \text{FIRST}(t) = \{t\}, \text{ therefore FOLLOW}(E) = \{t\}$$

ii) **Case 2:** α is iEt, B is S and β is S′. FOLLOW(S)= FIRST(S′) –ε.

$$\text{FIRST}(S') = \{e, \varepsilon\}, \text{ therefore FOLLOW}(S) = \{e\}$$

$S \rightarrow iEtSS'$ is in the form $A \rightarrow \alpha B$, where iEtS is α and S′ is B. So, FOLLOW(S) is a subset of FOLLOW(S′).

$S' \rightarrow eS$ is in the form $A \rightarrow \alpha B$. Therefore FOLLOW(S′) is a subset of FOLLOW(S).

Hence FOLLOW(S) = FOLLOW(S′)

$$\text{FOLLOW}(S) = \{e, \$\}$$
$$\text{FOLLOW}(S') = \{e, \$\}$$
$$\text{FOLLOW}(E) = \{t\}$$

Example 13.7 Construct FOLLOW of the non-terminal symbols of the given grammar.

$$E \rightarrow TE'$$
$$E' \rightarrow +TE' \mid \varepsilon$$
$$T \rightarrow FT'$$
$$T' \rightarrow -FT' \mid \varepsilon$$
$$F \rightarrow (E) \mid id$$

Solution: E is the start symbol. Hence FOLLOW(E) = {\$}.

There is a production E → TE′ [in the form A → αB]. So FOLLOW(E) is a subset of FOLLOW(E′).

There is a production E′ → +TE′ [in the form A → αBβ]. Everything FIRST(E′) is included in FOLLOW(T) except ε. FIRST(E′) = {+, ε}, so FOLLOW(T) = {+}.

As ε is in FIRST(E′), so FOLLOW(E′) is a subset of FOLLOW(T).

There is a production T → FT′ [in the form A → αB]. FOLLOW(T) is a subset of FOLLOW(T′).

There is a production T′ → –FT′ [in the form A → αBβ]. Everything FIRST(T′) is included in FOLLOW(F) except ε. FIRST(T′) = {–, ε} , so FOLLOW(F) = {+}.

As ε is in FIRST(T′), so FOLLOW(T′) is a subset of FOLLOW(F).

Consider the production F → (E). It is in the form A → αBβ. So everything in FIRST(')') except ∈ is in FOLLOW(E). Hence FOLLOW(E) ={')'}

Finally

FOLLOW(E) = {\$,)}
FOLLOW(E′) = {\$,)} FOLLOW(E) is a subset of FOLLOW(E′)
FOLLOW(T) = {+,), \$} FOLLOW(E′) is a subset of FOLLOW(T)
FOLLOW(T′) = {+,), \$} FOLLOW(T) is a subset of FOLLOW(T′)
FOLLOW(F) = {+, –,), \$} FOLLOW(T′) is a subset of FOLLOW(F)

Rules for LL(1) Parsing Table Construction: We know that the parsing table is a table whose top most row contains the terminal symbols including \$ and left most column contains the non-terminal symbols.

Let the parsing table is M[NT, T], where NT denotes non-terminals and T denotes terminals. The rules for constructing the parsing table are given below.

For each production rule in the form A → α of the given grammar G

{

i) for each terminal T_i in FIRST(α) add A → α to M[A, T_i]

ii) a) If FIRST(α) contains ε then for each terminal T in FOLLOW(A) the production rule A → α is added to M[A,T]

b) If FIRST(α) contains ε and FOLLOW(A) contains \$ then the production rule A → α is added to M[A, \$].

}

iii) All other undefined entries of the parsing table are ERROR.

Example 13.8 Construct LL(1) parsing table for the grammar.

S → iEtSS′/a
S′ → eS/ε
E → b

Solution: The FIRST of the Terminals and Non-Terminals and FOLLOW of the Non-Terminals are already calculated. These are given below.

FIRST(i) = {i}	FIRST(S) = {i, a}	FOLLOW(S) = {e, \$}
FIRST(t) = {t}	FIRST(S′) = {e, ε}	FOLLOW(S′) = {e, \$}
FIRST(a) = {a}	FIRST(E) = {b}	FOLLOW(E) = {t}
FIRST(e) = {e}		
FIRST(b) = {b}		
FIRST(ε) = {ε}		

i) $S \to iEtSS'$ is in the form $A \to \alpha$. FIRST(iEtSS′) = {i}. The production rule $S \to iEtSS'$ is added in the parsing table label [S, i].

ii) $S \to a$ is added in the parsing table label [S, a] as FIRST(a) = {a}.

iii) $S' \to eS$ is added in the parsing table label [S′, e] as FIRST(eS) = {e}.

iv) For the production $S' \to \$$ as FIRST(ε) = {ε} look for FOLLOW(S′). It is {e, \$}. Hence $S' \to \varepsilon$ is added in the parsing table label [S′, e] and [S′, \$].

v) $E \to b$ is added in the parsing table label [E, b] as FIRST(b) = {b}.

Therefore the constructed parsing table is

	i	a	b	e	t	\$
S	$S \to iEtSS'$	$S \to a$				
S′				$S' \to eS$ $S' \to \varepsilon$		$S' \to \varepsilon$
E			$E \to b$			

Example 13.9 Construct LL(1) parsing table for the grammar.

$$E \to TE'$$
$$E' \to +TE' \mid \varepsilon$$
$$T \to FT'$$
$$T' \to -FT' \mid \varepsilon$$
$$F \to (E) \mid id$$

Solution: The FIRST of the terminals and non-terminals and FOLLOW of the non-terminals are already calculated. These are given below.

FIRST(+) = {+}	FIRST(E) = {(, id}	FOLLOW(E) = {\$,)}
FIRST(() = {(}	FIRST(E′) = {+, ε}	FOLLOW(E′) = {\$,)}
FIRST(‘(‘) = {(}	FIRST(T) = {(, id}	FOLLOW(T) = {+,), \$}
FIRST(‘)’) = {)}	FIRST(T′) = {–, ε}	FOLLOW(T′) = {+,), \$}
FIRST(id) = {id}	FIRST(F) = {(, id}	FOLLOW(F) = {+, –,), \$}
FIRST(ε) = {ε}		

i) $E \to TE'$ is in the form $A \to \alpha$. FIRST(TE′) = {(, id}. $E \to TE'$ is added in the parsing table label [E, (] and [E, id].

ii) $E' \to +TE'$ is added in the parsing table label {E′, +}.

iii) FIRST(ε) = {ε} in $E' \to \varepsilon$. FOLLOW(E′) = {\$,)}. Therefore $E' \to \varepsilon$ is added in the parsing table label [E′,)]. As FOLLOW(E′) contains \$, so $E' \to \varepsilon$ is added in the parsing table label [E′, \$].

iv) FIRST(F) = {(, id} in T (FT′. Therefore the production $T \to FT'$ is added in the parsing table label [F,( ] and [F, id].

v) The production rule $T' \to -FT'$ is added in the parsing table label [T′, –].

vi) FIRST(ε) = {ε} in T′ → ε. FOLLOW(T′) = {+,), $}. Therefore T′ → ε is added in the parsing table label [T′, +] and [T′,)]. As FOLLOW(T′) contains $, so T′ → ε is added in the parsing table label [T′, $].

vii) The production rule F → (E) is added in the parsing table label [F, ( ].

viii) The production rule F→ id is added in the parsing table label [F, id].

Therefore the constructed parsing table is

	id	+	−	(	)	$
E	E → TE′			E → TE′		
E′		E′ → +TE′			E′ → ε	E′ → ε
T	T → FT′			T → FT′		
T′		T′ → (	T′ → −FT′		T′ → ε	T′ → ε
F	F → id			F → (E)		

Checking Whether a Grammar is LL(1) or Not: A grammar is LL(1) if parsing table of it does not contain multiple entry in a single label [NT, T].Look the parsing table for the grammar for If-Else statement. It contains two production rules S′ → eS and S′ → ε in the parsing table label [S′, e]. Hence the grammar is not LL(1).

But it become difficult to say whether a grammar is LL(1) or not by constructing its parsing table. If we are able to mark those conditions for which multiple entry appears in a single label [NT, T] of the parsing table, then it becomes easy to check whether a grammar is LL(1) or not. The conditions are as follows.

- A grammar which is left recursive is not LL(1).

  Explanation: Let a grammar contains a production rule A → Aα/β, where α and β are string of terminal and/or non terminals.

  $$A \overset{+}{\Rightarrow} \beta\alpha$$

  Hence FIRST(Aα) contains all the terminal symbols those appear in FIRST(β). Therefore the parsing table block label [A, T ∈ FIRST(β)] contains the productions A → Aα and A→ β.

- A grammar which is not left factored is not LL(1).

  Explanation: Let a grammar contains a productions $A \rightarrow \alpha S_1/\alpha S_2$, where α is a terminal and S_1, S_2 are string of terminal and/or non terminals. FIRST(A) contains the terminal α. Therefore the parsing table block label [A, α] contains the productions $A \rightarrow \alpha S_1$ and $A \rightarrow \alpha S_2$.

- An ambiguous grammar is not LL(1) grammar.

- If a grammar contain productions A → a and A → β.

  - If both of α and β generate ε, the grammar is not LL(1).

    Explanation: The parsing table block label [A, ε] contain A → α and A → β.

  - If β produce ε and α derive any string starting with a terminal belongs to FOLLOW(A), the grammar is not LL(1).

    Explanation: Let there is a production $NT_i \rightarrow T_i$<String>/ε and FOLLOW(NT_i) contains T_i. For the production $NT_i \rightarrow T_i$<String> the parsing table block label [NT_i, T_i] contains the production $NT_i \rightarrow T_i$<String>, as FIRST(T_i<String>) contains T_i.

Now consider the production $NT_i \rightarrow \varepsilon$. As FOLLOW($NT_i$) contains T_i, the production rule $NT_i \rightarrow \varepsilon$ is place in parsing table block label [NT_i, T_i].

It means parsing table block label [NT_i, T_i] contains two productions $NT_i \rightarrow T_i$<String> and $NT_i \rightarrow \varepsilon$, thus the grammar is not LL(1).

Example 13.10 Check whether the grammar is LL(1) or not.

$$S \rightarrow iEtSS'/a$$
$$S' \rightarrow eS/\varepsilon$$
$$E \rightarrow b$$

Solution:

i) The grammar is not left recursive. There is no option to make it left factored.
ii) The production rule $S \rightarrow iEtSS'/a$ and $S' \rightarrow eS/\varepsilon$ contain two productions in a group. Any of the first group of productions do not produce ε. The production $S' \rightarrow eS$ do not generate ε.
iii) For the group of production $S' \rightarrow eS/\varepsilon$, $S' \rightarrow \varepsilon$ produce ε. FOLLOW(S′) = {e, \$}. But the production $S' \rightarrow eS$ starts with $e \in$ FOLLOW(S′).

Therefore the grammar is not LL(1).

13.3.3.3 Bottom Up Parsing

Operation of bottom up parsing is the reverse of top down parsing. A bottom-up parsing starts traversing with the string of terminals from the leaves and by applying the productions in reverse it moves upward until it gets the start symbol. In this process, a bottom-up parser searches for substrings of the working string that match the right side of some production. When such a substring is found, parser reduces it i.e. the parser substitutes the left side non-terminal of the matching production rule with the substring (Which is the right hand side of the production rule). By this process at the last stage a substring is replaced by the start symbol reporting successful parsing. There are different types of bottom up parsing. Among them we shall discuss mainly shift reduce parsing and LR parsing.

Shift Reduce Parsing: A shift reduce parser is one type of bottom up parsing. It means the parser starts traversing a string from the string itself and reaches to the start symbol. A shift reduce parser consists of an input tape and a stack. The bottom of the stack is marked by \$, the stack bottom symbol. The end of the string in the input tape is marked by \$ as end marker.

The block diagram of a shift reduces parser is shown in Fig. 13.6.

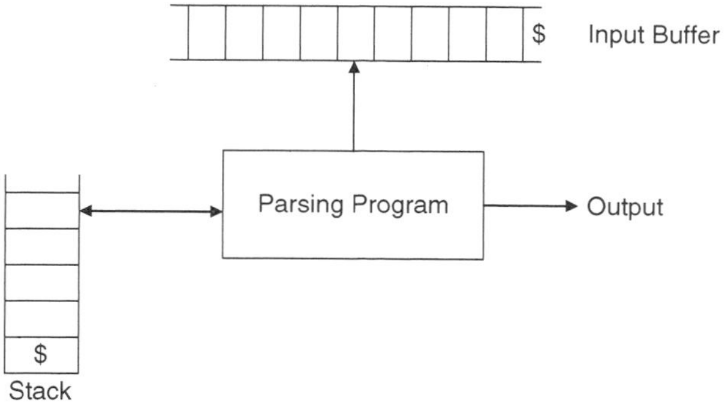

Fig. 13.6 *Shift Reduce Parser*

As the name suggests, the two operations attached with this parsing are shift and reduce.

- **Shift:** This operation traverses the input alphabets from input tape to the stack one by one in each step.
- **Reduce:** This operation reduces the handle by some non-terminal. A handle is a right sentential sequence of terminal and/or non-terminal within the string being parsed and same as the rightmost derivation of a production rule of the grammar.

The following are some examples of shift reduce parsing.

Example 13.11 Parse the string (a + b)*c using shift reduce parsing from the given grammar.

E → E + E
E → E * E
E → (E)
E → a/b/c

Solution:

Stack	Input Tape	Action
$	(a + b)*c $	Shift '('
$ (	a + b)*c $	Shift 'a'
$ (a	+b)*c $	Reduce by E → a
$ (E	+b)*c $	Shift +
$ (E +	b)*c $	Shift 'b'
$ (E + b	)*c $	Reduce by E → b
$ (E + E	)*c $	Reduce by E → E + E
$ (E	)*c $	Shift '('
$ (E)	*c $	Reduce by E → (E)
$E	*c $	Shift *
$E*	c $	Shift c
$E*c	$	Reduce by E → c
$E*E	$	Reduce by E *E
$E	$	Accept

Example 13.12 Parse the string ibtictaeac using shift reduce parsing from the given grammar.

S → iEtSE/a
E → eS/ε
E → b/c

Solution:

Stack	Input Tape	Action
$	ibtictaeac$	Shift i
$i	btictaeac$	Shift 'b'
$ib	tictaeac$	Reduce by E → b
$iE	tictaeac$	Shift t
$iEt	ictaeac$	Shift i
$iEti	ctaeac$	Shift c
$iEtic	taeac$	Shift t
$iEtict	aeac$	Shift a
$iEticta	aeac$	Reduce by S → a
$iEtictS	eac$	Shift e
$iEtictSe	ac$	Shift a
$iEtictSea	c$	Reduce by S → a
$iEtictSeS	c$	Reduce E → eS
$iEtictSE	c$	Reduce by S → iEtSE
$iEtS	c$	Shift c
$iEtSc	$	Reduce by E → c
$iEtSE	$	Reduce by S → iEtSE
$E	$	accept

Two types of problems that found in a shift reduce parsing are shift reduce conflict and reduce reduce conflict. There is no scope of discussing them details in this chapter.

LR Parsing: LR parsing is one type of bottom up parsing. It is also shift reduce parsing. LR parsing is a backtrack-free deterministic parsing technique performed in linear time. LR recognizes virtually all programming languages. For this reason LR is widely used parsing technique. In LR, L represents left to right scanning of the input string, R represents rightmost derivation in reverse. In LR(1) , the number '1' represents one symbol look ahead.

LR parsing is also a table driven parsing like LL(1) parsing. Here the parser table is divided into two parts—Action and Goto. The parser is attached with an input tape and a stack. The input tape contains the string to be parsed ended by a $ as end marker and the stack contains the grammar symbols (terminal and non-terminal both). The rows of the LR parsing table are marked by state number. The columns of the ACTION part are marked by the terminal symbol of the grammar and the GOTO part are marked by the non-terminal symbols of the grammar. Already it is discussed that LR parser is nothing but a shift reduce parser. The fields [State and Terminal] of the ACTION part contains SHIFT (the states) and REDUCE (by the given productions). The field [State and Non-Terminal] of the GOTO part contains the next state number to go. The block diagram of a LR parser is given in Fig. 13.7.

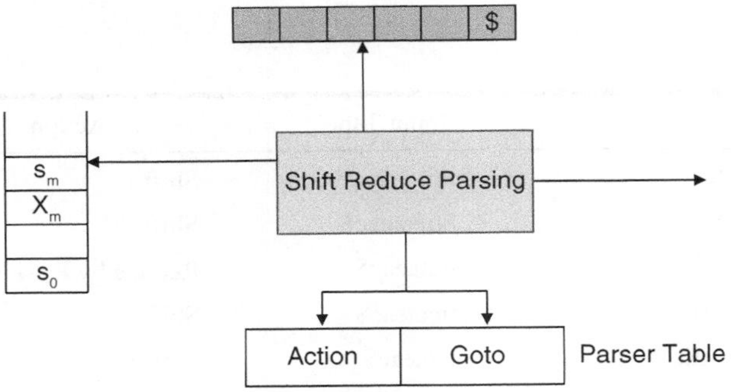

Fig. 13.7 *Block Diagram of LR Parser*

13.3.3.4 Parsing Algorithm

Input: Grammar with n number of productions, A LR parsing table.

A LR parser works as follows.

- It checks the state S_i currently on the top of the stack and the current input symbol I_i in the input tape.
- The parser consults in the Action part of the parsing table with label $[S_i, I_i]$. This returns one of the following.
  - Shift S_j. This places the current input in the stack followed by S_j as the top of the stack
  - Reduce by production number k. k = 1 to n. [Let the production k is $A \rightarrow B$. The parser pushes A by popping 2*| B | elements from the stack, consult with the goto part with label (A, previous input) and places state S_l on the top of the stack. S_l is the entry in (A, previous input) in the goto part of the parsing table.]
  - Accept
  - Error

Example 13.13 Parse the string id + id * id by the following grammar and the LR parsing table.

1. $E \rightarrow E + T$
2. $E \rightarrow T$
3. $T \rightarrow T * F$
4. $T \rightarrow F$
5. $F \rightarrow (E)$
6. $F \rightarrow id$

State	Action						Goto		
	id	+	*	(	)	$	E	T	F
0	S_5			S_4			g_1	g_2	g_3
1		S_6				acc			
2		r_2	S_7		r_2	r_2			
3		r_4	r_4		r_4	r_4			
4	S_5			S_4			g_8	g_2	g_3
5		r_6	r_6		r_6	r_6			
6	S_5			S_4				g_9	g_3
7	S_5			S_4					g_{10}
8		S_6			S_{11}				
9		r_1	S_7		r_1	r_1			
10		r_3	r_3		r_3	r_3			
11		r_5	r_5		r_5	r_5			

Solution:

Stack	Input	Action
0	id + id * id$	Shift 5
0id5	+ id * id$	Reduce by production no 6 (| id | = 1, so 2 symbols are popped from the stack, F is placed. For [0, F] there is g_3, thus 3 is top of the stack)
0F3	+ id * id$	Reduce by production no 4 (| F | = 1, so 2 symbols are popped from the stack, T is placed. For [0, T] there is g_2, thus 2 is top of the stack)
0T2	+ id * id$	Reduce by production no 2, then g_1. Top of the stack is 1.
0E1	+ id * id$	Shift 6
0E1 + 6	id * id$	Shift 5
0E1 + 6id5	*id$	Reduce by production no 6, then g_3. Top of the stack is 3.
0E1 + 6F3	*id$	Reduce by production no 4, then g_9. Top of the stack is 9.
0E1 + 6T9	*id$	Shift 7
0E1 + 6T9*7	id$	Shift 5
0E1 + 6T9*7id5	$	Reduce by production no 6, then g_{10}. Top of the stack is 10.
0E1 + 6T9*7F10	$	Reduce by production no 3. | T * F | = 3. Six symbols are popped. T is placed. Entry for [6, T] = g_9, thus 9 is the top of the stack.
0E1 + 6T9	$	Reduce by production number 1. Then g_1, thus 1 is top of the stack.
0E1	$	accept

LR Parsing Table Construction: LR parsing is an efficient parsing technique in compiler design. Whole parsing technique is dependent on the parsing table. But the most difficult part is the LR parsing table construction. In this section we shall discuss the construction process of Simple LR parsing table. The construction process consists of several steps as follows.

- Augmented grammar formation
- State construction of FA using item grammar and closure operation from the augmented grammar:

Item Grammar Construction: This is done by item grammar construction. An item of a grammar G is a production rule in G with a dot at some position of the right hand side of the production. Item (The presence of . [dot]) indicates how much of the production is parsed so far.

As an example If E → E + T is a production rule of a grammar, then the items of the grammar are

$$E \rightarrow .E + T$$
$$E \rightarrow E. + T$$
$$E \rightarrow E + .T$$
$$E \rightarrow E + T.$$

If there is a production T → ε in the grammar then T → . is an item of the grammar.

Closure Operation: If I is the set of items of a grammar G then Closure of I is a set of item constructed by the following rules.

- Closure(I) contains every item in I.
- If $A \rightarrow \alpha . B \beta$ exist in Closure(I) and $B \rightarrow \gamma$ is a production, then $B \rightarrow .\gamma$ is also added in the Closure(I).

Goto Operation: If a set of item is I and B is a grammar symbol, the operation goto(I, B) is constructed as

$$\text{goto}(I, B) = \text{closure } (A \rightarrow \alpha B. \beta) \text{ if } A \rightarrow \alpha . B \beta \text{ is in I}$$

- In FA if goto(I_i, T) $\rightarrow I_j$ then Action[I_i, T] = S_j. If goto(I_i, NT) $\rightarrow I_j$ then Goto[I_i, NT] = g_j
- To construct the reduce operations in Action table formulate the FOLLOW of the non-terminals. If I_i contains a production in the form NT $\rightarrow \alpha$., and FOLLOW(NT) = {T_j}, then Action [I_i, T_j] = r_k. [Where k is the production number of the production NT $\rightarrow \alpha$] If I_i contains a production S $\rightarrow \alpha$. , where S is the new start symbol and FOLLOW(S) = \$, then Action[I_i, \$] = Accept.

The process is described by the following example.

Example 13.14 Construct SLR parsing table for the following grammar

$E \rightarrow E + T$
$E \rightarrow T$
$T \rightarrow T * F$
$T \rightarrow F$
$F \rightarrow (E)$
$F \rightarrow id$

Solution:

a) The augmented grammar is

(0) $E' \rightarrow E$
(1) $E \rightarrow E + T$
(2) $E \rightarrow T$
(3) $T \rightarrow T * F$
(4) $T \rightarrow F$
(5) $F \rightarrow (E)$
(6) $F \rightarrow id$

b) **Construction of state of NFA:** This is done by two steps

i) **Item grammar construction:** Let take $E' \rightarrow . E$ in I. The closure(I) is the set as follow.

{
$E' \rightarrow .E$
$E \rightarrow .E + T$
$E \rightarrow .T$
$T \rightarrow .T * F$

$T \rightarrow .F$
$F \rightarrow .(E)$
$F \rightarrow .\,id$
}

Mark it I_0.

goto(I_0, E) = closure($E' \rightarrow E\,.$) = { $E' \rightarrow E.$, $E \rightarrow E. + T$} = $\mathbf{I_1}$
goto(I_0, T) = closure($E \rightarrow T\,.$) = { $E \rightarrow T.$, $T \rightarrow T. * F$} = $\mathbf{I_2}$
goto(I_0, F) = closure($T \rightarrow F\,.$) = {$T \rightarrow F.$} = $\mathbf{I_3}$
goto(I_0, +) = φ [There is no production in the form NT → .+ in I_0]
goto(I_0, *) = φ
goto(I_0, () = closure($F \rightarrow (.\,E)$) = {$F \rightarrow (.E)$, $E \rightarrow .\,E + T$, $E \rightarrow .T$, $T \rightarrow .T * F$, $T \rightarrow .F$, $F \rightarrow .(E)$
$F \rightarrow .id$} = $\mathbf{I_4}$
goto(I_0,)) = φ
goto(I_0, id) = closure($F \rightarrow id\,.$) = {$F \rightarrow id.$} = $\mathbf{I_5}$

goto(I_1, E) = φ
goto(I_1, T) = φ
goto(I_1, F) = φ
goto(I_1, +) = closure($E \rightarrow E + .T$) = {$E \rightarrow E + .T$, $T \rightarrow .T * F$, $T \rightarrow .F$, $F \rightarrow .(E)$, $F \rightarrow .id$} = $\mathbf{I_6}$
goto(I_1, *) = φ
goto(I_1, () = φ
goto(I_1,)) = φ
goto(I_1, id) = φ

goto(I_2, E) = φ
goto(I_2, T) = φ
goto(I_2, F) = φ
goto(I_2, +) = φ
goto(I_2, *) = closure($T \rightarrow T *.F$) = { $T \rightarrow T * .F$, $F \rightarrow .(E)$, $F \rightarrow .id$} = $\mathbf{I_7}$
goto(I_2, () = φ
goto(I_2,)) = φ
goto(I_2, id) = φ

goto(I_3, E) = φ
goto(I_3, T) = φ
goto(I_3, F) = φ
goto(I_3, +) = φ
goto(I_3, *) = φ
goto(I_3, () = φ
goto(I_3,)) = φ
goto(I_3, id) = φ

$I_3 = \{T \rightarrow F.\}$ i.e. $\{T \rightarrow id.\}$. So I_3 is the final state.

$goto(I_4, E) = closure(F \rightarrow (E.)) = \{F \rightarrow (E.), E \rightarrow E. + T\} = \mathbf{I_8}$
$goto(I_4, T) = closure(E \rightarrow T.) = \{E \rightarrow T., T \rightarrow T. * F\} = I_2$
$goto(I_4, F) = closure(T \rightarrow F.) = \{T \rightarrow F.\} = I_3$
$goto(I_4, +) = \varphi$
$goto(I_4, *) = \varphi$
$goto(I_4, () = closure(F \rightarrow (.E)) = \{F \rightarrow (.E), E \rightarrow .E + T, E \rightarrow .T, T \rightarrow .T * F, T \rightarrow .F, F \rightarrow .(E)$
$F \rightarrow .id\} = I_4$
$goto(I_4,)) = \varphi$
$goto(I_4, id) = closure(F \rightarrow id.) = \{F \rightarrow id.\} = I_5$

I_5 is the final state.

$goto(I_5, E) = \varphi$
$goto(I_5, T) = \varphi$
$goto(I_5, F) = \varphi$
$goto(I_5, +) = \varphi$
$goto(I_5, *) = \varphi$
$goto(I_5, () = \varphi$
$goto(I_5,)) = \varphi$
$goto(I_5, id) = \varphi$
$goto(I_6, E) = \varphi$
$goto(I_6, T) = closure(E \rightarrow E + T.) = \{E \rightarrow E + T., T \rightarrow T. * F\} = \mathbf{I_9}$

I_9 is final state as the parse ends at $E \rightarrow E + T$.

$goto(I_6, F) = closure(T \rightarrow F.) = \{T \rightarrow F.\} = I_3$
$goto(I_6, +) = \varphi$
$goto(I_6, *) = \varphi$
$goto(I_6, () = closure(F \rightarrow (.E)) = \{F \rightarrow (.E), E \rightarrow .E + T, E \rightarrow .T, T \rightarrow .T * F, T \rightarrow . F, F \rightarrow .(E)$
$F \rightarrow . id\} = I_4$
$goto(I_6,)) = \varphi$
$goto(I_6, id) = closure(F \rightarrow id.) = \{F \rightarrow id.\} = I_5$
$goto(I_7, E) = \varphi$
$goto(I_7, T) = \varphi$
$goto(I_7, F) = closure(T \rightarrow T * F.) = \{T \rightarrow T * F.\} = \mathbf{I_{10}}$

I_{10} is final state as parse ends at $T \rightarrow T * F$.

$goto(I_7, +) = \varphi$
$goto(I_7, *) = \varphi$
$goto(I_7, () = closure(F \rightarrow (.E)) = \{F \rightarrow (.E), E \rightarrow .E + T, E \rightarrow .T, T \rightarrow .T * F, T \rightarrow .F, F \rightarrow .(E)$
$F \rightarrow .id\} = I_4$
$goto(I_7,)) = \varphi$

goto(I_7, id) = closure(F → id.) = {F → id.} = I_5
goto(I_8, E) = φ
goto(I_8, T) = φ
goto(I_8, F) = φ
goto(I_8, +) = closure(E → E +. T) = {E → E + .T, T → . T * F, T → .F, F → .(E), F → .id } = I_6
goto(I_8, *) = φ
goto(I_8, ( ) = φ
goto(I_8,)) = closure(F→ (E).) = {F → (E). } = $\mathbf{I_{11}}$

I_{11} is also final state.

goto(I_8, id) = φ
goto(I_9, E) = φ
goto(I_9, T) = φ
goto(I_9, F) = φ
goto(I_9, +) = φ
goto(I_9, *) = closure(T → T * .F) = {T → T * .F, F → .(E), F → .id} = I_7
goto(I_9, ( ) = φ
goto(I_9,)) = φ
goto(I_9, id) = φ

For I_{10} and I_{11} all are null.

The final FA is

State	Input, Next State								
	id	+	*	(	)	$	E	T	F
I_0	I_5			I_4			I_1	I_2	I_3
I_1		I_6							
I_2			I_7						
I_3									
I_4	I_5			I_4			I_8	I_2	I_3
I_5									
I_6	I_5			I_4				I_9	I_3
I_7	I_5			I_4					I_{10}
I_8		I_6			I_{11}				
I_9			I_7						
I_{10}									
I_{11}									

c) Shift in the Action part and Goto Table is as follow.

State	Action						Goto		
	id	+	*	(	)	$	E	T	F
I_0	S_5			S_4			g_1	g_2	g_3
I_1		S_6							
I_2			S_7						
I_3									
I_4	S_5			S_4			g_8	g_2	g_3
I_5									
I_6	S_5			S_4				g_9	g_3
I_7	S_5			S_4					g_{10}
I_8		S_6			S_{11}				
I_9			S_7						
I_{10}									
I_{11}									

d) FOLLOW(E′) = {$}
FOLLOW(E) = {+,), $}
FOLLOW(T) = {+,), *, $}
FOLLOW(F) = {+,), *, $}

I_1 contains a production $E' \rightarrow E.$, and FOLLOW(E′) = {$}. Thus Action [$I_1$, $] = Accept.
I_2 contains a production $E \rightarrow T.$, and FOLLOW(E) = {+,), $}. Hence

Action[I_2, +] = r_2, Action[I_2,)] = r_2, Action[I_2, $] = r_2 [$E \rightarrow T$ is production number 2]

For I_3: Action[I_3, +] = r_4, Action[I_3,)] = r_4, Action[I_3, *] = r_4, Action[I_3, $] = r_4,
By this process the final table is

State	Action						Goto		
	id	+	*	(	)	$	E	T	F
I_0	S_5			S_4			g_1	g_2	g_3
I_1		S_6				acc			
I_2		r_2	S_7		r_2	r_2			
I_3		r_4	r_4		r_4	r_4			
I_4	S_5			S_4			g_8	g_2	g_3
I_5		r_6	r_6		r_6	r_6			
I_6	S_5			S_4				g_9	g_3

State	Action						Goto		
	id	+	*	(	)	$	E	T	F
I_7	S_5			S_4					g_{10}
I_8		S_6			S_{11}				
I_9		r_1	S_7		r_1	r_1			
I_{10}		r_3	r_3		r_3	r_3			
I_{11}		r_5	r_5		r_5	r_5			

What We Have Learned So Far

1. Compiler is a program that takes a program written in a source language and translates it into an equivalent program in a target language.
2. The knowledge of automata theory is required in different phases of compiler design.
3. Compiler are of three types—single pass, multi-pass and load and go.
4. Single pass compiler is known as narrow compiler and multi-pass is known as wide compiler.
5. Phases of Compiler are divided into two phases—analysis and synthesis phase.
6. Analysis part can be divided into three parts—linear or lexical analysis, hierarchical analysis or syntax analysis and semantic analysis.
7. Synthesis part is divided into three parts—intermediate code generator, code optimizer and code generator.
8. In lexical analysis the source program is read from left to right, character by character and generates token.
9. In syntax analysis the rules associated with the context free grammar for the programming language are checked.
10. Parsing is of two types—top down and bottom up parsing.
11. Recursive decent parsing is a top down parsing and it is most straight forward form of parsing. The major drawback of it is backtracking.
12. LL(1) parsing is a backtrack free table driven parsing.
13. LL(1) parsing table is constructed by FIRST and FOLLOW operation.
14. If any cell of the table constructed from a grammar contains multiple entry then the grammar is not LL(1).
15. Shift reduce and LR are bottom up parsing where the parsing starts from the leaf nodes and ends on the root.
16. LR parsing table consists of two parts Action and Goto.
17. LR parsing table is constructed by item grammar construction, closure operation, finite automata construction etc.

Multiple Choice Questions

1. Choose the role of compiler
   a) Converting high level language to machine level language
   b) Converting low level language to high level language
   c) Converting a source language to a target language
   d) Only checking the errors in the written program.
2. Compiler is a
   a) Software b) Operating system
   c) Program d) Translator
3. Symbol table keeps record of
   a) Keywords b) Identifier
   c) Constant d) Loop
4. Which is true for lexical analysis?
   a) It removes comments and white spaces from the source program.
   b) It breaks the source program into tokens
   c) Its operation is dependent on syntax analysis
   d) All of these
5. Pattern of an identifier is
   a) letter (letter | digit)*-
   b) digit (letter | digit)*
   c) (letter | digit)* letter
   d) (letter | digit)*digit
6. Which is not included in LL(1) parsing
   a) Input buffer b) Queue
   c) Parsing table d) Stack
7. Which may be a cause for a grammar not to be LL(1) ?
   a) Grammar is left recursive
   b) Grammar is left factored
   c) Parsing table does not contain multiple entry
   d) Grammar is not ambiguous.
8. Identify the invalid drawback of Shift Reduce conflict
   a) Shift-Reduce Conflict
   b) Reduce-Reduce Conflict
   c) Shift-shift Conflict
   d) Shift-Reduce-Shift Conflict

Answers:

1. c 2. d 3. b 4. d 5. a 6. b 7. a 8. c

GATE Questions

(Common for Question 1 and 2)

For the grammar below, a partial LL(1) parsing table is also presented along with grammar. Entries that need to be filled are indicated as E_1, E_2, E_3. ε is the empty string, \$ indicates the end of input, and, | separates alternate right hand side productions.

$$S \rightarrow aAbB \mid bAab \mid \varepsilon$$
$$A \rightarrow S$$
$$B \rightarrow S$$

	a	b	\$
S	E1	E2	$S \rightarrow \varepsilon$
A	$A \rightarrow S$	$A \rightarrow S$	Error
B	$B \rightarrow S$	$B \rightarrow S$	E3

1. The FIRST and FOLLOW sets for the non-terminals A and B are

   a) FIRST (A) = {a, b, ε} = FIRST(B)
   FOLLOW(A) = {a, b}
   FOLLOW (B) = {a, b, $}

   b) FIRST (A) = {a, b, $}
   FIRST(B) = {a, b, ε}
   FOLLOW(A) = {a, b}
   FOLLOW (B) = {$}

   c) FIRST (A) = {a, b, ε} = FIRST(B)
   FOLLOW(A) = {a, b}
   FOLLOW (B) = ∅

   d) FIRST (A) = {a, b} = FIRST(B)
   FOLLOW(A) = {a, b}
   FOLLOW (B) = {a, b}

2. The appropriate entries for E_1, E_2 and E_3 are

   a) E_1: S → aAbB, A → S
   E_2: S → bAaB, B → S
   E_3: B → S

   b) E_1: S → aAbB, S → ε
   E_2: S → bAaB, S → ε
   E_3: S → ε

   c) E_1: S → aAbB, S → ε
   E_2: S → bAaB, S → ε
   E_3: B → S

   d) E_1: A → S, S → ε
   E_2: B → S, S → ε
   E_3: B → S

3. In compiler keywords of a language are recognized during

   a) parsing of a program
   b) the code generation
   c) the lexical analysis of the program
   d) dataflow analysis

4. The grammar S → aSa | bS | c is

   a) LL(1) but not LR(1)
   b) LR(1) but not LL(1)
   c) both LL(1) and LR(1)
   d) neither LL(1) nor LR(1)

Answers:

1. a 2. b 3. c 4. c

Hints:

1. FIRST (S) = {a, b, ε} = FIRST(A) = FIRST(B)
   FOLLOW(S) = {$}

   Take S → aAbB
   α B β

   FIRST(bB) – ε is included in included in FOLLOW(A)

   FOLLOW(A) = {b}

   From S → bAab considering ab as β using same rule FOLLOW(A) = {a}.

   FOLLOW(A) = {a, b}

   Consider A → S and B → S. These two productions are in the form A → αB, where α is null.

   $$\text{FOLLOW}(A) \subset \text{FOLLOW}(S)$$
   $$\text{FOLLOW}(B) \subset \text{FOLLOW}(S)$$

Consider S → aAbB, with aAb as α. Using the rule A → αB, we get

$$\text{FOLLOW(S)} \subset \text{FOLLOW(B)}.$$

Thus FOLLOW(S) = {a, b, $}= FOLLOW(B)
FOLLOW(A) = {a, b}

Exercise

1. Parse the string S = abbfgg using the following LL(1) parsing table.

	a	b	f	g	d
S	S → A				
A	A → aS			A → d	
B		B → bBC	B → f		
C				C → g	

2. Construct LL(1) parsing table for the following grammar.

   S → E
   E → aF | Ed
   F → bFC | f
   C → g

3. Check whether the following languages are LL(1) or not.
   a) $L = a^n b^n, n \geq 1$
   b) $L = \{a^m b^n, n, m \geq 1\}$

4. Parse the string S = abbfgg using the following grammar by shift reduce parsing

   S → A
   A → aB |Ad
   B → bBC |f
   C → g

5. Parse the string id/id + id * id ↑ id by the following grammar using shift reduce parsing.

   E → E + E
   E → E * E
   E → E ↑ E
   E → E/E
   E → id/id/id

   In this relation describe shift-reduce and reduce-reduce conflict.

6. Construct LR parsing table for the grammar

   E → E + T/T
   T → T ↑ F/F
   F → id

Advance Topics Related to Automata

14

Introduction

Throughout 13 chapters many topics related to automata are discussed. Those are not the ultimate. Researches are going on in the field of automata like in other fields. In this chapter we shall discuss some new topics related to automata. From these we can accumulate some ideas about the current trend of researches in the field of automata. In this chapter we shall mainly discuss matrix grammar, probabilistic finite automata, cellular automata.

14.1 Matrix Grammar

Matrix grammar is an extension of context free grammar. The difference with CFG is that, here in matrix grammar the production rules are grouped into finite sequences and it is applied in the sequence instead of applying it separately. These sequences are called matrix. The name of the grammar came from this sequence (matrix).

Definition: A matrix grammar is denoted by

$$G = \{V_N, \Sigma, M, S\}$$

where

V_N : Set of non-terminals
Σ : Set of terminals
S : Start symbol
M : Finite non empty set of sequence of production rules known as matrices.

It can be written as

$$m_i = [P_1 \rightarrow Q_1, P_2 \rightarrow Q_2, \ldots\ldots P_r \rightarrow Q_r]$$

where $P_i \in V_N$ and $Q_i \in V_N \cup \Sigma$.

Example 14.1 Design a matrix grammar for $a^nb^nc^n$, n > 0.

Solution: It is already discussed that the grammar is context sensitive. The problem is in designing the productions in such a way that it adds one 'a', one 'b' and one 'c' in each replacement. This is not possible for context free grammar. But if the productions are designed in a group that when 'a' and 'b' are placed at the same time 'c' is also placed, then the problem will be solved.

$S \to XY$ is the starting production. Then $[X \to aXb, Y \to cY]$ are added in a group. It meant when in XY, X is replaced by aXb, at the same time Y will also be replaced by cY. For final replacement make the sequence as $[X \to ab, Y \to c]$.

Thus the grammar is

$$G = \{V_N, \Sigma, M, S\}$$

where

$V_N : \{S, X, Y\}$
$\Sigma : \{a, b, c\}$
S : Start symbol
M: $[S \to XY], [X \to aXb, Y \to cY], [X \to ab, Y \to c]$

Example 14.2 Design a matrix grammar for L = WW, where $W \in (a, b)^*$.

Solution: The length of the string is either 0 or 2n, if $|W| = n$. For any string $\in$ L the ith symbol and the $n + i^{th}$ symbol are same as the string is WW. Already it is discussed that the grammar is not context free. If the start production is $S \to XY$, where X generates first W and Y generates last W, then the problem lies in the number of replacement of X and Y by their corresponding productions. For generating W the productions are

$$S \to aS \mid bS \mid \varepsilon$$

If it can be made confirm that when X will be replaced by a production, Y must be replaced by the same production then the grammar can be constructed. It can only be done using matrix grammar, where we can group the productions as matrices as $[X \to aX, Y \to aY], [X \to bX, Y \to bY], [X \to \varepsilon, Y \to \varepsilon]$.

Thus the grammar is

$$G = \{V_N, \Sigma, M, S\}$$

where

$V_N : \{S, X, Y\}$
$\Sigma : \{a, b\}$
S : Start symbol
M : $[S \to XY], [X \to aX, Y \to aY], [X \to bX, Y \to bY], [X \to \varepsilon, Y \to \varepsilon]$

Example 14.3 Design a matrix grammar for $a^nb^nc^nd^n$, $n > 0$.

Solution:

$$G = \{V_N, \Sigma, M, S\}$$

where

$V_N : \{S, X, Y\}$
$\Sigma : \{a, b, c, d\}$
S : Start symbol
M : $[S \to XY], [X \to aXb, Y \to cYd], [X \to ab, Y \to cd]$

14.2 Probabilistic Finite Automata

Probabilistic finite automata was first introduced by O. Rabin in 1963. It is an extension of non-deterministic finite automata where the transition functions are assigned to some probability. The transitional function with probability forms a matrix called stochastic matrix. Probabilistic finite automata generalize the concept of Markov model.

Consider the following diagram in Fig. 14.1.

Here the number in bracket in each transition represents the probability for the input alphabet. Here we can construct a probability matrix M for each of the alphabets for the state transition.

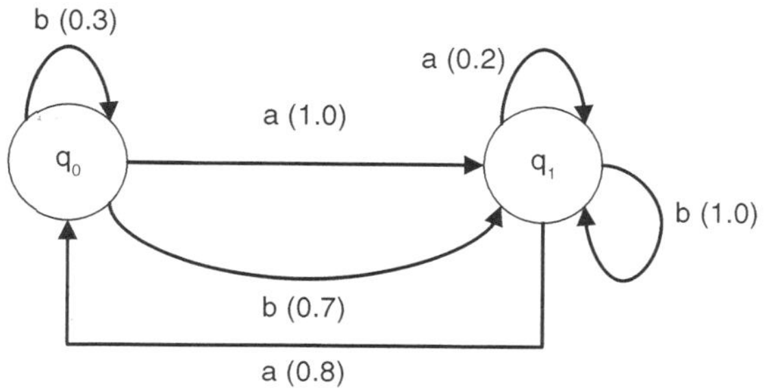

Fig. 14.1 *Probabilistic Finite Automata*

M(a) =		q_0	q_f
	q_0	0	1.0
	q_f	0.8	0.2

M(b) =		q_0	q_f
	q_0	0.3	0.7
	q_f	1.0	0

Now which is the initial and which is the final state? Probability is also assigned here in two forms:

1. Initial State probability (I)
2. Final State probability (F)

Let in the previous diagram $I(q_0) = 1$, $I(q_1) = 0$ and $F(q_0) = 0$, $F(q_1) = 1$.
So, the initial state probability I = {1, 0} and final state probability F= {0, 1}.
This is an example of probabilistic automata.

Definition: *A probabilistic finite automata is denoted as {Q, Σ, δ, I, F, M} with a probability matrix M with size | Q | × | Q | for each input alphabet. Here I and F denote the initial and final state probability.*

14.2.1 String Accepted by a PA

Let {Q, Σ, δ, I, F, M} be a probabilistic finite automata, η be a real number such that $0 \leq \eta \leq 1$ and $\overline{F}$ is the n dimensional column vector whose components are either 0 or 1. The language w is accepted by the probabilistic finite automata on $\overline{F}$ with the intersection point η if $(I\ M(w)\ \overline{F}) \geq \eta$.

The language accepted by probabilistic finite automata is one type of regular language known as stochastic language.

For a finite probabilistic automata the domain of M can be increased by including the following operations.

$$M(\varepsilon) = M$$
$$M(wx) = M(w)\,M(x)$$

Example 14.4 Consider the PFA given below.

Q : $\{q_0, q_2\}$
Σ : {a, b}
M :

		q_0	q_f
M(a) =	q_0	.25	.5
	q_f	0	1

		q_0	q_f
M(b) =	q_0	.5	0.75
	q_f	1.0	0

I : {1 0}
F : {0 1}

a) Calculate IM(x) $\overline{F}$, for each x = {ab, ba}.
b) Check whether x = {aa and bb} are accepted by the PFA with intersection point η = .5.

Solution:

a) i) x = ab

$$IM(x)\,\overline{F} = [1 \quad 0] \times M(ab) \times \begin{bmatrix} 0 \\ 1 \end{bmatrix} = [1 \quad 0]\,M(a) \times M(b) \times \begin{bmatrix} 0 \\ 1 \end{bmatrix}$$

$$= [1 \quad 0] \times \begin{bmatrix} 0.25 & 0.5 \\ 0 & 1 \end{bmatrix} \times \begin{bmatrix} .5 & .75 \\ 1 & 0 \end{bmatrix} \times \begin{bmatrix} 0 \\ 1 \end{bmatrix}$$

$$= .1875$$

ii) x = baa

$$IM(x)\,\overline{F} = [1 \quad 0] \times M(ba) \times \begin{bmatrix} 0 \\ 1 \end{bmatrix} = [1 \quad 0]\,M(b) \times M(a) \times \begin{bmatrix} 0 \\ 1 \end{bmatrix}$$

$$= [1 \quad 0] \times \begin{bmatrix} .5 & .75 \\ 1 & 0 \end{bmatrix} \times \begin{bmatrix} 0.25 & 0.5 \\ 0 & 1 \end{bmatrix} \times \begin{bmatrix} 0 \\ 1 \end{bmatrix}$$

$$= 1$$

b) i) x = aa

$$IM(x)\ \overline{F} = [1 \quad 0] \times M(aa) \times \begin{bmatrix} 0 \\ 1 \end{bmatrix} = [1 \quad 0]\, M(a) \times M(a) \times \begin{bmatrix} 0 \\ 1 \end{bmatrix}$$

$$= [1 \quad 0] \times \begin{bmatrix} 0.25 & 0.5 \\ 0 & 1 \end{bmatrix} \times \begin{bmatrix} 0.25 & 0.5 \\ 0 & 1 \end{bmatrix} \times \begin{bmatrix} 0 \\ 1 \end{bmatrix}$$

$$= .625$$

As .625 > .5, thus x = aa is accepted by the PFA on intersection point .5.

ii) x = bb

$$IM(x)\ \overline{F} = [1 \quad 0] \times M(bb) \times \begin{bmatrix} 0 \\ 1 \end{bmatrix} = [1 \quad 0]\, M(b) \times M(b) \times \begin{bmatrix} 0 \\ 1 \end{bmatrix}$$

$$= [1 \quad 0] \times \begin{bmatrix} .5 & .75 \\ 1 & 0 \end{bmatrix} \times \begin{bmatrix} .5 & .75 \\ 1 & 0 \end{bmatrix} \times \begin{bmatrix} 0 \\ 1 \end{bmatrix}$$

$$= .375$$

As .325 < .5, thus x = bb is not accepted by the PFA on intersection point .5.

14.3 Cellular Automata

Cellular automata in short CA was proposed by Ulam and Neumann in 1940 at Los Alamos National Laboratory. This model is used widely in different areas of computer science, mathematics, physics, theoretical biology etc. It is called ‘cellular’ as it consists of regular grid known as cell, where each cell is either ‘0’ or ‘1’. The cells are organized in the form of lattice. The grid may be one dimensional or multi dimensional. For each cell, its surrounding cells (including the cell itself) are called neighbourhood. For a one-dimensional array each cell has (2 + 1) neighbour within radius 1. In general for a cell in a one-dimensional array has (2r + 1) neighbours where radius is r.

In this section we shall mainly discuss about one-dimensional cellular automata.

CA evolves in discrete space and time, and can be viewed as an autonomous finite state machine (FSM). Each cell stores a discrete variable at time that refers to the present state (PS) of the cell. The next state (NS) of the cell at (t + 1) is affected by its state and the states of its neighbours at time t.

Definition: *A cellular automata (CA) is a collection of cells arranged in the form of lattice, such that each cell changes state as a function of time according to a defined set of rules that includes the states of neighbouring cells.*

14.3.1 Characteristics of Cellular Automata

- It performs *synchronous computation.*
- Each cell can be in any *finite number of state.*
- *Neighbourhood* describes how the cells are connected with its surrounding cells.

- *Update rule* computes the change in the state of a cell based on the states of its neighbours. State of a cell at t_{i+1} is dependent on the states of its neighbourhood cell at t_i.

Let us consider a one-dimensional CA. There are 256 rules [From 0 to 255] are defined for updatation.

Update Rule: A 3-neighbourhood CA (self, left and right neighbours), where a CA cell is having two states –0 or 1 and the next state of ith CA cell is

$$S^{t+1}_i = f_i\,(S^t_{i-1}, S^t_i, S^t_{i+1})$$

where S^t_{i-1}, S^t_i and S^t_{i+1} are the present states of the left neighbour, self and right neighbor of the ith cell at time t and f_i is the next state function. The states of the cells $S^t = (S^t_1, S^t_2, \ldots, S^t_n)$ at t is the present state of the CA. Therefore, the next state of the n-cell CA is

$$S^{t+1} = (f_1(S^t_0, S^t_1, S^t_2), f_2(S^t_1, S^t_2, S^t_3), \ldots, f_n(S^t_{n-1}, S^t_n, S^t_{n+1}))$$

Rules can be constructed in two ways—warp around or by null stuffing. In wrap around the sequence is considered as circular as given in Fig. 14.2(a).

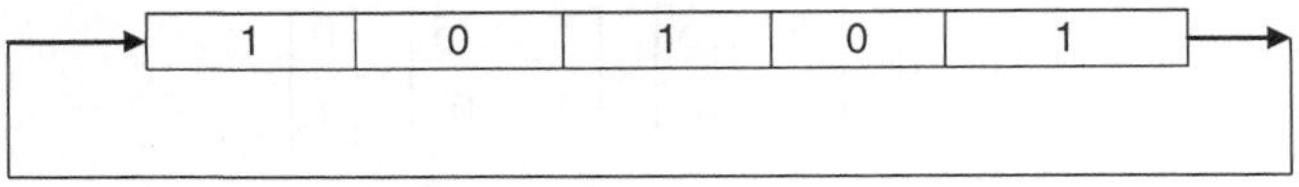

Fig. 14.2 (a) *Wrap Around*

In null stuffing '0' is stuffed at both sides of the original sequence. This is described in Fig. 14.2(b).

Original		1	0	1	0	1	
Null Stuffed	0	1	0	1	0	1	0

Fig. 14.2 (b) *Null Stuffed*

Let us consider the value as 210 (< 256). The binary equivalent of 210 is 11010010. Consider a one-dimensional cellular automata. Now construct the rules for it.

Decimal	7	6	5	4	3	2	1	0
Binary	111	110	101	100	011	010	001	000
210	1	1	0	1	0	0	1	0

The cell representations of 111 and 010 is given below.

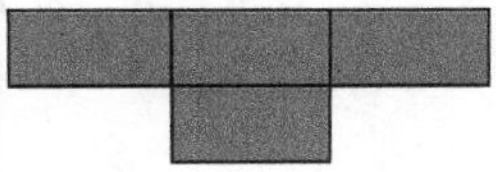

111

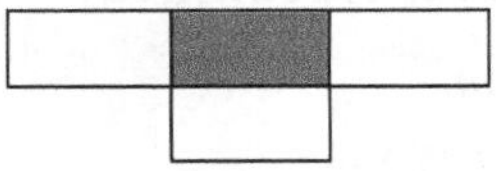

010

Now take a sequence 1001. Let us construct the sequence to t_3 using wrap around and null stuffing technique.

Wrap Around

t_0	1	0	0	1
t_1	1	1	1	0
t_2	0	1	1	0
t_3	1	0	1	1

The cells representation are given below

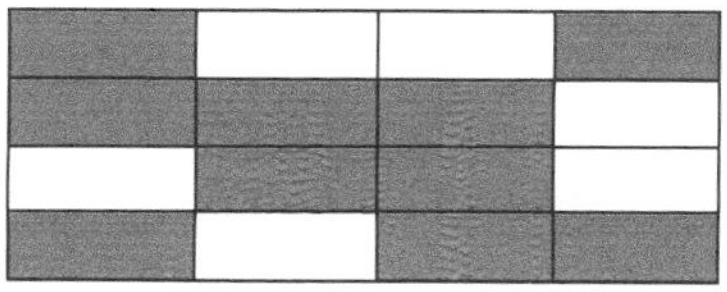

Null Stuffed: In null stuffed two '0' are added at both sides of the original string.

t_0	0	1	0	0	1	0
t_1	0	0	1	1	0	0
t_2	0	1	0	1	1	0
t_3	0	0	0	0	1	0

The cells representation are given below

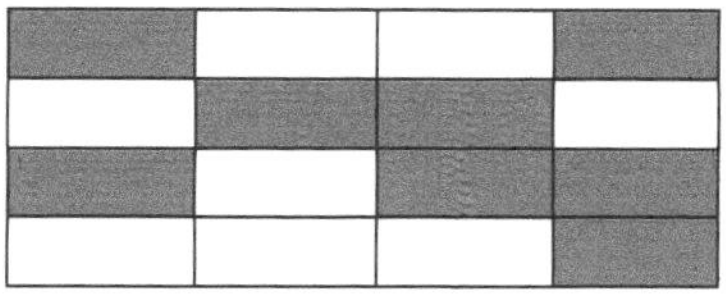

Example 14.5 Find the update of a one-dimensional CA rules for 212.

Solution: The binary equivalent of 212 is 11010100.
The rules are

Decimal	7	6	5	4	3	2	1	0
Binary	111	110	101	100	011	010	001	000
212	1	1	0	1	0	1	0	0

14.3.2 Applications of Cellular Automata

Cellular automata is not only used in Computer Science field, it also has different applications in other fields.

In the Field of Computer Science

- Cryptography
- Detecting fault tolerance in digital circuit
- Simulation of complex system

Beyond the Field of Computer Science

- Simulation of gas behavior
- Simulation of forest fire propagation
- Simulation of bone erosion.

Cellular automata are used to identify fault in some digital circuits.
Let consider an OR gate
Its truth table is

X	Y	O/P
0	0	0
0	1	1
1	0	1
1	1	1

It may happen that the input Y is faulty. It takes '1' for any input applied to it. Thus the output that we get will always be '1'. This is called 'Struck at 1'. Similarly the problem 'Struck at 0' can occur for a digital circuit.

Rule 192: Binary of 192 is 11000000
The rule is

	7	6	5	4	3	2	1	0
	111	110	101	100	011	010	001	000
192	1	1	0	0	0	0	0	0

Let the initial sequence is 1111.
By null stuffing at both sides the sequence become 011110.
So

t_0	0	1	1	1	1	0
t_1	0	0	1	1	1	0

Let it t_0 the bits are labeled as S_0, S_1, S_2 .... S_5.

State S_1 in t_1 depends on S_0, S_1 and S_2 of t_0. The bit pattern is 011, so the value of S_1 in t_1 is 0 (according to rule 192). By the same way all the other bits are placed. By the same way the patterns for t_2 and t_3 are generated.

t_2	0	0	0	1	1	0

t_3	0	0	0	0	1	0

Let a digital circuit has four inputs. To identify the 'Struck at 0' problem all bits are taken as '1' and patterns are generated for next (n–1) steps, where n is the number of input line. If the LSB is other than '1' in t_{n-1} step then we can say that the circuit is faulty.

References

Aho, Hopcroft and Ullman, *The Design and Analysis of Computer Algorithms,* Addison Wesley.

Aho, Sethi, Ullman, *Compilers Principles, Techniques and Tools,* Pearson Education, 2003.

Hopcroft and Ullman, *Introduction to Automata Theory, Languages and Computation,* Addison Wesley.

Kohavi, ZVI, *Switching And Finite Automata Theory*, Tata McGraw-Hill, 2006.

Lewis and Papadimitriou, *Elements of the Theory of Computation,* Prentice-Hall.

Martin, *Introduction to Languages and the Theory of Computation,* McGraw-Hill, 2nd edition,1996.

Mishra, KLP, Chandrasekaran, N. *Theory of Computer Science,* (Automata, Languages and Computation) PHI, 2002.

Pandey, *An Introduction to Automata Theory and Formal Languages*, S. K. Kataria & Sons, 2010.

Pandey, *Concept of Compiler Design*, S. K. Kataria & Sons, 2010.

Moore, Edward F. "Gedanken-experiments on Sequential Machines". *Automata Studies, Annals of Mathematical Studies* (Princeton, N.J.: Princeton University Press) (34): 129–153.

Mealy, George H. *A Method to Synthesizing Sequential Circuits*. Bell Systems Technical Journal. 1955, 1045–1079.

Chomsky, Noam. "Three models for the description of language". *IRE Transactions on Information Theory*, 1956, 113–124.

Davis et al. *Computability, Complexity, and Languages: Fundamentals of Theoretical Computer Science*. Boston: Academic Press, Harcourt, Brace, 1994, 327.

S. Y. Kuroda, "Classes of languages and linear-bounded automata", *Information and Control*, 7(2), 1964, 207–223.

"Two way Finite Automata" http://www.cs.cornell.edu/courses/cs682/2008sp/Handouts/2DFA.pdf

Cole Richard "Converting CFGs to CNF" http:// www.cs.nyu.edu/courses/fall07/V22.0453-001/cnf.pdf.

"Greibach Normal Form" *http://www.iitg.ernet.in/gkd/ma513/oct/oct18/note.pdf*.

Seungjin Choi, "Context-Free Grammars: Normal Forms" http://www.postech.ac.kr/~seungjin/courses/automata/2006/handouts/handout07.pdf.

"Simplification of CFG" http://www.cs.cmu.edu/~lblum/flac/Homeworks/hw3.pdf.

"Backus-Naur Form" http://www.math.grin.edu/~stone/courses/languages/Backus-Naur-form.pdf.

Scott, Dana and Rabin, Michael "Finite Automata and Their Decision Problems". *IBM Journal of Research and Development* 3(2), 1959, 114–125.

Ogden, W. "A helpful result for proving inherent ambiguity". *Mathematical Systems Theory* 2(3), 1968, 191–194.

"Ogden's Lemma" http://www.cs.binghamton.edu/~lander/cs573/pdf/cs573print424.pdf.

Helmbold David, "Example of Converting a PDA into a CFG", http://classes.soe.ucsc.edu/cmps130/Fall10/Handouts/PDA-CFGconv.pdf.
Church, Alonzo, "An Unsolvable Problem of Elementary Number Theory". *American Journal of Mathematics* 58(58), 1936, 345–363.
Church, Alonzo, *The Calculi of Lambda-Conversion*. Princeton: Princeton University Press, 1941.
E. Horowitz and Shani, "Fundamentals of Computer Algorithms" Galgotia Publication.
Weiss, Weiss Mark Allen, "Data Structures And Algorithm Analysis In C++, 3/E", Pearson Education.
Puntambekar A. A, "Principle of Compiler Design", Technical Publication, Pune.
Robin Hunter, "The Essence of Compiler Design", Pearson Education.

Index